T0276172

Planning, Writing and Reviewing Medical Device Clinical and Performance Evaluation Reports (CERs/PERs)

A Practical Guide for the European Union and Other Countries

Planning, Writing and Reviewing Medical Device Clinical and Performance Evaluation Reports (CERs/PERs)

A Practical Guide for the European Union and Other Countries

Joy L. Frestedt
Frestedt Incorporated, St. Louis Park, MN, United States

Academic Press is an imprint of Elsevier
125 London Wall, London EC2Y 5AS, United Kingdom
525 B Street, Suite 1650, San Diego, CA 92101, United States
50 Hampshire Street, 5th Floor, Cambridge, MA 02139, United States

ISBN: 978-0-443-22063-0

For Information on all Academic Press publications
visit our website at https://www.elsevier.com/books-and-journals

Publisher: Mica Haley
Acquisitions Editor: Mara Conner
Editorial Project Manager: Soumya Yadav
Production Project Manager: Maria Bernard
Cover Designer: Miles Hitchen

Typeset by MPS Limited, Chennai, India

Contents

List of figures

List of tables

List of boxes

Preface

1 Historical context

Massive changes occurred in the medical device industry under the European Union (EU) Medical Device Regulation (MDR) EU Reg. 2017/745 [1] and the EU *In Vitro* Diagnostic Device Regulation (IVDR) EU Reg. 2017/746 [2]. These 2017 laws required more and better clinical/performance data to support medical device and *In Vitro* diagnostic (IVD) device safety and performance (S&P) claims, and to document precisely how device benefits outweighed patient risks. The timelines for transitioning to these new regulations were interrupted multiple times by device manufacturer resistance to change, the global COVID-19 emergency, and notified body (NB) inability to keep up with demand for CE mark recertification. While MDR transitional timelines were severely delayed, IVDR requirements moved forward more rapidly.

The urgent need and exponential increase in global IVD technologies during the COVID-19 emergency likely contributed to the rapid speed to market for new IVD devices and the more rapid IVDR acceptance. The European Commission adopted common specifications for high-risk IVD tests including SARS-CoV-2 for COVID-19 on July 4, 2022 [3] and, on March 15, 2023, EU Reg. 2023/607 [4] conditionally extended the transition period until 2027/2028 depending on the device class. The EC website [5] provided additional information about the transition, and the MDR and IVDR were updated multiple times including a March 20, 2023 "consolidated" version [6,7]. In addition, a January 2023 guideline issued by the Medical Device Coordination Group (MDCG) entitled "MDCG 2023-2 List of Standard Fees" [8] provided greater clarity on NB fees to be made publicly available on each NB website.

As each NB became certified, they enforced MDR and IVDR requirements; however, NBs were not able to review all clinical evaluation reports (CERs, for medical devices) and performance evaluation report (PERs, for IVD medical devices) in a timely fashion. The review lag time combined with globally insufficient and inexperienced staff created an unparalleled scientific vacuum in the medical device industry. In addition, the greatly expanded CE mark certification requirements created an era of exponentially amplified NB regulatory enforcement activity which directly translated into increased manufacturer business risk and medical device costs.

NBs demanded more detailed and higher-quality clinical data collection in postmarket clinical follow-up (PMCF) and postmarket performance follow-up (PMPF) activities than the manufacturer previously determined was necessary. If a manufacturer failed or refused to provide a PMCF/PMPF plan (PMCFP/PMPFP) to gather additional clinical/performance data and a PMCF/PMPF evaluation report (PMCFER/PMPFER) to evaluate the added clinical data, the NB threatened to withdraw the CE mark, or not to issue it. Often, companies decided to remove the device from the EU market (or not offer it in the first place) as a result.

During the early years, many future clinical data requirement decisions happened in the absence of clear regulatory guidance on the PMCF/PMPF data type, quality, or amount required to keep whole classes of medical devices on the EU market. For example, PMCF guidance from January 2012 entitled "MEDDEV 2.12/2 rev. 2 Guidelines on Medical Devices: Post Market Clinical Follow-Up Studies" [9] clearly defined controlled clinical trials, registries, or retrospective data collections as acceptable PMCF studies; however, NBs were allowing surveys and other less rigorous tools like advisory boards and clinician anecdotal discussions when the clinical risk level did not require a rigorous clinical trial. The requirement for high-quality clinical/performance data was unclear for retrospective data collections, and real-world data and NBs were highly variable with their enforcement actions. In addition, NBs often changed their PMCF expectations over time. What was acceptable for a PMCFP/PMPFP at one time was not acceptable in future years. NBs invariably expected more clinical/performance data to fill CER/PER-identified clinical data gaps over time. Unfortunately, sometimes NB staff required clinical/performance data for devices without a clear need for the clinical data.

NB oversight was changing while the MDCG and expert panels came into existence. The expert panels began to review implantable, high-risk medical/IVD device CERs/PERs to ensure the devices conformed to the general safety and performance requirements (GSPRs) under MDR/IVDR. Manufacturers and NBs struggled to understand all the changes. For example, "MDCG 2022-2 Guidance on general principles of clinical evidence for *In Vitro* Diagnostic Medical Devices (IVDs)" [10] and MDCG 2021-21 rev.1 "Guidance on Performance evaluation of SARS-CoV-2 *In Vitro* diagnostic medical devices" [11] were released in January/February 2022, and the guidelines were expected to be implemented immediately. In summary, many new regulatory requirements were created, and all operational systems were having trouble keeping pace with these.

Although this textbook was created while EU regulations were rapidly evolving, excellent clinical evaluation (CE) and performance evaluation (PE) practices had been developed centuries earlier. Basic principles of good clinical and performance data evaluation were "spurred on" by similar regulatory changes in other geographies over time. Just as geographies around the world demanded more clinical data from manufacturers to keep their medical products safe and performing as intended, the EU Member States began to require more clinical data and better clinical data evaluation documentation for high-risk medical products than ever before.

The increased emphasis on medical device S&P may be good for patients and for business; however, these changes can be expensive and may lead companies to remove needed products from the market. In addition, whole companies and departments were created to support or manage the new regulatory requirements, so medical device costs unavoidably increase. For manufacturers navigating these EU MDR and IVDR changes, the CER and PER became greatly expanded components in the EU technical files/design dossiers. Driven by NB assessments and expert panel advice, all implantable and high-risk devices currently on the EU market should be supported by high-quality CERs/PERs. These European regulatory changes and diverse clinical evaluation quality concerns lead to the creation of this textbook as a "one-stop" resource to support CE/PE development to meet EU Regulations.

2 Intended audience

This textbook was designed to help develop stronger, more robust clinical and performance evaluations, and to support training for clinical/performance evaluation scientists and experts. To begin, we need to be able to talk to each other about clinical/performance data found in clinical trial, clinical literature, and clinical experience data in order to become better at evaluating device-related clinical and performance data as well as the medical device benefit-risk profile for the patient. This textbook was designed to help develop a common language and better communication between different team members from many different disciplines including, but not limited to:

- clinical affairs specialists and managers;
- clinical education specialists;
- clinical engineers;
- lawyers;
- corporate librarians;
- regulatory affairs specialists and managers;
- device vigilance specialists;
- CER team members;
- physicians, dentists, and nurse specialists;
- directors of clinical performance and evaluation;
- project managers;
- clinical evaluators and clinical researchers;
- postmarket surveillance specialists;
- risk management specialists;
- clinical data specialists;
- technical communication team members;
- medical and technical writers;
- notified body assessors and auditors;
- regulatory authority personnel;
- medical device designers;
- medical device regulatory affairs personnel;
- quality system managers and auditors;
- risk management specialists;
- usability specialists;
- biomedical engineers on graduate courses;
- physicians specializing in clinical evaluation;

- notified body personnel;
- vice-presidents/directors of clinical/regulatory affairs;
- chief medical officers
- chief executive officers
- directors of quality management for medical device, IVD, and CRO companies;
- staff on pharmaceutical companies with combination products (i.e., drug/device); and
- professors of clinical, regulatory, or quality sciences, biomedical engineering, and technology assessment.

Since starting this textbook, many jobs have been created including clinical evaluation specialists, clinical evaluation leads and managers, clinical reviewers, and clinical evaluation report writing services directors, and thousands of open jobs have been listed for these positions. In general, far fewer jobs were found for performance evaluator roles within the IVD device industry than for clinical evaluators in the medical device industry. Even so, this textbook was designed to help an individual reader transition into more advanced roles and responsibilities within a medical device manufacturing company needing more and better support for high-quality clinical/performance data evaluation.

3 Purpose, objectives, and goals

This textbook was intentionally written to discuss international good CE/PE practices, skills, and techniques. The overall goal was to develop further a CE/PE training curriculum and to expand instruction about how to evaluate and think critically about clinical data regardless of the medical device type. The critical thinking skills and techniques discussed in this textbook have been time-tested and should remain in place regardless of future regulatory changes. Simply stated: critical thinking is always a good practice in all industries and all clinical settings.

After reading this textbook, the reader should be able to

1. find and read MDR (EU Reg. 2017/745) and IVDR (EU Reg. 2017/746);
2. list resources to help with clinical/performance data evaluations;
3. describe a medical/IVD device and a CER/PER using MDR/IVDR terminology;
4. identify "suitably qualified" CER/PER team members and roles;
5. list key CER/PER stages;
6. discuss NB roles during CER/PER assessments and give CER/PER deficiency examples;
7. use MDCG and other guidance documents;
8. explore expert panel findings to improve CERs/PERs;
9. apply critical thinking concepts to specific clinical/performance data evaluations; and
10. give practical advice and tips to others when writing CE/PE documents.

The overarching goal for this textbook was to develop specific ways to evaluate product-related clinical/performance data from the patient/user perspective. This critical thinking and evaluation strategy should be useful for all medical products, not only medical devices.

This textbook follows the age-old five-step data analysis process (plan, identify, appraise, analyze, report). This textbook is based on details and examples from customized training services offered over 15 years within the Frestedt Learning Center (FLC). Creating this textbook was consistent with the Frestedt Incorporated mission to continuously improve and support clinical, regulatory, quality, and biomedical engineering affairs for our clients, to adhere to strong business ethics, and to work with good people on good projects using good processes. The FLC used an experimental and pragmatic approach by continuously learning and teaching others as projects were completed for pharmaceutical, device, and food company staff members in clinical, regulatory, quality, and engineering affairs. FLC teams authored hundreds of CERs/PERs for many different devices. This experience and expertise gained from working with many different NBs during these CE/PE processes is intended to "breathe" in this textbook. The added perspectives drawn from clinical safety teams in the pharmaceutical and food industries outside of the medical device community are intended to give a broader interpretation and use for the textbook. The goal is to help the learner understand how to apply critical thinking skills when evaluating clinical/performance data across the board, not just for medical devices and *in vitro* diagnostic devices.

4 Main references and starting points

For all readers, the main reference documents to read along with this textbook include, but are not limited to:

- Medical Device Regulation (MDR) (EU Reg. 2017/745) [6];

- *In Vitro* Diagnostic Medical Device Regulation (IVDR) (EU Reg. 2017/746) [7]; and
- MDCG Guidance documents [12].

This textbook is not meant to replace a careful reading of the regulations and guidelines. As with all things, change is constant and keeping pace is a challenge.

This textbook was designed to develop CE/PE document writing concepts at multiple levels to enhance learning for all readers—for example:

- a beginner with 0–3 years of experience working on CERs/PERs should be able to get "exposure" to all the jargon used in this work;
- an intermediate reader with 3–10 years of experience working on CERs/PERs should be able to develop proficiency in clinical and performance data evaluations, to improve CER/PER quality, and to make the CE/PE work easier; and
- an advanced reader with 10 + years of experience working on CERs/PERs should be able to review and consolidate their learning, and to offer constructive critical suggestions for the next edition.

To use this textbook as a beginner: first, read through each chapter one at a time and become familiar with the terms and interrelated topics; next, practice doing clinical/performance data planning, identification, appraisal, analysis, and conclusion steps; and finally, discuss improvements to help master clinical/performance data evaluations. For experienced clinical data evaluators, this textbook is organized by CER/PER writing steps, and a comprehensive table of contents and complete index are included so information of interest can be easily located. Each chapter ends with a conclusion summarizing key points and a few questions for readers to assess knowledge and retention. The appendices include document templates, checklists, and exercises to reinforce learning and improve CE/PE competence.

Please send feedback

Thank you to each and every reader! I greatly appreciate your work while reading and writing appropriate risk-based CE/PE documents so we can collectively develop and support more medical research innovation. I love feedback, so please email me any suggestions for changes or improvements. I can be reached at info@frestedt.com. Thank you for your participation in this learning experiment, and thank you even more for your contributions to the next edition.

Acknowledgments

Writing this book took 6 years, and I am grateful to so many wonderful friends and colleagues who shared experiences and helped with various drafts. First, thank you to the many current and former Frestedt Incorporated employees and consultants for their contributions to this book over the years: Brad Argue, PhD, Matt Harris, Hayley Jacobs, Kelsey McFinkel, PhD, Sarah McFinkel, Mike Perez, Amanda Rogers, Kasey Sands, PhD, Linnea Schmidt, Ali Schulz, PhD, Lindsay Young, PhD, MBA, and Adam Zhang. These individuals helped to develop training slides and workbooks for many FLC courses which directly led to this textbook. In addition, thanks to Dr. Bassil Akra for sharing his experiences working on MedDev 2.7/1 rev. 4 and serving as TUV SUD global lead during so many manufacturer CE mark discussions. Thanks also to Steve Erickson for many lunch break discussions about postmarket surveillance and managing clinical data collection activities designed to fill clinical data gaps identified in various CERs. I greatly appreciate the many conversations we shared while discussing the CE/PE challenges to we have faced. They are the reason why this textbook exists!

References

[1] European Parliament and Council of the European Union. Consolidated text: Regulation (EU) 2017/745 of the European Parliament and of the Council of 5 April 2017 on medical devices, amending Directive 2001/83/EC, Regulation (EC) No 178/2002 and Regulation (EC) No 1223/2009 and repealing Council Directives 90/385/EEC and 93/42/EEC. Accessed on September 28, 2023 at http://orcid.org/https://eur-lex.europa.eu/legal-content/EN/TXT/?uri = CELEX%3A32017R0745.

[2] European Parliament and Council of the European Union. Consolidated text: Regulation (EU) 2017/746 of the European Parliament and of the Council of 5 April 2017 on in vitro diagnostic medical devices and repealing Directive 98/79/EC and Commission Decision 2010/227/EU. Accessed on September 28, 2023 at eur-lex.europa.eu/legal-content/EN/TXT/?uri = CELEX%3A32017R0746.

[3] European Parliament and Council of the European Union. Commission Implementing Regulation (EU) 2022/1107 of 4 July 2022 laying down common specifications for certain class D in vitro diagnostic medical devices in accordance with Regulation (EU) 2017/746 of the European Parliament and of the Council. Accessed on June 23, 2024 at eur-lex.europa.eu/eli/reg_impl/2022/1107/oj.

[4] European Parliament and Council of the European Union. Regulation (EU) 2023/607 of the European Parliament and of the Council of 15 March 2023 amending Regulations (EU) 2017/745 and (EU) 2017/746 as regards the transitional provisions for certain medical devices and in vitro diagnostic medical devices. Accessed on June 23, 2024 at https://eur-lex.europa.eu/legal-content/EN/TXT/?uri = uriserv:OJ.L_0.2023.080.01.0024.01.ENG.

[5] European Commission. New Regulations (website). Accessed on June 23, 2024 at health.ec.europa.eu/medical-devices-sector/new-regulations_en.

[6] European Parliament and Council of the European Union. Consolidated text: Regulation (EU) 2017/745 of the European Parliament and of the Council of 5 April 2017 on medical devices, amending Directive 2001/83/EC, Regulation (EC) No 178/2002 and Regulation (EC) No 1223/2009 and repealing Council Directives 90/385/EEC and 93/42/EEC. Accessed on June 23, 2024 at https://eur-lex.europa.eu/legal-content/EN/TXT/?uri = CELEX%3A02017R0745-20230320.

[7] European Parliament and Council of the European Union. Consolidated text: Regulation (EU) 2017/746 of the European Parliament and of the Council of 5 April 2017 on in vitro diagnostic medical devices and repealing Directive 98/79/EC and Commission Decision 2010/227/EU. Accessed on September 28, 2023 at eur-lex.europa.eu/legal-content/EN/TXT/?uri = CELEX%3A02017R0746-20230320.

[8] Medical Device Coordination Group. MDCG 2023-2. List of Standard Fees – January 2023. Accessed on June 23, 2024 at health.ec.europa.eu/system/files/2023-01/mdcg_2023-2_en.pdf.

[9] European Commission. MEDDEV 2.12/2, rev2 (January 2012). Guidelines on Medical Devices. Post Market Clinical Follow-Up Studies: A Guide for Manufacturers and Notified Bodies. Accessed on June 23, 2024 at meddev.info/_documents/2_12_2_ol_en_.pdf.

[10] Medical Device Coordination Group. MDCG 2022-2 (January 2022). Guidance on general principles of clinical evidence for in Vitro Diagnostic Medical Devices (IVDs) Accessed on June 23, 2024 at health.ec.europa.eu/system/files/2022-01/mdcg_2022-2_en.pdf.

[11] Medical Device Coordination Group. MDCG 2021-21, Rev.1 (February 2022). Guidance on performance evaluation of SARS-CoV-2 *in vitro* diagnostic medical devices. Accessed on June 23, 2024 at health.ec.europa.eu/system/files/2022-02/mdcg_2021-21_en.pdf.

[12] European Commission. Medical Device Coordination Group (website). Guidance—MDCG endorsed documents and other guidance Accessed on June 23, 2024 at health.ec.europa.eu/medical-devices-sector/new-regulations/guidance-mdcg-endorsed-documents-and-other-guidance_en.

Chapter 1

Introduction

In order to explore clinical/performance data evaluations in detail, this chapter describes clinical evaluation/ performance evaluation (CE/PE) origins, regulatory history and requirements; general safety and performance requirements; basic clinical evaluation report/performance evaluation report (CER/PER) definitions, value propositions and guidelines; notified body involvement; linkages between CE/PE, clinical trial, postmarket surveillance and risk management systems as well as CE/PE controls and change management. This chapter also describes a novel suggested system to classify different CER/PER types across the device lifetime, including appropriate documentation when no clinical data are deemed appropriate to collect in a CER/PER. The document cycle and five CE/PE output documents are defined, and some hints are provided about how to write CE/PE documents.

Consumers expect governments:

1. to ensure medical products are safe for use as intended; and
2. to act in situations when risks exceed benefits to users.

Global regulations often require medical products, including drugs, devices, and novel foods, to be supported by clinical data documenting product safety and performance (S&P). Unfortunately, these supporting data are not always provided.

Historically, in many areas around the world, pharmaceutical development required both nonclinical and clinical data evaluations, while device development required only nonclinical technical and engineering data with little to no human clinical data. For decades, medical device companies suggested that no clinical data were needed due to the "low-risk" nature of these devices. Over time, however, most regulatory authorities developed device regulations distinguishing between low- and high-risk medical devices, and required clinical and performance data evaluations to document device S&P especially for high-risk devices.

One regulatory example contributing to this perception of low- versus high-risk devices needing different amounts of clinical data was in the United States. The US Food and Drug Administration (FDA) regulations required premarket clinical data reviews to ensure significant risk (SR) devices were safe and performing as intended; however, nonsignificant risk (NSR) devices rarely needed clinical/performance data reviews by the US FDA before they were allowed onto the US market.

Currently, regulations about device clinical and performance data evaluations vary by location, device risk level, and time on market (e.g., premarket or postmarket). In addition, government regulations change over time. For example, Europe has a risk continuum as the regulatory framework rather than a binary SR versus NSR device decision, as in the United States. In addition, the European Union (EU) Medical Devices Regulation (EU MDR Reg. 2017/745) and *in vitro* diagnostic (IVD) regulation (EU IVDR Reg. 2017/746) were changed in 2017 to require clinical and performance data evaluations for all medical devices across all three clinical data sources: clinical trials, clinical literature, and clinical experiences. The regulations required medical device and IVD device evaluation results to be documented in clinical evaluation reports (CERs) and performance evaluation reports (PERs), respectively.

BOX 1.1 Clinical evaluation and performance evaluation details in one textbook

This CER/PER textbook takes a comprehensive approach by including both medical device regulation (MDR) and *in vitro* diagnostic regulation (IVDR) details wherever possible. To help emphasize the similarity between these two distinct regulations, this book abbreviates long titles for the two different processes and document types by combining terms into contractions. For example: CE/PE for clinical evaluation and performance evaluation and CER/PER for clinical evaluation report/performance evaluation report. The CE/PE process and CER/PER differences are not emphasized; however, these differences are described as it is important to study them carefully when they pertain to a particular device. Explaining and harmonizing regulations across all device types is challenging, and this book proposes that regulatory harmonization should make regulations more logical and manageable.

Planning, Writing and Reviewing Medical Device Clinical and Performance Evaluation Reports (CERs/PERs). DOI: https://doi.org/10.1016/B978-0-443-22063-0.00014-6

After this introduction, the subsequent chapters discuss:

- planning clinical/performance evaluations;
- identifying clinical/performance data;
- appraising clinical/performance data;
- analyzing clinical/performance data;
- establishing clinical benefit–risk ratios;
- writing clinical/performance evaluation documents;
- writing summaries of safety and clinical performance (SSCPs)/summaries of safety and performance (SSPs);
- reviewing CE/PE documents;
- integrating CE/PE, PMS, and RM systems;
- understanding CER/PER regulations outside of Europe; and
- forecasting CER/PER future directions.

After reading each chapter, the reader should reflect and consolidate their learning while remembering the overarching textbook goal, which is to enable the reader to evaluate completely and accurately product-related S&P clinical data from the patient/user perspective.

1.1 The start of clinical or performance evaluations

A good way to learn about CERs and PERs is to explore global medical device regulatory history (**Appendix A: Example Regulatory Evolution Timeline**). For example, CERs and PERs have a long history in Europe and the need for CE/PE is fundamental to regulatory developments worldwide, not only in the EU. Global medical device regulatory goals are intended to ensure medical devices are safe for human use and benefits outweigh risks to the patient when using the medical device.

An interesting book entitled *Designing Clinical Evaluation Tools: The State-of-the-art* [1] traced clinical and performance evaluations back to 1873 and the origins of US nursing schools. The authors found "little literature... assessing the application of knowledge in a hands-on clinical situation." Nursing clinical efficiency reports started in about 1921, during an "awakening to the concept of evaluation..." After more than a century of using increasingly complex and sophisticated methods, educators still struggled with subjectivity when teaching how to evaluate clinical/performance data for those practicing clinical/performance evaluations daily in real-world settings.

Another book, *Clinical Evaluation of Medical Devices: Principles and Case Studies* [2], emphasized CE complexity over several hundred years and suggested using similar standards for medical device and drug trials while recognizing differences. For example, the 1976 US Medical Device Amendments compared devices to drugs and suggested that medical devices:

- were sometimes implanted;
- were more often used by healthcare professionals (HCPs);
- had more frequent incremental innovations;
- were more often developed by small companies with limited development, testing, and sales; and
- often used existing clinical data (rather than new clinical trial data) to evaluate medical device safety and effectiveness.

The authors identified international medical device use databases and follow-up studies published in the literature as clinical data sources; they also encouraged more rigorous medical device clinical data evaluations by adding well-designed, specific clinical trials when needed.

A third book, *Understanding Clinical Data Analysis: Learning Statistical Principles from Published Clinical Research* [3], provided advice about managing statistical software programs and why various statistical tests were used when evaluating clinical/performance data. Evaluating trial, literature, and experience data requires a critical "statistical quality" review. Individual experience reports have no statistical value. In addition, poor statistical methods often contribute to research design failures and insufficiently supported data analyses. The evaluator must be able to identify and explain these statistical quality concerns during the CE/PE appraisal and analysis steps so that the conclusions about the clinical data are scientifically sound. Often, authors of published literature are not statisticians and they may not understand the need for a good statistical design to ensure analytic quality and study validity. The evaluator should ensure the clinical data included in the CE/PE are appropriate and have been analyzed appropriately and accurately using robust and valid statistical analysis plans.

Current regulations in most areas require both well-designed statistical analysis plans and a rigorous CE/PE process (i.e., special effort was undertaken to ensure accuracy). Of interest, EU laws and regulations acknowledge individual country (i.e., Member State) sovereignty to govern drug and device regulations and marketing. The Competent Authority (an organization with the legal authority to perform a particular function—usually monitoring compliance with national statutes and regulations) for each EU Member State designates regulatory enforcement by a third-party NB to ensure manufacturers meet minimum S&P requirements for risk-based classes of medical products. Each EU Member State notifies the European Commission (EC) about each NB designation and is responsible for additional NB requirements as well as enforcement actions against underperforming NBs. The EC created the Medical Device Coordination Group (MDCG) to coordinate NB functions including NB uses of clinical evaluation assessment reports (CEARs) and performance evaluation assessment reports (PEARs). The EC also created standing expert panels to review NB assessments of novel, implantable, high-risk device types without common specifications using clinical evaluation consultation procedures (CECPs) for CERs and performance evaluation consultation procedures (PECPs) for PERs [4]. EU manufacturers must apply to the NB for a new CE mark process every 1–5 years, depending on the device risk classification.

Many manufacturers are not used to this diversity of EU opinions and multiparty, interactional complexity when making regulatory decisions. For comparison, the US FDA requires manufacturers to apply for and to submit to a federal assessment process prior to device approval or clearance only for medium- to high-risk devices. The absence of a federal-level process in the EU can be confusing for US manufacturers when trying to learn the EU regulatory process, and vice versa. In addition, US manufacturers are only required to update the US FDA about device changes occurring after initial device approval or clearance (i.e., not on a regular update schedule, as in the EU). The US FDA also creates expert panels (advisory committees) (21CFR814.44), but these are typically ad hoc and created only as needed (e.g., an expert panel or public meeting to discuss a "first-of-a-kind," high-risk device). The FDA does not require a CER/PER as defined in the EU regulations; however, it does require a report of prior clinical investigations for any and all Investigational Device Exemption (IDE) submissions as follows:

> *The report of prior investigations shall include reports of all prior clinical, animal, and laboratory testing of the device and shall be comprehensive and adequate to justify the proposed investigation.* (21CFR812.27)

Some device manufacturers use the CER/PER to provide this "report of priors" justification in their IDE submissions to the US FDA. Unfortunately, the EU and US regulations are not harmonized, so manufacturers must beware and must pay careful attention to the constantly changing, area-specific regulations surrounding critical clinical evidence evaluations.

Many regulatory authorities embrace international standards to ensure high-quality clinical/performance testing and to reduce regulatory review time. For example, the US FDA encourages using standards from groups like the American National Standards Institute/Association for the Advancement of Medical Instrumentation (ANSI/AAMI), National Fire Protection Association (NFPA), or other nongovernmental organizations (NGOs) to facilitate regulatory oversight during specific data review steps. Internationally, groups like the Global Harmonization Task Force (GHTF), which became the International Medical Device Regulators Forum (IMDRF), and the International Organization for Standardization (ISO) among others are driving global regulatory harmonization.

To put this area-specific regulatory evolution in historical context, the first National Standards Body was formed in 1901 as the British Standards Institute (BSI), which standardized many different products such as steel sections for tramways, furniture, and toys [5]. What is now the EU was formed in the 1950s, and the first EC "standardized" drug approval regulation was passed in 1965 (65/65/EEC). The EU single market was completed in 1993, the European Medicines Agency (EMA) was formed in 1995, and the Clinical Trials Directive 2001/20/EC (for drugs) was repealed and replaced by EU Reg. 536/2014 in 2014 [6], providing four routes for drug approval in Europe: (1) centralized (controlled by EMA with a single license valid in all Member States; mandatory for certain drug classes); (2) single Member State; (3) mutual recognition via EMA after approval in a single state; or (4) decentralized (simultaneous application in multiple Member States) [7]. The EC did not initially define or require special laws for medical devices separate from drugs; however, from 1990 to 2017, 27 EU Member States and many NBs increased their involvement in medical devices, especially in terms of CER/PER oversight and review.

Like the EU, most countries had no medical device regulations until recently. For example, China created medical device regulations in 2000 and the US FDA started regulating devices under the 1976 US Medical Device Amendments to the 1938 Food Drug and Cosmetic Act. Although the EU created multiple MDRs in the 1990s, the updates in 2017 focused effort and accountability on clinical data evaluations specifically.

1.2 European Union clinical and performance evaluation regulatory requirements

Prior to 2017, EU MDRs were detailed in three separate documents: Active Implantable Medical Device Directive (AIMDD) (90/385/EEC) adopted in 1990 [8], Medical Device Directive (MDD) (93/42/EEC) in 1993 [9], and In Vitro Diagnostic Medical Device Directive (IVDD) (98/79/EC) in 1998 [10]. All three EU Council Directives required clinical data prior to CE marking. Unfortunately, poor clinical data review and enforcement, by the responsible third-party NBs, was common. This poor NB performance continued even after the clinical data requirements were strengthened in the 2007 MDD and AIMDD amendments (2007/47/EC) to emphasize CE, conformity assessment and requirements for separate RM, PMS, and CE systems. In particular, all medical devices were required to have a CE as follows:

> *Demonstration of conformity with the essential requirements must include a clinical evaluation in accordance with Annex X* (2007/47/EC amendment: Annex I, 6a [11])

The IVDD also remained in force, with amendments through 2012 requiring:

> *the clinical condition or the safety of the patients... users... or... other persons and adequate performance evaluation data... should originate from studies in a clinical... environment...* (IVDD Essential Requirement 1 and Annex III, respectively) [10]

The AIMDD and MDD were repealed in 2017 and replaced by the merged MDR (EU Reg. 2017/745) [12]; however, the EU did not repeal Council Directive 98/79/EC (IVDD) [10] even though IVDR (EU Reg. 2017/746) [13] stated that "Directive 98/79/EC should be repealed..." More work will be needed in the future if the EU decides to merge the MDR and IVDR into a more cohesive regulatory framework.

This regulatory history is important because the MDR CER and IVDR PER requirements are similar but not identical. For example, the EU regulations consistently and universally require clinical evidence, including clinical investigations for both medical devices and IVD devices, and the clinical investigations are required to follow international standards like ISO 14155 [14]. Uniquely, the IVDR clinical evidence must specifically address IVD device scientific validity, analytical performance, and clinical performance. Essentially, the IVD device analytic sensitivity and specificity differs from the more generic specifications used for other medical devices, and the IVDR uses distinct terminology (e.g., PE and performance studies in the IVDR versus CE and clinical studies in the MDR). In addition, IVDR Article 64 states that IVD device clinical benefits are "fundamentally different" from drugs and other medical devices since the IVD device provides "accurate medical information on patients" but may not directly change the patient benefit or "final clinical outcome." This last point is perhaps most important, since the measures of success are different (i.e., how well the IVD device "diagnosed" versus how well the medical device "affected" the patient outcome).

Note the different "clinical benefit" definitions in IVDR and MDR:

> *... the positive impact of a device related to its function... screening, monitoring, diagnosis or aid to diagnosis of patients, or a positive impact on patient management or public health* (IVDR Article 2 (37))

> *... the positive impact of a device on the health of an individual, expressed in terms of a meaningful, measurable, patient-relevant clinical outcome/s, including outcome/s related to diagnosis, or a positive impact on patient management or public health* (MDR Article 2 (53))

Be careful to review specific laws and regulations applicable to the specific CER/PER devices. In addition, remember CE/PE systems require safety reporting, clinical trial, RM, and PMS/postmarket performance follow-up plan (PMPF) systems to be fully functional, aligned, and integrated with the CE/PE system processes.

BOX 1.2 CER/PER purpose

The CER/PER purpose is to evaluate all clinical data in a single, stand-alone document to determine if the device is safe and performs as intended and if the benefits outweigh all clinical risks to the patient when the device is used as intended in the indicated population. The clinical benefit–risk ratio evaluation should be a main CER/PER focus, and the CER/PER must have sufficient clinical data to meet these regulatory requirements.

CE/PE-specific regulatory changes are becoming more frequent and more harmonized throughout the world. Robust medical and *in vitro* diagnostic device clinical/performance data evaluations are now required in most areas (e.g., the EU, the United Kingdom, the United States, Canada, Japan, Australia, and China).

1.3 Relevant general safety and performance requirements

All device manufacturers with a product on the EU market must ensure their CE/PE processes accommodate the relevant GSPRs in the MDR and IVDR. For MDR 2017/745 Annex I, GSPRs 1, 6, and 8 require clinical data from patients, users, and other persons to document S&P data, benefit−risk ratios, etc. as follows:

> *GSPR 1: Performance and safety—"Devices shall achieve the performance intended by their manufacturer and shall be designed and manufactured in such a way that, during normal conditions of use, they are suitable for their intended purpose. They shall be safe and effective and shall not compromise the clinical condition or the safety of* ***patients****, or the safety and health of* ***users*** *or, where applicable,* ***other persons****, provided that any risks which may be associated with their use constitute acceptable risks when weighed against the benefits to the* ***patient*** *and are compatible with a high level of protection of health and safety, taking into account the generally acknowledged state-of-the-art."* (Emphasis added to highlight need for human clinical data.)

> *GSPR 6: Lifetime of device—"The characteristics and performance of a device shall not be adversely affected to such a degree that the health or safety of the* ***patient*** *or the* ***user*** *and, where applicable, of* ***other persons*** *are compromised during the lifetime of the device, as indicated by the manufacturer, when the device is subjected to the stresses which can occur during normal conditions of use and has been properly maintained in accordance with the manufacturer's instructions."* (Emphasis added to highlight need for human clinical data.)

> *GSPR 8: Benefit−risk ratio—"All known and foreseeable risks, and any undesirable side-effects, shall be minimized and be acceptable when weighed against the evaluated benefits to the* ***patient*** *and/or* ***user*** *arising from the achieved performance of the device during normal conditions of use."* (Emphasis added to highlight need for human clinical data.)

IVDR 2017/746 had identical GSPRs 1 and 6; however, GSPR 8 varied only slightly as follows:

> *GSPR 8: Benefit−risk ratio—"All known and foreseeable risks, and any undesirable* ***effects*** *shall be minimised and be acceptable when weighed against the evaluated* ***potential*** *benefits to the patients and/or the user arising from the* ***intended*** *performance of the device during normal conditions of use."* (Emphasis added to highlight differences compared to MDR GSPR 8.)

The IVDR versus MDR GSPR 8 differences were: (1) "undesirable" effects (not side effects); (2) "potential" benefits (not benefits); and (3) "intended" performance (not "achieved" performance). These minor differences could be construed as inconsequential. For example, simple edits would align GSPR 8: (1) change "side effects" and "effects" to "findings"; (2) use "potential benefits" in both; and (3) remove the adjectives "intended" or "achieved" from "performance." This GSPR alignment between the MDR and IVDR would be helpful since human safety data are already uniformly required and these medical device requirements do not vary for "nonactive," "active," or "*in vitro* diagnostic" medical devices any more.

BOX 1.3 MDR and IVDR GSPRs

Annex I in EU MDR includes 23 GSPRs, IVDR has 20 GSPRs, and the evaluator should fully understand what is and what is *not* relevant to the CER/PER. Specifically, the CER/PER must demonstrate conformity with GSPRs 1, 6, and 8 to ensure the device performs as intended, is safe over the device lifetime, and the clinical benefits outweigh the clinical risks (including residual risks to the patient/user/others when used as indicated in the labeling). The other GSPRs do not generally require human clinical data and would therefore be covered in other technical documentation *outside* the CER/PER. These other documents are defined in MDR/IVDR Annex II as the Technical Documentation required for CE mark (e.g., within design drawings, RM reports, device design verification and validation details, bench and animal test reports). In addition, any unanswered clinical CER/PER questions or clinical data gaps (especially any gaps in data supporting device S&P or preventing an assessment of the benefit−risk ratio) must be addressed in the postmarket clinical follow-up plan/postmarket performance follow-up plan (PMCFP/PMPFP) and the resulting evaluation in the PMCF/PMPF evaluation report (PMCFER/PMPFER).

Prior to 2021, medical device CER/PER requirements were specified under MDD Essential Requirements (ERs) 1, 3, and 6 in Council Directive 93/42/EEC of June 14, 1993, AIMDD ERs 1, 2, and 5 in Council Directive 90/385/EEC of June 20, 1990, and IVDD ERs 1, 3, and 4 in Council Directive 98/79/EC of October 27, 1998 (Table 1.1).

TABLE 1.1 Past MDD, AIMDD, and IVDD ERs related to human clinical data.

MDD (93/42/EEC)	AIMDD (90/385/EEC)	IVDD (98/79/EC)
ER 1: Performance and safety: *"The devices must be designed and manufactured in such a way that, when used under the conditions and for the purposes intended, they will not compromise the clinical condition or the safety of patients, or the safety and health of users or, where applicable, other persons, provided that any risks which may be associated with their intended use constitute acceptable risks when weighed against the benefits to the patient and are compatible with a high level of protection of health and safety. This shall include:—reducing, as far as possible, the risk of use error due to the ergonomic features of the device and the environment in which the device is intended to be used (design for patient safety), and—consideration of the technical knowledge, experience, education and training and where applicable the medical and physical conditions of intended users (design for lay, professional, disabled or other users)."*	ER 1: Performance and safety: *The devices must be designed and manufactured in such a way that, when implanted under the conditions and for the purposes laid down, their use does not compromise the clinical condition or the safety of patients. They must not present any risk to the persons implanting them or, where applicable, to other persons.*	ER 1: Performance and safety: *"The devices must be designed and manufactured in such a way that, when used under the conditions and for the purposes intended, they will not compromise, directly or indirectly, the clinical condition or the safety of the patients, the safety or health of users or, where applicable, other persons, or the safety of property. Any risks which may be associated with their use must be acceptable when weighed against the benefits to the patient and be compatible with a high level of protection of health and safety."*
ER 3: Performance: *"The devices must achieve the performances intended by the manufacturer and be designed, manufactured and packaged in such a way that they are suitable for one or more of the functions referred to in Article 1 (2) (a), as specified by the manufacturer."*	ER2: Performance: *"The devices must achieve the performances intended by the manufacturer, viz. be designed and manufactured in such a way that they are suitable for one or more of the functions referred to in Article 1 (2) (a) as specified by him."*	ER 3: Performance: *The devices must be designed and manufactured in such a way that they are suitable for the purposes referred to in Article 1(2)(b), as specified by the manufacturer, taking account of the generally acknowledged state-of-the-art. They must achieve the performances, in particular, where appropriate, in terms of analytical sensitivity, diagnostic sensitivity, analytical specificity, diagnostic specificity, accuracy, repeatability, reproducibility, including control of known relevant interference, and limits of detection, stated by the manufacturer. The traceability of values assigned to calibrators and/or control materials must be assured through available reference measurement procedures and/or available reference materials of a higher order.*
ER 6: Risk-performance ratio: *"Any undesirable side-effect must constitute an acceptable risk when weighed against the performances intended."*	ER 5: Risk-performance ratio: *"Any side effects or undesirable conditions must constitute acceptable risks when weighed against the performances intended."*	ER 4: Risk-performance ratio: *"The characteristics and performances referred to in sections 1 and 3 must not be adversely affected to such a degree that the health or the safety of the patient or the user and, where applicable, of other persons, are compromised during the lifetime of the device as indicated by the manufacturer, when the device is subjected to the stresses which can occur during normal conditions of use. When no lifetime is stated, the same applies for the lifetime reasonably to be expected of a device of that kind, having regard to the intended purpose and the anticipated use of the device."*

AIMDD, active implantable medical devices directive; *ER*, essential requirement; *IVDD*, *in vitro* diagnostic medical devices directive; *MDD*, medical device directive. Similarities are bolded.

The MDR and IVDR GSPRs 1, 6, and 8 were more harmonized and more focused on patient/user safety over device lifetime (i.e., acceptable device S&P and benefit–risk ratios) than simply the device performing as intended as required in 93/42/EEC, 90/385/EEC, and 98/79/EC ERs. NBs insisted that manufacturers develop "acceptance criteria" for prespecified human clinical S&P measures and benefit–risk ratios in the clinical evaluation plan/performance evaluation plan (CEP/PEP), and evaluators were required to evaluate the S&P measures and benefit–risk ratios against the "acceptance criteria." Clinical risks identified in the CER/PER should be minimized through well-defined and separate RM and PMS system processes.

During this transition, the available NB number fell to zero, then increased slowly as NBs became MDR and IVDR certified. The initial MDR transition period closed in May 2020 after the initial 3-year grace period; however, a 1-year extension was added due to the coronavirus disease (COVID-19) pandemic (EU Reg. 2020/561) [15], and another extension was granted in 2023 "as a matter of urgency... related to insufficient capacity of notified bodies to certify medical devices in accordance with the MDR within the remaining transition period that ends on 26 May 2024 and the [insufficient] level of preparedness of manufacturers" [16]. More deadline extensions to 2026 (end of derogation for class III custom-made implantable devices), 2027 (end of transition for class III and class IIb implantable devices) and 2028 (end of transition period for other class IIb, IIb, I, sterile/measuring and new devices) are in place for manufacturers meeting specific regulatory preparation requirements (e.g., MDR application on file, MDR QMS in place, written surveillance agreement with MDR NB, continuous compliance with applicable MDD/AIMDD, no significant changes in design or intended purpose, acceptable risk).

1.4 Defining a clinical evaluation report/performance evaluation report

The CER/PER evaluates clinical data to ensure the "subject device" performed safely when used as indicated with benefits outweighing risks to the patient.

BOX 1.4 "Subject" device and "clinical data" terminology

This textbook uses the term "subject device" to refer to the specific device under evaluation as made by the manufacturer. This detail is important because many different "clinical data" types are evaluated in the CER/PER, and clinical data from the subject device are most valuable. The term "clinical data" refers strictly to data derived from humans (i.e., not bench data, not animal data, but human "clinical data" specifically). This detail is important because many different data sources will be evaluated in the CER/PER, and human clinical data are most valuable.

The CER/PER evaluator must examine the nonclinical bench and animal testing to ensure these required data were acceptable before human clinical testing began. The nonclinical data typically reside outside the CER/PER in technical reports within the technical file or design dossier. Nonclinical data are evaluated by technical evaluators outside the CER/PER whereas human clinical data are evaluated by clinical evaluators in the CER/PER. Nonclinical data are generally only included in the CER/PER when the evaluator has a particular need to explain a human clinical data point or a specific design-related human benefit or risk. Nonclinical data may also be referenced in the CER/PER to address specific safety concerns in compliance with international standards.

The CER/PER also includes clinical data from other devices including a wide variety of "similar" devices made by the manufacturer and competitors. The CER/PER requires a discussion of background knowledge, alternative therapies, and the state of the art (SOTA). As a result, the CER/PER also includes clinical data from benchmark devices, which help to put subject device data in context.

The manufacturer must determine if clinical data from an "equivalent" device will be included. Equivalence can *only* be claimed if the two devices are documented as clinically, biologically, and technically equivalent as defined in the MDR/IVDR and the guidance document entitled "MDCG 2020-5 Clinical evaluation—Equivalence. A guide for manufacturers and notified bodies" [17]. The equivalent device must be CE marked and similar to the subject device.

As stated in the MDR Annex XIV, Section 4:

The results of the clinical evaluation and the clinical evidence... shall be documented in a clinical evaluation report which shall support the assessment of the conformity of the device.

Similarly, as stated in IVDR Annex XIII, Section 1:

Clinical performance shall be demonstrated and documented in the clinical performance report... manufacturer shall assess all relevant scientific validity, analytical and clinical performance data to verify the conformity of its device with the general safety and performance requirements...

In other words, the CER/PER is the main document required from medical device companies to demonstrate compliance/conformity with MDR/IVDR clinical data requirements. The CER/PER is also required to scientifically substantiate all clinical claims made in product labeling, including all marketing materials, websites, brochures, ads, etc.

As stated in MDR Annex XIV, Section 2:

The clinical evaluation shall be thorough and objective, and take into account both favourable and unfavourable data. Its depth and extent shall be proportionate and appropriate to the nature, classification, intended purpose and risks of the device in question, as well as to the manufacturer's claims in respect of the device.

Similarly, as stated in IVDR Annex XIII, Section 1:

The performance evaluation shall be thorough and objective, considering both favourable and unfavourable data. Its depth and extent shall be proportionate and appropriate to the characteristics of the device including the risks, risk class, performance and its intended purpose... The performance evaluation report shall in particular include... any claims made about the device's safety and performance.

A CER/PER is required for each medical device (or appropriate medical device group) to be marketed in the EU. It is an important document used to meet many specific regulatory requirements including, but not limited to, the following:

1. To document clinical/performance data from clinical trials, clinical literature, and clinical experiences about device S&P, as well as benefits and risks when devices were used as intended.
2. To evaluate clinical/performance data fairly and without bias in order to determine if S&P were acceptable and if the benefits outweighed the risks for devices used as indicated in appropriate patient populations.
3. To require more clinical/performance data whenever needed to provide sufficient scientific substantiation for all clinical claims regarding medical device uses.
4. To remove products from the market if the S&P were not acceptable or if the benefits did not sufficiently outweigh the risks to the patient.

For example, a CER/PER may conclude sufficient clinical/performance data were evaluated and no further clinical/performance data were required for CE/PE; however, the CER/PER may also conclude more clinical/performance data are needed and should be collected to address specific clinical/performance data gaps found during CE/PE. Sometimes these clinical/performance data gaps may be so overwhelming and the CER/PER may conclude the product should not remain on the market. In these cases, the product would be removed from the European market and, ethically, all other markets as well, while the device is redesigned, relabeled, and retested to ensure risks are reduced and/or benefits increased to an acceptable benefit–risk ratio. At other times, the risks are less serious and additional clinical/performance data should be gathered in a PMCF/PMPF study as the manufacturer contemplates changes to address the risks.

The CER/PER must document specific clinical data collection requirements for the PMCFP/PMPFP whenever CER/PER data were insufficient or unclear about clinical/performance benefits outweighing the clinical risks for medical device S&P.

1.5 Comparing clinical evaluation reports and performance evaluation reports

The MDR and IVDR have many similarities and a few important differences for CERs and PERs (Table 1.2).

TABLE 1.2 High-level clinical and performance evaluation report similarities and differences.

Item	CER	PER
Regulation	EU MDR Article 61, Annex XIV	EU IVDR Article 56, Annex XIII
Relevant GSPRs	1, 6, and 8	1, 6, and 8
Device definition	"'medical device' means any instrument, apparatus, appliance, software, implant, reagent, material, or other article intended by the manufacturer to be used, alone or in combination, for human beings for one or more of the following specific medical purposes: – diagnosis, prevention, monitoring, prediction, prognosis, treatment or alleviation of disease; – diagnosis, monitoring, treatment, alleviation of, or compensation for, an injury or disability; – investigation, replacement or modification of the anatomy or of a physiological or pathological process or state; or – providing information by means of *in vitro* examination of specimens derived from the human body, including organ, blood, and tissue donations, and which does not achieve its principal intended action by pharmacological, immunological or metabolic means, in or on the human body, but which may be assisted in its function by such means. The following products shall also be deemed to be medical devices: – devices for the control or support of conception; – products specifically intended for the cleaning, disinfection or sterilisation of devices..." [12]	"'*in vitro* diagnostic medical device' means any medical device which is a reagent, reagent product, calibrator, control material, kit, instrument, apparatus, piece of equipment, software, or system, whether used alone or in combination, intended by the manufacturer to be used *in vitro* for the examination of specimens, including blood and tissue donations, derived from the human body, solely or principally for the purpose of providing information on one or more of the following: (a) concerning a physiological or pathological process or state; (b) concerning congenital physical or mental impairments; (c) concerning the predisposition to a medical condition or a disease; (d) to determine the safety and compatibility with potential recipients; (e) to predict treatment response or reactions; (f) to define or monitor therapeutic measures." [13] Specimen receptacles shall also be deemed to be *in vitro* diagnostic medical devices.
Guidance documents *(*four are in common; however, in the absence of IVDR guidance, meaning* is often *inferred from MDR guidance)*	MEDDEV 2.7/1, Rev. 4 (2016, *mostly outdated*) MDCG 2019-3, Rev. 1 (expert panel) MDCG 2019-9, Rev. 1 (SSCP template) *MDCG 2019-11 (software) MDCG 2019-15, Rev. 1 (Class I) *MDCG 2019-16 (cybersecurity) *MDCG 2020-1 (CE/PE software) MDCG 2020-5 (equivalence) MDCG 2020-6 (sufficient evidence, legacy) MDCG 2020-7 (PMCFP template) MDCG 2020-8 (PMCFER template) MDCG 2020-9 (ventilators) MDCG 2020-12 (medicinal product) MDCG 2020-13 (CEAR template for NBs) MDCG 2021-1, Rev. 1 (EUDAMED delay) MDCG 2021-3 (custom-made devices) MDCG 2021-6 (clinical trials Q&A) MDCG 2021-24 (device classification) MDCG 2021-25 (old/legacy devices) *MDCG 2023-1 (health inst. exemption)	*MDCG 2019-11 (software) *MDCG 2019-16 (cybersecurity) *MDCG 2020-1 (CE/PE software) MDCG 2020-16, Rev. 2 (IVD class. rules) MDCG 2021-2 (COVID-19 tests) MDCG 2021-4 (Class D certification) MDCG 2021-21 (SARS CoV-2) MDCG 2021-22, Rev. 1 (expert panel) MDCG 2022-1 (third party SARS CoV2) MDCG 2022-2 (IVD clinical evidence) MDCG 2022-8 (IVD old/legacy devices) MDCG 2022-9 (SSP template) MDCG 2022-10 (IVD clinical trials) MDCG 2022-15 (IVDR Surveillance) *MDCG 2023-1 (health inst. exemption)

(*Continued*)

TABLE 1.2 (Continued)

Item	CER	PER
Clinical/performance data sources	Clinical trials/investigations Clinical literature Clinical experiences	Clinical trials/investigations Clinical literature Clinical experiences
Intended purpose requirements	Indications for use Contextual information about indication Device function (e.g., to diagnose, prevent, monitor, treat, mitigate disease) Patient target group/s Intended users Disease treated Contraindications Associated medicinal products	Indications for use Contextual information about indication IVD function (e.g., examine specimens and inform about physiology/pathology, prediction, safety) Testing population Intended users Analyte detected/measured (disease) Contraindications Specimen type Automated or not Qualitative, semiquantitative, quantitative *International nonproprietary name of associated medicinal products (companion diagnostics only)*
Plan	CEP – Developmental phases – Device characteristics – Relevant GSPRs – Intended purpose/use – Target patient group (indication, limitations, contraindications) – Means to determine benefit–risk ratio acceptability based on SOTA, CS – Clinical development plan – PMCF planning – Intended clinical benefits with specified clinical outcome parameters – Clinical safety evaluation methods including methods to identify risks and side effects – Benefit-risk issues related to pharmaceuticals, nonviable animal or human tissues	PEP – Developmental phases – Device characteristics – Relevant GSPRs – Intended purpose/use – Target patient group (indication, limitations, contraindications, test needs) – Means to determine benefit–risk ratio acceptability based on SOTA, CS – Performance development plan – PMPF planning – Scientific validity, analytical, and clinical performance methods – Analyte or marker device determinates/measures – Reference materials/measurement procedures
Evaluation basis	Clinical S&P data Benefit–risk ratio Available alternative treatment options	Scientific validity Analytical performance Clinical performance Clinical impact on patient safety
Equivalence	MDR Annex XIV, Part A, 3 Clinical, Biological and Technical Equivalence *MDCG 2020-5 (equivalence)*	IVDR Annex IX, Chapter I, 2.2(c) Scientific Validity, Analytical Performance and Clinical Performance *MDCG 2022-2 (IVD clinical evidence)*
Report	CER – Justification for clinical evidence gathering approach – Identify and appraise all relevant clinical data based on suitability – Literature search methodology, protocol, and report – Analyze all relevant clinical data to reach conclusions about device safety, performance, and clinical benefits	PER – Justification for clinical evidence gathering approach – Identify and appraise all relevant clinical data based on suitability – Literature search methodology, protocol, and report – Analyze all relevant clinical data to reach conclusions about device safety, performance, and clinical benefits

(*Continued*)

TABLE 1.2 (Continued)

Item	CER	PER
	– Technological basis, intended purpose, claims about performance or safety – Nature and extent of evaluated data – Clinical evidence of acceptable performance against SOTA – Any new PMCF conclusions benefits – Identify relevant clinical data gaps – Expert panel views (if consulted)	– Technological basis, intended purpose, claims about performance or safety – Nature and extent of evaluated data – Clinical evidence of acceptable performance against SOTA – Any new PMPF conclusions benefits – Identify relevant clinical data gaps – Expert panel views (if consulted)
Expert panel	MDR Articles 55, 61(2) and 106 Annexes IX(5.1) and X(6) • Class III/IIb to administer/remove drug • MDCG/European Commission may seek advice from expert panel • Manufacturer may consult expert panel before CE/investigation, document in CER • NB transmits CEAR and CER to EC and EC transmits documents to expert panel	IVDR Article 48 (6) • Class D • If first certification for device type and no CS, NB must submit PER to expert panel within 5 days of receipt from manufacturer
Public summary	SSCP Class III and implantable medical devices MDR Article 32 *MDCG 2019-9, Rev. 1 (SSCP template)*	SSP Class C and Class D IVD devices IVDR Article 29 *MDCG 2022-9 (SSP template)*
Postmarket clinical follow-up	PMCFP/PMCFER MDR Article 10, 74(1), Annex XIV, Part B *MDCG 2020-7 (PMCFP template)* *MDCG 2020-8 (PMCFER template)*	PMPFP/PMPFER IVDR Article 10, 70(1), Annex XIII, Part B
Postmarket surveillance	PMS, PMSP, PMSR MDR, Article 83, 84, 85 Class I devices updated as necessary, available on request	PMS, PMSP, PMSR IVDR Article 78, 79, 80 Class A and B devices, updated as necessary, available on request *MDCG 2022-15 (IVDR Surveillance)*
PSUR	MDR, Article 86 Class IIa: at least every 2 years Class IIb, III: at least annually	IVDR Article 81 Class C, D devices, at least annually
CE certificate term	5-year maximum 1 year for significant risk, implantable or not well-established device	5-year maximum 1 year for significant risk or not well-established device
Justification (JUS)	MDR, Article 61: "The manufacturer shall specify and justify the level of clinical evidence necessary to demonstrate conformity with the relevant general safety and performance requirements. That level of clinical evidence shall be appropriate in view of the characteristics of the device and its intended purpose.... where the demonstration of conformity with general safety and performance requirements based on clinical data is not deemed appropriate, adequate justification for any such exception shall be given based on	IVDR Article 56: "The manufacturer shall specify and justify the level of the clinical evidence necessary to demonstrate conformity with the relevant general safety and performance requirements. That level of clinical evidence shall be appropriate in view of the characteristics of the device and its intended purpose.... Clinical performance studies... shall be carried out unless it is duly justified to rely on other sources of clinical performance data." (65) "Where specific devices have no analytical or clinical performance or

(*Continued*)

TABLE 1.2 (Continued)

Item	CER	PER
	the results of the manufacturer's risk management and on consideration of the specifics of the interaction between the device and the human body, the clinical performance intended and the claims of the manufacturer." (Annex II, 6.1(d)) "the PMCF plan and PMCF evaluation report... or a justification why a PMCF is not applicable." (Annex III, 1.1(b)) "a PMCF... or a justification as to why a PMCF is not applicable."	specific performance requirements are not applicable, it is appropriate to justify in the performance evaluation plan, and related reports, omissions relating to such requirements." (Annex III) "a PMPF plan... or a justification as to why a PMPF is not applicable." (Annex XIII) "Where any... elements are not deemed appropriate in the Performance Evaluation Plan due to the specific device characteristics a justification shall be provided in the plan... The manufacturer shall demonstrate the analytical performance of the device... unless any omission can be justified as not applicable... The manufacturer shall demonstrate the analytical performance of the device... unless any omission can be justified as not applicable... Clinical performance studies shall be performed unless due justification is provided for relying on other sources of clinical performance data... If PMPF is not deemed appropriate for a specific device then a justification shall be provided and documented within the performance evaluation report."

CE, conformité européenne (European conformity); *CEAR*, clinical evaluation assessment report; *CEP*, clinical evaluation plan; *CER*, clinical evaluation report; *CS*, common specification; *EC*, European Commission; *EU*, European Union; *GSPR*, general safety and performance requirements; *IVD*, *in vitro* diagnostic medical device; *IVDR*, *in vitro* diagnostic medical device regulation; *JUS*, justification; *MDCG*, Medical Device Coordination Group; *MDR*, medical device regulation; *MEDDEV*, medical devices documents; *NB*, notified body; *PEP*, performance evaluation plan; *PER*, performance evaluation report; *PMCF*, postmarket clinical follow-up; *PMCFER*, postmarket clinical follow-up evaluation report; *PMCFP*, postmarket clinical follow-up plan; *PMPF*, postmarket performance follow-up; *PMPFER*, postmarket performance follow-up evaluation report; *PMPFP*, postmarket performance follow-up plan; *PMS*, postmarket surveillance; *PMSP*, postmarket surveillance plan; *PMSR*, postmarket surveillance report; *PSUR*, periodic safety update report; *S&P*, safety and performance; *SOTA*, state of the art; *SSCP*, summary of safety and clinical performance; *SSP*, summary of safety and performance. *Emphasis is added in bold and notes are added in italics.*

In addition to similarities between MDR and IVDR requirements, many regulatory process details are similar, if not identical, across ALL clinical/performance data evaluations for all product types. For example, the clinical data sources (i.e., clinical trials, literature, and experiences) are the same and are typically evaluated in similar ways whether a human is exposed to a drug, food, or medical/IVD device. Clinical safety and performance or efficacy evaluations include dissecting human adverse events (AEs) and human body reactions to product uses as intended as well as product evaluations to ensure the product is not contaminated, adulterated, out of specification, dysfunctional, or otherwise substandard.

1.6 The clinical evaluation report/performance evaluation report value proposition

When well-crafted, the CER/PER is an extremely valuable asset. The CER/PER is designed to evaluate all device clinical data known by the company. As such, the CER/PER is a marvelous teaching and marketing tool and knowledge repository.

The CER/PER includes (but is not limited to) the following details:

- device description including parts, components, and accessories;
- device background knowledge and SOTA;
- similar devices, competitor devices and alternative therapies;
- marketing claims substantiated with solid clinical data;
- developmental context with device regulatory history;
- product labeling details supported by clinical data;
- specific clinical data from trials, literature, and experiences using the device as indicated;
- S&P data including notable adverse events, and analytical and clinical performance details;
- rationale and documentation to support scientific validity for all data evaluated;
- rigorous clinical data evaluation methods and results details; and
- conclusions documenting acceptable device S&P and benefit–risk ratio.

Every medical device company team member working on the device may benefit from reading the CER/PER including everyone from the CEO to the assembly line worker. The CEO would be able to read a rather concise record about device "success" and the assembly line worker would see how the device benefits each patient while also having risks. In addition, the MDR requires the CER/PER to be summarized and made publicly available to the layperson in an SSCP for Class III and implantable medical devices and SSP for Class C and D IVD devices.

1.7 Clinical evaluation/performance evaluation guidance documents

Guidance documents have been available for decades to assist with EU CE/PE work including guidance documents aligned with the historical MDR (**Appendix B: MEDDEV Guidance List**). More recently, the MDCG was empowered to ensure MDR/IVDR implementation, and the MDCG maintains a website entitled "Guidance—MDCG endorsed documents and other guidance" [18] with appropriate guidance documents and tasks completed as required by the MDR and IVDR (Table 1.3).

TABLE 1.3 MDCG tasks for MDR and IVDR.

MDR—EU Reg. 2017/745 Article 105	IVDR—EU Reg. 2017/746 Article 99
(a) to contribute to the assessment of applicant conformity assessment bodies and notified bodies pursuant to the provisions set out in Chapter IV;	(a) to contribute to the assessment of applicant conformity assessment bodies and notified bodies pursuant to the provisions set out in Chapter IV;
(b) to advise the Commission, at its request, in matters concerning the coordination group of notified bodies as established pursuant to Article 49;	(b) to advise the Commission, at its request, in matters concerning the coordination group of notified bodies as established pursuant to Article 45;
(c) to contribute to the development of guidance aimed at ensuring effective and harmonised implementation of this Regulation, in particular regarding the designation and monitoring of notified bodies, application of the general safety and performance requirements and conduct of clinical evaluations and investigations by manufacturers, assessment by notified bodies and vigilance activities;	(c) to contribute to the development of guidance aimed at ensuring effective and harmonised implementation of this Regulation, in particular regarding the designation and monitoring of notified bodies, application of the general safety and performance requirements and conduct of performance evaluations by manufacturers, assessment by notified bodies and vigilance activities;
(d) to contribute to the continuous monitoring of technical progress and assessment of whether the general safety and performance requirements laid down in this Regulation and Regulation (EU) 2017/746 are adequate to ensure safety and performance of devices, and thereby contribute to identifying whether there is a need to amend Annex I to this Regulation;	(d) to contribute to the continuous monitoring of technical progress and assessment of whether the general safety and performance requirements laid down in this Regulation and Regulation (EU) 2017/745 are adequate to ensure safety and performance of devices, and thereby contribute to identifying whether there is a need to amend Annex I to this Regulation;
(e) to contribute to the development of device standards, of CS* and of scientific guidelines, including product specific guidelines, on clinical investigation of certain devices in particular implantable devices and class III devices;	(e) to contribute to the development of device standards and of CS;
(f) to assist the competent authorities of the Member States in their coordination activities in particular in the fields of classification and the determination of the regulatory status of devices, clinical investigations, vigilance and market surveillance including the development and maintenance of a framework for a European market surveillance programme with the objective of achieving efficiency and harmonisation of market surveillance in the Union, in accordance with Article 93;	(f) to assist the competent authorities of the Member States in their coordination activities in particular in the fields of classification and the determination of the regulatory status of devices, performance studies, vigilance and market surveillance including the development and maintenance of a framework for a European market surveillance programme with the objective of achieving efficiency and harmonisation of market surveillance in the Union, in accordance with Article 88;
(g) to provide advice, either on its own initiative or at request of the Commission, in the assessment of any issue related to the implementation of this Regulation;	(g) to provide advice, either on its own initiative or at request of the Commission, in the assessment of any issue related to the implementation of this Regulation;
(h) to contribute to harmonised administrative practice with regard to devices in the Member States.	(h) to contribute to harmonised administrative practice with regard to devices in the Member States.

CS, common specification; *EU*, European Union; *IVDR*, *in vitro* diagnostic regulation; *MDCG*, Medical Device Coordination Group; *MDR*, medical device regulation; *Reg*, regulation. Differences are bolded.

The MDCG tasks for MDR and IVDR are strikingly similar. The guidance documents on the MDCG website support the regulations, and are especially useful because the regulation change process created some internally conflicting requirements and guidance (see **Appendix C: MDCG Endorsed Guidance Documents**). Initially, during the MDR/IVDR transition, little guidance was provided (especially for IVDs needing careful attention to diagnostic sensitivity and specificity); however, both the overall document number and release rate increased rapidly with 122 guidance documents being issued from January 2018 to September 2023 (~6 years) (Fig. 1.1).

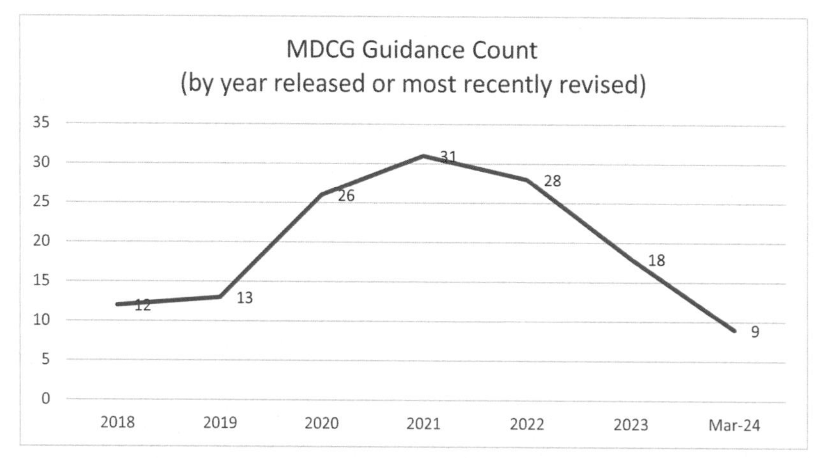

FIGURE 1.1 Annual guidance document volume January 2018-March 2024.

Guidance documents are not legally binding; however, following these wide-ranging guidance documents should help ensure CE/PE documents will be reviewed easily and quickly by the contracted NB.

> **BOX 1.5 Remember to keep a copy for later on**
>
> Tip: some guidance documents are removed from the MDCG website over time, so remember to keep a file copy, since the document may not be available to download later on.

Unfortunately, as many new documents flooded the Internet and as these documents included massive changes, manufacturers and NBs had significant challenges keeping up. The Europeans seem to be engaged in a massive operation to rewrite EU regulatory compliance one guidance document at a time, in real time. To understand the scale of these changes, the following subsections trace EU regulatory clinical evaluation guidance history back more than 25 years to 1998.

1.7.1 Notified Bodies, Medical Devices/2.7/Rec1 (June 1998) clinical investigations for clinical evaluation (CE) marking

The co-ordination of Notified Bodies, Medical Devices (NB-MED) issued a four-page "Guidance on when a clinical investigation is needed for CE marking," which stated "The purpose of clinical data is to provide clinical evidence of the compliance with essential requirements." The guideline defined clinical investigations, safety, performance, and clinical evidence, and specified the following:

> *Devices must be designed and manufactured in such a way that, when used under the conditions and for the purposes intended by the manufacturer, they will not compromise the clinical condition or the safety of patients, or the safety an [sic] health of users or, where applicable, other persons, provided that any risks which my [sic] be associated with their use constitute acceptable risks when weighed against the benefits to the patient and are compatible with a high level of protection of health and safety* (annex 1.1 MDD) [19]

This guideline requires clinical evidence "in all cases" as demonstrated by trials, literature, relevant data (i.e., experiences), or any combination of these. Clinical trials are likely for a completely new device with unknown "components, features and/or method of action"; a device with "modification" significantly affecting the clinical S&P

or device with a new indication; a device with new, unknown materials in contact with the body; or a device with materials contacting new locations in the body or for a longer time in the body with no prior clinical experience. Although documented as Rev. 1, this "old" recommendation was accepted at a January 1996 NB-MED meeting, and Rev. 2 was discussed in February, April, and June 1998 by the Medical Devices Experts Group [19].

1.7.2 MEDDEV 2.10/2, Rev. 1 (April 2001) Notified Bodies designation and monitoring

The EU assigned responsibility for conformity assessments to NB team members with training and experience to assess "design documentation and clinical evaluation data to determine that all aspects of the design are in compliance with the... regulations..." [20]. Shortly after this guideline was released, the EU issued a "Report on the functioning of the Medical Device Directive MDD 93/42/EEC" in 2002, and most device manufacturers did not have adequate clinical data to withstand a robust evaluation to answer clinical device S&P questions [21].

1.7.3 MEDDEV 2.7.1 (April 2003) evaluation of clinical data

A 19-page MEDDEV 2.7.1 guidance on clinical evaluation stated the purpose was "to provide guidance to Manufacturers on reviewing and analysing clinical data and to Notify Bodies when reviewing the manufacturers evaluation of clinical data as part of the conformity assessment procedures" relevant to AIMD/MDD regulations (p. 2) [22]. In addition, "The manufacturer must demonstrate that his intended purpose/s and claim/s made in relation to safety and performance are achieved..." This early version of MEDDEV 2.7.1 defined only two terms:

Clinical data is data which is relevant to the various aspects of the clinical safety and performance of the device. This must include data obtained from: (i) published and/or unpublished data on market experience of the device in question; or a similar device for which equivalence to the device in question can be demonstrated; or (ii) a prospective clinical investigation/s of the device concerned; or (iii) results from a clinical investigation/s or other studies reported in the scientific literature of a similar device for which equivalence to the device in question can be demonstrated.

The evaluation of clinical data is the process by which clinical data from all selected sources (literature, results of clinical investigations and other) is assessed, analysed and deemed appropriate and adequate to establish conformity of the device with the pertinent essential requirements of the Directive as they relate to safety and performance, and to demonstrate that the device performs as intended by the manufacturer. The outcome of this process is a report which includes a conclusion on the acceptability of risks and side effects when weighed against the intended benefits of the device.

Note the narrower definition of clinical data sources than in the more recent MDR/IVDR (i.e., without "unpublished data" and "other" included clinical experience data). In addition, three clinical evaluation "routes" to establish conformity with the regulations used "sufficient" data available: (1) the "literature route"; (2) the "clinical investigations route"; or (3) the "combination" route using both literature and clinical investigations, and including "market experience." Note how MDR/IVDR now envisions using clinical data sources including clinical trials and literature along with experience data. The MDR/IVDR separates the analysis into a subject/equivalent device section and a background/SOTA/alternative therapy section while using background/SOTA/alternative therapy data to set expectations (i.e., acceptance criteria) for the subject/equivalent device S&P expectations.

A risk analysis was envisioned to help specify clinical data needs and to decide if a clinical investigation would be required because the clinical data were insufficient.

The literature route required a "protocol" defining "the identification, selection, collation and review of relevant studies" based on a recognized systematic literature review process. Literature was required to be relevant to the device under consideration or a similar device shown to have clinical, biological, and technical equivalence to the subject device. The clinical data assessment (i.e., appraisal) focused on identifying literature "significance" when:

- the author has "expertise" with device and medical procedures;
- conclusions are "substantiated" by data presented;
- the article "reflects... current medical practice" and SOTA technologies;
- references are from "recognized scientific publications" and "peer reviewed journals"; and
- reported studies follow "scientific principles" for study design including "appropriate endpoints, inclusion and exclusion criteria," statistical power, treatment duration and safety analyses for "all adverse incidents, deaths, exclusions, withdrawals and subjects lost to follow-up" using "an appropriate" statistical analysis plan.

Ideal evidence was expected from controlled clinical trials, cohort/case-controlled study, etc., but not case reports, opinions, or reports lacking sufficient detail for analysis. The literature report structure required:

1. a "short" device description with intended functions, intended purpose, and application;
2. literature data analysis;
3. literature relationship to device features;
4. a description about how the data addressed device risks and showed device performance and intended purpose;
5. a hazards, risks, and safety measures analysis including PMS details;
6. a description of weighting and statistical methods used;
7. listed and cross-referenced articles;
8. a clear equivalence demonstration, if claimed;
9. conclusions including a benefit-risk assessment; and
10. the author's dated signature.

The "clinical investigations route" required submission of the protocol with Competent Authority and Ethics Committee approvals, and the signed and dated final report. The evaluator was required to confirm all clinical trials used in the clinical evaluation complied with the regulations and guidelines including EN 540: Clinical Investigations of Medical Devices for Human Subjects, 1993 (this guideline eventually became ISO 14155). Studies conducted outside the EU were required to demonstrate the device use and population studied were equivalent to the planned EU device use and population. The final clinical study report (CSR) was required to have specific detailed information including a "safety report" with a "summary of all adverse events and adverse device events... including... severity, treatment required, resolution and assessment by the investigator of relation to treatment; performance or efficacy analysis; any sub group analysis for special population; a description of how missing data, including patients lost to follow up or withdrawn, were dealt with in the analysis."

The guidance listed specific trial documents for the evaluator to request and review, along with specific information to be "checked in all cases," and information that the final report "should contain." The guidance also stated that "an assessment and analysis carried out by an independent and unbiased expert in the field should always be considered, particularly if in-house expertise is not available." Although the literature review details remained quite similar from 2003 to 2023, the ability to seek a CE mark based solely on a "literature route" as envisioned in this guidance became obsolete and the drive to include a greater number of clinical investigations increased exponentially.

1.7.4 Global Harmonization Task Force Study Group 5 clinical evidence history (January 2005)

Part of the history behind the current MDR and the required details needed in the CER/PER can be found in the International Medical Device Regulators Forum (IMDRF) website [23], which includes an archive of earlier Global Harmonization Task Force (GHTF) documents (Table 1.4).

TABLE 1.4 GHTF SG5 documents.

GHTF code	Title	Link to PDF
GHTF/SG5/N6:2012	GHTF SG5 Clinical Evidence for IVD Medical Devices—November 2012	https://www.imdrf.org/sites/default/files/docs/ghtf/final/sg5/technical-docs/ghtf-sg5-n6-2012-clinical-evidence-ivd-medical-devices-121102.pdf
GHTF/SG5/N7:2012	GHTF SG5 Scientific Validity Determination and Performance Evaluation—November 2012	https://www.imdrf.org/sites/default/files/docs/ghtf/final/sg5/technical-docs/ghtf-sg5-n7-2012-scientific-validity-determination-evaluation-121102.pdf
GHTF/SG5/N8:2012	GHTF SG5 Clinical Performance Studies for IVD Medical Devices—November 2012	https://www.imdrf.org/sites/default/files/docs/ghtf/final/sg5/technical-docs/ghtf-sg5-n8-2012-clinical-performance-studies-ivd-medical-devices-121102.pdf
GHTF/SG5/N5:2012	Reportable Events During Pre-Market Clinical Investigations	https://www.imdrf.org/sites/default/files/docs/ghtf/final/sg5/technical-docs/ghtf-sg5-n5-2012-reportable-events-120810.pdf
GHTF/SG5/N4:2010	GHTF SG5—Post-Market Clinical Follow-Up Studies—November 2009	https://www.imdrf.org/sites/default/files/2021-12/ghtf-sg5-n4-post-market-clinical-studies-100218.pdf

(Continued)

TABLE 1.4 (Continued)

GHTF code	Title	Link to PDF
Archived documents		
GHTF/SG5/N3:2010	GHTF SG5—Clinical Investigations—February 2010	https://www.imdrf.org/sites/default/files/docs/ghtf/archived/sg5/technical-docs/ghtf-sg5-n3-clinical-investigations-100212.pdf
SG5-N2R8:2007	GHTF SG5—Clinical Evaluation—May 2007	https://www.imdrf.org/sites/default/files/docs/ghtf/archived/sg5/technical-docs/ghtf-sg5-n2r8-2007-clinical-evaluation-070501.pdf
SG5-N1R8	GHTF SG5—Clinical Evidence—Key Definitions and Concepts—May 2007	https://www.imdrf.org/sites/default/files/docs/ghtf/archived/sg5/technical-docs/ghtf-sg5-n1r8-clinical-evaluation-key-definitions-070501.pdf
GHTF SG5 (PD)/N2R7 (2006)	GHTF SG5—Clinical Evaluation	https://www.imdrf.org/sites/default/files/docs/ghtf/archived/sg5/technical-docs/ghtf-sg5-n2r7-guidance-clinical-060426.pdf
GHTF SG5 (PD)/N1R7 (2006)	GHTF SG5—Clinical Evidence—Key Definitions and Concepts	https://www.imdrf.org/sites/default/files/docs/ghtf/archived/sg5/technical-docs/ghtf-sg5-n1r7-guidance-definitions-060426.pdf

GHTF, Global Harmonization Task Force; *SG*, study group.

The GHTF-SG5 meeting minutes from January 17, 2005 [24] reported: "Susanne Ludgate volunteered to draft a document on how to conduct a clinical evaluation and compile an evaluation report." Members of the various delegations were asked to provide definitions and to review their jurisdictional/guidance documents to assist in the development of these clinical evaluation guidelines. In addition, the initial draft guidance "Clinical Evidence—Key Definitions, Concepts and Principles" [25] offered by the Chair (Larry Kessler) was edited and finalized over the next 2 years. This guidance defined clinical data as "Safety and/or performance information... generated from the clinical use of a medical device," and clinical data sources included clinical trials, literature, and experiences.

Study Group 5 (SG5) records provide an excellent resource to review historical discussions about medical device "Clinical Safety and Performance" evaluations. In particular, SG5 developed the basic MEDDEV 2.7/1, Rev. 4 (2016) concepts from their first meeting in Canberra, Australia on January 17, 2005 and first slide show in 2005 to the GHTF/SG5 final document in 2007 (SG5/N1R8 [25]) and the last GHTF/SG5 document in 2012 (SG5/N6:2012) about IVD medical device clinical evidence.

The IMDRF is a "voluntary group of medical device regulators from around the world" started in 2011 to build on the work of the GHTF. The IMDRF Management Committee comprises regulatory authority representatives from the following jurisdictions:

- Australia—Therapeutic Goods Administration;
- Brazil—Brazilian Health Regulatory Agency (ANVISA);
- Canada—Health Canada;
- China—National Medical Products Administration;
- European Union—European Commission—Directorate-General for Health and Food Safety;
- Japan—Pharmaceutical and Medical Devices Agency (PMDA);
- Russia—Russian Ministry of Health;
- Singapore—Health Sciences Authority;
- South Korea—Ministry of Food and Drug Safety;
- United Kingdom—Medicines and Healthcare products Regulatory Agency; and
- United States of America—US FDA.

As stated on the IMDRF website (http://www.imdrf.org/about/about.asp, accessed on September 29, 2023), Argentina's National Administration of Drugs, Food and Medical Devices, the World Health Organization (WHO), and Swissmedic are "Official Observers," the South African Health Products Regulatory Authority (SAHPRA) is an "Affiliate Member," and the APEC LSIF Regulatory Harmonization Steering Committee, Global Harmonization Working Party (GHWP), and Pan American Health Organization (PAHO) have IMDRF Regional Harmonization Initiatives.

1.7.5 Global Harmonization Task Force Study Group 5/N2R8:2007 clinical evaluation (May 2007)

The Global Harmonization Task Force (GHTF) published a clinical evaluation guidance in May 2007, preceding the September 2007 EU MDD/AIMDD amendments. GHTF defined clinical evaluation as "an ongoing process conducted throughout the life cycle of a medical device... first performed during the conformity assessment process... then repeated periodically as new clinical safety and performance information about the device is obtained during its use. This information is fed into the ongoing risk analysis and may result in changes to the Instructions for Use" [26].

In order to conduct a clinical evaluation, the evaluator must "evaluate data in terms of its suitability for establishing the safety and performance of the device." The guidance included "data generated through clinical experience" as a data source in addition to data from literature searching and clinical trials. Further, the guidance defined clinical data as "Safety and/or performance information... generated from the clinical use of a medical device." The guidance also specified the clinical evaluation scope was "based on a comprehensive analysis of available pre- and post-market clinical data relevant to the intended use of the device in question." This guideline also provided details about clinical data identification, appraisal (relevance, applicability, quality, and clinical significance), and analysis (to allow conclusions about S&P and product information). Specifically, this guidance included IVDs as follows:

> *Clinical evaluation should be performed for* in vitro *diagnostic devices as part of conformity assessment to the Essential Principles in a manner similar to other devices. The basic principles of objective review of clinical data will apply as described in this guidance document. However, IVDDs offer some unique challenges which will be addressed in a future document.*

The clinical data identification step noted how the literature may "represent the greater part (if not all) of the clinical evidence" and "Papers considered unsuitable for demonstration of performance because of poor study design or inadequate analysis may still contain data suitable for assessing the safety of the device."

Two important points to remember: (1) literature may be the primary focus of the CE/PE; and (2) even poor-quality literature should be included for safety details.

Clinical experience data were identified as data "generated through clinical use... outside the conduct of clinical investigations..." including clinical data in PMS reports, AE databases, compassionate use programs, and field corrective actions (e.g., recalls). This guidance provided a clear rationale for the "value of clinical experience data" because:

> *... it provides real world experience obtained in larger, heterogeneous and more complex populations, with a broader (and potentially less experienced) range of end-users than is usually the case with clinical investigations... most useful for identifying less common but serious device-related adverse events; providing long term information about safety and performance, including durability data and information about failure modes; and elucidating the end-user "learning curve"... a particularly useful source of clinical data for low risk devices that are based on long standing, well-characterized technology and, therefore, unlikely to be the subject of either reporting in the scientific literature or clinical investigation.*

Appraisal is a discrete CE/PE stage assessing "each individual data set" for data quality, applicability/relevance to the device and intended use and clinical significance "to determine the contribution of each data subset to establishing the safety and performance of the device." This guidance specified "no single, well established method for appraising clinical data" exists; however, these ill-defined appraisal "criteria should be applied consistently." Creating appraisal criteria can be challenging, so this guidance also included some examples in a checklist for randomized controlled trials (RCTs), cohort studies, case-control studies, and case series (see **Appendix D: Appraisal Question Examples**).

Two things to keep in mind when preparing to appraise clinical data are: (1) keep safety and performance in separate categories to be analyzed separately; and (2) weigh each data point weight based on the relative data point contribution.

One potential appraisal scoring method was described in multiple international guidelines (GHTF SG5/N2R8 2007 [26], MEDDEV 2.7.1, Rev. 3 [27], and IMDRF 2019 [28]) with scoring and weighting (grades) based on "suitability" and "data contribution" (Table 1.5).

TABLE 1.5 Sample appraisal criteria for suitability and data contribution.

Suitability	Description	Grading system
Appropriate device	Were the data generated from the device in question?	D1 Actual device D2 Comparable/equivalent device D3 Other medical/device
Appropriate device application	Was the device used for the same intended use (e.g., methods of deployment, application)?	A1 Same use A2 Minor deviation A3 Major deviation
Appropriate patient group	Were the data generated from a patient group representative of the intended treatment population (e.g., age, sex) and clinical condition (i.e., disease, including state and severity)?	P1 Applicable P2 Limited P3 Different population
Acceptable report/data collation	Do the reports or collations of data contain sufficient information to be able to undertake a rational and objective assessment?	R1 High quality R2 Minor deficiencies R3 Insufficient information
Data contribution	**Description**	**Grading system**
Data source type	Was the design of the study appropriate?	T1 Yes T2 No
Outcome measures	Do the outcome measures reported reflect the intended performance of the medical/device?	O1 Yes O2 No
Follow-up	Is the duration of follow-up long enough to assess treatment effects and identify complications?	F1 Yes F2 No
Statistical significance	Has a statistical analysis of the data been provided and is the analysis appropriate?	S1 Yes S2 No
Clinical significance	Was the magnitude of the treatment effect observed clinically significant?	C1 Yes C2 No

*This table has the questions, criteria, and grading system across all three documents; however, in GHTF SG5/N2R8 2017 [26] and MEDDEV 2.7.1, Rev. 3 [27], there is one minor language change: "comparable device" becomes "Equivalent device" and IMDRF 2019 [28] "device" becomes "medical device."

The GHTF SG5/N2R8:2007 [26] stated the following:

The criteria may be worked through in sequence and a weighting assigned for each dataset. The data suitability criteria can be considered generic to all medical devices... actual method used will vary according to the device considered... To assess the data contribution criteria of the suitable data, the evaluator should sort the data sets according to source type and then systematically consider those aspects... most likely to impact on the interpretation of the results... what types of issues are most important in relation to the nature, history and intended clinical application of the device... characteristics of the sample, methods of assessing the outcomes, the completeness and duration of follow-up, as well as the statistical and clinical significance of any results... the weightings would be used to assess the strength of the datasets' contribution to demonstrating overall performance and safety of the device... As a general guide... the more level 1 grades, the greater the weight of evidence provided... in comparison to other datasets, however, it is not intended that the relative weightings from each category be added into a total score.

Thus, here is the problem for this particular guideline-suggested appraisal method: how can this method be used to prioritize data? Unfortunately, the process often becomes alphanumeric soup without an actual pragmatic use. First, not all clinical data fit into these specified questions, and the authors envisioned this as a framework only, since the comments stated that this table was an "example" and the "method used will vary." Unfortunately many users missed this "offered only as an example" concept, and did not understand the inherent need to develop a new, specific appraisal criterion process for each device. In addition, many individual users fail to grasp the inherent problem when using this "suitability and data contribution tool": even if the data can be "weighted" under the "suggested" suitability and data contribution code for each of the "SAMPLE" questions, what happens next?

The suitability criteria speak to relevance and data contribution criteria assess quality; however, this method is difficult to use because the questions are often not relevant to the clinical trial, literature, or experience data under evaluation. The scores require a significant amount of time to apply to each data record, and the method is unclear about how to sort the clinical data consistently based on the scores. As such, this "suitability and data contribution tool" often fails to yield a useful output. Suitability and contribution scoring was removed from MEDDEV 2.7/1, Rev. 4 (June 2016) [29] but included in the IMDRF Clinical Evaluation guideline (October 2019) [28], even though this grading system may not be particularly useful for making appraisal decisions. Table 1.6 provides a hypothetical example. Consider four articles weighted using only the "suitability and

TABLE 1.6 Article suitability and data contribution example.

#	Suitability criteria				Data contribution criteria				
	Appropriate device	Device application	Patient group	Data collation	Data source	Outcome measures	Follow-up	Statistical significance	Clinical significance
1	D1	A2	P2	R2	T1	O1	F2	S1	C2
2	D3	A2	P3	R1	T1	O1	F1	S1	C1
3	D2	A1	P1	R2	T2	O2	F1	S1	C1
4	D1	A1	P1	R3	T1	O1	F1	S1	C2

data contribution tool" grades as suggested. How should I&E appraisal criteria be assigned, and how should each article be weighted or ranked compared to the others?

This grading system alone does not provide enough information or context to appraise these articles adequately. For example, no information is included about sample size or study type—two important considerations for study quality. In addition, the criteria do not distinguish between background/SOTA/alternative therapy information and device-specific information. Now, compare this to a much simpler scoring process: are the clinical data relevant and of sufficient quality to include?

Often, medical device clinical data are limited and a qualitative analysis will be more likely than a quantitative analysis. Separating exploratory and pilot clinical trials from pivotal clinical trials may help to isolate the higher-quality data. Finding consistency across results from different datasets will increase certainty. In addition, understanding reasons for differences will be helpful. In the end, the evaluator needs to determine if the device S&P are as intended and if the benefit–risk ratio remains acceptable. This guidance [26] suggested careful consideration for:

> *... the number of patients exposed to the device, the type and adequacy of patient monitoring, the number and severity of adverse events, the adequacy of the estimation of associated risk for each identified hazard, the severity and natural history of the condition being diagnosed or treated. The availability of alternative diagnostic modalities or treatments and current standard of care should also be taken into consideration. The product literature and instructions for use should be reviewed to ensure they are consistent with the data and that all the hazards and other clinically relevant information have been identified appropriately.*

This last point is quite important: remember to review the product information to ensure consistency with the CER/PER clinical data.

A possible CER format was included with seven sections:

1. general details;
2. devices description and intended application;
3. intended therapeutic and/or diagnostic indications and claims;
4. evaluation context and choice of clinical data types;
5. summary of the clinical data and appraisal;
6. data analysis (performance, safety and product literature); and
7. conclusions.

The format does not include any clarity about how to present the various data sources (trials, literature, and experiences), does not separate out the background/SOTA/alternative therapy information from the subject device clinical data analyses, and does not integrate the proposed "suitability and data contribution tool" scoring (i.e., this was only a suggested tool).

1.7.6 MEDDEV 2.7.1, Rev. 3 (December 2009) clinical evaluation

MEDDEV 2.7.1 (2003) [22] was updated in 2009 [27] to align with the 2007 MDD/AIMDD amendments (Directive 2007/47/EC) [11] and the 2007 GHTF clinical evaluation guidelines (SG5/N2R8:2007) [26], which included suitability, contribution, and CER format details. The CE scope was broadened to be "based on a comprehensive analysis of available pre- and post-market clinical data relevant to the intended use of the device", appraisal was defined as a "distinct [CE] stage" and clinical experience was explicitly specified as a data source including but not limited to:

- manufacturer-generated postmarket surveillance reports, registries, or cohort studies (which may contain unpublished long-term safety and performance data);
- adverse events databases (held by either the manufacturer or Regulatory Authorities);
- data for the device in question generated from individual patients under compassionate usage programs prior to marketing of the device; and
- details of clinically relevant field corrective actions (e.g., recalls, notifications, hazard alerts).

As mentioned in the earlier GHTF SG5/N2R8:2007 clinical evaluation (May 2007) discussion above, this 2009 MedDev document presented idential text about real world experiences which are not likely to be found in clinial trials or clinical literature. This is particularly relevant when writing CERs/PERs, especially during clinical data analysis. Often, real-world evidence drawn from individual human experiences is the only clinical data source available. The data collected and reported by the manufacturer is crucial to CERs/PERs for all devices, but especially so for long-standing, low-risk devices where these experiences may be the entire dataset under evaluation for the subject/equivalent device in some CERs/PERs.

BOX 1.6 CER/PER must avoid bias and document conflicts of interest

MEDDEV 2.7.1, Rev. 3 [27] defined bias as a *"systematic deviation of an outcome measure from its true value, leading to either an overestimation or underestimation of a treatment's effect. It can originate from, for example, the way patients are allocated to treatment, the way treatment outcomes are measured and interpreted, and the recording and reporting of data"* (p. 15).

Bias can also occur when a company employee is the CER/PER evaluator for any CE/PE steps. The employee has a known conflict of interest, since the employee may feel motivated to help the company succeed so they can remain employed by the company. This conflict of interest may introduce bias, which can influence clinical data selection, inclusion/exclusion, weighting and analysis decisions, etc. Sometimes the employee's influence is unintentional and other times the influence may be intentional; however, either way, the potential for bias needs to be documented. Often companies hire external consultants to help avoid this known conflict of interest and potential unacceptable bias. Even external consultants can be biased, so careful descriptions of bias and recording of any and all conflicts of interest are required. Attaching the evaluator conflict of interest details is a required CER/PER component. Company quality control systems should be set up to document and to avoid bias as much as possible.

The guidance [27] defines data as "suitable for appraisal" when the clinical data contains "sufficient information for the evaluator to... undertake a rational and objective assessment... and make a conclusion about" subject device S&P. The appraisal purpose is to document clinical data "merits and limitations," "suitability to address questions about the device" and to prioritize data "contribution" to device S&P.

After clinical data are appraised for quality (a difficult parameter to measure and quantify) and relevance (i.e., "the data must be either generated for the device in question or for an equivalent device"), "further appraisal" should "determine the data subset contribution to establishing" device S&P. The clinical data should be "categorised to allow for separate [S&P] analysis," the "appraisal should be justified by the evaluator" and the analysis should "determine if the appraised data sets... demonstrate the clinical performance and safety of the device in relation to its intended use."

Note: this guidance does not consider background/SOTA/alternative therapy clinical data appraisal and analysis. Only subject device clinical data were envisioned for the CE/PE process in 2009.

The NB will determine if the manufacturer's proposed benefit–risk ratio was "unacceptable... broadly acceptable... or... acceptable under specified conditions." In general, the NB is seeking to understand if "valid" decision-making criteria were employed (e.g., whether the manufacturer used applicable guidance documents and international standards like ISO 14971 for medical device risk management activities outside the CER/PER). The NB also decides if the CER provided a sufficient device clinical evaluation to demonstrate conformity to the regulatory requirements (i.e., the Essential Requirements at the time of this guideline version). This guidance aligned best NB and Competent Authority practices as the NB began using the CEAR to document the NB CER assessment and the NB conclusions. Using a quality system approach, the NB will assess the CE/PE procedure including a sample to ensure the Class IIa or IIb CER meets the regulatory requirements; however, under the design dossier (or "type") examination, the NB assesses the data and processes used to identify, appraise, analyze, and draw conclusions about the data. See the NB checklist provided in **Appendix E: NB Clinical Evaluation Checklist**.

1.7.7 Notified Bodies Operations Group guidance documents before and after 2009

The NB Operations Group (NBOG) was formed in July 2000 with meetings twice a year including experts from the EC and Competent Authorities relating to NB designation and control. The NBOG identified and created best practice examples to guide NBs and Competent Authorities responsible for NBs. Prior to the establishment of the MDCG, the NBOG issued a "Designating Authorities Handbook" [30] to describe the NBOG, Designating Authority, Designation Process and NB Monitoring, and multiple documents related to EU regulatory requirements (Table 1.7).

TABLE 1.7 NBOG documents.

Number	Title	Date
NBOG documents for Regulation (EU) 2017/745 (MDR) and Regulation (EU) 2017/746 (IVDR)		
NBOG BPG 2017-1, Rev. 1	Designation and notification of conformity assessment bodies	Nov 2017
NBOG FORMS		
NBOG F 2017-1, Rev. 2	Application form to be submitted by a conformity assessment body when applying for designation as notified body under the medical device regulation (MDR)	Nov 2017
NBOG F 2017-2, Rev. 2	Application form to be submitted by a conformity assessment body when applying for designation as notified body under the in vitro diagnostic devices regulation (IVDR)	Nov 2017
NBOG F 2017-3	Applied-for scope of designation and notification of a Conformity Assessment Body—Regulation (EU) 2017/745 (MDR)	Nov 2017
NBOG F 2017-4	Applied-for scope of designation and notification of a Conformity Assessment Body—Regulation (EU) 2017/746 (IVDR)	Nov 2017
NBOG F 2017-5	Preliminary assessment report form—Regulation (EU) 2017/745	NA
NBOG F 2017-6	Preliminary assessment report form—Regulation (EU) 2017/746	NA
NBOG documents for medical device directives 90/385/EEC, 93/42/EEC, and 98/79/EC		
NBOG BPG 2016-1	(Re-)Designation of Notified Bodies: Process for Joint Assessments	Jun 2016
NBOG BPG 2014-3	Guidance for manufacturers and Notified Bodies on reporting of Design Changes and Changes of the Quality System	Nov 2014
NBOG BPG 2014-2	Guidance on the Information Required for Notified Body Medical Device Personnel Involved in Conformity Assessment Activities	Nov 2014
NBOG BPG 2014-1	Renewal of EC Design-Examination and Type-Examination Certificates: Conformity Assessment Procedures and General Rules	Nov 2014
NBOG BPG 2010-3	Certificates Issued by Notified Bodies with Reference to Council Directives 93/42/EEC, 98/79/EC, and 90/385/EEC	Mar 2010
NBOG BPG 2010-2	Guidance on Audit Report Content	Mar 2010
NBOG BPG 2010-1	Guidance for Notified Bodies Auditing Suppliers to Medical Device Manufacturers	Mar 2010
NBOG BPG 2009-4	Guidance on Notified Body's Tasks of Technical Documentation Assessment on a Representative Basis	Jul 2009
NBOG BPG 2009-3	Guideline for Designating Authorities to Define the Notification Scope of a Notified Body Conducting Medical Devices Assessment	Mar 2009
NBOG BPG 2009-2	Role of Notified Bodies in the Medical Device Vigilance System	Mar 2009
NBOG BPG 2009-1	Guidance on Design-Dossier Examination and Report Content	Mar 2009
NBOG BPG 2006-1	Change of Notified Body	Nov 2008

(Continued)

TABLE 1.7 (Continued)

Number	Title	Date
NBOG Checklists		
NBOG CL 2010-1	Checklist for Audit of Notified Body's Review of Clinical Data/Clinical Evaluation	Mar 2010
NBOG Forms		
NBOG F 2014-1	Application Form to be Submitted When Applying for Designation as a Notified Body	Nov 2014
NBOG F 2014-2	Qualification of Personnel (see NBOG BPG 2014-2)	Jun 2016
NBOG F 2012-1	Notification Form—Directive 93/42/EEC	Jan 2013
NBOG F 2012-2	Notification Form—Directive 90/385/EEC	Jan 2013
NBOG F 2012-3	Notification Form—Directive 98/79/EC	Jan 2013
NBOG F 2010-1	Certificate Notification to the Commission and other Member States	Mar 2010

BPG, best practice guide; *CL*, checklist; *EC*, European Commission; *EU*, European Union; *F*, form; *IVDR*, in vitro diagnostic regulation; *MDR*, medical device regulation; *NBOG*, notified body operations group. *Bolded documents guide NBs as they write their CEAR (clinical evaluation assessment report).

These guidelines now fall under MDCG oversight, and minimum NB CEAR contents were discussed briefly in two NBOG BPG documents. NBOG BPG 2009-1 [31] stated: "Any particular performance claims or product benefits claimed should be verified by adequate clinical data or design testing during the NB assessment of the technical documentation." Note how preclinical ("deisgn testing") and clinical evaluation were described separately. This is important to understand, because preclinical data will reside in other technical file locations separate from the CER/PER unless the clinical data specifically needs support from the preclinical data within the CER/PER for a particular reason. The CE/PE NB assessment was described as follows:

The NB should assess the validity of the clinical evaluation and should verify that the device has met the claimed performance as outlined. The documentation provided by the manufacturer should contain all necessary data according to MEDDEV 2.7.1... and / or GHTF SG 5 documents... to allow for a proper review of the clinical evaluation done by the manufacturer. The NB reviewer should assess the clinical investigation data and / or the literature review assembled and the validity of conclusions drawn by the manufacturer. The Design Dossier Report should include... the clinical safety and performance [assessment and the] conclusion (the NB should justify and document each step of the decision making process...)... The NB should review the manufacturer's proposed Post Market Surveillance programme specified for this device, including details of any post market clinical follow up... [31]

NBOG BPG 2009-4 [32] specifies the NB's technical file assessment must include the clinical evaluation.

1.7.8 Poly Implant Prothèse breast implant scandal (2010)

Although not a guidance document, the Poly Implant Prothèse (PIP) scandal offers guidance for those who have learned not to repeat past mistakes. The NB, TÜV Rheinland, issued CE mark (safety and performance) certificates to PIP after the company was founded in 1991. The breast implant materials were changed in 2001 to save costs after the US FDA banned silicone implants in 2000. A cheaper nonmedical grade silicone (i.e., ~10% cost) was secretly used without bench, animal or human safety testing as required. The substandard implants were five times more likely to rupture or leak than the approved models, the devices were implicated in several deaths due to systemic toxicity, and several breast cancer cases were reported.

PIP sold more than 2 million breast implants including sales to ~400,000 women between 2001 and 2010 who received breast implants made with substandard, industrial-grade silicone rather than medical-grade silicone gel as approved for human use initially. The French Health Ministry issued a full recall in 2010 after the company was defunct and the founder was imprisoned for 4 years. In 2011, the French government recommended removal of the faulty implants, and in 2021 (10 years later) the French court found TÜV Rheinland negligent in not detecting the fraud and thus partially liable for damages. The British Association of Aesthetic Plastic Surgeons identified poor PMS as the root cause and requested more stringent, compulsory monitoring. This case is often cited as the reason for the 2017 EU regulatory changes [33].

1.7.9 International Standards Organization 14155:2003, 2011, and 2020 medical device clinical trials

At the time of the inaugural GHTF-SG5 meeting in 2005, SG5 members were also attending the Berlin meeting of ISO TC 194 WG4 to discuss revisions to ISO 14155-1 and 14155-2 regarding medical device clinical trial conduct. The SG5 agreed to develop a Memo of Understanding (MoU), with the team putting together the standard for clinical investigation of medical devices. Patterned after ICH E6 documents used for pharmaceutical research, ISO 14155 optimized scientific validity and ability to reproduce clinical investigation results; however, this international standard did not apply to IVD medical devices. The ISO 14155 update was released in 2011 [34], replacing the first edition which was in two parts (ISO 14155-1:2003 General Requirements, https://www.iso.org/standard/31723.html; and ISO 14155-2:2003 Clinical Investigation Plans, https://www.iso.org/standard/32217.html). The standard was updated again in 2020 [14] and is undergoing additional updates in 2024. This standard included the following sections:

1. scope;
2. normative references;
3. terms and definitions;
4. summary of good clinical practice (GCP) principles (new in 2020 version);
5. ethical considerations (e.g., improper influence, compensation, responsibilities, registration in publicly accessible database, responsibilities, ethics committees, vulnerable populations and informed consent);
6. clinical trial planning (e.g., risk evaluation, design justification, the protocol, investigator's brochure, case report forms, monitoring plan, site selection, agreements, labeling, data monitoring committee);
7. clinical trial conduct (e.g., site initiation and monitoring, adverse events and device deficiencies, documents and documentation, site team members, privacy and confidentiality, document and data control, investigational device accountability, subject accounting and auditing);
8. suspension, termination, and close out;
9. sponsor responsibilities (quality assurance/control, planning, and conduct including monitoring and safety reporting, outsourcing, and regulatory communications); and
10. principal investigator responsibilities (qualifications, ethics committee communication, informed consent process, following the protocol, medical care for participants and safety reporting).

In addition, ISO 14155 included details about the protocol or Clinical Investigational Plan (CIP), Investigator's Brochure (IB), Case Report Forms (CRFs), *Clinical Investigation Report*, Essential Documents and Adverse Event Categorization, *EC responsibilities*, *Application of ISO 14971 to Clinical Investigations*, *Clinical Development Stages*, and *Clinical Investigation Audits* (items in italics were new in 2020 version).

1.7.10 Global Harmonization Task Force Study Group 5/N6, N7, N8 clinical evidence for *in vitro* diagnostic medical devices (2012)

Three 2012 guidance documents [35–37] supported gathering clinical evidence for IVD medical devices, as follows: "Generally, from a clinical evidence perspective, ... the manufacturer... [is expected to demonstrate] the device achieves its intended performance during normal conditions of use in the intended environment (e.g., laboratories, physician's offices, healthcare centers, home environments) and in the intended use population..." IVD devices are designed to examine specimens taken from the human body.

The GHTF/SG5/N6:2012 "Clinical Evidence for IVD Medical Devices—Key Definitions and Concepts" [35] guidance defined terms including "clinical evidence... scientific validity, analytical performance... clinical performance... and clinical utility." The "Illustrative Example" discussed an HBV mutation genotype test intended to determine mutation treatment failure likelihood as follows:

- Scientific validity (i.e., analyte associated with clinical condition or physiological state)
  - Example associated HBV mutation with a clinical state and asked if mutation in clinical setting was associated with drug resistance or worse outcome.
- Analytical performance (i.e., IVD device detects or measures a particular analyte)
  - Example referred to detecting HBV mutation with genotype test and asked if test would identify HBV mutation in spiked or native samples.
  - "Analytical performance may include analytical sensitivity (e.g., limit of detection), analytical specificity (e.g., interference, cross-reactivity), accuracy (derived from trueness and precision), and linearity."
- Clinical performance (i.e., IVD device yields results correlated with clinical condition or physiological state in target population and intended user; supports intended use)
  - Example referred to detecting HBV mutation in intended population and asked if test would identify HBV mutation in patient samples including HBV mutant, and how well HBV genotype test results correlated with treatment failure.
  - "Clinical performance may include expected values, diagnostic sensitivity and diagnostic specificity" based on individual clinical/physiological state and predictive values based on disease prevalence; derived from clinical performance studies, literature, diagnostic testing experience, etc.
- Clinical utility (i.e., IVD device test results are "useful" and information has "value" to the "individual being tested and/or the broader population"; supports clinical patient management decisions about effective treatment and prevention strategies)
  - Explored if patient benefits from genotype test information, asked if treatments manage HBV infections and if HBV test results lead to improved patient outcome.
  - Included "acceptability, appropriateness; availability of treatments/ interventions, and health economics... Aside from scientific validity and clinical performance, a manufacturer is not required to demonstrate any other elements of clinical utility for premarket conformity assessment..."

Performance is measured by the IVD medical device achieving the intended use/purpose as "reflected in the specifications, instructions and information provided by the manufacturer...' The intended use/purpose "should... define the IVD medical device test purpose such as diagnosis, aid to diagnosis, screening, monitoring, predisposition, prognosis, prediction, and physiological status determination. Other relevant aspects include the specific disorder, condition or risk factor of interest, the testing population and, where applicable, the intended user. This information will determine the type and depth of the clinical evidence..." needed for the performance evaluation.

The GHTF/SG5/N7:2012 "Clinical Evidence for IVD Medical Devices—Scientific Validity Determination and Performance Evaluation" [36] guidance discussed how to collect, appraise, analyze, and document clinical evidence in the CER/PER. The definitions included details about the following:

- Sensitivity (i.e., diagnostic or clinical sensitivity to "identify the presence of a target marker associated with a particular disease or condition... defined by criteria independent of the IVD medical device under consideration")
  - "... percent positivity in samples from subjects where the target disease or condition is known to be present... "
  - "... calculated as true positive values divided by the sum of true positive plus false negative values..."
- Specificity (i.e., diagnostic or clinical specificity to "recognize the absence of a target marker associated with a particular disease or condition... defined by criteria independent of the IVD medical device under consideration"
  - "... percent negativity in samples where the target analyte (measurand) is... absent..."
  - "... calculated as true negative values divided by the sum of true negative plus false positive values..."
- Predictive value (i.e., likelihood a "positive IVD medical device test result" identifies the "given condition..., or... negative... test result does not" identify the condition.
  - "... determined by the diagnostic sensitivity and diagnostic specificity... and by the prevalence of the condition..."
  - "Prevalence means the proportion of persons with a particular disease within a given population at a given time."
  - "positive predictive value indicates how effectively an IVD medical device separates true positive test results from false positive test results..."
  - "negative predictive value indicates how effectively an IVD medical device separates true negative test results from false negative test results..."

This guidance describes how IVD medical device risks and benefits are related to test results aiding patient diagnosis and management rather than the medical device directly contacting or affecting human cells and tissues. The IVD medical device benefit–risk ratio is based on risks of incorrect test results rather than direct device-related adverse events or patient harms, and these risks increase related to the novelty, innovation, disease/patient/user variability, and technological difficulty when identifying the analyte. Even so, the clinical data sources are identical for all medical devices (i.e., clinical trials, literature, and experience data), and all intended uses must be supported with suitable clinical evidence. Similarly, PMS and RM systems as well as proactive PMCF plans are required for all medical devices including IVD devices.

The clinical evidence stages (Fig. 1.2) were similar to future guidance documents using the same steps to identify, appraise, analyze, and write up CER/PER clinical data.

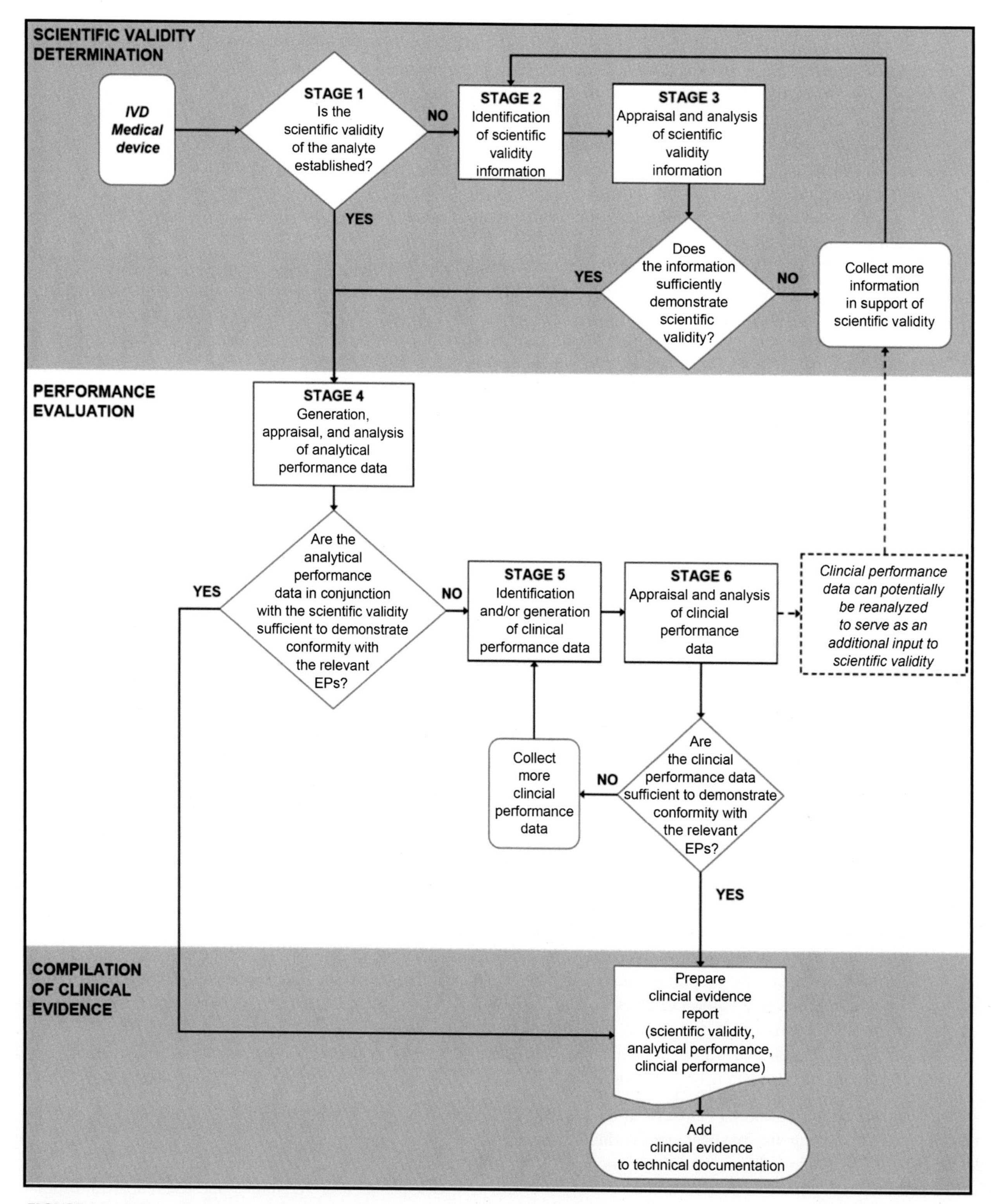

FIGURE 1.2 IVD medical device clinical evidence stages. *Image is from GHTF SG5/N7:2012 [36]. Essential Performances (EPs) of safety and performance of medical devices have been updated to become general safety and performance requirements (GSPRs).*

Analyte scientific validity research is followed by IVD medical device performance evaluation testing of human samples; however, this sequence of events is not required. Performance data can be used in the scientific validity determination and scientific validity steps are not required when the analyte is a well-known marker for the clinical condition/physiological state (i.e., only a "brief rationale" is needed with key references). All medical devices require the mechanism of action (i.e., scientific validity) and the device S&P to be supported with sufficient clinical data (i.e., clinical trials, literature, and experience data).

For performance testing, established/standardized tests were separated from established/nonstandardized tests and from novel tests. Technical (i.e., analytical) performance was described to include potentially: accuracy testing (i.e., trueness and precision) and "sensitivity (e.g., limit of detection, limit of quantitation)... specificity, linearity, cut-off, measuring interval (range), carry-over... appropriate specimen collection and handling, and endogenous and exogenous interference..." Novel and nonstandardized IVD tests may not allow accuracy testing even through comparison to other well-documented methods; a clinical performance study may be required, and sometimes "spiked" human samples may be acceptable. Clinical performance may test specificity (i.e., does IVD correctly classify patients without the disease or condition?), sensitivity (i.e., is device effective to identify patients with the disease or condition?), positive and negative predictive values, likelihood ratios, expected values, etc. Novel and nonstandardized IVD tests may require clinical performance testing; however, Class A IVD medical devices and simple reagent migration between instruments with the same technology often do not require performance testing.

When an IVD clinical performance study is required, the test is typically compared to "routine diagnostic testing... for patient management... [and all] patient management decisions are based on the routine standard test" (i.e., not the subject IVD medical device needing the clinical performance testing). The protocol should reflect the subject IVD medical device intended use, benefits, and risks, and should include a well-documented "reference test method" as the "routine diagnostic test." In general, premarket studies without a clinical performance test design (e.g., customer feedback, external analytical performance, research studies) are not considered clinical performance studies.

Clinical literature and clinical experience data gained by routine diagnostic testing and real world evidence are valuable especially since these data can represent large, heterogeneous, complex populations with a wide variety of interfering substances. Clinical experience data like this can help to identify rare events and may provide a rich data resource for low risk devices including those based on long standing, well-characterized technologies which are rarely studied in clinical trials or published in the medical literature. These PMS/real world data may be sufficient to demonstrate performance; however, high-risk (Class C and D) devices may require a clinical performance study to demonstrate clinical performance.

The clinical data are appraised based on relevance (i.e., both analyte and clinical condition/disease state data should be evaluated and clearly linked together) and quality (i.e., sufficient, high quality data are required to carry out a rational and objective scientific validity and clinical performance assessments). The clinical data (trials, literature, and experiences) are also appraised for weight and significance. Clinical trial data are typically weighted higher than clinical literature data, which in turn are weighted higher than clinical experience data based on the data quality and quality control expected for each of these three different clinical data sources. The combined clinical data should be evaluated to see if sufficient clinical data are available to document the IVD medical device performs as indicated/intended, the benefit–risk ratio is acceptable, and the instructions for use are consistent with the evaluated clinical data. If not, clinical data gaps should be identified and remediated prior to CER/PER completion including the compiled scientific validity, clinical, and analytical performance, or any justification about missing or planned future clinical data collection.

The CER/PER should outline the device technology (intended use, claims about device S&P), overall approach, and methods used to identify, appraise, analyze, and report the clinical evidence. As expected based on the regulations, this guidance states "the level of detail in the report content will vary according to the risk class of the IVD medical device and its intended use." Although Appendixes A and B in this guidance provide a "Possible format for the Literature Search Report" and a method for literature screening and selection, a separate literature search report is not recommended, since the literature is just one form of clinical data to be included seamlessly in the CER/PER. A better solution may be to use the suggested format for the literature search report within the CER/PER structure without creating a separate report.

The GHTF/SG5/N8:2012 "Clinical Evidence for IVD Medical Devices—Clinical Performance Studies for In Vitro Diagnostic Medical Devices" [37] guidance provided suggestions about clinical performance study design, conduct, and reporting for both premarket and postmarket studies as follows:

... clinical performance studies must be designed according to, and take into account scientific principles underlying the collection of clinical performance data along with accepted operational and ethical standards surrounding the use of human subjects. The clinical performance study objectives and design should be documented in a clinical performance study protocol. The data collection process must ensure patient safety and data integrity along the entire process of the study... The purpose of a clinical performance study is to establish or confirm aspects of IVD medical device performance which cannot be determined by analytical performance studies, literature and/or previous experience gained by routine diagnostic testing... When a clinical performance study is conducted, the data obtained is used in the performance evaluation process and is part of the clinical evidence for the IVD medical device...

IVD clinical performance studies can be observational (e.g., cross-sectional, longitudinal, retrospective, prospective, or randomized controlled trials for therapeutic agents) or interventional (e.g., when the IVD impacts patient clinical outcomes), and these designs are discussed briefly, including various design considerations as follows:

- objectives;
- intended use—purpose; population to include age, race, sex, location, disease state, and clinical condition; sample type such as serum, plasma, or urine; and intended healthcare professional or layperson user;
- common test purposes (to diagnose, screen, monitor, predict, etc.) and examples;
- test characteristics like analytical precision, interference, range, and cutoff;
- clinical performance including sensitivity and specificity;
- novelty;
- reference method to establish patient clinical status;
- sample collection and handling;
- study site;
- statistical design;
- risks; and
- ethics (i.e., follow Declaration of Helsinki ethical principles to protect study subject rights, safety, well-being using informed consent, ethics committee review, reporting).

Sample size was linked to test purpose and selection criteria including sample collection timing, follow-up, and simultaneous testing for multiple purposes. Sample collection types included intentional collection, leftovers, and archived test samples. The needs to avoid selection or treatment bias and to ensure appropriate handling and storage were stressed. Selection of study sites and collection methods should mimic the intended use environment and intended user, and the carefully designed statistical analysis plan should have an appropriate sample size, subject and sample inclusion/exclusion criteria, methods to minimize bias, re-examination procedures, missing/excluded data criteria, analysis methods, and details about IVD medical device performance characteristics. Sample collection using invasive procedures must be disclosed as risks to the study subjects, and adverse events must be monitored and reported in interventional studies. The study must also have a study protocol defining study conduct; invesgtigators must follow the protocol and comply with all applicable laws and regulations. The clinical study report imust document the study methods, results, and conclusions.

1.7.11 MEDDEV 2.7/1, Rev. 4 clinical evaluation guideline (July 2016)

The EC updated CER Guideline MEDDEV 2.7.1, Rev. 3 (December 2009) [27] to MEDDEV 2.7/1, Rev. 4 (June 2016) [29] entitled "Clinical Evaluation: A Guide for Manufacturers and Notified Bodies under Directives 93/42/EEC and 90/385/EEC" just prior to releasing MDR/IVDR in 2017; unfortunately, many MEDDEV 2.7/1, Rev. 4 sections conflicted with the MDR.

Clinical evaluation was defined as a "methodologically sound ongoing procedure to collect, appraise and analyse clinical data pertaining to a medical device and evaluate whether there is sufficient clinical evidence to confirm compliance with relevant essential requirements [sic] for safety and performance when using the device according to the manufacturer's Instructions for Use." Postmarket clinical follow-up (PMCF) was an ongoing CE component, and sufficient clinical evidence was defined as "an amount and quality of clinical evidence to guarantee the scientific validity of the conclusions". After planning and identifying clinical data, appraisal was again defined as a distinct CE stage where "each individual data set" is considered for "scientific validity, relevance and weighting" to determine the data value. Weighting was discussed with respect to "uncertainty [arising] from two sources": data methodological quality and data relevance to the device evaluation "in relation to the different aspects of its intended purpose." Evaluators were instructed to:

- *"identify information contained in each document,*
- *evaluate the methodological quality of work done by the authors and from that, the scientific validity of the information,*
- *determine the relevance of the information to the clinical evaluation, and*
- *systematically weight the contribution of each data set to the clinical evaluation."*

Evaluators were also instructed to "set up an appraisal plan that describes the procedure and criteria to be used" and ensure appraisal is "thorough and objective," in part, by documenting and justifying appraisal criteria based on "current knowledge / the state-of-the-art, applying accepted scientific standards". The guidance stated:
"evaluators should:

- *follow the... appraisal plan strictly and apply... criteria consistently throughout the appraisal;*
- *base their appraisal on the full text of publications... (not abstracts or summaries)... to review all... contents, the methodology... results... validity of conclusions drawn... and evaluate any limitations and potential sources of error in the data;*
- *document the appraisal in the... [CER] to... be critically reviewed by others."*

The suitability and data contribution matrixes were removed and only referenced in a footnote as one of "many acceptable ways, both qualitative and quantitative, by which the appraisal can be carried out...". The guidance also stated "For many well established devices and lower-risk devices, qualitative data may be adequate."

When evaluating quality and validity, evaluators were advised to critically "evaluate the extent to which the observed effect (performance or safety outcomes) can be considered to be due to intervention with the device" versus confounding factors such as underlying disease, concomitant treatments, bias, random error, inadequate information disclosure, or misinterpretation. The guideline noted "some papers considered unsuitable for demonstration of adequate performance... may still contain data suitable for safety analysis or vice versa"; the guideline also identified study design, quality issues, differences between data sources, data processing, statistics, quality assurance, and report quality to consider.

Additional guidance was provided for clinical trial appraisal, as well as a distinction between pivotal data (i.e., data quality needed to demonstrate S&P for subject/equivalent device used as indicated) and other data (e.g., not pivotal data, appraised and weighted for the background, SOTA) for determining relevance. The nuance involved in determining clinical data weight was also discussed and the identification of "appropriate criteria to be applied for a scientific evaluation" and strict adherence to the "pre-defined criteria" were stressed. Examples of criteria to determine relevance included, but were not limited to, data about subject/equivalent device, claims, risks, background knowledge/SOTA, intended purpose, specific models/sizes, user group, indication, age group, sex, medical condition, duration of use, etc. The highest weight for clinical data should be reserved for "well designed and monitored randomized controlled clinical investigation... with the device under evaluation in its intended purpose, with patients and users that are representative of the target population." In addition, the guidance noted "When rejecting evidence, the evaluators should document the reasons."

MEDDEV 2.7/1, Rev. 4 listed important analysis concerns and stated: "In order to demonstrate compliance, the evaluator should... use sound methods... make a comprehensive analysis... determine if additional clinical investigations or other measures are necessary... determine PMCF needs...". The CER needs to outline the CE stages/steps, and the NB is responsible for generating the CEAR guided by the NBOG documents including "best practice guides, checklists and forms." Appendixes provided details about equivalence, when to carry out a clinical trial, typical device descriptions, literature sources, key literature search, and review protocol elements. Appendix A6 included study examples lacking scientific validity (e.g., reports without methods, device identity, outcomes, etc.; numbers too small to be statistically significant; missing details about deaths and serious adverse events, misinterpreted data; illegal activities).

MEDDEV 2.7/1, Rev. 4 warned appraisers to be on the lookout for issues not limited to: "Improper statistical methods" and "Lack of adequate controls":

- *"results obtained after multiple subgroup testing, when no corrections have been applied for multiple comparisons.*
- *calculations and tests based on a certain type of distribution of data (e.g., Gaussian distribution with its calculations of mean values, standard deviations, confidence intervals, t-tests, others tests), while the type of distribution is not tested, the type of distribution is not plausible, or the data have not been transformed. Data such as survival curves, e.g., implant survival, patient survival, symptom-free survival, are generally unlikely to follow a Gaussian distribution."*

"bias or confounding are probable in single arm studies and in other studies that do not include appropriate controls...

- *when results are based on subjective endpoint assessments (e.g., pain assessment).*
- *when the endpoints or symptoms assessed are subject to natural fluctuations (e.g., regression to the mean when observing patients with chronic diseases and fluctuating symptoms, when natural improvement occurs, when the natural course of the disease in a patient is not clearly predictable).*
- *when effectiveness studies are conducted with subjects... likely to take or are foreseen to receive effective co-interventions (including over-the-counter medication and other therapies).*
- *... other influencing factors (e.g., outcomes... affected by variability of the patient population, ... disease... user skills... infrastructure available for planning/ intervention/ aftercare, use of prophylactic medication, other factors).*
- *... significant differences between the results of existing publications, pointing to variable and ill controlled influencing factors."*

An important quality marker is whether the study has a comparator group to provide adequate control. Traditionally, the control group receives a placebo—a nonactive drug or treatment. Placebos are less common in device than drug trials. Device trials often use a standard of care control arm for the disease or condition being treated. For example, oral anticoagulants are the standard of care to treat blood clots. A study comparing how well two thrombectomy devices treat blood clots without an arm where patients only get oral anticoagulants may not be well controlled. Appropriate controls help prevent bias and mitigate confounding effects. Testing controls are also common in IVD testing (i.e., the test control test sample may be missing the analytic test chemistry of interest and the test is run in the same manner as the sample with the correct analytic test chemistry).

A suggested CER Table of Contents and checklist were provided and compared to earlier GHTF CER Table of Contents suggestions (Table 1.8).

TABLE 1.8 CER Table of Contents changed from 2007 to 2016.

GHTF SG5/N2R8:2007 [26]	MEDDEV 2.7/1, Rev. 4 (2016) [29]	Differences
1. General details	1. Summary	Changed to executive summary format rather than general details.
	2. Scope	Not mentioned in 2007
	3. Clinical background, current knowledge and SOTA	Not mentioned in 2007
2. Devices description and intended application	4. Device under evaluation	Less detailed in 2016 than 2007
3. Intended therapeutic and/or diagnostic indications and claims		Less detailed in 2016 than 2007
4. Evaluation context and choice of clinical data types		Less detailed in 2016 than 2007
5. Summary of the clinical data and appraisal		Less detailed in 2016 than 2007
6. Data analysis (performance, safety and product literature)		Less detailed in 2016 than 2007
7. Conclusions	5. Conclusions	Same
	6. Date of next evaluation	Not mentioned in 2007
	7. Dates and signatures	Not mentioned in 2007
	8. Evaluator qualifications	Not mentioned in 2007
	9. References	Not mentioned in 2007

GHTF, Global Harmonization Task Force; *MEDDEV*, medical devices documents; *SG*, study group.

Note how inadequate the MEDDEV 2.7/1, Rev. 4 Table of Contents is compared to the earlier MDCG Table of Contents, especially for claims, clinical data types, appraisal, and analysis. At least in the 2007 Table of Contents, data analysis had a home in Section 6. This was left as a gaping hole nearly a decade later, in the 2016 guideline, even though clinical data analysis (i.e., evaluation) is the critical "work" required in all clinical/performance evaluations.

BOX 1.7 CER checklist modified from MEDDEV 2.7/1, Rev. 4 (2016)

Check the following before releasing the CER:

- Can the report be read and understood by a third party?
- Does the report provide sufficient detail to understand all clinical data, assumptions, and conclusions?
- Are all clinical data generated and held by the manufacturer mentioned and adequately summarized?
- Is demonstration of equivalence included?
  - If no, are all clinical data requirements met to ensure no clinical data are needed from an equivalent device?
  - If yes, are all differences between device under evaluation and equivalent device disclosed and justified, to explain why each and every difference is not expected to affect device S&P?
- Is the product already in the EU or other global market?
  - If yes, have the latest PMS/PMCF data been considered, summarized, and referenced?
- Are current knowledge/SOTA details up to date, summarized, and adequately substantiated by literature?
- Are the report's contents fully aligned with current knowledge/SOTA?
- Does the report explain benefit, risk, and undesirable side-effect acceptability based on current knowledge/SOTA?
- Does the report align with the risk management file, with appropriate discussions about "residual risk"?
- Is sufficient clinical evidence provided for all devices, models/sizes/settings, and/or different clinical situations?
- Are conclusions correct and inclusive for all devices/models/sizes/settings/clinical situations as stated in the instructions for use (IFU)?
  - Smallest/largest size, highest/lowest dose, etc.?
  - Every medical indication and including an understanding of all contraindications and other concerns?
  - Entire target population (e.g., preterm infants to geriatrics, males and females)?
  - Every applicable medical condition form, stage, severity (e.g., severe/benign, acute/chronic)?
  - All intended users (e.g., healthcare professionals, laypersons, others)?
  - Whole use duration (e.g., maximal number of repeated exposures)?
  - Any and all IFU discrepancies?
- Does the report clearly state conformity to each relevant essential requirements* (AIMDD ER 1, 2, 5/MDD ER 1, 3, 6), and are all discrepancies identified in the report's conclusions?
- Are the report's contents aligned with information materials supplied by the manufacturer, and are all discrepancies identified in the report's conclusions?
- Are all residual risks, uncertainties, and unanswered questions identified in the report's conclusions and assigned to be addressed with PMS/PMCF studies?
- Is the report signed, dated, and version controlled?
- Are evaluator qualifications included and correct in the report?
- Does the manufacturer hold up-to-date signed and dated CVs and declaration of interests for each evaluator?
- Do the declarations of interest include the needed information?
  - relevant financial interests outside current work as an evaluator
  - statements clarifying declaration extent (e.g., time span for grants, revenue sources, benefits paid or promised over 36 months prior to evaluation, are family member financial interests included or not—namely spouse or partner living in same residence as evaluator—children and adults for whom evaluator is legally responsible)
  - employment by manufacturer
  - participation as device preclinical or clinical study investigator
  - ownership/shareholding possibly affected by evaluation outcome
  - grants sponsored by manufacturer
  - benefits like travel or hospitality beyond reasonable work necessity as employee or external evaluator
  - interests related to device/constituent manufacturing
  - interests related to intellectual property (e.g., patents, copyrights, royalties; pending, issued or licensed) possibly affected by evaluation outcome
  - other interests or revenue sources possibly affected by evaluation result
  - manufacturer signature and date.

Checklist modified from MEDDEV 2.7/1, Rev. 4 (2016) [29]

**Note essential requirements (ERs) were changed to general safety and performance requirements (GSPRs) in the MDR/IVDR.*

In addition, Appendix A12 described NB CE quality system assessment by the "conformity assessment route" where the NB audits a sample of procedures for CE, PMS, and PMCF plans and reports. Alternatively, the NB may examine a submitted CER in a "type" examination for identity, appraisal, analysis, and conclusions related to regulatory requirement conformity, to verify and determine if the manufacturer completed the required activities adequately (Table 1.9).

TABLE 1.9 Changes in Notified Bodies (NB) decision-making guidelines from 2009 to 2016.

MEDDEV 2.7.1, Rev. 3 (2009) [27]	MEDDEV 2.7/1, Rev. 4 (2016) [29]
supplied clinical evaluation documentation...	supplied clinical evaluation documentation...
followed relevant procedures...	followed relevant procedures...
described and verified the intended characteristics and performances related to clinical aspect	described and verified the intended characteristics and performances related to clinical aspects
performed an appropriate risk analysis and estimated the undesirable side effects	performed an appropriate risk analysis and estimated the undesirable side-effects which are aligned with the clinical evaluation
involved appropriate clinical expertise in the compilation of the risk analysis to ensure risks and benefits associated with real clinical use are adequately defined	involved appropriate clinical expertise in the clinical evaluation and in the compilation of the risk analysis to ensure risks and benefits associated with real clinical use are adequately defined
	provided a solid justification as the basis for their estimations of benefits, risks, undesirable side-effects, indications and contraindications of the device in question
justified the chosen route/s of clinical data retrieval...	justified the chosen route/s of clinical data retrieval...
identified, appraised, analysed and assessed the clinical data... and demonstrated the relevance and any limitations of the clinical data identified in demonstrating compliance with particular requirements of the Directive or cited in particular aspects of the risk analysis	identified, appraised, analysed and assessed the clinical data (according to previous sections) and demonstrated the relevance and any limitations of the clinical data identified in demonstrating compliance with particular requirements of the Directive or cited in particular aspects of the risk analysis
	identified all clinical data, favourable and unfavourable, that is relevant to the device and using an appropriately robust, reproducible and systematic search strategy
provided sufficient clinical data relating to the safety, performance, design characteristics and intended purpose of the device in order to demonstrate conformity with each of the relevant essential requirements	provided sufficient clinical evidence relating to the safety, including benefits to the patients, the clinical performance intended by the manufacturer (including any clinical claims for the device the manufacturer intends to use), design characteristics and intended purpose of the device, in order to demonstrate conformity with each of the relevant essential requirements
if a critical evaluation of relevant scientific literature is provided, the notified body verifies that this data relates to... [the] safety, performance, design characteristics and intended purpose of the device;	conducted and provided a critical evaluation of relevant scientific literature and data relating to the safety, benefits, performance, design characteristics and intended purpose of the device
if a critical evaluation of relevant scientific literature is provided the notified body verifies that the device under assessment is demonstrated as equivalent to the device to which the data relates in all necessary areas (i.e. clinical, design, biological etc.)	demonstrated the equivalence of the device under evaluation to the device to which the data relates in all necessary areas, i.e. clinical, technical, biological and that the data available adequately addresses conformity to each of the relevant essential requirements (if a critical evaluation of relevant scientific literature is provided as the only source of clinical data)
if a critical evaluation of relevant scientific literature is provided the notified body verifies that the data presented for equivalent devices adequately addresses each of the relevant essential requirements	designed appropriate clinical investigations, when necessary, to address specific questions arising from the critical review of the scientific literature and address each of the relevant essential requirements
provided specific justification if a specific clinical investigation was not performed for class III or implantable devices	provided specific justification if a specific clinical investigation was not performed for class III or implantable devices
Note: A clinical evaluation is required for all classes of medical devices, the relevance of the data or the need for clinical investigation data should always be assessed and documented by the notified body	
provided evidence that clinical investigations presented are in compliance with applicable regulatory and ethical requirements, e.g., ethics committee approval, competent authority approval	provided evidence that clinical investigations presented are in compliance with applicable regulatory and ethical requirements, e.g., scientific validity, ethics committee approval, competent authority approval
	- provided detail of the PMS plan in place for the particular device and justified the appropriateness and adequacy of this plan
	- clearly identified which areas in the clinical evaluation and related data need to be further addressed and confirmed in the postmarket phase, with specific alignment to the PMCF

(*Continued*)

TABLE 1.9 (Continued)

MEDDEV 2.7.1, Rev. 3 (2009) [27]	MEDDEV 2.7/1, Rev. 4 (2016) [29]
justified the appropriateness of the planned PMCF	justified the appropriateness of the planned PMCF
justified and documented if PMCF is not planned as part of the PMS plan for the device	justified and documented if PMCF is not planned as part of the PMS plan for the device
	identified the sources of clinical data which will be gathered from the manufacturer's PMS system and PMCF
	concluded that the contents of the IFU are supported by clinical evidence (description of the intended purpose, handling instructions, type and frequency of risks, warnings, precautions, contraindications, others) and are in line with the risk analysis and clinical evaluation
concluded on the basis of documented justification that the risks are acceptable when weighed against the intended benefits and the relevant Essential Requirements are met.	concluded on the basis of documented *evidence* (1) that the risks are acceptable when weighed against the intended benefits and are compatible with a high level of protection of health and safety
	(2) that the intended clinical performances described by the manufacturer are achieved by the device, and
	(3) that any undesirable side-effect constitutes an acceptable risk when weighed against the performances intended.
The assessment carried out by the Notified Body will typically cover the following aspects of the manufacturer's clinical evaluation:	The assessment carried out by the notified body will *in addition* typically confirm the following aspects of the manufacturer's clinical evaluation:
appraisal to determine suitability and any limitations of the data presented to address the essential requirements in particular relating to the safety and performance of the device as outlined...	appraisal to determine suitability and any limitations of the data presented to address the essential requirements in particular relating to the safety, and performance of the device as outlined...
complete and adequate documentation...	
adequate procedures...	
the validity of any justification given	the validity of any justification given
the listing, characterisation and proof of the clinical performance of the device intended by the manufacturer and the expected benefits for the defined patient group/s	characterisation and evidence-based proof of the clinical performance of the device intended by the manufacturer and the expected benefits for the defined patient group/s
the use of harmonised standards	the application of all relevant harmonised standards or appropriate justifications if not
the use of the list of identified hazards to be addressed through evaluation of clinical data as described	identified hazards to be addressed through *analysis* of clinical data as described
the adequate estimation of the associated risks for each identified hazard by:	the adequate estimation of the associated risks for each identified hazard by:
a) characterising the severity of the hazard;	characterising the severity of the hazard;
b) estimating and characterising the probability of occurrence of harm, health impairment or loss of benefit of the treatment (document with rationale).	estimating and characterising the probability of occurrence of harm, impairment of health
	or loss of benefit of the treatment (documented and discussed based on scientifically valid clinical data)
	the adequate description and estimation of the current state-of-the-art in the corresponding medical field
	a justifiable and reasoned basis for estimation of risks and hazards.
	Where a device incorporates, as an integral part, a substance which, if used separately, may be considered to be a medicinal product, the notified body is responsible for verifying the usefulness of the medicinal substance as part of the device prior to the submission of an application for scientific opinion from a medicines authority.
The decision on the acceptability of risks... in relation to each identified hazard, and characterisation of the corresponding risk/benefit ratio as:—unacceptable; or—broadly acceptable; or—acceptable under specified conditions.	

(Continued)

TABLE 1.9 (Continued)

MEDDEV 2.7.1, Rev. 3 (2009) [27]	MEDDEV 2.7/1, Rev. 4 (2016) [29]
For drug-device combination products where a scientific opinion from a medicinal competent authority or from the EMEA has been sought, the notified body should consider any comments or considerations raised in the medicinal clinical assessment when making its final decision on the device. In the case of devices with a human blood derivative the notified body may not deliver a positive decision to issue a certificate if the EMEA's scientific opinion is unfavourable.	For drug-device combination products and products incorporating stable human blood derivatives, where a scientific opinion from a medicinal competent authority or from the European Medicines Agency (EMA) has been sought, the notified body should consider any comments or considerations raised in the medicinal clinical assessment when making its final decision on the device. In the case of devices with a human blood derivative the notified body may not deliver a positive decision to issue a certificate if the EMA's scientific opinion is unfavourable

Bolded items represent MEDDEV 2.7/1 guideline changes from Rev. 3 (2009) [27] to Rev. 4 (2016) [29].
EMA, European Medicines Agency; *EMEA*, Europe, Middle East, and Africa; *MEDDEV*, medical devices documents; *PMCF*, postmarket clinical follow-up; *PMS*, postmarket surveillance.

In general, the NB seeks to understand if "valid" decision-making criteria were employed (e.g., using applicable guidance and international standards like ISO 14971 for medical device risk management activities outside of the CER/PER). The NB will document the CER assessment and conclusions in their CEAR. Over time, the NB CE quality system assessment and decision-making guidelines have expanded the NB responsibilities.

Unfortunately, the historical MEDDEV 2.7/1 guidelines have not kept pace with the regulatory changes, and two more recent documents have added some context. MDCG 2019-6, Rev. 4 [38] explained that the NB internal/integrated clinician role was to "clinically judge the opinion provided by any external expert" and to "be responsible to make a recommendation to the decision maker on the adequacy of the clinical evaluation." This clinician should identify when specialist input from one or more clinical experts is required for CE/PE assessment (as defined in MDR/IVDR Annex VII, Section 3.2.4, MDR Annex IX, Section 4.3, and NBOG BPG 2017-2, Section 5.4) regardless of whether the device conformity to Annex I requirements are based on clinical data or not (i.e., when clinical data are not deemed appropriate, the clinician will determine if the justification for this claim is adequate before requesting clinical expert support for the CE/PE assessment). MDR/IVDR Annex VII, Section 3.2.4 requires the clinician to be internal (i.e., an employee) or a clinician "otherwise integrated" into the NB assessment and decision-making process, but who cannot be the final reviewer or decision maker per Annex VII, Section 3.2.7.

In addition, MDCG 2019-13 [39] discussed the MDR/IVDR sampling requirements including the "depth and extent of the technical documentation assessment" for Class IIa/IIb and Class B/C devices which "will be the same as the depth of assessment carried out for Class III and Class IIb implantable and Class D devices." In other words, device technical documentation will be assessed against all GSPRs, as specified in Annex I as well as the Annex II and III requirements. The NB assessment records must allow "a third party to understand the functionality of the device and all aspects of the assessment including judgements made by the assessor." The assessment should account for "every device (i.e. Basic UDI-DI)" including "different variants, models or sizes." The technical document review will also assess "how the differences among these have been addressed in the technical documentation and whether all of them are in line with the relevant requirements." Specifically, "According to Article 54, Class IIb active devices intended to administer and/or remove a medicinal product falling into rule 12 of Annex VIII are subject to the clinical evaluation consultation procedure prior to issuing of the certificate. These devices can be subject to sampling but according to Articles 54(3) and 55 the notified body must ensure... at least the clinical evaluation assessment report (CEAR) for each device is uploaded in Eudamed prior to issuing the QMS certificate. This means... sampling will not apply to the clinical evaluation as it has to be assessed for every device."

In general, the MDCG appears to be replacing MEDDEV 2.7/1, Rev. 4 (June 2016) [29] a little bit at a time, with hundreds of new guidance documents gradually being issued.

1.7.12 International Medical Device Regulators Forum/Good Regulatory Review Practices Group WG/N47 essential principles guidance (October 2018)

The IMDRF/GRRP WG/N47 "Essential Principles of Safety and Performance of Medical Devices and IVD Medical Devices" [40] uses new MDR language to put the GHTF SG5 documents into context and supersedes the similarly named GHTF/SG1/N68:2012 [41]. The "Essential Principles" are described for worldwide adoption, and guidance is offered about using the standard to meet harmonized S&P Essential Principles for medical devices and IVD medical device design and manufacture. The global regulatory harmonization goal is ongoing, with the intent to decrease cost and to allow earlier access to new devices. The references and standards lists are extensive and helpful with information about quality systems (ISO 13485), risk management (ISO 14971), sterilization (ISO 11135), packaging (ISO 11607), biocompatibility (ISO 10993), and many others.

The definitions generally align with the GHTF SG5 details about clinical data, evaluation, evidence, investigation, and performance. This 2018 IMDRF document [40] defines "Indications for Use" as separate from Intended Use and similar to the US FDA definition.

Indications for Use: A general description of the disease or condition the medical device or IVD medical device will diagnose, treat, prevent, cure, or mitigate, including a description of the patient population for which the medical device or IVD medical device is intended.

Intended Use/Intended Purpose: The objective intent regarding the use of a product, process or service as reflected in the specifications, instructions and information provided by the manufacturer. (Modified from GHTF/SG1/N77:2012)... NOTE: The intended use can include the indications for use. [40]

The indication for use is now being used more globally to limit medical device sales and marketing to specific disease, conditions, and populations where the medical device manufacturer has clear and convincing S&P clinical data accumulated for the device.

In addition, this guidance included terms from GHTF/SG5/N6:2012 [35] for IVD medical devices as follows:

Analytical Performance of an IVD Medical Device: The ability of an IVD medical device to detect or measure a particular analyte.

Clinical Evidence for an IVD Medical Device: All the information that supports the scientific validity and performance for its use as intended by the manufacturer.

Clinical Performance of an IVD Medical Device: The ability of an IVD medical device to yield results... correlated with a particular clinical condition/physiological state in accordance with target population and intended user... Clinical performance can include diagnostic sensitivity and diagnostic specificity based on the known clinical/physiological state of the individual, and negative and positive predictive values based on the prevalence of the disease.

Effective: The ability of a medical device or IVD medical device to provide clinically significant results in a significant portion of the target population... This ability is assessed in situations where the medical device or IVD medical device is used for its intended uses and conditions of use and accompanied by adequate directions for use and warnings against unsafe use.

In Vitro Diagnostic (IVD) Medical Device: ... means a medical device, whether used alone or in combination, intended by the manufacturer for the in-vitro *examination of specimens derived from the human body solely or principally to provide information for diagnostic, monitoring or compatibility purposes. NOTE 1: IVD medical devices include reagents, calibrators, control materials, specimen receptacles, software, and related instruments or apparatus or other articles and are used, for example, for the following test purposes: diagnosis, aid to diagnosis, screening, monitoring, predisposition, prognosis, prediction, determination of physiological status...*

Performance Evaluation of an IVD Medical Device: Assessment and analysis of data to establish or verify the scientific validity, the analytical and, where applicable, the clinical performance of an IVD medical device.

The manufacturer must ensure each device is safe and effective for the entire device life cycle. The device must perform as intended and must meet "broad, high-level, criteria for design, production and post-production." Devices should perform as intended during normal conditions of use. The risk management plan (RMP) should analyze hazards, estimate/evaluate risks, reduce/eliminate/control risks, and evaluate overall information, and should be updated regularly to address any new risk details. Safety principles should guide risk control measures to ensure residual risks remain "acceptable," and solutions should be prioritized to remove or reduce risks, develop protection measures (e.g., alarms), and provide safety information and training for users when needed. The risk processes should include a means to inform users of residual risks and designs to reduce user risks (e.g., ergonomic/environmental features, consideration of user knowledge, experience, physical condition). The device should not be adversely affected by the stresses of normal use when properly maintained and calibrated, should maintain integrity and cleanliness when used as intended, and should not be negatively affected by expected transportation or storage conditions for the entire shelf life of the device. Stability should be acceptable for the entire device shelf life, and undesirable side effects should be minimized.

In particular, this 2018 guideline [40] re-iterates two basic clinical evaluation components:

1. Clinical data should be used to establish and evaluate the benefit–risk ratio based on clinical trials, clinical literature, and clinical experience.
2. Clinical trials should be ethical and should protect the rights, safety, and well-being of all human subjects.

Outside the CE/PE processes, the device's chemical, physical, and biological details should ensure the device is biocompatible, minimize risks posed by contaminants/residues, reduce risks posed by leaking/leaching and by ingress of substances into the device, and reduce risks of infection. Sterilization details should be optimized, microbial contamination should be reduced, and the device's impact on the environment as well as any restrictions on use conditions should be carefully indicated in the labeling to ensure safe and effective use. Users should be protected from electrical, mechanical, and thermal risks. Active medical devices should have appropriate testing and features (e.g., warnings, alarms, monitors) to ensure safe operations, and devices including software (or software as a medical device) or diagnostic/measuring functions should be accurate, reliable, precise, SOTA, and safe, and should perform as intended. Appropriate information should accompany the device and patients should be protected from any radiation emitted from the device. Devices used by laypersons, devices incorporating drugs or materials of biological origin, and implantable and *in vitro* diagnostic devices have additional requirements.

1.7.13 International Medical Device Regulators Forum Medical Device Clinical Evaluation Working Group (MDCE WG)/N56FINAL 2019 clinical evaluation (October 2019)

The IMDRF updated the GHTF SG5/N2R8:2007 [26] clinical evaluation guidance in 2019 [28] and widened the CE scope to include "safety, clinical performance and/or effectiveness data" from "pre- and post-market clinical data relevant to the intended use of the device in question" (p. 8). The overall guidance was similar to GHTF SG5/N2R8:2007 and the recommendations for appraising data did not change substantially. The main changes were a recommendation for evaluators to "identify, in advance, the appropriate criteria to be applied for a specific circumstance" (p. 17) and to "assess whether clinical data are collected in conformance with the applicable regulatory requirements or other relevant standards (ISO 14155:2011) and whether clinical data are applicable to the population for which the marketing authorization is being sought" (p. 16). The suitability and data contribution details were included along with the formulation criteria checklists developed in the earlier 2007 guidance.

1.7.14 Medical Device Coordinating Group 2020-1 clinical evaluation/performance evaluation for medical device software

Medical device software (MDSW) used to inform or drive clinical management including treatment or diagnosis requires clinical and performance evidence to support safety, performance, and claims just like any other medical device type. The MDCG 2020-1 document entitled "Guidance on Clinical Evaluation (MDR)/Performance Evaluation (IVDR) of Medical Device Software (Mar-20)" [42] provided "a framework" to determine the "appropriate level of CLINICAL EVIDENCE required" for MDSW to fulfill the MDR/IVDR requirements. MDSW was defined as "software... intended to be used, alone or in combination, for a purpose as specified in the definition of a "medical device" in the" MDR or IVDR. The MDCG 2020-1 accounted for concepts outlined in the IMDRF/SaMD WG/N41FINAL:2017 Software as a Medical Device (SaMD): Clinical Evaluation [43].

The MDCG 2020-1 explained how MDSW was distinct from medical device component/accessory software because "the manufacturer claims a specific medical intended purpose" and "has a CLINICAL BENEFIT." As a result, MDSW required CE/PE "within its own conformity assessment" (Table 1.10).

TABLE 1.10 MDSW CE/PE requirements.

Software model	Software description/function	CE/PE scope
Software driving or influencing medical device use	No independent (software-specific) intended purpose or independent claimed clinical benefit (i.e., not MDSW)	Medical device driven or influenced by included software (i.e., a medical device component or accessory)
MDSW	Independent (software-specific) intended purpose and claimed clinical benefit	MDSW only
MDSW	Intended purpose and claimed clinical benefit (software-specific) related to driving or influencing a medical device for a medical purpose	MDSW and the driven or influenced medical device

CE, clinical evaluation; *MDCG*, Medical Device Coordination Group; *MDSW*, medical device software; *PE*, performance evaluation.
Source: Adapted from MDCG 2020-1.

In other words, only software features claiming a specific intended purpose and clinical benefit required MDSW-specific clinical performance data typically reported in source data documents including technical standards, medical society guidelines, systematic literature reviews, clinical trials, clinical experience data, real-world data analysis, etc. In particular, MDSW clinical benefits can be distinct from clinical benefits derived from traditional medical devices or pharmaceuticals. For example, "providing accurate medical information on patients" is considered a clinical benefit and performance should be evaluated "against medical information obtained through the use of other diagnostic options and technologies."

As for all medical devices and IVDs, MDCG 2020-1 also explains if the MDSW has multiple indications, "each indication and claimed clinical benefit" in the intended purpose "should be assessed individually and have the supporting clinical evidence." In addition, MDSW CE/PE follows the same steps as CE/PE for traditional medical devices and IVDs, including planning, data identification, appraisal, analysis, and documentation; however, compiling MDSW clinical evidence required three "key components" (similar to IVD key components) as follows:

- Clinical association (MDR) and scientific (IVDR) validations about "MDSW's output (e.g., concept, conclusions, calculations) based on the inputs and algorithms selected… associated with the targeted physiological state or clinical condition." For example, if an MDSW analyzes certain analytes in blood to detect an underlying disease, the evaluator must demonstrate a link between the analytes and the disease.
- Technical performance (MDR) and analytical performance (IVDR) validations about "MDSW's ability to accurately, reliably and precisely generate the intended output, from the input data."
- Clinical performance validation about "MDSW's ability to yield clinically relevant output in accordance with the intended purpose" where the clinical relevance was a demonstrable positive impact on patient health, patient management, public health or "screening, monitoring, diagnosis or aid to diagnosis." If MDSW clinical benefits cannot "be specified through measurable, patient-relevant clinical outcome/s," clinically relevant outputs can be related to "demonstrated predictable and reliable use and usability."

The MDCG 2020-1 guidance provided specific considerations for collecting clinical/performance data to validate these three key MDSW CE/PE components and several examples for strategies to establish MDSW validity and performance.

Like all medical device CE/PE processes, the MDCG 2020-1 guidance reiterated how the MDSW CE/PE process "must consider the Benefit-Risk Ratio in light of the state-of-the-art related to practice of medicine for diagnosis, treatment or patient management," and must include all CE/PE components. For example, the clinical data amount and quality must be appraised in the same manner as other medical devices and MDSW CERs/PERs should not differ fundamentally from other medical device CERs/PERs.

Also, the software "driving or influencing [another medical device] is covered by the medical device regulations either as a part/component of a device or as an accessory for a medical device." More guidance about classifying and qualifying software is provided in MDCG 2019-11 "Guidance on Qualification and Classification of Software in Regulation (EU) 2017/745—MDR and Regulation (EU) 2017/746—IVDR (Oct-2019)" [44]. This guidance describes how MDSW can act as a stand-alone product or as a component of a device system, and devices with MDSW require special qualification and classification.

1.7.15 Medical Device Coordinating Group 2020-6 clinical evidence for legacy devices

Three years after the MDR and IVDR were released, MDCG 2020-6, entitled "Regulation (EU) 2017/745: Clinical Evidence Needed for Medical Devices Previously CE Marked under Directives 93/42/EEC or 90/385/EEC: A Guide for Manufacturers and Notified Bodies (April 2020)," [45] reiterated and defined eight important terms not yet defined in MDR/IVDR Article 2:

> *"indication" and "indication for use" are "the clinical condition… to be diagnosed, prevented, monitored, treated, alleviated, compensated for, replaced, modified or controlled by the medical device…" This is not the same as the "intended purpose/intended use" "which describes the device effect… (e.g., … an intended purpose of disinfection or sterilisation of devices)."*
>
> *"intended use" and "intended purpose" "have the same meaning"*
>
> *"legacy devices" are "devices previously CE marked under… MDD… or… AIMDD…"* [presumably IVDD].

"level of clinical evidence" was applied "to requirements for demonstration of conformity with the relevant GSPR and overall benefit-risk..." including the medical device "evidence... quality, quantity, completeness and statistical validity, etc.... required to demonstrate safety, performance and the benefit-risk conclusion... not... 'levels of evidence' (as used in evidence-based medicine) which is used to rank study designs, and is only a part of the concept..."

"scientific validity" and "scientifically valid" are often associated with "clinical data planning, evaluation and conclusions" when clinical evaluations follow a "defined and methodologically sound procedure" with "implicit" "expectations of scientific validity..." and fully consider the "adequacy of study design and controls for bias, appropriateness and relevance of research questions, adequacy of sample sizes and statistical analyses, completeness of data, adequacy of follow-up period, and appropriateness of conclusions on the basis of objective evidence."

And "'scientific validity of an analyte' means the association of an analyte with a clinical condition or a physiological state" [13] (IVDR Article 2 (38)).

similar device" is defined as "devices belonging to the same generic device group... [or] devices having the same or similar intended purposes or a commonality of technology allowing them to be classified in a generic manner not reflecting specific characteristics..."

state-of-the-art" (SOTA) was defined by IMDRF/GRRP WG/N47 [40] as a "Developed stage of current technical capability and/or accepted clinical practice in regard to products, processes and patient management, based on... relevant... findings of science, technology and experience..." SOTA "embodies... good practice in technology and medicine... [and] does not necessarily imply the most technologically advanced solution..."

"well-established technology" (WET) devices meet common features including "relatively simple, common and stable designs with little evolution... well-known safety... not... associated with safety issues... well-known clinical performance characteristics... standard of care devices...[with] little evolution in indications and the state-of-the-art... [and] a long history on the market."

Note: WET does not require a previous CE mark or equivalence and was previously limited to "sutures, staples, dental fillings, dental braces, tooth crowns, screws, wedges, plates, wires, pins, clips or connectors for which the clinical evaluation is based on sufficient clinical data and is in compliance with the relevant product-specific CS, where such a CS is available" [2] (MDR Article 61(6)(b)). CS, common specification.

Understanding these terms is important when working on CERs and PERs. In addition, this guidance on clinical evaluation of legacy devices noted the MEDDEV 2.7/1, Rev. 4 sections still relevant under MDR. Oddly, MDCG 2020-6 simply embedded the entire MEDDEV 2.7/1, Rev. 4 in a Background Note [46], even though many details conflicted directly with the MDR. In addition, MEDDEV 2.7/1, Rev. 4 did "not concern *in vitro* diagnostic devices" and MDCG 2020-6 did not mention IVD devices at all.

Manufacturers followed MEDDEV 2.7/1, Rev. 4 [29] general principles for CE and PE even though the documents were not for IVD devices until the MDCG 2022-2 entitled "Guidance on General Principles of Clinical Evidence for In Vitro Diagnostic Medical Devices (IVDs) (Jan-22)" [47] summarized the PE requirements from EU IVDR Article 56 and Annex XIII, Part A as "An assessment and analysis of data to establish or verify the scientific validity, the analytical and, where applicable, the clinical performance of a device." Essentially, the guidelines for medical devices and IVD medical devices are saying the same thing: CE/PE clinical data must be scientifically valid to document device safety and performance including IVD device clinical and analytical performance. Both documents also outlined similar evaluation processes including RM, scoping/planning, data identification/establishment, appraisal, analysis, conclusions reporting, continuous updates, and follow-up with more data gathered in the PMS system.

MEDDEV 2.7/1, Rev. 4 [29] indicated medical device evaluators:

"should address if the following points are adequately supported by sufficient clinical evidence:

- *the intended purpose described in the information materials supplied by the manufacturer (including for all medical indications);*
- *the clinical performance and benefits described in the information materials supplied by the manufacturer (including, for example, any claims on product performance and safety);*
- *measures for risk avoidance and risk mitigation described in the information materials supplied by the manufacturer (including, for example the declaration of the residual risks, contraindications, precautions, warnings, instructions for managing foreseeable unwanted situations);*
- *the usability of the device for the intended users and the suitability of the information materials supplied by the manufacturer for the intended users (including, if applicable, for lay or disabled persons);*
- *instructions for target population groups (including, for example, pregnant women, paediatric populations)."*

Similar to the assessment made for medical devices by the medical device evaluator, MDCG 2022-2 [47] guides the IVD evaluator to "make a qualified assessment whether the IVD will achieve the intended clinical benefit/s and safety, when used as intended by the manufacturer," taking into account:

- *"the intended users,*
- *the state-of-the-art,*
- *the nature, severity and the evolution of the condition being diagnosed or treated,*
- *the adequacy of the estimation of associated risks for each identified hazard,*
- *the number and severity of adverse events,*
- *the availability of alternative diagnostic devices and current standard of care."*

The MDCG 2020-6 [45] guidance about legacy devices "previously CE marked under the European Medical Devices Directive 93/42/EEC (MDD) or Active Implantable Medical Devices Directive 90/385/EEC (AIMDD)" stated "Legacy devices which have been placed on the market have been subjected to conformity assessment and therefore are presumed to have been supported by clinical data." Unfortunately, the guidance also stated: "As requirements and guidance developed over time, it is not necessarily the case that the clinical data used for conformity assessment under the Directives is clinical data providing sufficient clinical evidence for the purpose of MDR requirements." In this case, "Post-market clinical data together with the clinical data generated for the conformity assessment under the MDD/AIMDD will be the basis of the clinical evaluation process for legacy devices under the MDR... The MDR compliant clinical evaluation for a legacy device must contain the identification of available clinical data as well as their appraisal / analysis / evaluation and shall lead to a demonstration of conformity to the MDR GSPR based on clinical data providing sufficient clinical evidence as part of a lifecycle approach."

1.7.16 Medical Device Coordinating Group 2022-10 clinical trials and *in vitro* diagnostics

MDCG 2022-10, entitled "Q&A on the interface between Regulation (EU) 536/2014 on clinical trials for medicinal products for human use (CTR) and Regulation (EU) 2017/746 on *in vitro* diagnostic medical devices (IVDR) (May-22)" [48], clarified clinical trial assay requirements by exploring interactions between EU Reg. 536/2014 [6] clinical trial regulation (CTR) about "clinical trials for medicinal products for human use" and EU Reg. 2017/746 IVDR. The overall aim of this guidance was "to support the conduct of clinical trials using diagnostic assays, including combined trials for the development of companion diagnostics." Companion diagnostics were defined as devices "essential for the safe and effective use of a corresponding medicinal product to... Identify... patients... likely to benefit from the corresponding medicinal product; or... patients... at increased risk of serious adverse reactions as a result of treatment with the corresponding medicinal product."

This guidance [48] included 16 questions and answers including but not limited to the following:

- transitioning devices to IVDR conformity to remain on the market;
- clinical-trial IVD devices must be CE marked, an in-house IVD, or an IVD for a performance study;
- clinical-trial IVD assays with a medical purpose during the trial (e.g., guiding medical management decisions, selecting subjects, assigning subjects to a treatment arm or monitoring care) are subject to the IVDR;
- IVD devices with a clinical trial medical purpose are not research use only devices, they become IVDs subject to the IVDR; and
- in-house IVD devices may not be "transferred to another legal entity"; however, samples can be transported and analyzed by an in-house device at another location.

Clinical performance studies including IVD devices need to comply with traditional clinical study regulations and the specific IVD performance evaluation needs.

1.7.17 Medical Device Coordinating Group 2022-2 clinical evidence for *in vitro* diagnostics

MDCG 2022-2 [47] begins by linking clinical evidence to the IVD medical device "intended purpose" definition per IVDR Annex I, 20.4.1(c) as follows:

"the device's intended purpose:

(i) *what is detected and/or measured;*
(ii) *its function (e.g., screening, monitoring, diagnosis or aid to diagnosis, prognosis, prediction, companion diagnostic);*

(iii) *the specific information that is intended to be provided in the context of:*
- *a physiological or pathological state;*
- *congenital physical or mental impairments;*
- *the predisposition to a medical condition or a disease;*
- *the determination of the safety and compatibility with potential recipients;*
- *the prediction of treatment response or reactions;*
- *the definition or monitoring of therapeutic measures;*

(iv) *whether it is automated or not;*
(v) *whether it is qualitative, semiquantitative or quantitative;*
(vi) *the type of specimen/s required;*
(vii) *where applicable, the testing population; and*
(viii) *for companion diagnostics, the International Nonproprietary Name (INN) of the associated medicinal product for which it is a companion test."*

The guidance defines "Metrological Traceability' as a "Property of a measurement result whereby the result can be related to a reference through a documented unbroken chain of calibrations, each contributing to the measurement uncertainty. The metrological traceability chain is a sequence of measurement standards and calibrations... used to relate a measurement result to a reference." In addition, the guidance reminds the reader how the IVD "clinical benefit... is fundamentally different from... pharmaceuticals or of therapeutic medical devices, since the benefit of IVDs lies in providing accurate medical information on patients... assessed against... other diagnostic options and technologies."

The IVDR and this guidance stated "the necessary clinical evidence should be based on a sufficient amount and quality of data in order to allow a qualified assessment of whether the IVD is safe, performant and achieves the intended clinical benefit/s, when used as intended." The IVDR analyte clinical context should be captured and some "non-exhaustive questions" were provided to assist with IVD PE (these can be used for medical device CEs as well; bold emphasis added below to identify IVD focus).

- *"Does the data support the intended purpose, intended users indications, device specifications, target groups, clinical claims and the relevant safety and performance requirements?*
- *Has the novelty and level of innovation/history on the market been evaluated and considered?*
- *Have the risks been identified, mitigated and the effectiveness of the risk control measures been verified?*
- *Have for example environmental conditions, interference factors, exogenous factors and endogenous factors been evaluated?*
- *Has the quality of literature retrieved and reviewed been evaluated and has a rationale for the selection process been provided?*
- *Has there been a sufficient number of observations to draw scientifically valid conclusions?*
- *Have any limitations within the observations been appropriately justified?*
- *Was the statistical approach including sample size appropriate to reach a scientifically valid conclusion?*
- *Have the scientific validity, analytical and clinical performances been demonstrated?*
- *Is data from performance studies or other sources sufficient to verify the safety and performance, including clinical benefits (where applicable) of the devices when used as intended with respect to the state-of-the-art?*
- *Does the design and results of the performance studies support the clinical evidence?*
- *Have all deviations from and all planned changes to the performance evaluation plan been justified?*
- *Has the relevance of the information of the performance evaluation been assessed and documented?*
- *Has the contribution of each data set to the performance evaluation been weighted according to systematic criteria?*
- *Is the data set appropriate and takes into account the state-of-the-art of the device?*
- *Is all supporting data fully traceable, documented and is integrity assured?*
- *Were all ethical, legal and regulatory considerations/ requirements taken into account?*
- *Have all omissions been clearly outlined and justified?"*

MDCG 2022-2 [47] described three IVD-specific PE themes: scientific validity, analytical performance, clinical performance, and clinical performance studies; specific IVD IFU content details; and how to apply harmonized standards and/or common specifications (CSs) and IVD-specific definitions. The PE process was defined as "a structured, transparent, iterative and continuous process which is part of the quality management system and is conducted throughout the lifecycle of an IVD." Guidelines were offered for PEP, identifying data specific to each PE theme, PER writing, and updating and establishing a PMPF plan.

This guidance described IVD PMS and PMPF processes for "continuous update of the performance evaluation" to detect signals and ensure performance. The PMPF plan must include "appropriate methods, procedures and product specific appropriate triggers for proactively collecting and evaluating safety, performance and scientific data." Appendix I provided a flow chart to generate IVD CE starting with the PEP to ensure scientific validity before moving into analytical and clinical performance analyses, followed by conclusions, and continuous PE. Appendix II provided PER update requirements based on device class.

MEDDEV 2.7/1, Rev. 4 and MDCG 2022-2 have many similarities even though MEDDEV 2.7/1, Rev. 4 focuses on CE (i.e., CEP, CER, PMCF, SSCP, updates, and NB review processes) and MDCG 2022-2 guides PE (i.e., PEP, PER, PMPF, SSP, updates, and NB processes).

1.8 Designation of Notified Bodies under medical device regulation and *in vitro* diagnostic regulation

The EC website for NBs [49] lists NBs designated to assess conformity under MDR [50] and IVDR [51] (Table 1.11).

TABLE 1.11 Fifty-six NBs designated under MDR ($N = 44$) and IVDR ($N = 12$) as of 01 Apr 2024.

Body Type	Name	Country
MDR (EU Reg. 2107/745) $N = 44$		
NB 0044	TÜV NORD CERT GmbH	Germany
NB 0050*	National Standards Authority of Ireland (NSAI)	Ireland
NB 0051*	IMQ Istituto Italiano del Marchio di Qualità S.P.A.	Italy
NB 0123*	TÜV SÜD Product Service GmbH	Germany
NB 0124*	DEKRA Certification GmbH	Germany
NB 0197*	TÜV Rheinland LGA Products GmbH	Germany
NB 0297*	DQS Medizinprodukte GmbH	Germany
NB 0318	Centro Nacional de Certificacion de Productos Sanitarios	Spain
NB 0344*	DEKRA Certification B.V.	Netherlands
NB 0373	Istituto Superiore di Sanita'	Italy
NB 0425	ICIM S.P.A.	Italy
NB 0426	ITALCERT SRL	Italy
NB 0459*	GMED SAS	France
NB 0476	Kiwa Cermet Italia S.P.A.	Italy
NB 0477	Eurofins Product Testing Italy S.r.l.	Italy
NB 0482*	DNV MEDCERT GmbH	Germany
NB 0483*	MDC Medical Device Certification GmbH	Germany
NB 0494	SLG Prüf und Zertifizierungs GmbH	Germany
NB 0537	Eurofins Electric & Electronics Finland Oy	Finland
NB 0546	Certiquality S.r.l.	Italy
NB 0598 (ex-0403)	SGS Fimko Oy	Finland
NB 0633	Berlin Cert Prüf- und Zertifizierstelle für Medizinprodukte GmbH	Germany
NB 1023	Institut pro Testování a Certifikaci, a. s. (Institute for Testing and Certification) merged with ex-NB 1390	Czech Republic

(Continued)

TABLE 1.11 (Continued)

Body Type	Name	Country
NB 1282	Ente Certificazione Macchine S.r.l.	Italy
NB 1304	Slovenian Institute of Quality and Metrology—SIQ	Slovenia
NB 1370	Bureau Veritas Italia S.P.A.	Italy
NB 1383	Cesky Metrologicky Institut	Czech Republic
NB 1434	Polskie Centrum Badan i Certyfikacji S.A.	Poland
NB 1639	SGS Belgium NV	Belgium
NB 1912	Kiwa Dare B.V.	Netherlands
NB 1936	TÜV Rheinland Italia S.r.l.	Italy
NB 2265	3EC International a.s.	Slovakia
NB 2274	TÜV NORD Polska Sp. z o.o.	Poland
NB 2292	UDEM Uluslararasi Belgelendirme Denetim Egitim Merkezi San. ve Tic. A.Ş.	Türkiye
NB 2409*	CE Certiso Orvos- és Kórháztechnikai Ellenőrző és Tanúsító Kft.	Hungary
NB 2460*	DNV Product Assurance AS	Norway
NB 2696	UDEM Adriatic d.o.o.	Croatia
NB 2764	Notice Belgelendirme, Muayene ve Denetim Hizmetleri Anonim ŞirketiTürkiySI Group The Netherlands B.V.	Netherlands
NB 2797*	BSI Group The Netherlands B.V.	Netherlands
NB 2803	HTCert (Health Technology Certification Ltd)	Cyprus
NB 2862*	Intertek Medical Notified Body AB	Sweden
NB 2975	SZUTEST Konformitätsbewertungsstelle GmbH	Germany
NB 3022	Scarlet NB B.V.	Netherlands
NB 3033	RISE Medical Notified Body AB	Sweden
IVDR (EU Reg. 2107/746) *N* = 12		
NB 0050	National Standards Authority of Ireland (NSAI)	Ireland
NB 0123*	TÜV SÜD Product Service GmbH	Germany
NB 0124*	DEKRA Certification GmbH	Germany
NB 0197	TÜV Rheinland LGA Products GmbH	Germany
NB 0344	DEKRA Certification B.V.	Netherlands
NB 0459	GMED SAS	France
NB 0483	MDC Medical Device Certification GmbH	Germany
NB 0537	Eurofins Electric & Electronics Finland Oy	Finland
NB 2265	3EC International a.s.	Slovakia
NB 2797*	BSI Group The Netherlands B.V.	Netherlands
NB 2962	QMD Services GmbH	Austria
NB 3018	Sertio Oy	Finland

* 17 were among the first accredited NBs for conformity assessments (14 for MDR and 3 for IVDR) by September 7, 2020.

Each NB designation may have conditions restricting or specifying the products or procedures that the NB can review. For example, Eurofins Electric & Electronics Finland Oy has many conditions for their EU Reg. 2017/745 MDR "Tasks performed by the Body," including (but not limited to) the following:

Devices that directly contact central nervous system or central circulatory system, active therapeutic devices with an integrated or incorporated diagnostic function which significantly determines the patient management by the device (e.g., closed loop systems or automated external defibrillators), and devices that are intended for controlling, monitoring or directly influencing the performance of active implantable devices are excluded.

Active prostheses and exoskeletons are excluded.

Devices for sterilization are excluded.

Prior to 2020, many NBs were available to support medical device CE marking; however, the changes in EU and Competent Authority oversight and NB designation and certification severely limited the NBs qualified and designated to perform CER/PER assessments. Device manufacturers struggled because fewer accredited NBs, with insufficient staffing, were available to meet CER/PER assessment needs.

BOX 1.8 BREXIT and BSI

Accredited NBs can be removed from the accredited NBs list to conduct conformity assessments to EU Reg. 2017/745 and 2017/746. For example, NB 0086 in the United Kingdom was "BSI Assurance UK LTD" and their certification was "withdrawn" on December 31, 2020. After the United Kingdom withdrew from the EU on January 31, 2020, the BSI (NB 0086) began UK Conformity Assessed (UKCA) marking as a *UK approved body* for products on the market in England, Scotland, and Wales, and UKNI marking for products on the market in *Northern Ireland*.

In addition, NBs sometimes changed names. For example, an NB in the Netherlands (NB 1912) named DARE!! Services B. V. became Kiwa Dare B.V.

Companies and products in the hundreds and thousands (respectively) were waiting for NB reviews, and the wait times typically exceeded 9 months. At the same time, thousands of workers came into this workflow without the necessary training or experience. Companies and NBs were trying to meet the regulatory requirements; however, NBs were finding extensive CER/PER deficiencies and lengthy remediation processes, especially in large medical device companies, and lengthy, multiyear timelines were required to accommodate such sweeping changes throughout the organization.

1.9 Linking clinical or performance evaluation, postmarket surveillance, and risk management systems in the quality management system

Device manufacturers must have a reliable and well-documented quality management system (QMS) for medical device manufacturing. This QMS must document how multiple independent systems work together to accomplish clinical trial, clinical/performance, PMS, and RM evaluations. This overarching framework with multiple interdependent systems is required for conformity to the MDR/IVDR GSPRs in order to receive an EU CE mark (Fig. 1.3).

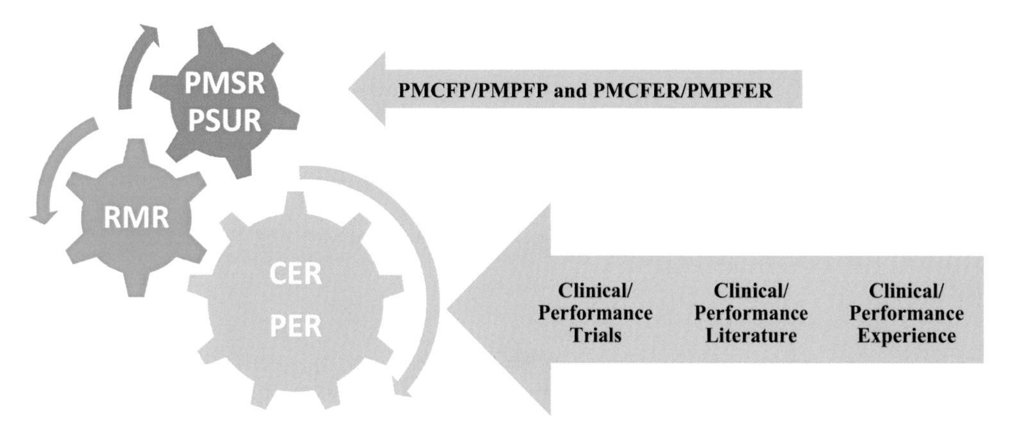

FIGURE 1.3 Manufacturing QMS interrelated reporting processes. *CER*, clinical evaluation report; *PER*, performance evaluation report; *PMCFER*, postmarket clinical follow-up evaluation report; *PMCFP*, postmarket clinical follow-up plan; *PMPFER*, postmarket performance follow-up evaluation report; *PMPFP*, postmarket performance follow-up plan; *PMSR*, postmarket surveillance report; *PSUR*, periodic safety update report; *QMS*, quality management system; *RMR*, risk management report.

Both RM and PMS systems require clinical data inputs, and they need to integrate with CER/PER findings just as the CER/PER needs to evaluate the clinical data considered in the RM and PMS systems. A good way to think about this is as a "check and balance" between different systems with detailed areas of overlap. Each system should inform the other and each system should include an independent risk evaluation about the patient using the device. The RM system will evaluate any/all theoretical or real risks to the device and patient; the PMS will evaluate any/all PMS and safety reports received by the company and needing to go out to the regulatory authorities; and the CER/PER will evaluate all clinical data from trials, literature, and experience data including the clinical data housed in the separate clinical trial, RM, and PMS systems.

To put this into perspective, the RM system is often developed in compliance with the international standard ISO 14971 [52], which requires a risk management report (RMR) documenting the subject device evaluation, including device design, process, and use failure modes and effects analyses (FMEAs). The RMR's purpose is to evaluate theoretical (and actual) risks and to create new designs, processes, and uses to manage and reduce "device" risks as low as possible until the risks are "acceptable" when balanced against the user benefits (Fig. 1.4).

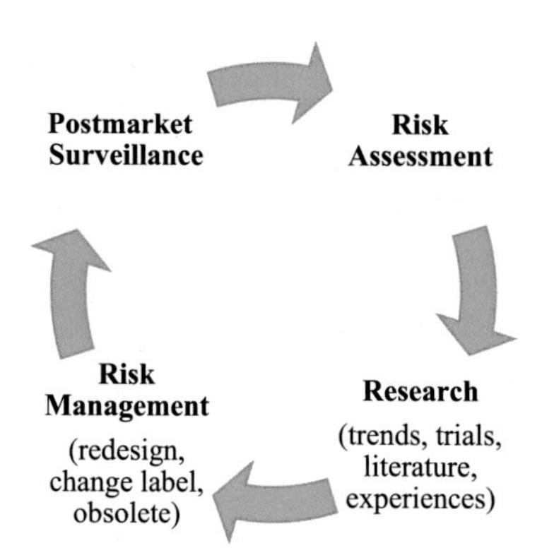

FIGURE 1.4 A risk management system works to reduce risks as much as possible.

At the same time, medical device clinical trials are often developed in compliance with ISO 14155 [14], and PMS systems are often developed in compliance with a relatively new international standard, ISO/TR 20416 [53]. If the device is already on the market, a Postmarket Surveillance Report (PMSR) for Class I devices or a Periodic Safety Update Report (PSUR) for Class II and III devices is required. MDCG 2023-3 [54] provides guidance on vigilance terms including some medical device vigilance terms updated from MEDDEV 2.12/1, Rev. 8 (January 2013) [55], which is no longer applicable under MDR 2017/745. MDCG 2023-3 is pending updates to include IVDR terminology; however, "serious incidents" were already aligned between the MDR Article 2 (65) and IVDR Article 2 (68) as:

the subset of incidents that directly or indirectly led, might have led or might lead to the death or the temporary or permanent serious deterioration in the state of health of a patient, user or other person or posed a serious public health threat.

Under MDR Article 87 and IVDR Article 82, serious incidents must be reported to the Competent Authority within the medical device vigilance system (i.e., the PMS system) (Fig. 1.5).

FIGURE 1.5 A postmarket surveillance system works to report incidents.

MDCG 2023-3 links to the Manufacturer Incident Report (MIR) and lists appropriate timelines for reporting within 2–15 calendar days after the manufacturer becomes aware and depending on the serious incident type (i.e., 2 days for serious public health threat, 10 days for death or unanticipated serious deterioration in health, or 15 days for any device-related serious incident). Having insufficient information is not an acceptable excuse for not filing, since the report should contain all information that was available at the time. Eudamed was designed to house device vigilance and PMS information, and MDCG 2021-1 [56] provided guidance while Eudamed was under development.

Although the clinical trial, RM, and PMS systems are outside the CE/PE scope for this textbook, the reader is advised to learn more about each of these interrelated, independent, and comprehensive systems. For example, the serious incident must have at least a suspected causal relationship (even indirectly) with the device before reporting is required, and indirect harms may include a missed, delayed, or inappropriate diagnosis or treatment. All available guidance documents should be used; for example, MDCG 2023-2 reiterates definitions useful for both MDR and IVDR and gives examples of:

... malfunction... or deterioration... in the characteristics or performance of a device... as a situation where a device fails to achieve, or is unable to uphold, the performance... intended by the manufacturer when used in accordance with the information supplied with the device.

... user... is any healthcare institution, healthcare professional or lay person (e.g., caregiver, patient) who uses the device, or persons installing or maintaining the device... the operator...

... use error... when the user's action or lack thereof while using the device leads to a different result or outcome than that expected by the user or intended by the manufacturer

Ergonomic features... as the physical features of a device... designed to facilitate and ensure... interaction between the user and the device is safe, effective and efficient...

... undesirable side-effect... as any unintended and unwanted medical manifestation in the human body, as a consequence of the normal use of a device... not the result of a malfunction, deterioration in the device's characteristics or performance, or an inadequacy in the information supplied by the manufacturer... (can be expected or unexpected, related or unrelated, serious or not serious—serious, unexpected, and related side effects must be reported in expedited fashion)

... manufacturer awareness date... is the date when the first employee or representative of the manufacturer's organisation, receives information (e.g., a complaint) regarding the potentially serious incident

... Periodic summary report... is an alternative reporting regime by which the manufacturer... can report similar serious incidents with the same device or device type in a consolidated way

In addition, "field safety corrective action (FSCA)" and "field safety notice (FSN)" are also already aligned between the MDR Article 2 (68, 69) and IVDR Article 2 (71, 72) as:

... corrective action taken by a manufacturer for technical or medical reasons to either prevent or reduce the risk of a serious incident... [associated with a device] made available on the market

... a communication sent by a manufacturer to users or customers in relation to a field safety corrective action

Clinical data from device-related serious incidents, FSCAs, and FSNs are important clinical experience data to include in the CE/PE process. The input for the CE/PE process should be the PMSR or PSUR with the full and complete analysis of all complaints and incidents, including root cause analyses, FSCAs, FSNs, etc. as required for the PMS system.

For medical devices needing a PSUR (MDR Article 86 and IVDR Article 81), MDCG 2022-21 [57] states two specific PMS objectives and can be used to guide PMSR for Class I devices or PSUR for Class II and III devices:

Identification and evaluation of changes of the benefit-Risk Ratio

The main objective of a PSUR is to present a summary of the results and conclusions of the analyses of post-market surveillance data relating to a device or a device group, thus allowing the reporting of any possible changes to the benefit-Risk Ratio of the medical device/s, considering new or emerging information in the context of cumulative information on benefits and risk... The PSUR should summarize the results and conclusions of the analysis of the data that the manufacturer has systematically and actively gathered in post-market surveillance with its device/s and, where relevant, with similar devices.

Information on Preventive or Corrective Actions (CAPA)...

The PSUR may also... provide information about Corrective Action/s or Preventive Action/s (CAPA)... [and] can be related to: ... Devices... on the EU market... Issues that might... impact product safety, performance or quality and... Any action related to a voluntary and non-temporary suspension of marketing... [including] Evaluation of benefits and risks identified through post-market activities... records referring to non-serious incidents and data on any undesirable side-effects... relevant specialist or technical literature, databases and/or registers... feedbacks and complaints... and... publicly available information about similar medical devices... [A]ll safety related CAPA should be part of the PSUR...

The PMSR/PSUR should be a clear, stand-alone document summarizing all PMS activities including all serious incidents (including FSCAs and FSNs), nonserious incidents, and undesirable side effects. The PMSR/PSUR should report any PMS data trends found among all complaints and feedback reported from users, distributors, and importers, as well as any findings from ongoing literature and registry data surveillance including information about similar medical devices. The PMSR/PSUR should also include the benefit–risk ratio conclusions, main PMCF/PMPF findings, device sales volume, etc. The manufacturer "shall proactively collect and evaluate clinical data" from devices used as intended (MDR 2017/745) or "shall proactively collect and evaluate performance and relevant scientific data" from *in vitro* diagnostic device uses (IVDR 2017/746). PMS device performance monitoring and clinical safety data surveillance analyses and conclusions provide useful inputs into the CE/PE and RM processes, just as the CE/PE and RM process outputs provide useful inputs into the PMS processes.

1.10 Quality management system control of clinical or performance evaluation

The QMS should state how CE/PE, PMS, and RM systems are related, integrated, and controlled, and how the systems interact, yet react independently and interrelatedly to new data. The reaction to high-risk clinical data needs to be swift and adequately controlled. Each system should have clear and distinct standard operating procedures (SOPs) and step-by-step work instructions (WIs) for normal operations and for emergent responses using appropriate forms (i.e., templates) to document data in discrete source documents (i.e., records). These records are the documents needed to withstand regulatory and legal scrutiny at any point in time. According to the MDR and IVDR, manufacturers should use EU Harmonized Standards to support claims of conformity to EU regulations, and these standards should be specifically included in the QMS.

1.10.1 Use harmonized international standards

Specifically, the EC provided separate MDR and IVDR Harmonized Standards lists (Table 1.12) as previously published in the *Official Journal of the European Union*, including a list of EC decisions made about formal objections to the standards [58].

TABLE 1.12 MDR and IVDR harmonized standards.

MDR harmonized standards	IVDR harmonized standards
IDENTICAL Standards	
EN ISO 13485:2016; EN ISO 13485:2016/A11:2021; EN ISO 13485:2016/AC:2018 Medical devices—Quality management systems—Requirements for regulatory purposes (ISO 13485:2016)	*EN ISO 13485:2016; EN ISO 13485:2016/A11:2021; EN ISO 13485:2016/AC:2018* Medical devices—Quality management systems—Requirements for regulatory purposes (ISO 13485:2016)
EN ISO 14971:2019; EN ISO 14971:2019/A11:2021 Medical devices—Application of risk management to medical devices (ISO 14971:2019)	*EN ISO 14971:2019; EN ISO 14971:2019/A11:2021* Medical devices—Application of risk management to medical devices (ISO 14971:2019)
EN ISO 15223-1:2021 Medical devices—Symbols to be used with information to be supplied by the manufacturer—Part 1: General requirements (ISO 15223-1:2021)	*EN ISO 15223-1:2021* Medical devices—Symbols to be used with information to be supplied by the manufacturer—Part 1: General requirements (ISO 15223-1:2021)
EN ISO 13408-6:2021 Aseptic processing of health care products—Part 6: Isolator systems (ISO 13408-6:2021)	*EN ISO 13408-6:2021* Aseptic processing of health care products—Part 6: Isolator systems (ISO 13408-6:2021)

(*Continued*)

TABLE 1.12 (Continued)

MDR harmonized standards	IVDR harmonized standards
EN ISO 11135:2014; EN ISO 11135:2014/A1:2019 Sterilization of health-care products—Ethylene oxide—Requirements for the development, validation and routine control of a sterilization process for medical devices (ISO 11135:2014)	*EN ISO 11135:2014; EN ISO 11135:2014/A1:2019* Sterilization of health-care products—Ethylene oxide—Requirements for the development, validation and routine control of a sterilization process for medical devices (ISO 11135:2014)
EN ISO 11137-1:2015; EN ISO 11137-1:2015/A2:2019 Sterilization of health care products—Radiation—Part 1: Requirements for development, validation and routine control of a sterilization process for medical devices (ISO 11137-1:2006, including Amd 1:2013)	*EN ISO 11137-1:2015; EN ISO 11137-1:2015/A2:2019* Sterilization of health care products—Radiation—Part 1: Requirements for development, validation and routine control of a sterilization process for medical devices (ISO 11137-1:2006, including Amd 1:2013)
EN ISO 11737-1:2018; EN ISO 11737-1:2018/A1:2021 Sterilization of health care products—Microbiological methods—Part 1: Determination of a population of microorganisms on products (ISO 11737-1:2018)	*EN ISO 11737-1:2018; EN ISO 11737-1:2018/A1:2021* Sterilization of health care products—Microbiological methods—Part 1: Determination of a population of microorganisms on products (ISO 11737-1:2018)
EN ISO 11737-2:2020 Sterilization of health care products—Microbiological methods—Part 2: Tests of sterility performed in the definition, validation and maintenance of a sterilization process (ISO 11737-2:2019)	*EN ISO 11737-2:2020* Sterilization of health care products—Microbiological methods—Part 2: Tests of sterility performed in the definition, validation and maintenance of a sterilization process (ISO 11737-2:2019)
EN ISO 25424:2019/A1: 2022 Sterilization of health care products—Low temperature steam and formaldehyde—Requirements for development, validation and routine control of a sterilization process for medical devices (ISO 25424:2018)	*EN ISO 25424:2019* Sterilization of health care products—Low temperature steam and formaldehyde—Requirements for development, validation and routine control of a sterilization process for medical devices (ISO 25424:2018)
Standards Unique to MDR or IVDR	
NA	*EN ISO 17511:2021* In vitro diagnostic medical devices—Requirements for establishing metrological traceability of values assigned to calibrators, trueness control materials and human samples (ISO 17511:2020)
EN ISO 14160:2021 Sterilization of health care products—Liquid chemical sterilizing agents for single-use medical devices utilizing animal tissues and their derivatives—Requirements for characterization, development, validation and routine control of a sterilization process for medical devices (ISO 14160:2020)	NA
EN ISO 17664-1:2021 Processing of health care products—Information to be provided by the medical device manufacturer for the processing of medical devices—Part 1: Critical and semi-critical medical devices (ISO 17664-1:2021)	NA
**EN IEC 60601-2-83:2020; EN IEC 60601-2-83:2020/A11:2021* Medical electrical equipment—Part 2-83: Particular requirements for the basic safety and essential performance of home light therapy equipment	NA
EN 285:2015 + A1:2021 Sterilization—Steam sterilizers—Large sterilizers	NA
EN ISO 10993-9:2021 Biological evaluation of medical devices—Part 9: Framework for identification and quantification of potential degradation products (ISO 10993-9:2019)	NA
EN ISO 10993-10:2023 Biological evaluation of medical devices—Part 10: Tests for skin sensitization (ISO 10993-10:2021)	NA
EN ISO 10993-12:2021 Biological evaluation of medical devices—Part 12: Sample preparation and reference materials (ISO 10993-12:2021)	NA
EN ISO 10993-23:2021 Biological evaluation of medical devices—Part 23: Tests for irritation (ISO 10993-23:2021)	NA

EN, English; *IEC*, International Electrotechnical Commission; *ISO*, International Organization for Standardization; *IVDR*, in vitro diagnostic regulation; *MDR*, medical device regulation; *MDSW*, medical device software; *PE*, performance evaluation. *Cenelec; all others are developed by CEN as the European Standards Organization. Bolded items are the main QMS and RM system standards.

Nine internationally recognized standards were identical for both MDR and IVDR, including standards about QMS and RM systems as well as methods for labeling symbols, aseptic processing, and sterilization. Nine unique standards included one for IVDR devices about calibration and eight for MDR devices about processing animal tissues, device processing, electrical equipment, large sterilizers, and biocompatibility (degradation, sensitization, sample prep and irritation testing).

When using international standards, the manufacturer should have an overall plan for teams working on all the interrelated, stand-alone, reports (i.e., the CER/PER process will need to review all reports based on human clinical data, often including the clinical trial reports, PMSR/PSUR, RMR, CER/PER, PMCFER/PMPFER, etc.).

Globally, medical device manufacturers should be operating under an ISO 13485 certified QMS or an ISO 14971 RMS, while reports including clinical trial reports, PMSR/PSUR, and CER/PER, etc. are aligned with the MDR/IVDR and other regulations as well as other relevant guidelines and documents. The critical observer will note that clinical trial standards like ISO 14155, PMS standards like ISO/TR 20416, and others are missing from this list. Thus, the MDR and IVDR Harmonized Standards lists provided by the EC are a good starting point, but should not replace additional due diligence and critical thinking about standards not listed but clearly relevant to the device. Good practices are required to manage actively and accurately all Harmonized Standards needed to support a given product.

1.10.2 Require data integrity and ALCOA + compliance

In order to meet regulatory requirements in most areas around the world, the QMS must ensure each data record maintains integrity. Data integrity should be monitored frequently and audited appropriately in real time to ensure "ALCOA + " compliance (Fig. 1.6).

Attributable	•Name person generating the data, action performed, and when •Initial and date document or use electronic audit trail
Legible	•Text and data must be readable and permanent •Review for clarity before storing paper and electronic data
Contemporaneous	•Record data in real time when the event occurs •Date and time stamps, ensure sequence flow (NO pre- or back dating)
Original	•The first time data are written down or documented electronically •Preserve content and meaning in initial records
Accurate	•Without errors, truthful and consistent with the observation •Edit with audit trail or single line through, correction, initals, and date
Complete	•Recorded clearly, included details for all related data (i.e., repeat test) •Include person performing tests, date and time performed, etc.
Consistent	•Aligns with historical observations as expected based on prior work •Monitor and document changes over time
Enduring	•Records kept in laboratory notebooks or validated software systems •Ensure storage media are reviewed often to ensure usability
Available	•Able to access for review, audits, and inspections •Clearly index, label, and file so records can be found easily

FIGURE 1.6 ALCOA + requires careful recordkeeping.

All CE/PE records should be managed with good data practices (GDPs) and good recordkeeping incorporating all nine ALCOA + principles. If CE/PE documents are used in US FDA regulatory submissions, ALCOA + compliance is expected and required.

1.10.3 Use integrated standard operating procedure workflows

SOPs should be high-level descriptions of each required and standardized process flow, and WIs should explain stepwise actions suggested during normal operations, including what should happen if deviations are needed or changes are identified. This is especially critical if the deviations or changes affect the benefit–risk ratio.

BOX 1.9 Patient versus subject terminology

This textbook uses common terminology to distinguish clearly between human "patients" who come to a healthcare professional (HCP) for medical care and "study subjects" or "volunteers" or "participants" who come to a study center (the terms study site, investigational site, and investigational center are used interchangeably in the literature) to participate in a clinical trial (the terms clinical study, clinical investigation, and research project terms are also used interchangeably).

For the CE/PE system procedure, the sequence of events for each CER/PER should be defined by CER/PER TYPE, and the approximate timeline for CER/PER completion should be specified.

1.10.4 Include detailed STEPS in work instructions

Work instructions need to define specific steps to be completed within each system. The documents from the CE/PE processes are meant to record the entire clinical dataset at a particular moment in time. When writing WIs, use a STEPS approach (Fig. 1.7).

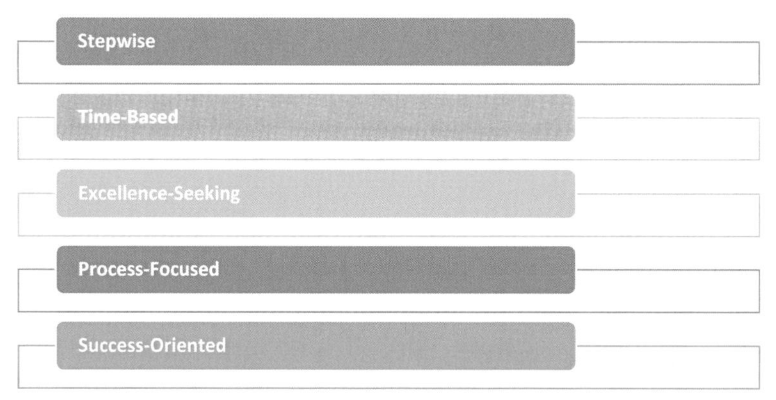

FIGURE 1.7 Steps to follow when writing WIs. *WI*, work instruction.

WIs should have steps to follow. Among the worst WIs are those without any steps to follow. Use basic, logical steps and write each instruction clearly so others can follow the instructions. Use images and test out your WIs to see if people can actually do the process before issuing the document to be followed.

Timeliness is essential. The regulations and guidelines explain the appropriate CER/PER update frequency (every 1–5 years depending on the device risk level) and the need for a more frequent update when new clinical data emerges and changes the benefit–risk ratio. The method used to update a CER/PER earlier than planned should be discussed in the CE/PE WI.

Quality controls should drive evaluators toward excellence. The CE/PE process should encourage all staff to suggest improvements. Each revision to the WI should strive toward a higher-quality document than past or any other documents and aim for overall excellence.

Clear CE/PE processes save time and money. The CE/PE WI must specify how CER/PER writing begins by ensuring a clear CE/PE strategy within a version-controlled CEP/PEP. The CE/PE strategy should consider all CE/PE development needs from beginning to end. The CE/PE process must include writing all CE/PE required documents as well as the negotiation process to respond to critical reviewer comments.

Success is obvious when work is well done. The regulations required a robust and well-substantiated CER/PER accounting for all S&P details and clearly defining the benefit–risk ratio. CE/PE WI statements about expected data quality, relevance, and sufficiency are critical to ensure the clinical evidence demonstrates GSPR conformity.

Explore the boundaries in the CE/PE WI with care. When transitioning work between persons or teams, spell out details about the transitions (forward and back), expectations (who is doing what, how the work is to be done, and why this work is needed), and timelines (when each draft is needed and what happens if a deliverable is early or late).

1.11 Clinical data

Clinical data are defined in the MDR Article 2(48) [12] as follows:

(48) *"'clinical data' means information concerning safety or performance... generated from the use of a device and is sourced from the following:*

- *clinical investigation/s of the device concerned,*
- *clinical investigation/s or other studies reported in scientific literature, of a device for which equivalence to the device in question can be demonstrated,*
- *reports published in peer reviewed scientific literature on other clinical experience of either the device in question or a device for which equivalence to the device in question can be demonstrated,*
- *clinically relevant information coming from post-market surveillance, in particular the post-market clinical follow-up"*

Although not defined clearly, clinical data are discussed in IVDR Annex XIV, Section 2.4 [13]:
"2.4. Existing clinical data, in particular:

- *from relevant peer-reviewed scientific literature and available consensus expert opinions or positions from relevant professional associations relating to the safety, performance, clinical benefits to patients, design characteristics, scientific validity, clinical performance and intended purpose of the device and/or of equivalent or similar devices;*
- *other relevant clinical data available relating to the safety, scientific validity, clinical performance, clinical benefits to patients, design characteristics and intended purpose of similar devices, including details of their similarities and differences with the device in question."*

In general, three clinical data sources exist for use in clinical/performance evaluations: clinical trials, clinical literature, and clinical experiences.

1.12 How clinical data sources change over time

The CE/PE process will have significant differences at time of launch and during early release compared to when the subject device has been on the EU market for decades. Typically, early release devices have multiple, significant design and use changes early on, while more mature devices tend to have fewer changes, especially after the device has been on the market for many decades. For IVD devices and those incorporating information technology or computerized systems, the change frequency may be much more frequent, and devices may have significantly reduced longevity compared to other types of medical devices.

Clinical data are expected to accumulate over time, and device changes may need to reset the clinical data collections clock and the evaluated clinical data must remain relevant to the specific subject device designs and uses. In these device change situations, clinical data from older device versions are potentially evaluated separately from clinical data from new device versions. Each updated CER/PER should carefully differentiate the device versions used in each dataset, and the conclusions must fully substantiate the S&P for each device version used in humans. If the clinical data are insufficient for each device version on the market, a clinical data gap should be documented and the appropriate clinical data should be recommended for collection in the next PMCFP/PMPFP.

BOX 1.10 Do not pool data before/after device changes impact benefit–risk ratio

A key point to remember when evaluating clinical data is to pool data only when scientifically appropriate. In other words, avoid pooling data from an old device version together with clinical data from a new subject device version when the two device versions have different benefit–risk ratios.

Reliance on clinical data from similar and benchmark devices may lessen over time after launch, but will not disappear in the postmarket phase or even after the device has been obsoleted (when devices are still in use). Similarly, the need for clinical data from the equivalent device is expected to decline over time as clinical data from the subject device accumulate. Even in the absence of a premarket clinical trial, device sales will accumulate, authors will publish literature, and users will file complaints and use reports. The clinical data about subject device uses are more relevant than clinical data about equivalent device uses, and the need for the equivalent device data should eventually disappear once sufficient clinical data are evaluated about the subject device itself. This cycle of data collection will begin anew after any device changes alter the benefit–risk ratio and create new requirements for clinical data collection.

Background knowledge, SOTA, alternative therapies, and clinical data from similar and benchmark devices are important throughout the device lifetime; however, clinical data should be included from the actual subject and/or equivalent device used as intended in the appropriate patient populations. If clinical data from the subject device are not available, the CER/PER will likely be found to have insufficient clinical data to show MDR/IVDR conformity.

Understanding clinical data sources and the relative clinical data abundance is helpful when writing the CER/PER. Hypothetically, before launch in the premarket phase, a clinical trial may be implemented to collect clinical data. Clinical literature about the subject/equivalent device may be increasing and the manufacturer may begin collecting human clinical trial experience data. At the time of launch, clinical trial activity may peak, since novel devices generate intense interest in the medical community and many authors may compete to be the "first" to report new device use in various settings in the clinical literature. This clinical interest drives trials, literature, and experience data (Fig. 1.8).

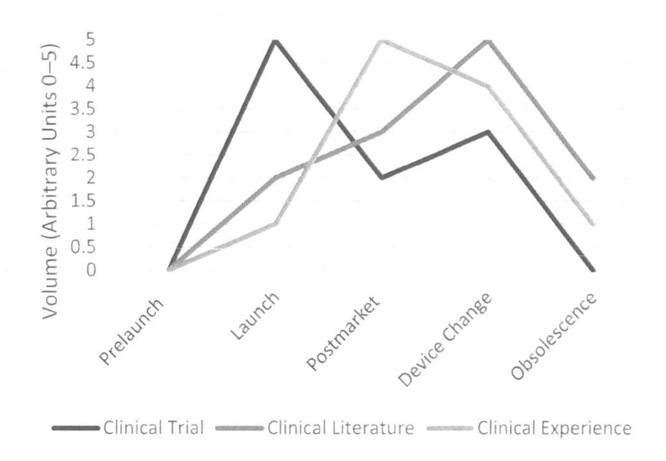

FIGURE 1.8 Theoretical clinical data available over time by development phase.

During the postmarket phase, clinical literature and clinical experience data will tend to increase, while clinical trial activities are expected to decrease as the device becomes more commonly used (and theoretically less "interesting" as the subject of a published article in a reputable medical journal). This medical device "aging" effect is sometimes mitigated by a device change or improvement to increase use, which may require a clinical trial and should maintain the clinical literature volume and the clinical experiences volume at maximal levels.

BOX 1.11 Determine if device is "NOVEL"

When defining a medical device, questions often arise about how "novel" the device may be. Novelty is a complex concept in this context. The answer is not simply whether or not the device is "new." Specifically, in order to determine if the device is novel, the manufacturer must consider the design, construction materials, components, manufacturing process, application site, device-patient interface, interaction and control, operational principles, medical purpose, intended use, indication for use, etc.

The EU MDR Annex IX, Chapter II, 5.1(c) described how the expert panel decides whether to offer an opinion based in part on device novelty, as follows:

The expert panel shall decide, under the supervision of the Commission, on the basis of...: (i) the novelty of the device or of the related clinical procedure involved, and the possible major clinical or health impact thereof (ii) a significantly adverse change in the benefit-Risk Ratio of a specific category or group of devices due to scientifically valid health concerns in respect of components or source material or in respect of the impact on health in the case of failure of the device; (iii) a significantly increased rate of serious incidents..., whether to provide a scientific opinion on the clinical evaluation assessment report of the notified body based on the clinical evidence provided by the manufacturer, in particular concerning the benefit-risk determination, the consistency of that evidence with the medical indication or indications and the PMCF plan... Where the information submitted is not sufficient for the expert panel to reach a conclusion, this shall be stated in the scientific opinion.

Separately, the EU IVDR defined the certification period and device technical documentation sample selected for Class B and C devices based in part on novelty, as follows:

The NB shall "decide, based on the novelty, risk classification, performance evaluation and conclusions from the risk analysis of the device, on a period of certification not exceeding five years." (IVDR Annex VII, 4.8).

(Continued)

BOX 1.11 (Continued)

For "class B and C devices, the quality management system assessment shall be accompanied by the assessment of the technical documentation for devices selected... [based on] the novelty of the technology, the potential impact on the patient and standard medical practice, similarities in design, technology, manufacturing and, where applicable, sterilisation methods, the intended purpose and the results of any previous relevant assessments... The notified body in question shall document its rationale for the samples taken." (IVDR Annex IX, 2.3).

Expert panels reviewing Class III implantable and Class IIb active devices (and presumably Class D IVD devices) follow a specific eight-page guidance document [59] to interpret clinical evaluations consistently regarding when to offer an opinion about the NB CEAR:

... in particular... the benefit-risk determination, the consistency of that evidence with the medical indication or indications and the PMCF plan.... on the basis of the following criteria: (i) the novelty of the device or of the related clinical procedure involved, and the possible major clinical or health impact... (ii) a significantly adverse change in the benefit-Risk Ratio of a specific category or group of devices due to scientifically valid health concerns in respect of components or source material or in respect of the impact on health in the case of failure of the device; (iii) a significantly increased rate of serious incidents reported... [59]

The guidance describes how to assess device/procedural novelty and clinical/health impact. In this setting, novelty means no device/procedural S&P experience exists and no or insufficient experience is available regarding similar devices to appraise future device S&P when used. When a product is modified, careful review of PMS is needed, and the expert panel must *"estimate the clinical impact or health impact in conjunction with the novelty. Novelty alone is insufficient to trigger a scientific opinion."*

The major novelty dimensions include clinical procedure-related (e.g., clinical/surgical procedure with a changed application/use/treatment, device-user interface, device interaction/control or deployment) and device-related (e.g., new purpose, design, mechanism, materials, application, components, manufacturing) dimensions. For decision-making purposes, the expert panel rates the novelty level as high, medium, or low, and *"no scientific opinion is required when the level of novelty is not high... [with] no major potential negative clinical and/or health impact"*; however, *"a scientific opinion is required when a major negative clinical or health impact is anticipated"* for all novelty levels.

After the novelty level is determined, the expert panel assesses the following.

Clinical impact

- *"Clinical outcomes leading to changes in mortality, morbidity, health-related quality of life, burden of treatment, duration of hospitalization, mode of administration, severity, intensity, duration and timing of effects, need for medical or surgical re-intervention, and consideration of preferences, acceptability, usability as well as patient compliance where relevant;*
- *Clinical benefit or major contribution to care of individual patients or specific groups of patients, i.e. changes in clinical performance, and/or safety profile resulting in clinical advantages compared to existing state-of-the-art methods;*
- *Negative clinical outcomes, clinical hazards and related risks;*
- *Risks related to potential severity, types, number and rates of adverse events, incidents, probability and duration of serious adverse events and of serious incidents* (Article 2(57), (58), (64) and (65) of the MDR);
- *Risks related to incompatibility with the use of other medical devices;*
- *Risks related to medical device dysfunction due to reasonably foreseeable inappropriate conditions of use and misuse."*

Health (population) impact ("net potential benefits and risks... in real-world conditions")

- *"Effects on an individual level cumulatively expressed on a population level..."* (i.e. effect size, population exposed and effect duration)
- *"Probability of serious public health threats* (Article 2(66) of the MDR);
- *Anticipating a justifiable high market penetration due to innovation, leading to a greater uptake of the device and subsequently to a higher number of patients being exposed to the device and, as a result, a higher overall probability of harms to occur leading to a higher net risk. Risk is defined in Article 2(23) of the MDR as the combination of the probability of occurrence of harm and the severity of that harm."*

The expert panel assessment considers "hazards, harms and risks," including the NB conclusions about the manufacturer benefit–risk ratio, "possible impacts" and uncertainties, including possible outcomes without direct clinical evidence, and an impact severity estimation "based on their clinical experience and knowledge from the published literature."

In addition, the expert panel considers "scientifically valid health concerns" initially predicted by the expert panel based on reliable (high-quality, sufficient number, scientifically plausibility with robust causal links) and relevant (same or similar device materials, specific to device under evaluation, device use in similar clinical procedure linked to the health concern) evidence before issuing a scientific opinion. The data under evaluation may also come to the expert panel from the expert panel Commission Secretariat (i.e., from review of manufacturer PMS and vigilance activities, PSURs, trend reports, etc. or from a "significantly increased rate of serious incidents"). The scientific opinions may be updated after the devices are actually used in real-world conditions.

As the device reaches obsolescence and removal from the market, all forms of clinical data are expected to fall to minimal levels. This leads to a recommended CER/PER writing approach designed to leverage all available historical clinical data by CER/PER Type.

1.13 Clinical evaluation report/performance evaluation report TYPES

All devices on the EU market must have a current CER/PER, regardless of the medical device class or life cycle stage. As a result, CERs/PERs vary widely and evolve over time as the clinical/performance data change. This textbook takes a novel approach to begin NAMING and CLASSIFYING five specific CER/PER TYPES (Fig. 1.9).

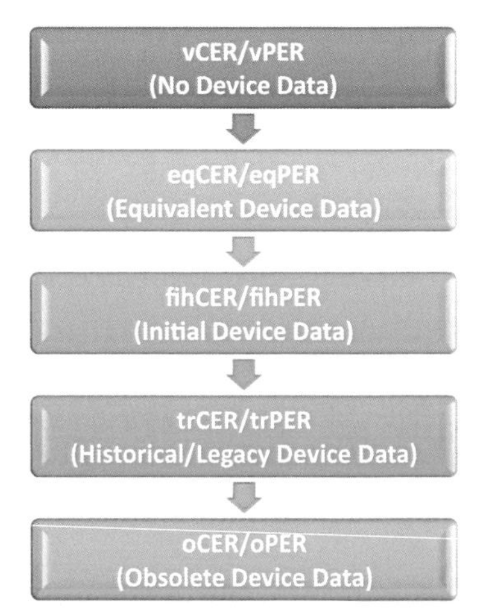

FIGURE 1.9 CER/PER types aligned with device life cycle stages. *CER*, clinical evaluation report; *eq*, equivalent device; *fih*, first in human; *o*, obsolete device; *PER*, performance evaluation report; *tr*, traditional; *v*, viability.

The unique clinical data situations at different device life cycle times generated the idea for creating CER/PER TYPES (or categories) based on the expected clinical data amounts available for the CER/PER, which change over the device lifetime. Although the medical device industry has not yet adopted these proposed CER/PER TYPE names, having this distinct naming convention is helpful to clarify and differentiate the specific regulatory, analytic, and quality requirements for each CER/PER TYPE, which change as devices progress through the device lifetime from ideation and prototyping through launch, marketing, and sales to obsolescence.

1.13.1 Viability clinical evaluation report/performance evaluation report

During the earliest premarket development phase, early stage "prototype" device designs are typically regulated as "investigational use only" devices because they are not yet CE marked for sales and marketing in the EU. During this prelaunch period, manufacturers will likely be collecting nonclinical bench and animal data; therefore, a full CE/PE may be premature because little to no actual clinical data are available for subject device uses in humans as intended. The lack of clinical trial, literature, and experience data for the subject device naturally makes writing this CER/PER type particularly challenging (if not impossible). This is the case for all medical devices before the first human use, and is especially true for low-risk devices, which may not require clinical trial data at any point throughout the entire device lifetime.

Regulatory requirements: Although a CER/PER is not required during the premarket phase, a fully developed CER/PER is required prior to CE marking. In other words, the CER/PER must be developed during the premarket phase to be available as required prior to CE marking. The need for a CER/PER in this setting is unique, since most devices in early development have never been used in a human before and the CER/PER is needed before the specific medical device is launched or used anywhere in the world. No clinical data are available to be evaluated for this new device in the CER/PER. This is classified as a viability CER/PER (vCER/vPER).

Analytic requirements: In this premarket setting, the medical device manufacturer needs to evaluate clinical data from similar and benchmark devices to determine if the new medical device design is viable. As a result, the vCER/vPER is based entirely on theoretical clinical data and is unlike any other CER/PER type. The vCER/vPER often provides an entirely hypothetical clinical data evaluation, and sometimes device manufacturers use the vCER/vPER to help determine if the device should even be created. Obviously, without a strong clinical/performance rationale, the planned medical device may be difficult to market and sell.

Quality requirements: The vCER/vPER requires a well-written rationale to explain why the vCER/vPER exists without human clinical data from the subject device. If no clinical data from the device are evaluated in the vCER/vPER, then a CER/PER claiming that a clinical/performance evaluation was carried out for the device is untruthful and misleading. The writer must make the clinical data sources clear, and specific technical writing requirements will be helpful to differentiate the vCER/vPER from other CER/PER types that are based on actual clinical/performance data evaluations from humans using the subject device as intended.

The vCER/vPER is often based on clinical data from similar/benchmark devices exploring the subject device viability before any clinical data are collected or when clinical data collection is deemed inappropriate (MDR Article 61, clause 10). Writing an effective vCER/vPER will require strong development of a rationale about why the clinical data from the similar/benchmark devices clearly indicate that the new subject device should be created.

1.13.2 Equivalence clinical evaluation report/performance evaluation report

Sometimes, a manufacturer creates a device to mimic another device already on the market. This regulatory strategy allows the manufacturer to claim equivalence (eq) to the device already on the market with a CE mark before any clinical data collection is completed for the subject device. Using nonclinical test data to show the similarity between the two devices, the manufacturer may not need to collect clinical/performance data for the subject device. By claiming equivalence, the subject device will be allowed to enter the market based on the clinical/performance data from the equivalent CE-marked device before sufficient data are available for the subject device.

Regulatory requirements: Manufacturers can choose to claim equivalence to other devices on the market when their devices are not novel (i.e., another device has already used the technical innovation embodied in the subject devices). The equivalence scientific rationale is supported by documenting clinical/performance (e.g., same clinical purpose, condition severity, disease stage, population), biological (e.g., same materials of construction or substances in contact with human cells and tissues), and technical (e.g., similar design, specifications, properties) equivalence between the subject device and the chosen "equivalent" device. If the two devices are deemed equivalent, then they are assumed to have the same S&P profile with the same benefit–risk ratio in the CER/PER. This is classified as an equivalent CER/PER (eqCER/eqPER).

Analytic requirements: Manufacturers must demonstrate and justify all claims of clinical/performance, biological, and technical equivalence using test data. The equivalence data must document the differences between the device under evaluation and the "equivalent" CE-marked device already on the EU market. Each difference must be analyzed for clinical/performance impact on the benefit–risk ratio, and a written justification must be provided to establish equivalence. The device under evaluation must have data similar enough to the claimed equivalent so the clinical/performance data from the equivalent device can be used to demonstrate conformity with GSPRs for the device under evaluation.

Quality requirements: The eqCER/eqPER requires a clear justification and rationale to allow the claim of equivalence. Equivalence testing does not require the devices to be identical; however, the devices need to be "similar enough" in all three considered areas (clinical, technical, biological), and the eqCER/eqPER must justify why the clinical impacts for any clinical, technological or biological differences will still allow the devices to be considered equivalent.

Writing an effective eqCER/eqPER will require strong attention to the details of the clinical, biological, and technical similarities and differences between the subject and the equivalent devices. Analyzing and documenting the clinical impacts for all differences will also require excellent organizational skills. Subject device clinical/performance data will begin to accumulate after product launch (e.g., uses published by others in the literature and incidents/experiences documented in the PMS system), and the equivalence claim can be replaced when sufficient clinical and performance data are accumulated from subject device uses as indicated.

1.13.3 First in human clinical evaluation report/performance evaluation report

Sometimes, a human clinical/performance trial is required for high-risk or implantable devices; however, most device manufacturers only conduct a single, relatively small, clinical/performance trial before securing a CE mark and placing the high-risk device on the EU market. Well-controlled clinical/performance trial data from human subjects using the subject device are critical for the CER/PER; however, the accumulation of clinical/performance data takes time, so the manufacturer will probably want to leverage the first in human (fih) data from a single, small clinical trial or a limited launch of the subject device to maximal benefit.

Regulatory requirements: Manufacturers are required to collect PMS data from any reports of human experiences with the device. In addition, manufacturers can choose to collect additional clinical/performance information about the medical device in many ways (e.g., a clinical trial, survey, registry, patient information collected at the time of use). Often, manufacturers capture a good deal of clinical/performance data during the first medical device uses. Typically, a manufacturer captures as much fih data as possible to be used in the CER/PER process. This is classified as a first in human CER/PER (fihCER/fihPER).

Analytic requirements: Manufacturers must analyze all clinical/performance and PMS data, even when the clinical/performance data are limited during the initial launch of a new product. The data limitations (e.g., small number of patients with data collected over a short time) become clinical data gaps to be filled by additional data collected later—for example, during PMCF/PMPF activities and continuously during PMS when users file incident reports, complaints, etc.

Quality requirements: The fihCER/fihPER requires data collection during the first device uses. Rare events may not have occurred before, and the manufacturer must be prepared to monitor patients for rare clinical data details in the clinical experience data and case reports found in the clinical literature and PMS data over the long term. The fihCER/fihPER must justify why this evaluation, with such limited clinical/performance data is appropriate and conforms to the GSPRs. The PMCFP/PMPFP is expected to contain more detailed clinical/performance data in the future to document how the "device is safe and achieves the intended clinical benefit/s" (MDR [12]) and to ensure "a high level of safety and performance" (IVDR [13]) for GSPR compliance.

The fihCER/fihPER is especially focused on the limited clinical data amount collected from the initial device uses in humans. It often has a specific PMCFP consideration where a manufacturer may need to collect clinical data from a limited product launch to address all clinical S&P concerns in a small human patient sample before a wider product launch is completed in many more patients.

1.13.4 Traditional clinical evaluation report /performance evaluation report (trCER/trPER)

After being on the EU market, the legacy device will be updated according to the required update frequency or to address changes reducing the benefit–risk ratio.

Regulatory requirements: Over time, manufacturers are expected to evaluate all clinical/performance data collected for a given device used as intended in the appropriate human population. Clinical/performance data accumulates over time and often include substantial amounts of safety and performance data. Typically, previously marketed medical devices are no longer "novel" and should cease to rely upon "equivalence" because the active PMCF process will have gathered sufficient clinical/performance data over time to satisfy the GSPRs directly in the CER/PER. This is classified as a traditional CER/PER (trCER/trPER).

Analytic requirements: The trCER/trPER is supported directly by subject device clinical/performance data accumulated over many years in documented clinical/performance trials, literature, and experiences. The evaluator will likely need to identify which subject device clinical/performance data to exclude since too much data may be available.

Quality requirements: The trCER/trPER requires the data evaluation to accommodate large amounts of subject device data and to document how the data relate to all the subject device changes over time. In addition, the S&P acceptance criteria should be much more advanced, more specific, and better-informed than for CERs/PERs earlier in the device life cycle, since more data are available and more evaluations have been completed for the subject device in this trCER/trPER setting.

The trCER/trPER is the most common CER/PER TYPE and is most likely to have sufficient clinical data to demonstrate conformity with the GSPRs. More refined and appropriate S&P acceptance criteria should be developed over time to improve the benefit and risk documentation as much as possible in concert with the integrated CT, PMS, and RM systems.

1.13.5 Obsolete clinical evaluation report /performance evaluation report (oCER/oPER)

The device life cycle ends when the device is obsoleted and removed from the EU market, even though some device versions may still be in active use. For example, implanted devices may continue to function within the patient even though the device is no longer available for purchase on the EU market. Planning for obsolescence is a part of the medical device life cycle.

Regulatory requirements: The MDR and IVDR require CERs/PERs throughout the device lifetime. When a device is obsoleted and not actively marketed but is still used, device clinical/performance data should still be analyzed in the CER/PER; however, the manufacturer may not wish to support this work due to cost in updating the CER/PER when device sales may be minimal. The process for obsoleting the CER/PER should be carefully considered. Although the regulations may not require continuation of the CER/PER work once the device has been removed from the EU market, good clinical practice may dictate CER/PER continuation while all devices are being obsoleted. This is classified as an obsolete device CER/PER (oCER/oPER).

Analytic requirements: Manufacturers should account for all devices to determine the relative amount still in use. The PMS system should track how the device sales decrease over time as the device reaches end of life. During the obsolescence process, safety signals should still be monitored, risks should still be managed, and the CER/PER should still be updated to protect all subjects still using the device.

Quality requirements: The oCER/oPER requires data evaluation to look specifically for device-related end-of-life concerns and risks encountered by the final few users. In addition, the CER/PER for all future versions of the same or similar device should be considered carefully before a device oCER/oPER is no longer updated.

The oCER/oPER requires clinical/performance data evaluation plans until no subject devices remain in use.

1.13.6 Changing clinical evaluation report /performance evaluation report TYPES over time

The ability to move between certain CER/PER TYPES over time is critical. The CER/PER evaluator will need to merge in new data and to modify old clinical data evaluations in light of the changing background/SOTA/alternative therapy clinical data and the growing subject device clinical data. Understanding CER/PER TYPE features and needs may be helpful (Table 1.13).

TABLE 1.13 CER types: unique feature examples.

Name	vCER	eqCER	fihCER	trCER	oCER
	vPER	eqPER	fihPER	trPER	oPER
Definition	Viability (justification)	Equivalence	First in human	Traditional	Obsolescence
Life cycle	Prelaunch	Launch	First use	Mid-life	End of life
Purpose during device life cycle	Viability should be maintained throughout life cycle, to be replaced with subject device data as soon as feasible	Equivalence maintained until no longer needed; to be replaced with subject device data as soon as feasible	New subject device data should accumulate and continues until no longer novel	Subject device data accumulates throughout device lifetime	Ends once last subject ends device use
Justification/ rationale	Nonclinical viability rationale to document why clinical data were not needed/not deemed appropriate	Comparative rationale to document clinical impacts for each clinical, biological, and technical difference when equivalent device represents subject device	Ongoing data collection plan to collect clinical data at first subject device use and beyond	Change management plan to ensure careful monitoring and reporting over time during product changes as subject device clinical data increases	Aging device data reviews to ensure product remains safe and performing as indicated for patients/ users during entire life cycle to end of life
How much subject device clinical data?	None	None	Little	Lots	Historical
Clinical data substitutes?	When no subject device clinical data, use bench and animal test results, and similar device human use data	If clinically, biologically, and technically equivalent, use equivalent device clinical data	When little to no subject device clinical data, use all initial (first in human, fih) subject device clinical data	When subject device clinical data is abundant, use only highest weighted, relevant subject device clinical data	For expired subject device, use historical subject device clinical data and monitor until last use
Next steps?	PMCFP—can this move to eqCER/ eqPER?	PMCFP—can this move to fihCER/fihPER?	PMCFP—can this move to trCER/ trPER?	PMCFP—can this move to oCER/oPER?	File once last subject device use confirmed

(Continued)

TABLE 1.13 (Continued)

Name	vCER vPER	eqCER eqPER	fihCER fihPER	trCER trPER	oCER oPER
PMS needs	PMS system needs to collect subject device data as soon as any are available	PMS system must account for both eq and subject devices	PMS system must scale up to house commercial data on larger scale than before	PMS system efficiencies required (especially for links to clinical trial, CE, and RM systems)	Shut down and archive PMS system
RM needs	RM system needs to anticipate actual clinical data arrival	RM system must account for both eq and subject devices	RM system must scale up to house commercial data on larger scale than before	RM system efficiencies required (especially for links to clinical trial, CE, and PMS systems)	Careful RM system shut down required
Amount of iteration expected (frequency of updates)	May not be able to iterate if no clinical data available and device is not used in humans	Significant updates expected when subject device clinical data become available	Significant updates expected as subject device clinical data accumulates	Significant updates, weighted and relevant as abundant subject device clinical data accumulates	No updates expected after last subject device is no longer in use
Regulatory requirements	Justify rationale (no clinical data from subject or equivalent device are available)	Document equivalence clinical, biological, and technical (differences and clinical impacts inconsequential)	Specify how CT, CE/PE, RM, and PMS systems are integrated (consider data from limited market release)	Describe clinical and performance data flows between operational units (document and report any S&P issues)	Plan for obsolescence (device S&P accounted for until device end of life and removal from market)
Regulatory references	MDR Article 61, clause 10	MDR Article 6, Annex XIV, Pt A, MDCG 2020-5 (equivalence)	MDR Article 61	MDR Article 61	MDR Article 61(6a)
	IVDR Article 56	IVDR Annex IX 2.2(c); Article 56; MDCG 2022-01 (Notice to 3rd country mfr.) MDCG 2022-02	IVDR Article 56	IVDR Article 56	MDCG 2020-6 (sufficient evidence)
					MDCG 2021-25 (old devices)
					IVDR Article 56
					MDCG 2022-8 (legacy devices)

CE, clinical evaluation; *CER*, clinical evaluation report; *CT*, clinical trial; *eq[document type]*, equivalence; *fih[document type]*, first in human; *IVDR*, in vitro diagnostic regulation; *MDCG*, Medical Device Coordination Group; *MDR*, medical device regulation; *o[document type]*, obsolete (legacy); *PER*, performance evaluation report; *PMCFP*, postmarket clinical follow-up plan; *PMS*, postmarket surveillance; *RM*, risk management; *S&P*, safety and performance; *tr[document type]*, traditional; *v[document type]*, viability (justification document).

This CER/PER TYPES classification scheme is intended to help both clinical/performance evaluators and NB reviewers, since clinical/performance data expectations will differ between these CER/PER TYPES. For example, some CER/PER TYPES have no subject device data (i.e., vCER/vPER), some have little clinical/performance data from the subject device (i.e., fihCER/fihPER), and some have the benefit of all the clinical/performance data collected and analyzed from the entire device lifetime (i.e., oCER/oPER). The regulations, data analyses, and even the data quality will differ between the five CER/PER TYPES. For example, the vCER/vPER and fihCER/fihPER clinical/performance data quality may be poor since all clinical/performance data will be theoretical and/or predominantly nonclinical, respectively. Conversely, the trCER/trPER and the oCER/oPER clinical/performance data quality are expected to be of the highest quality among these CER/PER TYPES, since they will have the largest amount of relevant clinical/performance data after the subject device was on the EU market for the longest time.

In general, as real clinical data accumulate from human subject uses, the need for theoretical clinical/performance and nonclinical data in the CER/PER declines. The CER/PER TYPE classification system was intended to foster more concrete and focused learning opportunities about sorting out and standardizing details specific to each CER/PER TYPE.

1.13.7 The clinical evaluation/performance evaluation document cycle

The CE/PE document cycle starts with good planning documented in the CEP/PEP (Fig. 1.10).

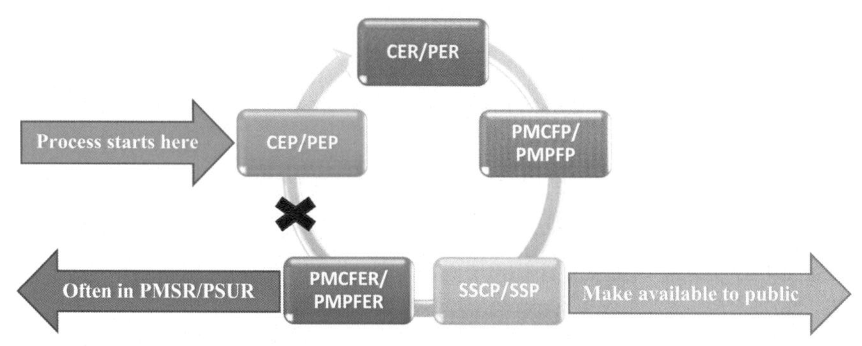

FIGURE 1.10 The CE/PE document cycle. *CEP*, clinical evaluation plan; *CER*, clinical evaluation report; *PEP*, performance evaluation plan; *PER*, performance evaluation report; *PMCFER*, postmarket clinical follow-up evaluation report; *PMCFP*, postmarket clinical follow-up plan; *PMPFER*, postmarket performance follow-up evaluation report; *PMPFP*, postmarket performance follow-up plan; *PMSR*, postmarket surveillance report; *PSUR*, periodic safety update report; *SSCP*, summary of safety and clinical performance; *SSP*, summary of safety and performance.

The red X in the figure indicates where the process completely ends and fully restarts each time the cycle is engaged. The cycle is continuous like the CT, PMS, and RM system processes, and each updated report is a snapshot in time. Setting an appropriate data cutoff date for the clinical/performance data to be evaluated in each report is a best practice.

The CEP/PEP defines the actions to be taken in the CER/PER. The CER/PER evaluates all prior subject device clinical data, and the clinical evaluator draws conclusions based on these clinical data to determine if the device should be issued into the EU market for the first time, stay on the market, or be withdrawn from the market. The CER/PER also documents any and all clinical data gaps and directs the PMCFP/PMPFP to collect specific clinical/performance data for future CER/PER evaluations. The PMCFP/PMPFP guides the additional data collection and reporting within the PMCFER/PMPFER, and the PMCFER/PMPFER may become a PMSR/PSUR component. The PMCFER/PMPFER is required as a clinical data input for the next CER/PER update, and will be documented as a specific clinical data source in the CEP/PEP for evaluation in the next CER/PER update.

A written process should ensure appropriate and sufficient clinical/performance data have been gathered and evaluated in the CER/PER, which must document how the clinical benefits outweigh clinical risks to the patient when the device is used as indicated in the appropriate population.

The CE/PE document cycle has five CE/PE output documents (Table 1.14).

TABLE 1.14 Five CE/PE output documents.

Number	Name	Description*	Regulation
1	CEP/PEP	Defines scope and plan before evaluation begins	MDR Annex XIV and IVDR Annex XIII
2	CER/PER	Documents entire evaluation and lists clinical/performance data gaps for PMCFP/PMPFP	MDR Annex XIV and IVDR Annex XIII
3	PMCFP/ PMPFP	Defines scope and plan to address clinical/performance data gaps found in CER/PER or justifies why no PMCF was required	MDR Annex XIV and IVDR Annex XIII
4	SSCP/SSP	Summarizes high-risk device CER/PER findings for public review	MDR Article 32 and IVDR Article 29
5	PMCFER/ PMPFER	Documents entire PMCF process and how CER/PER clinical data gaps were addressed	MDR Annex XIV and IVDR Annex XIII

CE, clinical evaluation; *CEP*, clinical evaluation plan; *CER*, clinical evaluation report; *IVDR*, in vitro diagnostic regulation; *MDR*, medical device regulation; *PE*, performance evaluation; *PEP*, performance evaluation plan; *PER*, performance evaluation report; *PMCFER*, postmarket clinical follow-up evaluation report; *PMCFP*, postmarket clinical follow-up plan; *PMPFER*, postmarket performance follow-up evaluation report; *PMPFP*, postmarket performance follow-up plan; *SSCP*, summary of safety and clinical performance; *SSP*, summary of safety and performance. *Note: All documents are written, signed, dated, version controlled.

These five CE/PE output documents are single, stand-alone documents, and the most current version should be the most complete and comprehensive document available. Similar to the CE/PE system process, the CT, RM, and PMS systems each have their own protocols, plans, and reports. In addition, each CER/PER must have sufficient clinical/performance data from human device exposures to support the conclusions and to ensure all clinical/performance data points are covered. Any clinical/performance data gaps should be stated explicitly in the CER/PER, and PMCF/PMPF plans will need to gather the missing data in the future, using the appropriate clinical/performance data collection tools for the next CER/PER update (e.g., The CER/PER should determine if a clinical investigation is needed? If an animal study could provide the needed clinical/performance data? Should a retrospective review article be written?).

Just meeting the lower limit with enough clinical/performance data needed to claim conformity to the basic GSPRs 1, 6, and 8 is often not "sufficient," since companies may make many claims about device clinical/performance, and these claims must also be supported in the CER/PER.

1.14 Writing clinical evaluation/performance evaluation documents

Although CE and PE writing processes are not identical, they both include planning, identifying, appraising, analyzing, and reporting CE/PE data (Fig. 1.11).

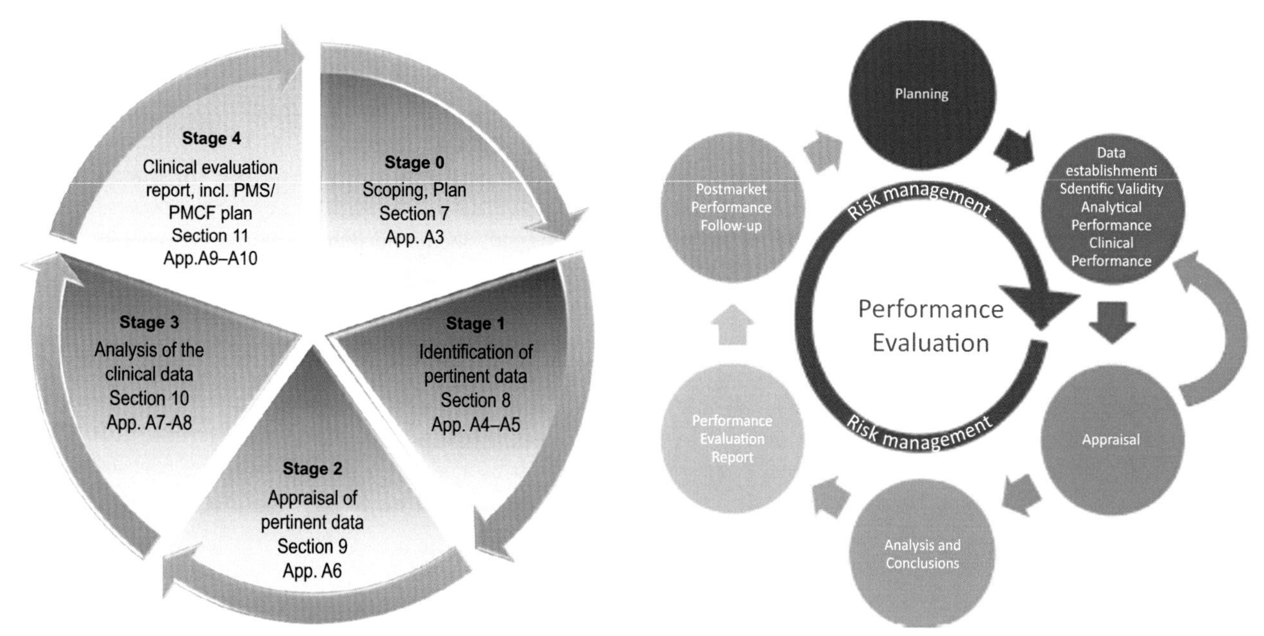

FIGURE 1.11 CE (left) and PE (right) writing processes. *App.*, appendix; *PMCF*, postmarket clinical follow-up; *PMS*, postmarket surveillance. *Taken from MEDDEV 2.7/1, Rev. 4 [29] and MDCG 2022-2 [46], respectively.*

The main difference between the MDR CE and IVDR PE writing process diagrams is the identification step, which is called identification of pertinent data in the CE process and data establishment in the PE process. Although these have different names, they define the methods used to identify or "establish" the data to be evaluated in the CE/PE process (Table 1.15).

TABLE 1.15 CE/PE writing stages.

Stage	CE	PE	Differences
0	Scoping, plan	Planning	NA
1	Identification of pertinent data	Data establishment (scientific validity, analytical performance, clinical performance)	Main PE difference about analyte: associated with target, correct detection, clinical correlation
2	Appraisal of pertinent data	Appraisal	NA
3	Analysis of the clinical data	Analysis and conclusions	NA
4	Clinical evaluation report including PMS/PMCF plan	Performance evaluation report including PMPF (as a separate step)	NA

CE, clinical evaluation; *NA*, not applicable; *PE*, performance evaluation; *PMCF*, postmarket clinical follow-up; *PMPF*, postmarket performance follow-up; *PMS*, postmarket surveillance. Bolded information is the only difference.

These five stages are quite similar and relatively easy to follow at a high level; however, the details can be complicated and specific for each evaluation type. In addition, MDR/IVDR guidelines do not explain how to link various systems interacting with CE/PE processes, and they do not provide much detail or support specifying the actual work needed to create a CER/PER. Furthermore, these five simple stages give a false impression, suggesting CER/PER writing is simple when CER/PER writing is quite complex. This complexity is due to the clinical/performance data and the need for the evaluator to understand precisely *how* to identify relevant clinical/performance data, *how* to appraise clinical/performance data for inclusion or exclusion, *how* to analyze clinical/performance data fully for S&P criteria, risks, and benefits, and *how* to draw conclusions documenting how the clinical/performance data demonstrates compliance with MDR/IVDR GSPRs 1, 6, and 8.

Understanding and using specific, rigorous process steps for clinical/performance data evaluation should help. For example, the evaluator must do the following:

Plan: Follow the process and specify details including required, prespecified, and prequantified CER/PER S&P acceptance criteria.
Identify: Find and gather together all clinical/performance data for the background/SOTA/alternative therapies and the subject device—good and bad, recent and long past.
Appraise: Determine if specific clinical/performance data have sufficient relevance (e.g., by product, therapy, disease state), quality (e.g., large enough population, well-designed study, high-quality data), and weight (e.g., randomized controlled trials, large observational trials) to include or exclude for further evaluation. Include rare cases and real-world experience (RWE) data.
Analyze: Develop data summary tables and benefit−risk ratios based on the included clinical/performance data and specify reliable conclusions for product safety, efficacy, and performance.
Write: Document details in CER/PER based on data analyzed and decide if data meet the prespecified S&P acceptance criteria documented in the CEP/PEP. Specify any clinical/performance data gaps for the PMCFP/PMPFP to gather more data for the next evaluation.

The MDCG 2020-1 [42] about software as a medical device (SaMD) put this together for both MDR and IVDR (Fig. 1.12).

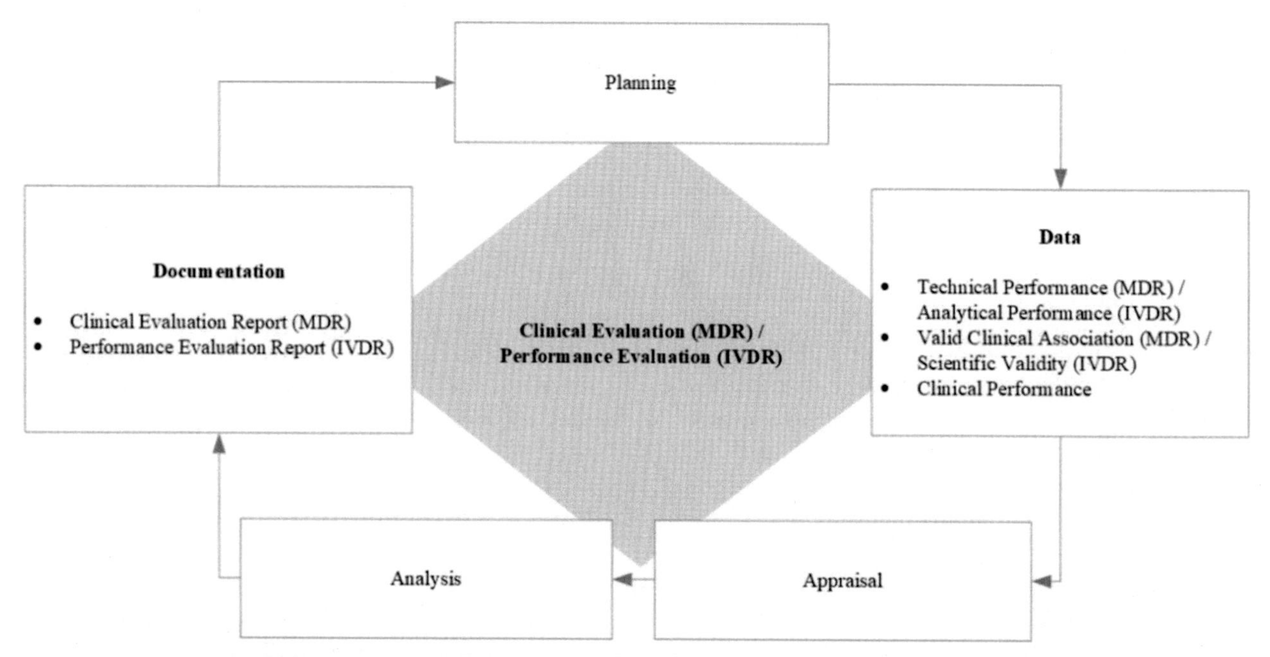

FIGURE 1.12 Merged CE/PE stages for SaMD—Overview. *Taken from MDCG 2020-1 [42].*

The planning stage resulted in the CEP/PEP and the identifying (data), appraising, analyzing, and writing (documentation) stages resulted in the CER/PER; this process also generated the clinical data gaps resulting in the PMCFP/PMPFP. These three document types are dependent on each other, and should be completely separate documents under version control. The process expects the CEP/PEP to be written before the CER/PER, which in turn is written before the PMCFP/PMPFP, which is written before the PMCFER/PMPFER, which documents the new clinical data for the next CER/PER.

1.15 Conclusions

This chapter described CE/PE history, EU regulatory requirements, and CE/PE definitions, and highlighted some similarities and differences between CERs and PERs. The CER/PER can have a broad use to train manufacturing teams and interested others.

This chapter also highlighted and discussed many clinical evidence, clinical trials, and medical device software guidance documents, and summarized the EU regulatory system, NB CE/PE review processes, and expert panel details when providing advice specifically about high-risk device CERs/PERs already reviewed by the NB in their CEAR/PEAR documents. Critically important details about integrating manufacturer clinical/performance trials, evaluations, vigilance reporting (i.e., PMS), and risk reduction (i.e., RMS) systems and processes were described.

Process controls included the potential to follow international standards and to use common specifications like ALCOA + to ensure data quality and integrity. Many changes are anticipated over time, so understanding change management is imperative. As devices mature, the data sources and amounts change. A new naming convention was introduced to align with the medical device life cycle and to accomodate different detailed process steps when working on different CER/PER TYPES including the viability, equivalence, first in human, traditional, and obsolete CER/PER TYPEs.

The CE/PE process involves multiple documents linked to many other technical file or design dossier documents, and this chapter ended with a brief discussion about writing CE/PE documents. When writing CERs/PERs, the reader is strongly encouraged to establish good practices as follows:

1. Review past CE/PE work.
2. Prepare completely new submission on appropriate frequency to renew CE mark.
3. Evaluate prior reviewer comments and NB deficiencies.
4. Understand decentralized EU regulatory authority structure.
5. Expect different legislation in different EU countries.
6. Be fully trained on MDR/IVDR regulations and guidelines.
7. Employ competent quality and regulatory experts early in the CE/PE process.
8. Learn about NB and expert panel work processes.
9. Communicate CE/PE story clearly and completely.
10. Anticipate CER/PER reviewer, approver, NB, and regulatory authority needs.
11. Meet minimum CE/PE requirements.
12. View compliance cost as an investment in future market access.

The next few chapters will review each step in the CE/PE process in detail for planning, identifying, appraising, and analyzing clinical/performance data, establishing benefit–risk ratios for the subject device uses in humans, and writing CERs/PERs. The next chapters review SSCP/SSP writing, CE/PE document review, integrating the PMS and RM systems with the CE/PE system, and how to understand and relate this EU regulatory work to CER/PER requirements outside of Europe, with some thoughts about future directions for CER/PER writing globally.

1.16 Review questions

After reading this chapter, the reader should be able to answer the following questions.

1. What do the following acronyms mean? CEP/PEP, CER/PER, MDR/IVDR, PMCF/PMPF, and SSCP/SSP.
2. Which devices require a CER in the EU?
3. When is a CER required?
4. At what stage in product development should a vCER or a fihCER be written?
5. Why is a CER required?
6. What is the purpose of a CER?
7. Is a CER alone sufficient to apply the CE mark and keep a device marketed in the EU?
8. What five stages are defined for the clinical evaluation in MEDDEV 2.7/1, Rev. 4 (2016)?
9. What is the most important part of the CER?
10. When are the clinical/performance data evaluations considered sufficient in the CER/PER?
11. What happens after the CER/PER is written?
12. True or false: all CERs/PERs must be updated and submitted annually.

13. What are the five CER types, and why is this CER classification system helpful?
14. How does a manufacturer choose a NB?
15. How many EU notified bodies were designated under MDR (EU Reg. 2017/745) and IVDR (EU Reg. 2017/726)?
16. How does a notified body become accredited?
17. How long does the NB take to issue a response after the CER is submitted?
18. How is a CER evaluated?
19. What are the five CE/PE output documents?
20. What is the document sequence in the CE/PE document writing cycle?

References

[1] Battenfield B.L. Designing clinical evaluation tools: the state-of-the-art. National League for Nursing. 1986. Pub. No. 32-2160. ISBN 0-88737-290-2.

[2] K.M. Becker, J.J. Whyte (Eds.), Clinical Evaluation of Medical Devices, 1st ed., Humana Press, 1997. 2nd ed. 2006/2010.

[3] T.J. Cleophas, A.H. Zwinderman, Understanding Clinical Data Analysis: Learning Statistical Principles from Published Clinical Research, 1st ed, Springer, 2017.

[4] European Commission. Medical devices—expert panels. Accessed January 8, 2023. https://health.ec.europa.eu/medical-devices-expert-panels_en.

[5] BSI. Our history—1901 to present day standards. Accessed January 15, 2023. https://www.bsigroup.com/en-GB/about-bsi/our-history/#chapter5.

[6] European Parliament and Council of the European Union. Consolidated text: Regulation (EU) No 536/2014 of the European Parliament and of the Council of 16 April 2014 on clinical trials on medicinal products for human use, and repealing Directive 2001/20/EC. Accessed March 1, 2023. https://eur-lex.europa.eu/legal-content/EN/TXT/?uri = CELEX%3A02014R0536-20221205.

[7] G.A. Van Norman, Drugs and devices: comparison of European and U.S. approval processes, JACC: Basic. Transl. Sci. 1 (5) (2016) 399–412. Available from: https://www.ncbi.nlm.nih.gov/pmc/articles/PMC6113412/pdf/main.pdf.

[8] European Parliament and Council of the European Union. Directive 90/385/EEC of 20 June 1990, also known as the Active Implantable Medical Devices Directive (AIMDD). https://eur-lex.europa.eu/legal-content/EN/TXT/?uri = CELEX:31990L0385/ and consolidated version in 2007 at https://eur-lex.europa.eu/legal-content/EN/TXT/?uri = CELEX/3A01990L0385-20071011.

[9] European Parliament and Council of the European Union. Directive 93/42/EEC of 14 June 1993 also known as the Medical Device Directive (MDD). https://eur-lex.europa.eu/legal-content/EN/TXT/?uri = CELEX:31993L0042.

[10] European Parliament and Council of the European Union. Consolidated text: Directive 98/79/EC of the European Parliament and of the Council of 27 October 1998 on in vitro diagnostic medical devices. Accessed February 25, 2023. https://eur-lex.europa.eu/legal-content/EN/TXT/?uri = CELEX%3A01998L0079-20120111.

[11] European Parliament and Council of the European Union. Directive 2007/47/EC of the European Parliament and of the Council of 5 September 2007 amending Council Directive 90/385/EEC on the approximation of the laws of the Member States relating to active implantable medical devices, Council Directive 93/42/EEC concerning medical devices and Directive 98/8/EC concerning the placing of biocidal products on the market. Accessed January 8, 2023. https://eur-lex.europa.eu/legal-content/EN/TXT/?uri = CELEX%3A32007L0047.

[12] European Parliament and Council of the European Union. Consolidated text: Regulation (EU) 2017/745 of the European Parliament and of the Council of 5 April 2017 on medical devices, amending Directive 2001/83/EC, Regulation (EC) No 178/2002 and Regulation (EC) No 1223/2009 and repealing Council Directives 90/385/EEC and 93/42/EEC. Accessed September 29, 2023. https://eur-lex.europa.eu/legal-content/EN/TXT/?uri = CELEX%3A02017R0745-20230320.

[13] European Parliament and Council of the European Union. Consolidated text: Regulation (EU) 2017/746 of the European Parliament and of the Council of 5 April 2017 on in vitro diagnostic medical devices and repealing Directive 98/79/EC and Commission Decision 2010/227/EU. Accessed September 29, 2023. https://eur-lex.europa.eu/legal-content/EN/TXT/?uri = CELEX%3A02017R0746-20230320.

[14] ISO. ISO 14155:2020. Clinical investigation of medical devices for human subjects—good clinical practice. Accessed April 8, 2023. https://www.iso.org/standard/71690.html.

[15] EU REG 2020/561 of 23 April 2020 amending EU Reg 2017/745 on medical devices dates of application. https://eur-lex.europa.eu/legal-content/EN/TXT/PDF/?uri = CELEX:32020R0561&from = EN.

[16] Council of the European Union. Update regarding the state of play on the implementation of the Medical Device Regulations. December 6, 2022. https://data.consilium.europa.eu/doc/document/ST-15520-2022-INIT/en/pdf.

[17] MDCG 2020-5 Clinical evaluation—Equivalence. A guide for manufacturers and notified bodies. Accessed February 10, 2023. https://health.ec.europa.eu/system/files/2020-09/md_mdcg_2020_5_guidance_clinical_evaluation_equivalence_en_0.pdf.

[18] European Commission. Guidance—MDCG endorsed documents and other guidance. Accessed February 18, 2023. https://health.ec.europa.eu/medical-devices-sector/new-regulations/guidance-mdcg-endorsed-documents-and-other-guidance_en.

[19] Co-ordination of Notified Bodies Medical Devices (NB-MED). 1998. Guidance on clinicals. NB-MED/2.7/Rec1. Accessed September 29, 2023. https://www.team-nb.org/wp-content/uploads/2015/05/nbmeddocuments/Recommendation-NB-MED-2_7-1_rev2_Guidance_on_clinicals.pdf.

[20] European Commission. 2001. DG enterprise. Directorate G. Unit 4—Pressure equipment, medical devices, metrology. Medical devices guidance document. MEDDEV 2.10-2 Rev.1. *Designation and Monitoring of Notified Bodies within the Framework of EC Directives on Medical*

Devices. Accessed April 22, 2023. https://www.medical-device-regulation.eu/wp-content/uploads/2019/05/2_10_2date04_2001_en.pdf or https://ec.europa.eu/docsroom/documents/10291/attachments/1/translations/en/renditions/pdf.

[21] Donawa M. European medical device regulation: a new era? Medical Device Technology. 2004. pp. 36–37. Accessed April 22, 2023. https://citeseerx.ist.psu.edu/document?repid = rep1&type = pdf&doi = dbc18f44020169a7135ff7178ddfcf5ac9ce6599 or at https://www.donawa.com/european-cro/files/8%20Clinical%20Evaluation%20Nov2006%20MDT%20issue.pdf.

[22] European Commission. 2003. Enterprise and Industry Directorate-General. Single Market: regulatory environment, standardization and new approach. Pressure equipment, medical devices, metrology. MEDDEV 2.7.1 (version 2003-03-27). *Guidelines on Medical Devices. Evaluation of Clinical Data: A Guide for Manufacturers and Notified Bodies*. Accessed September 29, 2023. http://meddev.info/_documents/2_7.pdf.

[23] International Medical Device Regulators (IMDRF). Global Harmonization Task Force Study Group 5—clinical safety/performance documents. Accessed April 2, 2023. https://www.imdrf.org/documents/ghtf-final-documents/ghtf-study-group-5-clinical-safetyperformance.

[24] Global Harmonization Task Force Study Group 5. GHTF SG5 clinical evidence minutes of meeting Monday 17 January and Tuesday 18 January 2005. Accessed April 2, 2023. https://www.imdrf.org/sites/default/files/docs/ghtf/final/sg5/meetings/ghtf-sg5-meeting-minutes-050117-australia-canberra.pdf.

[25] Global Harmonization Task Force Study Group 5. GHTF SG5/N1R8:2007. Clinical evidence—key definitions and concepts. May 2007. Accessed April 2, 2023. https://www.imdrf.org/sites/default/files/docs/ghtf/archived/sg5/technical-docs/ghtf-sg5-n1r8-clinical-evaluation-key-definitions-070501.pdf.

[26] Global Harmonization Task Force Study Group 5. 2007. Clinical evaluation. GHTF SG5/N2R8:2007. Accessed September 29, 2023. https://www.imdrf.org/sites/default/files/docs/ghtf/archived/sg5/technical-docs/ghtf-sg5-n2r8-2007-clinical-evaluation-070501.pdf.

[27] European Commission. 2009. MEDDEV 2.7.1 Rev. 3. Guidelines on medical devices. Clinical evaluation: a guide for manufacturers and notified bodies. Enterprise and Industry Directorate General. Consumer Goods. Cosmetics and Medical Devices. Accessed September 29, 2023. http://meddev.info/_documents/2_7_1rev_3_en.pdf.

[28] IMDRF Medical Device Clinical Evaluation Working Group. Clinical evaluation (IMDRF MDCE WG/N56FINAL:2019) (10OCT2019). Accessed September 29, 2023. https://www.imdrf.org/sites/default/files/docs/imdrf/final/technical/imdrf-tech-191010-mdce-n56.pdf.

[29] European Commission. 2016. Guidelines on medical devices; MEDDEV 2.7/1 Rev 4. Clinical evaluation: a guide for manufacturers and notified bodies under directives 93/42/EEC and 90/385/EEC. Accessed September 29, 2023. https://www.medical-device-regulation.eu/wp-content/uploads/2019/05/2_7_1_rev4_en.pdf.

[30] NBOG. *Designating Authorities Handbook*. Accessed September 29, 2023. https://www.nbog.eu/app/download/421905/da_handbook.pdf.

[31] NBOG BPG 2009-1. 2009. *Guidance on design-dossier examination and report content*. Accessed April 9, 2023. http://www.doks.nbog.eu/Doks/NBOG_BPG_2009_1.pdf.

[32] NBOG BPG 2009-4. 2009. *Guidance on notified body's tasks of technical documentation assessment on a representative basis*. Accessed April 9, 2023. http://www.doks.nbog.eu/Doks/NBOG_BPG_2009_4_EN.pdf.

[33] L. Keutzer, U.S.H. Simonson, Medical device apps: an introduction to regulatory affairs for developers, JMIR Mhealth Uhealth 8 (6) (2020) e17567. Available from: https://mhealth.jmir.org/2020/6/e17567/PDF.

[34] ISO 14155. 2011. *Clinical investigation of medical devices for human subjects—Good clinical practice*. Accessed July 17, 2023. https://www.iso.org/standard/45557.html.

[35] Global Harmonization Task Force Study Group 5. 2012. GHTF/SG5/N6:2012. *Clinical evidence for IVD medical devices—key definitions and concepts*. Accessed May 12, 2023. https://www.imdrf.org/sites/default/files/docs/ghtf/final/sg5/technical-docs/ghtf-sg5-n6-2012-clinical-evidence-ivd-medical-devices-121102.pdf.

[36] Global Harmonization Task Force Study Group 5. 2012. GHTF/SG5/N7:2012. Clinical evidence for IVD medical devices—scientific validity determination and performance evaluation. Accessed May 12, 2023. https://www.imdrf.org/sites/default/files/docs/ghtf/final/sg5/technical-docs/ghtf-sg5-n7-2012-scientific-validity-determination-evaluation-121102.pdf.

[37] Global Harmonization Task Force Study Group 5. GHTF/SG5/N8:2012. Clinical performance studies for in vitro diagnostic medical devices. Accessed May 12, 2023. https://www.imdrf.org/sites/default/files/docs/ghtf/final/sg5/technical-docs/ghtf-sg5-n8-2012-clinical-performance-studies-ivd-medical-devices-121102.pdf.

[38] MDCG 2019-6, Rev 4. Questions and answers: requirements relating to notified bodies. Accessed April 9, 2023. https://health.ec.europa.eu/system/files/2022-10/md_mdcg_qa_requirements_notified_bodies_en.pdf.

[39] MDCG 2019-13. Guidance on sampling of devices for the assessment of the technical documentation. December 2019. Accessed April 9, 2023. https://health.ec.europa.eu/system/files/2020-09/md_mdcg_2019_13_sampling_mdr_ivdr_en_0.pdf.

[40] IMDRF/GRRP WG/N47 FINAL:2018. Essential principles of safety and performance of medical devices and IVD medical devices. October 31, 2018. Accessed November 18, 2018. http://www.imdrf.org/docs/imdrf/final/technical/imdrf-tech-181031-grrp-essential-principles-n47.pdf.

[41] Global Harmonization Task Force Study Group 1. GHTF/SG1/N68:2012. Essential principles of safety and performance of medical devices. November 2, 2012. Accessed April 9, 2023. https://www.imdrf.org/sites/default/files/docs/ghtf/archived/sg1/technical-docs/ghtf-sg1-n68-2012-safety-performance-medical-devices-121102.pdf.

[42] MDCG 2020-1. Guidance on clinical evaluation (MDR)/performance evaluation (IVDR) of medical device software. March 2020. Accessed February 23, 2023. https://health.ec.europa.eu/system/files/2020-09/md_mdcg_2020_1_guidance_clinic_eva_md_software_en_0.pdf.

[43] IMDRF/SaMD WG/N41FINAL:2017. Software as a medical device (SaMD): Clinical evaluation. Accessed May 12, 2023. https://www.imdrf.org/sites/default/files/docs/imdrf/final/technical/imdrf-tech-170921-samd-n41-clinical-evaluation_1.pdf.

[44] MDCG 2019-11. Guidance on qualification and classification of software in regulation (EU) 2017/745—MDR and Regulation (EU) 2017/746—IVDR. October 2019. Accessed February 26, 2023. https://health.ec.europa.eu/system/files/2020-09/md_mdcg_2019_11_guidance_qualification_classification_software_en_0.pdf.
[45] MDCG 2020-6. Regulation (EU) 2017/745: clinical evidence needed for medical devices previously CE marked under directives 93/42/EEC or 90/385/EEC. A guide for manufacturers and notified bodies. April 2020. Accessed February 18, 2023. https://health.ec.europa.eu/system/files/2020-09/md_mdcg_2020_6_guidance_sufficient_clinical_evidence_en_0.pdf.
[46] MDCG. Background note on the relationship between MDCG 2020-6 and MEDDEV 2.7/1 rev.4 on clinical evaluation. Accessed April 21, 2023. https://health.ec.europa.eu/system/files/2022-09/md_borderline_bckgr-note-manual-bc-dir_en_1.pdf.
[47] MDCG 2022-2 Guidance on general principles of clinical evidence for in vitro diagnostic medical devices (IVDs). January 2022. Accessed April 21, 2023. https://health.ec.europa.eu/system/files/2022-01/mdcg_2022-2_en.pdf.
[48] MDCG 2022-10. Q&A on the interface between Regulation (EU) 536/2014 on clinical trials for medicinal products for human use (CTR) and Regulation (EU) 2017/746 on in vitro diagnostic medical devices (IVDR). May 2022. Accessed February 23, 2023. https://health.ec.europa.eu/system/files/2022-05/mdcg_2022-10_en.pdf.
[49] European Commission. Notified bodies. Accessed February 19, 2023. https://ec.europa.eu/health/md_topics-interest/notified_bodies_en.
[50] European Commission. Notified bodies—NANDO for EU Reg 2017/745 on medical devices. Accessed February 19, 2023. https://ec.europa.eu/growth/tools-databases/nando/index.cfm?fuseaction = directive.notifiedbody&dir_id = 34.
[51] European Commission. Notified bodies—NANDO for EU Reg 2017/746 on in vitro diagnostic medical devices. Accessed February 19, 2023. https://ec.europa.eu/growth/tools-databases/nando/index.cfm?fuseaction = directive.notifiedbody&dir_id = 35.
[52] International Organization for Standardization. ISO 14971: 2019 Medical Devices—application of risk management to medical devices. Accessed February 19, 2023. https://www.iso.org/standard/72704.html (Note: access fee).
[53] International Organization for Standardization. ISO/TR 20416:2020 Medical devices—post-market surveillance for manufacturers. Accessed October 1, 2023. https://www.iso.org/standard/67942.html (Note: access fee).
[54] MDCG 2023-3. Questions and Answers on vigilance terms and concepts as outlined in the Regulation (EU) 2017/745 on medical devices. February 2023. Accessed February 19, 2023. https://health.ec.europa.eu/system/files/2023-02/mdcg_2023-3_en_0.pdf.
[55] European Commission. 2013. MedDev 2.12/1 rev 8. Guidelines on a medical devices vigilance system. Accessed March 4, 2023. http://meddev.info/_documents/2_12_1_rev8.pdf.
[56] MDCG 2021-1 Rev.1. *Guidance on harmonised administrative practices and alternative technical solutions until EUDAMED is fully functional.* May 2021. Accessed February 19, 2023. https://health.ec.europa.eu/system/files/2021-05/2021-1_guidance-administrative-practices_en_0.pdf.
[57] MDCG 2022-21. Guidance on Periodic Safety Update Report (PSUR) according to Regulation (EU) 2017/745. December 2022. Accessed February 19, 2023. https://health.ec.europa.eu/system/files/2023-01/mdcg_2022-21_en.pdf.
[58] European Commission. Harmonized standards. Accessed October 1, 2023. https://single-market-economy.ec.europa.eu/single-market/european-standards/harmonised-standards_en including specific standards for MDR https://single-market-economy.ec.europa.eu/single-market/european-standards/harmonised-standards/medical-devices_en and for IVDR https://single-market-economy.ec.europa.eu/single-market/european-standards/harmonised-standards/iv-diagnostic-medical-devices_en.
[59] European Commission. 2020/C 259/02. Commission guidance for the medical devices expert panels on the consistent interpretation of the decision criteria in the clinical evaluation consultation procedure. Accessed May 11, 2023. https://eur-lex.europa.eu/legal-content/EN/TXT/?uri = uriserv%3AOJ.C_0.2020.259.01.0002.01.ENG&toc = OJ%3AC%3A2020%3A259%3ATOC.

Chapter 2

Planning clinical/performance evaluations

This chapter explores the planning steps required in the clinical evaluation plan/performance evaluation plan (CEP/PEP). The product must be a medical device and the CEP/PEP process must follow the EU regulatory requirements. Strategies and process controls are discussed, including ideas about monitoring, auditing, and controling CEP/PEP development. Device groups and families are defined in detail, along with specifics about making an equivalent device claim. The plan must introduce the developmental context for the subject device, as well as a plan for the background and state-of-the-art section in the CEP/PEP. A "sandboxing" method (trying various search terms) is recommended while writing the CEP/PEP to solidify the best database search string strategies to identify appropriate, high-quality clinical data in the many different databases to be searched for clinical data to add to the CER/PER. The five clinical evaluation/performance evaluation types are discussed, and details are presented about the different planning concerns encountered for each type.

Planning will set the tone for the entire clinical evaluation/performance evaluation (CE/PE) process; therefore, it is important to plan well.

2.1 Product must be a medical device

When starting the clinical evaluation plan/performance evaluation plan (CEP/PEP), the product to be marketed for sale in the European Union (EU) must be a medical device defined in the EU regulations, as follows:

a 'medical device' means any instrument, apparatus, appliance, software, implant, reagent, material or other article intended by the manufacturer to be used, alone or in combination, for human beings for one or more of the following specific medical purposes:

- diagnosis, prevention, monitoring, prediction, prognosis, treatment or alleviation of disease,
- diagnosis, monitoring, treatment, alleviation of, or compensation for, an injury or disability,
- investigation, replacement or modification of the anatomy or of a physiological or pathological process or state,
- providing information by means of *in vitro* examination of specimens derived from the human body, including organ, blood and tissue donations, and which does not achieve its principal intended action by pharmacological, immunological or metabolic means, in or on the human body, but which may be assisted in its function by such means.

The following products shall also be deemed to be medical devices:

- devices for the control or support of conception;
- products specifically intended for the cleaning, disinfection or sterilisation of devices... Article 1(4) and... Article 2 (1), EU medical device regulation (MDR)(EU REG 2017/745) [1]

an 'in vitro diagnostic medical device' means any medical device which is a reagent, reagent product, calibrator, control material, kit, instrument, apparatus, piece of equipment, software or system, whether used alone or in combination, intended by the manufacturer to be used* in vitro *for the examination of specimens, including blood and tissue donations, derived from the human body, solely or principally for the purpose of providing information on one or more of the following:*

(a) concerning a physiological or pathological process or state;
(b) concerning congenital physical or mental impairments;
(c) concerning the predisposition to a medical condition or a disease;
(d) to determine the safety and compatibility with potential recipients;
(e) to predict treatment response or reactions;
(f) to define or monitoring therapeutic measures.

Planning, Writing and Reviewing Medical Device Clinical and Performance Evaluation Reports (CERs/PERs). DOI: https://doi.org/10.1016/B978-0-443-22063-0.00013-4

Specimen receptacles shall also be deemed to be in vitro *diagnostic medical devices... Article 2 (2) EU* in vitro *diagnostic regulation* (IVDR)(EU REG 2017/746) [2]

**Note: An* in vitro *diagnostic (IVD) medical device is just one of many medical device types (e.g., active/inactive, implantable/not implantable, sterile/nonsterile, software/no software medical devices).*

These EU MDR and EU IVDR medical device and *in vitro* diagnostic medical definitions apply to the clinical evaluation report/performance evaluation report (CER/PER). If the product is not a medical device, then use a different report name to avoid confusion. If the product is a medical device, then the CER/PER is always about the clinical data evaluation and not other types of evaluation needed for device design and testing, etc. The report should be focused on the patient receiving the benefit and risk when using the medical device and the report should always describe the safety and performance (S&P) from the patient perspective.

2.2 Follow the regulatory requirements

Specifically, the MDR included 7 references to "clinical evaluation plan" while the IVDR included 10 references to "performance evaluation plan." These CEP/PEP regulatory requirements were quite similar (Table 2.1) including requirements for clinical studies and postmarket clinical/performance follow-up (PMCF/PMPF).

TABLE 2.1 MDR and IVDR CEP/PEP statements.

UID	MDR [1]	IVDR [2]	Differences
1	"the clinical evaluation plan referred to in Article 61(12) and Part A of Annex XIV" shall be in the technical V&V documentation (Annex II, 6.1c)	"The clinical evidence shall support the intended purpose of the device as stated by the manufacturer and be based on a continuous process of performance evaluation, following a performance evaluation plan." (Article 56, 2)	Similar need for CEP/PEP
2	"documentation on the clinical evaluation plan, and... description of the procedures in place to keep up to date the clinical evaluation plan, taking into account the state-of-the-art" (Annex IX, 2.1)	"documentation on the performance evaluation plan, and... description of the procedures in place to keep up to date the performance evaluation plan, taking into account the state-of-the-art" (Annex IX, 2.1)	Only name changed
3	"Clinical investigations[a] shall be in line with the clinical evaluation plan as referred to in Part A of Annex XIV." (Annex XV, 2.4)	"To allow for a structured and transparent process, generating reliable and robust data, sourcing and assessment of available scientific information and data generated in performance studies should be based on a performance evaluation plan." (61)	Similar need for CE/PE study and plan to align
4	*NA; however, appropriate justifications are required when clinical investigations are exempted or "not deemed appropriate" (Article 61)*	"Where specific devices have no analytical or clinical performance or specific performance requirements are not applicable, it is appropriate to justify in the performance evaluation plan, and related reports, omissions relating to such requirements." (65)	Similar need to justify omissions
5	Clinical investigation application shall contain "details and/or reference to the clinical evaluation plan" (Annex XV, 1.5)	*NA; however, clinical performance studies require an application even though IVDR does not mention a link to the PEP in Annex XIV.*	CEP in clinical trial application
6	Clinical investigation plan shall include "General information such as type of investigation[a] with rationale for choosing it, for its endpoints and for its variables as set out in the clinical evaluation plan." (Annex XV. 3.6.1.)	*NA; however, IVDR required data obtained from clinical performance studies to "be used in the performance evaluation process and be part of the clinical evidence for the device." A clinical performance study plan (CPSP) was required but no direct link to the PEP was stated in Annex XIII,* Section 2.2.	CEP in clinical investigation plan and PEP use implied for CPSP

(*Continued*)

TABLE 2.1 (Continued)

UID	MDR [1]	IVDR [2]	Differences
	Annex XIV CE and PMCF Below	*Annex XIII PE, Performance Studies, and PMPF Below*	
7	*"... manufacturers shall... establish and update a clinical evaluation plan, which shall include at least: an identification of the general safety and performance requirements that require support from relevant clinical data;"*	*"... manufacturer shall establish and update a performance evaluation plan. The performance evaluation plan shall specify the characteristics and the performance of the device and the process and criteria applied to generate the necessary clinical evidence...1.1. Performance evaluation plan. As a general rule, the performance evaluation plan shall include at least: ... an identification of the general safety and performance requirements... that require support from relevant scientific validity and analytical and clinical performance data;"*	Similar need for general safety and performance requirements (GSPR) detail
8	"a specification of the intended purpose of the device;"	"a specification of the intended purpose of the device;"	Identical
9	NA	*"a specification of the characteristics of the device... a specification of the analyte or marker to be determined by the device... a specification of the intended use of the device; identification of certified reference materials or reference measurement procedures to allow for metrological traceability;"*	PEP needs, analyte/marker, intended use and reference materials
10	*"a clear specification of intended target groups with clear indications and contra-indications;"*	*"a clear identification of specified target patient groups with clear indications, limitations and contra-indications;"*	Similar target group needs
11	*"a specification of methods to be used for examination of qualitative and quantitative aspects of clinical safety with clear reference to the determination of residual risks and side-effects;"*	*"a specification of methods, including the appropriate statistical tools, used for the examination of the analytical and clinical performance of the device and of the limitations of the device and information provided by it;"*	Similar details needed for CEP/PEP methods
12	*"an indicative list and specification of parameters to be used to determine, based on the state-of-the-art in medicine, the acceptability of the benefit-risk ratio for the various indications and for the intended purpose or purposes of the device;"*	*"a description of the state-of-the-art, including an identification of existing relevant standards, CS, guidance or best practices documents... an indication and specification of parameters to be used to determine, based on the state-of-the-art in medicine, the acceptability of the benefit-risk ratio for the intended purpose or purposes and for the analytical and clinical performance of the device;"*	Similar need for state-of-the-art (SOTA) and benefit-risk ratio
13	*"a detailed description of intended clinical benefits to patients with relevant and specified clinical outcome parameters... an indication how benefit-risk issues relating to specific components such as use of pharmaceutical, non-viable animal or human tissues, are to be addressed;"*	*NA*	CEP needs patient benefits and outcomes
14	*NA*	*"for software qualified as a device, an identification and specification of reference databases and other sources of data used as the basis for its decision making;"*	PEP identifies software needs

(Continued)

TABLE 2.1 (Continued)

UID	MDR [1]	IVDR [2]	Differences
15	*"a clinical development plan indicating progression from exploratory investigations, such as first-in-man studies, feasibility and pilot studies, to confirmatory investigations, such as pivotal clinical investigations, and a PMCF... with an indication of milestones and a description of potential acceptance criteria..."*	*"an outline of the different development phases including the sequence and means of determination of the scientific validity, the analytical and clinical performance, including an indication of milestones and a description of potential acceptance criteria; the PMPF planning... Where any of the above mentioned elements are not deemed appropriate in the Performance Evaluation Plan due to the specific device characteristics a justification shall be provided in the plan."*	Similar development stages and postmarket surveillance (PMS) needs for PMCF/PMPF

CE, clinical evaluation; *CEP*, clinical evaluation plan; *CPSP*, clinical performance study plan; *CS*, common specification; *GSPR*, general safety and performance requirements; *IVDR*, *in vitro* diagnostic regulation; *MDR*, medical device regulation; *PE*, performance evaluation; *PEP*, performance evaluation plan; *PMCF*, postmarket clinical follow-up; *PMPF*, postmarket performance follow-up; *PMS*, postmarket surveillance; *SOTA*, state of the art. Bolded text emphasizes the limited differences between the MDR and IVDR in specific CEP/PEP discussion sections while italic text denotes when MDR and IVDR are more broadly different in other quoted CEP/PEP discussion sections.
[a]*A strong link was made between planned clinical investigations and the CEP but not the PEP.*

The MDR and IVDR regulations required a comprehensive, scientifically valid plan (i.e., the CEP/PEP) preceding the report (i.e., the CER/PER). Differences included the special analyte/marker characterization and specification needs for *in vitro* products and the MDR required the CEP to have a tighter, more specific relationship to the "clinical investigation application" and "clinical investigation plan" than the IVDR required for the relationship between PEP and the "clinical performance studies." Also, for the MDR, the CEP must discuss device components and whether any pharmaceutical, nonviable animal/human tissues were included, whlie the PEP must discuss any software used in decision-making for the IVDR.

The MDR and IVDR similarities were used to create a common checklist for plan development (**Appendix F: Clinical evaluation plan/performance evaluation plan requirements checklist**), including but not limited to the required CEP/PEP functions to:

1. support device intended use;
2. describe procedures to keep the plan up to date;
3. include across-the-board SOTA considerations;
4. align with clinical/performance studies using structured and transparent processes to generate reliable and robust data and scientific information;
5. justify in writing any/all exceptions from clinical/performance requirements;
6. link to the clinical investigations and performance studies, required regulatory applications and the investigation/study plans;
7. identify GSPR needing support from relevant clinical/performance data;
8. specify device intended purpose, target groups, indications, contraindications and methods used to examine device S&P;
9. indicate and specify the SOTA parameters used to determine the acceptability of the benefit-risk ratio for the device intended purpose/s; and
10. include details about the device developmental phase and PMCF planning with milestones and potential acceptance criteria.

These MDR and IVDR foundational similarities encouraged including both CE and PE in this one textbook; however, focusing on the specific device under evaluation should always be the strategy.

2.3 Use international regulatory standards

Many common specifications and international standards are related to medical device clinical/performance trials and other evaluations (Table 2.2). These will be helpful when working on clinical trial (CT), risk management (RM), or postmarket surveillance (PMS) operations.

TABLE 2.2 Example international standards for clinical/performance trials.

Standard [ref[a]]	Title	Notes
ISO 14155 [3]	Clinical Investigation of Medical Devices for Human Subjects—Good Clinical Practice	• Guides ethical conduct, clinical quality, audits, and stats during clinical trials • Includes risk-based monitoring and trial registration references • Clarifies requirements for each stage of clinical development • Describes clinical trial (CT) risk management with links to ISO 14971 • Indicates IVD medical devices have additional requirements
ISO 14971 [4]	Medical Devices—Application of Risk Management to Medical Devices	• Defines procedure for risk management including software as a medical device (SaMD) • Requires objective risk acceptability criteria • Does not apply to clinical decisions although clinical risk processes are developing
ISO/TR 20416:2020 [5]	Medical Devices—Post-Market Surveillance for Manufacturers	• Aligns with relevant international standards, in particular ISO 14155, ISO 13485 and ISO 14971 • Describes a proactive and systematic process for manufacturers ◦ to collect and analyze appropriate data ◦ to provide feedback information for other processes ◦ to meet applicable regulatory requirements ◦ to gain post-production experience data
[a]BS EN 13612 [6]	Performance Evaluation of *In Vitro* Diagnostic Medical Devices	• Offers ethical guidance on CTs and responsibilities • Describes PE requirements including study planning and organization, sites, designs, records, observations, unexpected findings, and reports
ISO 20916 [7]	*In Vitro* Diagnostic Medical Devices—Clinical Performance Studies Using Specimens from Human Subjects—Good Study Practice	• Offers ethical guidance on CTs and responsibilities • Defines how to plan, design, conduct, record, and report clinical performance studies for regulatory purposes • Describes site visits, training, documentation • Does not assess technical specifications, analytical performance or re-imbursement issues
ISO 17511 [8]	*In Vitro* Diagnostic Medical Devices—Requirements for Establishing Metrological Traceability of Values Assigned to Calibrators, Trueness Control Materials and Human Samples	• Specifies technical requirements and documentation needed • Clarifies roles and responsibilities • Applies to IVD medical devices with measurement results
[a]GHTF/SG5/N6 [9]	Clinical Evidence for IVD Medical Devices—Key Definitions and Concepts	• Defines clinical evidence for IVD Medical Devices • Clarifies scientific validity, analytical, clinical performance • Guides generation of clinical evidence to support marketing
[a]GHTF/SG5/N7 [10]	Clinical Evidence for IVD Medical Devices—Scientific Validity Determination and Performance Evaluation	• Guides how to collect and document clinical evidence • Guides regulators when assessing clinical evidence • Defines when scientific validity for an analyte is expected • Explains how to demonstrate clinical performance • Guides how to appraise and analyze the data

(Continued)

TABLE 2.2 (Continued)

Standard [ref[a]]	Title	Notes
[a]GHTF/SG5/N8 [11]	Clinical Evidence for IVD Medical Devices—Clinical Performance Studies for *In Vitro* Diagnostic Medical Devices	• Defines clinical performance study purpose • Guides clinical performance study design selection (observational or interventional) • Describes considerations prior to clinical performance study • Describes clinical performance study protocol and conduct • Explains details vary in the report based on device risk class
[a]GHTF/SG1/N068 [12]	[a]Essential Principles of Safety and Performance of Medical Devices	• Describes general principles of device safety and performance • Explains how [a]Essential Principles are applicable to all devices • Reviews [a]Essential Principles for medical and IVD devices
[a]GHTF/SG1/N078 [13]	Principles of Conformity Assessment for Medical Devices	• Defines quality management system (QMS) • Describes PMS system and technical documentation • Describes declaration of conformity and registration • Guides harmonized conformity assessment system
[a]GHTF/SG1/N046 [14]	Principles of Conformity Assessment for IVD Medical Devices	• Defines QMS, PMS, and technical documentation • Describes declaration of conformity and registration • Guides conformity assessment
[a]GHTF/SG1/N063 [15]	Summary Technical Documentation (STED) for Demonstrating Conformity to the Essential Principles of Safety and Performance of *In Vitro* Diagnostic Medical Devices	• Defines STED preparation and format • Describes device description, variants, accessories • Refers to prior device generations, similar devices • Discuses risk analysis and control • Describes design and manufacturing information • Defines device verification and verification and labeling

BS, British Standard; *CT*, clinical trial; *EN*, English; *GHTF*, Global Harmonization Task Force; *ISO*, International Organization for Standardization; *IVD*, *in vitro* diagnostic; *PE*, performance evaluation; *PMS*, postmarket surveillance; *QMS*, quality management system; *SaMD*, software as a medical device; *SG*, study group; *STED*, Summary Technical Documentation.
[a]*CAUTION—OUTDATED INFORMATION: Use only for historical context, consider all relevant regulatory changes since date of international standard release.*

This list is not comprehensive and is only intended to help put current regulations in context. *Note*: The new "ISO/AWI 18969: Clinical Evaluation of Medical Devices" [16] is under development by the ISO Technical Committee 194 with a description as follows:

> This document specifies terminology, principles and a process for the clinical evaluation of medical devices. The process described in this document aims to assist manufacturers of medical devices to estimate the clinical risks associated with a medical device and evaluate the acceptability of those risks in the light of the clinical benefits achieved when the device is used as intended. The requirements of this document are applicable throughout the life cycle of a medical device. The process described in this document applies to the assessment of risks and benefits from clinical data obtained from the use of medical devices in humans. This document specifies general requirements intended to:
>
> - verify the safety of medical devices when used in accordance with their instructions for use;
> - verify the clinical performance or effectiveness of a medical device meet the claims of the manufacturer in relation to its intended use;
> - verify sufficient clinical evidence exist to demonstrate the achievement of a positive benefit/risk balance when a medical device is used as intended in the appropriate patient population;
> - ensure the scientific conduct of a clinical evaluation and the credibility of conclusions drawn on the safety and performance of a medical device;
> - define the responsibilities of the manufacturer and those conducting or contributing to a clinical evaluation; and
> - assist manufacturers, clinicians, regulatory authorities and other bodies involved in the conformity assessment of medical devices.
>
> Note 1 This standard can be used for regulatory purposes.
>
> Note 2 This document does not apply to *in vitro* diagnostic medical devices. However, there may be situations, dependent on the device and national or regional requirements, where sections and/or requirements of this document might be applicable.

Medical device regulations and standards are developed over many decades, and sometimes they are updated but not aligned with each other. Specifically, many available international standards are no longer fully aligned with the EU MDR and IVDR. In addition, unfortunately, many international standards are out of touch with current local geography regulatory changes, so use this listing with caution. The new ISO standard should help to harmonize CE (but not PE) regulations globally.

2.4 Specify how to document the plan in the quality management system

Using regulatory requirements, guidelines, and international standards together is the expected process for conducting clinical evaluations. Starting with the plan and progressing through each evaluation stage, the CE/PE processes must be rigorous and well controlled.

The CEP/PEP needs to be:

1. documented in writing before CE/PE begins; and
2. version controlled (signed and dated).

The quality management system (QMS) should carefully control the CEP/PEP before, during, and after use to create the required clinical evaluation report/performance evaluation report (CER/PER) each time the CER/PER is written.

BOX 2.1 Plan carefully: each CER/PER is a comprehensive revision

Some people think an update to a document is easier than writing the first document. This makes sense if the work is simply adding in bits of data. Unfortunately, this is not the case for the CE/PE documents since each version is intended to be a single, stand-alone, comprehensive telling of the "device story."

New clinical/performance data added into the story may actually change the whole CER/PER storyline and structure. For example, a previously unknown risk may require a complete device redesign, or a new competitor product or regulatory requirement may lead the device into obsolescence. Conversely, new clinical data may support a clinically significant, new, patient benefit leading to an improved benefit-risk ratio. The evaluator and CE/PE team need to exercise due diligence during CE/PE planning to ensure each rewrite and update is a complete and fully annotated tale.

The CE/PE team must include all details and the team must rearrange the details to ensure the story makes sense. Simply inserting new data bits and fragments into an otherwise well-written CER/PER is a bad practice. Isolated data bits should not be inserted without an overall re-evaluation of the clinical data in total. Evaluation inconsistencies like these could trigger a lack of clarity or a series of misunderstood details leading to serious deficiencies, which can prevent the assignment of CE Mark or might require device removal from the EU market until the clinical data evaluation issues can be addressed.

The QMS should house the most current CEP/PEP template and the CE/PE process should be detailed with step-by-step instructions. Draft documents should be controlled to ensure the drafting process is well-organized and not confusing. The CEP/PEP is intended to direct the CER/PER development and should not change during use. If an amendment to the CEP/PEP is needed during the CER/PER development, this CEP/PEP amendment should be documented with sufficient detail to show careful control of the planning process and the resulting reporting process.

2.5 Clinical evaluation/performance evaluation strategies and process controls

Do not leave strategy to chance. Plan and "bake" the clinical strategy into all manufacturing systems.

First, explore and spell out the goal and then the process to achieve the goal (i.e., the strategy). Design all systems for the win. Make the role and function within each system unique and not duplicative. Think carefully about each operational system within the company: how does the CE/PE system work? What about the CT, PMS, and RM systems?

One strategic goal should be to create synergies between company systems and another strategic goal should be to ensure confidence and competence within each system (Fig. 2.1).

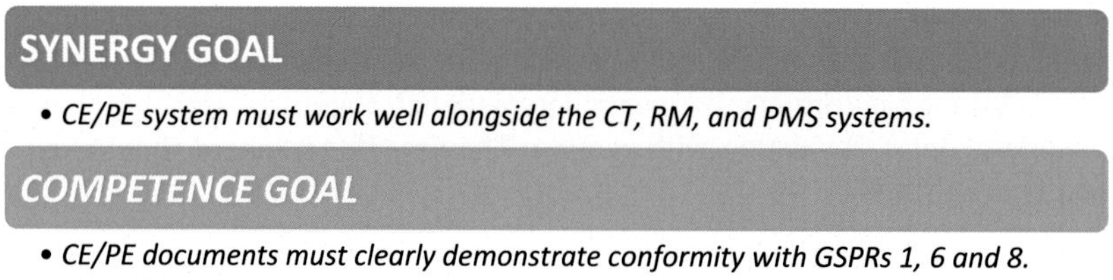

SYNERGY GOAL

- *CE/PE system must work well alongside the CT, RM, and PMS systems.*

COMPETENCE GOAL

- *CE/PE documents must clearly demonstrate conformity with GSPRs 1, 6 and 8.*

FIGURE 2.1 Setting strategic CE/PE goals. *CE*, clinical evaluation; *CT*, clinical trial; *GSPR*, general safety and performance requirements; *PE*, performance evaluation; *PMS*, postmarket surveillance; *RM*, risk management.

Adding in this new expanded CE/PE system alongside the previously existing and "well-oiled" CT, RM, and PMS systems was a not a small task. Companies should "watch out for the obvious pitfalls" when integrating these systems. For example, employees may say things like:

- work was always done this way and should not change (*not true*);
- one literature search will work for all company systems (*not true*);
- device risks are the same as patient safety risks (*not true*);
- field engineers have sufficient clinical training to address all CE/PE issues (*not true*);
- clinical staff have sufficient clinical training to address all PMS and RM issues (*not true*);
- data analyzed previously do not need to be updated (*not true*); or
- solved complaints and problems are not listed as complaints during PMS (*not true*).

A skilled leader can avoid these pitfalls and traps. Here are some suggestions to help:

1. Make each integrated process specific to the purpose for each process (note differences and use care when linking and potentially integrating processes together). The CE/PE, CT, RM, and PMS system processes must meet company-specific needs.
2. Train staff to perform CE/PE, CT, RM, and PMS system processes well. This will reduce notified body (NB) deficiencies, which in turn will reduce time and cost for this work in the future.
3. Have strong system differentiation at the core (i.e., be unique for each of the CE/PE, CT, RM, and PMS systems). This differentiation will often be a good indicator for future financial success.
4. Strategize ways to map out device design and development processes uniquely and to integrate the processes vertically within each CE/PE, CT, RM, and PMS system. This will create value which is often recognized by each team member, repeatedly.
5. Ensure staff understand terms vary and may have specific, but different, definitions and uses in the CE/PE, CT, RM, and PMS systems. Ensure all team members know EU regulatory terms like clinical data, performance data, safety data, benefit-risk ratio, equivalence, background knowledge, alternative therapy/device, and SOTA. Lay terms or regulatory terms from other countries may conflict and cloud the ability for individual team members to work well together. For example, the term "equivalence" in the EU is not the same as "equivalence" in the US, just as the term "risk" has a different meaning to an engineer than to a physician.
6. Know how NBs work by reading the NB sections in the MDR and getting to know your NB team (especially the clinical reviewer). Understanding the NB role and responsibility will help manage expectations.
7. Know if your CER/PER will be sent to an expert panel for advice, and plan how expert panel feedback will flow back into your CE/PE, CT, RM, and PMS integrated systems.

A great strategy defines what to expect, how to identify problems early and how to control those problems to ensure success built on insights derived from combining past and new process needs. As CE/PE planning starts, the evaluator/leader should have a well-defined strategy and should plan to use strong, yet flexible tactics to get this complex, multifactorial work done within a reasonable timeline and at a reasonable cost.

2.5.1 Strategies for success in clinical evaluation/performance evaluation work

Focus on the CE/PE process has been increasing. The MDR/IVDR clarified CE/PE expectations and NBs are enforcing these EU regulations more vigorously than ever before because they are being held accountable by the regulatory authorities overseeing them. In this setting, the CE/PE process must ensure the GSPR are met. For the planning part, the CEP/PEP should:

- identify each GSPR to be addressed in the CER/PER by number and description;
- provide a clear device-specific justification documenting why each GSPR was included;
- specify S&P acceptance criteria/threshold values based on SOTA comparisons;
- describe specific clinical/performance data needed to conform to each GSPR;
- list relevant device benefits and risks and define the historical benefit-risk ratio; and
- require the CER/PER to use specific CLINICAL data to document GSPR conformity.

Learning all of these bolded terms is required (see the *Index and Definition of Terms* at the end of this textbook to study specific CER/PER terms). Although this list is not complete, the point should be clear: providing specific descriptions and detailed explanations and lists should provide the NB with clear information to assess both the GSPR conformity rationale and the clinical/performance data supporting the GSPR conformity (Fig. 2.2).

FIGURE 2.2 PLAN: Gather ALL clinical data and write single, stand-alone CER/PER.

Each company must interpret GSPR conformity specifically given the device and the QMS surrounding the device. Being able to document safety, performance, risks, benefits, side effects, and other details throughout the device lifetime and across all indications for use has become critical work required before the CE mark can be applied for a medical device to be marketed in Europe or to stay on the market in Europe. In particular, the S&P acceptance criteria threshold values and the benefit-risk ratio need to be spelled out, prespecified, and prequantified before the CEP/PEP drafting process begins. These details were previously underdeveloped or simply not present in most existing medical device manufacturing processes; therefore, companies needed to develop these detailed S&P acceptance criteria threshold values and the benefit-risk ratio specifications to meet NB requirements.

Although seemingly simple, some companies experienced significant challenges when asked to express patient-specific clinical benefits. This was a somewhat surprising challenge since risk evaluations had been well developed previously. One suggestion to overcome this potential challenge was to approach benefits by considering how the device improves patient health and to ensure the CEP/PEP states precisely how the specific device-related benefits will be accurately measured. For example, the CEP/PEP for a cardiac pacemaker might state a performance benefit is "fewer cardiac deaths" measured in acute cardiac patients over 18 years of age who received the pacemaker versus similar patients who did not receive the pacemaker. The PMS data would then need to collect S&P data from patients with and without the pacemaker. Details like time after implantation, concomitant medications, additional procedures, and comorbidities would also need to be considered and addressed in the CEP/PEP.

The new regulations and guidelines also required the PMS system to address the postmarket clinical follow-up plan/ postmarket performance follow-up plan (PMCFP/PMPFP) and postmarket clinical follow-up evaluation report/ postmarket performance follow-up evaluation report (PMCFER/PMPFER) covering the clinical data gaps identified in the CER/PER. Often, the new CE/PE process uncovered serious clinical data gaps and suggested the company had insufficient clinical/performance data to complete the CER/PER. In addition, the clinical/performance evaluator often struggled to determine exactly what the S&P acceptance criteria should be in the CEP/PEP and how to determine if the benefits outweighed the risks in the CE/PE benefit-risk ratio.

Resource management is a critical CE/PE strategy component and the manufacturer will probably need to hire and empower new clinical employees or consultants to build a stronger "clinical voice" for the company. Previously, medical device companies may not have had a Chief Medical Officer (CMO) or any clinical team members. Today, the expectation should be quite clear; the 2017 MDR/IVDR requirements to evaluate clinical data are nonnegotiable. As a result, the manufacturer must anticipate the need for a trained and experienced clinical person or team to address clinical issues and gaps. For example, many small companies hired a medical advisor to serve as a fractional CMO or clinical team member to develop and write a clinically accurate CER/PER representing the patient benefits and risks appropriately. The strategy lesson here is to staff the clinical human resources appropriately at the right time, with the right talents, and fitting within the available budget to ensure conformity with the clinical/performance evaluation regulations.

In addition, the company QMS needs to show people how to spend their time, talents, and budgets to achieve and maintain conformity with the regulations in the required geographies where the device is sold. Management and teams need to be disciplined and diligent in meeting the requirements especially in any remote working environments and when working with fractional time equivalent employees like a part-time evaluator or CMO. Aligning people to goals and timelines is important, and having team leaders who do this well is an excellent strategy.

The company can also anticipate QMS changes needed to embrace the new "clinical voice" while creating and updating the CE/PE documents. Company management can employ tactics like improving templates, work instructions (WIs) and standard operating procedures (SOPs) to be more clinically relevant while planning training sessions for clinical team leaders to lead internal and external teams working on these CE/PE documents.

2.5.2 Improve quality thorugh clinical evaluation/performance evaluation monitoring and auditing

Having a well-structured and well-planned monitoring and auditing process to review past CE/PE documents should help improve the CE/PE process. For example, careful review of past S&P acceptance criteria and benefit-risk ratios should help to better define each round of CEP/PEP development. Engaging an external expert consultant for some specific monitoring and auditing functions may bring added value by engaging a different point of view when reviewing documents and processes. This should lead to fewer NB deficiencies, which will save cost and time involved in dealing with many deficiencies after the fact. The monitoring and auditing process should also help the manufacturer develop and set company CE/PE process expectations as each CER/PER iteration is developed and released. Employing a CE/PE check and balance system is good practice. Monitoring and auditing functions are common to many other manufacturing areas, so the CE/PE monitoring and auditing should not be difficult to understand or to implement; however, having qualified and competent clinical affairs staff to do the monitoring and auditing is required. Having this strong CE/PE monitoring and auditing process specifically for the CER/PER and other CE/PE-related documents should help to support continuous CE/PE quality improvements in a good and appropriate manner.

2.5.3 Identify clinical evaluation/performance evaluation problems early

Monitoring and auditing functions may help to identify problems early; however, another strategy is to observe the work being done. Sometimes, after a new strategy is put into motion, great leaders may simply watch and wait. This watchful waiting is meant to support the team by identifying problems early, which may not be obvious to the people executing the strategy for the first time. A great leader might ask: "Does the CE/PE system need other tools to identify problems early?" One good tactic might be to create a place to store identified problems or issues in a "parking lot" or knowledge center for triage, prioritization, and potentially for immediate adjudication (Fig. 2.3).

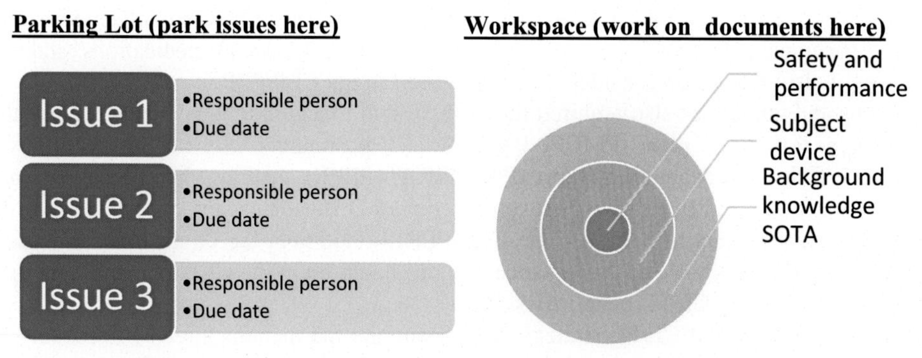

FIGURE 2.3 Create a "parking lot" for unsolved issues to stay on TARGET.

The leader should encourage all evaluators to raise CE/PE system problems. This "parking lot" should take the focus off the problem and keep the focus on meeting the goal to complete the CE/PE process. Some problems must be addressed in real time because they may present an unpassable obstruction to the process, but most problems are not critical issues and can be recorded to be addressed later. The "parking lot" process should track each problem, the responsible person and the timeline for each problem to be resolved in the future, but only if the problem actually requires resolution. Correction of problems in the future might be managed by establishing a stage gate process and assigning the problem to a particular stage gate function in order to allow stepwise process measurements to see if the team can navigate successfully through the defined process-related stage gates and all of the identified problems in a timely fashion.

Trouble-shooting the CE/PE process is multilayered. Since clinical/performance data are derived from three sources (trials, literature, and experiences), the method to identify problems early should be within the evaluation process for each clinical data source. For example, routine surveillance of ongoing clinical trials should be tracked by the CE/PE team so team members can see when new clinical safety and performance issues arise during relevant clinical trials. Similarly, as the PMS team looks for reportable events in routine literature, experiences, and complaints surveillance, they can share any new and relevant S&P data with the CE/PE team as they meet the PMS reporting timelines.

An early hurdle in the CE/PE race might relate to ensuring S&P acceptance criteria/thresholds and benefit-risk ratio details are defined clearly in the CEP/PEP. Understanding these requirements and the device-specific details should help team members to identify and escalate problems efficiently and effectively in real time (e.g., what happens when an acceptance criterion is not met? How is the benefit-risk ratio revised when new benefits and risks are identified?).

Parking lots, stage gates, S&P acceptance criteria/thresholds, and benefit-risk ratio details should be clearly understood. All team members should be able to monitor and escalate any and all identified problems swiftly so the CE/PE team can determine if a new or modified CE/PE process is needed.

2.5.4 Controlling clinical evaluation/performance evaluation problems to ensure success

Strategically moving resources to address a new problem is required. Controlling CE/PE problems are not the same as typical device manufacturing PMS problems. Simply answering the phone and recording a complaint for a marketed medical device is no longer sufficient. The MDR/IVDR now requires proactive PMS. This means the manufacturer must have a strategic solution to pull together all new CE/PE needs with past manufacturer details. For example, creating the PMCFP/PMPFP requires a careful assessment of the "out of control" clinical/performance problems identified in each CER/PER update.

Getting these CE/PE problems under control requires time and resources. The company may be required to run a clinical trial or to collect new clinical/performance data in other ways. When the PMS or RM system identifies a new clinical problem, how will the integrated systems react? What do these systems need to do to control newly identified CE/PE problems?

The most common problem seems to be when clinical data are insufficient to meet the CE/PE requirements to claim conformity to GSPR 1, 6, and 8. In this case, the manufacturer must determine how new clinical data will be gathered to address these regulatory requirements.

Synergy is important to leverage when deciding how to control problems strategically and ensure success. How do the CT, RM, PMS, and CE/PE systems work together and how does each individual system control problems? In areas of overlap (i.e., device clinical/performance problems), how can the company design controls to ensure the systems react synergistically to control each identified problem?

For example, consider the problem of insufficient clinical data. When the clinical/performance data are insufficient to support a thorough CE/PE process, then, by definition, the data were insufficient to support fully operational RM and PMS systems as well, so any occurrence of insufficient clinical data should be an identified clinical/performance problem in all clinical/performance systems. The lack of a "clinical voice" at the company may have contributed to a decision suggesting the data collected were "good enough" or, the number of problems was "small enough" to keep the product on the market while ignoring the need to analyze carefully each problem/complaint clinically to document how the patient was affected and treated. This past practice of sweeping the lack of clinical data under the proverbial rug should be avoided with proactive interacting systems designed to optimize the way clinical data is identified, appraised, analyzed, and reported. This needs to happen in the most time-efficient, effective, and comprehensive manner possible.

Based on typical scenarios observed from 2017 to 2024, a more appropriate control might be to ensure the CE/PE team has weighed in on the clinical/performance data reviewed by the CT, RM, and PMS teams. Were the clinical/performance data sufficient for the RM, PMS, and CE/PE purposes individually and as a whole? If not, what should be done at each step in each system to ensure sufficient clinical/performance data are collected by the company in the future? How do the CE/PE needs fit into this new more integrated corporate structure?

Controlling CE/PE processes requires a strong and clear "clinical voice." Typically, starting with the CMO and leading down to the persons collecting the clinical/performance data, this clinical voice needs to encourage the CT, CE/PE, PMS, and RM teams to help each other collect and analyze all available clinical/performance data confidently in real time.

2.5.5 Controlling integration of clinical evaluation/performance evaluation, clinical trial, postmarket surveillance, and risk management systems

For CT team members collecting data from clinical trials, the CE/PE process should integrated the clinical trial data with all the other clinical/performance data (i.e., clinical literature and experience data). These clinical teams should define the appropriate S&P acceptance criteria and benefit-risk ratio details for the company. The CE/PE evaluator and team members should meet with clinical trial staff members to discuss how the clinical trial data fit within the S&P acceptance criteria and the how the clinical trial outcomes affect the benefit-risk ratio. Changes should be carefully controlled in a rationale and pragmatic process to ensure the S&P acceptance criteria and benefit-risk ratio details are based on scientifically sound methods of clinical data analysis in all nonclinical and clinical research settings.

For the PMS and RM staff persons surveilling the literature and complaints/incidents, the CE/PE process requires not only the number of incidents found but also the clinical/performance details for each incident. The need to collect as much clinical/performance data for each and every complaint/incident should be clear.

Key clinical/performance data elements include, but are not limited to:

- specific patient details (e.g., age, sex, disease state, why did patient need the device?)
- device details (e.g., device name, manufacturer, size, components, did any device part have a problem?)
- device clinical/performance issues encountered by patient (e.g., what happened to the person, to the device, was this unexpected, life-threating, related to the device?)
- device clinical/performance issue remediation by patient (e.g., how was patient treated when device problem occurred? Was a routine treatment given or did the device issue require extreme measures?)
- device clinical/performance issue outcomes by patient (e.g., what happened to the device and to the patient? Is the patient fully recovered? Does the patient have lingering problems or sequelae? What ongoing therapy is needed for the patient, if any?)

The CE/PE evaluator and team members may also want to meet with the PMS call center staff to discuss how complaints/incidents relate to the S&P acceptance criteria and the benefit-risk ratio. In addition to documenting all reportable incidents in the postmarket surveillance report (PMSR)/periodic safety update report (PSUR), the PMS team can also identify and forward any relevant literature or incidents to the CE/PE staff to triage and store for the next CE/PE update. For devices on a 5-year update frequency, the complaints data storage library recording the individual reported patient issues or devise problems will need to be routinely monitored and cleaned so the data are most useful for the CE/PE update. A CE/PE update may be required at any time, so the PMS and RM systems need to be maintained in a constant state of reporting readiness. The CE/PE evaluator and team members may want to meet with the PMS and RM staff members to discuss how newly identified S&P data will impact the next CE/PE. The CE/PE, CT, PMS, and RM system team members should work together to identify appropriate QMS details, individual steps and process owners to ensure all systems operate efficiently together, data can be easily accessed and frequently updated.

Checks and balances are required and typically include a hierarchical structure leading up to the CMO who should make important clinical/performance data-based decisions about the S&P acceptance criteria and benefit-risk ratio. The QMS should describe how incoming clinical/performance data will be evaluated in the CE/PE, CT, PMS, and RM processes both separately and together. This is a critical senior leadership team review function.

2.5.6 Clinical evaluation/performance evaluation success

CE/PE success happens when the clinical/performance data were sufficient and the CT, PMS, and RM system processes were integrated enough to determine the device was safe and performed as indicated in the appropriate patient population with an acceptable S&P and an acceptable benefit-risk ratio. In addition, the manufacturer completed a high-quality CE/PE within the appropriate update frequency so the CE mark could be applied to the device and the device could be marketed and sold in the EU.

The CE/PE functions must fit well within the QMS and must integrate well with the CT, PMS, and RM systems. The "clinical voice" must help manufacturing teams keep an appropriate focus on the patient clinical data as well as the device data. All clinical team members must have sufficient training, experience, and expertise in the medical fields addressed by the device, and clinical team members need to be empowered to lead and change device designs when required by unacceptable S&P clinical data and/or benefit-risk ratios. The CE/PE team members also need to be empowered to design appropriate PMCF/PMPF studies and clinical/performance data-gathering activities to drive product success, not only in the EU but worldwide.

The patient needs a safe device performing as indicated with an appropriate benefit-risk ratio. The device must deliver the promised benefits and those benefits must outweigh the risks to the patient. Device CE/PE success results from good planning and teams must be aligned to work together well before, during, and after each CER/PER update cycle.

2.5.7 A good process timeline for clinical evaluation report/performance evaluation report writing

This textbook is based on a detailed stage-gated process used successfully to complete hundreds of CE/PE documents over many decades.The CE/PE process typically lasted ~12–16 weeks after the kickoff meeting (KOM) with timelines overlapping as follows: 2 weeks for CEP/PEP; 10 weeks for CER/PER; and 2 weeks each for subsequent PMCFP/PMPFP and SSCP/SSP (if required) document development. The PMCFER/PMPFER was drafted separately after all the new clinical/performance data were gathered according to the PMCFP/PMPFP. Similarly, the PMCFP/PMPFP was created after the CER/PER was drafted in order to address all of the identified clinical data gaps. The completed PMCFER/PMPFER was used as an input for next CER/PER and the CE/PE update cycle begins again with an updated CEP/PEP to include all the new clinical data collected and discussed in the updated PMCFER/PMPFER, PMSR/PSUR, RMR, labeling, etc.

Of interest, the EU MDR/IVDR mandated timeframes for CE/PE updates (i.e., within 1 year for high risk/implantable devices and within 2–5 years for lower risk devices or earlier if new information or device changes might impact the subject device benefit-risk ratio). These timelines were not met before, during or after the 2017 MDR/IVDR and many changes, updates, enforcement delays, and additional guidance documents were released. Most companies were not able to control their processes well enough to meet these timelines. In addition, most companies needed significant training support to develop the span and control required to arrive at a reasonable timeline to complete the CE/PE process deliverables, especially for high-risk devices, which needed updates every year. One of the goals for this textbook is to help deliver this training.

2.6 TEN stage gates

The CE process should be standardized in the company QMS using an SOP (**Appendix G: Clinical evaluation standard operating procedure example 1 (short)** and **Appendix H: Clinical evaluation standard operating procedure example 2 (long)**) to specify the policies, process flows, work requirements and a CE WI (**Appendix I: Clinical evaluation work instruction example**) to detail step-by-step instructions with CE output document forms and templates (Table 2.3).

TABLE 2.3 10 stage gate steps

SG #	SG name	SG deliverable	SG templates or forms
SG1	Initiating	CE kickoff meeting (KOM)	**Appendix J: Kick off meeting agenda/minutes/slides**
SG2	Planning	CE PLAN drafted/ reviewed/approved[a]	**Appendix K: Clinical evaluation plan template**
SG3	Searching	All clinical/performance data identified/listed	CER/PER Workbook
SG4	Coding	Clinical/performance data appraised/coded	Workbook: data given appropriate codes
SG5	Analyzing	Clinical/performance data analyzed/abstracted	Workbook: data details abstracted
SG6	Writing	CER/PER drafted/ reviewed/approved[a]	**Appendix L: Clinical evaluation report template**, **Appendix M: Clinical data requirements checklists**[b], **Appendix N: Declaration of interest template**[c]
SG7	Integrating[d]	PMCFP/PMPFP drafted/ reviewed/approved[a]	MDCG 2020-7 Guidance on PMCF plan template (*requires modification for IVD device use*) [17]
SG8	Summarizing	SSCP/SSP drafted/ reviewed/approved[a]	**Appendix O: Summary of safety and clinical performance/summary of safety and performance template**[e]
SG9	New data	New clinical data collected to fill gaps	Clinical trial system database
SG10	Follow-up	PMCFER/PMPFER drafted/ reviewed/approved[e]	MDCG 2020-8 Guidance on PMCF evaluation report template (*requires modification for IVD device use*) [18]

CEP, clinical evaluation plan; *CER*, clinical evaluation report; *CT*, clinical trial; *DOI*, declaration of interest; *KOM*, kickoff meeting; *PEP*, performance evaluation plan; *PER*, performance evaluation report; *PMCFER*, postmarket clinical follow-up evaluation report; *PMPFER*, postmarket performance follow-up evaluation report; *PMCFP*, postmarket clinical follow-up plan; *PMPFP*, postmarket performance follow-up plan; *PMS*, postmarket surveillance; *RMS*, risk management system; *SG*, stage gate; *SSCP*, summary of safety and clinical performance; *SSP*, summary of safety and performance. Bolded items identify locations of additional information.
[a]*Approved document must be signed, dated, and version controlled within a reliable QMS.*
[b]*CEP/PEP requirements checklists (see also MEDDEV 2.7/1 CER [19], MDCG 2020-1 medical device software [20], and MDCG 2022-2 PER Checklists [21]).*
[c]*DOI is required for each CER/PER evaluator.*
[d]*Integrating refers to CT, PMS, and RM system alignment in the PMCFP/PMPFP.*
[e]*SSCP/SSP documents are quite similar; template merged info from MDCG 2019-9, Rev. 1 Summary of safety and clinical performance [22] and MDCG 2022-9 Summary of safety and performance template [23] into one document.*

For example, the SOP should indicate how specific assigned staff members "will follow" the WI when completing CE activities controlled and monitored by management. The WI should include the following steps:

1. Management will assign CE/PE activities to appropriate staff members as needed.
2. Assigned staff members should review CE/PE history to ensure CE/PE scope is clear.
3. Clinical staff will interact with other team members to complete CE/PE assignments.
4. Assigned staff members should conduct the CE/PE using 10 stage gate (SG) steps
5. An independent person should conduct quality checks before document approval.
6. Completed documents should be stored in the appropriate master file within the QMS.

Templates should include all the required elements in an organized presentation to meet all international regulatory requirements for the geographies where the device is sold.

The 10 SG steps process will allow project managers to track CER/PER data flows through the complicated CE process. For example, the sequential, stepwise flow ensures the CEP/PEP must be done before the CER/PER work begins; the clinical/performance data are identified, appraised, and analyzed before the CER/PER is completed; and the related plans, processes, and reports are completed sequentially and in an orderly fashion. In addition, the 10 SG steps are expected to:

- formalize CE timelines (start and end dates);
- improve quality and efficiency by verifying a document was ready to move to the next gate before working toward the next document/stage gate;
- increase right-first-time rates;
- allow appropriate progress assessment;
- ensure plans are completed before reports are developed;
- control processes;
- encourage regular "check in" points for team members to ensure documents are "on track";
- identify process delays;
- anticipate impact of process delays;
- reduce scope changes;
- reduce document updates needed due to scope changes; and
- encourage well-planned integration between document types.

During this process, the CE/PE leader must help the team to identify all clinical data gaps in the CER/PER which will need additional clinical data collection in the future. These clinical data gaps must not cause undue delays in completing the assigned processes. Specifically, the PMCFP/PMPFP is required to document and address all of the clinical data gaps identified in the CER/PER using appropriate clinical data collection tools (e.g., clinical trial, retrospective chart review, survey, registry). The PMCFP/PMPFP must also document a required timeline commitment to complete the new data collection. The PMCFP/PMPFP may be a stand-alone document, or this specific plan may reside within the PMSP. The PMCFP/PMPFP location should be stated clearly in the QMS and in the CER/PER. In addition, for Class III/implantable and Class C or D IVD devices, the SSCP/SSP must be developed to summarize the CER/PER for the lay public.

After the new PMS/PMCFP/PMPFP data collection is completed, the collected and fully analyzed data must be documented in the PMCFER/PMPFER which will be evaluated in the next CE/PE process cycle. To begin the next cycle, the CEP/PEP must be updated based on the continuous cumulative data collection and assessment in the ongoing CT, PMS, and RM systems. The updated CEP/PEP must be issued and version controlled before the next CE/PE clinical data identification, appraisal, and analysis cycle begins. In other words, the PMCFER/PMPFER contents should be reviewed and fully considered in the new CEP/PEP. The CE/PE team leader will use the KOM checkpoint to ensure all necessary clinical/performance data are available before starting the CE/PE process. The goal for each CE/PE cycle is to keep the team focused and delivering new content on time.

Since clinical trial, literature, and experience data are complex and scattered sources of S&P evidence, having a stage gate around searching and sorting the clinical/performance data should ensure all available clinical/performance data are gathered and organized to prevent missing information as much as possible. In addition, the stage gates about appraising and analyzing the clinical/performance data should organize the clinical data contents to compare the clinical data from the subject device to the background/SOTA/alternative therapies. The five stage gates about writing CE/PE documents (i.e., 50% of the 10 SG steps process) occur after the combined clinical data are analyzed to ensure the writing process is completed in a timely manner, the data are correct, the conclusions are derived directly from the collected information, and the analyzed clinical data are fully aligned and correct in all CE/PE documents.

The stage gate after writing the full CE/PE draft should ensure inconsistencies, missing information and inappropriate conclusions are cleared up before each final version is signed off. Similarly, stage gates requiring PMS planning completion, SSCP/SSP writing, new data collection, and evaluation should ensure these important processes are completed before the next CE/PE process. Some steps are much more complex than others (Fig. 2.4).

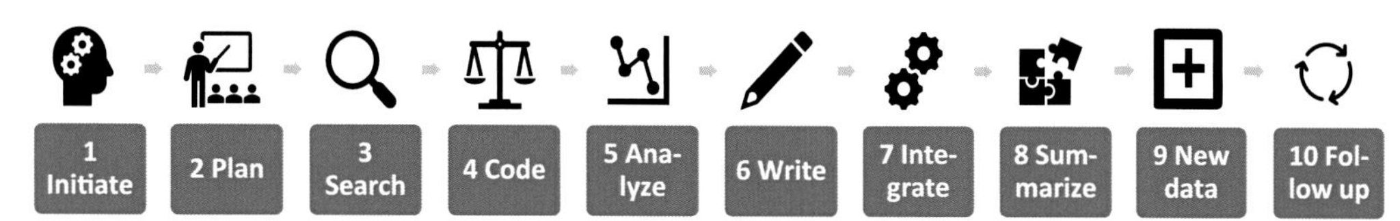

FIGURE 2.4 CE/PE road map.

Team leaders should recognize the transition from one SG to the next in order to measure the momentum and progress and to ensure the individual SG and overall project timelines are met.

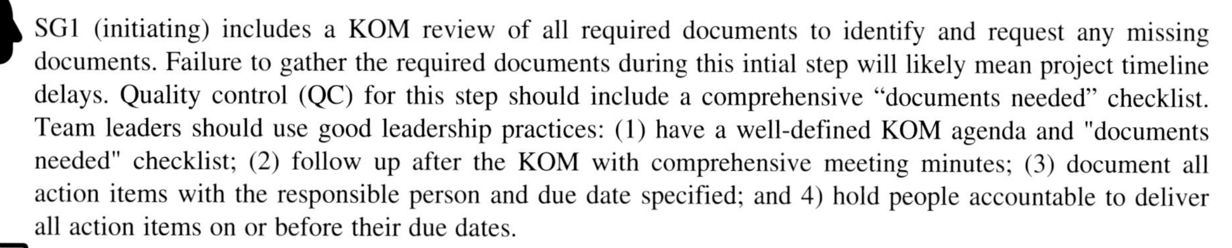

SG1 (initiating) includes a KOM review of all required documents to identify and request any missing documents. Failure to gather the required documents during this intial step will likely mean project timeline delays. Quality control (QC) for this step should include a comprehensive “documents needed” checklist. Team leaders should use good leadership practices: (1) have a well-defined KOM agenda and "documents needed" checklist; (2) follow up after the KOM with comprehensive meeting minutes; (3) document all action items with the responsible person and due date specified; and 4) hold people accountable to deliver all action items on or before their due dates.

SG2 (planning) requires the CEP/PEP to describe the relevant device, CER/PER scope, clinical claims, S&P acceptance criteria, methods to be used during CE/PE for identification, search strategies, appraisals, clinical data analyses, CER/PER writing, etc. Often the CEP/PEP device description is used to begin populating the CER/PER, and the PMCFER/PMPFER is reviewed to describe the new clinical/performance data collected to address all past CER/PER clinical/performance data gaps. QC for this step involves detailed CEP/PEP review and adjudication by senior, experienced persons. The evaluator should always review the device design history for any device changes potentially affecting the S&P data and/or the benefit-risk ratio.

SG3 (searching) identifies and collects all clinical data so CE/PE can begin. This stage includes numerous clinical data searches performed as defined in the CEP/PEP. All search terms and processes are subjected to a rigorous QC checking for accuracy and objectivity so resulting clinical data are robust and represent an unbiased clinical data collection for the CE/PE. QC often involves a second person performing identical searches to ensure all appropriate data were identified and no data were missed. If new searches are needed, a third, more experienced person may need to adjudicate any proposed search strategy changes, to ensure as much relevant data are captured as possible. Additional searches beyond those initially planned in the CEP/PEP should be discussed in the CER/PER.

SG4 (coding and weighting) requires all identified clinical data to be reviewed and grouped. Background/SOTA clinical data are grouped separately from subject/equivalent device clinical data and the appropriate inclusion and exclusion criteria are applied in a process often called "coding". The coding must “weight” each identified and included clinical data item to separate the higher-quality from the lower-quality clinical data items. QC typically involves three steps: (1) enter data; (2) check data; and (3) adjudicate data changes. For example, one person “codes” each clinical data item for inclusion or exclusion. A second, more senior/experienced person reviews and either agrees or suggests a change to the proposed code. Finally, a third, more senior/experienced person adjudicates any changed codes and assigns the appropriate codes with justifications. This coding process is often modified as the evaluator acquires more knowledge and is able to refine the groups and priorities among the included and excluded clinical/performance data. This process step is iterative and should be repeated as the CER/PER develops.

SG5 (analyzing) involves analyzing the included clinical/performance data and is the longest stage. For the CER/PER, the background and SOTA clinical data are analyzed separately from the subject/equivalent device clinical data. Clinical trial data are typically summarized from the final, clinical study report (including safety and efficacy data, adverse event (AE) listings, and any important deviations or findings). Clinical literature data are analyzed by abstracting and tabulating relevant details from each article including relevant S&P and benefit-risk tables and writing the literature analyses in narrative form. Clinical experience data are also tabulated and analyzed separately for patient problems and device deficiencies and then described in the CER/PER in the context of device sales volumes and use numbers. Like SG4, the QC for SG5 often involves a second, more senior/experienced person reviewing the data abstractions, tables and analyses and a third (more senior/experienced) person adjudicating all data changes. The product literature (i.e., device labeling) must also be analyzed to document in the CER/PER if all clinical data, S&P criteria, and benefit-risk ratio details are appropriately represented in the product literature.

SG6 (writing) requires each CE/PE document to be developed and edited section by section including the CER/PER S&P acceptance criteria tabulations in the background/SOTA as well as the subject device sections. The CER/PER clinical data analyses, benefit-risk ratio description, product literature evaluation, executive summary, and conclusions must be drawn directly from the clinical data. All references must be cited in text, detailed in the reference section and included in an appendix (if critical) to the CER/PER. Details about the evaluators are included in CER/PER appendixes (e.g., the curriculm vitae [CV] and declaration of interests [DOI] are required for each qualified evaluator). Again, a three-step QC adjudication process is helpful: (1) one person drafts each CER/PER section; (2) a second, more senior/experienced person reviews and corrects any errors; and (3) a third, more senior/experienced person adjudicates all writing changes to ensure the messages and analyses are accurate, clear, and complete, to demonstrate MDR/IVDR conformity. QC often relies on a CER/PER checklist to ensure all CER/PER parts are present and complete. If MDR/IVDR conformity is not supported by the clinical data; then, SG6 is the time when management will be alerted to the impending CE/PE failure.

SG7 (integrating) links CT, PMS, and RM systems together specifically and especially when clinical data gaps were identified in the CER/PER. This step requires drafting, reviewing, and approving the PMCFP/PMPFP to collect the additional clinical/performance data needed to fill the clinical data gaps identified in the CER/PER. The CE/PE process needs to integrate seamlessly with the separate and independently operating CT, PMS, and RM systems, even when no clinical data gaps were identified.

SG8 (summarizing) requires drafting, reviewing, approving and translating the SSCP/SSP into multiple languages and then making this CER/PER summary document available to the public.

SG9 (new data) includes collecting new itemized clinical/performance data listed in the PMCFP/PMPFP for the next CER/PER.

SG10 (follow-up) evaluates and records the new clinical data collection in the PMCFER/PMPFER within the PMSR (Class I devices) or PSUR (Class II and III devices) before this 10-step CE/PE cycle repeats.

This chapter discusses the CE/PE initiation (SG1) and planning (SG2) stages in detail including CEP/PEP development. The remaining stages will be discussed further in subsequent chapters.

2.7 Stage Gate 1: Kick-off meeting defines clinical evaluation/performance evaluation scope

Gathering all relevant information and team members together for the KOM is meant to be the tactical planning session to detail the CE/PE scope in the CEP/PEP as the CER/PER writing process begins. A good practice is to have a strategy meeting with senior management prior to the KOM to articulate the CE/PE strategy, history, and team member responsibilities.

2.7.1 Clinical evaluation/performance evaluation strategy session checklist

At the highest level, each manufacturer should have a well-developed and well-documented CE/PE strategy for all devices on the EU market. The manufacturer will want to manage all existing EU-marketed devices needing additional clinical/performance data to keep their CERs/PERs compliant with the MDR/IVDR so those devices can retain their CE mark and stay on the EU market.

For devices with little additional clinical data needed (e.g., traditional (trCERs) and obsolete (oCERs)), planning the CE/PE activities should be executed on a least burdensome path to completion. Alternatively, for devices needing a lot of new clinical/performance data (e.g., viability (vCERs), equivalent device (eqCERs), and first-in-human (fihCERs), the manufacturer may first want to determine which ones are worth the cost to collect and analyze the new data and which ones are not. The EU end-of-device-life strategy is important to plan. Removing medical devices from the EU market or never entering new devices into the EU market have been major consdierations during this EU MDR/IVDR comliance process globally.

Business decisions about which devices to invest in and which devices to obsolete are required. For example, continuing to spend time and money to solve problems for devices fraught with design issues and clinical/performance data deficiencies could lead a company to bankruptcy, while focusing on devices which do not have these problems might improve the chances of company-wide commercial success. Knowing the difference between clinically insignificant clinical data problems and business-critical clinical data problems is helpful during these strategy meetings, since the goal will be to decide where to focus effort, time, and money from a clinical data development perspective.

The CE/PE strategy session should consider the following:

- devices needing CE/PE (what is the "book of business"—how many devices? what type?);
- past NB deficiencies (which CE/PE documents had deficiencies?);
- team members (who can work on CE/PE processes going forward?);
- NB required CE/PE document delivery timelines for each document (what are the specific CEP/PEP, CER/PER, PMCFP/PMPFP, PMCFER/PMPFER, SSCP/SSP due dates?);
- meeting plans (number of meetings, topics, agendas, and minutes for each meeting?);
- assignments (who is doing eachCE/PE process?);
- due dates (when are deliverables required from each team member?);
- coordinator functions (who is collecting deliverables and updating team on progress?);
- data sourcing details (where are data housed and are all needed clinical data present?);
- roadblocks (what is missing and where have problems arisen in the past?);
- who can be called for help (when, why, and how soon should outside support be called?);
- links to staff (who will support PMSR/PSUR, RMR, senior management reviews from the scientific, medical affairs, statistics departments, etc.);
- what clinical data should be requested on a regular basis (can clinical data sharing be streamlined to make these clinical data analysis requests more efficient?);
- what will happen if required clinical data are not found;
- what will happen if CE/PE reports insufficient clinical data were available for analysis;
- PMCFP/PMPFP overview (who needs to sign off for the future clinical data collection plan for all devices in this CE/PE process);
- design details (how are changes made based on the CE/PE findings?);
- labeling (how are changes made based on the CE/PE findings?);
- NB requirements (how will NB requirements be addressed if company feels the product is safe and performs well with an appropriate benefit-risk ratio but the NB disagrees?, and how does the NB weigh risks/benefits?) *Note: Some NBs suggest the device must be "superior" to all other devices on the market or data must include RCTs as opposed to other data types. The company CE/PE strategy should be clear if any of these issues arise because these are not regulatory requirements*; and
- the "parking lot" (place to write down future CE/PE negotiations and needs).

Setting the right strategy in motion helps to develop an efficient and effective CE/PE process. Key strategic goals should be shared and reminders should be sent to ensure team members stay aligned with the strategy. When leading the CE/PE team, a well-defined strategic approach will help to avoid confusion among team members (especially in a large group).

The leader should state strategy components clearly, including, but not limited to, the following:

- All manufacturer held clinical data will be gathered and discussed at the KOM.
- Device and device development will be carefully described and reviewed at the KOM.
- Decisions about equivalence, device groupings, and scope will be verified at the KOM.
- Each team member will have a defined project-related role and responsibility.
- Each deliverable document will have an assigned team member with a deadline.
- Each document will be reviewed and edited by a more senior/experienced writer/evaluator.
- Changes made during the review/edit process will be adjudicated by a more senior/experienced writer/evaluator
- Each CE/PE document will be consistent and clear.
- Details about how to do the assigned work will be written down and doable.
- Training sessions will be provided.
- The CEP/PEP will be signed off prior to CE/PE starting.
- Clinical data gaps will be anticipated, planned and documented in PMCFP/PMPFP.
- Company will share device goals with NB.
- Guidance from NB and expert panel (if received) will be followed.

Management should understand the various CE/PE business risks. For example, if the CER/PER is not acceptable by the NB, then the device will not gain a CE mark and will not be allowed on the EU market. Reducing CE/PE business risk requires good leadership, strategy setting, and management throughout the CE/PE process, including NB negotiations. NBs no longer issue CE mark certificates without any form of assessment. Negotiations are the norm, so plan to talk to the NB. Building a good relationship with NB team members by understanding their processes and the services they provide will be helpful. Remember all NBs are changing and learning to adapt to regulatory changes just like everyone else.

BOX 2.2 Case study—benefit-risk ratio

Case study situation

A company CER concluded device S&P acceptance criteria were met and the documented benefit-risk ratio was appropriate; however, no clinical trial data were available for the Class IIb device, and only limited clinical literature and experience data were available from equivalent and similar devices.

Case study issues identified

The clinical evaluatordocumented in the CER how the clinical data were insufficient for the viability (v)CER, even when including all clinical data from the equivalent device. The clinical evaluator also reported the equivalent device requirements were not met due to clinically significant differences between the subject and equivalent devices. The clinical evaluator recommended collecting real-world evidence (RWE) in an observational study during initial market release. Collecting PMCF data during early PMS was expected to satisfy the NB in this vCER setting.

Case study outcomes

The vCER evaluated clinical data from similar devices as well as the background knowledge and SOTA. The vCER identified clinical data gaps and recommended a PMCF study to collect additional clinical data. Because the NB was likely to require a trial or at least some clinical data to address specific clinical data gaps identified in the CER, the company agreed to do an observational study in addition to more bench work to strengthen the equivalence claim and to support the S&P details.

A better, case-study solution enacted

Some people want a clinical trial for every device; however, others agree clinical trials should only be conducted when appropriate, since clinical trials are expensive and will drive the cost of devices up without providing much value. For example, the S&P criteria and benefit-risk ratio may already be known without running a clinical trial. In addition, many devices have been allowed on the market without clinical trials and users have found the devices were safe and the benefits outweighed the risks. In addition, legacy devices had sufficient clinical experience data to document the acceptable S&P and benefit-risk profile when used as indicated prior to 2017. In this setting, the collection of additional specific clinical data during the initial limited device launch was able to satisfy the GSPR conformity requirements.

Reducing CE/PE risk by generating high quality clinical data is helpful, getting all required documents delivered and stored in one place before the CE/PE process begins improves efficiency, and describing the CEP/PEP writing strategy, setting the tone for next steps, and following up on assignments are required CE/PE leadership functions during the KOM. Do not underestimate the value of this leadership - spend the time to make the KOM worthwhile.

2.7.2 Clinical data amounts vary

CE/PE document writing requires sufficient clinical data and the amount of clincal data available should be discussed at the KOM. The CE/PE process should begin before the KOM by the CE/PE leader requesting and securing all relevant documents (Table 2.4).

TABLE 2.4 CER/PER relevant documents checklist.

UID	Document description	Reason needed
1	CE/PE SOP	To ensure compliance with company policies and QMS
2	CE/PE WI	To describe steps in CE/PE process
3	CE/PE forms/templates	To encourage similar format/contents for all CE/PE plans, reports, etc.
4	CE/PE reviewers and approvers list (names, credentials, titles, contact info, CV and DOI with conflict of interest management plan when needed)	To schedule times for review and approval and to document education, training and experience credentials as well as any potential conflicts of interest
5	Competitor/equivalent device list (device name, manufacturer, city, state) and relevant product literature or publications	To build background/SOTA/alternative therapy section
6	CE/PE past comments/deficiencies from NB, expert panel, or other reviewers	To ensure past problems do not recur
7	Plans and reports: a. CEP/PEP b. CER/PER c. PMCFP/PMPFP d. PMCFER/PMPFER e. SSCP/SSP f. Risk Management Plan (RMP) and report (RMR) (per EN ISO 14971:2019) g. PMS Plan (PMSP) and PMSR/PSUR (e.g., experience data including all prior PMCF/PMPF data, complaints and issues from Manufacturer and User Facility Device Experience (MAUDE), Database of Adverse Event Notifications (DAEN), HealthCanada, SwissMedic)	To ensure all related documents are well integrated and aligned
8	Device and use description	To ensure consistent device and use description across all documents
9	Novelty decision document	To document if device is "novel"
10	Instructions for use (IFU), operating manuals, patient brochures (all relevant labeling)	To ensure consistent labeling
11	Clinical trial data: a. company-sponsored, trial protocols, reports, publications b. investigator-initiated trial protocols, reports, publications c. other clinical "trial" data relevant to subject device or background knowledge/SOTA	To evaluate all clinical data from this highest-value clinical trial data source
12	Clinical literature data: a. literature searches b. articles/manuscripts (published or unpublished)	To evaluate all clinical data from this medium-valued clinical data source

(*Continued*)

TABLE 2.4 (Continued)

UID	Document description	Reason needed
13	Clinical experience data: a. complaints (with details per human patient: what happened? Serious? Unexpected? Related? Treated? Resolved?) and device deficiencies (DD) (e.g., what malfunctioned? Expected? Root cause? Recall? Resolved?) b. registries c. surveys d. advisory boards e. patient information forms f. http://www.clinicaltrials.gov and other listings (e.g., https://www.australianclinicaltrials.gov.au, https://health-products.canada.ca/ctdb-bdec/index-eng.jsp, https://trialsearch.who.int) g. other experience data	To evaluate all clinical data from this lowest-valued clinical data source
14	Product information/marketing materials: a. marketing history (list of countries with sales, date or year first on market in each country/geography) b. sales numbers (broken out by year to align with complaints) c. marketing materials (e.g., brochures, ads, promo pieces.) d. claims matrix (e.g., including S&P claims, website claims and claims in press releases and promotional documents)	To ensure all product information is aligned with the clinical data evaluation (all clinical claims must be fully substantiated in the CER/PER)
15	EU regulatory documents: a. CE mark approval date, renewal, history b. NB name and number for CE mark certification c. ISO 13485 certificates d. declaration of conformity e. audit reports f. deficiencies lists and remediation report g. EU regulatory history h. other EU regulatory information	To ensure device developmental context is clearly stated in the CER/PER
16	US regulatory documents: a. approval date b. approval/clearance type (e.g., 510k, PMA, *de novo*, HDE) c. audit/inspection reports (e.g., 483s, warning letters, EIRs) d. deficiencies e. US regulatory history f. other regulatory information from any country	To provide context for EU MDR/IVDR filing
17	ISO certifications	To ensure standards were followed
18	UDI, GUIDID, and other identification information	To verify UDI info for EUDAMED
19	Design history file and/or technical files (concise/available): a. labels b. specifications c. sterilization information d. manufacturing information	To ensure device design changes over time are properly cataloged and linked to clinical/performance data for evaluation purposes

CE, clinical evaluation; *CE mark*, conformité européenne (European conformity) mark; *CER*, clinical evaluation report; *DAEN*, Database of Adverse Event Notifications; *DD*, device deficiency; *EIR*, establishment inspection report; *EN*, English; *EU*, European Union; *EUDAMED*, European Database on Medical Devices; *GUDID*, Global Unique Device Identification Database; *HDE*, humanitarian device exemption; *IFU*, instructions for use; *ISO*, International Organization for Standardization; *IVDR*, *in vitro* diagnostic regulation; *MAUDE*, Manufacturer and User Facility Device Experience; *MDR*, medical device regulation; *NB*, notified body; *PE*, performance evaluation; *PER*, performance evaluation report; *PMA*, premarket approval; *PMCFER*, postmarket clinical follow-up evaluation report; *PMPFER*, postmarket performance follow-up evaluation report; *PMCFP*, postmarket clinical follow-up plan; *PMPFP*, postmarket performance follow-up plan; *PMS*, postmarket surveillance; *PMSP*, postmarket surveillance plan; *PMSR*, postmarket surveillance report; *PSUR*, periodic safety update report; *QMS*, quality management system; *RMP*, risk management plan; *RMR*, risk management report; *S&P*, safety and performance; *SOP*, standard operating procedure; *SOTA*, state of the art; *SSCP*, summary of safety and clinical performance; *SSP*, summary of safety and performance; *UDI*, unique device identifier; *WI*, work instruction.

This document request serves a dual purpose. Not only are all necessary clinical/performance data gathered from the manufacturer in one location for CE/PE work, but the process helps the evaluator to identify who within the company is responsible for relevant clinical/performance data. These people should be invited to the KOM and the evaluator should call on them as resources during the CE/PE document writing process.

The team attending the KOM should review the information gathered including past and planned device designs, mechanisms of action, background, SOTA, alternative therapies and product information including all prior CE/PE documents, clinical/performance trial reports, clinical/performance literature and clinical/performance experience data including all complaints and sales data, and all PMS, PMCFER/PMPFER, and RM reports. The timeline expectations require the evaluator to analyze large amounts of information quickly from diverse areas. In order to facilitate cross-functional team interactions, the KOM leader should answer the following questions before the KOM:

- Which details are required and which are not?
- Which device issues are most problematic?
- Which problems are allowed to continue and for how long if the product is safe and performing as intended?
- What is the CE/PE scope?
- What clinical data will be included or excluded during the CE/PE process?

The KOM agenda should include:

- introductions for all attendees;
- reviewing specific MDR/IVDR requirements;
- requesting missing documents from each responsible party;
- defining the due date for missing documents;
- reviewing documents and sharing CE/PE insights and concerns;
- creating document development timeline and action item list;
- assigning responsible persons to each action item;
- ensuring meeting minutes will be drafted and circulated within 2 business days for review/sign-off; and
- scheduling the next meeting.

Each KOM department expert (i.e., each team member from the engineering, regulatory, clinical, and quality departments) should briefly present the relevant clinical S&P data from each CE/PE document identified and provided from their department. The KOM should expect the evaluator to understand the available, subject device clinical/performance data from clinical trials, literature, and experiences. Each team member should also offer ideas for a smooth CE/PE document writing process. The CE/PE team leader should react to problems as quickly as possible (e.g., some devices will have no clinical/performance data while other devices will have detailed clinical/performance data from 100,000 subjects or more and both extremes can be difficult to manage during the CE/PE document writing process; so, the CE/PE team leader should be well versed in the specific CE/PE document types required for the particular device under review in the CE/PE process).

Although clinical data are required to establish conformity to EU MDR/IVDR GSPR, the clinical data type and amount vary greatly by device risk. More clinical trial data are required for high risk, implantable devices and for devices with a questionable benefit-risk ratio. For example, a left ventricular assist device (LVAD) will require clinical trial data to determine if benefits outweigh risks; however, a surgical drape typically has no clinical trial data collected and must still document device benefits and risks (Fig. 2.5).

More Clinical Data	Less Clinical Data
Example: LVAD	*Example: surgical drape*
high risk, limited experience/use data, invasive, implantable, S&P varies across manufacturers (complex design, materials, shapes, sizes, uses, etc.)	low risk, extensive experience/use data, noninvasive, S&P uniform across manufacturers (common design, materials, shapes, sizes, uses, etc.)

FIGURE 2.5 Clinical data requirements vary.

The clinical evaluator must consider the clinical data type, amount, and quality specified in the CEP/PEP to ensure sufficient clinical data will be evaluated in the CER/PER.

BOX 2.3 Case study—Book of business?

Case study situation

A mid-sized medical device company had hundreds of "legacy" devices on the EU market for many decades. The engineering team had dozens of device groups with dozens of CERs to cover all devices. A project was created to see if work flows could be improved and if CER quality was sufficient for NB review for the first time under MDR. The company was also updating the CER work instruction and developing a new GSPR checklist to ensure all devices were covered with specific evidence for each applicable GSPR.

Case study issues identified

This company needed to define their "book of business" so they could clearly define their products and their CE Mark strategy for the CERs to cover all of their devices on the EU market. The review process identified CERs being written for nonmedical devices, numerous CERs being written when fewer CERs would suffice, and high-risk device CERs without sufficient clinical data to ensure the benefits outweighed the risks. In addition, no existing CERs had the required prespecified and prequantified S&P acceptance criteria, and the highest-risk devices (i.e., Class IIb devices) had CERs with similar content and quality as the CERs developed for the lowest-risk devices.

Case study outcomes

After reviewing the CERs, some devices were identified and reclassified because they did not meet the EU MDR definition of a medical device. Other devices were classified as low-risk, Class I devices and these were grouped together in a single CER by device function. This one CER replaced four previously created and routinely updated CERs, leading to an 4-fold increased efficiency and 4-fold reduced cost. Specific S&P criteria were developed and tested in the CERs, and the company QMS was updated to include SOPs, WIs, templates (e.g., including CER development and GSPR checklists in addition to CEP, CER, PMCFP and PMCFER as well as PMSR/PSUR templates). Regulatory files were updated to include classification documents for products no longer considered medical devices, with adequate justification to support these decisions. Devices were grouped into similar device families, and justifications were written about why specific devices were grouped together for coverage in fewer CERs than in the past. CERs were written for Class I, IIa, and IIb device groups, including appropriate clinical data details based on the device benefits and risks with increasing detail based on the increasing device risk.

A better, case-study solution enacted

The regulatory files were updated to reclassify products that were not medical devices and to remove the CER writing burden when no CER was recommended. In addition, a complete "book of business" was created and all medical devices were grouped with clinical rationales and justifications, which resulted in fewer CERs being written. Required CERs were improved with additional detail based on the level of risk associated with each "higher-risk" device family.

Starting CE/PE planning with a KOM should ensure needed materials are assembled and shared while missing clinical data pieces are assigned to responsible team members to collect and share by the deliverable dates so the CEP/PEP can be completed on time. The CE/PE scope should be well-defined in the CEP/PEP, including all anticipated CE/PE challenges (e.g., when clinical data are scant but more are being collected in the PMCFP/PMPFP or when clinical data have been sufficient over time and no new changes have occurred). The CE/PE team should document scope decisions including which devices will be covered in each CE/PE document.

2.7.3 Other meetings beyond the kick-off meeting

In addition to the KOM, several additional CE/PE update meetings may be helpful, especially if many people need to submit clinical/performance data elements not available at the KOM. This lack of adequate KOM preparation is a common but entirely undesirable situation. The planning challenges grow exponentially when CE/PE starting materials are not ready for the KOM. Companies needing dozens of CE/PE planning meetings for one CE/PE document set are likely to have a QMS in need of significant repair and remediation. One KOM should be sufficient for the entire suite of CE/PE documents in any device class.

2.8 Stage Gate 2: Clinical evaluation plan/performance evaluation plan development and version control

CEP/PEP template version control within the QMS is helpful; however, too much version control during the drafting process can be problematic. This note is important because so many examples are available when inexperienced leaders use available software solutions for CE/PE document writing with strict version controls in place during document drafting and development.

A rigid drafing process often wastes time and costs more than less rigid systems. Of course, the opposite is also true: a complete lack of control for each draft version will also waste time and be costly if writers need to repeat steps previously addressed. The CE/PE document drafting process needs clear and experienced leadership to avoid these pitfalls.

BOX 2.4 Tracking every change may "break the bank"

Case study situation

A large medical device manufacturer assigned a team to conduct the CE and write the CE documents. The team had never done a CE before. The clinical data searches for trials, literature, and experiences were done by separate groups and the literature review was written as a separate document to be incorporated into the CER. All CE writing was done by filling in templates for each of the five required CE documents. The team leader employed a strict policy about recording every edit and every comment within a document stored in a SharePoint site. Document sharing was simultaneous, and conventions were discussed about including initials for each person who entered each comment. Dozens of weekly team meetings were scheduled to review updates.

Case study issues identified

The team may have misunderstood the team meeting purpose and team member roles and responsibilities. The literature review process identified more than 1,000 articles for appraisal and evaluation. The SOP and WI did not specify and the CEP did not provide any guidance about how the separate documents were to be incorporated into the CE process—for example, when writing the CER and identifying clinical data gaps. The separate literature review was difficult to use in the CER because the focus was not on the device under evaluation but was focused on the broad device type and development history rather than the S&P surrouding the device, the background konwledge and SOTA and the alternative therapies. Hundreds of hours were spent by multiple team members reviewing and re-reviewing work to ensure each comment was fully addressed and documented. Frequently, a new document needed to be created because individual comments were numerous, confusing, and often conflicting with the other comments and document text. Dozens of meetings were held while attempting to understand each reviewer comment. Each draft version for each document took many weeks and often months to produce and/or update.

Case study outcomes

The team was unable to complete the project with the initial 6-month timeline and needed to extend the project timeline to more than 12 months. The QC process for each document was interrupted multiple times to adjust expectations. After more than a year in development, the CER did not meet NB reviewer expectations and dozens of deficiencies were issued by the NB. The poorly constructed CE documents needed extensive remediation effort by the company.

A better, case-study solution

Writing CE documents by a committee of undereducated, untrained and inexperienced staff is not a best practice. A better solution might include having a more qualified leader as the responsible, individual evaluator/writer managing expectations and leading the CE process from end to end. Leadership by an educated, trained and epxerienced clinical evaluator is especially important during the document review cycle to manage reviewer inputs and to keep the project moving forward. This responsible CE team leader and/or clinical evaluator should oversee planning, searching, appraising, analyzing, and CE reporting. For example, the leader/evaluator might require a better literature search strategy to limit the clinical literature search output to fewer, more relevant/appropriate, and higher-quality articles. The details about the search strategy and literature review integration should have been detailed in the CEP, but flexible enough to allow revision when too many articles were returned to the team prior to the data appraisal step. The literature appraisal step should control the volume of high-quality articles returned to ensure CE process. The literature review writing should be integrated with the whole CE document set to reduce team member confusion when reviewing documents and entering conflicting suggestions for others to fix. A better solution would be to have stronger leader and a more experienced, smaller team with a clear focus on the CE purpose and less focus on each individual reviewer comment.

Writing by committee is not easy and the leader needs to understand how to balance QC in order to move forward quickly and cost effectively. The right balance is needed between clearly controlling each document change and allowing significant change without tracking pointless details during the initial drafting process. A good practice is to start the CEP/PEP update by reviewing past CE/PE documents (i.e., CEP/PEP, CER/PER, PMCFP/PMPFP, PMCFER/PMPFER, SSCP/SSP) and using all data from past documents to begin to transform the latest CEP/PEP template into the next CEP/PEP draft with all details up to date.

During CEP/PEP writing, ensure appropriate comments and changes are included as the document matures. The leader must understand how to differentiate high-value edits, comments and suggestions for change from low value, time-wasting minutia. The leader must also have the skill to help the weakest team members to contribute constructively to the process rather than wasting time (e.g., assign specific roles and responsibilities to specific team members based on their education, training and experience). Once the document is signed-off, the final version should be uploaded into the version-controlled record storage area defined in the QMS.

2.8.1 Use a quality management system clinical evaluation plan/performance evaluation plan template

The CEP/PEP template should be updated to include all required elements (**Appendix K: Clinical evaluation plan template**) and should provide reasonable CE/PE process quality assurance (QA) by helping writers to avoid simple omission errors. For example, the CEP/PEP template should include S&P acceptance criteria at the beginning so the writer and reader both understand the plan to ensure clinical/performance data clearly meet or exceed these prespecified and prequantified thresholds for S&P acceptance criteria.

Note: If S&P acceptance criteria were overlooked in earlier CEP/PEP documents, this error should be corrected by including this section the redesigned CEP/PEP template.

The CEP/PEP template should also help evaluators and writers to avoid style mistakes by following the company style guide (i.e., using standard type font, spacing, figure and table layouts, etc.) and including specific templated language (as appropriate) to help jump-start CEP/PEP writing.

A good practice is to start the CEP/PEP draft document before the KOM. This is possible because the CEP/PEP is largely background information for the CER/PER and other CE/PE documents. For example, the device description may change only slightly over the years for a mature device, so the device description can sometimes be included from prior documents and simply updated during the KOM.

CEP/PEP information must be updated to ensure the current CE/PE scope and plan are clear, correct, and version controlled to avoid confusion about which version is the most up to date while drafting, reviewing, and finalizing the CEP/PEP.

Note: The "early" draft can be used to help focus the KOM discussion and to speed the CEP/PEP completion timeline (i.e., ideally the CEP/PEP should be signed off within a time frame of less than 2 weeks after the KOM).

2.8.2 Follow the clinical evaluation plan/performance evaluation plan table of contents

The CEP/PEP table of contents orients the reader and illustrates how important the device description, background/SOTA, CE/PE, and PMCFP/PMPFP methods are for planning and scoping the entire CE/PE process. The device description includes important details like device intended use, indications for use, device lifetime, clinical claims, device developmental context and any equivalence claims with appropriate justifications to be supported with clinical data in the CER/PER.

The example CEP/PEP table of contents template (Table 2.5) aligns with the MDR [1] and IVDR [2], outdated international standard, EN 13612:2002 [6], the outdated EU guidance in MEDDEV 2.7/1, Rev 4 [19], MDCG 2020-1 [20] for medical device software and MDCG 2022-2 [21] for IVD medical devices to include general device characteristic information (e.g., device description, design features, indications, equivalence, risks, background/SOTA), data sources planned for clinical/performance evaluation (e.g., clinical trial, literature, and experience sources, both internal and external), and details from devices already on the EU market including design changes, new clinical concerns, and needed PMS updates and activities.

TABLE 2.5 Example CEP/PEP table of contents.

Title/header: CEP/PEP
Table of contents
Requirements map to each document section (especially if following an international standard)
Abbreviations list
Relevant documents list – SOP clinical/performance evaluation – CEP/PEP work instruction – CEP/PEP template
Responsibilities and resources – Overall coordinator – Scientific validity (assess available test results/data) – Performance (examine/confirm claims) – Document CEP/PEP (define procedures/tests) – Other (e.g., software, hardware, physical location)
1. Objectives and scope

(Continued)

TABLE 2.5 (Continued)

Title/header: CEP/PEP

2. Specifications and acceptance criteria
   - Safety: define one or more specific safety claim/s with quantified threshold/s
   - Performance: define one or more specific performance claim/s with quantified threshold/s
   - Explain process to be used if clinical data meets or exceeds the threshold values stated in CEP/PEP.
   - Explain process to be used if clinical data does NOT meet the threshold values stated in CEP/PEP
   - Describe any available common specifications, international standards, etc.
   - Link to SOTA benefit-risk ratio

3. Device description
3.1 Device name/version
   - Basic UDI-DI
   - UMDNS, GMDN, EMDN code
3.2 Manufacturer
3.3 Device description and principles of operation
   - features, images, and part numbers as needed
   - assay technology, certified reference materials/methods available
   - stability of specimens and reagents for testing
3.4 Device lifetime
3.5 Intended purpose/use
   - Intended user/s, variability of study subject population
   - Disease state, condition, risk factor of interest
   - Disease state prevalence
   - Target population
3.6 Indication for use
3.7 Contraindications

3.8. Clinical/performance claims to be substantiated

Claim (specify and quantify each claim)	Applicable (y/n)?	Explanation
Safety 1		
Safety 2		
Performance 1		
Performance 2		
Analytical sensitivity		
Diagnostic sensitivity		
Analytical specificity		
Diagnostic specificity		
Accuracy		
Repeatability		
Reproducibility		
Marketing claims		

3.9. Equivalent device
   - Name, description
   - Justification

3.10 Developmental context
   - Novelty, degree of innovation
   - Risks (e.g., direct patient harm from medical device or incorrect/delayed IVD result) and overall risk level
   - Regulatory history (classification, etc.)

(Continued)

TABLE 2.5 (Continued)

Title/header: CEP/PEP
4. Background and state of the art (SOTA) 4.1 Medical context, SOTA, standard of care 4.2 Similar devices 4.3 Safety and performance considerations 4.4 Benefits and risks considerations
5. Evaluation process 5.1 Data selection rationale (trial, literature, experience) 5.2 Identification (describe methods and justify databases used)—for example: – scientific databases—bibliographic (e.g., MEDLINE, EMBASE) – specialized databases (e.g., MEDION) – systematic review databases (e.g., Cochrane Collaboration) – clinical trial registers (e.g., CENTRAL, NIH) – adverse event report databases (e.g., MAUDE, IRIS) – reference texts 5.3 Appraisal (inclusion/exclusion criteria, weighting) 5.4 Analysis (specify how results will be analyzed to identify sufficient data versus clinical data gap conclusions)
FOR IVD DEVICE PEP ANALYSIS DESCRIPTION *Scientific validity—typically uses a literature search to specify the "valid" reason why the IVD can measure the analyte and why the test should be done to detect the analyte. (Art. 2(38), Annex XIII, A1.2)* *Analytical performance—requires metrics (methods and test results) for IVD medical device detecting what device should detect (e.g., sensitivity and specificity values; e.g., if used in 100 patients, what results are expected?). (Art. 2(40), Annex I.II.9.1a, XIII, A1.2)* *Clinical performance—typically uses literature search to describe metrics (methods and test results) from IVD medical device use in the intended patient population (e.g., sensitivity and specificity values when tested in humans); however, may need clinical performance study results. (Art. 2(41), Annex I.II.9.1b, XIII, A1.2)* *Each item may generate a separate report (SVR, APR, CPR) to be included in the PER (Annex XIII, A1.3)* 5.5 If new trial planned, define: – human subject protection – sites (laboratories/institutions in study) – timelines (plan, PE testing, write PER) – investigator briefing – software validation – study records – results (observations and unexpected outcomes) 5.4 Writing the CER/PER
6. Postmarket follow-up plan 6.1 PMSR/PSUR 6.2 Device changes 6.3 Newly emerged clinical/performance concerns 6.4 SSCP/SSP
7. CER/PER conclusion – Data gaps to be addressed in PMCFP/PMPFP – Re-evaluation/change management procedure
8. References
9. Signatures and dates by function (e.g., creation, review, approval)

APR, analytic performance report; *CPR*, clinical performance report; *CE*, clinical evaluation; *CE mark*, conformité européenne (European conformity) mark; *CENTRAL*, Cochrane Central Register of Controlled Trials; *CEP*, clinical evaluation plan; *CER*, clinical evaluation report; *Embase*, Elsevier's unique medical literature database; *EMDN*, European Medical Device Nomenclature; *GMDN*, Global Medical Device Nomenclature; *IRIS*, integrated risk information system; *IVD*, *in vitro* diagnostic; *MAUDE*, Manufacturer and User Facility Device Experience; *MEDION*, medical diagnostic studies database; *MEDLINE*, National Library of Medicine's bibliographic database; *NIH*, National Institutes of Health; *PE*, performance evaluation; *PEP*, performance evaluation plan; *PER*, performance evaluation report; *PMCFP*, postmarket clinical follow-up plan; *PMPFP*, postmarket performance follow-up plan; *PMSR*, postmarket surveillance report; *PSUR*, periodic safety update report; *QMS*, quality management system; *SOP*, standard operating procedure; *SOTA*, state of the art; *SSCP*, summary of safety and clinical performance; *SSP*, summary of safety and performance; *SVR*, scientific validity report; *UDI-DI*, unique device identifier-device identifier; *UMDNS*, Universal Medical Device Nomenclature System. Bolded text adds emphasis and italic text provides additional detail.

Some topics are relatively hidden in the CEP/PEP table of contents, including where to address concepts linking CE/PE system processes to CT, RM, and PMS system processes. In addition, the CEP/PEP table of contents purposely does not include other technical file details (e.g., nonclinical animal and bench work, biocompatibility, verification, and validation testing), as they are not required. Be careful not to include nonclinical details in the CEP/PEP unless absolutely necessary (e.g., certain bench or animal data may be helpful when required to examine safety problems which were inappropriate to test in humans; however, including a lot of non-clinical data in an attempt to hide the lack of human clinical data available is unacceptable - these situations would be better served by transparent documentation about identifying the missing clinical data and documenting how the clinical data gaps will be addressed in the PMCFP/PMPFP).

2.8.3 State the clinical evaluation/performance evaluation objective and scope

Having a clear objective and scope is a key differentiating feature between efficient and less effective CE/PE processes. One goal should be to describe the objective and scope as concisely as possible, and another goal should be to describe how the CE/PE scope changed over the device lifetime. Start with device labeling to define the intended purpose/use and indications for use. These details will shape the CE/PE process. As the device changed or had components added/removed or when new risks or benefits were discovered, the labeling and the scope was required to change to encompass the new details. Being clear on device changes over time is important. Documenting all objective and scope changes clearly and concisely will expedite the CE/PE process.

BOX 2.5 CE/PE scope

Each CER/PER needs the right "scope," and the scope is expected to change over time. Consider the CER strategy for lifesaving Class III devices like a pacemaker and LVAD device. Then compare these to a low-risk device like a patient drape.

Example 1: Pacemaker

The first pacemaker in the early 1900s used external needles to deliver an electrical current directly to the heart. One early device used a spring motor turning a magneto-generator to generate electricity and a neon lamp to show when the electrical surge passed through the needle to the heart [24]. Unfortunately, this design could not identify pacing defects and could not shock more than once, which led to cardiac arrhythmias and death.

In the late 1950s, an implantable pacemaker used silicon transistors to create a pulse in transvenous leads to deliver cardiac pacing. Pacemakers with extended battery life used a mercuric oxide-zinc and then a lithium battery encapsulated in epoxy to treat atria-ventricular heart block with a pulse delivered through implanted leads and repeated 60 times a minute to mimic a normal heart beat. Unfortunately, battery replacement surgery as well as lead fractures, lead dislodgements and fluid collecting in the pulse generator were typical problems.

Dual chamber pacemakers were developed along with improved sensing and stimulation leads and complex software to deliver customized cardiac stimulation. Modern pacemakers included a titanium battery case and steroid eluting leads to improve longevity and to reduce foreign body responses to implanted materials. In addition, leadless, intracardiac pacing devices were created and the future may include nanotechnology solutions to create "miniaturized cardiac implantable electronic devices... including leadless pacemakers and piezoelectric nanogenerators to self-power symbiotic cardiac devices" [25].

CER scope differs over time

If all "pacemaker" advancements had occurred within one "pacemaker" CER, then many design changes would need to be incorporated and updated in the one CER over time (Fig. 2.6).

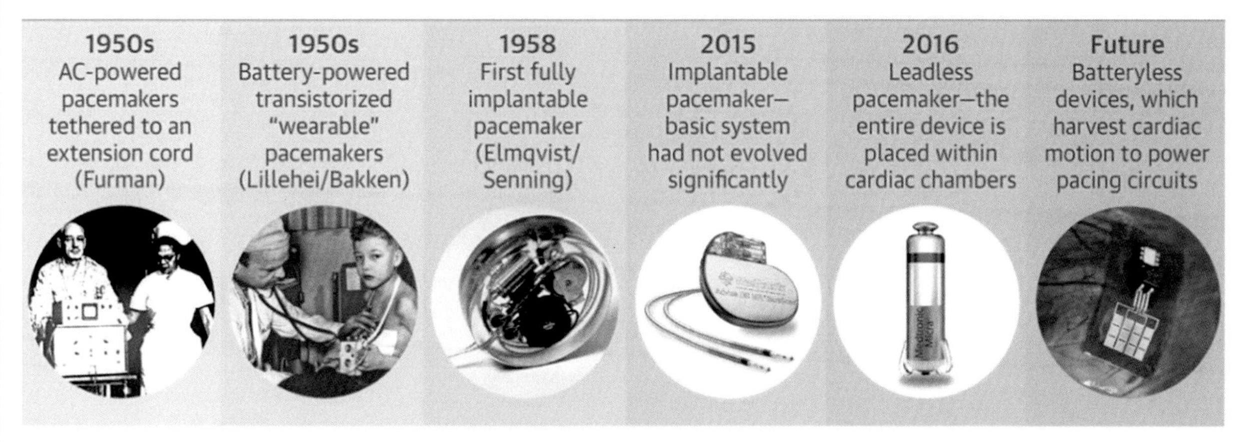
1950s
AC-powered pacemakers tethered to an extension cord (Furman)
1950s
Battery-powered transistorized "wearable" pacemakers (Lillehei/Bakken)
1958
First fully implantable pacemaker (Elmqvist/ Senning)
2015
Implantable pacemaker– basic system had not evolved significantly
2016
Leadless pacemaker–the entire device is placed within cardiac chambers
Future
Batteryless devices, which harvest cardiac motion to power pacing circuits

FIGURE 2.6 Cardiac pacing history. *Image taken from published work [26].*

(Continued)

BOX 2.5 (Continued)

The technology changed significantly over time, from the patient being tethered to a shopping cart battery in the 1950s to having a small battery implanted within the heart in 2016. In addition, while the clinical application stayed the same (i.e., to pace the heart), the biological interactions with the human body (e.g., external communicating device vs. fully implanted device) and the technical features changed dramatically (e.g., smaller battery size) to improve pacemaker S&P and benefit-risk ratio. Currently, the leadless and implantable pacemakers both still exist on global markets with significantly different risks (e.g., related to the leads, battery, overall size) and each unique device should now have a separate CER to focus on device-specific details (e.g., leads vs. no leads).

Example 2: Left ventricular assist device (LVAD)

Currently, few clinical studies support LVAD use even though the technology, as well as the medical and surgical management of heart failure patients, have changed rapidly over time since the first LVAD was approved in the US in 1994 [27]. For comparison to the pacemaker, the LVAD development timeline covered the same timespan from 1950 to present day (Fig. 2.7).

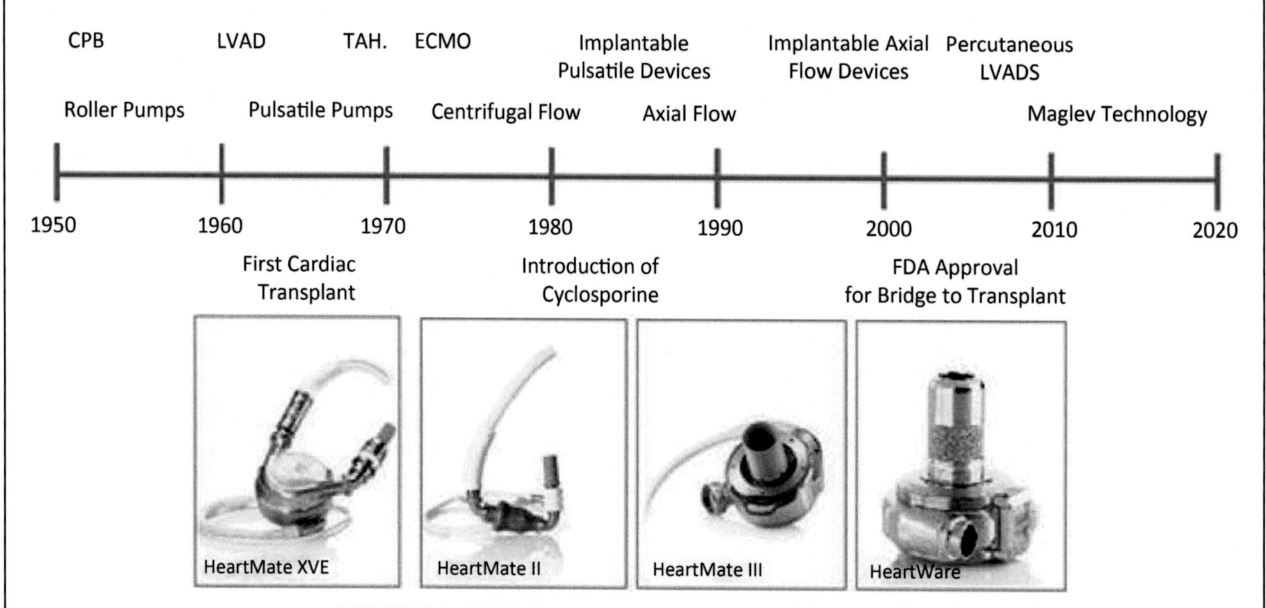

FIGURE 2.7 LVAD history. *Image taken from published work [29].*

Unfortunately, current LVAD technology is not optimal. Device complications, poor quality of life, and comparisons to heart transplantation as an alternative therapy/device make writing a LVAD CER challenging, even though LVAD devices are also available in the EU market.

Current LVAD CERs should comment on future LVAD devices. Many new device designs are in planning stages to enhance blood-pump compatibility and reliability, potential dynamic pump adaptations to simulate cardiac contractility responses to physiologic demands for changes in blood flow and LVAD computer software designs with artificial intelligence and machine learning to monitor and react to LVAD complications like impending pump failures or physiologic changes like cardiac arrhythmias or blood volume fluctuations. The future may even include an implantable and fully autonomous artificial heart to optimize cardiac physiology.

CER scope requires clinical data

When limited LVAD CER clinical data are available, the CER scope must be sufficiently broad to accommodate all available clinical data and to generate new clinical data for the small heart failure population who might undergo an LVAD implant including those who may need a heart transplant (Fig. 2.8).

(*Continued*)

BOX 2.5 (Continued)

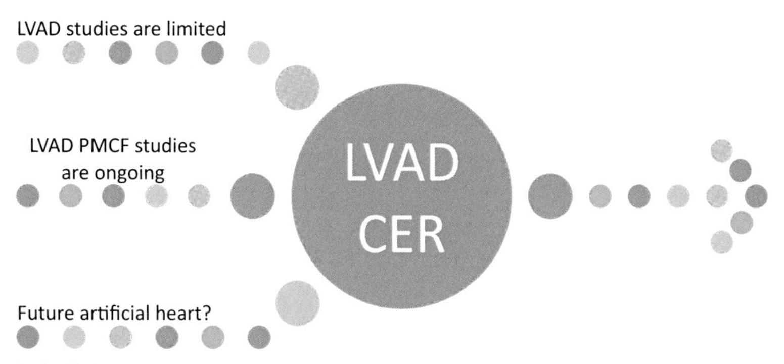

FIGURE 2.8 LVAD CER must account for current SOTA and cover clinical data gaps.

In this example, the available clinical data amount is limited and the clinical data quality is poor; therefore, clinical data gaps must be addressed in ongoing PMS clinical studies as detailed in the corporate PMCFP. The PMCF clinical studies must address all CER S&P and benefit-risk ratio concerns to demonstrate conformity with MDR GSPR 1, 6, and 8 for safety, performance, lifetime of device, benefit-risk ratio, etc.

Example 3: Patient drape (a "low-risk" and "well-established" device)

Patient drapes have been used since the 19th century to help maintain a sterile field and to create a barrier to protect the patient from dust and particulates as well as microbial contamination (e.g., by the patient's own flora or from nonsterile areas of the operating room table) or environmental contamination (e.g., from surgical fluids used during the procedure). Patient drapes also provide privacy, warmth/comfort and may protect vulnerable skin including wounds, surgical sites, and neurologically impaired areas. The Centers for Disease Control (CDC) and Prevention and the Association of periOperative Registered Nurses (AORN) recommend draping the patient to prevent surgical site infections; however, clinical evidence is limited for this low-risk and well-established technology.

Limited CER scope for low-risk, well-established devices like patient drapes

Over more than 100 years, from the 1920s to the 2020s, patient drapes have evolved from all white cloth drapes to all disposable multicolored drapes with many shapes and sizes including some drapes with adhesive backing and/or "windows" for wound isolation during surgery, and more (Fig. 2.9).

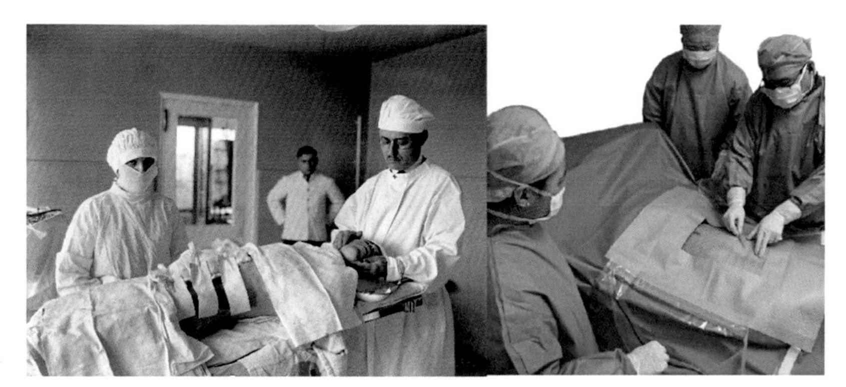

FIGURE 2.9 Operating room patient drapes: 1920–2020. *Images from https://i.pinimg.com/originals/6e/22/5a/6e225a125b3ee4da7b9363a18361be50.jpg (left) and https://www.stevimed.com/images/product/10/drap-2.jpg (right).*

The lack of clinical trials focused on patient drapes makes the clinical evidence rather weak for patient drapes; however, the regulations allow an exemption for CER/PER when the collection of clinical data is not deemed appropriate. In this case, the low-risk combined with the extensive, long-term requirements to use patient drapes in standard practice makes the collection of new clinical data in an expensive clinical trial rather inappropriate since the benefit-risk ratio is stable, many millions of surgical and nonsurgical procedures use patient drapes each year and no risk "signal" has been associated with patient drapes used for more than 100 years.

(*Continued*)

BOX 2.5 (Continued)

Conclusions

Knowing and defining the CE/PE scope is a critical evaluator function. The evaluator must address any data relevance and quality challenges while developing the objective and scope. When clinical data are abundant (as in the pacemaker example), splitting clinical data into groups will be required; however, when little clinical data are available (as in the LVAD example), a broad data collection scope will be required to bring in enough clinical data for the evaluation. In addition, when the collection of clinical data is deemed "not appropriate" the evaluator should be prepared to write the justification (JUS) explaining the rationale to forgo writing the CER because doing so is not appropriate.

The evaluator should also be ready for the clinical data relevance challenges when the clinical data evaluated are from an equivalent or similar device and not the actual device under evaluation (as is common in the LVAD example above). Ultimately, the evaluator also controls the PMCFP/PMPFP scope by identifying the clinical data gaps in the CER/PER based on the scope defined in the CEP/PEP. The evaluator must require additional, future clinical data collection to address clinical data gaps and the PMCF/PMPF clinical data generating activities must be incorporated into the PMS system.

Although the objective and scope may have many parts, always start by evaluating the intended purpose/use and indications for use in the labeling. Creating one- or two-sentence CEP/PEP objective and scope statement/s from the intended purpose/use and indications for use may help team members decide where to focus effort and these brief CEP/PEP objective and scope statement/s may even help to defuse team member disputes. These sentences are worth the time to create. The device and device use must be clear. Resist the temptation to use multiple sentences to define the objective and scope. Practice writing concise and clear objective and scope statements. CEP/PEP objectives and scope statements are sourced from the device labeling and should be found in the prior CER/PER. The details need to be linked from the CER/PER to the labeling and back again. Ultimately, the intended purpose/use and indications for use may need to change based on the clinical data evaluated in the CE/PE process. This link to labeling is also important to keep in mind when writing the CER/CEP.

The key to developing an executable CEP/PEP is to understand the available technical files and CE/PE data prior to drafting the CEP/PEP. Explore every new postmarket, equivalent device, clinical trial, or other clinical data available for the CER/PER. If documentation or process points are missing, create a plan to gather them or be prepared to justify the current clinical data status (e.g., the clinical trial has only just begun or animal trials are currently underway to address a particular performance concern). The CEP/PEP should provide enough detail to execute the CER/PER and should limit the CE/PE to a relevant and high-quality focus on the clinical data for the specific device use. Anyone reading the CEP/PEP clinical/performance trial section should understand exactly which clinical/performance trials will be evaluated in the CER/PER. The clinical literature and experience search details should be detailed enough (e.g., databases searched, search terms used, dates included, appraisal inclusion and exclusion criteria, analysis methods) for another evaluator to reproduce the results and draw the same/similar conclusions.

Note: The evaluator will identify clinical data gaps while executing each planned CE/PE stage. This does not mean the CEP/PEP data source process descriptions were incorrect; rather, the ability to find and address additional clinical data gaps during PMS is a healthy and iterative process. The CEP/PEP scope should accommodate the future PMCFP/PMPFP to address clinical data gaps, both old and new.

2.8.4 Ensure safety and performance acceptance criteria are "well-developed"

When S&P acceptance criteria are defined for the first time, a good rule of thumb is to start small and stay focused. Identify the top two or three safety and the top two or three performance concerns for the device. Consider how each concern can be measured and review all clinical/performance data from the background knowledge, SOTA, and alternative device therapies to estimate and set thresholds specifically for the subject/equivalent device under evaluation. Be careful to set appropriate threshold values (i.e., not too high and not too low) so device data can be appropriately evaluated in the CE/PE process. The goal is to use these threshold values to ensure subject/equivalent device clinical/performance data are at least as safe and perform at least as well as other devices and therapies (including no therapy at all, if appropriate).

The CMO should have a clear and specific list of actual clinical/performance risks and benefits, and the RM team should have a theoretical risk list developed during the RM process. The CEP/PEP should consider each item on both lists as well as the PMS system patient and device problems lists. The evaluator should also identify the most common, clinically significant PMS clinical complaints when isolating safety issue acceptance criteria thresholds. For performance, the evaluator should consider the greatest clinical benefits provided by the subject/equivalent device when used as indicated and when the device performs as intended. The goal is to identify clear, quantifiable subject device S&P details to measure and then to set the threshold requirements for each measure. Focus on indicated uses only and understand all findings in best and worst clinical/performance cases before setting the threshold number for each S&P criterion. As time goes on, expand and refine the S&P acceptance criteria to have an appropriate and sufficient list of criteria to cover all appropriate S&P concerns with enough sensitivity and specificity to detect changes in S&P over the entire device lifetime.

BOX 2.6 S&P acceptance criteria development process

Read the most relevant, highest-quality, and most recent metaanalyses and systematic reviews discussing the subject/equivalent device, and list the S&P and benefit-risk ratio details discussed in these reviews. If no meta-analyses or systematic reviews are available about the subject/equivalent device, then review all available device-related clinical data and develop the list and occurrence rate for each risk based on the available data. For each item listed, document the occurrence rate separately for the subject/equivalent device and for the similar devices or alternative therapies. Prioritize the most clinically relevant items and compare the S&P and benefit-risk ratio details for the subject/equivalent device to the S&P and benefit-risk ratio details for the similar devices or alternative therapies. In addition, compare the new data to the most recent CER/PER, RMR and PMSR/PSUR data. If the S&P acceptance criteria and benefit-risk ratio differ from those found in the most recent CER/PER, you should update the acceptance criteria as appropriate, and explain the changes in the CEP/PEP for the next CER/PER.

S&P acceptance criteria and benefit-risk ratios must be specific and measurable. For example, one safety criterion for an LVAD could be the proportion of patients with device-related infections after implantation. The focused research on LVAD background knowledge and SOTA literature including other bridge-to-transplant therapies and standard of care treatments for heart failure indicated infections occurred in 1.5 + 0.5% transplanted patients and a safety acceptance threshold criterion was set at less than 2% infection. Any observation of safety data for the subject device exceeding this value would be subject to careful evaluation and potential future adjustment of the S&P acceptance criteria in the next CEP/PEP, as the company reacts to all out-of-specification situations evolving over time.

When developing S&P acceptance criteria and benefit-risk ratios, evaluate similar device S&P data carefully. Primary and secondary study endpoints in clinical trials can demonstrate how similar devices were evaluated and what metrics were considered reasonable comparators. What were typical values and units measured? How was clinical success defined?

High-quality systematic reviews and metaanalyses about both the subject/equivalent device and the similar/alternative therapy devices are helpful for defining S&P acceptance criteria. Unfortunately, many devices do not have any metaanalyses or systematic reviews and having only low-quality clinical data will make the S&P acceptance criteria difficult to establish. In this setting, evaluators should practice defining "reasonable" S&P acceptance criteria based on the available clinical data and institutional knowledge about the device. Initial S&P acceptance criteria should evolve and improve over time as more S&P data are evaluated during CE/PE updates.

2.8.5 Describe the device

Like a clinical trial protocol, the CEP/PEP starts by defining the medical/IVD device to be evaluated in the CER/PER. The device description includes many parts and a good CEP/PEP template should help lead the CE/PE process to have no deficiencies in the NB clinical evaluation assessment report (CEAR) after the NB completes the CE/PE document review. For example, the CEP template "Device description" heading should cover all device details needed to describe a full and complete CE/PE process scope for the entire CE/PE process (Table 2.6).

TABLE 2.6 CEP/PEP device description details

UID	Section name	What to include
1	Device name	Include brand names and generic names for each included device; refer to Declaration of Conformity (DoC) document with all devices and part numbers. If list is long, consider listing only names for each device group in CEP/PEP text and details (e.g., part numbers, descriptions, features) in an appendix.
2	Device manufacturer	Include a table with all essential manufacturing locations (e.g., including packaging, labeling, sterilization) and provide text specifying the certifications available for each location to ensure manufacturing is compliant with ISO 13485 and other relevant sterilization standard certifications.
3	Device description and principles of operation	Include device group/type, images, and start by describing entire system before describing individual components with a concise physical and chemical description (especially for materials contacting human tissues). Explain clearly how device functions and achieves the intended purpose. List all devices, catalog numbers, product names, starting components, sizes, uses, and accessories in a table. Specify if the device includes a medicinal substance (already marketed or new) and if the device includes any animal or human tissues. Provide all details important to the device-specific principles of operation, S&P features, and benefit-risk ratio [28].
4	Device lifetime	Include testing used to verify device lifetime. *Note: Do not confuse device lifetime with shelf life, patient lifetime, or indefinite use as none of those details will be acceptable here.* Expected lifetime is the "time-period… during which the medical device or IVD… is expected to maintain safe and effective use" [29] (e.g., battery life may determine pacemaker lifetime).
5	Intended purpose/use[a]	Use specific wording from labeling/instructions for use/promotional materials to describe critical performance and expected clinical effect for the specific intended purpose. Include labeling source document as a CEP/PEP appendix. *Note: Intended purpose/use describes device effect, not indication for use* [30].
6	Indication for use	Include disease name/state/clinical condition/s including disease stage/severity/symptoms or aspects to be treated, managed, diagnosed, prevented, monitored, alleviated, compensated for, replaced, modified, or controlled by the medical device, patient population (e.g., age: adults, children, anatomy, physiology), target group/intended user/s (e.g., healthcare professionals, lay persons), use environment (hospital, clinic, home), anatomic locations contacted by device, duration of use, repeat applications/restrictions, single/multiuse, and invasiveness (i.e., contact with mucosal membranes, implantation, etc.) [21,30] *Note: Devices intended to disinfect or sterilize etc. may not have an indication and, in general, higher-risk devices require greater specificity and detail in the indications for use than lower-risk devices.*
7	Contraindications and precautions	Include contraindications and precautions listed on device label/IFU. Note contraindications are defined by the European Medicines Agency (EMA) as "situations where the product must not be given for safety reasons, e.g., concomitant disease, demographic factor or predisposition, concomitant use" with another product [31].
8	S&P claims	Include intended safety and technical performance details, intended clinical benefits and all clinical S&P claims. List product literature including claim source, when relevant (e.g., device labeling: IFU, Operation Manuals, UDI info). *Note: The CEP/PEP should include the "current version number or date of the information materials supplied by the manufacturer (label, IFU, available promotional materials and accompanying documents possibly foreseen by the manufacturer)" [21].*
9	Equivalence	Specify if the device is considered equivalent to an existing CE-marked device on the market in the EU. If equivalence is being claimed, list equivalent device name, models, sizes, settings, components, software, etc. and if equivalence was used for CE/PE previously. Include a table comparing equivalent and subject devices for clinical, biological, and technical equivalence, specify any/all differences and document clinical impacts with scientific justifications for each difference.

(Continued)

TABLE 2.6 (Continued)

UID	Section name	What to include
10	Developmental context	Include regulatory details for device evolution over time from prototype to most recent model/s. Include predecessor device names, model, sizes, and whether the predecessor is still on the market with a description of modifications and dates for those modifications. Describe differences between current model and previous models/generations along with current device clinical development plan. Identify any unmet medical needs to be addressed by the device and any alternative therapies available for the device. Describe if device is undergoing initial CE marking or was already CE marked. Describe if the device is currently on the market in the EU or other countries, since when, and state the number of devices placed on the market.

CE, conformité européenne (European conformity); *CEP*, clinical evaluation plan; *DoC*, declaration of conformity; *EMA*, European Medicines Agency; *EU*, European Union; *IFU*, instructions for use; *ISO*, International Organization for Standardization; *IVD*, *in vitro* diagnostic; *MDCG*, Medical Device Coordination Group; *MDR*, medical device regulation; *PE*, performance evaluation; *PEP*, performance evaluation plan; *S&P*, safety and performance; *UDI*, unique device identifier. *Notes are in italics and emphasis is added in bold text.*
[a]*MDR defines "intended purpose" not "intended use"; MDCG 2020-6 [30] specifies "intended use" corresponds to "intended purpose."*

Sometimes the device description and principles of operation can be complicated. Some devices are systems with many components, some devices have changed over time, other devices use accessories not provided with the device, and still other devices are grouped into device families. All details need to be documented and described clearly in the device description.

2.8.6 Group low-risk devices into device families

The common high-level strategy about when to group low-risk devices together into a "device family" for the CE/PE process and when to split devices into separate CE/PE groups is critical to CER/PER success during EU CE marking. No formal guidance is available for grouping devices in technical files; however, NBs have routinely accepted grouped devices in CERs/PERs. Class I medical devices and Class A IVD devices do not require NB CER/PER assessment (i.e., unless they are sterile or have a measuring function when NB oversight is expected for selected product samples), and good planning and documentation for appropriate device family CERs/PERs is a helpful strategy when evaluating large, low-risk device groups.

The MDR/IVDR allows "generic device groups" to be defined for conformity assessment purposes for Class I/IIa/IIb WET/IIb Rule 12 devices (MDR) or Class A/B/C IVDs (IVDR). A generic device group is "a set of devices having the same or similar intended purposes or a commonality of technology" (MDR Article 2(7) [1]; IVDR Article 2(8) [2]). When assessing clinical evidence needed for generic device groups [30], NBs use a sampling strategy to meet MDR Article 52(4) and (6) [1] and IVDR Article 48(7) and (9) [2] where technical documentation for "at least one representative device" is reviewed and used to determine conformity per "generic device group" (for Class IIb and Class C devices) or "category of devices" (for Class IIa and Class B devices) [32]. "At least one technical documentation must be reviewed each year" for each group or category for surveillance assessment. Under this rule, manufacturers prepare a single CER/PER for each device family. The device family should be well defined, devices in the family should use the same technology, have the same benefit-risk ratio and the same risk classification. For example, a manufacturer can group multiple surgical clamp devices into a single surgical clamp device family CER provided they meet the details discussed above.

When writing a device family CEP/PEP, the process must identify clinical data covering the entire device family by running searches for all grouped devices. Trials, literature, and experience data, in particular, should be identified for every device within the device family to ensure the S&P details and the benefit-risk ratios for each device are evaluated.

BOX 2.7 Clinical evaluation report/performance evaluation report device family examples

To understand the CEP/PEP strategy for grouping devices together into device families, consider the following three scenarios for low-, medium-, and high-risk devices. Each scenario describes the company, devices, and factors considered during the device family decisions.

Scenario #1: A large global manufacturer was marketing drapes and gloves including different sizes, shapes, types, and uses. Before grouping devices into multiple CE device families, the manufacturer considered the following factors: surgical and nonsurgical uses, benefits and risks, materials, antibiotic incorporation or not, manufacturing site location, sterile or nonsterile packaging, individual unit or bulk sales, etc. Although drapes and gloves were considered well-established, relatively low-risk, single use, barrier-providing medical devices for short-term use, some products had a higher risk ratio than others. Drapes and gloves were placed into separate device families based on their differing uses, and drapes with antibiotics added versus those without antibiotics were also placed into separate device families due to the potential for antibiotic-specific allergic reactions or antibiotic drug resistance. Within the device family CERs, sterile and nonsterile devices as well as devices made with latex or nonlatex materials were discussed together, including comments about sterility and latex-allergy risks. In this example, three CERs were developed: drapes without antibiotic, drapes with antibiotics, and gloves. Eventually, the drapes with antibiotics were obsoleted and only two device families remained.

These low-risk drapes and gloves had many separate devices even though they all provided a similar type of the same intended barrier protection. This large, diversified company decided to obsolete the devices with antibiotics added rather than to continue the expensive process of supporting a separate CE process *for this added (higher-risk) antibacterial intended function/use.*

Scenario #2: A medium-sized manufacturer was marketing orthopedic plates and screws including different sizes, types, and uses. Before grouping devices into multiple CE device families, the following factors were considered including (but not limited to): benefits and risks, materials, device size and anatomic location, sterile or nonsterile packaging, individual unit or bulk sales, uses to repair broken or diseased bones/tendons, etc. Devices were grouped and discussed based on implant anatomic location because the benefits and risks were dependent on the materials and the size and shape of the individual components (e.g., the forefoot and toes used smaller, more delicate plates and screws than the mid or hind foot, or the knee or hip; similar to the hand and fingers compared to the wrist, elbow, or shoulder plates and screws). In this example, six CE processes were initially developed: (1) forefoot/toes, mid-foot, hind foot, and ankle; (2) knee; (3) hip; (4) hand, fingers, and wrist; (5) elbow; and (6) shoulder plates and screws.

The medium-risk, implantable, orthopedic plates and screws were grouped into six *device families based on the implant anatomic locations. Hundreds of devices were grouped together into each CER because so many different sizes and shapes of plates and screws existed for each anatomic location. Having only six CERs saved* time *and money since the company had several dozen CERs previously.*

Scenario #3: A medium-sized device company was manufacturing surgical tissue repair products in different sizes and using different animal tissue types. Many factors were considered before grouping devices into multiple CE device families based on the different tissue repair types used in different anatomical locations (e.g., bovine and porcine skin and bone used for dental, hernia or dura repair). These high-risk devices were grouped individually by material type (i.e., different sizes of the same tissue material were grouped within the same indication; however, different tissues types were split into separate CERs) and indication for use (i.e., devices were grouped by clinical benefit-risk ratio given similar indications for use and each group had a separate CER).

Grouping these devices by animal tissue type helped to focus the biological evaluation and grouping by anatomical location helped to focus CE on the functional features for these high-risk animal tissue repair devices. In general, far fewer devices were grouped in each device family compared to the other examples above.

Scenario #4: One IVD device manufacturer grouped devices with similar indications for use, the same analytes/dry chemistry and similar technologies (i.e., test strip and small automated device were grouped with a larger device supporting a common platform) into one PER.

Even though IVD device family members had different features, devices from one manufacturer can be grouped together if they have at least one common intended purpose.

CERs/PERs can cover "generic device groups" with different technical features as long as the intended uses and benefit-risk ratios are similar. When writing the rationale for the device grouping in the CER/PER, consider the European Medical Device Nomenclature (EMDN). According to MDCG 2019-13 [32], the MDR groups are defined at the fourth EMDN level and the IVDR groups are defined at the third EMDN level using the appropriate "IVP Code" as defined in MDCG 2021-14 [33].

Note: Although similar, grouping devices does not involve the same decision-making process as used when determining if products are equivalent.

When grouping devices into device families, start with the device indications for use and review past NB history back to when the CE mark was first applied. Any differences in the indications for use or in the clinical data deficiencies should be reviewed. In addition, any clinical data concerns for device groups from any source within the company should be considered. Any benefit-risk ratio differences between devices should also be considered when making decisions about CE/PE device family decisions. Once the NB deficiencies, clinical data concerns, all prior CERs/PERs, and benefit-risk ratio differences have been reviewed, the device family decisions can be applied. The company will want to ensure all associated technical files or design dossiers are aligned with the CE/PE device families.

2.8.7 Define "equivalence" if required

Equivalence generally means equal or virtually identical in effect or function. Under MDR and IVDR, device manufacturers were allowed to use clinical data from a claimed "equivalent device" to represent the expected clinical data from the subject device during the CE/PE process. Equivalence was documented when characteristics were sufficiently similar (i.e., devices had "no clinically significant difference in the safety and clinical performance of the device"), proper scientific justification and substantiation were provided for the equivalence claim and sufficient technical data were evaluated to justify the equivalence claim (i.e., direct access to equivalent device data was documented to justify the equivalence claim) (MDR, Annex XIV) [1].

Specifically, the manufacturer was required to demonstrate clinical, biological, and technical equivalence (Fig. 2.10) between the subject device and an equivalent device before the equivalent device clinical data could represent the subject device for CE marking without new clinical data collection.

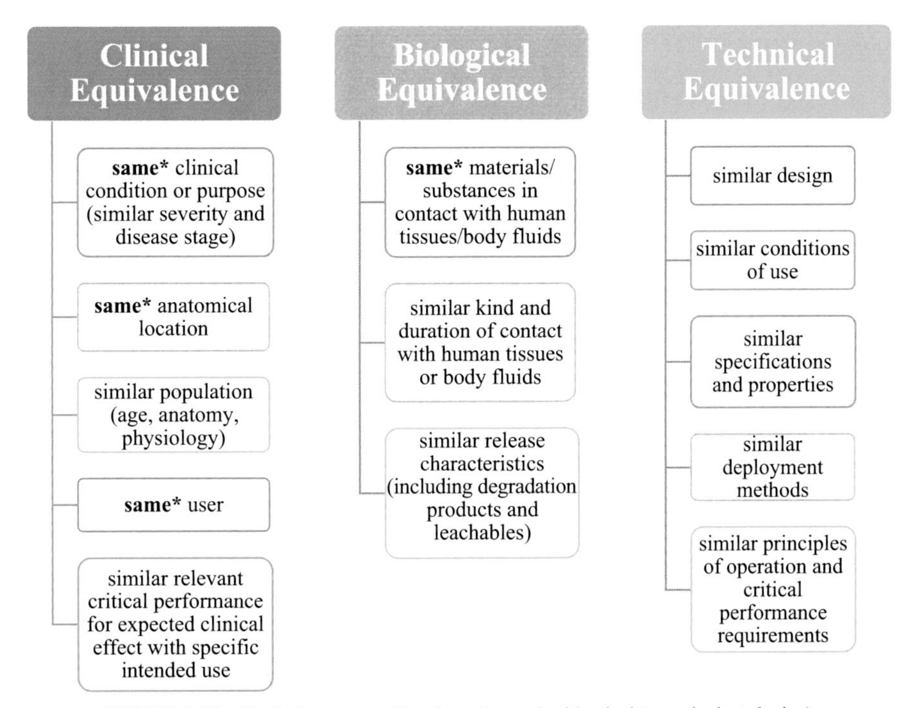

FIGURE 2.10 Equivalence considerations (*must be identical to equivalent device).

Unlike the MDR, the IVDR [2] only mentioned equivalence briefly, as follows:

"The notified body's assessment of performance evaluations... shall cover... validity of equivalence claimed in relation to other devices, the demonstration of equivalence, the suitability and conclusions data from equivalent and similar devices..." IVDR, Annex VII, Requirements to be met by NBs, Process requirements, Section 4.5.4

The QMS should include "the strategy for regulatory compliance, including processes for identification of relevant legal requirements, qualification, classification, handling of equivalence, choice of, and compliance with, conformity assessment procedures..." IVDR, Annex IX, QMS conformity assessment, Section 2.2(c)

"The notified body shall, in circumstances in which the clinical evidence is based partly or totally on data from devices which are claimed to be equivalent to the device under assessment, assess the suitability of using such data, taking into account factors such as new indications and innovation. The notified body shall clearly document its conclusions on the claimed equivalence, and on the relevance and adequacy of the data for demonstrating conformity." IVDR, Annex IX, QMS conformity assessment based on QMS and assessment of technical documentation, Section 4.5

"The notified body shall: ... in circumstances in which the clinical evidence is partly or totally based on data from devices which are claimed to be similar or equivalent to the device under assessment, assess the suitability of using such data, taking into account factors such as new indications and innovation. The notified body shall clearly document its conclusions on the claimed equivalence, and on the relevance and adequacy of the data for demonstrating conformity" IVDR, Annex X, Conformity assessment based on type-examination, Section 3(d)

"The PMPF plan shall include at least: ... an evaluation of the performance data relating to equivalent or similar devices, and the current state-of-the-art" (Annex XIII, Performance evaluation, performance studies and postmarket performance follow-up, Section 5.2 f)

"Investigator's Brochure... Existing clinical data, in particular: ... from relevant peer-reviewed scientific literature and available consensus expert opinions or positions from relevant professional associations relating to the safety, performance, clinical benefits to patients, design characteristics, scientific validity, clinical performance and intended purpose of the device and/or of equivalent or similar devices." Annex XIV, Interventional clinical performance studies and certain other performance studies, Section 2.4

This limited IVDR equivalence discussion created some confusion since the exact basis for IVD device equivalence was unclear; however, many equivalent device requirements were similar if not exactly the same between the MDR and IVDR.

The MDCG 2022-2 [19] guidance about IVD device clinical evidence was intended to help resolve confusion about CE general principles; however, this guidance did not detail IVD device performance studies or equivalence beyond the following:

evidence for an IVD's conformity is established by demonstrating and substantiating the scientific validity, analytical performance and clinical performance... clinical evidence should be based on a sufficient amount and quality of data in order to allow a qualified assessment of whether the IVD is safe, performant and achieves the intended clinical benefit/s, when used as intended... clinical evidence may include data from devices which are claimed to be equivalent to the device under assessment. The handling of equivalence should be defined in the manufacturers QMS (Annex IX 2.2.c). *Where data from equivalent devices is used, a justification should be provided... scientifically substantiated conclusions should be reached through... a systematic and explicit appraisal of data...* MDCG 2022-2 Section 2.3, Introduction [19]

Added to this MDR versus IVDR confusion, the "equivalent device" used during CE marking (so the device can be sold on the EU market) was often confused with the "predicate device" used during 510(k) clearance by the United States (US) Food and Drug Administration (FDA) [34] (so the device can be sold on the US market).

BOX 2.8 US "substantial equivalence" is not the same as EU "equivalence"

The MDR and IVDR equivalent device processes are similar to, but not the same as, the US FDA "substantial equivalence" (SE) versus "non-substantial equivalence" (NSE) process used to determine if a medical device can be cleared to be sold on the US market under the premarket notification (PMN, 510(k)) pathway when a premarket approval (PMA) was not required. The US 510(k) pathway is often (but not always) completed without new clinical data collection required prior to marketing the device in the US. This is similar to the desired outcome when claiming an "equivalent" device in the EU CE/PE process; however, many other details between these US and EU regulatory processes are dissimilar.

For example, in the EU, a device in any device class can be considered an equivalent device when specific requirements are met and the equivalent device clinical data can represent the subject device S&P and benefit-risk ratio in the CE/PE process so no new clinical data are required prior to marketing in the EU. In the US, no CER/PER is used and Class III devices cannot use the 510(k) pathway; therefore, all new Class III devices must gather new clinical trial data (with certain limited exceptions).

The US regulations in "Section 513(i) of the FD&C Act and 21 CFR 807.100(b) stated... to be considered substantially equivalent to a predicate device, the new device must have the same intended use as the predicate device and the same technological characteristics or different technological characteristics that do not raise different questions of safety and effectiveness than the predicate device."

The FDA considers a "significant change in the materials, design, energy source, or other features of the device from those of the predicate device" to be "different technological characteristics." The FDA reviews each 510(k) submission to determine if the new device is controlled sufficiently based specifically on the prior FDA determination of "reasonable assurance of safety and effectiveness" for the predicate device [34].

(*Continued*)

BOX 2.8 (Continued)

In addition, unlike the EU, the FDA uses a specific flow chart with six decisions to determine substantial equivalence:

"Decision 1: Is the predicate device legally marketed?"
"Decision 2: Does the device have the same intended use?"
"Decision 3: Do the devices have the same technological characteristics?"

If these answers are all "yes," then the devices are deemed substantially equivalent (SE). If the answer to Decision 3 is "no," the answers to the next three questions must be "no," "yes," and "yes," respectively, for a SE determination.

"Decision 4: Do the different technological characteristics of the devices raise different questions of safety and effectiveness?"
"Decision 5a: Are the methods acceptable?"
"Decision 5b: Do the data demonstrate substantial equivalence?"

These FDA decisions are the same for all medical devices including IVD devices: however, this PMN/510(k) regulatory pathway is distinctly different when compared to EU MDR and IVDR. For example, the EU regulations do not envision a "predicate device" and the NB does not make a SE/NSE (not substantially equivalent) determination—these decisions are unique to the US. In addition, the EU allows Class III, implantable devices to be considered equivalent, but this is not generally allowed in the US.

The EU MDR and IVDR focus is specifically on clinical data "equivalence" for CE/PE and not for overall regulatory classification or specific decision to clear a device for marketing. EU regulations require clinical, biological, and technical equivalence before clinical data from the "equivalent" device can be used as clinical data for the device under evaluation in the CER/PER.

The US regulations require specific bench testing comparing the subject and "predicate device" head-to-head for "technological characteristics" to demonstrate SE while the EU regulations require "clinical data" from the "equivalent device" to be used in the CER/PER.

Additional differences between the US and EU regulations exist, and due diligence is required when navigating regulatory pathways across global geographies. The CER/PER evaluator should avoid using terms like "predicate" or "substantial equivalence" in the CE/PE documents unless making a specific point about differences between US and EU regulatory information in the device description or background describing global regulatory approvals.

EU regulations specifically allowed equivalent device use for all medical device types and the MDCG 2020-5 [35] guidance described how to demonstrate equivalence between devices including devices with an "ancillary medical product" or "without an intended medical purpose..." This guidance envisioned a harmonized equivalence approach across the EU. Even though a clinical trial was described as "the most direct way to generate clinical data" so the medical device S&P could be rigorously documented for CE marking, clinical data could also be sourced from clinical literature and experiences using the "device in question or a device for which equivalence to the device in question can be demonstrated." This guidance described how NBs judge equivalence and highlighted differences between MEDDEV 2.7/1 rev 4 [21], MDR [1] and IVDR [2] for CE-mark decisions using an equivalent device (Table 2.7).

TABLE 2.7 Comparing equivalence in MDR/IVDR[a], MedDev 2.7/1, and MDCG2020-5.

MDR, Annex XIV Part A (3) [1]	MEDDEV 2.7/1 rev 4, App. A1 [21]	MDCG 2020-5 [35]
	Clinical equivalence	
"The device is used for the same clinical condition or purpose, including similar severity and stage of disease, at the same site in the body, in a similar population, including as regards age, anatomy and physiology; has the same kind of user; has similar relevant critical performance in view of the expected clinical effect for a specific intended purpose."	"... used for the same clinical condition (including when applicable similar severity and stage of disease, same medical indication), and... used for the same intended purpose, and... used at the same site in the body, and... used in a similar population (this may relate to age, gender, anatomy, physiology, possibly other aspects), and... not foreseen to deliver significantly different performances (in the relevant critical performances such as the expected clinical effect, the specific intended purpose, the duration of use, etc.)"	The devices need "the same kind of user." (i.e., "healthcare professional or lay person who uses a device... a lay person means an individual who does not have formal education in a relevant field of healthcare..."). Manufacturers must consider if "the intended user's competence or knowledge" might impact "the safety, clinical performance and outcome when considering equivalence... For example, a device intended for professional use and a device intended for home use, but for the same clinical condition or purpose, may have a

(Continued)

TABLE 2.7 (Continued)

MDR, Annex XIV Part A (3) [1]	MEDDEV 2.7/1 rev 4, App. A1 [21]	MDCG 2020-5 [35]
		different safety and performance profile due to the environment in which they are... used... MDR does not explicitly state... the medical device needs to be used for the same medical indication, gender, and duration of use as the equivalent device..."; however, "both devices should be used for the same clinical condition or purpose including similar severity and stage of disease... [with a] similar relevant critical performance..." *HINT: see intended purpose, clinical performance and clinical benefit definitions in the MDR and IVDR*
	Biological equivalence	
"The device uses the same materials or substances in contact with the same human tissues or body fluids for a similar kind and duration of contact and similar release characteristics of substances, including degradation products and leachables"	"Use the same materials or substances in contact with the same human tissues or body fluids. Exceptions can be foreseen for devices in contact with intact skin and minor components of devices; in these cases risk analysis results may allow the use of similar materials taking into account the role and nature of the similar material."	MedDev 2.7/1 "exceptions" are not acceptable since MDR requires materials or substances contacting body tissues/fluids to be the "same." ISO 10993 principles can be adopted including ISO 10993-1 for risk-based approach to biological evaluation [36], -18 for chemical characterization [37], -17 for leachable toxicological risk [38], as well as -13 [39], -14 [40], and -15 [41] about degradation products, etc. MDR requires the "same" device "substances" to meet "relevant requirements laid down in Annex I to Directive 2001/83/EC..." [42] "for... absorption, distribution, metabolism, excretion, local tolerance, toxicity, interaction with other devices, medicinal products or other substances and potential for adverse reactions." Devices with ancillary medicinal substances are Class III devices (e.g., drug-eluting, chemically bonded) and must consider biological characteristics. The NB will verify substance is useful when used as indicated. *Note: The NB must seek a scientific opinion from the competent authority or EMA (European Medicines Agency) for ancillary medicinal substances and anything systemically absorbed to achieve the intended purpose.*
	Technical equivalence	
"The device is of similar design; is used under similar conditions of use; has similar specifications and properties including physicochemical properties such as intensity of energy, tensile strength, viscosity, surface characteristics, wavelength and software algorithms; uses similar deployment methods, where relevant; has similar principles of operation and critical performance requirements."	– "be of similar design, and – used under the same conditions of use, and – have similar specifications and properties (e.g., physicochemical properties such as type and intensity of energy, tensile strength, viscosity, surface characteristics, wavelength, surface texture, porosity, particle size, nanotechnology, specific mass, atomic inclusions such as nitrocarburising, oxidability), and	"The conditions of use shall be similar... [with] no clinically significant difference in the safety and clinical performance between the device in question and the device presumed to be equivalent... the MDR specifically points out that software algorithms shall be similar in the device presumed to be equivalent... the functional principle of the software algorithm, as well as the clinical performance/s and intended purpose/s... shall be considered when demonstrating the equivalence of a software algorithm..." Equivalence is not required "for the software code... developed in line

(*Continued*)

TABLE 2.7 (Continued)

MDR, Annex XIV Part A (3) [1]	MEDDEV 2.7/1 rev 4, App. A1 [21]	MDCG 2020-5 [35]
	– use similar deployment methods (if relevant), and – have similar principles of operation and critical performance requirements"	with international standards for safe design and validation… of medical device software. Software solely intended for the configuration of the device… and not related to medical purpose… does not need to be similar…[when] justified to not negatively affect the usability, safety or clinical performance."

EC, European Commission; *EMA*, European Medicines Agency; *EU*, European Union; *IFU*, instructions for use; *ISO*, International Organization for Standardization; *IVDR*, *in vitro* diagnostic regulation; *MDCG*, Medical Device Coordination Group; *MDR*, medical device regulation; *MEDDEV*, medical devices documents; *NB*, notified body. Bold text emphasizes key concepts in each type of equivalence and italic text provides additional information. Table is modified from MDCG2020-5 [35] and includes IVDR and software as defined in MDCG 2019-11 [43].
[a]Note: *Equivalence may not be claimed "to a product without an intended medical purpose" [35].*

MDCG 2020-5 [35] describes how all prerequisite requirements must be fulfilled to demonstrate equivalence as follows:

1. Clinical, biological, and technical characteristics are similar enough, so no clinically significant difference in device S&P exists (some characteristics must be the "same" not only "similar" and a gap analysis must be done for any clinically significant differences; modifications/intended improvements must meet the same standard with "no additional risks" or "negatively altered performance").
2. When claiming more than one equivalent device, "each device shall be equivalent in all the listed technical, biological and clinical characteristics" and "shall be fully investigated, described and demonstrated" in the CER/PER. Different parts from different devices may not be claimed equivalent; however, different devices within different systems may be claimed equivalent provided device interactions with the system are fully considered and all other equivalence requirements are met.
3. The CER/PER shall "specify and justify the level of clinical evidence" required to demonstrate GSPR conformity, why this level is "appropriate" and the "proper scientific justification" for the equivalence claimed. "The manufacturer is expected to fully identify and disclose any differences between the two devices." Technical equivalence and biological equivalence are expected to be based on sound scientific evaluations (e.g., nonclinical data from animal and bench tests, literature, common/technical specifications, harmonized standards, using the actual subject device version compared to the specific equivalent device version).
4. Implantable and Class III devices without a clinical investigation must be either (a) a modified device marketed by the same manufacturer and claimed as an MDR equivalent device (i.e., with a valid CE mark, updated CER, favorable benefit-risk ratio, etc.) or (b) a device where the manufacturers have a contract allowing "full access to the technical documentation on an ongoing basis," and the CER must be MDR compliant (i.e., certified under the MDR) for the equivalent device.
5. Non-implantable, non-Class-III devices may claim equivalence to non-CE marked devices and no contract is required; however, the regulatory status should be disclosed and sufficient data must be available to claim equivalence satisfying all the MDR requirements. The clinical data must comply with international guidelines (e.g., ISO 14155 [3] and Declaration of Helsinki [42]), and a justification should define the data transferability and applicability to the EU population.
6. Clinical investigations are expected for "products without an intended medical purpose" unless existing clinical data are justified and available from an "analogous medical device" with a medical purpose (i.e., a medical device with a similar function, risk ratio and medical purpose, see MDR [1] Recital (12)). In this case, clinical benefit is understood as device performance and the product and "presumed analogous medical device" must have no clinically significant differences in safety or performance.

The MDR recommended obtaining direct access to equivalent device technical files; however, access is often not possible for a competitor device since most manufacturers do not share proprietary and confidential product information with their competitors. In most cases, public information can be located and used to claim equivalence with competitor devices; however, Class III implantable devices require a contract with the equivalent device manufacturer to allow long-term access to the equivalent device technical files.

Equivalence is a key CER/PER concept because clinical and performance information from other devices can be used to demonstrate GSPR conformity, which would otherwise not be permitted (i.e., equivalence is claimed in order to use clinical data from another device to represent the subject device safety and performance data). The MDCG 2020-5 guidance [35] described the NB method to assess equivalence claimed by the manufacturer including a required equivalence table to be used in an eqCER/eqPER. When claiming equivalence, the guidance suggested the manufacturer must document the comparison between the device under evaluation and the claimed equivalent device line-by-line in the proposed table. All clinically significant differences must be identified and the clinical impacts of those differences must be justified in the table.

The table is provided here for the purpose of claiming equivalence, and this table was modified to start with clinical equivalence, then biological equivalence and finally technical equivalence (Table 2.8).

TABLE 2.8 Equivalence table.

Characteristic[a] Expand and replace details as needed for specific device	**Device 1 (under clinical evaluation)** Description of characteristics and reference to specifying documents	**Device 2 (marketed device)** Description of characteristics and reference to specifying documents	**Identified differences or conclusion of no differences in the characteristic**
1. Clinical characteristics			
Same clinical condition or purpose, including similar severity and stage of disease			1.1
Same site in the body			(Characteristic must be same for demonstration of equivalence) 1.2
Similar population, including as regards age, anatomy and physiology			1.3
Same kind of user			(Characteristic must be same for demonstration of equivalence) 1.4
Similar relevant critical performance in view of the expected clinical effect for a specific intended purpose			1.5
Scientific justification for no clinically significant difference in the safety and clinical performance of the device, OR a description of the impact on safety and or clinical performance (use one row for each identified difference in characteristics, and add references to documentation as applicable)			Clinically significant difference Yes/No
1.1			
1.2			
1.3			
1.4			
1.5			
2. Biological characteristics			
Uses same materials or substances in contact with the same human tissues or body fluids			(Characteristic must be the same for the demonstration of equivalence) 2.1
Similar kind and duration of contact with same human tissues or body fluids			2.2
Similar release characteristics of substances including degradation products and leachables			2.3

(Continued)

TABLE 2.8 (Continued)

Characteristic[a] Expand and replace details as needed for specific device	**Device 1 (under clinical evaluation)** Description of characteristics and reference to specifying documents	**Device 2 (marketed device)** Description of characteristics and reference to specifying documents	**Identified differences or conclusion of no differences in the characteristic**
Scientific justification for no clinically significant difference in the safety and clinical performance of device, OR a description of the impact on safety and or clinical performance (use one row for each identified difference in characteristics, and add references to documentation as applicable)			Clinically significant difference Yes/No
2.1			
2.2			
2.3			
3. Technical characteristics			
Device is of similar design			3.1
Used under similar conditions of use			3.2
Similar specifications and properties including physiochemical properties such as intensity of energy, tensile strength, viscosity, surface characteristics, wavelength and software algorithms			3.3
Uses similar deployment methods where relevant			3.4
Has similar principles of operation and critical performance requirements			3.5
Scientific justification for no clinically significant difference in the safety and clinical performance of the device, OR a description of the impact on safety and or clinical performance (use one row for each of the identified differences in characteristics, and add references to documentation as applicable)			Clinically significant difference Yes/No
3.1			
3.2			
3.3			
3.4			
3.5			
Summary			
In the circumstance that more than one nonsignificant difference is identified, provide a justification whether the sum of differences may affect the safety and clinical performance of the device.			

CER, clinical evaluation report; *MDCG*, Medical Device Coordination Group; *PER*, performance evaluation report. Bold text emphasizes important information.

[a]*Table is modified from MDCG 2020-5 [34] to list clinical trial data first, then biological and finally technical (in order of relevance and quality for CER/PER).*

This modified order is preferred, since no further work should be done if the clinical applications are not the same between the subject and equivalent devices. In addition, if the materials contacting the human tissues are different, then biological equivalence may not be possible. Finally, technical details must only be similar in all aspects for technical equivalence. This prioritized order tends to help focus energy on the highest risk areas in order: clinical, then biological and lastly technical equivalence.

All details in the equivalence determination require "proper scientific justification" and the evaluator must document the appropriate citations and arguments in the CEP/PEP equivalence section to allow investigation and traceability for the clinical impact determinations for all differences. Using a single equivalent device is expected; however, multiple equivalent devices are allowed when each device is completely evaluated and documented to be equivalent.

MDCG 2020-5 [35] stated:

the MDR provided "a possibility to use clinical data related to an equivalent device in the clinical evaluation required for a device under conformity assessment" in lieu of conducting a clinical trial

"to generate clinical data concerning the safety and performance of medical devices for the purpose of CE marking, clinical data can also be sourced from... clinical investigation/s or other studies reported in scientific literature, of a device for which equivalence to the device in question can be demonstrated... reports published in peer reviewed scientific literature on other clinical experience of either the device in question or a device for which equivalence to the device in question can be demonstrated..."

"The demonstration of equivalence does not remove the requirement to always conduct a clinical evaluation in accordance with the MDR.... the demonstration of equivalence... allows the manufacturer to let clinical data from an equivalent device enter the clinical evaluation process of the device in question for the purpose of confirmation of conformity with relevant general safety and performance requirements..."

"A manufacturer of implantable devices and class III devices shall perform clinical investigations except if the device has been designed by modifications of a device already marketed by the same manufacturer and equivalence can be demonstrated according to the MDR... In this context, a marketed device is considered to be a device already placed on the market and CE marked with respect to either the MDR or the directives 93/42/EEC or 90/385/EEC. The CE marking should still be valid, should be based on an updated clinical evaluation, and the benefit/risk ratio for this device should be favourable." (emphasis added in bold)

For an equivalent, implantable, and Class III device from a different manufacturer; however, the equivalent device CER must also be in full MDR compliance. MDCG 2020-5 [35] also described using clinical data from similar devices defined as "devices belonging to the same generic device group" (i.e., "having the same or similar intended purposes or a commonality of technology allowing them to be classified in a generic manner not reflecting specific characteristics"). When equivalence was not demonstrated under MDR/IVDR, "similar" device clinical data use was encouraged to identify relevant risks and hazards for the comprehensive risk management system, to understand SOTA, natural disease history or alternative therapies, to support CE/PE scope designs by evaluating similar device design features posing special safety or performance concerns, to offer possible clinical trial, PMS or PMCFP/PMPFP design ideas, or to identify potential clinical benefit outcomes and minimum clinically relevant benefit and risk occurrence rates.

Sometimes no clinical data should be gathered in a trial (e.g., exposure to virus, radiation, cancer causing agents) and sometimes no equivalent device data are available. In these cases, the lack of clinical data should be justified with a full explanation in the CEP/PEP.

2.8.8 Describe device developmental context

Device developmental context is not defined as such in the regulations, guidelines, or international templates; however, this concept has been helpful to house and organize many different required details during the CE/PE process. For example, the "developmental context" section of the CEP/PEP and CER/PER documents was used to great effect to provide the story behind the device development over the years. This section should help the reader understand the device background knowledge, SOTA, and context for the S&P acceptance criteria as well as the benefit-risk ratio.

The "developmental context" section is the place to describe the device novelty and degree of innovation, any unmet medical needs to be addressed by the device and any alternative therapies available for the device. The lack of any medicinal substances or human/animal tissues can be documented here along with the key risks (e.g., direct patient harm from medical device or incorrect/delayed IVD result) and overall device risk level. In addition, regulatory details (e.g., classification, countries where the device is marketed, sales numbers, different device versions approved or cleared over time) should be included here because this regulatory history will help the reader to understand how the device evolved from prototype to the most recent model.

This is the place to describe all predecessor device names, models, and sizes, differences between the current model and previous models/generations, and whether the predecessor is still on the market with a description of modifications and modification dates along with the current device clinical development plan. This is also the place to document if the device is undergoing initial CE marking or was already CE marked. Describe if the device is currently on the market in the EU or other countries, since when, and the number of devices placed on the market worldwide as well as in the EU along with future marketing plans. This will provide the context for the rest of the CER/PER (e.g., to put into perspective the relative ratio of complaints to number of units on the market).

The SOTA refers to the current technical capability and/or accepted clinical practice but does not necessarily mean the device under evaluation is the newest, best, or most advanced clinical intervention. The device must be placed within the SOTA and the SOTA essentially defines the current device environment including the disease and conditions treated/diagnosed, standard practices, alternative therapies and other acceptable treatments/diagnostic processes/devices to treat/diagnose the disease, etc.

Reviewing the technical files and previous CE/PE documents understand the device prior to starting the CE/PE process. For example, the evaluator may need to know the following: were PMS activities updated recently with new clinical data? Did the PMCFER/PMPFER provide new clinical data results for evaluation? Were risk management changes made based on the new clinical data collected since the last CER/PER? If this is the first CEP/PEP, will an equivalent device be needed or are clinical trial data present for analysis in the CER/PER? If clinical data, documents, or CE/PE process points are missing, create a plan to gather the missing information or be prepared to justify the current clinical data status (e.g., animal trials are currently underway or clinical data collection was not deemed appropriate).

2.8.9 Discuss background knowledge, state-of-the-art, and alternative therapies

The CEP/PEP includes a place to begin the background knowledge and SOTA description including alternative therapies and CE/PE process strategies including the S&P acceptance criteria definition based on the historical background knowledge and SOTA/alternative therapy clinical data. This section should discuss the disease state, medical context, and standard of care for treating or diagnosing the disease, as appropriate for the device under evaluation. This broad discussion should include a high-level description of equivalent and similar devices to be used in the CE/PE process. The prior CE/PE, risk management and PMS documents should be reviewed, and any relevant details should be included and updated here as needed.

The CEP/PEP background knowledge, SOTA, and alternative therapy section should explain how the appropriate clinical data will be identified, appraised, analyzed, and documented in the CER/PER during the CE/PE process. Comparisons between the subject device clinical/performance data and the background/SOTA/alternative therapy clinical data will be required to verify the subject device is at least as safe and performs at least as well as other devices and therapies for this disease and treatment paradigm (including no therapy, if appropriate). Sometimes benchmark devices are included in this section for comparison.

Often the CER/PER background knowledge/SOTA/alternative therapy section may be segmented by the disease, condition, or indication treated, prevented, or diagnosed by the subject device. Sometimes, the hardest part in the CE/PE process is to manage the scope so the background/SOTA/alternative therapy does not become a never-ending research project. The objective is to set the context for the subject device. When writing the plan for the search criteria to identify the clinical data from trials, literature, and experiences for the background knowledge/SOTA/alternative therapy section, first ensure the clinical data searches identify a clear and relevant body of clinical evidence across the SOTA for the subject device including relevant and high-quality S&P parameters of interest. Then, remove extraneous information to focus the story on the device under evaluation.

The CEP/PEP should allow iterative approaches to refining the background/SOTA/alternative therapies and the CEP/PEP should be documented in a CER/PER appendix. In particular, the evaluator needs to build the CER/PER background/SOTA/alternative therapy section to define the device S&P. The clinical data in this background/SOTA/alternative therapy section must be appraised for inclusion and analyzed separately from subject device data. The goal of the CER/PER background/SOTA/alternative therapy section is to place the subject device S&P data in the proper clinical context among all available medical treatments and alternatives.

2.8.10 Document clinical evaluation/performance evaluation planning process

The CEP/PEP should specify how clinical data will be collected from the manufacturer (i.e., manufacturer-held clinical data) and others (e.g., externally located clinical data identified by contracted agreement or available in the public domain), as well as how various databases will be searched using specific search terms. The evaluator must be familiar with clinical data searching and retrieval processes and clinical data identification requirements. The evaluator must also be expert in separating out high-quality clinical data from low-quality, unfounded work while not losing any important, relevant data for the CE/PE process.

Clinical data from trials, literature, and experiences can be identified in separate searches for the background/SOTA and subject device; however, multiple searches can also be combined to generate a matrix of clinical datasets for appraisal and analysis. The ultimate goal in the clinical data identification step is to identify sufficient clinical data for S&P and benefit-risk ratio analyses and the CE/PE plan should aim to capture as much relevant and high-quality clinical data as possible. Another goal for the CE/PE process is to critically evaluate all the clinical data to ensure a robust and accurate CER/PER with minimal clinical data gaps.

The steps for planning, identifying/accessing, appraising, analyzing, and writing the CER/PER should be documented in the internal QMS and the CEP/PEP should define a specific, scientifically valid clinical data identification process (Fig. 2.11).

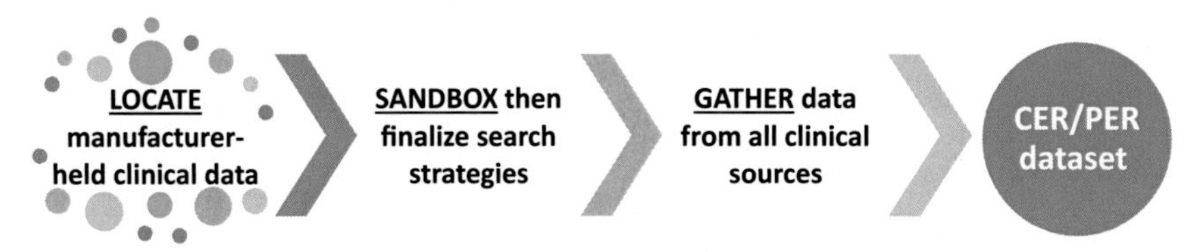

FIGURE 2.11 Clinical data identification process.

Clinical data identification must identify just the right clinical data. Any data missed in this step will not be included in the final evaluation and any extraneous data included will make the job harder. Consider the following steps:

1. First, locate all relevant manufacturer-held clinical data (e.g., clinical trials, literature, and publications, and all clinical experiences including complaints and PMCF/PMPF data).
2. Identify and document in the CEP/PEP all internal and external clinical data repositories for all three clinical data sources:
   a. Trials include premarket trials, PMCF study data, device registry data, and trials registered in international clinical trial databases (e.g., clinicaltrials.gov).
   b. Literature includes internal documents as well as published articles in literature databases (e.g., PubMed, Embase, Cochrane).
   c. Experiences include internal complaints as well as complaints stored in device problem databases (e.g., MAUDE, TPLC) [44].
3. Design each search strategy using relevant search terms:
   a. subject/equivalent device searches (e.g., the more unique the subject/equivalent device brand names, the more useful the "brand name" searches will be to identify relevant data using the subject/equivalent/competitor device);
   b. competitor device searches; and
   c. background/SOTA similar/generic devices and alternative therapies (i.e., in addition to specific brand name searches, the background/SOTA searches will need to include similar and generic devices and alternative therapies and may need to include disease specific searches to better understand relevant alternative therapies).
   d. Consider using specific PICO [45] terms to help structure background and alternative therapy/device database search strategies, as follows:
      P = Population/disease (e.g., age, gender, ethnicity, with a certain disorder)
      I = Intervention/variable of interest (e.g., disease exposure, risk behavior, prognostic factor)
      C = Comparison (e.g., placebo or standard of care/"business as usual" as in no disease, absence of risk factor, prognostic factor)
      O = Outcome (e.g., disease, risk of disease, diagnosis accuracy, adverse outcome occurrence rate, disease recurrence rate)
4. "Sandbox" searches (i.e., execute searches; test search terms) and select most appropriate search terms to generate appropriate clinical data for CE/PE process.
5. Document appropriate "sandbox" identified search terms and details in CEP/PEP and template workbook (e.g., Excel document or proprietary database/software)
6. Download search outputs into workbook/database
7. Conduct QC steps for the CEP/PEP and workbook:
   a. One person does sandboxes and downloads search results.
   b. Another, more senior/experienced person verifies all data are accurate and sufficient.
   c. A third, more senior person adjudicates any discrepancies (often the evaluator is the adjudicator).
8. Secure sign-off and file CEP/PEP in document control with strict version control.

Search strategies need to be developed and documented for each relevant external database. The CEP/PEP should fully document how workbook or database outputs will be identified, presented, and stored in the CER/PER appendix. Evaluators must document all search strategies used to generate the included clinical data. Another person should be able to check and/or replicate the searches and their outputs whenever needed after the planned appraisal, analysis, and writing steps begin.

Meeting regulatory requirements requires careful documentation across the entire CE/PE process. For example, CEPs must include "a specification of methods to be used for examination of qualitative and quantitative aspects of clinical safety with clear reference to the determination of residual risks and side-effects" (MDR 2017/745, Annex XIV, Part A, 1) [1] and PEPs must "specify the... process and criteria applied to generate the necessary clinical evidence" (IVDR 2017/746, Annex XIII, Part A, 1) [2].

The CEP/PEP must describe how the three different data sources (i.e, clinical trials, literature, and experiences) will be identified, appraised, analyzed, and presented consistently in CE/PE records to reflect the rank order of statistical quality and importance: company-sponsored, subject-device clinical trials, then clinical literature and clinical experiences. Being clear about the appropriate, relevant weight for various data types is important in the plan and in the report.

2.8.11 Use a good clinical data identification strategy

The proposed identification process should begin by first identifying all clinical data relevant to the subject device (and/or the equivalent device, if any), then the competitor device/s and finally the broader generic devices and alternative therapies for the background knowledge and SOTA discussion. This process should help the evaluator understand the most relevant data about the subject/equivalent/competitor devices first and then the context of the background knowledge and the SOTA/alternative therapies.

BOX 2.9 Exercise: Identifying "appropriate" pacemaker clinical data

When identifying clinical data for a cardiac implantable pacemaker CER, first search for and identify all clinical data about the subject device using the brand name. If the brand name is too generic, then use the manufacturer name along with the generic device name. Next, search for the equivalent cardiac implantable pacemaker by brand name (or manufacturer + generic device name). Lastly search for similar pacemakers and alternative therapies. The most important clinical/performance data in the CER/PER will be the clinical/performance data about the subject device, so don't lose focus.

The CEP/PEP typically requires searching for clinical/performance data using a method like PICO. The "similar device" and "alternative therapy/device" searches for the "background/SOTA" in the CER/PER may include search terms focused on the patient population and disease, intervention, comparator groups, and outcomes when searching rigorously for all clinical data. Using this PICO search strategy, the device name (i.e., the intervention) may need to be combined with a patient and disease type (i.e., the population) to help limit the search results. For example, pacemaker and heart failure might be a good search strategy for a new pacemaker device designed solely to treat patients with heart failure.

In order to identify relevant data, specific search terms must be constructed. The best search terms may be simple (e.g., searching device name) or more complex (e.g., search similar devices types with target population and disease state). Identify the right search terms capturing the most specific results without being too broad (e.g., results in the thousands) or too specific (e.g., fewer than 10 results). Testing search terms to identify the best volume results is called "sandboxing" in this textbook. The goal of sandboxing is to identify *clinical* data relevant to the device and intended purpose.

A good clinical data identification process requires appropriate search terms explicitly selected to comprehensively, yet succinctly, cover (1) all S&P issues as well as all risks and benefits for the subject or equivalent device, and (2) all needed background knowledge and SOTA data including competitor devices and alternative therapies.

2.8.12 Use "sandbox" to select appropriate search terms

Database searching requires using only the best search terms. Search term design is an art. Even well-designed search terms considering appropriate and relevant industry terminology does not necessarily lead to reasonable results. For example, a well-considered search term string may result in 25,000 results; however, attempting to evaluate this many articles is simply unreasonable and a change in the search term/string is necessary to arrive at an appropriately sized dataset to appraise and analyze. The "sandbox" technique allows the evaluator to:

- test Boolean operator effectiveness when connecting two terms
  - "AND" limits data to include BOTH terms,
  - "OR" includes ALL data from both terms,
  - "NOT" excludes all data from the specified term,
  - "" (quotation marks) returns all data with the exact phrase,
  - () (parentheses) groups data together to control the order, and
  - * (asterisk) allows variation of the key word;
- test filter uses to identify most relevant data (e.g., filter by year, article type);
- test capitalization, hyphens and special characters in manufacturer and device names.
- determine best search terms generating most relevant results; and
- identify best search strings with most appropriate results.

Using the "sandbox" technique allows the evaluator to test possible search terms and combination strategies and then to review the quantitative and qualitative results. For example, note capitalization, hyphens, and special characters in manufacturer and device names. The evaluator ensurest a thorough search is conducted by "testing" searches with and without special characters to ensure all device or manufacturer data are identified.

Review all product literature (e.g., instructions for use, brochures, marketing material) to identify specific device and disease terms for potential use during these clinical data searches. Review background studies and consult with experts to ensure the device and device uses are well understood before starting to write the clinical data identification plan. Terms used in trials, literature, and experiences are likely to be highly variable and the clinical data search plan should identify the best terms to use while searching. The plan should be flexible to allow the addition of more or different search terms as the evaluator learns more about the device and the intended use during the CE/PE process. For example, a device may be defined as a "hydrodynamic" catheter in all device marketing materials; however, the evaluator may learn physicians refer to the same specific device using many different names (e.g., "rheolytic," "pressurized," or "thrombectomy" catheters). These terminology variations are important to understand when designing and trying out different search strategies.

The evaluator must adjust the search terms and combination strategies in an iterative manner until sufficient clinical data coverage and appropriate data volume are achieved. Informally, noting "sandbox" searches can help the evaluator to track search outputs to develop the right sequence of search terms, combination strategies and filters to identify the best results. The best search term combinations can then be listed in the CEP/PEP as the plan to be executed during the CE/PE clinical data identification stage.

2.8.13 Select the right trials

Review all company-sponsored, manufacturer-driven trials as well as all investigator-initiated and any other types of clinical trials using the device under evaluation and search clinicaltrials.gov as well as other international clinical trial databases to identify additional relevant clinical trials. Manufacturers may need to have a contract with the equivalent device manufacturer to allow technical file access (especially for Class III and implantable devices). In this identification step, any anecdotal trials can be added at any time and for any reason and prior discussions with regulatory authorities about the docuemnted trials may be important to document in the CEP/PEP.

BOX 2.10 Expert panel advice

The CEP/PEP should document clinical trial advice received from the NB or expert panel. EU MDR Article 61(2) [1] states:

For all class III devices and for the class IIb devices[*]..., the manufacturer may, prior to its clinical evaluation and/or investigation, consult an expert panel... with the aim of reviewing the manufacturer's intended clinical development strategy and proposals for clinical investigation. The manufacturer shall give due consideration to the views expressed by the expert panel. Such consideration shall be documented in the clinical evaluation report... The manufacturer may not invoke any rights to the views expressed by the expert panel with regard to any future conformity assessment procedure. *Class IIb devices include only those designed to deliver or remove a medicinal product per Article 54(1)(b)

If the manufacturer uses advice from an expert panel, the advice must be documented in the CER/PER; however, the expert panel views are not binding in any future conformity assessments.

Unlike the EU MDR, where "expert" is mentioned 122 times, the same "expert" term is mentioned only 71 times in the IVDR. The most conspicuous absence of this requirement for expert advice is when the IVDR fails to provide access to expert panel advice for the IVD device manufacturer. This expert panel service is not mentioned in the EU IVDR even though extensive discussion of consultation with experts and sharing of expertise is present for the NB, MDCG, Member States, and EU Commission.

Once all relevant and high-quality clinical trials have been identified, list the trial names in the CEP/PEP and gather all relevant clinical trial technical file details (e.g., protocol, informed consent form, statistical analysis plan, risk management plan, case report forms) so the trials can be appraised and analyzed during the CE/PE process.

2.8.14 Select the right literature

In this identification step, anecdotal literature can be added at any time and for any reason as a "manual" search, meaning the data were identified outside of any database search. Sometimes, the most valuable articles are those from historical knowledge. Once added during identification, these "manual" search articles will be treated the same as all the other literature. When writing the CEP/PEP, the evaluator should ensure the clinical literature are sufficient. An example "sandbox" strategy is presented below for a literature search in PubMed based on an artificial human heart device called "BiVACOR" from a company also called BiVACOR (Table 2.9).

TABLE 2.9 Example Sandbox searches using specific and generic search terms.

Search #	Search term/string	Results	Notes: search date March 16, 2023
1	Manual searching	5	Keep
	Subject device search		
2	BiVACOR	35	Keep; subject device and manufacturer
	Competitor device search		
3	CardioWest	73	Keep; Competitor device by SynCardia
4	Aeson	7	Keep; Competitor device by Carmat
	Generic searches		
5	heart failure	307,117	Too broad
6	"heart failure"	254,036	Too broad; did not reduce enough
7	Total artificial heart	7,446	Too broad, Pulling results for each individual word; not relevant
8	"total artificial heart"	1,302	Too broad
9	"heart failure" AND "total artificial heart"	443	Too broad
10	"heart failure" AND "total artificial heart"*filters: guideline, practice guideline*	1	Not about total artificial heart; No guidelines; may be because fairly new technology with low patient numbers implanted
11	"heart failure" AND "total artificial heart"*filters: metaanalysis, systematic review*	3	Keep and expand by adding additional filters to include more clinical trials
12	"heart failure" AND "total artificial heart"*filters: clinical study, clinical trial, comparative study, controlled clinical trial, evaluation study, multicenter study, observational study, pragmatic clinical trial, randomized controlled trial, validation study*	38	Keep; relevant for background/SOTA

CEP, clinical evaluation plan; *PEP*, performance evaluation plan. Italic text indicates filters used during literature searches.

The device name is searched first, then competitor device names and finally broad generic terms are searched to gather information for the background knowledge and SOTA section and to include alternative therapies.

PICO [45] was used to "sandbox" the generic searches. Specifically, the patient disease was specified as "heart failure" and the intervention was the "total artificial heart." The search filters were applied in sequential order using the highest value filter first and then adding more filters until sufficient data were identified for each of the following three types of articles:

1. metaanalyses, systematic review, review (note review is rarely, if ever, used);
2. practice guideline, guideline; and
3. randomized controlled trial, controlled clinical trial, multicenter study, pragmatic clinical trial, clinical trial, clinical study, observational study, comparative study, evaluation studies.

In this example, the device type (total artificial heart) was only manufactured by two other companies (SynCardia and Carmat) and the competitor searches were reasonable to review manually. Searches 1, 2, 3, 10, and 11 yielded 156 articles to be kept, so these will be downloaded from PubMed and appraised for inclusion or exclusion. The CEP can leave out details for searches 4–9 and a simple explanation can be added to the CEP for any additional clinical data sources to be used in the CER. The report can always house more data than was envisioned in the plan. For example, if the CEP/PEP was signed off and the evaluator identified and searched an additional database, the data from the additional database search could be included in the CER/PER with a simple justification and explanation about the added data.

BOX 2.11 Clinical data search time frame

CERs/PERs were assigned to be updated every 2 years, and the searches were conducted as follows:

CER version A was released in 2020 with searches conducted without time limits.
CER version B was released in 2022 with updated searches limited to 2020-2.

In this setting, the clinical data for Rev. B were only reviewed for the last 2 years; however, this logic is flawed because the data identification, appraisal, and analysis steps were intended to be comprehensive for all clinical/performance data relevant to the device over all time (if relevant).

Tshi is important because articles listed in the various databases may be added (i.e., sometime articles are added to the database even if published years earlier), commented on (rebutted), edited or removed (retracted) due to issues, complaints, or misinformation at any time. Choosing to limit the time frame may cause important articles or updates to be missed by assuming only newly published articles should be identified.

If possible, always search without limits, even in CER/PER revisions and updates. Historical literature is often critically important for the CER/PER background knowledge/SOTA section.

Here is a second, more generic, Mölnlycke gown literature search example for comparison. As in the first example, always begin with most relevant, specific search terms (e.g., manufacturer name and device name), since these are the most critical clinical data to gather for the CE/PE process. Using a manufacturer or device name alone or both connected with Boolean operator "AND" will generate different results so explore many combinations when "sandboxing" to get the right search terms. Use filters to explore narrowing results (e.g., try study methodology filter types) (Table 2.10).

TABLE 2.10 Specific and generic search term sandboxing.

Search type	Search term/string	Results	Comments
1. Specific	Mölnlycke	217	Too broad: includes all products for this company; not relevant to subject device product
1. Specific	Mölnlycke AND gown	1	KEEP, relevant to subject device product; product information refers to product generically as "gown"
1. Specific	Mölnlycke AND barrier	8	KEEP, relevant to subject device product; some providers may refer to gown as "barrier"
2. Generic	gown	956	Too broad
2. Generic	barrier	314,562	Too broad
2. Generic	gown AND barrier*filters: guideline, metaanalysis, practice guideline, systematic review*	31	KEEP: isolated high quality article methods to narrow scope; relevant
2. Generic	"surgical gown"	56	KEEP Used quotes to isolate specific phrases noted in other search result article titles; relevant

The evaluator began with "specific" target-device-name searches (specific search type 1) and noted the manufacturer term alone generated results about the many different product types and devices irrelevant to the target gown. Adding in the requirement for the search to return only manufacturer gown or barrier device types limited the data results to be both more relevant and a more manageable size. The evaluator then used generic, nonspecific search terms (generic search type 2) to identify relevant, high-quality articles for background/SOTA/alternative therapies. Searches for all articles including the word "gown" or "barrier" were too broad (i.e., results were in the hundreds and thousands). These generic search results needed to be narrowed in order to conduct a literature appraisal and analysis effectively. Addtional branded competitor devices can be added to this search strategy.

No specific article number is required for the CE/PE process; however, the regulations require the clinical data must be "sufficient" to demonstrate conformity with the GSPR (e.g., device safety, performance, benefit-risk ratio). Searches which result in a few hundred articles for appraisal and analysis are appropriate. In some cases, even with high-quality study-type filters applied, thousands of articles may be identified. In this case, combine device names with terms about the intended use (e.g., "surgical gown"), disease state, preferred comparator or outcome measurement to deliver a more reasonable result (i.e., use PICO—population, intervention, comparator, outcome—to help focus results).

After sandboxing for this example, four searches were retained, including two searches relevant to the subject device and two searches relevant to background/SOTA (Table 2.11).

TABLE 2.11 Final searches after sandboxing

Search		Search Term/String	Results	Comments
~~1. Specific*~~		~~Mölnlycke~~	~~217~~	~~All products for this brand; not relevant to subject device product~~
1. Specific		Mölnlycke AND gown	1	Relevant to subject device product; product material refers to product generically as 'gown'
1. Specific		Mölnlycke AND barrier	8	Relevant to subject device product; some providers may refer to gown as 'barrier'
~~2. Generic~~		~~gown~~	~~956~~	~~Too broad~~
~~2. Generic~~		~~barrier~~	~~314,562~~	~~Too broad~~
2. Generic		gown *Filters: Guideline, Meta-Analysis, Practice Guideline, Systematic Review*	31	Try isolating by article methodological quality to narrow scope; relevant
2. Generic		"surgical gown"	56	Try quotes to isolate specific phrases noted in other search result article titles; relevant

*Crossed-out items were omitted and the others were retained in the final search strategy.

The general article information (e.g., title, authors, year published) should be downloaded and stored in an Excel workbook or custom software solution.

2.8.15 Select the right experiences

Experience data volume may be higher than other data sources; however, the data quality will be poor. All anecdotal experiences should be included and they will be appraised and analyzed later in the CE/PE process. Use the sandboxing technique to determine the precise searches to list in the CEP/PEP for the CE/PE process. Typically, using the product name in all searches will be sufficient. The challenge with using real-world data (RWD) like complaints and issues database data to generate real-world evidence (RWE) in the CER/PER is related to the poor experience data quality and lmited reliability [46]. Experience data are not typically used in the CER/PER background and SOTA section; however, subject device experience data are required and must be identified. If the subject device experience data are insufficient, additional searches will be required (i.e., search for the equivalent device, competitor devices or similar generic product types but only if needed, since these RWD have limited relevance and extremely poor data quality).

Experience data are particularly important to monitor for any safety signal found solely in experience data outside of any clinical trial and clinical literature report data. These anecdotal experience data should not be used to set the S&P threshold values in the CER/PER background and SOTA section, since neither the individual data reports nor the number of reports in the database are reliable report volume indicators. These complaints and database reports are anecdotal and subjective. The complaint author may be a healthcare professional or lay person, and the report may be mandatory or voluntary. As such, the reported issue volumes in these databases do not represent the incidence (i.e., the occurrence rate or frequency) or prevalence (i.e., the percentage affected in a population) of an issue in the population. To gather this type of important volume information, the data need to be better controlled than a random sample from an experience data repository (i.e., registry, issues/recall database, product complaint listing).

BOX 2.12 Experience data statistical analyses may be unfounded

Experience data are not generally used for statistical analyses because the report numbers are uncontrolled. For example, a typical CEP/PEP will recommend depicting the experience data using a "complaint rate" ratio to approximate the overall complaints over the "sales" estimated number used as follows:

Complaint Rate $= \frac{x}{y}$ **[complaint number (*x*) over units sold number(*y*)]**

Unfortunately, the numerator (i.e., complaints) underreports the actual complaints from all patients since many patients did not report their complaints into the complaint listing used and the denominator (units sold) overreports the actual units used since some units were sold but not used in any patient. As a result, these data do not provide reliable estimates and should not be overinterpreted to represent more than a simple unreliable estimate.

Experience data has a lower statistical quality then most clinical trial and clinical literature data Highest statistical quality is often found in metaanalyses and randomized, controlled clinical trial data, followed by clinical literature discussing other trials, and lastly the experience data, which generally lack any statistical relevance.

The CEP/PEP should spell out the experience data analysis plan to include sorting and grouping experience reports by patient safety issue and separately by device deficiency performance issue. The plan needs to allow each report to be accounted for separately in each of those two analyses. Each analysis (patient safety and device performance) should be comprehensive for the entire experience dataset.

2.8.16 Define the appraisal process

When sandboxing clinical data, remember to develop the plan for appraising and analyzing these data in the CEP/PEP. One appraisal strategy will be to sort all identified experience data by relevance:

- high priority and high weight for subject device reports;
- middle priority/weight for equivalent device reports; or
- lowest priority/weight for similar or generic device reports.

When completing the appraisal process, the relative weight becomes clearer as one learns from the included data. The CEP/PEP must be flexible enough to allow this learning from the data and to allow refinements to the inclusion and exclusion rationales in real time during the appraisal process.

The first clinical data appraisal step is to understand the CEP/PEP-defined inclusion and exclusion (I&E) criteria. These I&E criteria are expected to change over time as the evaluator learns more about the clinical data, so do not fall for the trap of simply using the prior CE/PE appraisal criteria from the last CER/PER. In addition, document in the CER/PER "exactly" how the clinical/performance data from each clinical data source were actually appraised during the CE/PE process.

2.8.17 Explain plans to develop related clinical evaluation/performance evaluation documents

The CEP/PEP needs to document the plan to develop the other related CE/PE documents, including not only the CER/PER, but also the PMS activities in the PMSR/PSUR as well as the PMCFP/PMPFP, the design records showing changes over time, risk management files showing newly emerged S&P concerns, and the SSCP/SSP. The QMS must detail how various documents interact and should support the CEP/PEP writing to ensure theinterrelated document systems are integrated and functioning, with appropriate checks and balances.

2.8.18 Specify postmarket surveillance activities to cover clinical data gaps

The CEP/PEP must define how the PMSR/PSUR will be reviewed and analyzed during the CE/PE process. The CEP/PEP also needs to specify how the CE/PE process will resolve old and identify new clinical data gaps. The PMCFP/PMPFP process description should explain clearly how new clinical data collection will resolve the identified clinical data gaps from the previous CER/PER. Specifically, the CEP/PEP must define (at a high level) how the CE/PE process will identify, appraise, and analyze available clinical data, and set out a plan to identify missing or insufficient clinical/performance data in the CER/PER. In other words, the CEP/PEP must explain how to determine if "sufficient" clinical data are present to complete the evaluation or not. If clinical data are missing or insufficient, the CEP/PEP must specify how the CER/PER will document the clinical data gaps and insufficiencies and how the PMCFP/PMPFP will enable the collection of the required new clinical data to address these clinical data gaps and insufficiencies.

BOX 2.13 Address PMCFP/PMPFP needs in the CEP/PEP

The CEP/PEP defines the evaluation plan and should address poor planning issues uncovered during past NB inspections. For example, the NB may have listed "deficiencies" preventing the CE mark process from moving forward previously and all of these deficienciesshould be reviewed prior to each CEP/PEP revision.

Many deficiencies from different companies were analyzed during the writing of this textbook and some were related to deficiencies in appropriate planning to collect additional data after the CE/PE process was completed. Here are just a few example NB deficiencies and the resulting actions which needed to be accounted for in the next CEP/PEP revision:

1. Not all "indication for use" aspects had clinical data evaluated in the CER/PER (e.g., insufficient CER/PER clinical data were evaluated to explore all S&P criteria for all indicated patient ages, device sizes, disease states, severity ratings, safety signals). Result: manufacturer changed all labeling to restrict the indication for use to details supported by evaluated and sufficient clinical data (e.g., labeling was updated to include terms like "to be used in adults over 18 years of age" or "for use in mild to moderate" disease state and this new restricted use population must be considered in the next CEP/PEP).
2. A safety signal was reported but was not present in device labeling or was not yet sufficiently evaluated in the CER/PER to determine device causality or relationship. Result: manufacturer updated all labeling to include the missing AE details and collected additional clinical data as required in the PMCFP/PMPFP. The new clinical data were, reported in the PMCFER/PMPFER and these new, clinically-relevant, data must be considered in the next CEP/PEP.
3. A performance issue was identified but was not fully evaluated with clinical data. Result: additional bench testing and clinical data collection were added to the PMCFP/PMPFP, reported in the PMCFER/PMPFER and these new, clinically-relevant These changesmust be considered in the next CEp/PEP, data must be considered in the next CEP/PEP.
4. PMCFER/PMPFER were not completed on time or did not capture clinical data identified as clinical data gaps in the prior CER/PER. Result: new timeline and commitment were negotiated with NB and CE mark and market launch were delayed. These new timelines and commitments must be considered in the next CEP/PEP.

The sufficient clinical data determination is subjective by nature and depends on the device in question. This determination should not be left to a novice or an evaluator-in-training. The CEP/PEP needs to be poised for success. The CEP/PEP must clearly and completely document how the clinical data generated by the PMCPF/PMPFP and reported in the PMCFER/PMPFER will be evaluated rigorously with all the other clinical trial, literature and experience data to determine if sufficient clinical data will be present in the next CER/PER.

The CEP/PEP requires early and comprehensive planning including PMCFP/PMPFP details to address previously identified clinical data gaps. Good CEP/PEP writing requires an expert ability to describe how postmarket clinical data collection was and will be performed in an ongoing fashion. This is especially important to address the clinical data gaps uncovered during the CE/PE process. Sometimes the clinical data are insufficient but the benefit-risk ratio is reasonable to allow awarding of a CE mark, as long as the manufacturer makes a written commitment to gather specific additional clinical data in the postmarket setting. The next CE/PE must ensure the past commitments were met or the manufacturer must justify the postmarket data collection changes in the agreed plans. In particular, the PMCFP/PMPFP is meant to document the specific clinical data to be collected for the purpose of covering the clinical data gaps identified in the CER/PER and the CEP/PEP is meant to review the changes since the last CE/PE.

2.8.19 Document device changes over time

The CE/PE process should describe the past clinically relevant design changes over the device lifetime. Capturing this subject device background knowledge in the CEP/PEP is important so the CER/PER will be comprehensive and have the appropriate scope to explain the device developmental context over time. This is a critical point since the clinical data identified, appraised, and analyzed during the CE/PE process may come from many different points in time. Telling the device story in the CEP/PEP requires a clear understanding about how and why the device changed over time. The CEP/PEP should clearly state why some clinical data will or will not be considered relevant to the subject device for the CE/PE process. Remember to stay focused on the "clinically relevant" changes. Other changes will be recorded in the technical documentation and may not be relevant for the CE/PE discussed in the CER/PER (e.g., a change in internal device materials with no patient contact and no impact on device S&P or benefit-risk ratio does not need to be highlighted in the CER/PER). All technical documentation will be reviewed by NB engineering experts; however, the CER/PER will be reviewed by NB clinical experts and is just one small part of the technical documentation so technical details are not needed in the CER/PER unless they have a clinical impact.

2.8.20 Document risk management including newly emerged safety and performance concerns

Similar to understanding the device changes over time, the CEP/PEP should also document all newly emerged clinical S&P concerns and how these concerns will be evaluated. The QMS should define the review processes for the risk management files prior to writing the CEP/PEP, and the CEP/PEP should only document the clinically relevant findings or concerns, if any.

2.8.21 Specify summary of safety and clinical performance/summary of safety and performance template development, if required

For Class III and implantable devices and Class C and D *in vitro* diagnostic devices, the CEP/PEP specify how the SSCP/SSP will be developed after the CER/PER and PMCFP/PMPFP are completed. The CEP/PEP should document the cross-linked information, including where to find the related PMCFP/PMPFP and SSCP/SSP. Furthermore, the CEP/PEP should specify where the SSCP/SSP will be released into the public domain and the appropriate revision control for each version of the related documents.

2.8.22 Write clinical evaluation plan/performance evaluation plan conclusions

The CEP/PEP conclusions must be based on the plan contents. Full consideration must be given to the objective, scope, S&P acceptance criteria, and device description details including the device class and benefit-risk ratio. The plan conclusions must consider if any equivalent device was appropriately justified in the plan and if the device developmental context and evaluation process were expected to generate enough relevant, high-quality data to allow CE marking. The conclusion should specifically state any concerns about potential clinical data gaps and the need for postmarket clinical data collection to address the documented clinical/performance data gaps. The conclusion should also review clearly how all the integrated documents will be monitored based on the plan.

2.9 Clinical evaluation/performance evaluation planning by clinical evaluation/ performance evaluation type

CE/PE planning will differ by CE/PE type. The CE/PE type will suggest specific needs and specific processes to be used during the evaluation. Not all required clinical/performance data will be available for each CER/PER type, and the CER/PER success often depends on how well the CEP/PEP planned the details to address the insufficient clinical/ performance data in the PMCFP/PMPFP and how well the CER/PER was written previously to convey all the details in a clear and compelling story about the benefit-risk ratio and the device S&P.

2.9.1 Viability clinical evaluation plan/performance evaluation plan

The viability evaluation plan relies heavily on the vPMCFP/vPMPFP because no human clinical data were available in this early device life cycle stage. The vCE/vPE is often based on bench and animal data with a justification to include similar clinical data whenever available. When no equivalent device is possible, this stage will progress until the first in human experience has occurred. The vCEP/vPEP needs to make this clear to the reader.

2.9.2 Key viability clinical evaluation plan/performance evaluation plan contents

- Rationale explaining why the proposed device should be considered safe enough to perform as indicated in proposed human population (not to be confused with design control ideation inputs and outputs).
- Nonclinical data and similar device clinical data evaluation (not to be confused with device verification and validation activities).
- Plan to collect clinical data about S&P during device use in humans (not to be confused with the PMS system).
- Plan to capture and weigh benefits and risks during device use (not to be confused with the RM system).
- FOR IVD devices: clearly state the methods to establish scientific validity, analytical performance and clinical performance for the IVD device (not to be confused with other technical requirements).
- Plan for evolution to eqCE/eqPE as an equivalent device is established to allow clinical data are collected for the device under evaluation.

2.9.3 Equivalence clinical evaluation plan/performance evaluation plan

Like the viability evaluation plan, the equivalent device evaluation plan has little to no human clinical data derived from subject device use, but instead uses clinical data from the equivalent device to represent the subject device clinical data. The equivalent device evaluation plan should require an equivalence table with full clinical, biological, and technical equivalence documentation in the eqCER/eqPER. The plan to complete the equivalence table must describe the need to determine the clinical impact for each specific difference noted in the table. The plan must also specify the comparison between subject and equivalent devices must not show any clinically significant differences in the S&P between devices.

Many manufacturers attempt to complete the equivalence table to show no differences between the subject and equivalent devices, but this is a mistake. The art here is to be truthful and not misleading about key clinical, biological, and technical differences between the subject and equivalent devices. All clinically significant differences should be listed and the impact on clinical S&P should be described for each clinically significant difference. The eqCEP/eqPEP can state the equivalence evaluation will be done in the eqCER/eqPER. Unlike devices in the US where nearly 80% of devices are qualified by a "substantial equivalence" route, the EU regulatory path using equivalence may be used in far fewer devices (maybe 20%). The differences between these two regulatory paths are complex.

If the clinically significant differences have no or only a minimal impact on device-related clinical S&P, then a justification should be written about why the differences will not negatively affect the benefit-risk ratio and the difference can be deemed acceptable. For any clinically significant difference with a negative impact on the benefit-risk ratio, the decision to use the equivalent device should be reconsidered.

2.9.4 Key equivalence clinical evaluation plan/performance evaluation plan contents

- Device descriptions *side-by-side* subject and equivalent device descriptions
- Indications table *side-by-side* subject and equivalent device indications for use
  - must show they are used for same clinical condition/purpose
  - use direct quotes to document "identical" clinical application (e.g., clinical condition, disease severity, intended user, use environment)
- Evaluation rationale equivalent device clinical data used for subject device CE/PE
  - Both devices have same indication for use
  - Clinical data must demonstrate CE-marked equivalent device was safe enough to perform as indicated in proposed human population
  - Not to be confused with design control ideation inputs and outputs
  - Contract may be required with different manufacturer to ensure access to equivalent device technical information (especially Class III and implantable devices or Class C and D IVD devices).
- Equivalence table with all clinically-relevant differences documented and all clinical impacts of those clinically-relevant differences described including determination if clinical impact will negatively affect the benefit-risk ratio or not.
  - If the benefit-risk ratio is negatively impacted by any individual, clinically-significant difference, the devices may not be equivalent for CE/PE
  - High-risk, implantable devices typically require human clinical data, and the CE/PE equivalence path may not be allowed

- Nonclinical data showing head-to-head testing between subject and equivalent device
  - Bench tests should focus on key, clinically-relevant device S&P issues
  - Animal testing should evaluate key, clinically-relevant device S&P issues (*in vivo* animal test data may be required for medium-risk devices, but may not be required for low-risk devices)
- Plan to collect clinical data about S&P during device use in humans (not to be confused with PMS system)
- Plan to capture and weigh benefits and risks during device use in future (not to be confused with RM system)
- Plan for IVD devices must define scientific validity, analytical performance and clinical performance for subject and equivalent IVD devices (not to be confused with other technical requirements)
  - Equivalent device must have sufficient data to demonstrate well-established and scientifically valid association between analyte and clinical condition
  - Equivalent device must have sufficient clinical data to define clinical performance
  - Equivalent device must also demonstrate analytical performance
  - Equivalent device gap assessments from prior PERs should be clearly addressed to allow continued CE mark
  - Plan for evolution to fihCE/fihPE as the initial clinical data are collected

2.9.5 First-in-human clinical evaluation plan/performance evaluation plan

Clinical data should be collected from the initial human subject device uses. The fihCEP/fihPEP will need to identify and include those data to show the actual subject device S&P when the device is actually used as indicated in humans. At this point in the medical device life cycle, the clinical evaluation data source is changing from nonhuman or equivalent device data to human clinical data from individual human subject device uses as indicated.

The fihCER/fihPER is not likely to have sufficient human clinical data from subject device uses to demonstrate conformity to the specified GSPR, so the fihCEP/fihPEP needs to explain how the first in human clinical data using the subject device will be supplemented with other data (e.g., simliar to the viability and equivalence device clinical evaluations).

Sufficient clinical data are required for the clinical evaluation and the clinical data sources as well as the clinical data analyses need to make sense and be appropriate. The fihCEP/fihPEP has a specific requirement to merge disparate clinical data sources in order to determine if the subject device is safe enough and performs well enough to be on the EU market. In the right situations, even clinical data from a similar device may be helpful to allow time for the additional data specified in the PMCFP/PMPFP to be collected and analyzed in order to address all identified clinical data gaps.

2.9.6 Key first-in-human clinical evaluation plan/performance evaluation plan contents

- Plan to collect new human clinical data as completely and quickly as possible
- Plan to merge new human clinical data with data used in prior CE/PE
- Plan to reduce reliance on nonclinical and similar device clinical data for CE/PE
- Plan to update the benefit-risk ratio as clinical data are collected
- Integrate new human clinical use S&P data into PMS and RM systems
  - Update plan to gather all additional required clinical data in PMCFP/PMPFP
- Plan for IVD devices must update initial scientific validity, clinical and analytical performance data with a "sufficient" clinical data amount
- Plan for evolution to trCE/trPE as clinical data accumulates

2.9.7 Traditional clinical evaluation plan/performance evaluation plan

The traditional CE/PE often has a significant amount of human clinical data from subject device uses; however, the collected clinical data may not meet the analytical needs to address subject device S&P as well as the benefit-risk ratio.

Often a prior trCEP/trPEP may be available and the routine trCEP/trPEP update process can be managed by simply describing how the clinical data must be updated (i.e., new search strategies, new trials, literature, and experience data identified, appraised, analyzed, and reported, new clinical data included in the prior CER/PER, PMCFER/PMPFER and PMSR/PSUR, etc.).

Alternatively, the trCEP/trPEP may need to be updated "for cause" (e.g., after a new life-threatening risk was identified or afer the labeling was changed) before the regularly scheduled update. This "for cause" trCEP/trPEP update requires a timely and responsive plan and process to begin the next CE/PE process sooner than scheduled in prior plans.

BOX 2.14 Case study—ever-changing CEP during CER process is unwise

Case study situation

A large company marketing both drugs and devices in the EU had a CE team working on a CER/PMCFER at the same time as the CEP and PMCFP were being developed. During the ongoing CE development, the PMCF study started as suggested by the clinical data gaps uncovered during the ongoing CE work. The PMCF study identified a new, life-threatening risk for the device. This new risk was addressed by changing the labeling including the indication for use and retraining all physicians using the devices. The company issued the new labeling before any CE documents were issued. In this situation, the CE team had spent more than 9 months working on the clinical data evaluation under the original planned labeling and indication for use. The timeline was constantly expanding and changing due to the new clinical data and remediation activities being shared with the CE/PE team related to this new life-threatening device-related risk. The company required the CE team to absorb this new info into the CE process and to reconsider all clinical data given the new labeling and the new indication for use.

Case study issues identified

This case illustrates a mismanaged CE process with poor documentation practices and a complete lack of appropriate leadership and oversight. The CE team did not understand: (1) how to collect and contain the clinical data to be used in the CE process; (2) when to evaluate the planned clinical data; (3) how to follow a finite CE timeline based on a set CEP; or (4) how to complete the CE according to the CEP as specific gated processes within a specific and relevant time frame for each output document. This was a complete, multisystem-wide QMS failure related to the PMS, RM, and CE processes.

Case study outcomes

Frustrated team members, financial losses, and timeline delays were the outcomes. Misinformation, misdirection, and missing data were the norm during the CE process when these new clinical data bits should have been irrelevant to the CER work being completed. Specifically, the CER deadlines were extended from 3 months to multiple years, the costs were much greater than planned, and all aspects of the CE process were out of control.

Five better case study solutions suggested

First, the QMS must design functional QC processes. Although everyone was focused on hurrying up to protect patients exposed to this device, the issue escalation path was misdirected to the CE team before a clear plan was in place to deal with the newly identified risk by the appropriate PMS and RM teams. This company would be better served by a more appropriate clinical data escalation path to the team or teams designated to handle newly identified serious, unexpected clinical risks related to the device. The PMS team should have handled the initial AE report documentation and then escalated the issue to the RM team so a more appropriately controlled response plan could be initiated for risk mitigation outside the CE team, which was working on a different project, the CER. The CER evaluates past clinical data collections and past remediations within a well-controlled and predefined process. *Note: The PMS and RM processes should probably have required the device to be pulled from the market (at least temporarily during the redesign stage) when the benefit-risk ratio was found to be "not acceptable" and the device was not considered "safe" to perform as intended. Gathering more clinical data in this situation might be deemed "clinically inappropriate" due to the extreme, life-threatening patient risk and forcing changes to the CER in this setting might be considered impertinent since bigger issues needed to be addressed first.*

Second, this company would have been better served by the CE team if the QMS had required the CEP to be completed with the specific CE scope written down and agreed to before starting the CER. For example, the CEP should have clearly stated the specific clinical dataset to be evaluated with exact dates for clinical data collection. For example, the CEP could have stated: "The CER will present the clinical evaluation results for clinical data collected from <date device was first placed on the market> through <end date aligned for clinical data collection from all three interrelated but independent systems, PMS, RM, and CE, here>." If this had been documented in the signed and dated CEP before the CER work started, the new clinical data would have been easily excluded, the CER would have been completed on time, and the team would have easily explained how the NEW clinical data would be included in the next CER update (in response to the newly identified life-threatening risk and perhaps concurrently, with a new trCEP and trCER scope ahead of the routine trCER update frequency, as required in the regulations). This is important because the new trCEP scope was quite different from the past trCEP.

The new trCEP should be planned to align with the ability to evaluate changes evident in the clinical data collected during and after the time when this event occurred. The new trCEP should be defined carefully, with an adequate scope to ensure sufficient clinical data would be included to evaluate the changed device given the new benefit-risk ratio. The PMS system should have been urgently and immediately directed to add a specific clinical data collection category relevant to this new life-threatening risk so the experience data in this category could be evaluated in future trCERs. The RM system should have been urgently and immediately directed to add a specific risk (or series of risks) to be mitigated and accounted for in the ISO 14971 compliant RM process. The new occurrence and severity rates as well as the success of the risk mitigation strategies included in the RMR would need to be considered in future trCERs provided the device was able to remain on the market in conformity with all the required regulations, standards, and company policies.

(Continued)

BOX 2.14 (Continued)

Third, the company QMS should require the CEP to document in writing the specific labeling (i.e., with the specific, required indication for use) to be evaluated in the CER. A best practice is to attach a copy of the current IFU as an appendix to the CEP. This should explicitly avoid the problem depicted in this case study. The CEP needs to define clearly the device under evaluation including the specific labeling and indication for use and these details must be carefully considered during the clinical data evaluation in the CER. Another good practice is to tabulate the clinically relevant labeling changes that have occurred over time, so the clinical data can be aligned with these changes. For this case study, the new life-threatening device risk and the changes to the labeling would be an important item to include in this tabulation within the next CER.

Fourth, the company QMS would be further improved by setting up a required stage-gate process within the CE SOP or WI. For example, the CEP stage must result in a CEP with an appropriate end date for clinical data collection, the CEP gate would not allow the CER to start until the CEP was signed off and the CE process overall would not allow clinical data collection beyond the end date specified in the CEP for clinical data collection. In addition, the PMCF stage would require a completed signed-off CER and then a signed-off PMCFP before beginning data collection in any PMCF study, and the clinical data collected in the PMCF study would need to be fully evaluated in the PMCFER prior to review in the next CER. The PMCF stage gates would include both requirements for both the PMCFP to be signed off before the PMCF study could begin and the PMCFER to be written before the next CER would evaluate the new PMCFER clinical data.

Fifth, the QMS should have a detailed and precise process to be followed for each initiated CER and before any CER update is initiated "for cause" like the change in the device benefit-risk ratio seen in this case study.

In other words, effective trCEP/trPEP management requires the evaluator to understand the CE/PE scope and the developmental context in order to analyze all available subject-device clinical data according to the plan.

Note: The CER/PER planning and writing processes are not meant to be continuously ongoing like the data gathering processes must be (behind the scenes) for the other ongoing RM and PMS processes. The CE/PE scope must be set (i.e., documented in writing and not changing) so the version-controlled CER/PER can be completed according to the plan. An ability to generae a new CE/PE cycle in reaction to unforeseen situations is also required.

2.9.8 Key traditional clinical evaluation plan/performance evaluation plan contents

- Define how human clinical data are evolving (i.e., highlight changes over time)
- If legacy device, explain clinical data available.
  - Describe gaps in previous clinical data collection, if any
  - Describe required changes to collect clinical data, as required
  - Define rationale about how the available clinical data allow conformity to the GSPR to be unequivocally established
- Plan to merge new human clinical data with data used in prior CE/PE
- Explain how clinical data are sufficient to address all S&P questions or not
  - If clinical data are sufficient, explain why no additional clinical data are needed under MDR/IVDR
- Plan to update the benefit-risk ratio as clinical data are collected
- Integrate new human clinical use S&P data into PMS and RM systems
  - Update plan to gather all additional required clinical data in PMCFP/PMPFP
- Plan for IVD devices must update initial scientific validity, clinical and analytical performance data with a "sufficient" clinical data amount
  - Plan for evolution to oCE/oPE as device life cycle ends

2.9.9 Obsoleted clinical evaluation plan/performance evaluation plan

The obsoleted or (o)CEP/(o)PEP is meant to cover obsoleted medical devices at the end of the medical device life cycle. Some medical devices stay viable for hundreds of years (e.g., surgical scissors, clamps, sutures) while others may be obsoleted after less than 5 years on the market. Research suggests most medical devices are replaced within 18–24 months [47], and some are recalled due to serious health risks (e.g., the Medtronic HeartWare ventricular Assist Device HVAD system was obsoleted on June 3, 2021 after launching on October 23, 2017, with sales only through April 30, 2020 for about 2.5 years [48,49]). During the obsolescence phase, understanding the benefit-risk ratio and all the available clinical data will help the evaluator to develop the oCEP/oPEP from the previous trCEP/trPEP when the product was still on the market.

The oCEP/oPEP was increasing in importance during a time of increased transparency under the MDR/IVDR and other regulatory changes worldwide. The ability to share information via the Internet and the regulatory changes will eventually align to require clear, expanded information for the user when a device is removed from the market or obsoleted. For example, users with an implanted device will still want the latest and best clinical data available to guide their healthcare decisions about what to do in the case of an obsoleted Class III implanted device.

2.9.10 Key obsoleted clinical evaluation plan/performance evaluation plan contents

- Define why device was obsoleted.
  - If due to a recall, be clear about all clinically relevant issues
  - Specify S&P details
  - State the benefit-risk ratio
  - Specify if users must do anything (e.g., return obsoleted device, have implanted device removed/replaced)
- Define remaining user status (how many, when will the last use be documented?)
- Explain how remaining users should interact with obsoleted device
- Define when CE/PE updates will end
  - Describe gaps in clinical data, if any
  - Describe how clinical data may or may not continue to be collected
- Specify if any alternative is available, such as:
  - medical device with similar or better S&P and benefit-risk ratio
  - IVD device with similar or better scientific validity, clinical, and analytical performance

2.10 Check, sign, and date clinical evaluation plan/performance evaluation plan before clinical evaluation/performance evaluation begins

The CE/PE strategy is best articulated in a single, stand-alone, comprehensive, well-written, statistically sound, signed, dated, and version-controlled CEP/PEP. Much like a protocol for a clinical trial, the CEP/PEP spells out the scope of work for the CER/PER and provides the steps to be followed so the work can be completed as planned.

CE/PE success starts with a comprehensive yet flexible CEP/PEP to accommodate the clinical data accumulating over time. From the device idea to the end of device life during obsolescence, planning for clinical data evaluation is important and needs to change as the clinical data mature. Understanding clinical data qualities and differences across the medical device life cycle should help the evaluator understand how to write each CER/PER type. The plan should not be in flux during the evaluation period; instead, the plan should be clearly stated and cleanly executed within a reasonable time frame (e.g., 12 weeks) to allow for the next plan update and subsequent CE/PE process to occur when needed.

Every CE/PE assessment or quality audit (whether internal by the manufacturer or external by the NB) should find a clear roadmap before the CE/PE process began (i.e., the CEP/PEP) and a clean report about what happened as the CE/PE process was completed (i.e., the CER/PER). Obviously, changes may occur at any and all time points, so the CEP/PEP must be flexible enough to accommodate the changes and the CER/PER must document all the clinical data evaluation rationales with sufficient clinical data detail to demonstrate how the device S&P and benefit-risk ratio were acceptable as required under the regulations (i.e., MDR/IVDR).

To save time and ensure quality, the CE/PE writing group manager should check (QC) the following during the CEP/PEP writing process:

- CEP/PEP reads well as a whole document
- device description is clear and complete to define CE/PE scope
- clinical trial, literature, and experience search details are executable
- clinical trial, literature, and experience appraisal (i.e., coding) and analysis details are specific
- all CEP/PEP related "parking lot" issues are completed before final sign-off

The CE/PE writing group manager should review the CEP/PEP draft as each section is completed and this review should be flexible since sections are often written concurrently.

The completed CEP/PEP should be fully executed (i.e., signed and dated by all required parties) before moving on to conduct the evaluation. Not only is the CEP/PEP signature and date an MDR/IVDR requirement; but, ensuring this stage gate process is followed is an efficiency benefit during CER/PER development. A solid plan will explain how to gather CER/PER clinical data, how to group, appraise, weigh and analyze the clinical data, and should keep the entire CE/PE process on track.

2.11 Conclusions

This chapter reviewed CEP/PEP steps, starting by ensuring the product is a medical device and following the regulatory requirements when drafting and developing the CEP/PEP. Strategies and process controls included ideas about monitoring, auditing, and control functions during CEP/PEP development to improve quality and to ingrate clinical data into the CT, PMS, and RM systems. Having a 12- to 16-week timeline using 10 stage gates for the CE/PE process was discussed. Starting with the KOM (kick-off meeting), CE/PE team members should contribute to the timely completion of each process step. Using templates with a well-organized table of contents and details about how to state the objective and scope as well as the S&P acceptance criteria should help to get the CEP/PEP started.

The device description included many detailed subsections to define the device groups and families, any equivalent device claim, and the developmental context for the subject device, as well as a plan for the background and SOTA section and the CE/PE plan.

A key concept was discussed regarding the need to test search terms using a "sandboxing technique" before finalizing the search strategy in the CEP/PEP. The sandboxing method (trying various search terms) was recommended to solidify the best database search string strategies to identify appropriate, high-quality clinical data in many different databases. One goal was to leave some flexibility while specifying the specific device, competitor device, and generic background/SOTA terms required to identify relevant and high-quality clinical trial, literature, and experience data. The amount of clinical data needed to have "sufficient" clinical data may not be entirely clear in the planning stage; therefore, the plan should identify and collect as much relevant, high-quality clinical data as possible to ensure sufficient clinical data will be evaluated.

Most NBs require clinical data searches to be less than 6 months old during NB assessment to ensure current data have been reviewed, or the searches should be repeated before regulatory submission/review. A good practice is to align the CE/PE clinical trial, literature, and experience searches to be conducted on one date. In addition, searches should not be limited by time (e.g., only articles in last 1, 2, or 5 years is usually not appropriate) unless a clear justification is given (e.g., subject device only entered the market recently).

The plan needs to specify what happens if clinical data were not sufficient, how to list the clinical data gaps, and PMS activities including the PMCFP/PMPFP to collect new clinical data to address the documented clinical data gaps in the CER/PER. The plan should also specify how related PMSR/PSUR, design records, and RM documents will need to be updated to show any changes over time and any newly emerged S&P concerns. The plan should also specify if an SSCP/SSP will need to be written and shared publicly for certain high-risk devices.

The five CE/PE types were discussed and details were presented about the different planning concerns encountered for each type (e.g., from the lack of clinical data in the early viability and equivalent device stages all the way to device obsolescence with the need for continued patient support with clinical evidence until the last devices use by the last patient).

Clinical data quality must stay in focus throughout the CE/PE process. From initial planning to the final signed CER/PER, significant effort should be made to ensure a critical statistical quality review is carried out at all CE/PE process steps along the way.

2.12 Review questions

1. What do the following acronyms mean? CT, KOM, QMS, SOP, WI.
2. How would you determine if a product is a medical device in the EU?
3. What is a good CE/PE strategy?
4. What are the 10 stage gates and deliverables for each?
5. Which comes first: strategy meeting or kickoff meeting (KOM), and why?
6. What are three key things to audit to ensure a reliable CEP/PEP process?
7. Which stage gate involves clinical data inclusion and exclusion decisions?
8. How is the required clinical data amount determined for the CER/PER?
9. Why would a company need to gather new clinical data and how is this documented?
10. True or false: version control is required for draft documents.
11. True or false: clinical claims should be well-defined in the CEP/PEP.
12. How can 100 different Class I devices or Class A IVD devices be grouped for the CE/PE process?

13. Do NBs evaluate all medical devices for conformity to the GSPR?
14. What three equivalence types must be documented to claim an equivalent device?
15. What types of information belong in the developmental context?
16. How should a "sandbox" be used when writing a CEP/PEP?
17. What should happen when a new, life-threatening clinical risk is discovered?
18. Why do CT, PMS, RM, and CE/PE processes need to be integrated?
19. What QMS details are needed to control CEP/PEP processes?
20. What are unique plan details for each CEP/PEP type (v, eq, fih, tr, o)?

References

[1] European Parliament and Council of the European Union. Consolidated text: Regulation (EU) 2017/745 of the European Parliament and of the Council of 5 April 2017 on medical devices, amending Directive 2001/83/EC, Regulation (EC) No 178/2002 and Regulation (EC) No 1223/2009 and repealing Council Directives 90/385/EEC and 93/42/EEC. Accessed April 23, 2024. https://eur-lex.europa.eu/legal-content/EN/TXT/?uri = CELEX/3A02017R0745-0200424.

[2] European Parliament and Council of the European Union. Consolidated text: Regulation (EU) 2017/746 of the European Parliament and of the Council of 5 April 2017 on in vitro diagnostic medical devices and repealing Directive 98/79/EC and Commission Decision 2010/227/EU. Accessed April 23, 2024. https://eur-lex.europa.eu/legal-content/EN/TXT/?uri = CELEX/3A02017R0746-20220128.

[3] ISO 14155:2020. Clinical investigation of medical devices for human subjects—good clinical practice. 2020. Accessed April 23, 2024. https://www.iso.org/standard/71690.html.

[4] ISO 14971:2019. Medical devices—application of risk management to medical devices. 2019. Accessed April 23, 2024. https://www.iso.org/standard/72704.html.

[5] ISO/TR 20416:2020. Medical devices—post-market surveillance for manufacturers. 2020. Accessed April 23, 2024. https://www.iso.org/standard/67942.html#: ~ :text = This%20document%20provides%20guidance%20on%20the%20post-market%20surveillance,standards%2C%20in%20particular%20ISO%2013485%20and%20ISO%2014971.

[6] BS EN 13612:2002. Performance evaluation of in vitro diagnostic medical devices. 2002. Accessed April 23, 2024. https://knowledge.bsigroup.com/products/performance-evaluation-of-in-vitro-diagnostic-medical-devices/standard.

[7] ISO 20916:2019. In vitro diagnostic medical devices—clinical performance studies using specimens from human subjects—good study practice. 2019. Accessed April 23, 2024. https://www.iso.org/standard/69455.html.

[8] EN ISO17511:2020. In vitro diagnostic medical devices—requirements for establishing metrological traceability of values assigned to calibrators, trueness control materials and human samples. 2020. Accessed April 23, 2024. https://www.iso.org/standard/69984.html.

[9] GHTF/SG5/N6:2012. Clinical evidence for IVD medical devices—key definitions and concepts. 2012. Accessed April 23, 2024. https://www.imdrf.org/sites/default/files/docs/ghtf/final/sg5/technical-docs/ghtf-sg5-n6-2012-clinical-evidence-ivd-medical-devices-121102.pdf.

[10] GHTF/SG5/N7:2012. Clinical evidence for IVD medical devices—scientific validity determination and performance evaluation. 2012. Accessed April 23, 2024. https://www.imdrf.org/sites/default/files/docs/ghtf/final/sg5/technical-docs/ghtf-sg5-n7-2012-scientific-validity-determination-evaluation-121102.pdf.

[11] GHTF/SG5/N8:2012. Clinical evidence for IVD medical devices—clinical performance studies for in vitro diagnostic medical devices. 2012. Accessed April 23, 2024. https://www.imdrf.org/sites/default/files/docs/ghtf/final/sg5/technical-docs/ghtf-sg5-n8-2012-clinical-performance-studies-ivd-medical-devices-121102.pdf.

[12] GHTF/SG1/N68:2012. Essential principles of safety and performance of medical devices. 2012. Accessed April 23, 2024. https://www.imdrf.org/sites/default/files/docs/ghtf/archived/sg1/technical-docs/ghtf-sg1-n68-012-safety-performance-medical-devices-121102.pdf.

[13] GHTF/SG1/N078:2012. Principles of conformity assessment for medical devices. 2012. Accessed April 23, 2024. https://www.imdrf.org/sites/default/files/docs/ghtf/final/sg1/technical-docs/ghtf-sg1-n78-2012-conformity-assessment-medical-devices-121102.pdf.

[14] GHTF/SG1/N046:2008. Principles of conformity assessment for in vitro diagnostic (IVD) medical devices. 2008. Accessed April 23, 2024. https://www.imdrf.org/sites/default/files/docs/ghtf/final/sg1/procedural-docs/ghtf-sg1-n046-2008-principles-of-ca-for-ivd-medical-devices-080731.pdf.

[15] GHTF/SG1/N063:2011. Summary technical documentation (STED) for demonstrating conformity to the essential principles of safety and performance of in vitro diagnostic medical devices. 2011. Accessed April 23, 2024. https://www.imdrf.org/sites/default/files/docs/ghtf/archived/sg1/technical-docs/ghtf-sg1-n063-2011-summary-technical-documentation-ivd-safety-conformity-110317.pdf.

[16] ISO/AWI 18969. Clinical evaluation of medical devices. Accessed April 23, 2024. https://www.iso.org/standard/85514.html.

[17] MDCG 2020-7. Guidance on PMCF plan template. April 2020. Accessed April 23, 2024. https://health.ec.europa.eu/system/files/2020-09/md_mdcg_2020_7_guidance_pmcf_plan_template_en_0.pdf.

[18] MDCG 2020-8. Guidance on PMCF evaluation report template. April 2020. Accessed April 23, 2024. https://health.ec.europa.eu/system/files/2020-09/md_mdcg_2020_8_guidance_pmcf_evaluation_report_en_0.pdf.

[19] European Commission. Guidelines on medical devices; MEDDEV 2.7/1 Rev 4. June 2016. Clinical evaluation: a guide for manufacturers and notified bodies under directives 93/42/EEC and 90/385/EEC. Accessed April 23, 2024. https://www.medical-device-regulation.eu/wp-content/uploads/2019/05/2_7_1_rev4_en.pdf.

[20] MDCG 2020-1. Guidance on clinical evaluation (MDR)/performance evaluation (IVDR) of medical device software. March 2020. Accessed April 23, 2024. https://health.ec.europa.eu/system/files/2020-09/md_mdcg_2020_1_guidance_clinic_eva_md_software_en_0.pdf.

[21] MDCG 2022-2. Guidance on general principles of clinical evidence for in vitro diagnostic medical devices (IVDs). January 2022. Accessed April 23, 2024. https://health.ec.europa.eu/system/files/2022-01/mdcg_2022-2_en.pdf.

[22] MDCG 2019-9. Rev. 1. Summary of safety and clinical performance. March 2022. Accessed April 23, 2024. https://health.ec.europa.eu/system/files/2022-03/md_mdcg_2019_9_sscp_en.pdf.

[23] MDCG 2022-9. Summary of safety and performance template. May 2022. Accessed April 23, 2024. https://health.ec.europa.eu/document/download/b7cf356f-733f-4dce-9800-0933ff73622a_en?filename = mdcg_2022-9_en.pdf&trk = public_post_comment-text.

[24] AZO Sensors. History of pacemakers. Accessed April 23, 2024. https://www.azosensors.com/article.aspx?ArticleId = 10.

[25] T. Almas, R. Haider, J. Malik, A. Mehmood, A. Alvi, H. Naz, et al., Nanotechnology in interventional cardiology: a state-of-the-art review, Int J Cardiol Heart Vasc. 43 (2022) 101149. Available from: https://doi.org/10.1016/j.ijcha.2022.101149. PMID: 36425567; PMCID: PMC9678733.

[26] S.K. Mulpuru, M. Madhavan, C.J. McLeod, Y.-M. Cha, P.A. Friedman, Cardiac pacemakers: function, troubleshooting, and management: part 1 of a 2-part series, J Am Coll Cardiol. 69 (2) (2017) 189–210. Available from: https://doi.org/10.1016/j.jacc.2016.10.061.

[27] C. Berardi, C.A. Bravo, S. Li, M. Khorsandi, J.E. Keenan, J. Auld, et al., The history of durable left ventricular assist devices and comparison of outcomes: HeartWare, HeartMate II, HeartMate 3, and the future of mechanical circulatory support, J Clin Med. 11 (7) (2022). Available from: https://doi.org/10.3390/jcm11072022.

[28] Mazzei M., Keshavamurthy S., Kashem A., Toyoda Y. Heart transplantation in the era of the left ventricular assist devices. InTech. 2018. https://doi.org/10.5772/intechopen.76935. Accessed May 16, 2023. https://www.intechopen.com/chapters/61419.

[29] IMDRF/GRRP WG/N52 FINAL:2019. Principles of labelling for medical devices and IVD medical devices. 2019. Accessed May 14, 2023. https://www.imdrf.org/sites/default/files/docs/imdrf/final/technical/imdrf-tech-190321-pl-md-ivd.pdf.

[30] MDCG 2020-6. Regulation (EU) 2017/745: clinical evidence needed for medical devices previously CE marked under directives 93/42/EEC or 90/385/EEC. A guide for manufacturers and notified bodies. April 2020. Accessed April 23, 2024. https://health.ec.europa.eu/system/files/2020-09/md_mdcg_2020_6_guidance_sufficient_clinical_evidence_en_0.pdf.

[31] European Medicines Agency. Section 4.3 Contraindications (europa.eu). Accessed April 23, 2024. https://www.ema.europa.eu/en/documents/presentation/presentation-section-43-contra-indications_en.pdf#: ~ :text = Contraindications%20should%20be%20unambigiously%2C%20comprehensively%20and%20clearly%20outlined,another%20medicine%2C%E2%80%A6%20must%20be%20stated%20in%20section%204.3.

[32] MDCG 2019-13. Guidance on sampling of MDR Class IIa/Class IIb and IVDR Class B/Class C devices for the assessment of the technical documentation. December 2019. Accessed April 23, 2024. https://health.ec.europa.eu/system/files/2020-09/md_mdcg_2019_13_sampling_mdr_ivdr_en_0.pdf.

[33] MDCG 2021-14. Explanatory note on IVDR codes. July 2021. Accessed April 23, 2024. https://health.ec.europa.eu/system/files/2021-07/md_mdcg_2021-14-guidance-ivdr-codes_en_0.pdf.

[34] US FDA. The 510(k) program: Evaluating substantial equivalence in premarket notifications [510(k) Guidance]. July 28, 2014. Accessed April 23, 2024. https://www.fda.gov/media/82395/download.

[35] MDCG 2020-5. Clinical evaluation—equivalence. A guide for manufacturers and notified bodies. Accessed April 23, 2024. https://health.ec.europa.eu/system/files/2020-09/md_mdcg_2020_5_guidance_clinical_evaluation_equivalence_en_0.pdf.

[36] ISO 10993-1:2018. Biological evaluation of medical devices—part 1: evaluation and testing within a risk management process. 2018. Available for purchase on April 23, 2024. https://www.iso.org/standard/68936.html.

[37] ISO 10993-18:2020. Biological evaluation of medical devices—part 18: chemical characterization of medical device materials within a risk management process. 2020. Available for purchase April 23, 2024. https://www.iso.org/standard/64750.html.

[38] ISO 10993-17:2002. Biological evaluation of medical devices—part 17: establishment of allowable limits for leachable substances. 2002. Available for purchase on April 23, 2024. https://www.iso.org/standard/23955.html.

[39] ISO 10993-13:2010. Biological evaluation of medical devices—part 13: identification and quantification of degradation products from polymeric medical devices. 2010. Available for purchase on April 23, 2024. https://www.iso.org/standard/44050.html.

[40] ISO 10993-14:2001. Biological evaluation of medical devices—part 14: identification and quantification of degradation products from ceramics. 2001. Available for purchase on April 23, 2024. https://www.iso.org/standard/22693.html.

[41] ISO 10993-15:2019. Biological evaluation of medical devices—part 15: identification and quantification of degradation products from metals and alloys. 2019. Available for purchase on April 23, 2024. https://www.iso.org/standard/68937.html.

[42] European Parliament and European Council. Consolidated text: directive 2001/83/EC of the European Parliament and of the Council of 6 November 2001 on the Community code relating to medicinal products for human use. Available on April 23, 2024. https://eur-lex.europa.eu/legal-content/EN/TXT/?uri = CELEX/3A02001L0083-20220101.

[43] MDCG 2019-11. Guidance on qualification and classification of software in regulation (EU) 2017/745—MDR and regulation (EU) 2017/746—IVDR. October 2019. Accessed April 23, 2024. https://health.ec.europa.eu/system/files/2020-09/md_mdcg_2019_11_guidance_qualification_classification_software_en_0.pdf.

[44] World Medical Association. Declaration of Helsinki. 2013. Accessed May 14, 2023. https://www.wma.net/policies-post/wma-declaration-of-helsinki-ethical-principles-for-medical-research-involving-human-subjects/.

[45] Purdue University. Evidence based nursing. What is PICO? Accessed April 23, 2024. https://library.purdueglobal.edu/ebn/whatispico.

[46] K. Sands, J.L. Frestedt, Real-world evidence and postmarket surveillance data: are they the same thing? RF Quarterly 3 (1) (2023) 11–21. Accessed April 23, 2024. https://www.raps.org/getmedia/a6f18af3-e651-45c2-ac68-609cbefe5440/23-3_RFQ-1_Frestedt.pdf?ext = .pdf.

[47] Galle B. 45 medical device industry statistics, trends & analysis. January 28, 2020. Accessed April 23, 2024. https://brandongaille.com/45-medical-device-industry-statistics-trends-analysis/.

[48] DAIC. Medtronic recalls HVAD Pump implant kits after 2 deaths and 19 serious injuries. 2021. Accessed April 23, 2024. https://www.dicardiology.com/content/medtronic-recalls-hvad-pump-implant-kits-after-2-deaths-and-19-serious-injuries#: ~ :text = LVAD%20may%20have%20delayed%20or%20failed%20restart%20after,delay%20in%20restarting%20after%20the%20pump%20was%20stopped.

[49] Medtronic. Medtronic HVAD™ system urgent product update. 2021. Accessed April 23, 2024. https://www.medtronic.com/us-en/healthcare-professionals/products/cardiac-rhythm/ventricular-assist-devices/hvad-system.html.

[50] Mazzei M., Keshavamurthy S., Kashem A., Toyoda Y. Heart transplantation in the era of the left ventricular assist devices. *InTech*. 2018. https://doi.org/10.5772/intechopen.76935. Accessed May 16, 2023. https://www.intechopen.com/chapters/61419.

[51] World Medical Association. Declaration of Helsinki. 2013. Accessed May 14, 2023. https://www.wma.net/policies-post/wma-declaration-of-helsinki-ethical-principles-for-medical-research-involving-human-subjects/.

Chapter 3

Identifying clinical/performance data

Clinical/performance data are defined as information derived from a human using a device (under the EU medical device regulation, MDR) or from a device testing human samples (under the EU *in vitro* diagnostic device regulation, IVDR) with resulting positive and negative clinical effects and performances documented. Three clinical data sources include clinical trial, literature, and experience reports. Identifying "appropriate" and "sufficient" clinical data from clinical trial, literature, and experience reports is required for all clinical evaluation/performance evaluation (CE/PE) types. Appropriate clinical data are required to be high-quality and relevant to the subject device. Sufficient clinical data are required to represent the device with enough safety and performance (S&P) data as well as benefit-risk ratio details to allow a comprehensive clinical evaluation of general safety and performance requirement (GSPR) conformity including scientific substantiation of all manufacturer clinical claims.

The manufacturer must use "a defined and methodologically sound procedure" (European Union, EU, Medical Device Regulation [MDR] Article 61(3) [1] and In Vitro Diagnostic Regulation [IVDR] Article 56(3) [2]) for all clinical evaluation/performance evaluation (CE/PE) steps including the critically important clinical data identification step. Specifically, prior to starting the clinical data identification, the clinical evaluation plan/performance evaluation plan (CEP/PEP) must specify how to identify systematically and reproducibly all relevant, high-quality clinical/performance data from all available clinical data sources. All identified data must be properly attributed to the search terms used as well as the physical locations and databases searched. Properly identified data must be appraised and analyzed, and the results and conclusions documented in the clinical evaluation report/performance evaluation report (CER/PER).

Specifically, the in vitro diagnostic (IVD) device PER must identify and define scientific validity, analytical performance, and clinical performance when using the IVD device to test human samples. These terms are specifically defined in the IVDR [2] as follows:

- *"'scientific validity of an analyte' means the association of an analyte with a clinical condition or physiological state"* [Article 2 (38)]
- *"'analytical performance' means the ability of a device to correctly detect or measure a particular analyte"* [Article 2 (40)]
- *"'clinical performance' means the ability of a device to yield results that are correlated with a particular clinical condition or a physiological or pathological process or state in accordance with the target population and intended user"* [Article 2 (41)]
- *"'clinical evidence' means clinical data and performance evaluation results, pertaining to a device of a sufficient amount and quality to allow a qualified assessment of whether the device is safe and achieves the intended clinical benefit/s, when used as intended by the manufacturer"* [Article 2 (38)].

When identifying IVD device data, analytical performance is an exception to clinical data collection and may not require a human person (i.e., clinical) element. The objects of interest for analytical performance are specifically the analyte (i.e., chemical substance) and the chemical detection by the device, not the human (i.e., clinical) relationship *per se*.

3.1 Clinical data definition

The PE/CE process requires clinical data, and the nuances for this clinical data definition must be clear to identify all relevant clinical/performance data (Fig. 3.1).

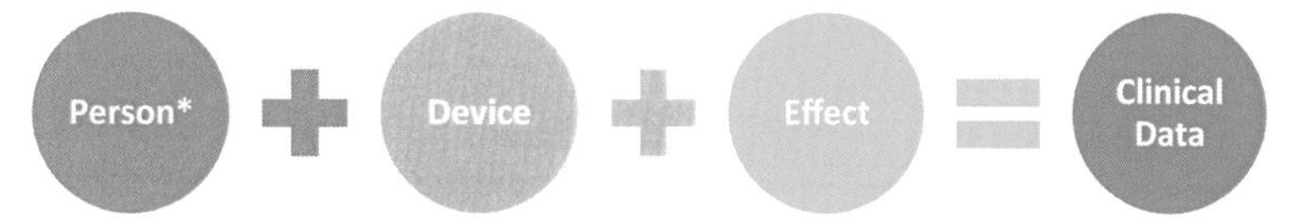

FIGURE 3.1 Clinical data needs to include a person, a device, and a clinical effect. *An analyte rather than a person is required to determine IVD analytical performance.

Planning, Writing and Reviewing Medical Device Clinical and Performance Evaluation Reports (CERs/PERs). DOI: https://doi.org/10.1016/B978-0-443-22063-0.00002-X

Clinical/performance data and clinical evidence are defined in EU regulatory documents (Table 3.1).

TABLE 3.1 Clinical/performance data and clinical experience definitions.

MDR [1]	IVDR [2]	MedDev 2.7/1, rev. 4 [3]	IVD CE MDCG 2022-02 [4]
Clinical data definitions			
"(48) 'clinical data' means information concerning safety or performance that is generated from the use of a device and is sourced from the following: – clinical investigation/s of the device concerned, – clinical investigation/s or other studies reported in scientific literature, of a device for which equivalence to the device in question can be demonstrated, – reports published in peer reviewed scientific literature on other clinical experience of either the device in question or a device for which equivalence to the device in question can be demonstrated, – clinically relevant information coming from [PMS]... in particular the post-market clinical follow-up";	"Article 56... conformity with relevant [GSPRs]... in particular those concerning the performance characteristics... under the normal conditions of the intended use of the device, and the evaluation of the interference/s and cross-reaction/s and of the acceptability of the benefit-risk ratio... shall be based on scientific validity, analytical and clinical performance data providing sufficient clinical evidence..." AND "Demonstration of the clinical performance of a device shall be based on one or a combination of the following sources: – clinical performance studies; – scientific peer-reviewed literature; – published experience gained by routine diagnostic testing."	"Clinical data: the safety and/or performance information that is generated from the clinical use of a device. Clinical data are sourced from: – clinical investigation/s of the device concerned; or – clinical investigation/s or other studies reported in the scientific literature, of a similar device for which equivalence to the device in question can be demonstrated; or – published and/or unpublished reports on other clinical experience of either the device in question or a similar device for which equivalence to the device in question can be demonstrated..."	"Examples of existing data (without any particular order): – appraised literature data, – peer-reviewed data – published clinical data (e.g., Summary of Safety and Performance (SSP), Registries[a] and databases from authorities), – relevant information on the scientific validity of devices measuring the same analyte or marker, – proof of concept studies... relevant consensus expert opinions/positions from relevant professional associations relating to the safety, performance, clinical benefits to patients, design characteristics, scientific validity, clinical performance and intended purpose of the device... Examples of generating new evidence (without any particular order) – Perform clinical performance study, – Other studies (e.g., analytical performance studies or PMPF studies)."
Clinical evidence definitions			
"(51) 'clinical evidence' means clinical data and clinical evaluation results pertaining to a device of a sufficient amount and quality to allow a qualified assessment of whether the device is safe and achieves the intended clinical benefit/s, when used as intended by the manufacturer";	"(36) 'clinical evidence' means clinical data and performance evaluation results, pertaining to a device of a sufficient amount and quality to allow a qualified assessment of whether the device is safe and achieves the intended clinical benefit/s, when used as intended by the manufacturer";	"Clinical Evidence: the clinical data and the clinical evaluation report pertaining to a medical device. [GHTF SG5/N2R8:2007]"	"Clinical Evidence: Clinical data and performance evaluation results pertaining to a device of a sufficient amount and quality to allow a qualified assessment of whether the device is safe and achieves the intended clinical benefit/s, when used as intended by the manufacturer. Source: EU 2017/746 (IVDR), Article 2 (36)" (i.e., identical to IVDR)

EU, European Union; *GHTF*, Global Harmonization Task Force; *GSPR*, general safety and performance requirements; *IVDR*, in vitro diagnostic regulation; *PMPF*, postmarket performance follow-up; *PMS*, postmarket surveillance.
[a]*Registries and databases are clinical experience data.*

The clinical/performance data definition has varied over time. For example, MedDev 2.7.1, rev. 3 explained clearly how to identify clinical data as follows:

> *Stage 1... Identify clinical data from... Literature searching &/or... Clinical experience &/or... Clinical investigation* (MedDev 2.7.1, rev. 3, Section 5.1, Figure 1) [5]

Clinical data identification details include three specific clinical data sources (Fig. 3.2). In practice, clinical trial data are ranked as the "top-tier" data source, followed by literature and finally experience data. Clinical data need to be identified and evaluated in the CER/PER to see if they demonstrate sufficient device safety and performance (MDR Article 61(1) [1] and IVDR Article 56(1) [2]).

Clinical Trials[1]

- Randomized, controlled, prospective, retrospective, and other trials
- Global clinical trials or trials outside the EU with justification on applicability
- Trials using subject device, equivalent device, and other[2] similar devices
- Curated trial databases and clinical study reports

Clinical Literature

- Metaanalyses/systematic reviews evaluating subject, equivalent, and other[2] devices
- Existing data from scientific studies using subject, equivalent, and other[2] devices
- Peer-reviewed publications

Clinical Experience

- Complaints and compliments
- Real-world S&P data (e.g., surveys, registries, adverse event databases)
- PMS data including PMCF/PMPF data
- Routine diagnostic testing

FIGURE 3.2 Clinical evaluation/performance evaluation data sources. [1]Note: Clinical trials, clinical investigations, and clinical studies are synonymous terms. [2]Clinical data for similar devices will be included in the CER/PER background/state-of-the-art (SOTA) section.

Unfortunately, the updated MedDev 2.7/1, rev. 4 guideline [3] moved away from this clear clinical data source definition and instead compared "manufacturer-held" data (i.e., data from manufacturer-sponsored trials, registries, surveys, publications, and complaints recorded within the clinical trial (CT), postmarket surveillance (PMS), and risk management (RM) systems) to data collected from nonmanufacturer sources. In addition, the clinical data sources were not distinguished based on premarket or postmarket data collection although postmarket data collection may be required for CE-marked devices whenever additional clinical data are required. These ambiguous clinical data identification details contributed to significant confusion, even though "manufacturer-held" data still include all three previously defined clinical data sources: trials, literature, and experiences.

Being clear about the clinical data needed for the CE/PE is important and understanding the different data volumes, qualities, and types for each of the three clinical data sources is helpful when trying to decide how much clinical data are sufficient for the CER/PER. For example, well-controlled and well-executed clinical/performance trials using the device under evaluation will result in relevant, high-quality data, if the trials were designed to look at subject device safety and performance (S&P) and benefits/risks specifically. Specifically, manufacturer-held clinical trials should include access to the study database with all the clinical data details for evaluation. Literature, on the other hand, often has less data volume and more data variation than company-sponsored clinical/performance trials. In addition, literature is often less focused on the subject device S&P or benefit:risk ratio than company-sponsored clinical/performance trial data. Lastly, experience data reports may be numerous but will have the lowest clinical/performance data quality since experiences are often documented as singular complaint records. Expereinces may be drawn from international databases or from surveys or registries without much detail available. These differences in clinical trial, literature, and experience data volume, quality, and type require different strategies to identify relevant and sufficient clinical data from each clinical data source for evaluation.

Ideally, all stakeholders involved in the CE/PE process will realize the clinical data identification process is not perfect and will not identify "every" clinical/performance data point (defined as an individual human using a device for a particular effect). For this reason, individual data points from outside the rigorous data searches should always be allowed into the identified dataset. The CEP/PEP must allow this "manual data entry" while also specifying processes to search and identify "appropriate" CE/PE data from clinical trials, literature, and experiences for the CE/PE process. The evaluator should carefully document data volumes, qualities, and types and the reviewer should verify and validate the specific data identification and collection processes used.

3.2 Clinical data identification process

"Sound" scientific process steps are required to identify all "appropriate" clinical data and to ensure "sufficient" clinical data are present for the CE/PE process. The CE/PE process must "scientifically substantiate" each clinical claim made by the manufacturer including the claim about the device meeting the EU MDR/IVDR GSPRs based on the clinical data available for the subject/equivalent device. The CE/PE process must also support the S&P acceptance criteria set by the manufacturer based on the background knowledge, competitor devices, and alternative therapies defining the state of the art (SOTA) for the subject/equivalent device. Any inappropriately included, missing, or insufficient clinical data at the clinical data identification step should be corrected. The regulations require all clinical data in the CE/PE process to be accurate, relevant and sufficient.

3.2.1 Clinical data identification should be standardized

Work instructions (WIs) within the manufacturer quality management system (QMS) should be as detailed as possible when outlining CE/PE clinical data identification. List each step in the WI chronologically for the clinical trial, literature, and experience data searches, and explain how the background/SOTA data will be presented separately from the device-specific data in the relevant CER/PER sections. The WI must comply with the MDR/IVDR requirements and should include sufficient details for a qualified person (i.e., the evaluator) to execute the steps. The WI for clinical data identification should include at least the following details:

1. The evaluator must follow the CEP/PEP (documentation should include CEP/PEP version/signature/date).
   a. CEP/PEP is often included as a CER/PER appendix
2. The evaluator must document actual search methods used in the CER/PER
   a. CEP/PEP exceptions/deviations
   b. exact search dates, databases, and search terms used for each clinical data type (i.e., clinical trials, literature, and experience data)
   c. rationale for search terms (e.g., PICO: population, intervention, comparator, outcome)
   d. output from each search (i.e., number of results)
3. The evaluator should document clinical data gaps observed while identifying clinical data.
4. The evaluator must document details about each data download, organization, and storage.
5. The evaluator must separate background/SOTA data from subject/equivalent device-specific data.
6. The quality control (QC) person must document the internal QC methods and results.

The WI should clearly state the steps needed for each specific CER/PER type—viability (v), equivalent (eq), first-in-human (fih), traditional (tr), and obsolete (o)—and who will conduct the clinical data identification steps including the searches themselves as well as the mechanisms to record search results in a workbook or database. The QC process should ensure all results are accurate and complete. The searches must identify clinical data to support all the various device applications allowed in the indication for use, all claims about S&P and all details dfining how the clinical benefits outweigh the risks to the patient when using the medical device as intended.

3.2.2 Sound scientific processes must be used to identify clinical data

Clinical evaluations must follow "sound" written processes and procedures, meaning the procedures have been tested and are in good condition. For example, the clinical data identification process must follow a rigorous "scientific method" with careful observation and skeptical evaluation at every step along the way, since assumptions and preconceived expectations can distort clinical/performance evaluations.

Using the "scientific method" often starts with a scientific question or observation phrased as a testable hypothesis. A hypothesis is a suggested explanation for a phenomenon based on limited knowledge and used as the basis for further scientific testing. Thus, a hypothesis is like an educated guess which is then researched and tested in experiments. Experiments use scientific procedures to evaluate the hypothesis and to produce data. The data are then analyzed and used to formulate a conclusion or an answer to the initial testable hypothesis question. Typically, answering the hypothesis question or completing the experiment leads to more questions. This test and retest process is the nature of science.

BOX 3.1 Hypothesis testing during clinical data identification

The methods used to identify CE/PE clinical data must be scientifically rigorous. Each step should be reviewed to ensure the scientific method has been followed. For example, an evaluator working on the initial clinical data identification step might hypothesize: "sufficient" clinical data are available to support the safety and performance of the subject device. Then, as the clinical data are identified, the evaluator would test the clinical data to see if the hypothesis is true. The evaluator might ask many critical questions not limited to the following:

- Was enough clinical trial data identified to show all parts of the indication for use were supported by actual human clinical trial data?
- Does the clinical literature data support all known S&P details?
- Do the clinical experience data "fit" the subject device S&P profile and benefit-risk ratio?

For any and all negative "test" results, the hypothesis will have failed and additional work must be triggered (e.g., the postmarket clinical follow-up plan (PMCFP), or the postmarket performance follow-up plan (PMPFP) must collect more clinical data to fill the identified gaps.

A well "defined and methodologically sound procedure" typically leads to reliable (consistent) answers and valid (accurate) findings (data). A "reliable" experiment has clear and comprehensible methods allowing other scientists to repeat the study and to arrive at the same result. A "valid" study collects data in an intentional and controlled manner specifically about the intended measurement in the original question and the data are found to be "true" or accurate when the repeated experiment has the same findings. A valid study directly and completely tests the hypothesis and has reproducile results.

The regulations also require the clinical data evaluator to use sound scientific processes when creating single, stand-alone CE/PE documents. Each evaluator must "critically evaluate" all relevant clinical trial, literature and expereince data using the subject device as well as all of the background knowledge and SOTA clinical data including clinical data from patients using alternative treatments. The CE/PE process requires a critical, skeptical, unbiased and comprehensive clinical data review for all human device uses whether the clinical data reflected positively or negatively on the device S&P. This identification step must keep all relevant clinical data included unless a clear and compelling reason to exclude the data is documented.

3.2.3 Quality control processes should be used during clinical data identification

The CE/PE goal is to include relevant and high-quality clinical/performance data while excluding irrelevant and low-quality data whenever possible. A clear and concise device S&P listing should clearly define the device benefit-risk ratio. In addition, the QMS should specify the required (and standardized) QC process during each CE/PE step including the clinical/ performance data identification step. For example, QC should occur after key search terms were sandboxed and the final terms were selected. QC should occur after the searches are completed (including all search modifications) and the identified clinical/performance data are entered into an Excel spreadsheet or proprietary software tool for use during the clinical data identification step (Fig. 3.3).

FIGURE 3.3 Quality management system must control the search terms use to identify clinical/performance data.

The QC step after sandboxing should help to ensure sufficient clinical data were identified during the clinical/ performance data identification step before moving forward. The QC step after storing the search outputs should help to ensure all searches used a sound strategy, were executed and documented as planned and had reasonable results before proceeding to the next CE/PE appraisal and analysis steps. In addition, these QC process steps are expected to save time, especially if the QC process identifies helpful improvements during the clinical/performance data identification step and not later in the process. For example, the QC process may identify missed clinical/performance data or different search terms or combinations to optimize the identified clinical/performance data results. Clinical/performance data identification problems will be amplified during downstream work since each clinical/performance data point needs careful consideration (often multiple times) during subsequent CE/PE process steps. Different issues arise when too many or too few data are identified for the CE/PE process, and the QC steps should identify these issues so the team can apply appropriate strategies to ensure, for example, additional data are generated when too little clinical data were found or additional exclusion criteria are applied when too much clinical data were found.

3.2.4 Qualified evaluators should oversee all clinical data identification efforts

Persons should have sufficient education, training, and experience to execute the clinical data searches to identify the relevant clinical trial, literature, and experience data, to organize the identified data from all three sources and to store the actual identified clinical data for the evaluation. Crucially, these persons must be able to follow the CEP/PEP by creating and executing the planned search strategy (i.e., accessing databases, setting up, executing, and storing the search terms and strings with their results). In addition, the persons doing the clinical data identification step must capture all favorable and unfavorable data (i.e., without introducing bias), should identify possible clinical data gaps, and should suggest ways to separate relevant background/SOTA and device-specific clinical data.

The clinical data identification process is challenging because multiple databases must be searched with many different search details. In addition, the identified clinical data are highly variable, which makes identifying the "right" clinical data difficult. For these reasons, persons doing this step are required to have a relevant advanced degree and 5 years of experience, or 10 years of experience without an advanced degree.

BOX 3.2 Case study—clinical data identified by a librarian

Case study situation

Librarians are helpful resources for clinical data identification searches. Often, manufacturers hire external librarians to construct literature search terms and strings, execute searches, and provide abstracts and/or other outputs from their searches.

Case study issues identified

One manufacturer hired an external librarian for a literature search and then put the search result directly into the CER. The librarian conducted an effective background/SOTA and targeted subject-device search; however, the librarian conducted only the literature search and this was done without a documented or planned methodology (i.e., no CEP was documented prior to search execution). Additionally, the librarian only completed the literature search without any clinical data appraisal or analysis. The manufacturer did not conduct any clinical data analysisin the CER.

Case study outcomes

The NB withheld the CE mark due to documented deviations from the regulatory requirements (i.e., no plan was documented before the searches were completed, the identified data were not appraised or analyzed and the rigor used in the clinical literature identification was not matched by the manufacturer for the clinical trial and clinical experience clinical data identification methods.

A better case study solution enacted

In this case, the clinical data identification step was incomplete because clinical trial and experience databases were not searched and the clinical literature searches did not ensure "sufficient" clinical data were identified. Clinical data identification is only one CE process step, but a failure at this early stage will impact all subsequent appraisal and analysis steps for all three clinical data sources in the CER/PER.

A good QC strategy during the CEP/PEP-directed clinical data identification step is to review results as they are generated from each search to ensure the clinical data are as expected. Even though the planning stage included a sandbox approach to test out the searches and to ensure the results generated the needed clinical data, the results may shift a little when the searches are actually completed. Be prepared to do additional work to modify search strategies as needed to ensure the clinical data identification step successfully identifies "sufficient," "high quality" clinical data needed for the CER/PER. For example, when a search strategy returns no results, expand the search terms to cover broader concepts. Alternatively, when a search strategy returns too many results, restrict the search terms to be more discrete or combine search terms to require more than one term.

Search methods and processes vary for each clinical data type (e.g., searches for clinical trials and experiences typically focus on the subject device while literature searches must include background/SOTA terms in addition to device-specific terms). All searches must capture all relevant, high-quality data and the clinical data identification rationale must be defensible.

3.2.5 Appropriate clinical data must be identified

The QC process should determine if the plan was appropriate to identify relevant and high-quality clinical data from a human using the subject device to be evaluated in the CE/PE process. The lack of any human clinical data often means the CER/PER may be a vCER/vPER (i.e., viability) type. In this case, the QC process should confirm the device has never been used in humans before. Alternatively, some "legacy" devices have little to no clinical data because they were low risk and CE marked prior to the regulations requiring clinical data for all medical devices. In this "legacy" device case, a vCER/vPER is simply NOT appropriate since clinical/performance data must reside somewhere but the clinical data identification process may have failed to identify the available PMS clinical experience data (for individual human experiences using the marketed device) or the manufacturer failed to prioritize the clinical data collection from the marketed and sold devices. Depending on the legacy device situation, the manufacturer should probably improve the PMS system to capture "sufficient" clinical data for the CE/PE process, and the PMCFP/PMPFP should document the clinical data gap and schedule additional clinical/performance data collection tools to collect the needed clinical data for the trCER/trPER (i.e., traditional type).

The CEP/PEP may be "appropriate" to write when specific failure modes are identified in the PMS experience data; however, the CE/PE process is only "successful" when "sufficient" clinical evidence (i.e., clinical data from device use in humans with clinical/performance evaluation results) allow CE/PE conclusions to be drawn in the CER/PER about the subject device S&P and benefit-risk ratio.

BOX 3.3 CE/PE failure mode: no "clinical" data in CER/PER

Prior to MDR/IVDR, manufacturers often had minimal or no clinical support within the company. As such, often engineers and other experts were tasked with writing the CER/PER after clinical data evaluation but they were not clinicians or clinical researchers and they were not actually evaluating "clinical" data.

Years of engineering experience provided high value and excellent skill development relevant to device data including device manufacturing, bench testing, design controls, engineering diagrams and records, etc. Unfortunately, these skills and data records were not entirely and directly relevant to the "clinical" data required in the CER/PER. The required clinical data are specifically seeking to understand actual human device use (i.e., what happened to the human using the device? What were the S&P details encountered by the human using the device?).

This inappropriate staffing issue for the CE/PE evaluator role often resulted in a CER/PER without clinical data but with lots of engineering data about the device. In other words, the clinical focus for the CER/PER was confused with irrelevant, nonclinical, engineering data, and the CER/PER was developed relevant to the engineering team only which did not meet the clinical evaluation requirement. Of course, bench/engineering data were relevant to the device, but not always to the human using the device and these data simply did not include the required human testing with the device in use.

Not all data are clinical data, and nonclinical data are generally excluded from CER/PER because nonclinical data (e.g., technical design verification and validation studies, human factors studies, animal studies and biocompatibility bench/animal studies) are generally expected to be stored in the technical files and to be replaced with actual clinical (i.e., human use) data.

For example, a biocompatibility technical report (TR) showing no cytotoxicity, sensitization, or irritation in cell and animal models does not need to be mentioned in the CER/PER, especially when human data demonstrate human biocompatibility directly in the more relevant, human use setting. These technical testing results are assumed to be the case for most CE/PE work. If the device had failed biocompatibility testing (or other required technical file testing) , the device would not likely be allowed in human use and ultimately would not be on the market because the device did not pass the technical requirements evaluated outside the CER before any clinical data collection could begin. Nonclinical data are only relevant in the CER/PER if the nonclinical data have a specific human clinical data impact.

Ultimately, clinical data will be reviewed by clinical experts and engineering data will be reviewed by engineering experts. The CER/PER writing process should clearly understand the clinical healthcare professional audience and the clinical reviewer because they will have expertise and interest in understanding human clinical data, unlike the engineering audience and the technical reviewer for the rest of the engineering data in the technical file.

As illustrated here, the term "appropriate" means the CE/PE clinical data identification step must include all relevant human clinical data, whether favorable or unfavorable, from device uses as indicated on the labeling. This understanding about appropriate human clinical data from each device use should help to focus the CE/PE scope which will include clinical data identified and collected from subject/equivalent device and competitor device uses in humans. Meanwhile, when identifying clinical data about general knowledge, clinical background, SOTA, and alternative therapies, the appropriate, identified, and collected human clinical data may not require a device to be used at all. In other words, many searches must be completed before the evaluator is satisfied all relevant and appropriate clinical data have been included during this identification step in order to adequately support the appraisal and analysis steps to follow.

3.2.6 Clinical data must be relevant to subject device

The clinical evaluation/performance evaluation must be about the subject device.

This point cannot be overstated. All too often, evaluators and writers go off on tangents about irrelevant information. The goal is to "tell the subject device story." Putting the subject device into the context of the background knowledge and SOTA and competitor device landscape seems to cause the most difficulty when training new CE/PE evaluators and writers. Writers often spend an inordinate amount of time writing about the background and SOTA, and the extensive background and SOTA contents are sometimes totally unrelated to the subject device. This is a mistake. The background knowledge, SOTA, and competitor device data are needed to put the subject/equivalent device into context; therefore, these data must be relevant to the subject device. The CE/PE is about the subject device, not the background knowledge, SOTA, and competitor device data. Keep the CE/PE process focused on the subject device and write a complete story with a global context.

In other words, a typical failure mode for clinical data identification is when a large amount of "irrelevant information" was identified and erroneously included for subsequent appraisal and analysis. For example, these "irrelevant" data typically include nonhuman (i.e., animal, bench data) or unsubstantiated (i.e., posters, abstracts, presentations, testimonials, etc.) or "off label" (i.e., human device use was not as described in the product labeling/ indication for use) data. Only in rare cases are nonhuman, unsubstantiated, and off-label use data relevant (e.g., animal data seeking to understand a particular safety concern, an abstract or off label use noting a rare safety concern applicable to on-label use). Search terms need to be refined to exclude as much irrelevant information as possible and the CEP/PEP should not pass the sandboxing step or the QC checks with extensive, irrelevant information identified for subsequent appraisal and analysis. A QC failure should be documented and appropriate corrections should be made when a large amount of irrelevant information are identified in the CE/PE identification step.

In addition, great care should be taken when the clinical data identified are mostly from "similar" devices without much clinical data for the subject/equivalent device. These "similar" device clinical data are somewhat irrelevant to the subject device, because the data may be different when using the subject device.

Highly relevant clinical data about the subject device should be distinguished from data with little to no relevance to the subject device. Including off-topic data may lead to incorrect conclusions about the subject device. Specifically, irrelevant data should be excluded from the CE/PE process during the identification step by choosing appropriate search terms (Fig. 3.4).

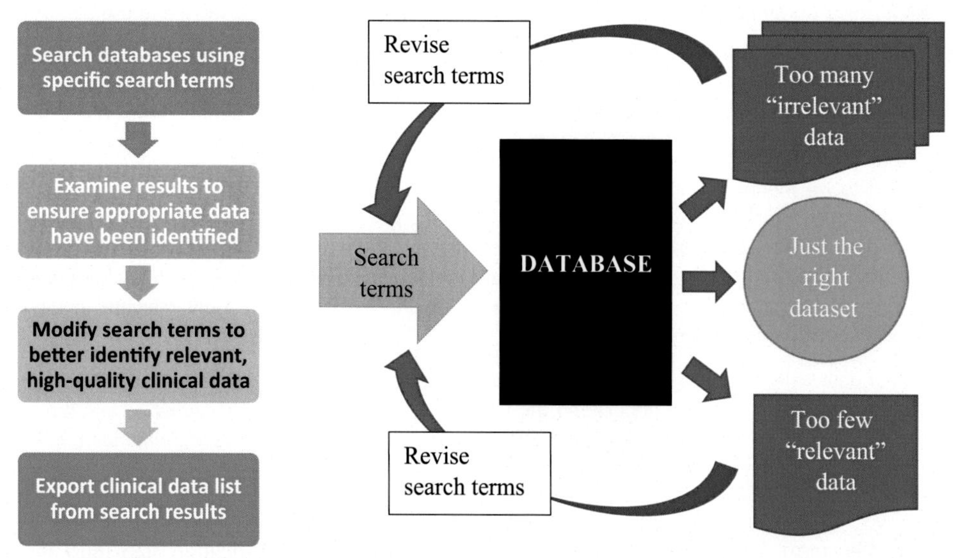

FIGURE 3.4 Modify database search terms to identify appropriate and sufficient clinical data.

During clinical data identification, avoid moving on to the appraisal step until after all searches, downloads, and data organizations have been completed. This will help to avoid premature clinical data identification acceptability decisions.

3.2.7 Clinical data must be high quality

In addition to clinical data relevancy, data quality is an important factor to consider in the clinical data identification step. "High-quality" clinical data have scientific rigor (i.e., special efforts were taken to ensure accuracy using validated methods) and require complete, methodologically sound, version-controlled, and quality-controlled documents based on the latest scientific evidence.

High-quality, manufacturer-held clinical trial data, for example, should include a version-controlled clinical trial protocol and final study report created by appropriately qualified personnel. Many different clinical trial types are possible and trial quality varies widely (e.g., randomized clinical trials are considered the gold standard while case reports are considered poor quality). Although clinical trial design considerations are outside the scope of this textbook, high-quality clinical trial data are produced using scientifically and methodologically "valid" research practices. Processes are considered valid when scientists agree the data presented were accurate and truthful. In addition, the full clinical trial dataset should be available for review and further analysis. This is an important consideration when identifying and including clinical trial data in the CE/PE process. Unfortunately, most low-risk devices do not have clinical trial data and this must be considered during the CE/PE process.

High-quality clinical literature data must also be scientifically rigorous, complete, methodologically sound, and quality-controlled. High-quality clinical literature data are often published in "top-tier" journals typically after the required peer-review process. Peer review is a form of QC before the journal editors will allow the article to be published in the journal. This peer-reviewed validation is a gold standard to require before clinical literature data are included in the CE/PE process; however, this standard is not always possible or appropriate. Sometimes clinical data from initial device uses are first accumulated in small non-peer-reviewed articles (i.e., case reports, testimonials, reviwe articles). Also, the CE/PE should include all clinical literature about safety concerns, even though the article may not be high-quality or peer-reviewed. The CE/PE process must consider manymany articles at different quality levels. For example, sometimes metaanalyses and systematic reviews are available and these are higher-quality than a case report because more data were reviewed in a more carefully planned manner than simply reporting an individual case without any comparator. Similarly, an article about clinical data collected in a clinical trial is of higher quality than a narrative review because the narrative review is a description of previously published information and generally does not contain primary clinical data.

When identifying clinical literature for CE/PE, the evaluator may want to develop key search terms from the "PICO" question including the patient/problem (e.g., main concern, disease, age), intervention (e.g., testing, treatments), comparison (e.g., control arm, standard of care), and outcome (e.g., measurable improved symptoms). Searching multiple databases with well-defined search terms is best practice. For example, these searches may include searches for systematic reviews in the Cochrane database. The Cochrane Handbook provides guidance on planning and conducting a systematic review, including methods to search and select studies, to gather data, and to interpret results among other topics (Fig. 3.5).

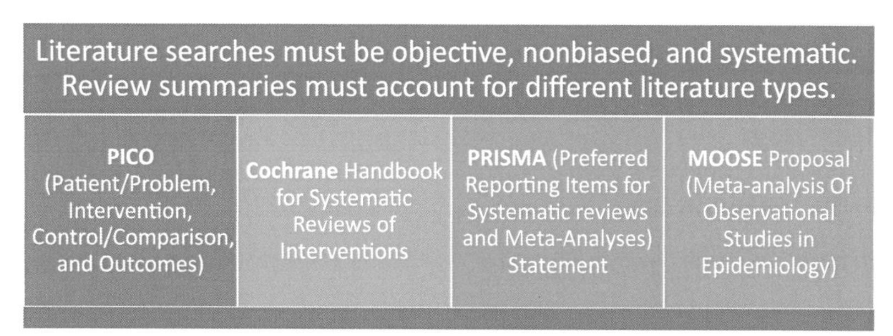

FIGURE 3.5 Literature searches should use standardized tools.

Systematic review articles evaluate two or more studies if they meet prespecified criteria to address a particular question. Systematic reviews use explicit methods to minimize bias including many judgments about the individual data findings being sensitive and robust enough to be included. Metaanalyses are special articles using statistics to evaluate two or more studies and to offer increased power especially when dealing with conflicting claims. Unfortunately, sometimes the metaanalaysis statistical tool may mislead the reader, so consider data variation between studies carefully when identifying metaanalyses for potential CE/PE inclusion (i.e., studies with greatly different effect sizes may reduce metaanalysis reliability). When identifying articles to include in the CE/PE, systematic reviews and metaanalyses can be quite helpful, especially if these articles followed specific guidelines (e.g., PRISMA and MOOSE provide details and checklists about processes to follow when creating a systematic review or metaanalysis).

BOX 3.4 Expert panel finding example: inadequate literature search

The expert panels on medical and IVD devices (EXPAMED) offer opinions and advice to NBs and regulatory authorities including findings related to clinical data identification for CE/PE. For example, one of the first 10 published Clinical Evaluation Consultation Procedure (CECP 2022-000232) opinions concluded in part:

...the literature search was not adequate as it only retrieved 3 articles, while at least 5 additional articles were found at the time of the literature search performed by the manufacturer. These articles were potentially relevant and would have increased the number of available publications to 8 instead of 3, significantly increasing the quality of available data on the device... NB should ensure that all published articles have been identified and presented by the manufacturer (in this case, at least 5 articles were missing...) A thorough and robust literature search is at the core of a reliable systematic review. Such literature search and study selection should be conducted by qualified experts according to internationally accepted guidelines (e.g., PRISMA or MOOSE) in order to maximize the validity of the report. [6]

The level of detail discussed by the expert panel (e.g., assigning "significant" value to the identification of eight versus three articles) should help put the importance of the clinical data identification process in perspective. Clinical data experts need to invest time and effort into identifying all the appropriate and relevant data, especially for the subject device.

Clinical experience data are the lowest-quality of the three clinical data sources. For example, experience data from complaints reported to the manufacturer or from database reports are generally of low quality due to the uncontrolled nature of the individual reports and the lack of education, training and experience for the individual reporters. Anyone can enter data into databases like the Manufacturer and User Facility Device Experience (MAUDE), and these reporters and users may also categorize their reports by applying tags related to device malfunction types (e.g., break, missing part) and patient problems (e.g., injury, death). Unlike clinical trial and clinical literature reports, which undergo extensive QC processes before the data are released, clinical experience data are simply entered into the database, and not investigated further. Often by the data are entered by the patient directly affected by the device. Sometimes these experience data are highly biased and they may even be misleading or untruthful. The analysis of experience data requires an experienced evaluator.

3.2.8 Clinical data must be sufficient

The MDR and IVDR require the manufacturer to determine the clinical evidence needed to meet the GSPRs as follows:

> *The manufacturer shall specify and justify the level of clinical evidence necessary to demonstrate conformity with the relevant general safety and performance requirements. That level of clinical evidence shall be appropriate in view of the characteristics of the device and its intended purpose.* (MDR Article 61(1) and IVDR Article 56(1) are identical) [1,2]

The manufacturer will generally specify GSPRs 1, 6, and 8 are relevant and to be supported by clinical evidence in the CER/PER. The CER/PER is required to provide "sufficient" clinical data to demonstrate conformity with the identified GSPRs. The quote above is clear about the clinical evidence needing to be "appropriate" based on the device characteristics and the intended purpose.

The regulations require sufficient clinical evidence in many areas including, but not limited to, assessing device safety, performance, benefit-risk ratio acceptability, intended purpose, indications, GSPR compliance, risk mitigation activities, device usability, product information, and instructions for use (especially for target populations). Unfortunately, the MDR/IVDR did not define the term "sufficient," and the MedDev 2.7/1, rev. 4 [3] guideline simply stated:

> *Sufficient clinical evidence: an amount and quality of clinical evidence to guarantee the scientific validity of the conclusions.*

Note the word "guarantee" is quite an unusual term in this definition since scientific endeavors do not offer "guarantees."

BOX 3.5 Setting the bar for "sufficient" clinical evidence

One of the most difficult CER/PER questions is: when do we have enough clinical data to claim the clinical evidence is sufficient? The answer depends on many issues, including the device risk level. A seasoned medical device and regulatory expert proposed a simplistic yet impactful view of sufficient clinical data in a CER/PER. The CER/PER should contain enough information to determine if an expert would use the device to treat their dearest family member.

The clinical data amount available often aligns with the device risk: more clinical trial, literature, and experience data are generally available for high-risk devices than low-risk devices. In particular, low-risk devices generally do not have clinical trials completed before the devices are trusted for use; they also tend to have less relevant and lower-quality clinical literature and experience data compared to high-risk devices.

A typical failure mode for the clinical data identification step is when little to no clinical data are identified or when the identified clinical/performance data are all poor quality. Insufficient clinical data examples may include:

- clinical data records with small sample sizes;
- low-quality, study design concerns;
- results reporting mistakes;
- lack of relevant clinical data;
- data skewed by significant bias;
- data collected outside the EU and not able to be generalized to an EU population;
- missing important background knowledge and SOTA clinical data; and
- literature data including only case reports, posters, presentations, anecdotal information, narrative reviews, etc.

In these cases, the CER/PER must specify the urgent and compelling need to generate new clinical data to meet the MDR/IVDR requirements to evaluate clinical data. For a Class III or implantable device, this typically means planning a clinical trial before the CER/PER can be signed off or the PMCFP/PMPFP must specify why a new clinical trial was not required and how the new clinical data to be generated in the PMS system will be sufficient to address this clinical data gap in a specified time frame.

Note: In general, PMS experience data are the largest clinical data amount available; however, these PMS experience data are also the lowest-quality data. As a result, PMS experience data alone are rarely sufficient for the CE/PEprocess.

Here are some questions to ask during the clinical data identification step to help ensure the clinical data being identified and collected are sufficient:

1. Were clinical data included from all three clinical data sources (i.e., were trial, literature, and experience data identified and included)?
2. Were both manufacturer-held and non-manufacturer-held clinical data included?
3. Were clinical data added from manual sources (outside the clinical data searches)?
4. Did the clinical data searches identify relevant information?
5. Were clinical data from both device-specific and background/SOTA data identified?
6. Were clinical data from competitor devices and alternative treatment options identified?
7. What clinical data gaps were identified during the clinical data searches?
8. Was a PMCFP/PMPFP proposed to cover the identified clinical data gaps (i.e., limited or no clinical data found) and to demonstrate conformity with applicable GSPRs.
9. Does the identified clinical data include both positive and negative information about the device under evaluation?
10. What is the minimal acceptable amount of clinical data required for this CE/PE and where is this acceptability standard documented?

Clinical data identification steps should be built to identify an acceptable device-specific and background/SOTA clinical/performance data amount and quality needed to ensure the clinical evidence is scientifically "valid" and scientists will agree the data presented are accurate and truthful when they are evaluated in the CER/PER.

3.2.9 Clinical trials may be exempted when clinical data are sufficient

MDCG 2020-6 [7] was developed "to support a harmonised approach with respect to clinical data providing sufficient clinical evidence necessary to demonstrate conformity with the relevant... GSPR[s]" across the EU Member States. This guide quoted MDR article 61, highlighting the difference between legacy devices (Article 61(6)) and well-established technologies (WET) (Article 61 (6b))and also stated:

> *MDR Article 61(4) states that clinical investigations shall be performed for Class III and implantable devices, but distinct exemptions from this requirement are identified... The common theme of these exempted devices is that they have been previously marketed (Article 61(6)) or have been demonstrated to be equivalent to devices previously marketed (Article 61(5))... all such exemptions from clinical investigations require that the clinical evaluation is based on "sufficient clinical data"... and compliance to common specifications where these exist... both the Directives and the MDR require the quantity and quality of clinical data to be sufficient to demonstrate safety, performance and the acceptability of the benefit-risk ratio: both the Directives and the MDR require clinical evidence to be sound and the conclusions derived from this evidence to be scientifically valid*

MDCG 2020-6 [7] defined sufficient evidence for legacy devices (i.e., devices previously CE marked before MDR and IVDR) and stated:

> *'sufficient clinical evidence' is understood as 'the present result of the qualified assessment which has reached the conclusion that the device is safe and achieves the intended benefits'... note that clinical evaluation is a process where this qualified assessment has to be done on a continuous basis.*

Furthermore:

> *In exceptional cases, particularly for low risk standard of care devices where there is little evolution in the state-of-the-art, and the device is identified as belonging to the group of 'well-established technologies'... a lower level of clinical evidence may be justified to be sufficient for the confirmation of conformity with relevant GSPRs. This may be supported by clinical data from the PMS provided... a quality management system [was] in place to systematically collect and analyse any complaints and incident reports, and that the collected data support the safety and performance of the device.*

In a similar manner, the IVDR Article 56(4) stated:

Clinical performance studies... shall be carried out unless... duly justified to rely on other sources of clinical performance data. [2]

The MDR/IVDR allows significant flexibility when constructing the clinical data collection plan and defining "sufficient" clinical data. For example, low-risk devices do not generally require a clinical trial, while Class III or implantable devices typically require specific clinical trial data.

3.2.10 Justification is allowed if clinical data are not deemed appropriate

All devices must have a CER/PER written and on file at the manufacturer; however, sometimes, the MDR/IVDR allows a device manufacturer to forgo collecting clinical data when the clinical data collection is not appropriate. A justification (JUS) is required when GSPR conformity based on clinical data collection is "not deemed appropriate," as stated in MDR 2017/745, Article 61(10):

> *10. Without prejudice to paragraph 4, where the demonstration of conformity with general safety and performance requirements based on clinical data is not deemed appropriate, adequate justification for any such exception shall be given based on the results of the manufacturer's risk management and on consideration of the specifics of the interaction between the device and the human body, the clinical performance intended and the claims of the manufacturer. In such a case, the manufacturer shall duly substantiate in the technical documentation... why it considers a demonstration of conformity with general safety and performance requirements that is based on the results of non-clinical testing methods alone, including performance evaluation, bench testing and pre-clinical evaluation, to be adequate.* [1]

The decision to forgo clinical data collection needs to be based on the RM details, device characteristics and details about how the device and the human body interact. In addition, the intended clinical performance and all clinical claims must be considered when making this decision. The manufacturer must adequately document "why" the collection of clinical data to fulfill the GSPRs is "not deemed appropriate" and the available nonclinical data are "adequate." This JUS document should include the relevant nonclinical performance evaluations, bench testing, and preclinical evaluations, explaining clearly why collection of clinical data is "not deemed appropriate."

This JUS documentation exception was not meant to represent situations where manufacturers simply have no clinical data available even though the clinical data can and should be collected. This JUS documentation should make a clear, rational, and scientifically sound argument about why clinical data are not appropriate to collect. For example, a device designed to protect a person from a lethal radiation dose should not be tested on otherwise healthy humans intentionally exposed to a lethal radiation dose to evaluate the device S&P and to see if the benefits outweigh the risks. The lethal radiation exposure is an unreasonable risk to the patient.

Specifically, low-risk and well-established devices as well as those where no clinical data are possible for a medical device may be able to use this MDR clause and a JUS should be fully considered by the evaluator before writing a full CER. Often, a JUS may serve instead of writing the CER or PER in specific situations. Similar to MDR Article 61, clause 10, IVDR Article 56, clause 4 states "Clinical performance studies... shall be carried out unless... duly justified to rely on other sources of clinical performance data" and ANNEX XIII, 1.1 states "Where any... elements are not deemed appropriate in the Performance Evaluation Plan due to the specific device characteristics a justification shall be provided in the plan." In other words, both the MDR and IVDR allow a JUS when clinical data collection is "not deemed appropriate."

BOX 3.6 When is enough, enough?

The vCER/vPER type was historically called a justification document (JUS) designed to explain why a CER/PER and other CE/PE documents were not required.

One NB, the British Standards Institute (BSI), objected to the JUS and required the CER/PER naming to be reserved to avoid confusion. The BSI required regular updates for all CE/PE documentation, even for the lowest-risk devices and devices where the collection of clinical data was deemed not appropriate. Other NBs (e.g., DEKRA, TUV Nord, MedCert) had no objection to the JUS name or to the details documented within the JUS about why no CER/PER was needed and why no CER/PER updates would be completed in the future unless a new safety signal or change in benefit-risk ratio was identified.

This conundrum about needing to write a CER/PER even though no clinical data were available, and when the need for collecting clinical data was not deemed appropriate as provided in MDR Article 61(10), is quite common. For example, a newly designed device will not have any human use until sufficient safety and performance are demonstrated on the bench and in animals, as appropriate. This is called a viability (v) vCER/vPER for this reason.

(Continued)

BOX 3.6 (Continued)

Historical devices (e.g., devices without specific clinical claims like equipment drapes, exam gloves, tongue depressors, band aids, suture, scissors, automated blood hematology analyzers, urine chemistry dipsticks) which have been used for decades without documented clinical data on file should not suddenly require additional S&P data evaluation when the PMS and RM systems are working well to keep the devices within their S&P specifications and when the benefit-risk ratio is acceptable. Additional clinical data collection and CE/PE report writing in these settings may be superfluous and likely to add significant cost without sufficient reciprocal benefit. Passing on these costs to the patient and completing clinically inappropriate clinical data evaluations with little to no hope of improving clinical benefit or reduing clinical risk may be unethical in certain situations.

In an effort to satisfy all NBs, the new vCER/vPER naming was created and a broader discussion was recommended to reduce unnecessary costs related to individual NBs requiring all CE/PE documents when no such burdensome documentation will improve the benefit-risk ratio for the device. As the regulations change to clarify clinical data needs, evaluators should be careful to make ethically appropriate decisions since CE/PE process is not a "one size fits all" situation. Each CER/PER should be clearly articulated and the conclusions should be scientifically and ethically sound based on the data analyzed. The evaluator must determine when identifying and evaluating more clinical data is just not appropriate.

MDCG 2020-6 [7] clarifies Article 61(10) allows nonclinical data to be used to demonstrate GSPR conformity in certain "exceptional" situations for all non-Class III and non-implantable devices when "demonstration of conformity" with GSPRs "based on clinical data is not deemed appropriate." In addition, MDCG 2020-13 [8] (clinical evaluation assessment report (CEAR) template) states a "clinical evaluation" is "still required" and an "evidence-based justification shall be presented in the clinical evaluation report." This CEAR template requires the NB to document the alternative data used instead of the clinical data (options included performance evaluation data, bench tests, preclinical evaluations, etc.). The CEAR template also included a number of bulleted items for the NB assessor to consider, as follows:

- Has any available clinical data for the device or an equivalent device been searched for and/or identified by the manufacturer? If yes −was the identified clinical data integrated in the clinical evaluation. This should include an evaluation of clinical data identified from the literature, and an appraisal of their relevance to the device under evaluation.
- Is clinical data available for similar devices, does this provide information with relevant to the safety and performance of the device under evaluation? Has the manufacturer conducted an appropriate search of scientific literature? If clinical data for similar devices is available – this should be included in the CER and evaluated and may be of particular relevance to post-market surveillance/PMCF planning.
- The results of the manufacturer's risk management – Are the results of the manufacturer's risk management supportive of the use of non-clinical testing methods?
- Consideration of the specifics of the interaction between the device and the human body—Is the device under assessment part of a system or stand-alone? Is there sufficient information regarding this interaction available from sources other than clinical data?
- The clinical performance intended – What is the intended performance? Is it reasonable to rely upon non-clinical data for the proposed intended performance?
- The claims of the manufacturer – The manufacturer should not make any claims which are not supported by clinical data.

Prior to this guide, many CER documents were named JUS documents in order to distinguish them from a CER/PER, which included clinical data.

BOX 3.7 Justifying exceptions is normal clinical evaluation/performance evaluation work

The term "just" with all word variants is present 67 times in MDR [1] and 58 times in IVDR [2], with consolidated updates on March 11, 2023. In addition to justifying exceptions to writing CERs when clinical data collection "is not deemed appropriate" (MDR Article 61(10)), justifications are often used in many other situations—for example:

MDR-specific justifications:

EC can add group to device list "without an intended medical purpose" (Article 1), device can be manufactured/used only in EU health institutions (Article 5), manufacturers can use S&P "solutions" equivalent to MDR common specifications (Article 9), systems/procedure packs can include other products (Article 22), NB designation/notification to follow MDCG recommendations (Article 42), NB can make choices about file sampling (Article 45), well-established technologies can be added or removed from class IIb (Article 52), NB/

(Continued)

BOX 3.7 (Continued)

expert panel can have divergent Class III & IIb conformity assessment views (Article 55), competent authority can place a device on the market... when MDR procedures are not met but use is "in the interest of public health or patient safety or health" or urgent "relating to the health and safety of humans" (Article 59).

In addition to the justification specified in clause 10, Article 61 also stated: "(1)... manufacturer shall... justify... clinical evidence necessary to demonstrate conformity with the relevant... [GSPRs and the]... level of clinical evidence shall be appropriate... (7)... Cases in which paragraph 4 [i.e., clinical investigations required for implantable and class III devices] is not applied by virtue of paragraph 6 [i.e., exceptions from clinical investigations when prior CE marked devices have CERs with sufficient clinical data, etc.] shall be justified in the clinical evaluation report by the manufacturer and in the clinical evaluation assessment report by the notified body... (8)... Where justified in view of well-established technologies similar to those used in the exempted devices... or where justified in order to protect the health and safety of patients, users or other persons or other aspects of public health... (9)... products without an intended purpose... [when relying] on existing clinical data from an analogous medical device is duly justified."

Article 62, 4e, 5; Article 63 and Article 71 described clinical investigations to demonstrate MDR conformity shall ensure benefits "justify the foreseeable risks and inconveniences and compliance... is constantly monitored" and subjects do not need to "provide any justification" to "withdraw from the clinical investigation..." The clinical investigation information disclosure in electronic form is "justified" to be kept confidential and not released to the public (Article 73). Justification is needed when: a clinical investigation has been "temporarily halted" or "terminated" or when a "clinical investigation report" will be late (Article 77), one Member State "disagrees" with the "coordinating Member State" regarding the conduct of a clinical investigation; the Member State can "refuse to authorise a clinical investigation" in disagreement "with the... coordinating Member State..." (Article 78), one Member State has a "situation" where serious incidents and field safety corrective actions vary from the consistent content of the notice to all Member States (Article 89), competent authorities "may require... necessary samples of devices... free of charge" for market surveillance activities (Article 93), devices have "unacceptable risk to the health or safety of patients, users or other persons, or to other aspects of the protection of public health," the competent authority "shall... take all appropriate and duly justified corrective action..." (Article 95) and a Member State may object to "Union law" (Article 96) or when taking "Preventive health protection measures (Article 98).

The MDR annexes also have many different justifications, for example, when: "Design and manufacture of devices" includes specified substances (e.g., carcinogenic, mutagenic, toxic to reproduction, endocrine-disrupting substances) above the 0.1% weight by weight limit or "adjustment parameters" (Annex I, GSPRs 10, 14, 21); device is assigned to a particular risk class, when GSPR conformity assessment details including PMCFP, PMCFER were documented and when studies regarding substances added into the body are absent in Technical Documentation (Annex II); "a PMCF is not applicable" (Annex III), combinations are adjusted within the UDI configuration (Annex VI), "exceptional circumstances" occur and NB personnel qualifications "cannot be fully demonstrated" when those personnel are authorized by the NB to "carry out specific conformity assessment activities" (Annex VII, NB must justify this situation to the "authority responsible for notified bodies"), no preclinical evaluation procedures are performed (Annex VII, 4.5.4), no clinical investigations or PMCF are performed (Annex VII, 4.5.5); "Changes and modifications" occur to NB assessment conclusions (Annex VII, 4.9); NB "has not followed the advice of the expert panel" (Annex IX, 5g); claims of equivalence are made in the clinical evaluation (Annex XIV, 3); PMCF activities are planned within "a detailed... time schedule" (Annex XIV, 6.2h) and clinical investigations, expected clinical outcomes for device risks/benefits, statistical designs, summary and results are planned and reported (Annex XV).

IVDR-specific justifications:

Similar to the MDR, the IVDR has many different justifications allowed – for example, when: a "health institution justifies... the target patient group's specific needs cannot be met, health institution... shall include a justification of their manufacturing, modification and use; information on which requirements are not fully met with a reasoned justification" (Article 5); "Manufacturers shall comply with the CS... unless they can duly justify that they have adopted solutions that ensure a level of safety and performance that is at least equivalent thereto" (Article 9); "Where... Member State does not follow... MDCG, it shall provide a duly substantiated justification" (Article 38); "The sampling of files... shall be planned and representative... and be appropriately justified" (Article 41); for "divergent views between the notified body and the experts, a full justification shall also be included" (Article 50); "any competent authority may authorise, on a duly justified request, the placing on the market... a specific device for which the procedures... have not been carried out but use of which is in the interest of public health or patient safety or health... On duly justified imperative grounds of urgency relating to the health and safety of humans, the Commission shall adopt immediately applicable implementing acts..." (Article 54)

In addition, "The manufacturer shall specify and justify the level of the clinical evidence necessary... Clinical performance studies... shall be carried out unless... duly justified to rely on other sources of clinical performance data" (Article 56); anticipated benefits to the subjects or to public health justify the foreseeable risks and inconveniences" (Article 58); "Member States shall assess whether the performance study is designed in such a way that potential remaining risks... are justified, when weighed against the clinical benefits to be expected" (Article 67); "The information... shall be accessible to the public, unless... confidentiality of the information is justified" (Article 69); "specify when the results of the performance study are going to be available, together with a justification" (Article 73); "A Member State... shall refuse to authorise a performance study... on duly justified grounds" (Article 74); "Unless duly justified... the field safety notice shall be consistent in all Member States" (Article 84); "Where... the competent authorities find that the device presents an unacceptable risk... they shall... require the manufacturer... to take all appropriate and duly justified corrective action to bring the device into compliance" (Article 90); "Where a Member State or the Commission considers that the risk to health and safety emanating from a device cannot be mitigated satisfactorily... may take... necessary and duly justified measures to ensure the protection of health and safety..." (Article 91); "Where a Member State... indicates... device or category or group of devices should be withdrawn from the market or recalled, it may take any necessary and justified measures"

(Continued)

BOX 3.7 (Continued)

(Article 93); "in duly justified and exceptional cases instructions for use shall not be required or may be abbreviated..." (Annex 1, 20.1(d)); "a PMPF plan... or a justification as to why a PMPF is not applicable" (Annex III); "Where, in exceptional circumstances, the fulfilment of the qualification criteria... cannot be fully demonstrated, the notified body shall justify... the authorisation of those... personnel to carry out specific conformity assessment activities." (Annex VII, 3.3.1)

Where no new testing has been undertaken... or... (for) deviations from procedures, the notified body... shall critically examine the justification presented by the manufacturer... (including) justifications in relation to non-performance of performance studies or PMPF" (Annex VII, 4.5.4); "The medicinal products authority consulted shall provide its opinion, within 60 days... [which] may be extended once for a further 60 days on justified grounds" (Annex IX, 5.2(d)); "Where any... elements are not deemed appropriate in the Performance Evaluation Plan due to the specific device characteristics a justification shall be provided in the plan" (Annex XIII, 1.1); "manufacturer shall demonstrate (device) analytical performance... unless any omission can be justified as not applicable" (Annex XIII, 1.2.2); "manufacturer shall demonstrate the (device) clinical performance... unless any omission can be justified as not applicable... Clinical performance studies shall be performed unless due justification is provided for relying on other sources of clinical performance data" (Annex XIII, 1.2.3); "performance evaluation report shall... include... justification for the approach taken to gather the clinical evidence" (Annex XIII, 1.3.2); "Each step in the clinical performance study, from the initial consideration of the need for and justification of the study to the publication of the results, shall be carried out in accordance with recognised ethical principles" (Annex XIII, 2.2); "description of and justification for the design of the clinical performance study, its scientific robustness and validity, including the statistical design, and details of measures to be taken to minimise bias..." (Annex XIII, 2.3.2(j)); "the analytical performance... with justification for any omission" (Annex XIII, 2.3.2(k)); "parameters of clinical performance... justification for any omission; and with the exception of studies using left-over samples the specified clinical outcomes/ endpoints (primary/secondary) used with a justification" (Annex XIII, 2.3.2(l)); "Where any... elements... are not deemed appropriate for inclusion in the CPSP due to the specific study design chosen, such as use of left-over samples versus interventional clinical performance studies, a justification shall be provided" (Annex XIII, 2.3.2 end); "a detailed and adequately justified time schedule for PMPF activities" (Annex XIII, 5.2(h)); "If PMPF is not deemed appropriate for a specific device then a justification shall be provided and documented within the performance evaluation report." (Annex XIII, 8)

These MDR and IVDR justification exceptions when conducting CE/PE are meant to empower all medical device clinical data evaluators and others to justify exceptions to regulations as appropriate. Consider how the law is constructed, know the device benefit-risk ratio, document how the benefits outweigh the risks, and specify if additional clinical data are "not deemed appropriate" to collect. Do not hesitate to put forward an ethical, rational, and strong justification when appropriate clinical and performance evaluation strategies need to be justified.

The regulations are flexible enough to allow situations when no clinical data should be gathered (e.g., trials are inappropriate when a patient may need to be exposed to a virus or to radiation, or when the device does not provide a direct clinical benefit, like a sterilizer). In this setting, clinical data still needs to be identified and collected for appraisal and analysis even though clinical data will not be for the subject/equivalent device, but for the background knowledge and SOTA, and to describe alternate treatment options using similar devices to define the expected subject device S&P.

3.3 Using appropriate software to store identified clinical data

Creating a designated workspace is advisable to organize and store identified clinical data from trials, literature, and experiences. Having a user-friendly workspace will allow each decision about each clinical data point to be documented quickly, easily, and reliably so others can see how the data were interpreted. For example, a simple Excel workbook can include all relevant data (e.g., a separate worksheet can be created to house clinical data collected from each clinical data source: trial, literature, and experience data). Similarly, a simple Word document (**Appendix L: CER/PER Template) can house the developing CER/PER draft document** (Fig. 3.6).

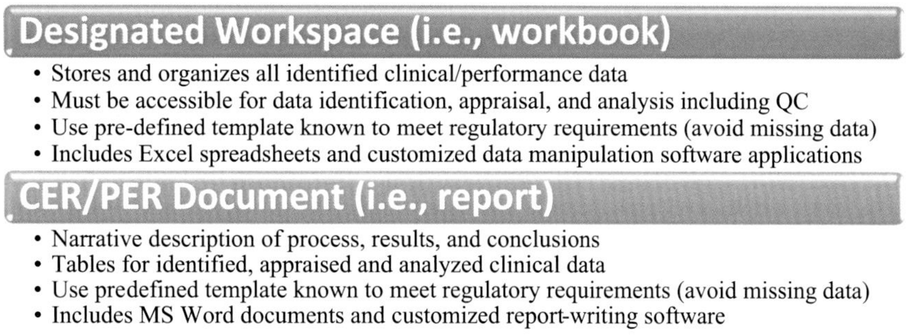

Designated Workspace (i.e., workbook)

- Stores and organizes all identified clinical/performance data
- Must be accessible for data identification, appraisal, and analysis including QC
- Use pre-defined template known to meet regulatory requirements (avoid missing data)
- Includes Excel spreadsheets and customized data manipulation software applications

CER/PER Document (i.e., report)

- Narrative description of process, results, and conclusions
- Tables for identified, appraised and analyzed clinical data
- Use predefined template known to meet regulatory requirements (avoid missing data)
- Includes MS Word documents and customized report-writing software

FIGURE 3.6 Workspace and clinical evaluation report document suggestions.

Avoid conducting "work" inside the CER/PER document itself. Using a separate spreadsheet or workbook outside the Word document report is critical to save time, to work most effectively, and to ensure accuracy. Entering data into a workbook in an organized fashion starting with the clinical data identification step will be most efficient for the clinical data identification, appraisal and analysis steps. Managing identified clinical data in the most appropriate data repository or software should facilitate creating and modifying data points, tables, and figures easily and repeatedly. The result can then be entered into the Word document after the data have been fully analyzed in the workbook or software fit for the analysis purpose.

For example, creating, modifying, and sorting a table of adverse events from multiple clinical trials in a table format within Excel is much easier and more reliable than the same table managed in an MS Word document. The Excel table can be auto-tabulated, graphed, and calculated using Excel functions which are not available in the MS Word document. In addition, the Excel workbook can contain all identified clinical data and the report can focus on the smaller "included" dataset after the data were manipulated to be in the proper format for presentation in the Word document. The workbook can simply be an appendix to the CER/PER so exhaustive lists do not need to be created in addition to the direct downloads from various databases into the Excel workbook.

In other words, the workbook should: (1) organize clinical data in one place (e.g., software, folder, spreadsheet) to prepare for an efficient appraisal process; (2) ensure all clinical data are visible since all information can be simply pulled over from different databases and listed in the workbook even if not used in the report (e.g., title, journal details, publications dates, full article location on web for literature and similar listings for clinical trials and experience reports can be directly exported into Excel); and (3) allow sorting for future identified clinical data appraisal and analysis.

3.4 Clinical trial data identification

Manufacturer-held or company-sponsored clinical trial data are often the most valuable clinical data, especially if the clinical trial was designed carefully to collect detailed clinical S&P data about the subject device. These well-designed clinical trials will often control as many variables as possible to answer specific questions about device benefits and risks. These trials often support a regulatory submission required before the subject device can be placed on the market and sold in a specific geography like the EU. PMCF/PMPF clinical trials are also conducted within the PMS system specifically to address clinical data gaps found during the CE/PE process or in the PMS or RM systems.

In general, when a clinical trial is completed by the manufacturer or others using the subject or equivalent device, the resulting clinical trial data are the most relevant and often the highest-quality clinical data available. For these reasons, as directed by the CEP/PEP, the first clinical trial data search should be conducted to see if the company holds any clinical trial data using the subject or equivalent device or any similar devices. The company may have sponsored a clinical trial or supported an investigator-initiated clinical trial completed by others. Alternatively, the company may have a relationship with another company having completed a clinical trial with the subject or equivalent device. All of these clinical trial data are needed for the CE/PE process, regardless of whether the clinical trial was completed during the premarket or postmarket phase of device development.

3.4.1 Clinical trial database searches

Unfortunately, clinical trial details are not always available to the manufacturer when trials are completed by others even when the manufacturer seeks a contract to allow direct access to the clinical trial data for CE/PE. The CE/PE process requires all the detailed measurements and outcomes, including all benefits, risks, and safety signals, as well as desirable and undesirable clinical outcomes. For clinical trials completed by others where direct data sharing is not possible, the manufacturer should follow clinical trial progress by checking publicly available clinical data in clinical trial, literature, and experience databases. Several international clinical trial databases and registries are freely and publicly available. These trial databases can be searched to identify relevant clinical trials; however, the listings and follow-up reports in these databases are often fragmented, incomplete and generally of low quality (Table 3.2).

TABLE 3.2 Clinical trial database/registry examples

Search engine	Geography	Description	Link
ClinicalTrials.gov	US	National Institutes of Health (NIH) US National Library of Medicine database with more than 4492,684 research studies in all 50 states and 223 countries (as of 27APR2024)	https://clinicaltrials.gov
EU Clinical Trials Register (CTR)	EU	Contains more than 43,865 clinical trials (as of 27APR2024)	https://www.clinicaltrialsregister.eu/ctr-search/search
World Health Organization: International Clinical Trials Registry Platform (ICTRP)	International	Voluntary trial registry with searchable portal of more than 469,858 trials (as of 27APR2024) including the following data providers: Australian New Zealand Clinical Trials Registry, last data file imported on 22 April 2024 Chinese Clinical Trial Registry, last data file imported on 22 April 2024 ClinicalTrials.gov, last data file imported on 22 April 2024 Clinical Trials Information System (CTIS), last data file imported on 2 April 2024 EU Clinical Trials Register (EU-CTR), last data file imported on 22 April 2024 ISRCTN, last data file imported on 22 April 2024 The Netherlands National Trial Register, last data file imported on 23 April 2024 Brazilian Clinical Trials Registry (ReBec), last data file imported on 1 April 2024 Clinical Trials Registry - India, last data file imported on 1 April 2024 Clinical Research Information Service - Republic of Korea, last data file imported on 2 April 2024 Cuban Public Registry of Clinical Trials, last data file imported on 1 April 2024 German Clinical Trials Register, last data file imported on 8 April 2024 Iranian Registry of Clinical Trials, last data file imported on 8 April 2024 Japan Registry of Clinical Trials (jRCT), last data file imported on 1 April 2024 Pan African Clinical Trial Registry, last data file imported on 8 April 2024 Sri Lanka Clinical Trials Registry, last data file imported on 1 April 2024 Thai Clinical Trials Registry (TCTR), last data file imported on 1 April 2024 Peruvian Clinical Trials Registry (REPEC), last data file imported on 12 March 2024 Lebanese Clinical Trials Registry (LBCTR), last data file imported on 1 April 2024 International Traditional Medicine Clinical Trial Registry (ITMCTR), last data file imported on 1 April 2024	https://trialsearch.who.int
International Standard Randomized Controlled Trial Number (ISRCTN) Registry	UK	Trial registry with 24,814 listings (as of 27APR2024)	https://www.isrctn.com

NOTE: While laws and policies around the world have set expectation to list clinical trials on public databases (esp. when the trial is supported by public funding or if the research is intended to be published in top-tier journals), most observational and many research studies aren't required to be listed on public databases by law or company policy.

For example, https://clinicaltrials.gov is the largest database listing clinical trials from more than 220 countries. As defined in the CEP/PEP, the clinical data identification searches in these clinical trial databases should use appropriate search terms and store results in an appropriate workbook (e.g., an Excel spreadsheet) or purchased CE/PE clinical data storage database. The evalutor must ensure the stored clinical data are reliable, valid, and can be reproduced easily with the same result. The details must be accurately recorded including the search methods (i.e., databases searched, terms used, search dates, etc.) along with the search outputs and results. The goal is to identify relevant, high-quality clinical trials for the CE/PE process (Table 3.3).

TABLE 3.3 Clinical trial search strategy summary

Where?	Manufacturer-held, clinical trial database or registry (e.g., https://clinicaltrials.gov).
What?	Planned search terms (e.g., *device name, *manufacturer name).
How?	Search database, download SELECTED outputs in CSV format and convert into Excel.
When?	Within 6 months of CER/PER completion; GOAL: same date as literature and experience searches.

A good strategy is to start with the subject device name and/or manufacturer name. This will work well for products with a unique name; however, when a product name is too generic, then a product and manufacturer name combination may help to focus the results (Table 3.4).

TABLE 3.4 Fictional clinical trial search

Search	Search term	Results[1]	Comment[2]
1	<Product name>	2493	Not specific enough, merged with company name.
2	<Manufacturer name>	255	Not specific enough, merged with product name.
3	1 AND 2	120	Keep.

[1]*Record total number of trials identified in each search output.*
[2]*Use a comment column in the spreadsheet or database to document which results were included and which were not included for the next step in the CE/PE process.*

Document the exact date, database (name and link) as well as the number of results from each search. Download all information about the trials to be kept in a "comma-separated values" (.csv) format for transport into the Excel workbook (or other software to be used for data appraisal and analysis). Ideally, all clinical trial data searches should be conducted on the same date. This single-date strategy offers a clean, uncomplicated search method description for documentation in the CER/PER.

BOX 3.8 Clinical trial data identification surprises

Case #1: During a CER training course, a manufacturer team conducted a clinical trial search on https://clinicaltrials.gov for their device. During the activity, the company team members noted a trial using the company product. The company was not aware of this non-company-sponsored trial currently recruiting participants with an on-label use of their device. These clinical data were highly valued and the company personnel were encouraged to reach out to the investigator to arrange to learn more about the clinical trial and the potential to gain access to the clinical trial data.

Case #2: Several company-sponsored clinical trials were found on https://clinicaltrials.gov during a CER update review; however, these trial data had not been analyzed in the CER update. The company CER team members were not aware the trials existed. After lengthy discussion and extensive searching within the company, the clinical trial data were retrieved and the CER was updated to include these highly valuable clinical trial data about the on-label device uses.

Avoid eliminating results during the clinical data identification step. Data inclusion/exclusion will be done in the appraisal step. To facilitate comprehensive identification of all relevant trials, the evaluator must combine terms appropriately during clinical trial searches. The evaluator also needs to review the output from each search to determine if the expected clinical trial data were identified. If not, a new search strategy may be required. Comments are helpful from the evaluator to documentwhy any search results were "not specific enough" and recording the strategy to merge searches in order to focus the results to identify as much relevant, high-quality data as possible without bringing forward too many trials to be appraised and analyzed.

Example 1: N95 face mask clinical trials

In this example, the evaluator used an Excel spreadsheet to store and organize identified clinical data about the "N95" face mask manufactured by "3M" (i.e., the subject device). The name of the device was generic and would not yield specific clinical trials so the search strategy included the device and the manufacturer names. The evaluator constructed a key to record search strategy and details for each search. The key was kept in a separate worksheet in the Excel workbook. The clinical trial key included the numbered search terms used and the outputs from each search with any Comments (Table 3.5).

TABLE 3.5 ClinicalTrials.gov search key: face masks.

Search no.	Search terms	Result <date>	Comments
1	N95	61	Generic search, use for background/SOTA.
2	"N95 and 3M"	2	Specific search, use for subject device.

First, the clinical trial data in the https://clinicaltrials.gov database were searched following the plan to use "N95" in the "other terms" search window. The first search identified 61 studies, which were downloaded directly into a .csv format (Fig. 3.7).

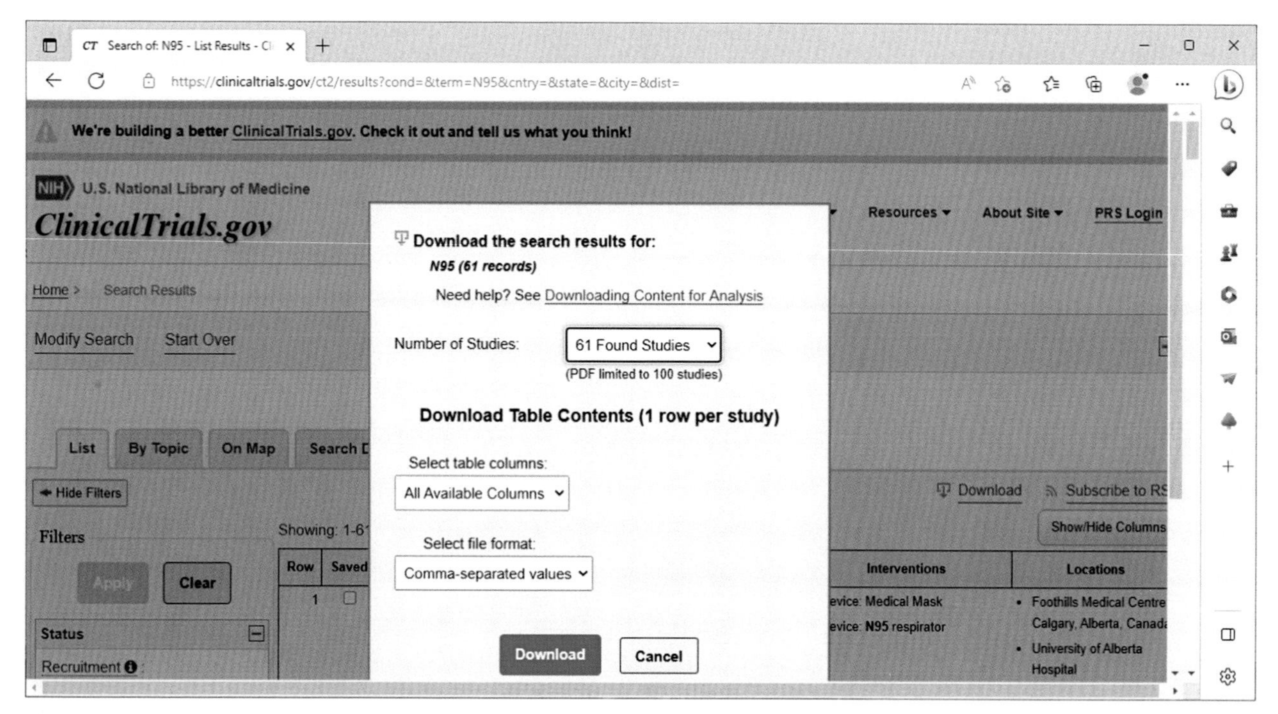

FIGURE 3.7 Clinical trial search download example: "N95".

The .csv file was opened and saved in an Excel workbook format (e.g., .xlsx) (Fig. 3.8).

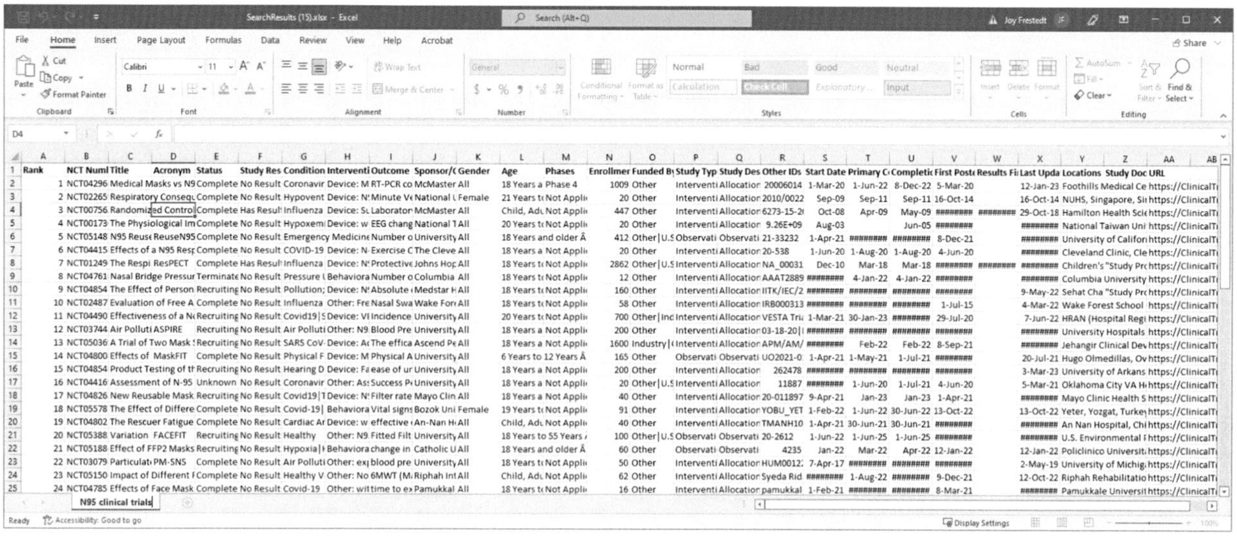

FIGURE 3.8 Clinical trial search export example: "N95".

The evaluator added columns to indicate a unique identifier (UID) number for each data point and search number for each search string used (Search 1, Search 2, etc.), and to allow the inclusion and exclusion (I&E) criteria to be entered during the appraisal step (Fig. 3.9).

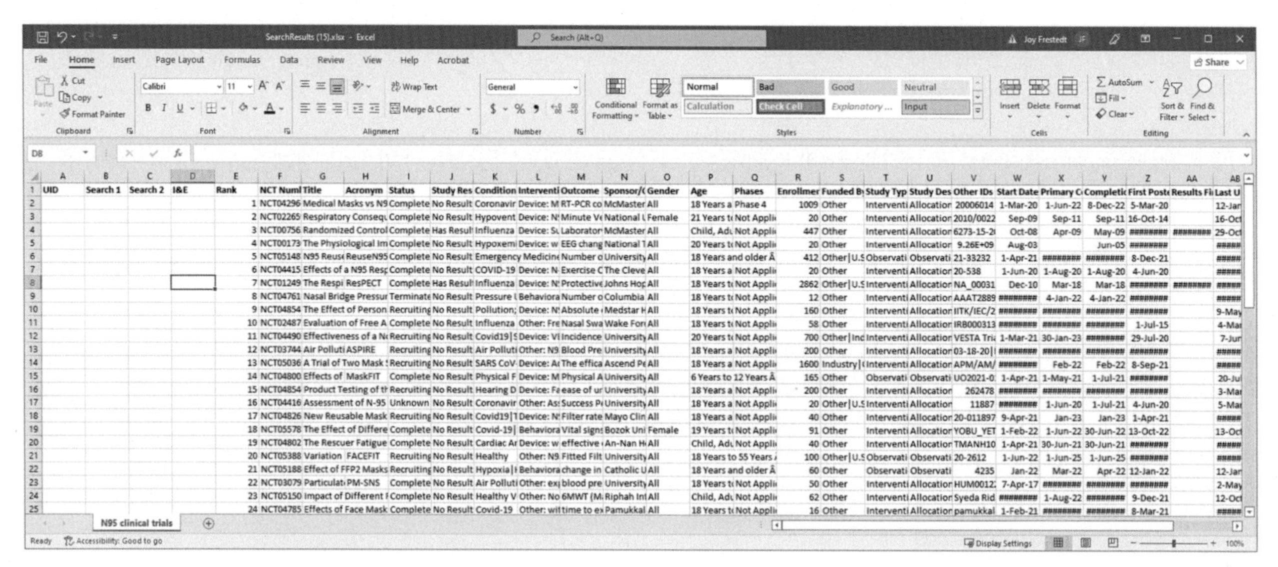

FIGURE 3.9 Adding spreadsheet columns.

To verify thes search identified appropriate clinical trials with the subject device, a separate search was done for "N95 and 3M," and this search resulted in two clinical trials (Fig. 3.10).

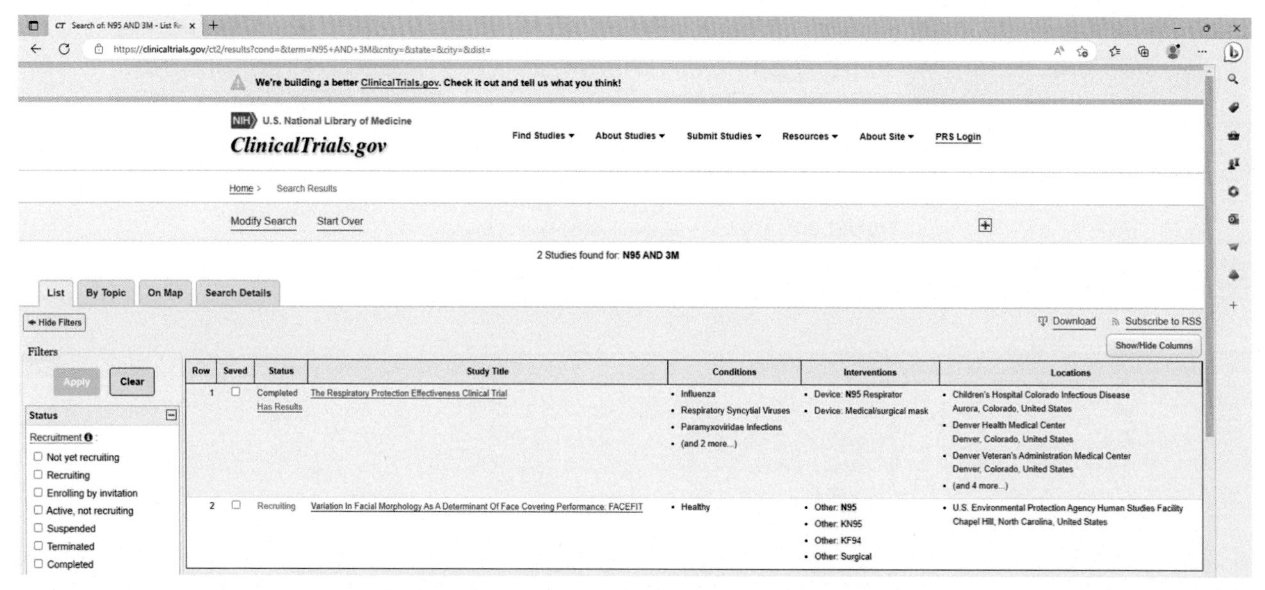

FIGURE 3.10 Clinical trial search result for "N95 and 3M".

The two trials were downloaded and highlighted as duplicates in the initial output for careful review and potential inclusion during the appraisal step as studies specifically about the subject 3M N95 face mask. TThe planned search identified 61 clinical trials and at least 2 trials appeared to be about the subject device. Searches may also be conducted for competitor devices or alternative therapies. After the clinical trial search methods, results and all downloads are documented, the workbook should be saved for furtherappraisal and analysis.

Example 2: Tongue depressor clinical trials

In this example, the evaluator used an Excel spreadsheet to store and organize identified clinical data about the "tongue depressor" manufactured by "Puritan" (i.e., the subject device). The evaluator documented the search strategy key and details for each search in an Excel worksheet pageincluding the numbered search terms used and the outputs, with any Comments (Table 3.6).

TABLE 3.6 ClinicalTrials.gov search key: tongue depressors.

Search no.	Search terms	Result <date>	Comments
1	"Tongue Depressor"	32	Generic search, use for background/SOTA.
2	Puritan	22	Generic search, use for background/SOTA.
3	Puritan AND "Tongue Depressor"	1	Specific search, use for subject device.

The output was converted from .csv to Excel (.xlsx) format including all contents available about the trials. and *NOTE: only the rank, NCT number, and title are shown in the three searches*

Search #1: https://clinicaltrials.gov search date: December 29, 2022; search term: tongue depressor Search Result: 32 (Table 3.7).

TABLE 3.7 "Tongue depressor" clinical trial search.

Rank	NCT number	Title
1	NCT03095183	Do Traditional or Flavored Tongue Depressors Make for Easier Posterior Oropharynx Exams in Pediatric Patients?
2	NCT04256590	Tongue Depressor-Related Tongue Swelling
3	NCT00507208	Dynasplint Therapy for Trismus in Head and Neck Cancer
4	NCT05464225	Colorado Oral Strengthening Device
5	NCT03443947	Modified Mallampati Scoring Technique for Airway Assessment
6	NCT05074784	Effects of IOPI on Swallowing Function and Functional Status in Geriatric Patients
7	NCT05656950	Assessment of Oral Health Status of Patients in Pediatric Intensive Care Units
8	NCT04205253	Tongue Depressor-Related Ischemia-Reperfusion Injury in Tongue
9	NCT05310019	Effect of Electroacupuncture in Patients Submitted Orthognathic Surgery and Mentoplasty
10	NCT04496960	Safety of Tofacitinib, an Oral Janus Kinase Inhibitor, in Primary Sjogren's Syndrome
11	NCT03605186	Prevention of Oral Mucositis After Using Chamomile Oral Cryotherapy Versus Oral Cryotherapy in Pediatric Cancer Patients Receiving Chemotherapy
12	NCT04648631	Hidradenitis Suppurativa (HS) Tunneling Wounds
13	NCT02144207	Optimal Laryngoscopic View to Enable GlideScope-Assisted Tracheal Intubation
14	NCT00094978	Depsipeptide/Flavopiridol Infusion for Cancers of the Lungs, Esophagus, Pleura, Thymus or Mediastinum
15	NCT03979924	Mouth Opening, Prevention, Education, Nutrition (OPEN)
16	NCT03889704	Neuromuscular Electrical Stimulation for Jaw-Closing Dystonia
17	NCT02673853	Assessment of Residual Paralysis in Patients Who Receive Mini-Dose Atracurium During Supraglottic Airway Insertion
18	NCT01695980	Laryngeal Mask Airway in Pediatric Adenotonsillectomy
19	NCT03253510	Measuring Biting Force of Poly-amide Complete Dentures
20	NCT03743805	Rapid Reversal of CNS-Depressant Drug Effect Prior to Brain Death Determination
21	NCT02318654	Ice Versus EMLA for Pain in Laser Hair Removal

(*Continued*)

TABLE 3.7 (Continued)

Rank	NCT number	Title
22	NCT05618938	Effects of Rocabado's Approach Versus Kraus Exercise Therapy
23	NCT04129229	LinguaFlex Tongue Retractor (LTR) for the Treatment of OSA and Snoring in Adults
24	NCT04694963	A Multi-Center Study to Determine the Prevalence and Influence of Pertussis on Subacute Cough in Shenzhen
25	NCT04694430	A Multi-Center Study to Determine the Prevalence and Influence of Pertussis on COPD Exacerbation in Shenzhen
26	NCT04606004	Perianal Maceration in Pediatric Ostomy Closure Patients
27	NCT01184118	DREAM: Does Inhaled Fluticasone Result in Obstructive Sleep Apnea Manifestations?
28	NCT01554488	Inhaled Fluticasone Effects on Upper Airway Patency in Obstructive Lung Disease
29	NCT00762177	Investigate Oral Bacteria in Adult Population
30	NCT02256280	A Randomized Double Blind Controlled Trial Comparing Sugammadex and Neostigmine After Thoracic Anesthesia
31	NCT05077865	Single Ascending and Multiple Dose Study to Evaluate Safety, Tolerability, and PK of MYMD1 in Healthy Male and Female Adult Subjects
32	NCT04349761	Single Ascending-Dose Study to Evaluate Safety, Tolerability, and PK of MYMD1 in Healthy Male Adult Subjects

Search #2: https://clinicaltrials.gov search date: December 29, 2022; search term: Puritan Search Result: 22 (Table 3.8).

TABLE 3.8 "Puritan" clinical trial search.

Rank	NCT number	Title
1	NCT02801994	Impact of Proportional Assisted Ventilation on Dyspnea and Asynchrony in Mechanically Ventilated Patients
2	NCT02639364	Airway Pressure Release Ventilation (APRV) Protocol Early Used in Acute Respiratory Distress Syndrome
3	NCT03095183	Do Traditional or Flavored Tongue Depressors Make for Easier Posterior Oropharynx Exams in Pediatric Patients?
4	NCT01403467	Impact of Continuous Positive Airway Pressure on the Treatment of Acute Asthma Exacerbation
5	NCT01479959	Impact of Patient Controlled Positive End-Expiratory Pressure on Speech in Tracheostomized Ventilated Patients
6	NCT05045443	Safety and Efficacy of Curcumin in Children with Acute Lymphoblastic Leukemia
7	NCT04140682	Proportional Assisted Ventilation and Pressure Support Ventilation in Adult Patients with Prolonged Ventilation
8	NCT05356299	Analysis of the Magnetic Tape Bandage on Respiratory Functional Effects
9	NCT05498454	Contribution of Online Stress and Pain Mindfulness Treatment to ACT Process Change and Outcomes in Chronic Pain
10	NCT02390024	Influence of Patient/Ventilador Decoupling in Neurocognitive and Psychopathological Sequelae in ICU Patients
11	NCT02071160	Melatonin for Neuroprotection Following Perinatal Asphyxia
12	NCT02085499	Flow-Synchronized Nasal IMV in Preterm Infants
13	NCT05570526	Effect of Melatonin in Pediatric Hemodialysis Patients
14	NCT01083277	Variable Ventilation during Acute Respiratory Failure
15	NCT02447692	Proportional Assist Ventilation for Minimizing the Duration of Mechanical Ventilation: The PROMIZING Study
16	NCT01944085	Percutaneous Pin Removal in Children—is Analgesia Necessary?
17	NCT04998383	HVNI Versus Noninvasive Ventilation for Acute Hypercapnic Respiratory Failure
18	NCT05499039	High Flow Nasal Cannula Versus Non-Invasive (NIV) in Both Hypoxemic and Hypercapnic Respiratory Failure
19	NCT03788304	High Flow Nasal Cannula Versus Non-Invasive Ventilation in Prevention of Escalation to Invasive Mechanical Ventilation in Patients With Acute Hypoxemic Respiratory Failure
20	NCT02822859	A Comparison of Three Nebulizers for Standard Clinical and Research Use in Methacholine Challenge Testing
21	NCT02404701	Effect of Over-the-Counter Dietary Supplements on Kidney Stone Risk
22	NCT01089010	A Study of CK-2017357 in Patients with Amyotrophic Lateral Sclerosis (ALS)

Search #3: https://clinicaltrials.gov search date: December 29, 2022; search term: Puritan AND tongue depressor Search Result: 1 (Table 3.9).

TABLE 3.9 "Puritan" AND "tongue depressor" clinical trial search.

Rank	NCT number	Title
1	NCT03095183	Do Traditional or Flavored Tongue Depressors Make for Easier Posterior Oropharynx Exams in Pediatric Patients?

Although some of these clinical trials may be interesting for the CER background/SOTA section, only one clinical trial was about the Puritan Tongue Depressor specifically and the concept of "easier" might be important to evaluate for tongue depressor safety and performance when used in children. Searches may also need to be conducted for competitor devices or alternative therapies. After the clinical trial search methods, results and all downloads are documented in the workbook, the workbook should be saved for further appraisal and analysis.

3.4.2 Controlling clinical trial identification and data gathering

All clinical trial identification steps should be conducted within an established a QMS including a standard operating procedure (SOP), WI, and an aligned checklist (Table 3.10).

TABLE 3.10 Example clinical trial data identification checklist

UID	Description
1	CEP/PEP detailed a comprehensive, objective, and justified clinical trial search method with appropriate rationales.
2	CEP/PEP clinical trial search strategy was followed (note deviations: ____________).
3	Manufacturer-held clinical trial data were compiled and entered into CER/PER workbook or database.
4	Search strategy was executed with search details documented in workbook (e.g., databases searched, date each search was conducted, specific search terms/strings used, number results).
4	Search terms were appropriate for subject device (based on product literature), equivalent device (if applicable) and generic device.
6	Search method was reproducible with same results.
7	Non-manufacturer-held clinical trial data were identified and entered into CER/PER workbook or database.
8	Clinical trial data were identified over all time (i.e., no time frame limits were applied).
9	Full clinical trial results and all identified clinical trial data were downloaded, gathered and stored for the next appraisal and analysis steps (i.e., final study protocols, reports, and data records were secured when available and/or all available clinical trial data were downloaded from international databases including article title, investigator, study details, links to any publications, etc.).
10	If no actual clinical trial data were available, mark the identified trial appropriately and suggest exclusion in the Comments because clinical/performance data were not available in the trial (include details about to any clinical literature referring to the identified trial).

QMS tools should be stored and accessible to all users to help ensure success when completing CE/PE clinical trial identification process steps. *Note how item #10 clarifies what to do when clinical trial data were not available for appraisal and analysis during the CE/PE (i.e., no clinical trial data can be sourced in a particular clinical trial).* This is typical for non-manufacturer-held data, since most companies do not share clinical trial details publicly. In many cases, some of the trial data may be available in the clinical literature, but clinical literature is not a full clinical trial report including all the clinical trial report details (e.g., protocols, informed consent forms, detailed patient clinical data, adverse event and device deficiency (DD) listings). A cross-reference to the published literature may be helpful for each identified clinical trial.

3.4.3 Challenges during clinical trial identification

Internal manufacturer-held and company-sponsored clinical trial data should be reviewed and assessed for potential bias as the clinical trial data are identified and collected. Unfortunately, the clinical trial data in clinical trial databases are often incomplete and out of date (e.g., ~50% of listed trials had missing data even though many trials were marked as completed). Follow-up information may be available by searching clinical literature databases for trial key words or publications by the principal investigator name. Sometimes, publications with complete study results are present prior to database listings being updated. Document this ancillary process to identify more information about the identified clinical trials clearly in the CER/PER workbook.

Different clinical trial databases store data in various formats. For example, clinicaltrials.gov uses standard columns when exporting clinical trial data into a .csv file which can then be converted into an Excel workbook including the study title, status, and results, interventions, study location, and study National Clinical Trial (NCT) number, etc.. Other clinical trial databases provide different data in different formats, so combining these data is time consuming.

NOTE: Like most regulatory authorities, the EU prefers clinical trials to be conducted in the country where the device will be marketed. The EU will accept external device clinical data as long as the clinical data have an appropriate justification about why the data can be considered relevant to the EU population expected to use the device.

Clinical trial data may be manufacturer-held or available in limited form within searchable databases or online registries. Search terms may include the subject device, equivalent device, alternative therapies, and generic device names as appropriate. Adding the manufacturer name along with the device name may be helpful to limit and focus clinical trial identification results (similarly, focusing on the subject device is most useful since little actual clinical data may be found in the international trial databases). The clinical trial identification search plan outlined in the CEP should be followed, and results must be documented (e.g., in a spreadsheet or commercial CE/PE database). Identifying clinical trials within international clinical trial databases and registries is not difficult. Simply search the appropriate terms and store the results in a user-friendly format for future appraisal and analysis. When searching to identify clinical trial data, always start with the manufacturer to see if they have any manufacturer-held or company-sponsored clinical trial data for the subject device. Then search all appropriate international clinical trial databases to see if any clinical trials have been conducted for the subject device and should be considered in the CE/PE process.

3.5 Clinical literature data identification

Scientists and clinicians (e.g., physicians, nurses) write and publish articles to share clinical data. Sometimes the article may be about a target medical device; however, more often, articles are written about a disease, condition, procedure, or treatment. This makes clinical literature identification, appraisal, and analysis complex. In nondevice-focused articles, the evaluator will need to dissect the available clinical data within the article carefully to see if the data are relevant to the subject/equivalent/other device or background/SOTA/alternative therapy/device. The CE/PE process must evaluate the clinical literature (i.e., published articles) to develop both the background knowledge/SOTA and alternative therapy/device discussion as well as the subject/equivalent device S&P and benefit-risk ratio evaluation.

The evaluator must understand the research intent as well as the clinical S&P data, which may not be the point of the article. Simply identifying articles because the device name is used in the article is often not "sufficient" to allow adequate CE/PE. The evaluator must apply their knowledge to identify all meaningful clinical literature, which may include clinical data about the device S&P when used as intended. The appraisal step will further refine the identified clinical literature and the analysis step will complete the decision-making process about whether the article should be included in the CE/PE or not. These ultimate decisions are not part of the identification step. Here, the goal is simply to identify relevant, high-quality literature for further appraisal and analysis.

Sometimes, the manufacturer may be responsible for publishing and/or sponsoring the publication of clinical literature by others (i.e., manufacturer-held clinical literature); at other times, the literature may be published without any connection to the manufacturer. Thus, the evaluator must search for clinical literature data within the manufacturer-held files and in clinical literature databases outside the company. For all clinical trials, documenting, obtaining and storing each identified full-length publication and all supplemental materials for the publication is important.

3.5.1 Clinical literature database searches

After gathering all manufacturer-held clinical literature (i.e., including all clinical literature used by the sales and marketing teams), a rigorous literature search should be conducted in several large literature databases (Fig. 3.11).

FIGURE 3.11 Images of literature search databases entry points. *(A) PubMed, (B) Embase, (C) Cochrane Library, (D) Google Scholar.*

For the EU, the CER/PER must include search results from Embase, even though this literature search engine is quite expensive to use. Many evaluators may choose to use free databases like PubMed, Cochrane Library, and Google Scholar first to search for relevant literature globally and then use EMBASE to save on costs (Table 3.11).

TABLE 3.11 Example databases for clinical literature identification

Search engine	Details	Link
PubMed	Free; 37 + million biomedical literature articles from MEDLINE, more than 7,000 journals, books, etc. in more than 50 languages, maintained by the US NIH	https://pubmed.ncbi.nlm.nih.gov
Embase	Fee-bearing; 45.6 million articles from 8,400 journals, 95 + countries, gray literature including 5,1 million conference abstracts; includes Medline; maintained by Elsevier	https://www.embase.com/landing?status = grey
Medline (OVID)	Free; component of PubMed, 5,200 global journals maintained by the US NIH	https://www.nlm.nih.gov/medline/medline_overview.html
Cochrane Library	Free databases; 9,000 + systematic reviews; a UK nonprofit organization maintained by Wiley	https://www.cochrane.org
Google Scholar	Free internet search engine; bibliographic database with160 + million documents, journal articles, patents, books, case law, etc.	https://scholar.google.com

Many literature databases allow search results to be exported into an Excel spreadsheet including title, authors, journal, etc. Multiple database searches are required. Downloaded literature search outputs should be combined into a single spreadsheet (similar to process used for the clinical trial database search outputs). The stored data must identify each of the separate searches discretely and must keep the same data from each database search in the same columns to avoid confusion (i.e., the data and the column order often differs between database downloads).

The detailed methods used for the completed literature searches must be written in the CER/PER, including any deviations from the CEP/PEP. These methods should describe exactly what happened during the clinical literature identification process (Table 3.12).

TABLE 3.12 Document literature identification strategy in CER/PER

Where?	Literature databases searched (e.g., Embase, PubMed, Cochrane, Google Scholar).
What?	Planned/added search terms (e.g., device-specific, background/SOTA, PICO).
How?	Boolean search strings (e.g., *manufacturer name AND *device name).
When?	Within 6 months of submission; same date as clinical trial and experience searches.

Literature searches are done in several databases which differ in important ways: (1) search engines/algorithms differ (i.e., searches may work in different ways after the search terms are entered); (2) publication types differ (e.g., Cochrane Library contains systematic reviews); and (3) study geographies differ (e.g., databases may contain more data from home regions). In addition, research practices and scientific rigor may vary depending on the locations and situations where the study occurs; therefore, the evaluator must document the published study geographic locations in the literature worksheet or database. The EU favors clinical literature data from EU countries, so EU databases must be searched and studies conducted in Europe should be highlighted in the CER/PER.

The literature search process needs to be iterative (i.e., repeated many times) since knowledge will be learned as the CE/PE process proceeds. The CEP/PEP can always be supplemented with more searches and more data as new knowledge is accumulated; however, the plan should never have fewer searches or less data than planned. Document additional searches with rationales for the added work. Clinical literature searches should identify all clinical/performance literature about the subject/equivalent device (and similar devices) including clinical data to substantiate all subject device indicated uses. Clinical literature searches should also identify appropriate background knowledge/SOTA/competitor/alternative therapy/device clinical literature to define the context for all the other patient treatment options (e.g., including relevant clinical application, target patient population, disease state or condition and treatments offered including treatment with similar devices, alternative treatments or no treatment at all, if medically appropriate) (Fig. 3.12).

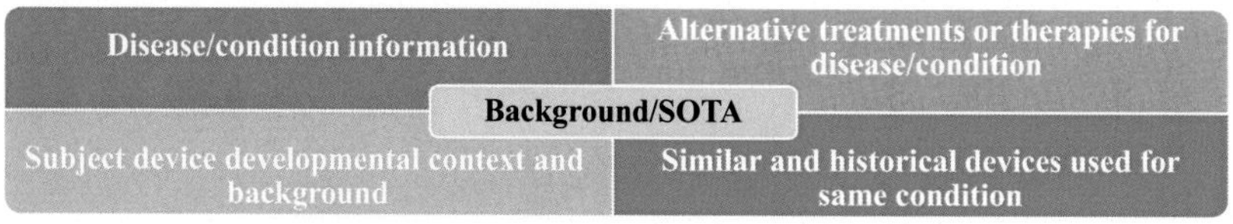

Disease/condition information
Alternative treatments or therapies for disease/condition
Background/SOTA
Subject device developmental context and background
Similar and historical devices used for same condition

FIGURE 3.12 Background/state-of-the-art contents.

Some search engines use Boolean search terms to expand or narrow search results (Table 3.13). Always define search methods to document search results while learning which operators are necessary to obtain the desired outcome for each CE/PE process.

TABLE 3.13 Example Boolean operators used in PubMed.

Boolean	Description	Action	Example in BOLD (description)
AND	All terms present and exclusive	Narrows	3M AND N95 (identifies only articles including both 3M and N95 terms present)
OR	Any term present and inclusive	Expands	mask OR respirator (identifies all articles with either mask or respirator terms present)
NOT	One term present but not other	Narrows	respirator NOT ventilator (identifies articles with respirator term but excludes articles with ventilator term)
""	Exact term order present	Narrows	"COVID-19 respirator" (identifies articles with exact phrase)
( )	Groups terms together	Narrows	(respirator OR mask OR N95) AND COVID (emphasizes and groups OR terms together separate from the AND term)
*	Allows various word endings	Expands	Respirat* OR pulmon* (identifies any word variant, e.g., respirator, respirators, respiration, respiratory, or pulmonary, pulmonology, pulmonologist)

When conducting literature searches, the evaluator must review selected database logic available on the website to ensure operators are used correctly. Some operators may be less valuable because databases contain logic superseding the operator. For example, PubMed contains logic automatically applied for word endings (in table as * use). If "respirator" is searched, then "respirators," "respirations," "respiratory," etc. are automatically applied to the search function. The evaluator must search the database, download the SELECTED outputs in .csv format, then convert into Excel and save for evaluation and storage.

Example 3: N95 Face Mask Clinical Literature.

The PubMed database main search window at https://pubmed.ncbi.nlm.nih.gov was searched following the plan for the device and company name "N95 and 3M," and 88 articles were identified. The evaluator constructed a clinical literature key to document the search strategy details for each search including the search number, search terms used, and outputs, with any notes or comments (Table 3.14).

TABLE 3.14 PubMed search key: N95 and 3M

Search no.	Search terms	Result <date>	Comments
1	N95	2308	Generic search, use for background/SOTA.
2	3M	39862	Generic search, use for background/SOTA.
3	N95 AND 3M	88	Specific search, use for subject device.

A quick review of the results indicated the first few articles were relevant to N95 face masks. For example, the first two articles explored in the N95 results reported1) device failures related to "off label" uses during an overlapping procedure using two face masks, and 2) the need for systematic QC when "reprocessing" and sterilizing N95 face masks (Fig. 3.13).

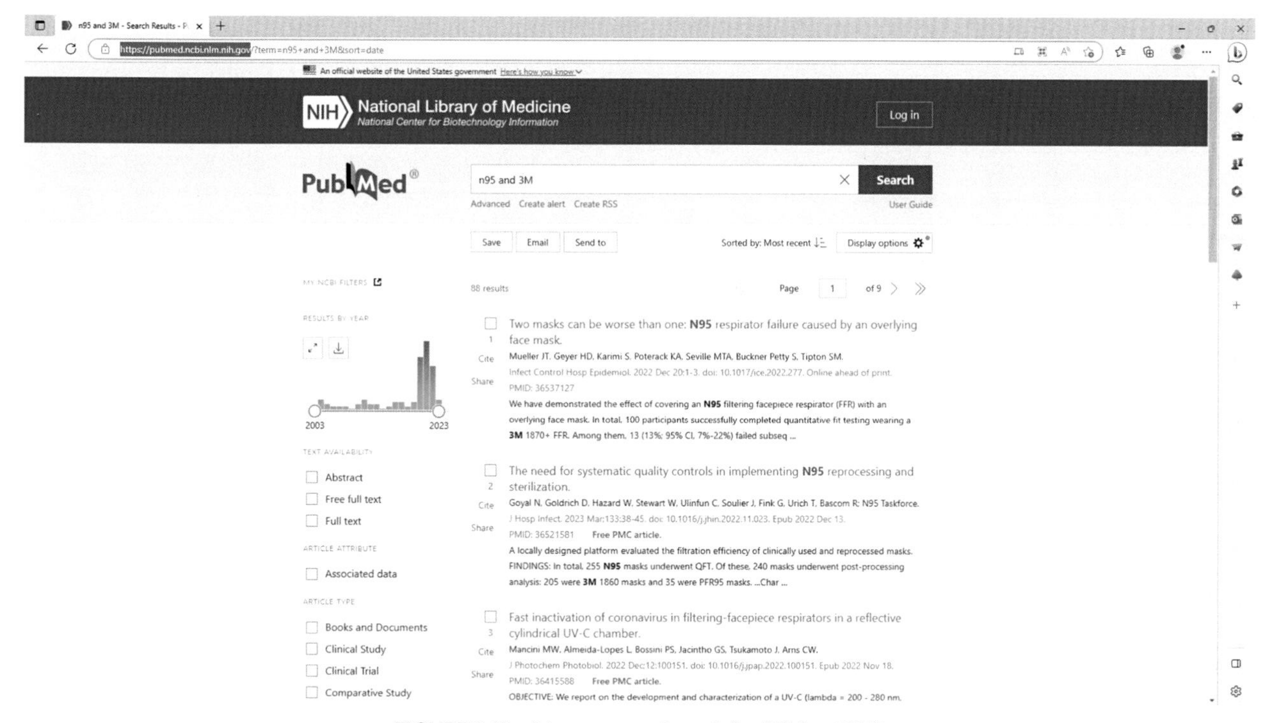

FIGURE 3.13 Literature search result for "N95 and 3M".

The literature search result was downloaded directly from the PubMed website by clicking "save," then selecting all results and the .csv format before clicking "create file." The .csv file was opened and the entire worksheet was moved to the Excel workbook in a worksheet separate from the worksheet holding the clinical trial data (Fig. 3.14).

	A	B	C	D	E	F	G	H	I	J	K	L	M	N	O
1	UID	Search 1	Search 2	I&E	PMID	Title	Authors	Citation	First Auth	Journal/B	Publicatio	Create Date	PMCID	NIHMS ID	DOI
2	1	1			36537127	Two mask	Mueller J1	Infect Con	Mueller J1	Infect Con	2022	12/20/2022			10.1017/ice.2022.277
3	2	1			36521581	The need	Goyal N, G	J Hosp Inf	Goyal N	J Hosp Inf	2023	12/15/2022	PMC9744483		10.1016/j.jhin.2022.11.023
4	3	1			36415588	Fast inacti	Mancini M	J Photoch	Mancini M	J Photoch	2022	11/23/2022	PMC9673160		10.1016/j.jpap.2022.100151
5	4	1			36402800	Plasma-ge	He J, Wan	Sci Rep. 20	He J	Sci Rep	2022	11/19/2022	PMC9675796		10.1038/s41598-022-23660-5
6	5	1			36388184	Case repo	Petrovic P	Front Neu	Petrovic P	Front Neu	2022	11/17/2022	PMC9649972		10.3389/fneur.2022.1003046
7	6	1			36386978	A qualitat	Moghimi E	Front Psyc	Moghimi E	Front Psyc	2022	11/17/2022	PMC9641701		10.3389/fpsyt.2022.1004143
8	7	1			36256682	Effectiven	Singh T, D	J Occup Er	Singh T	J Occup Er	2023	10/18/2022			10.1080/15459624.2022.2137299
9	8	1			36193586	Characteri	Wang Y, Z	Ecotoxicol	Wang Y	Ecotoxicol	2022	10/4/2022			10.1016/j.ecoenv.2022.114131
10	9	1			36189446	Electron b	Huang M,	Radiat Phy	Huang M	Radiat Phy	2023	10/3/2022	PMC9516357		10.1016/j.radphyschem.2022.110557
11	10	1			36168470	Comparat	Clark S, Ng	Antimicro	Clark S	Antimicro	2021	9/28/2022	PMC9495415		10.1017/ash.2021.183
12	11	1			35855909	Reusing F	Yan S, Sta	ACS Susta	Yan S	ACS Susta	2021	7/20/2022	PMC9284677		10.1021/acssuschemeng.1c04530
13	12	1			35800851	Effects of	Dacha S, C	Can J Resp	Dacha S	Can J Resp	2022	7/8/2022	PMC9212081		10.29390/cjrt-2022-014
14	13	1			35662553	Under-ma	Bhatia DD	J Hosp Inf	Bhatia DD	J Hosp Inf	2022	6/6/2022			10.1016/j.jhin.2022.05.015

FIGURE 3.14 Literature data export for "N95 and 3M".

As before, the evaluator added columns to indicate the unique identifier (UID) number for each data point, search number for each search string used (Search 1, Search 2, etc.), and to allow the inclusion and exclusion (I&E) criteria to be entered during the subsequent appraisal step.

Although many search iterations may be required to populate the background knowledge and SOTA (i.e., generic face mask) with "sufficient" clinical data, the example here included a literature search in PubMed only for the combined terms: "face mask" AND "safety" AND "performance," which identified 26 articles (Fig. 3.15).

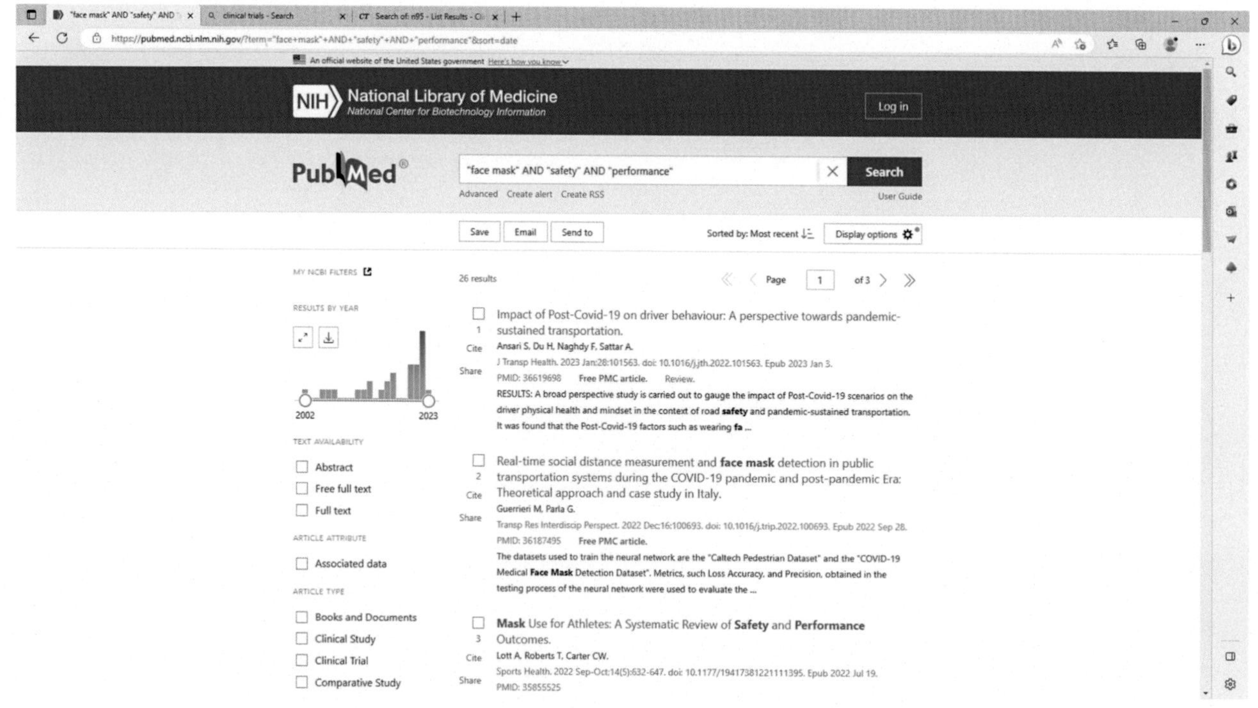

FIGURE 3.15 Literature search for "face mask" AND "safety" AND "performance".

The search output was downloaded into the literature worksheet as "Search 2." The "1" in the Search 1 column uniquely identified 88 articles from the first "N95" search and the "2" in the Search 2 column uniquely identified the 26 articles from the second "face mask" AND "safety" AND "performance" search, and the literature worksheet included 114 articles (Fig. 3.16).

1	UID	Search 1	Search 2	I&E	PMID	Title	Authors	Citation	First Auth	Journal/B	Publicatio	Create Date	PMCID	NIHMS ID	DOI
86	85	1			15340662	The physi	Kao TW, H	J Formos	Kao TW	J Formos	2004	9/2/2004			
87	86	1			15202153	Respirato	Lee K, Sla	J Occup Er	Lee K	J Occup Er	2004	6/19/2004			10.1080/15459620490250026
88	87	1			15202151	Release o	Kennedy	J Occup Er	Kennedy	J Occup Er	2004	6/19/2004			10.1080/15459620490250017
89	88	1			12908863	Evaluatior	Janssen L,	AIHA J (Fa	Janssen L	AIHA J (Fa	2003	8/12/2003			10.1202/477.1
90	89		2		36619698	Impact of	Ansari S, D	J Transp H	Ansari S	J Transp H	2023	1/9/2023	PMC9808417		10.1016/j.jth.2022.101563
91	90		2		36187495	Real-time	Guerrieri	Transp Re	Guerrieri	Transp Re	2022	10/3/2022	PMC9515336		10.1016/j.trip.2022.100693
92	91		2		35855525	Mask Use	Lott A, Ro	Sports He	Lott A	Sports He	2022	7/20/2022	PMC9460089		10.1177/19417381221111395
93	92		2		35578617	Multi-laye	Xu Y, Zhar	Nano Res.	Xu Y	Nano Res	2022	5/17/2022	PMC9094123		10.1007/s12274-022-4350-2
94	93		2		35402141	Performar	Htwe YZN	Arab J Sci	Htwe YZN	Arab J Sci	2022	4/11/2022	PMC8985389		10.1007/s13369-022-06801-w
95	94		2		35162087	Effects of	Steinhilbe	Int J Envir	Steinhilbe	Int J Envir	2022	2/15/2022	PMC8834111		10.3390/ijerph19031063
96	95		2		35090062	The effect	do Prado	Clin Exp P	do Prado	Clin Exp P	2022	1/28/2022			10.1111/1440-1681.13624
97	96		2		34956609	Functiona	Avinash P	Interface	Avinash P	Interface	2021	12/27/2021	PMC8662388		10.1098/rsfs.2021.0040
98	97		2		34924208	Face mask	Blachere F	Am J Infec	Blachere F	Am J Infec	2022	12/20/2021	PMC8674119		10.1016/j.ajic.2021.10.041
99	98		2		33797548	Comfort r	Maniaci A	Ann Ig. 20	Maniaci A	Ann Ig	2021	4/2/2021			10.7416/ai.2021.2439
100	99		2		33630951	Monitorin	Rahim A,	PLoS One.	Rahim A	PLoS One	2021	2/25/2021	PMC7906321		10.1371/journal.pone.0247440
101	100		2		33136621	Covid-19	Naylor G,	Ear Hear.	Naylor G	Ear Hear	2020	11/2/2020			10.1097/AUD.0000000000000948
102	101		2		32817806	Simulatio	Sanchez N	Adv Simul	Sanchez N	Adv Simul	2020	8/21/2020	PMC7424643		10.1186/s41077-020-00135-z
103	102		2		32780113	Filtration	Sickbert-B	JAMA Inte	Sickbert-B	JAMA Inte	2020	8/12/2020	PMC7420816		10.1001/jamainternmed.2020.4221
104	103		2		32720316	Recomme	Lawrence	Laryngosc	Lawrence	Laryngosc	2020	7/29/2020			10.1002/lary.29014
105	104		2		29073915	EMT-led l	Fiala A, Le	Scand J Tr	Fiala A	Scand J Tr	2017	10/28/2017	PMC5658918		10.1186/s13049-017-0446-1
106	105		2		28300710	Frameless	Fanous A	World Ne	Fanous A	World Ne	2017	3/17/2017			10.1016/j.wneu.2017.03.007
107	106		2		27553588	Initial stat	Donaldsso	Arch Dis C	Donaldsso	Arch Dis C	2017	8/25/2016			10.1136/archdischild-2016-310577
108	107		2		26092213	Nasal high	Fealy N, C	Aust Crit C	Fealy N	Aust Crit C	2016	6/21/2015			10.1016/j.aucc.2015.05.003
109	108		2		24257292	Preventio	Yan G, Mit	J Appl Clir	Yan G	J Appl Clir	2013	11/22/2013	PMC5714626		10.1120/jacmp.v14i6.4543
110	109		2		22753612	Determin	Wakui N,	J Oncol Ph	Wakui N	J Oncol Ph	2013	7/4/2012			10.1177/1078155212451196
111	110		2		21154104	Evaluatior	Gao P, Jac	J Occup Er	Gao P	J Occup Er	2011	12/15/2010			10.1080/15459624.2010.515554
112	111		2		17993787	Combined	Copeland	Clin J Spor	Copeland	Clin J Spor	2007	11/13/2007			10.1097/JSM.0b013e31815b187d
113	112		2		16426473	Comparis	Redfern D	Eur J Anae	Redfern D	Eur J Anae	2006	1/24/2006			10.1017/S0265021505002103
114	113		2		16301275	A multice	Hagberg C	Anesth Ar	Hagberg C	Anesth Ar	2005	11/23/2005			10.1213/01.ANE.0000184181.92140.7
115	114		2		12005132	Performar	Clayton M	Ann Occup	Clayton M	Ann Occup	2002	5/15/2002			10.1093/annhyg/mef020

FIGURE 3.16 Add search 2 output to workbook literature tab.

In this way, all search outputs can be combined, yet clearly and accurately identified. All results must be uniquely identified as originating from their specific search (e.g., this example uses the specific search number code to identify each search output).

3.5.2 Quality control for clinical literature identification

A good practice for the clinical literature search process is to have one person complete the clinical literature searches, a second person should check the clinical literature searches, and a third, more senior person should adjudicate any differences between the two literature search experts. Another good QC process is for the adjudicator to ensure all articles listed match the search criteria and the search intent defined in the CEP/PEP. The ability to complete clinical literature searches without inappropriate bias is critical. Keep the search terms as simple and broad-based as possible to avoid concerns about missing important clinical data and use a system of checks and balances with a goal to minimize bias. A clinical literature identification checklist may be helpful here to support company SOPs and WIs and to ensure the clinical literature identification procedure is completed as required by the QMS (Table 3.15).

TABLE 3.15 Literature identification checklist

No.	Description
1	Clinical literature searches followed the strategy defined in the CEP/PEP (e.g., all required searches were completed).
2	Clinical literature search strategy details were clearly documented (e.g., date/s search conducted, databases searched, search terms/strings used, results).
3	Clinical literature search method was comprehensive, objective, and justifiable.
4	Clinical literature search method was reproducible with same results.
5	Manufacturer held and third-party literature were identified.
6	Search terms identified device-appropriate clinical literature data based on product literature and field (e.g., terminology used by practitioners in field).
7	Brand names were used to identify relevant, high-quality data about subject, equivalent, similar device/s (as appropriate).
8	Generic terms were used to identify background, SOTA, and alternative therapy/device information.
9	Result numbers were reasonable (less than a few hundred).
10	Literature details (e.g., article title, authors, link, publication year) were downloaded into Excel workbook or software solution and full-length articles were also downloaded and stored in an accessible format.

The team leader may use the completed literature identification checklist to ensure stage gate completion. The checkilst should be stored in an accessible format and location to document the successful completion of the CE/PE clinical literature identification process.

3.5.3 Challenges during clinical literature identification

Having too many or too few data in the clinical literature identification process is a commonly discussed challenge. The clinical literature search strategy will need to be modified in both cases to get the right amount of relevant, high-quality clinical data in the CER/PER.

Several common misconceptions often cause confusion among those completing clinical literature identification steps including (but not limited to) the following false statements:

1. *Searches must not miss a single data point.*
2. *All included literature must be identified by literature searches.*
3. *Off-label data must not be included in identified clinical literature.*
4. *The evaluator must complete "X" searches.*
5. *Specific device and background/SOTA device searches must be completed separately.*
6. *A specific piece of literature must be found in the literature search.*
7. *If a specific piece of literature is not found in a literarture search the search is not valid and should be discarded.*
8. *Literature searches must be expansive and lengthy to cover all possible terms.*

None of these statements are true. The sheer number of articles available should explain why clinical literature data will be missed, no matter how much effort is extended to be comprehensive. In addition, the evaluator should understand how and why clinical literature added by manual means may be much more relevant and of much higher quality than the literature identified in the literature searches (e.g., sometimes an article is written which simply defies the selected search terms, yet is incredibly important to include—this is absolutely an acceptable article to add manually outside of any search strategy). Although the CE/PE literature search process is expected to focus on device use as indicated (i.e., on label), sometimes off-label uses are helpful to understand device usability and device safety and performance issues. These data can and should be included but use caution to document the off-label nature of the data in the report. No magic search number exists, and terms like "sufficient clinical data are needed" require the evaluator to use good judgment when deciding how many searches and how many articles are enough. In addition, the specifics about how to do clinical data searches are not defined in the regulatory requirements, and some combined searches may be of higher quality and more efficient than separate search strategies.

Sometimes, getting the search string strategy right may take a day or two. This is time well spent, since all of the downstream work depends on this step being done well. Poor clinical literature identification will result in irrelevant data needing to be sorted out and poor-quality data needing to be reviewed. Efficiency will increase if the clinical literature identification step removes the irrelevant and poor-quality data before the appraisal step. Unfortunately, good clinical literature identification processes require expert literature searching skills and a willingness to iterate the process to improve searches during the revision process (i.e., significant patience and perseverance are required).

The CE/PE clinical literature identification process must identify and gather relevant and high-quality clinical literature to be evaluated in the CER/PER. Start by gathering all relevant literature from the manufacturer clinical and regulatory as well as the sales and marketing teams. Then, search multiple clinical literature databases including databases covering literature from EU member states (e.g., Embase). Search terms should be designed using a sandboxing technique before the CEP/PEP is written. Clinical literature data should be identified using search terms about the subject, equivalent, competitor, and similar devices as well as for the background, SOTA, and alternative therapies for the patient population indicated to use the device. Store the results in a user-friendly format to facilitate appraisal and analysis.

3.6 Clinical experience data identification

Clinical experience data are typically recorded as *individual patient experiences* stored in company-held *complaints* records and in international databases documenting publicly available reports about individual adverse events, safety issues, recalls, and field actions.

Individual patient *experience* reports include diverse, poor-quality, highly variable, data (e.g., an individual device failure report, a device failure with a patient harm, or an adverse event/patient harm with a device use). These poor-quality clinical experience data have value in the CE/PE process because the experience data represent real-world experiences from individuals actually using the device. Often, legacy devices (i.e., those devices already marketed in the EU under a CE mark prior to the EU MDR/IVDR) may only have clinical experience data to be evaluated in the CE/PE process because no clinical trials were conducted and no clinical literature were published to explore the device S&P. This is a common situation needing careful and appropriate negotiation with the NB, since more clinical data may be helpful but may not be absolutely required if the device is safe and performs as intended within the MDR/IVDR GSPRs.

The CE/PE purpose is to evaluate clinical data about the subject device S&P, and this includes clinical experience data, which is less amenable to good scientific analysis than most clinical trial and clinical literature data.

3.6.1 Clinical experience database searches

Complaints are reported directly to the manufacturer (often via a phone call, email, or by direct data entry into a manufacturer held data collection system/database) or complaints may be entered into a complaint system existing outside the manufacturer. The manufacturer should analyze routinely and comprehensively all complaints entered in the many different databases. The manufacturer should document the complaint analysis the in the postmarket surveillance (PMS) system on a regular basis (e.g., quarterly PMS evaluations are a good standard practice) in either a PMS report (PMSR) for low-risk devices or a periodic safety update report (PSUR) for high risk, implantable devices. The CE/PE process should pull this experience data from the PMSR or PSUR into the CER/PER.

The regulatory requirements for analyzing experience data are quite similar under MDR and IVDR (Table 3.16). MDR specifies PMSRs and PSURs in Articles 85 and 86, respectively [1], and IVDR specifies PMSRs and PSURs in Articles 80 and 81, respectively [2].

TABLE 3.16 PMSR and PSUR regulatory requirements

Report	MDR [1]	IVDR [2]
PMSR (for low risk devices)	Article 85: "Manufacturers of class I devices shall prepare a post-market surveillance report summarising the results and conclusions of the analyses of the post-market surveillance data gathered as a result of the post-market surveillance plan… together with a rationale and description of any preventive and corrective actions taken. The report shall be updated when necessary and made available to the competent authority upon request."	Article 80: "Manufacturers of class A and B devices shall prepare a post-market surveillance report summarising the results and conclusions of the analyses of the post-market surveillance data gathered as a result of the post-market surveillance plan… together with a rationale and description of any preventive and corrective actions taken. The report shall be updated when necessary and made available to the notified body and the competent authority upon request."
PSUR (for high risk devices)	Article 86: "1. Manufacturers of class IIa, class IIb and class III devices shall prepare a periodic safety update report ('PSUR') for each device and where relevant for each category or group of devices summarising the results and conclusions of the analyses of the post-market surveillance data gathered as a result of the post-market surveillance plan… together with a rationale and description of any preventive and corrective actions taken. Throughout the lifetime of the device concerned, that PSUR shall set out: (a) the conclusions of the benefit-risk determination; (b) the main findings of the PMCF; and (c) the volume of sales of the device and an estimate evaluation of the size and other characteristics of the population using the device and, where practicable, the usage frequency of the device. Manufacturers of class IIb and class III devices shall update the PSUR at least annually. That PSUR shall, except in the case of custom-made devices, be part of the technical documentation…Manufacturers of class IIa devices shall update the PSUR when necessary and at least every two years. That PSUR shall, except in the case of custom-made devices, be part of the technical documentation…For custom-made devices, the PSUR shall be part of the documentation…2. For class III devices or implantable devices, manufacturers shall submit PSURs by means of the electronic system… to the notified body involved in the conformity assessment… The notified body shall review the report and add its evaluation to that electronic system with details of any action taken. Such PSURs and the evaluation by the notified body shall be made available to competent authorities through that electronic system.3. For devices other than those referred to in paragraph 2, manufacturers shall make PSURs available to the notified body involved in the conformity assessment and, upon request, to competent authorities."	Article 81: "1. Manufacturers of class C and class D devices shall prepare a periodic safety update report ('PSUR') for each device and where relevant for each category or group of devices summarising the results and conclusions of the analyses of the post-market surveillance data gathered as a result of the post-market surveillance plan… together with a rationale and description of any preventive and corrective actions taken. Throughout the lifetime of the device concerned, that PSUR shall set out: (a) the conclusions of the benefit-risk determination; (b) the main findings of the PMPF; and (c) the volume of sales of the device and an estimate of the size and other characteristics of the population using the device and, where practicable, the usage frequency of the device. Manufacturers of class C and D devices shall update the PSUR at least annually. That PSUR shall be part of the technical documentation…2. Manufacturers of class D devices shall submit PSUR by means of the electronic system… to the notified body involved in the conformity assessment of such devices… The notified body shall review the report and add its evaluation to that electronic system with details of any action taken. Such PSUR and the evaluation by the notified body shall be made available to competent authorities through that electronic system.3. For class C devices, manufacturers shall make PSURs available to the notified body involved in the conformity assessment and, upon request, to competent authorities."

Bold emphasizes the PMSR/PSUR references

The first step when identifying clinical experience data is to gather all the past PMSRs/PSURs to cover the entire device life cycle. After gathering all manufacturer-held clinical experience data, search all non-manufacturer-held clinical experience databases to identify device-specific experience data (both good and bad) for CE/PE. In each planned database search, the target and equivalent device names should be searched as outlined in the CEP (Table 3.17).

TABLE 3.17 Experience search database and search engine examples

Search Engine	Type	Geography	Link
Manufacturer and User Facility Device Experience (MAUDE)	Reports	US	https://www.accessdata.fda.gov/scripts/cdrh/cfdocs/cfmaude/search.cfm
Database of Adverse Event Notifications (DAEN)	Reports	Australia	https://apps.tga.gov.au/prod/DEVICES/daen-entry.aspx
European Union	Clinical and performance studies, PMS	EU	https://ec.europa.eu/tools/eudamed/#/screen/home
Medical Product Safety Network (MedSun)	Reports	US	https://www.fda.gov/medical-devices/medical-device-safety/medsun-medical-product-safety-network
Total Product Life Cycle (TPLC)	Reports[1]	US	https://www.fda.gov/about-fda/cdrh-transparency/cdrh-transparency-total-product-life-cycle-tplc
FDA Medical Device Recalls	Recalls	US	https://www.accessdata.fda.gov/scripts/cdrh/cfdocs/cfRES/res.cfm
Swissmedic	Recalls	Switzerland	https://fsca.swissmedic.ch/mep/#
Recalls and Safety Alerts	Recalls	Canada	https://recalls-rappels.canada.ca/en
BfArM	Recalls, reports, field corrective actions	Germany	https://www.bfarm.de/SiteGlobals/Forms/Suche/EN/Expertensuche_Formular.html;jsessionid = C666D57E79F097877268CB779F575C97.internet562?nn = 708434&cl2Categories_Format = kundeninfo
Medical Devices Regulation and Safety (MHRA)	Recalls	UK	https://www.gov.uk/drug-device-alerts
Health Products Regulatory Authority (HPRA)	Recalls	Ireland	https://www.hpra.ie/homepage/medicines/safety-notices

[1]*Contains full report about product lines in addition to complaint reports.*

Document and download all details from each search into the CER/PER Excel workbook or software to identify relevant, high-quality clinical experience data for the CE/PE process (Table 3.18).

TABLE 3.18 Clinical experience search strategy summary

Where?	Manufacturer-held complaints, external clinical experience databases (i.e., include EU database/s).
What?	Planned search terms (e.g., *manufacturer name, *device name).
How?	Search planned terms or <ctrl + f> to find in downloaded document or on web page.
When?	Complete searches within 6 months of submission; same date as clinical trial and literature searches.

Two commonly searched experience databases include the US Food and Drug Administration (FDA) Manufacturer and User Facility Device Experience database (MAUDE) and the Australian Therapeutic Goods Association (TGA) Database of Adverse Event Notifications (DAEN). In addition, the EUDAMED database is under development to provide access to surveillance data. Collectively, databases like these will need to be searched to identify clinical *experience* data not held by the manufacturer to be evaluated in the CE/PE process.

Example 4: N95 face mask clinical experiences

The MAUDE database (https://www.accessdata.fda.gov/scripts/cdrh/cfdocs/cfmaude/search.cfm) was searched using "N95" in the "brand name" search window. The evaluator constructed a key to record search strategy and details for each search including the search terms used and the outputs, with any notes. The device name was specific enough to identify relevant experience reports in MAUDE (Table 3.19).

TABLE 3.19 MAUDE search key: N95

Search no.	Search terms	Result <date>	Comments
1	N95	40	Specific search, use for subject device.

This search resulted in 40 records (Fig. 3.17).

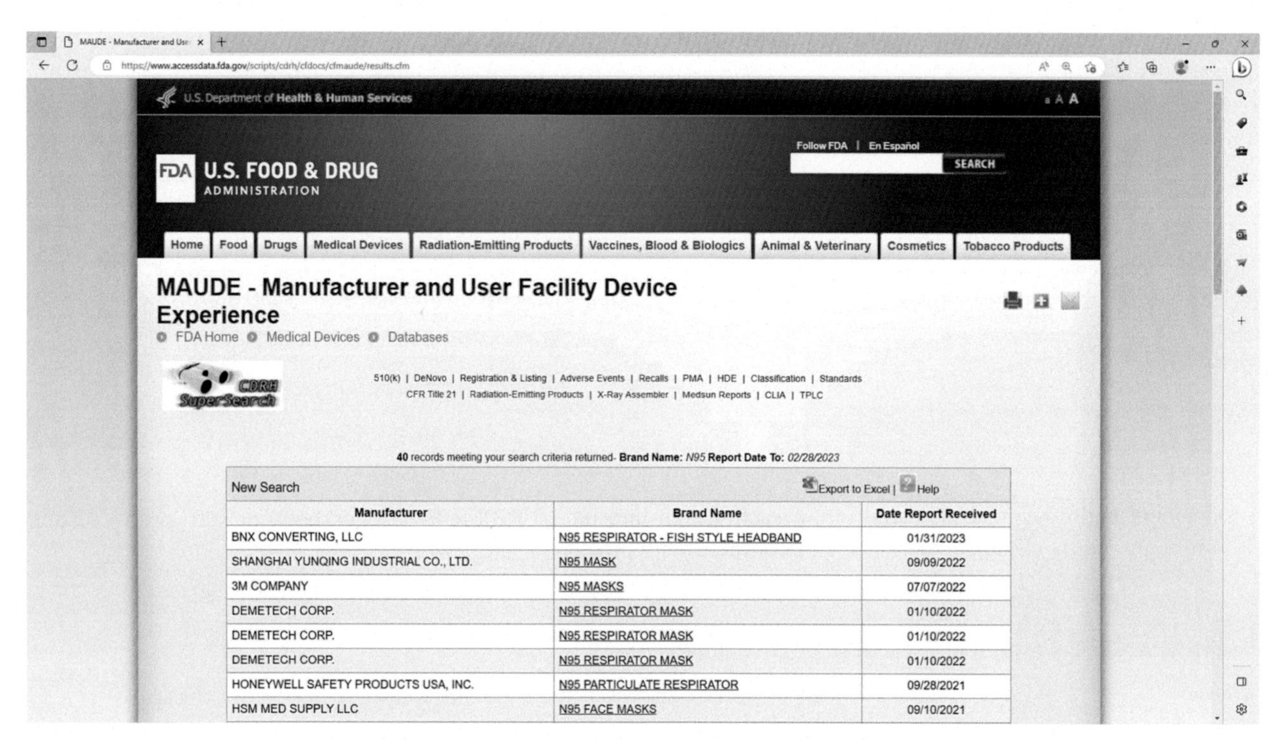

FIGURE 3.17 MAUDE database search identified 40 "N95" reports.

These 40 records were "exported to Excel" directly from the MAUDE database and added to the N95 CER workbook. The four columns for UID, Search 1, Search 2, and I&E columns were added, and extraneous text about the MAUDE database was removed (Fig. 3.18).

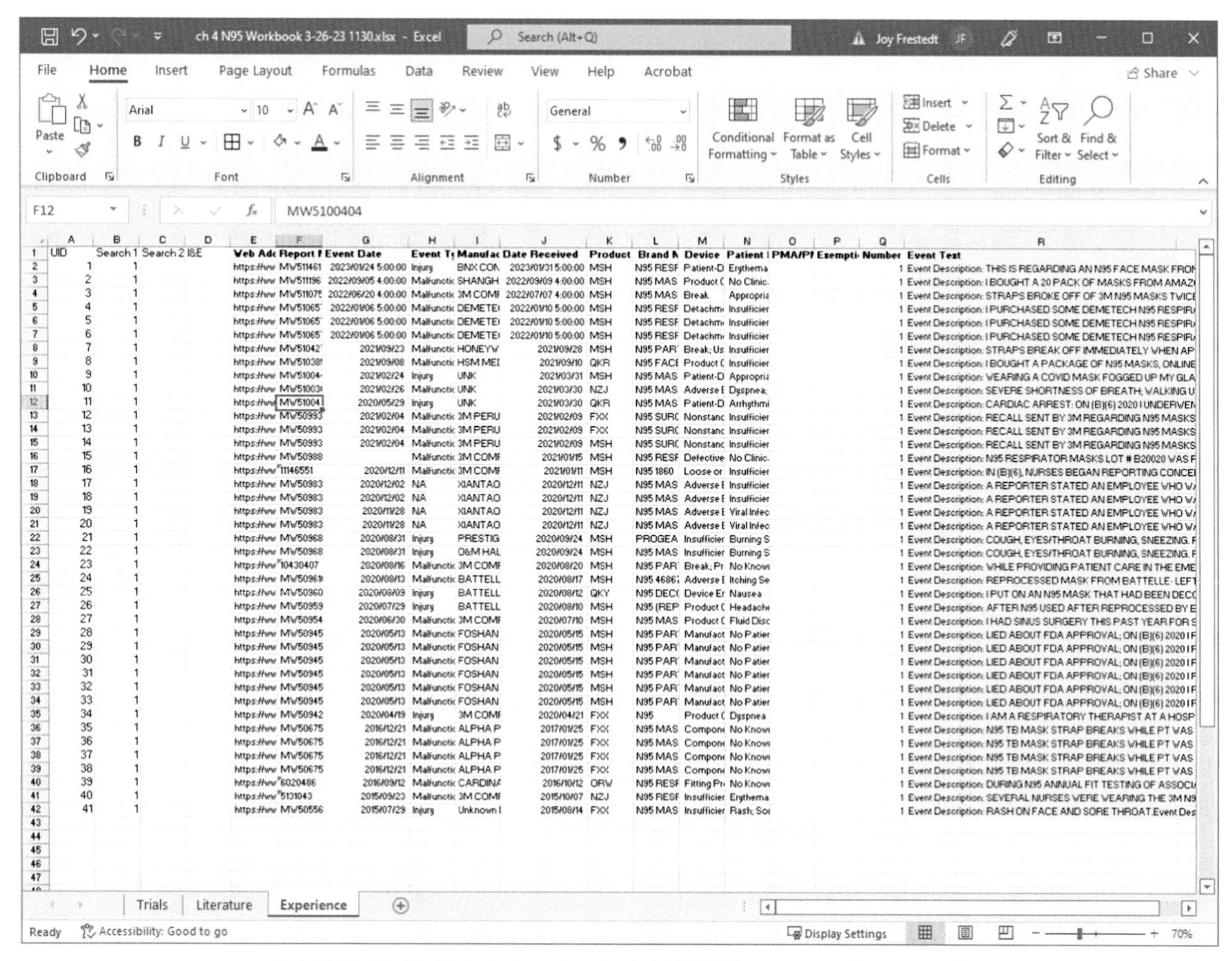

FIGURE 3.18 MAUDE database downloaded 41 "N95" reports (not 40 as expected).

Note how the MAUDE database exported more records than were counted in the database search (e.g., this search downloaded 41 results when only 40 were identified in the search output). This is a common and expected problem with the MAUDE database, so check and verify all results to ensure accuracy. The most important MAUDE data are in the "Event Text" (i.e., source data) column.

3.6.2 Quality control for clinical experience data identification

As required in all database searches, have one person complete the clinical experience searches, a second person should check the clinical experience searches and results, and a third, more senior person should adjudicate any differences between the two experience search experts. Do the experiences listed match the search criteria and the intent for the searches? If not, the searches may need to be refined. The ability to focus the clinical experience searches without inappropriate bias is critical. Keep the search terms as simple and broad-based as possible to avoid concerns about missing important clinical data.

Like clinical trial and literature identification data checklists, the clinical experience data identification checklist should ensure procedural success aligning with SOPs and WIs (Table 3.20).

TABLE 3.20 Experience data identification checklist

No.	Description
1	Manufacturer-held complaint data compiled and entered into CER workspace.
2	Search strategy follows CEP.
3	Search strategy details documented (e.g., date conducted, terms/strings, results, databases searched).
4	Search method was comprehensive, objective, and justifiable.
5	Search method reproducible with same results.
6	Non-manufacturer-held complaint/adverse event, recall data identified and entered into CER workspace.
7	Search terms appropriate for device (based on product literature).
8	No time frames applied (data identified over all time).
9	Identified data downloaded (e.g., event date. evemt text) and prepared appraisal.

Using a checklist may help to ensure appropriate clinical experience data are identified to meet industry standards. The checklist may also help evaluators and writers to work through the process. Checklists should be version controlled and updated whenever needed (e.g., if manufacturer processes or regulations change).

3.6.3 Challenges during clinical experience identification

The CE/PE clinical experience data identification processes should be designed specifically and comprehensively to accommodate the unique manufacturer situation. Often, clinical experience data are high volume, individual, uncontrolled data points with extremely high data variability and extremely low data quality. During the CE/PE process, the responsible PMS team member should explain any PMS actions taken by the manufacturer. All clinical experience data and all "potential" clinical experience data (i.e., complaints) should be maintained during the clinical experience data identification process for appraisal and analysis during the rest of the CE/PE process.

BOX 3.9 Case study—manufacturer-held complaints "analysis"

Case study situation

A manufacturer with a well-established, low-risk device reeported only five complaints over the device lifespan. Two complaints were identified as "user errors" causing safety concerns resulting in patient harm. Because the manufacturer determined the user "caused" the error, the manufacturer removed these complaints from the PMSR/PSUR and the complaints were "hidden" from view during the CE/PE process.

Case study issues identified

The PMSR/PSUR should report all complaints, especially those directly resulting in or potentially resulting in a patient harm. Device malfunction details were historically captured, but the details about what happened to the patient were not well documented.

Case study outcomes

The manufacturer QMS was updated to ensure all patient-harm complaints were reported in the PMSR/PSUR and CER/PER.

A better case study solution enacted

Although many complaints can be described as non-device related, any complaint where the device was used should be identified and included in the evaluation within the PMSR/PSUR. This includes complaints labeled as "user error," "unclear device relationship," and complaints where the manufacturer took action to resolve the issue. The PMSR/PSUR should detail exactly what happened to the patient who experienced each individual reported device complaint and the clinical data should be evaluated in the PMS, RM, and CE/PE systems, as required by the MDR/IVDR.

No database location exists for all medical device safety reports worldwide and each individual database uses a different method to collect and publicly share clinical experience data. For example, EUDAMED is designed to contain PMS data in the EU; however, this database is not yet functional for this purpose. In addition, unlike the MAUDE, DAEN, and EUDAMED databases, many "recall" databases exist, which may also include clinical experience data. Unfortunately, these recall databases suffer from significant overlap and redundancy with each other globally. Use caution when searching recall databases.

The clinical experience data reporting and recording landscape is changing. As real-world data and "real world evidence" are prioritized, many new clinical experience databases are becoming available. As a result, plan to search for new experience databases when experience data searches are conducted. Complete searches in as many databases as possible to identify all subject/equivalent device clinical experience data.

Clinical experience database searches work differently than clinical trial and clinical literature database searches. The inputs and the outputs vary in appearance, functionality, and data download/transfer abilities. Some clinical experience databases include search functions to help identify needed data (e.g., adverse event reports, recalls, corrective actions); however, most clinical experience databases do not provide search functions. In these cases, selecting <ctrl + f> on the keyboard will open search (i.e., "find") box for the experience document or web page. Enter search terms into the "find" box and record the date, database, search terms, results, etc. in the CER/PER workbook or software.

Clinical experience data should be identified both internally at the manufacturer and externally in reporting databases including complaints, recalls, and safety reports for the subject device. Identifying clinical experience data for other devices is not recommended because the clinical experience data are poor quality and difficult to analyze reliably in the CER/PER background/SOTA or alternative therapy/device section/s.

3.7 Clinical data identification by clinical evaluation/performance evaluation type

Clinical trials in patients using the subject device are of high value in CERs/PERs; however, clinical trial data are not often available for most medical devices. Clinical literature data are also high value when the article specifies patient details and the clinical data were rigorously collected and analyzed in the article. Unfortunately, clinical experience data hold minimal value because the clinical experience data are typically extremely poor-quality, anecdotal information.

The amount of clinical data available varies across devices, and sometimes clinical data cannot be identifed because no clinical data exist. This point is often missed when evaluators use nonclinical data to support a JUS about why clinical data may not be required. The evaluator should understand the clinical data volume changes throughout the device lifetime and the evaluator should leverage the accumulation of clinical data during the CE/PE process.

3.7.1 Viability clinical evaluation / viability performance evaluation

The viability (v), vCE/vPE, is a good example when no clinical data exist for the subject device. The vCE/vPE is typically generated before the device is placed on the EU market and before any humans have used the device. For the novel device, the vCE/vPE must identify appropriate and relevant nonclinical data (e.g., bench data, animal data, human factors testing), which are generally not included in established device CERs/PERs. The vCE/vPE may also identify similar device clinical data to defend the proposed and expected device S&P as well as the expected device benefit-risk ratio. The vCE/vPE is often used for a completely novel device, for a device quite early in development, or when an equivalence claim is not desired. The vCE/vPE requires a strong and well-planned vPMCFP/vPMPFP to define how sufficient, future clinical data will be collected to ensure GSPR conformity for the subject device.

3.7.2 Equivalent clinical evaluation / equivalent performance evaluation

The equivalent (eq), eqCE/eqPE, is another situation where little to no clinical data exist for the subject device. The eqCE/eqPE claims equivalence to a previously marketed device with sufficient clinical data to show S&P when the device is used as indicated. The MDR/IVDR require equivalent devices to be technically (similar design, conditions of use, specifications, properties, etc.), biologically (similar materials or substances in contact with human tissues/fluids, etc.), and clinically (same clinical application, same indication for use, similar condition, purpose, population, user, performance, etc.) similar without any clinically significant differences in device S&P. Equivalent device data are used to represent and demonstrate the subject device is safe, performs as intended and meets the relevant clinical GSPRs. When the two devices are equivalent, the clinical data from each can be used for the other device.

Claiming equivalence is a regulatory strategy especially for a new, but not novel, device. Often, subject device clinical data are lacking and the equivalent device clinical data are used instead. This strategy is often used as a transient claim during development where the equivalent device data will be replaced once sufficient clinical data have been collected from future subject device uses as indicated. The lack of clinical data for the subject device means the eqCE/eqPE requires a strong and well-planned eqPMCFP/eqPMPFP to define how sufficient clinical data will be identified and collected to ensure subject device GSPR conformity, once those data become available after the device has been on the market for some time.

3.7.3 First-in-human clinical evaluation / first-in-human performance evaluation

Clinical data from the first in human (fih) device uses should be collected and evaluated in the fihCE/fihPE. For high-risk or implantable devices, these fih uses may include premarket clinical trial data if a trial was completed for the device before CE marking. For low-risk devices, the fih data will likely be identified and collected as experience data when monitoring device sales. The limited clinical data evaluated in the fihCE/fihPE should expose the need for a detailed fihPMCFP/fihPMPFP to collect the future clinical data needed to fill in the clinical gaps identified in the fihCER/fihPER.

3.7.4 Traditional clinical evaluation / traditional performance evaluation

Subject device clinical data are most abundant in the traditional (tr) trCE/trPE type compared to all other CE/PE types. The manufacturer should have clinical data available; however, prior to the MDR/IVDR, the requirement for clinical data was not evenly enforced and clinical trials may not have been required even for Class III, implantable devices. In this setting, the minimal clinical trial and minimal clinical literature results should be documented and justified, especially if the lack of clinical data is related to a low-risk device or an infrequently studied device type. Prior to MDR/IVDR, clinical trials were rarely done on low-risk devices, and clinical literature rarely had low-risk device clinical S&P outputs studied, evaluated, and published.

BOX 3.10 Legacy devices

For subject devices released and CE marked prior to MDR/IVDR, requirements and enforcement regarding clinical data identification and use in CERs/PERs were inconsistent, so manufacturers most likely did not have any or had only minimal clinical data, yet these CE-marked devices were allowed onto the EU market.

As MDR/IVDR came into effect, transitional provisions were allowed for "legacy" devices which included devices placed on the market after MDR/IVDR was released and until 2024 to 2028 under certain conditions depending on device risk classification under MDR Article 120(3) [1], IVDR Article 110(3) [2]. Guidelines were also created. MDCG 2021-25 [9] stated "MDR requirements are in principle not applicable to 'old' devices." Old devices were those placed on the market before May 26, 2021. Similarly, MDCG 2022-8 [10] stated "IVDR requirements are in principle not applicable to 'old' devices." Old IVD devices were those placed on the market before May 26, 2022. A notable exception is the MDR/IVDR provisions for PMS surveillance activities, including PMSR/PSUR, are applicable to both old and new devices.

This is an important concept since legal and regulatory battles are defining how regulations will apply to devices marketed before the MDR/IVDR came into force, since "legacy devices" conformed to the existing regulations at the time they were placed on the EU market. In other words, certain parts of the MDR or IVDR were simply "not applicable" to devices CE marked prior to May 26, 2021 or May 26, 2022, respectively, during this transitional period.

Even in settings where no clinical trials and no clinical literature are available, trCE/trPE subject device clinical data should be easily identified as clinical experience data. The clinical experience data would include all complaints reported over time and all sales numbers recorded to document the units sold and, presumably, used in humans over the same time period. When clinical trial and literature data are minimal or not available, the available clinical experience data often become the highest-value clinical data in the trCE/trPE. The clinical experience data should be used to support the device S&P and benefit-risk ratio in the trCER/trPER.

3.7.5 Obsolete clinical evaluation / obsolete performance evaluation

Obsolete products are no longer actively marketed; however, they may still be used in patients. The MDR and good medical practices require device S&P to be supported throughout the device lifetime including this last device obsolescence stage. For example, some implantable medical devices will have lifetime extending for the entire patient lifetime while others may cease to function prior to the patient's end of life. Clinical data from trials, literature, and experiences should be identified and collected until no devices are in use any longer. Data should be continuously gathered and analyzed in PMS processes and included in obsolete (o) device oCERs/oPERs. This is crucial to identify new safety signals as the devices age, and these new S&P signals should be shared with all remaining device users. The QMS should define how to include oCER/oPER data in new device CE/PE processes when relevant and appropriate.

3.8 Conclusions

First, the evaluator must identify, collect, and gather together all manufacturer-held clinical data. The evaluator must follow the CEP/PEP to identify all clinical trial, literature, and experience data in various databases. The rigorous and methodical/systematic searches use manufacturer and device names. Literature database search strategies may also include additional generic terms for background/SOTA clinical data identification.

Clinical data vary greatly in quality depending on where the device is in the device life cycle. A helpful strategy is to consider specific CE/PE types aligned in five stages of the device life cycle: viability, equivalence, first in human, traditional, and obsolescence. Executing a clear, comprehensive, and systematic data identification search designed for the relevant CE/PE type will help the evaluator to collect all appropriate clinical data. Clinical trials are the best source of data for clinical evaluation; however, literature and experience data must also be evaluated. Specifically, IVD devices require documentation of analytical validity as well as clinical and analytical performance in the PER. The clinical data evaluation must accurately document the device S&P effects on the human, especially with regard to the potential for error or harm. This includes direct device effects in or on the human body for medical devices and lab result details for IVD devices.

Storing and organizing search details and all identified clinical data outputs in one Excel workbook or one proprietary software outside the CER/PER will facilitate organization of the identified clinical data, and should optimize clinical data appraisal and analysis later in the CE/PE process. The search key is the place to record details for each search (e.g., all database links, terms, and dates for each search) and to specify the meaning for all literature worksheet numbers and codes. A suggested goal is to conduct all clinical data identification searches (i.e., for all three data sources: trials, literature, and experiences) on one date. This should simplify and standardize the data identification methodology. Another best practice is to include QC and quality assurance details in the clinical data identification steps. Having a second (more senior) person check the data identified and a third (more senior) person adjudicate any discrepancies between the first person doing the clinical data identification and the second person checking the clinical data identification will help ensure the right clinical data were identified in the CE/PE process. Qualified and experienced personnel are required for this work to identify the appropriate and comprehensive clinical data for the CE/PE process. These clinical data identification steps should be detailed in an SOP or WI.

Unfortunately, in certain situations, identifying and collecting clinical data from some devices is impossible (e.g., when no clinical data exist). In addition, some devices may provide a benefit without a measurable or clinically meaningful end point, or some devices may be accessories without any measurable end point other than technical performance. For devices without any clinically meaningful or measurable end points (i.e., the clinical data for evaluation will not demonstrate conformity with the GSPRs 1, 6, and 8), a justification (JUS) should be clearly written with sufficient nonclinical data including literature and experience support for any and all clinical claims made about this device including S&P and benefit-risk ratio acceptability. For an accessory, one approach is to include the accessory within the main device CER/PER since the accessory cannot be used alone. Another approach would be to write a JUS for each group of accessories and to link the JUS documentation to the CER/PER for the main device using the accessory.

In other situations, a clinical trial may be required to address the clinical data gaps identified when collating the clinical data for the CER/PER. Ultimately, sufficient clinical data must be identified in this identification step to allow all requirements within GSPRS 1, 6, and 8 to be evaluated and conformity clearly documented in the CER/PER. After identifying these highly-relevant clinical trial data about the subject device, appraisal and analysis will be required because these trials may have highly variable data quality depending on the clinical trial details.

3.9 Review questions

1. For in vitro devices, specifically what three evaluations are required in the PER?
2. What are the three clinical data sources evaluated in CERs and PERs?
3. What three elements are needed to be considered "clinical data"?
4. What is an analyte?
5. What does "scientific validity of an analyte" mean?
6. What does "analytical performance" mean?
7. What does "clinical performance" mean?
8. What does "clinical evidence" mean?
9. What is the scientific method?

10. What clinical data identification steps should be followed to ensure a "scientifically sound" clinical evaluation?
11. What is the rationale for having a workspace/workbook separate from the CER/PER document (e.g., Word file)?
12. How does the CE/PE evaluator determine if the identified clinical data are "sufficient"?
13. What is a JUS document and when is a JUS appropriate?
14. What is the main limitation when using clinicaltrials.gov data in the CE/PE?
15. What two primary areas must the appropriately identified clinical literature focus on to be useful in the CER/PER?
16. Why is sandboxing recommended prior to search term finalization in the CEP/PEP?
17. What timespan should the literature search cover?
18. How are clinical experience database searches different from literature searches?
19. What should be in the SOP and WI for the clinical data identification step?
20. What are "legacy" devices and why is this regulatory designation important to know during the CE/PE identification step?

References

[1] European Parliament and Council of the European Union. Consolidated text: Regulation (EU) 2017/745 of the European Parliament and of the Council of 5 April 2017 on medical devices, amending Directive 2001/83/EC, Regulation (EC) No 178/2002 and Regulation (EC) No 1223/2009 and repealing Council Directives 90/385/EEC and 93/42/EEC. Accessed May 24, 2023. https://eur-lex.europa.eu/legal-content/EN/TXT/?uri = CELEX%3A02017R0745-20230320.

[2] European Parliament and Council of the European Union. Consolidated text: Regulation (EU) 2017/746 of the European Parliament and of the Council of 5 April 2017 on in vitro diagnostic medical devices and repealing Directive 98/79/EC and Commission Decision 2010/227/EU. Accessed May 24, 2023. https://eur-lex.europa.eu/legal-content/EN/TXT/?uri = CELEX%3A02017R0746-20230320.

[3] European Commission. Guidelines on medical devices; MEDDEV 2.7/1 Rev 4. June 2016. Clinical evaluation: a guide for manufacturers and notified bodies under directives 93/42/EEC and 90/385/EEC. Accessed May 24, 2023. https://www.medical-device-regulation.eu/wp-content/uploads/2019/05/2_7_1_rev4_en.pdf.

[4] MDCG 2022-02. Guidance on general principles of clinical evidence for In Vitro Diagnostic medical devices (IVDs). January 2022. Accessed May 24, 2023. https://health.ec.europa.eu/system/files/2022-01/mdcg_2022-2_en.pdf.

[5] European Commission. Enterprise and industry directorate general. Consumer goods. cosmetics and medical devices. MEDDEV 2.7.1 Rev. 3. December 2009. Guidelines on medical devices. *Clinical evaluation: a guide for manufacturers and notified bodies.* Accessed May 24, 2023. http://meddev.info/_documents/2_7_1rev_3_en.pdf.

[6] European Commission. Expert decision and opinion in the context of the clinical evaluation consultation procedure (CECP 2022-000232). Accessed May 25, 2023. cecp-2022-000232_opinion_en.pdf (europa.eu).

[7] MDCG 2020-6. Regulation (EU) 2017/745: clinical evidence needed for medical devices previously CE marked under Directives 93/42/EEC or 90/385/EEC A guide for manufacturers and notified bodies. April 2020. Accessed May 25, 2023. https://health.ec.europa.eu/system/files/2020-09/md_mdcg_2020_6_guidance_sufficient_clinical_evidence_en_0.pdf.

[8] MDCG 2020-13. Clinical evaluation assessment report template. July 2020. Accessed May 25, 2023. https://health.ec.europa.eu/system/files/2020-07/mdcg_clinical_evaluationtemplate_en_0.pdf.

[9] MDCG 2021-25. Regulation (EU) 2017/745—application of MDR requirements to 'legacy devices' and to devices placed on the market prior to 26 May 2021 in accordance with Directives 90/385/EEC or 93/42/EEC. October 2021. Accessed May 26, 2023. https://health.ec.europa.eu/system/files/2021-10/md_mdcg_2021_25_en_0.pdf.

[10] MDCG 2022-8. Regulation (EU) 2017/746—application of IVDR requirements to 'legacy devices' and to devices placed on the market prior to 26 May 2022 in accordance with Directive 98/79/EC. May 2022. Accessed May 26, 2023. https://health.ec.europa.eu/system/files/2022-05/mdcg_2022-8_en.pdf.

Chapter 4

Appraising clinical/performance data

The appraisal process includes making decisions about what clinical data to include or exclude from the clinical evaluation/performance evaluation (CE/PE). Appraisal includes sorting, categorizing, and organizing all identified clinical data. The appraisal step is vital to focus the CE/PE on the most relevant and high-quality clinical data without losing bits of novel clinical data or the device use "context" as a whole.

Appraisal includes assigning a relative weight to each included clinical/performance data point. The evaluator often starts to structure and complete the clinical evaluation report/performance evaluation report (CER/PER) by separating out the most relevant and highest quality clinical/performance data for two distinct CER/PER sections. One section is about the subject/equivalent device; and the other is about the background knowledge, state-of-the-art (SOTA), and alternative therapies including all competitor and similar devices. The CER/PER must define the overall context and how the subject/equivalent device clinical data fit into the background of all relevant SOTA data.

Clinical data vary greatly and including every data point from every identified data source is neither possible nor advisable. Often, identified clinical data are poor quality or are not relevant to the CE/PE even when the planning and identification steps were done to perfection.

The appraisal step is meant to remove irrelevant and unreliable clinical data while maintaining both positive and negative safety and performance (S&P) findings. This means the evaluator must complete an unbiased appraisal to keep all clinical data included until a clear, rational, and acceptable reason for inclusion or exclusion is documented for each identified data point.

These clear, rational, and acceptable reasons for inclusion or exclusion depend on the data. The appraisal must be completed for each unique dataset, each time the CE/PE begins. This means the clinical evaluation plan/performance evaluation plan (CEP/PEP) must be flexible enough to allow the evaluator to learn from the actual data appraisal process and to make changes in the inclusion and exclusion rationales during the appraisal process.

4.1 Good appraisal processes are iterative

As stated in MEDDEV 2.7/1 revision 4: "appraisal and analysis... may uncover new information and raise new questions, with a need to widen the scope of the evaluation, refine the clinical evaluation plan, and to retrieve, appraise and analyse additional data" [1]. Each appraisal round (i.e., when updating a CER/PER) must take into account the identified clinical data as a whole (i.e., all clinical data contents, volumes, qualities and weights) with a goal to limit and focus the CE/PE on relevant, high-quality clinical data in the existing clinical dataset. In other words, appraising a clinical data subset is not acceptable (i.e., do not have two separate appraisals for historical and new data). This is important because all new data must be integrated with all historical data to make the new appraisal decisions for the entire dataset including the new data. This is a crucial point.

BOX 4.1 Appraise all identified clinical data, not a subset of data

Setting: A large manufacturer had a large team working on a low-risk device CE.

Problem: The previous CER had appraised hundreds of articles and the new search was planned as a single, stand-alone 3-year update. Some team members recognized how a separate update was not appropriate because the prior CER had identified a potential "skin irritation" safety concern for the subject device; however, subsequent investigations had shown this safety issue was not possible for the subject device because the subject device did not adhere to the skin in any way. The skin irritation problem was seen only in other devices adhering to the skin by using an adhesive. The clinical application needed to be refined in the data identification step and carried throughout all CE steps including planning, identification, appraisal, analysis, and reporting.

(Continued)

Planning, Writing and Reviewing Medical Device Clinical and Performance Evaluation Reports (CERs/PERs). DOI: https://doi.org/10.1016/B978-0-443-22063-0.00015-8

BOX 4.1 (Continued)

Solution: The plan and entire search strategy was updated and completed to include clinical data from all time points, not just the 3 years since the last CER. In addition, the identified safety concern was specifically included in the background/SOTA as well as the subject device evaluation processes to ensure the safety concern was well documented in the background with the rationale about why this safety concern was not relevant to the subject/equivalent device which did not have any adhesive.

Clinical/performance data appraisal is more than a predefined include/exclude checklist. Inclusion and exclusion (I&E) reasons must be clearly defined, documented and consistently applied even though the reasons will vary based on the available clinical/performance data, overall clinical/performance data quality and specific clinical/performance data type being appraised. Specific justifications may be helpful to document for sensitive I&E criteria.

Overall, the clinical/performance data appraisal must be unbiased, rigorous and methodologically sound. In addition, the clinical/performance data appraisal must be actively updated and all included clinical/performance data must be weighted (i.e., rated or ranked) for CE/PE value or priority. As each document is coded for inclusion or exclusion, the included documents should be downloaded and stored so they can be further evaluated. For the clinical literature, many articles are available for free and can be directly downloaded; however, some articles require a copyright fee.

4.2 Quality management system must define clinical data appraisal quality

The European Union (EU) Medical Device Regulation (MDR) (EU REG 2017/745) [2] and the EU *In Vitro* Diagnostic Device Regulation (IVDR) (EU REG 2017/746) [3] require manufacturers to maintain a quality management system (QMS) addressing all CE/PE details. Standard operating procedures (SOPs) and work instructions (WIs) must document the rigorous, methodical, and required CE/PE appraisal steps in writing. The CE/PE WIs appraisal steps must be thorough and detailed enough to be reproducible while also being flexible enough to accommodate differences for each individual device. The CE/PE WIs should also include specific steps for quality control (QC) to ensure appropriate appraisal decisions were made.

Appraisal is a complex process and no two CE/PE appraisal processes will have the exact same I&E codes or appraisal considerations. In addition, the I&E criteria are expected to change over time, so avoid simply using the CE/PE appraisal criteria from the last CE/PE.

The CE/PE WI should provide specific step-by-step instructions to follow the CEP/PEP. The evaluator must apply the CEP/PEP-defined I&E criteria by documenting and justifying the specific I&E code assigned to each clinical/performance data point including details about the assigned scientific weight based on the clinical data relevance and quality. Data from all three clinical data sources (i.e., clinical trials, literature, and experiences) must be appraised and the included data must be presented in the appropriate CER/PER location (i.e., background/SOTA section or subject device section).

BOX 4.2 Use different appraisal methods for different clinical data types

Clinical trial, literature, and experience data typically require distinct I&E criteria based on clinical data volume and type . Often, few device clinical trials are available, while experience reports are typically reported in large numbers and literature reports are only reported in large numbers when the device has significant research interest in the medical community.

When appraising clinical trial data, virtually every clinical trial with the subject/equivalent device should be included as long as the clinical data are relevant, high quality, and available for evaluation. Most clinical/performance trials conducted by the subject/equivalent device manufacturer should be available for CE/PE including all patient details; however, most clinical/performance trial details are not generally made available to the competitor device manufacturer.

When appraising clinical literature data, the evaluator must separate subject/equivalent device literature from the background/SOTA/alternative therapy literature. These data will be analyzed separately and the background/SOTA/alternative therapy literature will be the predominant clinical data source used to establish the prespecified and prequantified S&P criteria for the CE/PE process.

When appraising clinical experience data, the subject/equivalent device experiences are highly valuable to document rare adverse events (AEs) and device deficiencies (DDs). Clinical experiences for other devices are difficult to use due to the extremely poor data quality, even if the clinical experience data volume is quite high.

Relevant and high-quality clinical trial, literature, and experience data must be included in the CE/PE. The appraisal step is crucial to include all relevant and high-quality data and to exclude the irrelevant and/or poor-quality clinical/performance data from the CE/PE process.

(Continued)

BOX 4.2 (Continued)

Often the lines between relevant and irrelevant and between poor and high quality are rather blurry and the boundary varies depending on the available data amount. Less relevant and less high-quality data will be included when little clinical/performance data are available and the clinical data will be more relevant and high-quality when clinical/performance data are available in abundance.

Both MDR and IVDR require a methodologically sound CE/PE appraisal to include relevant clinical/performance data:

"3. A clinical evaluation shall follow a defined and methodologically sound procedure based on the following:

(a) *a critical evaluation of the relevant scientific literature currently available relating to the safety, performance, design characteristics and intended purpose of the device, where the following conditions are satisfied:*
 – *it is demonstrated that the device subject to clinical evaluation for the intended purpose is equivalent to the device to which the data relate, in accordance with* Section 3*of Annex XIV, and*
 – *the data adequately demonstrate compliance with the relevant general safety and performance requirements;*

(b) *a critical evaluation of the results of all available clinical investigations, taking duly into consideration whether the investigations were performed under Articles 62 to 80, any acts adopted pursuant to Article 81, and Annex XV; and*

(c) *a consideration of currently available alternative treatment options for that purpose, if any*." [2] (EU REG 2017/745, Article 61(3), emphasis added)

and:

"3. A performance evaluation shall follow a defined and methodologically sound procedure for the demonstration of the following, in accordance with this Article and with Part A of Annex XIII:

(a) *scientific validity;*
(b) *analytical performance;*
(c) *clinical performance.*

The data and conclusions drawn from the assessment of those elements shall constitute the clinical evidence for the device. The clinical evidence shall be such as to scientifically demonstrate, by reference to the state-of-the-art in medicine, that the intended clinical benefit/s will be achieved and that the device is safe. The clinical evidence derived from the performance evaluation shall provide scientifically valid assurance, that the relevant *general safety and performance requirements set out in Annex I, are fulfilled, under normal conditions of use*." [3] (EU REG 2017/746 Article 56 (3), emphasis added).

As noted above, relevance is also determined by considering how the clinical/performance data fulfill the general safety and performance requirements (GSPRs), especially GSPRs 1, 6, and 8 related to human data showing device S&P throughout the device lifetime with an acceptable benefit-risk ratio.

4.2.1 General safety and performance requirement 1: suitable device, safety and performance, benefit-risk ratio

GSPR 1 requires clinical/performance data to be gathered and included from subject/equivalent devices used "during normal conditions of use" and the clinical/performance data evaluation must be included in the CER/PER to determine if the subject/equivalent device:

a. achieves "the performance intended by their manufacturer";
b. is "suitable for their intended purpose";
c. is "safe and effective";
d. "shall not compromise the clinical condition or the safety of the patients or the safety or health of users or, where applicable, other persons"; and
e. is known to have potential use risks which "constitute acceptable risks when weighed against the benefits to the patient" and offer "a high level of protection of health and safety, taking into account the generally acknowledged state-of-the-art." (MDR and IVDR, Annex I (1)) [2,3].

For IVD devices, the clinical/performance data must also specifically demonstrate scientific validity, analytical performance, and clinical performance [3] (EU IVDR Article 56 (3)).

BOX 4.3 Exclude data if no intended purpose, safety and performance, or benefit-risk data

Any clinical trial, literature, and experience data without details about the intended purpose, S&P, and/or benefit-risk ratio in human patients (i.e., study subjects) should be excluded during appraisal, since the data are generally not relevant for CE/PE. The clinical data included for CE/PE analysis needs to address all S&P and patient benefits and risks across the entire indication for use.

For example, if the device indication for use states the device may only be used in adults, then clinical data from device use in pediatric patients may not be relevant and could be excluded from the subject device CER/PER section, unless the clinical data raises S&P concerns relevant to adults and is not addressed in other clinical data.

If, however, the device indication for use statement does not specify an age range, then those same pediatric data are relevant and should be included in the CE/PE for the subject/equivalent device section because they provide support for use in pediatric patients who are included in the broad indication for use statement.

The CER/PER background/SOTA section must define the intended purpose, including the disease, condition, or indication to be treated, prevented, or diagnosed by the subject/equivalent device. In addition, the background knowledge/SOTA section needs to place the subject/equivalent device S&P data in the proper clinical context among all available alternative medical treatments (including competitor/similar devices, alternative therapies and no therapy at all, if appropriate) used for the same intended purpose. Specifically:

EU MDR requires *"a consideration of currently available alternative treatment options..." (Article 61 (3)(c)) and specifies S&P criteria used to "determine... the acceptability of the benefit-risk ratio for the various indications and for the intended purpose or purposes of the device" be "based on the state-of-the-art in medicine"* (Annex XIV 1(a)) [2].

EU IVDR requires *"clinical evidence... to scientifically demonstrate, by reference to the state-of-the-art in medicine, that the intended clinical benefit/s will be achieved and that the device is safe"* (Article 56 (3)) [3].

Specifically, the evaluator must define the device S&P and benefit-risk ratio based on the SOTA in medicine. In other words, the background/SOTA data needs to be included and analyzed separately from subject device data so subject device data can be evaluated and related to background/SOTA data. As a result, the first appraisal question for any given data point is usually "Is this clinical data point about the subject/equivalent device or the background/SOTA/alternative therapy/device, both or neither?" The answer to this question should be reflected in the appraisal coding for every data point.

In general, appraisal methods focus on ensuring relevant and high-quality clinical/performance data are included from all clinical/performance data sources across these two main datasets: subject device data and background/SOTA/alternative therapy/device data (Table 4.1).

TABLE 4.1 Clinical data appraisal matrix.

CER/PER appraisal	Background/SOTA section	Subject/equivalent device section
Clinical data relevance	About similar/benchmark/SOTA devices and alternative therapies About safety and performance outcomes	About subject/equivalent device specifically About safety and performance outcomes
Clinical data Quality[a]	Trials: Large, multinational RCTs using similar/ benchmark/SOTA devices and alternative therapies Literature: Articles about the background/SOTA/alternative therapy/device, especially including guidelines, meta-analyses, systematic reviews, etc. Experience: Experience data are typically anecdotal and extremely poor quality and are therefore, excluded when sufficient background/SOTA clinical trials and literature data are included	Trials: Manufacturer-held clinical trials, PMCF/PMPF studies, investigator-initiated studies, registries, surveys, etc. using the subject/equivalent device Literature: Articles about subject/equivalent device RCTs, controlled trials, retrospective reviews, observational trials and, rarely, case reports/series when safety signals are present, etc. Experience: Manufacturer-held complaints and sales data, international AE database search outputs

AE, adverse event; *PMCF*, postmarket clinical follow-up; *PMPF*, postmarket performance follow-up; *RCTs*, randomized controlled trials; *SOTA*, state of the art.
[a]*Other data types are often included, but the evaluator should include the highest-quality data available while commenting on all data available.*

Guidelines, metaanalyses, and systematic reviews including the subject/equivalent device should be included in the subject/equivalent device section rather than the background/SOTA section. All other guidelines, metaanalyses, and systematic reviews without the subject/equivalent device would remain in the background/SOTA section. The data type may need to expand into individual clinical trial data if little to no clinical/performance data are found in the guidelines, metaanalyses and systematic reviews for the background/SOTA section.

The CE/PE appraisal process must include clinical/performance data showing the background/SOTA/alternative therapies have been fully considered and the subject/equivalent device is used as intended, is safe and effective, and has an acceptable benefit-risk ratio for the entire indication for use over the entire device lifetime.

4.2.2 General safety and performance requirement 6: device lifetime

GSPR 6 requires the manufacturer to ensure any adverse device effects during the device-use lifetime do not compromise "the health or safety of the patient or the user and, where appropriate, of other persons... when the device is subjected to the stresses which can occur during the normal conditions of use" and when the device was "properly maintained in accordance with the manufacturer's instructions."

BOX 4.4 Include clinical data representing the entire device lifetime

All identified clinical data should be appraised to include clinical/performance data considering the entire device lifetime. For implanted devices or diagnostic devices with a long device lifetime, clinical/performance data should be included from all device lifetime phases (recently manufactured, initial/early use, mid-lifetime, end of life, and uses during/after obsolescence).

Do not confuse device lifetime (i.e., time needed for actual human device use or diagnostic test) with shelf life (i.e., time when the device remains stable on the shelf before use).

4.2.3 General safety and performance requirement 8: risks, undesirable side effects

GSPR 8 requires the manufacturer to minimize "All known and foreseeable risks, and any undesirable [side-]effects," and these risks must "be acceptable when weighed against the evaluated [potential] benefits to the patient and/or user arising from the achieved [intended] performance of the device during normal conditions of use." This work is typically captured in the risk management files, which are reviewed by the CE/PE evaluator. The evaluator must also review all clinical data from clinical trials, literature, and experiences to ensure all identified clinical risks are acceptable when weighed against the clinical benefits.

BOX 4.5 Include "normal" conditions of use and comment on "off-label" uses

Off-label uses (i.e., subject devices used outside the intended use and/or indication for use) are generally considered "irrelevant" to the subject/equivalent device section, but are relevant to include separately in the background knowledge and SOTA section. If off-label use is found to be quite common with the subject/equivalent device, the manufacturer may want to address this serious, potential, systematic misuse concern prior to other CE mark activities.

All identified clinical data should be appraised to include background/SOTA as well as subject/equivalent device clinical data regarding the benefit-risk ratio during normal uses.

4.2.4 Clinical/performance data must be relevant

Background/SOTA and subject/equivalent device clinical data (i.e., trials, literature, and experiences) are considered highly relevant and should be included when products are used in a human as indicated in the device labeling. In addition, high-quality and scientifically sound pivotal S&P data about the subject/equivalent device used as indicated are relevant and should be included in the CER/PER subject device section. High-quality and scientifically sound nonpivotal, indirect, supportive clinical data are also relevant and should be included in the CER/PER background/SOTA section to provide important context about S&P details, if needed. Even poor-quality clinical data may be highly relevant and should remain included if the clinical data provide additional context or novel information about subject/equivalent device S&P (i.e., benefits and risks) or the background/SOTA [1].

Subject devices vary widely and may be noninvasive or invasive, or active or nonactive; some may need to follow special rules, or may include accessories/components/systems/robotics/software. The evaluator must understand the subject device details, so relevant clinical data can be included or excluded.

BOX 4.6 Use numerical codes to track inclusion and exclusion decisions

Numerical codes are often used to track clinical data I&E decisions. Often, positive code numbers are assigned to represent specific inclusion reasons and negative code numbers are assigned to specific exclusion reasons. For example, I&E codes can be developed to include relevant and to exclude irrelevant information as follows:

2 = clinical/performance data about "other" devices/background/SOTA/alternative therapies
1 = clinical/performance data about the subject/equivalent device used as indicated
− 1 = not about subject/equivalent/other devices/background/SOTA/alternative therapies
− 2 = duplicate data

During appraisal, these broad, high-level I&E codes can be expanded functionally by adding additional specific numerical codes as required by the evaluator conducting the appraisal. Often these expanded criteria are directly related to the clinical data quality. For example, certain case reports may be included because they include rare safety concerns while other case reports may be excluded because the case report is poor quality and provides no novel information of interest for the CE/PE.

Relevant clinical data addresses subject/equivalent device S&P or provides clinical context for generic device S&P in the background/SOTA section. If clinical data is about subject device use in humans to address the stated indication for use, then, by definition, the clinical data is relevant to the subject/equivalent device section and should be included until potentially excluded for another reason (e.g., poor-quality data, duplicated data). Clinical data addressing background knowledge, standard of care, SOTA, or alternative therapy/device S&P for the same indication are also relevant for the background knowledge/SOTA section when treatment, prevention, or diagnosis S&P are measured.

Sometimes, relevance is questionable. An example is a combination product, when a drug and device are used together (e.g., drug eluting stent). For the CER, if needed, clinical data on the individual product parts (i.e., the device or drug alone) could be included in the background/SOTA section while the clinical data from the combination product use in humans would likely be required in the subject device section since the clinical evidence for the S&P, benefit-risk ratio, etc. requires the device and drug to be used together.

4.1.5 Clinical/performance data should be high-quality

No hard and fast rules exist for appraising clinical data quality. Typically, quality (like relevance) is appraised and documented on a case-by-case basis. The evaluator must understand what makes clinical data scientifically valid.

For example, a high-quality clinical trial typically has the following features:

- an appropriate scientific question/objective;
- a sufficiently large sample size to have statistical power;
- appropriate outcome measures;
- validated measurement methods;
- an appropriate statistical analysis plan;
- data aligned with the background knowledge and SOTA;
- data aligned with the historical benefit-risk ratio;
- methodological reliability, consistency and transparency;
- scientifically sound data analyses; and
- conclusions based on data presented.

To be most helpful during CE/PE, the clinical trial must have sufficient data to evaluate subject/equivalent device S&P for all benefits and risks across all indications for use. This type of RCT would be ideal; however, since clinical trials can vary widely (different study designs, sizes, inclusion/exclusion criteria, measures, etc.), the evaluator must take care to appraise each clinical trial for inclusion or exclusion.

Like clinical trials, the clinical literature also varies widely. Many different literature types are published. Sometimes articles are about individual clinical trials or summaries of many different clinical trials. Sometimes the articles are a single person's opinion without any clinical data provided. Some articles are peer-reviewed and others are not. Some articles are paid advertisements while others are rigorously evaluated to account for many different potential and actual biases before publication is allowed. The evaluator must understand how to assess quality for each literature type and for all types grouped together.

Fortunately, many guidelines, articles, checklists, and tools have been published to help in assessing published literature quality. These "methodologic quality tools" generally fall into three tool types: items (individual considerations relevant to clinical research), checklists (many items without scores), and scales (many items which are scored and combined into a summary score) [4]. Some websites list specific report appraisal guidelines, checklists, and other tools to help assess methodological quality (Table 4.2).

TABLE 4.2 Literature assessment tools.

No.	Name	Used to assess (details)	Link
1	Research Reporting Guidelines and Initiatives	Checklists and guidelines for many trial types	http://www.nlm.nih.gov/services/research_report_guide.html
2	Critical Appraisal Skills Programme (CASP) Checklists	Checklists for many trial types	https://casp-uk.net/casp-tools-checklists
3	Enhancing the Quality and Transparency of health Research (EQUATOR) network	Checklists for many trial types	http://www.equator-network.org
4	National Institute for Health and Care Excellence (NICE). The social care guidance manual: Process and methods [PMG10] (30 April 30, 2013, last updated 1 July 2016)	Appendixes have checklists for many trial types	https://www.nice.org.uk/process/pmg10/chapter/introduction
5	SIGN methodology tools	Checklists for many trial types	https://www.sign.ac.uk/what-we-do/methodology/checklists
6	Appraisal of Guidelines, Research and Evaluation (AGREE-II)	Clinical practice guidelines	https://www.agreetrust.org/agree-ii
7	Systematic Reviews & other Review Types (Temple University)	Checklists for many trial types	https://guides.temple.edu/systematicreviews/criticalappraisal
8	Preferred Reporting Items for Systematic reviews and Meta-Analyses (PRISMA) checklist	Systematic reviews/ metaanalyses	https://www.prisma-statement.org/
9	Meta-analysis of Observational Studies in Epidemiology (MOOSE) checklist	metaanalyses (observational trials)	https://www.elsevier.com/__data/promis_misc/ISSM_MOOSE_Checklist.pdf
10	Assessment of Multiple Systematic Reviews (AMSTAR)	Systematic reviews/ metaanalyses	https://amstar.ca/Amstar_Checklist.php
11	Checklist for Systematic Reviews and Research Syntheses (Joanna Biggs Institute, JBI)	Systematic reviews	https://jbi.global/sites/default/files/2019-05/JBI_Critical_Appraisal-Checklist_for_Systematic_Reviews2017_0.pdf
12	AHRQ Practical Tools and Guidance for Systematic Review of Complex Interventions	Complex intervention systematic reviews	https://effectivehealthcare.ahrq.gov/products/interventions-tools-guidance/overview
13	Cochrane collaboration tool for assessing risk of bias	RCT	https://sites.google.com/site/riskofbiastool/welcome/rob-2-0-tool
14	Physiotherapy Evidence Database (PEDro) Scale	RCT	https://pedro.org.au/english/resources/pedro-scale
15	Modified Jadad Scale	RCT	https://onlinelibrary.wiley.com/doi/pdf/10.1002/9780470988343.app1
16	Methodological Index for Non-Randomized Studies (MINORS)	Nonrandomized interventional studies (12 points: comparative and noncomparative studies)	http://cobe.paginas.ufsc.br/files/2014/10/MINORS.pdf

(Continued)

TABLE 4.2 (Continued)

No.	Name	Used to assess (details)	Link
17	Newcastle-Ottawa Scale (NOS)	Nonrandomized (cohort and case-control) studies in metaanalyses	https://www.ohri.ca/programs/clinical_epidemiology/oxford.asp
18	Crombie's items (adapted)	Controlled studies (12 items)	https://cebma.org/assets/Uploads/Critical-Appraisal-Questions-for-a-Controlled-Study-July-2014-1-v2.pdf
19	Quality Assessment of Diagnostic Accuracy Studies-2 (QUADAS-2)	Diagnostic test accuracy studies	http://www.bristol.ac.uk/population-health-sciences/projects/quadas
20	SYstematic Review Centre for Laboratory animal Experimentation (SYRCLE) risk of bias tool	Animal studies (10 items)	https://www.ncbi.nlm.nih.gov/pmc/articles/PMC4230647/pdf/1471-2288-14-43.pdf

RCT, randomized controlled trial.

The evaluator should be familiar with these tools before attempting to appraise the medical literature. In the ideal case, every clinical trial article would be about a "gold standard" large, randomized, controlled, multiarm, multisite clinical trial; however, in reality, few published articles about medical device studies are this type, so understanding how to appraise less than ideal clinical data is important.

4.2.6 Good statistical quality is required

Statistical quality is another important appraisal consideration. For most comparative clinical data, statistical significance is a key clinical data quality parameter to consider during appraisal. In most research settings, the difference between two numbers must have less than a 5% chance the difference occurred from chance alone (i.e., the probability, or *p*-value, must be less than 0.05 to suggest the finding was significant and the difference in the two compared data distributions was significantly different).

Good statistical quality requires proper statistical methods were used to determine significance accurately. Often, clinical/performance data are based on exceedingly poor statistical methods, or the data analyses have clear misunderstandings about the importance of sound statistical methods. In these cases, the clinical/performance data should be excluded for insufficient statistical quality. In other words, the evaluator must assess the statistical methods used and exclude clinical data analyzed with poor or improper statistical methods. For example, some researchers claim a $p > 0.05$ result is significant when a 95% significance level was chosen in the statistical analysis plan before the study started. The rules are clear: $p = 0.05$ or above is simply not significant, thus the *p*-value must be less than 0.05 to be considered significant. Other authors may use a term like "trend" to suggest a *p*-value closer to, but greater than, $p = 0.05$ has value; however, this is incorrect since a *p*-value at or above 0.05 is simply not significant; as defined in the plan, the *p*-value must be less than 0.05 to be considered significant.

Just because two number sets look different, this does not mean they are different. When comparing two number sets (e.g., two averages), a mathematical (i.e., statistical) test is required to determine the probability the two data sets are actually different. More accurately, to demonstrate at least a 95% chance for a real difference between number sets (i.e., to show the numbers did not come from the same underlying numerical data distribution), several statistical methods and tests are available, including the Student's T-Test (to compare averages of two samples) and the Fisher Exact Test (for 2x2 contingency tables), among many other possibilities. The statistical test outputs are reported as probabilities (i.e., *p*-values) estimating if the two number sets are actually different. For example, groups may have "a statistically significant difference" when the *p*-value is less than 0.05 (i.e., the statistical tests show a less than a 5% chance the two groups are from the same underlying number distribution).

When appraising clinical data for inclusion or exclusion, here are some helpful questions to ask when assessing data processing and statistical methods:

- Do the protocol methods include a detailed statistical analysis plan?
- What level of significance was chosen (i.e., was a 95% power chosen to detect a difference between two groups)?
- How are the clinical data presented for statistical analyses?
- Are clinical data presented in a consistent format for statistical analyses?
- How are missing clinical data handled?
- What clinical data were included or excluded from statistical analysis?
- How do included or excluded clinical data affect the conclusions?
- What statistical methods were used to analyze the clinical data?
- Is a comparison or control group included (i.e., are two or more groups compared to each other, compared over time or some other comparison)?
- Do comparisons result in *p*-values or some other measure of probability?
- Are statistical details missing?
- Do the statistical results align with the conclusions or have the statistics been misinterpreted?

One important clinical/performance data quality parameter to consider during appraisal is the clinical data sample size or subject number treated in the trial (i.e., the study will have greater power with larger sample sizes). To determine the sample size in the protocol (i.e., before the trial begins), the statistician needs to determine the effect size for the item being measured (e.g., the study will have greater power with larger effect sizes and as the sample size increases, smaller effect sizes can be detected), the measured response variability (i.e., signal-to-noise ratio or standard deviation which should be documented in prior research), and the statistical power to be used (i.e., clinical study ability to detect a difference, if a difference exists, between two groups) at a chosen significance level (i.e., most often, significance is defined as $p < 0.05$).

The effect size and variability are generally fixed by the disease state and treatment types; therefore, significance level and sample size are the main power determinants in clinical research. Everything else being equal, the larger the sample size the better, until sufficient power is achieved. Power is directly related to the number of subjects required in each comparator arm. Each arm needs to be large enough to power a comparison sufficiently across and between groups as defined in the study objective or hypothesis question. The more comparator arms a study has, the larger the overall sample size will need to be. Single-armed studies (e.g., studies without a control or comparator arm) do not have statistical power, as statistical power requires a comparison between two or more groups. Single-arm trials with no comparator groups or control arms are of much lower quality than multiarm trials with controlled, comparator data and at least one control arm.

A large study does not always have higher quality than a small study. For example, an RCT with 200 subjects may be of higher quality than a retrospective review with 500 patients due to the statistical quality used to determine (i.e., statistically test) the differences in the clinical data collected from subjects in the control arm to the clinical data collected from subjects in the test arm.

4.2.7 High-quality research should limit bias

Another important quality marker to appraise is bias. No research is free from bias, but high-quality research attempts to understand and to mitigate potential bias. For example, when appraising author bias, high-quality articles tend to include statements about potential conflicts of interest in the text (often right before the reference section). This conflict-of-interest statement typically lists each author's financial and other relationships with the device manufacturer, study-related interest groups, or others. During the appraisal process, the CE/PE evaluator should review conflict-of-interest statements in each clinical trial or published article and document concerns about any potential or real clinical data bias in the appraisal worksheet. Bias does not always disqualify the clinical data; however, a study performed by investigators with potential bias related to significant financial or other conflicts of interest should be documented and evaluated carefully for potential exclusion of the clinical data from the CE/PE.

Clinical/performance data appraisal must consider many quality factors and each CE/PE update presents unique—and sometimes hidden—appraisal challenges. The evaluator must assess both individual data point quality for each individual trial, article, or experience report and overall dataset quality for the entire clinical/performance data group.

The appraisal process is iterative as device-specific clinical trial knowledge increases and may change the relevance and quality assumptions made during the evaluation. Different research fields use different reporting standards and conventions, so familiarity with the clinical literature is necessary to assess adequately whether clinical trial data quality is consistent with device-specific data reporting standards in the published literature.

4.2.8 Document quality control and quality assurance details

Quality control (QC) is required when collecting clinical data during clinical trials, literature, and experiences. This means the clinical data have been inspected to ensure the clinical data are appropriate and accurate or "scientifically valid." The quality assurance process to ensure QC should be defined in the methods used to collect the clinical trial, literature, and experience data or the CE/PE. For example, clinical trial data could include a:

- control group;
- comparator arm;
- randomization process; or
- blinding process.

Some questions to help assess QC and quality assurance might include the following:

- Were data collected in compliance with good clinical practice (GCP) guidelines (i.e., did the clinical trial comply with ISO 14155, the international medical device clinical trial standard or equivalent standards for good clinical practice)?
- Did clinical trial activities follow the protocol or were major protocol deviations noted?
- Were independent monitoring and auditing completed without significant issues being identified?
- Did data collection comply with all applicable local, federal/national and international laws, regulations, and requirements (e.g., Declaration of Helsinki, Institutional Review Board ongoing oversight)?

The QC and quality assurance details are different for each clinical data source. Clinical trials require well-designed systems specified in the clinical trial protocol, documented in the clinical trial data, and analyzed in the clinical trial report. Clinical literature should be published in journals with a strong peer-review process to ensure data integrity, reporting honesty, and a careful consideration of author bias. Clinical experience are often collections of individual reports and these poor-quality clinical experience data points should never be overinterpreted to represent more than the data they actually include. When appraising clinical data, insist on actual clinical data (i.e., exclude reports with only a complaint number and without any actual clinical data describing what actually happened to each individual device user/patient).

Use a workbook or software application to store the appraisal details for each clinical data point. Some software applications for clinical data appraisal may provide a sort and organize function; however, the software should also support data appraisal as an iterative process - as more data are reviewed, inclusion and exclusion criteria are likely to change. In addition, the appraisal and analysis steps overlap, so the software must accept appraisal changes once the data are more fully analyzed. Good software applications and processes will allow for such changes before and during the clinical data appraisal process. The software should also keep an audit trail documenting the QC and quality assurance functions used to ensure data integrity.

4.2.9 Clinical/performance data must be sufficient

Both the EU MDR and IVDR require "sufficient" clinical data to demonstrate device S&P. How much data are sufficient, however, depends on the specific subject/equivalent device and background knowledge/SOTA/alternative therapy/device clinical/performance data. The manufacturer is required to "specify and justify the level of clinical evidence necessary to demonstrate conformity with the relevant general safety and performance requirements." Furthermore, the clinical data amount and type shall be sufficient and the overall "clinical evidence shall be appropriate in view of the characteristics of the device and its intended purpose" (MDR Article 61(1) [2] and IVDR Article 56(1) [3]).

In addition, according to MEDDEV 2.7/1 rev. 4 "When clinical data are required in order to draw conclusions as to the conformity of a device... the data need to be in line with current knowledge/the state of the art, be scientifically sound, cover all aspects of the intended purpose, and all products/models/sizes/settings foreseen by the manufacturer" [1].

The evaluator must read the regulations and guidance documents carefully. Documenting "sufficient clinical evidence" must be explicit and clear, especially because this concept was not clearly defined in the MDR or IVDR. Essentially, the clinical/performance data must be available in a "large-enough" quantity and a "high-enough" quality to allow a qualified evaluator to determine if the device conforms to GSPRs 1, 6, and 8. In other words, the device must have documented clinical evidence demonstrating (at a minimum) the device is safe, and achieves the intended performance and benefits when used as intended with an acceptable benefit-risk ratio throughout the device life cycle and across all indication for use details.

MDCG 2020-6 [5] is helpful here because Appendix 1 defines MEDDEV 2.7/1 rev. 4 sections still relevant under MDR, Appendix 2 defines minimum CEP requirements for a legacy device, and Appendix III provides a clinical evidence hierarchy "ranked... from strongest to weakest" including 12 "types of clinical data and evidence" (Table 4.3), which can be used for clinical data appraisal to confirm MDR/IVDR GSPR conformity. Basic MDR principles in this table may also be applicable to IVDR clinical evidence requirements.

TABLE 4.3 Clinical evidence hierarchy.

Rank[a]	Clinical data types/evidence	Considerations/comments
1	Results of high-quality clinical investigations covering all device variants, indications, patient populations, duration of treatment effect, etc.	This is the gold standard, but may not be feasible or necessary for certain well-established devices with broad indications (e.g., Class IIb legacy sutures, which could be used in every conceivable patient population).
2	Results of high-quality clinical investigations with some gaps	Gaps must be justified/addressed with other evidence in line with an appropriate risk assessment and clinical safety, performance, benefit and device claims and the appropriate PMCFP/PMPFP should address residual risks. Otherwise, manufacturers shall narrow the device indication for use until sufficient clinical data are generated.
3	Outcomes from high-quality clinical data collection systems such as registries	Same as #2 and is sufficient evidence of quality available in data collected by the registry? Are devices adequately represented? Are data appropriately stratified? Are end points appropriate to safety, performance, and end points identified in CEP/PEP? [6,7]
4	Outcomes from studies with potential methodological flaws but where data can still be quantified and acceptability justified	Same as #2 and many literature sources are at this level due to limitations such as missing data, author bias, publication bias, selection bias, time-lag bias, etc. This applies equally to publications in peer-reviewed scientific literature; however, for legacy devices where no safety or performance concerns have been identified, these sources can be sufficient for confirmation of conformity to relevant GSPRs if appropriately appraised and all clinical data gaps have been identified and handled. High quality surveys may also fall into this category.
Class III legacy devices and implantable legacy devices (and Class D IVD devices) which are not WET should have sufficient clinical data as a minimum at level 4. WET devices may be able to confirm GSPR conformity via an evaluation of cumulative evidence from additional sources as listed below. Reliance solely on complaints and vigilance is not sufficient.		
5	Equivalence data (reliable/quantifiable)	Equivalence must meet MDR/IVDR criteria. Manufacturers should gather data on their own devices in postmarket phase. Reliance on equivalence should be duly justified and linked to appropriate PMCFP/PMPFP and proactive PMS.
6	SOTA evaluation, including evaluation of clinical data from similar devices	These are not considered clinical data relevant to the subject device under MDR or IVDR, but for WET only, SOTA and background data can be considered supportive for confirmation of conformity to relevant GSPRs. Data from similar devices may be important to establish if the device under evaluation and similar devices can be considered WET. Data from similar devices may be used to demonstrate ubiquity of design, lack of novelty, known S&P profile of a generic group of devices, etc.
7	Complaints and vigilance data; curated data	These are poor-quality, often high-volume "real-world data" with reporting limitations even when collected in a robust quality system. These data are often used to identify safety trends or performance issues and may provide supportive device safety evidence.
8	Proactive PMS data, e.g., derived from surveys	These are poor-quality clinical data due to limitations associated with sources of bias and poor data collection practices. May be useful for identifying safety concerns or performance issues.
9	Individual case reports on the subject device	These are poor-quality clinical data due to inability to generalize findings to a wider patient population, selection bias, reporting bias, etc. May provide supportive or illustrative information for claims.
10	Compliance to nonclinical elements of common specifications considered relevant to device safety and performance	This is not clinical data, but common specifications may address clinically relevant end points through nonclinical evidence such as mechanical testing for strength and endurance, biological safety, usability, etc.
11	Simulated use/animal/cadaveric testing involving healthcare professionals or other end users	This is not clinical data, but may be considered evidence of confirmation of conformity to relevant GSPRs, particularly in terms of usability, such as for accessories or instruments. *May also be of interest when clinical data are deemed not appropriate to collect under MDR Article 61(10).*
12	Preclinical and bench testing/compliance to standards	This is not clinical data, but may be considered evidence of confirmation of conformity to relevant GSPRs, particularly in terms of usability, such as for accessories or instruments.

CEP, clinical evaluation plan; *GSPR*, general safety and performance requirements; *IVDR*, *in vitro* diagnostic device regulation; *MDR*, medical device regulation; *PEP*, performance evaluation plan; *PMCFP*, postmarket clinical follow-up plan; *PMPFP*, postmarket performance follow-up plan; *PMS*, postmarket surveillance; *S&P*, safety and performance; *SOTA*, state of the art; *WET*, well-established technology. Emphasis added in bold text.

[a]*Table contents modified from MDCG 2020-6, Appendix III [5].*

In practice, this hierarchy allows and encourages the evaluator to make decisions. If sufficient high-quality data are available, lower quality studies lacking novel safety or performance information can be excluded based on quality alone; however, if the subject device does not have sufficient high-quality data, lower-quality data need to be included until sufficiency is achieved.

In addition, in certain exceptional cases, the evaluator should be aware nonclinical evidence may be used without actual clinical data available for the CE/PE process. For example, nonclinical data may be used to support certain legacy devices or well-established technologies (WET) or when equivalence is claimed for devices in compliance with relevant common specifications and only if the CE/PE has "sufficient clinical data." The evaluator must clearly define how clinical data sufficiency has been achieved based on the existing, available clinical data. MDCG 2020-6 [5] states the need to "perform analysis of the methodological quality of data obtained from different sources to identify and assess the level of evidence, bias, other inherent weakness or other possible shortcomings" and defines the legacy devices potentially "exempt" from the "Clinical evaluation consultation procedure for certain class III and class IIb devices."

Identifying these and other clinical data gaps early during appraisal is helpful because the evaluator is trying to determine if the clinical data are likely to be sufficient. If the clinical data are insufficient to meet the MDR/IVDR GSPRs 1, 6, and 8 requirements then the need for the PMCFP/PMPFP to generate and collect new clinical trial, literature, and experience data should be escalated to the postmarket surveillance (PMS) team during the appraisal step. Additional clinical data generation should begin as soon as possible to address the clinical data gaps. Ultimately, the manufacturer may need to "narrow the intended purpose" during the CE/PE process until all indication details are supported by the "available clinical evidence." The labeling changes resulting from this change in indication will take time to execute for a device already on the market and the plan for this should be clearly stated in the CE/PE documents.

To avoid needing to narrow the intended use/indication for use, the manufacturer must ensure sufficient clinical data are available across the entire intended use population and the CE/PE process must be designed to appropriately appraise all relevant clinical trials, literature and experience data in compliance with the EU regulations using specific appraisal criteria especially designed for each clinical data type.

4.3 Clinical trial data appraisal

In general, Class III, implantable devices or Class D IVD devices must have clinical trial data documenting device S&P along with the specific benefit-risk ratio drawn from reliable clinical data. In some cases, the manufacturer may have sponsored multiple clinical trials during product development and regulatory approval. The evaluator must evaluate and appraise each trial protocol, study design and the study results to determine if the trial should be included or not. The trial should be included if the clinical trial is directly relevant to the subject/equivalent device S&P and the available clinical trial data are of high quality. The evaluator should exclude irrelevant and low-quality clinical trial data during the clinical trial appraisal step.

Manufacturer-held clinical trial data is often the highest-quality clinical data available; however, clinical trials completed by others may also be available to the manufacturer. The evaluator must take care not to duplicate data when both the clinical study report and a publication are available for a given clinical trial.

4.3.1 Clinical trial data must be relevant to the subject/equivalent device

All relevant trials should be included unless and until the evaluator determines a particular trial brings no value to the analysis or has no clinical trial S&P data available to evaluate. If two trials cover the same device use and patient population, these will probably both be included; however, a smaller "pilot" trial may be excluded if a larger pivotal trial is available and is much higher quality, especially if all S&P data from the smaller pilot trial are documented in the larger pivotal trial. Always to include both trials if the smaller study has any novel S&P data.

BOX 4.7 Clinical trial appraisal example

Situation: A subject device under evaluation has the following indication for use: "for use in adults with Indication A and/or Indication B" and the company ran five clinical trials using the subject device as follows:

- **a.** Phase I ($n = 15$) in healthy adults;
- **b.** Phase II ($N = 25$) in children with Indication A;
- **c.** Phase II ($N = 45$) in adults with Indication B;
- **d.** Phase II ($N = 30$) in adults with Indication A; and
- **e.** Phase III ($N = 5000$) in adults with Indication A.

(*Continued*)

BOX 4.7 (Continued)

Appraisal question: Which trials should you include in your evaluation, and why or why not?

Appraisal considerations: With only the information provided, all five trials would be included; however, trials A and B would be appraised for inclusion in the Background/SOTA section of the CER/PER due to the "off-label" use in "healthy adults" and "children" respectively when the indication for use specified use only in "adults" with "Indication A" or "Indication B."

Trials C and E would be appraised for inclusion in the subject device section of the CER/PER because each trial is relevant to a different device indication for use (i.e., Indication B and A, respectively).

The small sample size in Trial C should be noted and the PMS team should be alerted to a potential clinical data gap for Indication B where little clinical trial data were identified and where the potential to need additional clinical data gathering in the PMCFP/PMPFP is high.

If the clinical trial data from Trial D are completely included in Trial E because the two trial datasets were combined in a phase II/III trial design, then Trial D may be excluded if fully duplicative with Trial E.

In addition, if Trials A and D do not provide any novel subject/equivalent device S&P information compared to the large Trial E (i.e., all S&P issues in Trials A and D were also seen in Trial E), Trials A and D can be excluded because the data were not novel and all S&P concerns were clearly represented in the larger Trial E.

In general, smaller, lower-quality trials can be excluded when the smaller, lower-quality trial provides the same (or less) clinical data compared to an included larger, higher-quality trial. In this situation, however, the overall sample size and the individual trial sizes matter. If Trial E had only 50 subjects, then the additional 30 subjects from Trial D would warrant inclusion even if the S&P data were quite similar.

Clinical trial data appraisals often need to consider other clinical trial factors potentially including, but not limited to, trial design quality, primary and secondary end point data relevance, comparator device data, etc.

The appraisal is an iterative process. Start by appraising clinical trial relevance to the subject/equivalent device or the background/SOTA/alternative therapies before appraising clinical trial quality. If a trial is not relevant, no need to appraise clinical trial quality just exclude the trial because the clinical trial data are not relevant to the subject/equivalent device or the background/SOTA/alternative therapies.

The appraisal process typically involves three passes through the data to make inclusion/exclusion decisions:

- first pass—review trial title and include/exclude*;
- second pass—review trial synopsis and include/exclude*; and
- third pass—review all trial details and include/exclude*.

**Keep all trials included unless the exclusion rationale is completely clear and justified.*

During appraisal, remember to segregate trials for subject/equivalent device section or background/SOTA/alternative therapies section of the CER/PER. For example: using code 1 as the appraisal code might mean the trial should be included because the data are relevant to the subject/equivalent device and using code 2 as the appraisal code might mean the trial should be included because the data are relevant to the background/SOTA/alternative therapies section of the CER/PER.

If the trial is clearly irrelevant, assign an appropriate exclusion code. For example: using code −1 as the appraisal code might mean the trial should be excluded because the trial is not relevant to the subject/equivalent device or the background knowledge/SOTA/alternative therapies or similar devices. Using code −2 as the appraisal code might mean the trial should be excluded because the trial data are duplicated in another trial.

The clinical/performance trials identified and appraised for inclusion must have all records gathered together for the evaluator to complete the analysis next. Every effort should be made to access the full clinical trial report records from all potentially included trials. The trial records should include, but are not limited to, the following:

- Legal and regulatory details (i.e., contracts and regulatory applications/clearances/approvals, if any)
- Final study report (signed and dated with all data included)
- Protocol, informed consent forms and all amendments
- Primary data records on all measured details including disease states and outcomes
- Primary data listings for all adverse events (AEs) and device deficiencies (DD)
- Narrative details for all deaths and device-related serious and unexpected events

When these records are not available the trial should be excluded due to insufficient information. For example, the trial may have been done by a competitor who is unwilling to share access to the data details OR the manufacturer may have sponsored trials done by others where the data were kept at the trial location but were not shared with the manufacturer.

In some cases when clinical trial records are not available to the manufacturer, the trials may have some data published in the clinical literature. These clinical trial data should be excluded from the clinical trial data section because the clinical trial data are not available for appraisal or analysis; however, the clinical trial information available in the literature should be cross-referenced between the specific, related clinical trial and the clinical trial article evaluated in the literature section. This is quite common for articles identified in clinical trial database searches since the manufacturer may not have any right of access to those clinical trial data and the only possibility may be to wait for a publication about the clinical trials identified in the databases.

4.3.2 The clinical trial data must be high quality

Be careful when using preliminary or unpublished clinical trial data or data reported in clinical trial databases since these data are subject to change and may be poor quality. Clinical trial registry databases are not peer-reviewed and have little to no quality controls for the stored clinical trial report data.

Many different trial designs exist and many different questions should be asked when appraising the quality of a particular study design. For example, RCTs and cohort studies are just two examples of different clinical study designs. The RCT is the "gold standard" clinical trial design where the study subjects are prospectively assigned to a treatment arm at random. Often, one of the treatment arms is a control or placebo treatment. The RCT is typically "double blinded" which means neither the person giving the treatment nor the study subject knows if the treatment or the placebo were given. A cohort trial is one of many other clinical trial designs. The cohort trial takes place over time and the study subjects check-in periodically. A given study may be retrospective (i.e., events happened in the past) or prospective (i.e., events are planned and recorded as they occur). In addition, a study may be observational (i.e., no treatment is given) or interventional (i.e., a treatment is given).

Below are some example questions to ask about study quality when appraising RCTs and cohort trials (answers to appraisal questions should be stored in the spreadsheet or CE/PE software):

Randomized Controlled Trials (including interventional, comparative trials)

- Study design? (Were subjects enrolled prospectively or retrospectively?)
- Sample size? (Was trial statistically powered to show a difference between groups?)
- Number of arms? (Was trial controlled? Did each arm have sufficient sample size?)
- Number of sites? (Was trial completed at more than one EU site?)
- Subject device? (Was an equivalent or similar device used in the trial?)
- Which indications were treated? (Were any indications for use missing in the trial?)
- Are methods comparable across arms? (Were methods appropriate for the trial?)
- Were appropriate statistical methods used? (Were statistics appropriate for the trial?)
- Are detailed inclusion and exclusion criteria specified? (Was population appropriate?)
- Was the treatment group assigned at random? (How was randomization assured?)
- What are the prognostic factors by group? (Were groups appropriately balanced?)
- Were groups comparable at baseline? (Were any "demographics" significantly different)?
- How was blinding applied? (Were recruiters, outcome assessors, care providers and/or subjects blinded to treatment allocation?)
- Were all randomized subjects included? (How were intent-to-treat, modified intent-to-treat and per protocol groups constructed and verified?)
- What point estimates and variability measures were reported for primary outcomes? (Were appropriate significant figures details provided in all values?)
- What proportions of each group attended follow-up visits (What were the dropout rates and the specific reason for each dropout?)
- What were the S&P end points? (Did trial include the needed S&P measures?)
- Follow up? (Were study subjects followed long enough?)

Cohort studies (observational trial, subject group has common characteristics)

- Study design? (Were observed subjects selected prospectively or retrospectively?)
- What was the sample size? (Was population justified with an appropriate statistical power calculation?)
- Were any comparator groups included? (Was a control group included?)
- Number of sites? (How many sites were in the EU?)
- Subject device? (Was an equivalent or similar device used in the trial?)
- Which indications were treated? (Were any indications for use missing in the trial?)
- Are methods comparable across subgroups? (Were methods appropriate for the trial?)
- Were appropriate statistical methods used? (Were the statistics appropriate for the trial?)
- Are detailed inclusion and exclusion criteria specified? (Was population appropriate?)
- What are the prognostic factors by group? (Were groups appropriately balanced?)
- Were groups comparable at baseline? (Were any "demographics" significantly different?)
- Was adequate control applied for confounding factors?
- Were measurement outcomes unbiased?
- What proportions of each cohort attended follow-up visits? (What were the dropout rates and the specific reason for each dropout?)
- What were the S&P end points? (Did trial include the needed S&P measures?)
- Follow up? (Were study subjects followed for long enough?)

As illustrated, some questions are the same for the different types of trials; however, some questions differ depending on the clinical trial type. The evaluator consider all quality details specific to the clinical trial type. Sometimes these questions are part of the analysis step. Only some of the questions may need to be asked during the appraisal step so an initial inclusion/exclusion code can be applied. Detailed analysis will follow the appraisal for all included clinical/performance data.

4.4 Clinical literature data appraisal

The clinical literature appraisal purpose is to determine which articles to include or exclude from further clinical evaluation. Appraising clinical literature often involves winnowing down search results to focus on the most appropriate articles to analyze. This work can be done in multiple ways, so long as relevant, high-quality literature are included and unique S&P issues are not excluded.

No standard appraisal methods exist for clinical literature. Literature needs to be appraised both individually and as a whole and decisions about sufficiency, relevance and quality depend on the device risk, CER/PER type and overall literature depth and quality. No single appraisal algorithm applies to all, or even most, datasets; however, some broad questions can be applied to literature appraisal in general:

- Was article relevant to subject/equivalent device indication for use?
- Was article about a similar device with the same indication for use as the subject/equivalent device?
- Was article about disease state background/SOTA?
- What study type was described (e.g., metaanalysis, RCT, case series)?
- What study design elements were described (i.e., number of subjects? controls? other comparators, if any?)
- Were data analyzed appropriately?
- What were the study limitations (e.g., small sample size, no controls, biased investigator)?
- Does article describe something novel (e.g., an S&P issue not described elsewhere in the literature)?
- How does article fit into overall dataset (i.e., what value did the data bring to the CE/PE)?

Literature appraisal is an iterative process and an article appraisal (i.e., I&E code) may change during the process. For example, an article with a seemingly relevant title and abstract, upon reading the entire article, may not warrant inclusion due to quality, validity or relevance issues. Or a small case series, after reviewing identified literature, may be the only article addressing a particular patient population or otherwise unreported risk and therefore should be included. Appraisal is an ongoing process, and the exact literature included will change as the literature are analyzed, considered, and reappraised.

BOX 4.8 Does publication in a "top-tier" journal matter?

Some scientists believe top-tiered (higher-impact, highly rated) journals have higher-quality articles because the editors draw from a highly competitive and expansive pool of highly experienced authors; however, high-quality papers are also published in lower-impact journals and vice versa (i.e., sometimes low-quality articles are published in top-tiered journals).

In other words, where the paper was published should not matter and the journal impact factor (i.e., rating) should not be considered a reason to reduce any appraisal activities. MEDDEV 2.7/1 rev. 4 [1] states "While publication in a renowned peer reviewed scientific journal is generally accepted as an indicator of scientific quality, such publication is not considered an acceptable reason for bypassing or reducing appraisal activities." A journal's renown reflects a perception of scientific quality and prestige but does not guarantee quality. Some have reported the more citations a journal has (a proxy for journal quality), the higher the retraction rate [8].

Repetitive publication of the same data should be discouraged and peer review should be required. No matter where the article was published, the appraisal should be the same: the evaluator should confirm the article was published in an accredited, peer-reviewed journal with experience publishing in the applicable clinical field. Journals without a history of publishing in a particular clinical field may have trouble identifying appropriate peer-reviewers with the necessary expertise.

Each article must be appraised in the same scientific and analytical manner, without bias. Is the paper relevant to the device under evaluation or the background/SOTA and is the article high quality or not?

Once articles are identified, downloaded and organized in the workbook or software system, begin relevancy and quality determinations by first reading titles, then abstracts and finally full-length articles. Separate and prioritize the highest quality articles from the lower quality articles (e.g., metaanalyses, systematic reviews and RCTs are often higher quality than literature reviews, retrospective studies or studies solely about questionnaires, registries or case reports). The evaluator should determine the relevancy and quality based on the article clinical data details, not the category where the article is placed (e.g., a registry study with high-quality clinical data is appropriate to prioritize above a small RCT with poor-quality data).

4.4.1 Clinical literature inclusion and exclusion criteria and coding

Clearly irrelevant titles can be appraised or coded for exclusion and excluded immediately without further review. For example, a search for "N95" masks to mitigate COVID-19 spread in PubMed identified an article titled "Macular Blood Flow and Pattern Electroretinogram in Normal Tension Glaucoma" due to the search algorithm identifying "N95 amplitudes" within the article abstract. Based on title alone, the article could be immediately excluded as irrelevant to N95 face masks and unrelated to target condition COVID-19 or background/SOTA. We can code this with a " − 1" number where the minus means the article is excluded and the serialized number separates exclusion codes in a ranked order from most obvious to least obvious. This exclusion code is described as "Not about subject device, background knowledge or SOTA" (i.e., the article is irrelevant to N95 masks or the specific intended use and the article is not about a similar, equivalent, or alternative therapy/device) (Table 4.4).

TABLE 4.4 Example I&E coding key.

Code	Code description
1	About subject device, background knowledge or SOTA
−1	Not about subject device, background knowledge or SOTA

SOTA, state of the art.

The evaluator should work through the identified articles and they should not delete identified articles from the listing in the worksheet or software system. The process is to code each article for inclusion (e.g., +1) or exclusion (e.g., −1) in the worksheet or CER/PER software and to retain the coding decision in the documentation (Table 4.5).

TABLE 4.5 Example: literature appraisal (I&E exclusion coding for N95 masks in bold).

UID	I&E	Title	Author, year	Abstract
1	1	P2/N95 respirators & surgical masks to prevent SARS-CoV-2 infection: Effectiveness & adverse effects	Kunstler B, 2022	Healthcare workers (HCWs) are required to wear personal protective equipment (PPE), including surgical masks and P2/N95 respirators, to prevent infection while treating patients. However, the comparative effectiveness of respirators and masks in preventing SARS-CoV-2 infection and the likelihood of experiencing adverse events (AEs) with wear are unclear...
1	1	Comparative effectiveness of N95, surgical or medical, and nonmedical face masks in protection against respiratory virus infection: A systematic review and network meta-analysis	Kim Ms, 2022	The aim of this systematic review and network metaanalysis is to evaluate the comparative effectiveness of N95, surgical/medical and nonmedical face masks as personal protective equipment against respiratory virus infection. The study incorporated 35 published and unpublished randomized controlled trials and observational studies investigating specific mask effectiveness against influenza virus, SARS-CoV, MERS-CoV and SARS-CoV-2...
1	−1	Macular Blood Flow and Pattern Electroretinogram in Normal Tension Glaucoma	Jeon SJ, 2022	... Patients with higher VD had higher N95 amplitude ($p = 0.048$). Macular VD was significantly correlated with N95 amplitude, irrespective of disease severity ($r = 0.352$, $p = 0.002$) for the total...
2	−1	Diagnostic issues in second opinion consultations in prostate pathology	Jara-Lazaro AR, 2010	...The increase in early detection of prostate cancer in the Asian population has bolstered second opinion consultations in prostate pathology in this region. In this review, we aimed to identify...

This process generates a meaningful, organized historical record of all identified articles and the inclusion or exclusion decisions made for each article. Clinical literature describing metaanalyses, systematic reviews, randomized controlled trials about the subject device or the background/SOTA are typically included in the appraisal process. Observational trials are also commonly included (these may include prospective or retrospective reviews, cohort studies, case-control/cross-sectional studies and certain case reports/case series or other nonanalytic studies including unusual/new results or adverse events/safety signals).

Common inclusion and exclusion criteria are as follows (numerical code and description):

Typical inclusion criteria

1. Clinical practice guidelines, metaanalyses and systematic review articles characterizing prevalence and/or natural course of conditions affecting the intended patient population (i.e., background knowledge/SOTA/alternative therapies)
2. Articles reporting similar device S&P clinical data outcomes, background, SOTA and alternative therapies available to the intended patient population, including clinical practice guidelines
3. Articles reporting subject-device-specific S&P clinical data outcomes (i.e., subject device benefits and/or risks clinical data)

Typical exclusion criteria

1. Not about the subject or similar devices, background knowledge, SOTA or alternative therapies
2. Duplicate article
3. Published in languages other than English without novel information
4. Bench, animal, cadaver or simulation studies
5. Case reports/Case series without novel information
6. No author or abstract, or unable to retrieve article
7. Other medical areas/far removed

These broad appraisal criteria are initially stated in the CEP/PEP and they are further developed during the appraisal process into specific I&E codes using positive numbers for included data and negative numbers for excluded data. For example, appraisal for the "N95" masks used to mitigate COVID-19 spread required additional I&E codes (Table 4.6).

TABLE 4.6 Expanded I&E coding key (N95 mask example).

I&E	Description
6	Survey about N95 use
5	RCT or crossover study about N95 masks used to protect from pollution or other environmental contaminants or nanomaterials (i.e., background/SOTA information)
4	Guideline about N95 masks used for infection prevention (e.g., COVID-19, influenza)
3	Metaanalysis about N95 masks used for infection prevention (e.g., COVID-19, influenza)
2	RCT or systematic review about N95 masks used for infection prevention (2 about uses during pregnancy)
1	About subject device (i.e., Honeywell N95 Respirator) S&P
−1	Not about S&P of N95 Respirator or similar device (e.g., about drug trials, psychotherapy, injury rehab, disaster preparedness, visual processing/ophthalmic eval/retinal function/macular degeneration/amblyopia, muscle damage, ERG electrodes, goiter, aortic valve, Vocal Fold Leukoplakia, surgical performance/revascularization/liver transplant, physiological biomarkers)
−2	Duplicate
−3	Not in English and no novel information or article not available
−4	Animal, *in vitro*, bench, cadaver, fit factor testing and simulations/metaanalyses of simulation studies (e.g., training, surrogate exposure, airborne particle counts, protection factors)
−5	Article with more current version available (e.g., metaanalysis with updates, most current version included) OR study protocol only
−6	About decontamination of single use masks
−7	About modified N95 with silicone-based dressing underneath N95 to prevent facial skin injury or about surgical smoke
−8	About infection control program (e.g., compliance or behavior related to mask use)
−9	Narrative review or case series or low-quality Cluster RCT about N95 masks used for infection prevention without novel information

RCT, randomized controlled trial; *S&P*, safety and performance; *SOTA*, state of the art.

Numerical codes were used to appraise both data relevant to the device (code 1) and data relevant to the background and SOTA (codes 2, 3, and 4). The codes represent different quality and relevance levels. For example, articles about decontamination of single use masks were not relevant to the indication for use and were given a specific exclusion code (−6). Guidelines are given specific consideration in the CER/PER background/SOTA and were given a specific inclusion code (4). RCTs were given three separate codes: two for high quality studies or studies with novel information, split by indication, and one for low quality studies without novel information (−9). Developing I&E codes is an ongoing process. New codes may be included or codes may be combined if significant overlap is observed.

A different evaluator may develop different I&E codes, or may use the same I&E codes but may make slightly different I&E decisions. No single appraisal process is required and no single "correct" final dataset is known. The evaluator must use their best judgment for each step in the clinical/performance evaluation, including the appraisal step, where I&E criteria must be applied consistently to identify sufficient clinical data. Consistency and sound methodology are required. For example: if an article is excluded because a more current version is available (code −5 in the above example), then every article with a more current version available should be excluded or the code rationale should be changed to define the I&E code specifically.

A good three-step appraisal QC process is helpful (i.e., one person appraises, a second, more senior person reviews and documents any coding disagreements, and a third, more experienced person adjudicates and determines the final appraisal coding).

4.4.2 Literature appraisal process

Like the clinical trial data appraisal, an effective method for literature appraisal begins with the identified article list and uses three passes through this clinical literature listing while adding more data appraisal and analysis data during each subsequent pass. In other words, during appraisal, for any articles where relevance and I&E coding could not be determined based on the article title alone, then read the abstract followed by entire article, if necessary, to determine appropriate I&E Code (Fig. 4.1).

Title → Abstract/Summary → Entire Article/Document

FIGURE 4.1 Clinical data relevancy determination process.

Sometimes the appraisal process spills over into the analysis process step when the rationale for exclusion is unclear.

First pass: Read the article title and decide if the article should be included or excluded based on the title alone. If the article title clearly does not address the device under evaluation, background/SOTA, alternative therapy/device, or human clinical data, then exclude the article by documenting the appropriate negative exclusion code (number) in the spreadsheet I&E column or software system. Conversely, if the title states the article is about the device S&P or might be helpful in the background and SOTA section, then include the article by documenting the appropriate positive inclusion code (number) in the spreadsheet I&E column or software system. Keep each article included until the clear and appropriate I&E decision is documented.
Second pass: Be sure the abstract was copied into the workbook in the abstract column or upload the details into the software system, then read the abstract and decide if the article should be included or excluded based on the title and abstract. Again, if the abstract does not provide enough information to determine a clear I&E decision, keep the article included for further review. Abstracts typically define the study design/type (e.g., case report, retrospective review, RCT, metaanalysis), study objective (i.e., study question), sample size (i.e., number of subjects), device name, number of subjects treated with device under evaluation and other study details including results and author conclusions. During appraisal, when analyzing the abstract to determine I&E, these data should be copied into the appropriate spreadsheet data analysis columns or into the appropriate software data analysis sections to inform both appraisal decisions and subsequent clinical data analysis processes.
Third pass: Skim through the full-length article and decide if the article should be included or excluded. Record the appropriate I&E code in the spreadsheet or software.

All appraisal (I&E) decisions must be documented and the spreadsheet or software system serve to record search outputs and appraisal decisions which can be reviewed at any time to understand the literature search and appraisal results. Using numerical codes (i.e., negative numbers excluded and positive numbers included) allows the spreadsheet or database data to be sorted from largest to smallest and separating excluded articles from included ones. Do not delete or remove excluded or duplicate articles from the spreadsheet; simply assign them the appropriate exclude code and sort them out of the way as needed. Use good version control to track I&E changes, since the evaluator (or QC persons) may decide (on subsequent passes) to include a previously excluded article (or vice versa) based on the developing dataset.

4.4.3 Full-length articles are required!

As documented in MEDDEV 2.7/1 rev. 4 [1]:

> Abstracts lack sufficient detail to allow issues to be evaluated thoroughly and independently, but may be sufficient to allow a first evaluation of the relevance of a paper. Copies of the full text papers and documents should be obtained for the appraisal stage... literature search protocol/s, the literature search report/s, and full text copies of relevant documents, become part of the clinical evidence and, in turn, the technical documentation for the medical device.

Each included article must be available in the full-length form to complete the appraisal and analysis steps. Articles are typically saved as a PDF with an identifiable title. A common article pdf title is: <First Author Last Name> and <Year> with <a, b, c> after the year for instances where the same first author has published multiple papers in the same year.

BOX 4.9 Pay attention to copyright laws

Not all literature is available for free. Many publishers require a copyright payment to use the published article especially when using the article for a commercial interest like a CER/PER.

A good practice is to have the company sponsoring the CER/PER be responsible for paying the copyright fee for any included "paywalled" articles. The company must pay the appropriate price for their intended use of the article beyond the initial evaluation by the evaluator. In other words, academic library credentials and permissions for personal use by the

(*Continued*)

BOX 4.9 (Continued)

evaluator should not be used to access paywalled articles for a device manufacturer. The CER/PER is used in a for-profit business marketing application to gain a CE mark allowing the device to be marketed and sold in the EU. As such, academic uses are generally not appropriate during CER/PER development.

The number of articles available for free has been increasing over the decades; however, if the article is not free to access, then the article must be purchased. One source of free clinical literature is supported by the National Institutes of Health (NIH), where any article about a study completed with NIH funding must be made available for free via PubMed Central (PMC) [9]. This distinction can be identified by the PMCID number, which is distinct from the PMID number among PubMed results. This PMCID number means the article is publicly available through PubMed Central.

Many other articles are also available for free through the publisher. Some authors and groups actually pay a fee to make their work freely available on the Internet. PubMed listings often specify which articles are free and provide a direct link to the PDF; however, the "Free Article" tag may not be included in the exported search results and must be accessed directly through the PubMed website. For articles without a free link on PubMed, always search Google Scholar for the article to see if you can access the article for free on another website.

Be sure to enter data into the spreadsheet while skimming the article during the appraisal step. Every article msut be reviewed in full with the data abstracted, tabulated, analyzed, and used in the CER/PER.

Appraisal happens both at the individual article level (i.e., for relevance and quality) and at the group level (i.e., to determine the weight and value for each article among all included articles). If an article is not immediately excludable in the first three appraisal rounds, then the article should be appraised in the group context while the analysis phase begins. At the end of the appraisal step, the evaluator must consider how the quality and relevance for each individual included article compared to all the other included articles. What value does each article bring to the group? If the article adds no value, the article can be excluded and the evaluator should focus on higher-quality, more relevant, and more appropriate clinical data for the CER/PER.

As full-length articles are reviewed and data is entered into the spreadsheet, a more complete literature picture will start to come into focus. How many articles were available? How many articles were metaanalysis or systematic reviews or RCTs about the device under evaluation? How many used the subject device versus similar devices or alternative therapies? The questions then become: What value did each article bring to the evaluation? What were the relevance and quality for each particular article compared to all the others? The value each article brings to the evaluation can change depending on the overall literature quality.

Some devices are well-established with multiple high-quality RCTs completed using the subject device with clear S&P data describing and quantifying detailed benefits and risks to the study subjects. Other devices may have only one or two RCTs, or may never have been evaluated with an RCT. For devices claiming equivalence to another device, the subject device may have no clinical data whatsoever. All of these details need to be documented, and the device-specific "learning" is important during the appraisal step.

BOX 4.10 How to skim/read a journal article during appraisal/analysis

Only certain article sections need to be reviewed during appraisal. A good practice is to read the abstract carefully and to focus first on the study design, methodology, and results. Often, the introduction can be skipped altogether during the appraisal step since new clinical data are found in the results and conclusions not within the introduction. Although results are not present in the methods section, this section is important to understand the methodologic study quality. Similar to the introduction, the discussion and conclusions can be skimmed for device related information.

Important methodological questions to ask while skimming through an article during the appraisal step include the following:

What was the study design (e.g., retrospective review, randomization, control)?
What were the subject inclusion/exclusion criteria?
What primary and secondary study end points were studied?
What clinical data were collected and how were the data collected?
How were clinical data reported? (I.e., what metrics were used, were data reported on an individual subject level or only on a group level, etc.?)
How were safety information collected and reported? (E.g., what AEs definitions were used? Were nonserious AEs collected and reported? How were complications measured and analyzed?)
What performance data were collected and reported?

(*Continued*)

BOX 4.10 (Continued)

Some articles will use the subject device, but the research will not be about the device. In these cases, the CER/PER goals often differ from the study goals in the literature research.

The evaluator is not simply repeating the study findings. Instead, they are looking for clinical data pieces across a wide and varied abundance of clinical data, so they must use good judgment when summarizing the study findings related to the subject device. Even in the background, SOTA, and alternative therapy/device section of the CER/PER, a critical requirement is to keep the evaluation focused on the subject device used in the setting of a particular disease state, secondary treatment modality, etc. Some articles may not report data in a standard form or in a form useful for direct analysis. The evaluator must understand how to identify the important information for the CER/PER. The evaluator must read between the lines to close the gap between what the authors studied and the useful clinical data needed to evaluate the device S&P.

If the overall literature quality is high (i.e., several high-quality RCTs exist), lower-quality articles may be excluded as part of the appraisal step; however, if any article raises S&P questions not observed in the other included articles, the article should not be excluded. For example, case studies generally have poor quality and should not be included; however, a case study documenting a rare S&P issue (i.e., a safety concern related to the devices under evaluation but not seen in any other included article) should not be excluded. Additionally, if the only published articles about the device under evaluation are case studies, then the case studies must be included. In other words, if the overall literature quality is low (particularly with regard to literature about the device under evaluation), then these low-quality articles have increased value to the CER/PER. Also remember: low-quality literature with few high-quality RCTs is a clinical data gap to be documented and addressed in the PMCFP/PMPFP.

4.4.4 Weighting clinical literature using levels of evidence

Clinical practice guidelines, metaanalysis, and systematic reviews are considered the highest-quality clinical literature because these articles have systematically searched all available clinical literature and analyzed all identified clinical data in a consistent and statistically valid process (Table 4.7). The CER/PER evaluation should use a similar process for the background/SOTA/alternative therapy/device discussion. Clinical practice guidelines, metaanalysis, and systematic reviews are secondary data; however, primary data are also required especially for the subject device. Primary data tend to be much more detailed than secondary data sources, and the primary data source should always be referred to when questions arise about secondary data interpretations.

TABLE 4.7 Levels of evidence (LOE) example: primary and secondary data.

LOE	Description	Notes
1	Clinical practice guidelines	Secondary data, filtered, consensus
2	Metaanalyses/systematic reviews of RCTs with definitive results	Secondary data, filtered
3	RCTs with definitive results	Primary[a] data, interventional study
4	RCTs nondefinitive results, nonrandomized, controlled studies	Primary[a] data, interventional study
5	Cohort studies, longitudinal studies	Primary[a] data, observational study
6	Case-controlled studies	Primary[a] data, observational study
7	Cross-sectional surveys	Primary[a] data, observational study
8	Case series and reports, retrospective reviews	Primary[a], observational, no design
9	Textbooks, narrative reviews, expert opinions, editorials	Secondary data, no design
10	Animal or laboratory (*in vitro*[b]) studies	No human use data

RCT, randomized controlled trial.
[a]*Guidelines and metaanalyses (i.e., secondary data sources) are ranked highest for the CER/PER background/SOTA/alternative therapy/device section; however, "primary" human data sources (LOE 3–8) are typically required for the CER/PER subject device section.*
[b]*In vitro data are required for IVD devices scientific validity, analytical performance and clinical performance.*

Note the inverse relationship, the larger the LOE number, the lower the literature weight (e.g., animal or laboratory studies have the lowest clinical data weight with a "10" LOE number while clinical practice guidelines have the highest clinical data weight with a "1" LOE number). Rather like being first in line, the evaluator will prioritize analyses of the highly weighted articles first before evaluating the lower weighted clinical data. In addition, note the reduced LOE score for observational data (LOE 5, 6, 7) when compared to interventional data (LOE 3, 4).

The value of observational trials has been hotly debated over the years and certain devices may be well-supported with observational data only. Some authors even consider observational studies "more informative" than RCTs—for example, in hypertension research [10]. These authors traced the first RCTs back to 1948, and they listed the strengths and weaknesses of RCTs and observational trials including bias/selective reporting, exposure to confounding factors, methodological control, cost, time, sample sizes, etc.

This book was written during a time of great scientific research upheaval, including an increase in artificial intelligence and machine learning to mine massive amounts of "observed" individual patient data, the desire to have decentralized clinical trials and remote trial monitoring, reduced access reactions to devices due to the changing global regulatory requirements, and astronomically increasing costs. Evaluators must be aware and must carefully appraise data quality in any included clinical literature based on all the data available. One size or type of clinical data appraisal will certainly not fit all literature.

BOX 4.11 "Rank" excluded articles and "weigh" included articles

One key strategy when appraising a given piece of literature for inclusion or exclusion is not only to "rank" the excluded data type (i.e., not about the subject device/background/SOTA, duplicative data, animal study, case report without novel information), but also to assign a clinical data "weight" or value (i.e., quality) for each included article.

This relative weight assignment is restricted to the included articles only. As the evaluator weighs each included article, they must decide if the lowest-value (i.e., lowest-weighted) articles should be excluded from the CER/PER analysis and report writing. The inclusion appraisal process requires the evaluator to determine if each individual article adds value beyond the other included articles. This delicate balance requires the evaluator to write a clear justification for the specific exclusion code assigned to each article with otherwise "includable" clinical data.

The evaluator must apply due diligence without bias to exclude only those articles which do not alter the overall clinical data interpretation in any way. In other words, the CER/PER team must not exclude articles which offer differing opinions or rare safety signals—these articles must be included for full analysis even though they may be rare events or unpopular opinions.

Many methods are used to appraise, rank, and weigh the identified clinical data in order to determine if the clinical data should be included or excluded.

Weighting is performed using both quantitative (scoring) and qualitative (holistic) methods. Different scoring methods are used to weigh articles for inclusion or exclusion. One method is to first assign a "level of evidence" (LOE) to each article. For example, many evaluators use the Oxford Centre for Evidence-Based Medicine (OCEBM) levels of evidence (Table 4.8) to assign a score (level) to each article representing the specific study-type strength.

TABLE 4.8 OCEMB levels of evidence (LOE).

LOE	Study type
1a	SR/MA of RCTs
1b	Single RCT
2a	SR/MA of cohort studies
2b	Single cohort study (including low-quality RCT; e.g., <80% follow-up)
3a	SR/MA of case-control studies
3b	Single case-control study
4	Case study or case series (and poor-quality cohort and case-control studies)
5	Expert opinion without critical appraisal or based on physiology, bench research, or "first principles"

MA, metaanalysis; *RCT*, randomized controlled trial; *SR*, systematic review.
Source: Adapted from OCEBM Levels of Evidence Working Group 2009 [11].

Using the OCEBM LOE, a well-done metaanalysis is given higher coded weight than an expert opinion (i.e., a narrative review) and an RCT is given higher coded weight than a nonrandomized, uncontrolled trial which are both given a higher coded weight than a case study or expert opinion. Unfortunately, the OCEBM LOE failed to include guidelines in the evidence ranking. This is important because so many other types of literature exist, and the evaluator must apply the knowledge from these LOE and other ranking tools to the data under evaluation.

A well-done guideline supported by a professional society or established organization and using systematically reviewed clinical data to directly support the guideline would most likely be coded "1a" using the OCEBM system. Alternatively, a "consensus" standard might be coded "5" if the authors failed to use clinical data in the development of their standard (i.e., the consensus standard may be evaluated as an expert opinion if not derived from the clinical data directly and if clinical data are not shown as evidence to support the standard).

Article weighting is helpful to aide inclusion (highest-weighted articles) and exclusion (lowest-weighted article) decisions. For example, if several RCTs used a particular device, then case reports/series and narrative reviews about the device should be excluded, unless a particular article discussed a unique safety or performance concern. In this case, manual searching should be done to identify, appraise, and include specific clinical data related to this unique safety or performance concern for the analysis step.

Grouping similar article types together based on LOE scores is a useful way to appraise the clinical literature data amount and type available. If many high-quality articles are available (e.g., multiple metaanalyses and systematic reviews covering relevant RCTs), then lower-quality articles may not warrant inclusion unless they have novel clinical data (when they absolutely should be included). Similarly, if few or no high-quality articles are available, lower-quality articles must be included because they are the only clinical data available (Fig. 4.2).

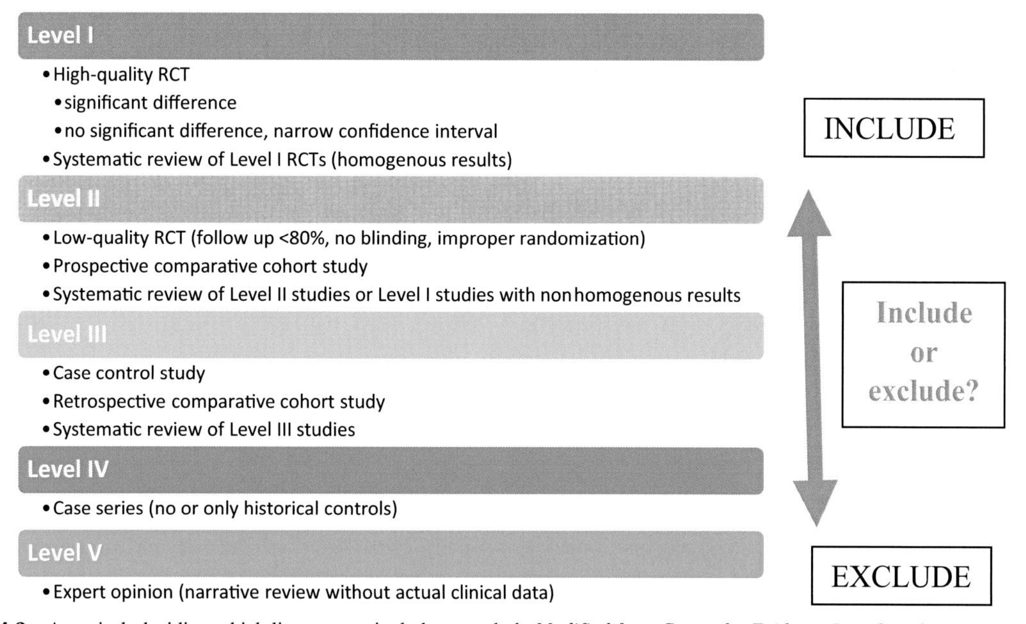

FIGURE 4.2 Appraisal: deciding which literature to include or exclude *Modified from Center for Evidence-Based Medicine LOE table [11].*

Many weighting methods may be appropriate depending on the device type. For example, orthopedic devices may benefit from the American Academy of Orthopaedic Surgeons (AAOS) levels of evidence [12,13]. AAOS listed seven domains (adapted from others) including strength of evidence, benefits and harms, outcome importance, cost-effectiveness/resource utilization, acceptability, feasibility, and future research for consideration before clinical data are included in their clinical practice guidelines.

In most cases, when designed and conducted appropriately, a clinical practice guideline, MA, SR, or RCT will be considered high-quality evidence and should be included; however, an expert opinion (i.e., a narrative review espousing an opinion), a case report or a case series article (with only a few study subjects) are considered low-quality clinical data and should be excluded except when reporting rare, novel safety or performance issues (i.e., risks or benefits) which should always be included.

Another holistic weighting process may use study type, size, author bias, data accuracy, or procedural correctness (e.g., errors, appropriate statistics) criteria to develop the overall article quality weight. Holistic weighting is based on many article factors rather than a single factor such as article type or LOE score. This is especially helpful when multiple articles are identified with the same LOE score and they have widely different qualities. For example, the OECBM [14] assigns LOE score "1b" to RCT articles; however, within the 1b LOE score, the RCTs with better QC, specific S&P measures relevant to the subject device, larger sample sizes, and less author bias should warrant a higher weight for quality and relevance.

LOE scores are useful for assessing article type, but are often not sufficient for inclusion and exclusion appraisal decisions. Depending on the clinical literature data volume and quality, inclusion and exclusion appraisal decisions will shift to accommodate the appropriate data needed for analysis in order to address fully the S&P, benefit-risk ratio, and GSPR conformity questions asked in the CER/PER.

4.4.5 Appraise background/state-of-the-art and specific-device literature

Quality assessment and weighting needs to be done for both background/SOTA literature and device-specific literature. The two groups can be searched, identified, and appraised at the same time but they also need to be grouped and appraised separately. For example, an RCT relevant to the background knowledge, SOTA, or alternative therapies should be given a different appraisal I&E code than an RCT relevant only to the evaluated device. When deciding whether to include low-quality literature, only the relevant literature group should be considered. I&E decisions about metaanalyses and systematic reviews relevant to the background knowledge, SOTA or alternative therapies should have no bearing on whether to include or exclude a case study or case series about the device under evaluation and vice versa.

Broad metaanalyses or systematic reviews may include the subject device (i.e., the device under evaluation) along with all many other devices or alternative therapies. This type of article may be best to include in the CER/PER discussion for the device under evaluation with other metaanalyses or systematic reviews which do not specifically discuss the subject device in the background/SOTA section.

The CER/PER background/SOTA section typically includes guidelines, metaanalyses, and/or systematic reviews and may not require any lower-quality articles; however, the background/SOTA data must include sufficient S&P as well as benefit-risk data for competitor devices and alternative therapies. The S&P acceptance criteria will need to be based on the background/SOTA clinical data for comparison to the subject/equivalent device clinical/performance S&P data; therefore, sometimes lower-quality articles are included in the background/SOTA section when detailed S&P data are required.

BOX 4.12 Appraising clinical guidelines

Clinical guidelines and recommendations provide important context for the CER/PER background/SOTA section; however, not every guideline needs to be included. In addition, guidelines tend to be updated over time, so the evaluator must continuously update the CER/PER literature to ensure the most recent guidelines are considered.

If multiple guidelines make the same recommendations regarding standard of care or device use, only the most recent should be included. If two guidelines from the same year make the same recommendations and one comes from a European group and the other does not, the European guideline should be included. The CER/PER includes device evaluations to meet the EU CE mark requirements.

Not every guideline addressing the disease or condition treated by the device under evaluation will be relevant to the background/SOTA. For example, a guideline about diagnosing heart disease without any information about heart rhythm disorder treatment is not relevant to the clinical evaluation of a pacemaker. Only guidelines addressing heart rhythm disorder treatment or management are relevant.

The evaluator should avoid writing long details about the disease state in the background/SOTA/alternative therapies section of the CER/PER. The CER/PER should focus on the device under evaluation after an appropriate introduction to the disease background/SOTA/alternative therapies.

The CER/PER subject/equivalent device section will also include MAs, SRs, and RCTs using the subject/equivalent device. These articles will need to report studies with sufficient sample sizes to address all S&P, benefit-risk ratio and clinical claim details and may not need to include lower quality articles; however, as already mentioned, if lower-quality data document novel S&P issues (i.e., issues not documented elsewhere) they should be included, regardless of quality. Specifically, articles with novel safety concerns should remain included regardless of article quality.

BOX 4.13 Literature appraisal example: hysteroscope

A recent hysteroscope CER literature search identified 464 articles and coded 59 for inclusion based on five indications for use (Fig. 4.3).

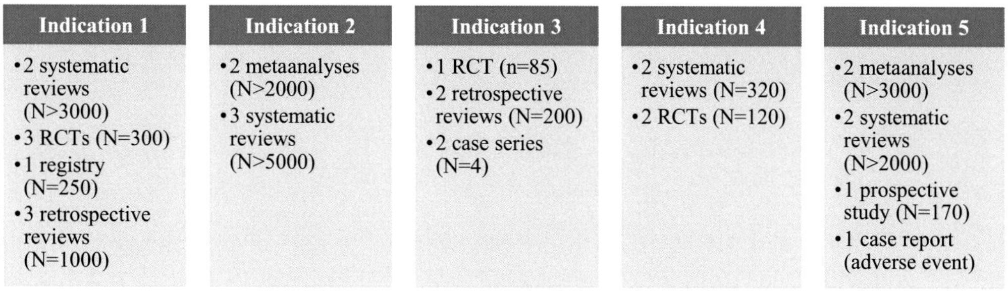

Indication 1	Indication 2	Indication 3	Indication 4	Indication 5
•2 systematic reviews (N>3000) •3 RCTs (N=300) •1 registry (N=250) •3 retrospective reviews (N=1000)	•2 metaanalyses (N>2000) •3 systematic reviews (N>5000)	•1 RCT (n=85) •2 retrospective reviews (N=200) •2 case series (N=4)	•2 systematic reviews (N=320) •2 RCTs (N=120)	•2 metaanalyses (N>3000) •2 systematic reviews (N>2000) •1 prospective study (N=170) •1 case report (adverse event)

FIGURE 4.3 Included hysteroscope articles by indication.

Appraisal coding for literature inclusion was influenced by relevant article number and quality for each indication. Some indications (e.g., indications 2 and 5) had relevant, high quality metaanalyses and systematic reviews covering subject/equivalent device S&P and other, lower quality articles did not need to be included. Indication 1 had two systematic reviews covering more than 3000 subjects, however, lower quality articles (RCTs, a registry and retrospective reviews) identified S&P issues not addressed in the systematic reviews and were therefore included. Indication 4 had systematic review and RCT data; however, no metaanalyses were available and lower quality articles were excluded. Indication 3 had no metaanalyses or systematic reviews and very few articles in general. Because high-quality data were not found, two low-quality case series, were included.

Appraising clinical literature is complex. Literature volume and quality are variable and the evaluation details are multiple, multifaceted, and often subtle. Sometimes, the clinical data within the clinical literature may not be entirely obvious. For example, sometimes articles are published after using a device, but the article does not offer any device-specific information. In this case, if the data were collected after the device was successfully used, the clinical data can be used to support safe use because the article did not mention any safety concerns associated with the device use during the study. This is a common situation when a device is used during research about topics unrelate to the device itself.

In addition, the background, SOTA, and alternative therapy/device data must be used to set "acceptable" S&P goals for the subject device. These data are typically found in the literature and must be abstracted in a manner similar to the subject device data abstraction in order to determine if the subject device S&P are acceptable. The subject device data will need to meet the S&P acceptance criteria based on the background, SOTA and alternative therapy/device clinical data.

When appraising clinical literature data, many different types of problems can arise. For example, too little or too much clinical data may be available, or the literature may have only poor-quality data. The evaluator should avoid common literature appraisal errors (Table 4.9).

TABLE 4.9 Common literature appraisal errors.

Excluding all articles relevant to background/SOTA or alternative therapy/device.
Excluding all low-quality articles without reviewing for novel S&P information
Excluding articles with unfavorable results
Including low-quality articles without novel S&P information (e.g., small retrospective studies may cloud the details from higher-quality articles addressing the same S&P information)
Not evaluating articles for potential bias
Not assigning a weight to each article
Including articles about off-label uses in the subject device section (i.e., since these data are not relevant to subject device indication for use, they should be included in the background knowledge/SOTA section)

(Continued)

TABLE 4.9 (Continued)
Including articles without clinical data (e.g., narrative reviews, preclinical data)
Incomplete or unclear justification for each included or excluded article
Including too many or too few articles
Only including free, publicly available articles
Excluding all foreign language articles
S&P, safety and performance; *SOTA*, state of the art.

In general, clinical literature data are intermediate weight compared to clinical trial and clinical experience data. Clinical trial data are typically weighted as the highest-quality data when the detailed clinical trial data are relevant to the device used as intended, and clinical experience data are typically weighted as the poorest-quality data because so little detail is available for these anecdotal reports. Careful appraisal methods must ensure all relevant clinical literature data are included in the CER/PER evaluation.

4.5 Clinical experience data

MEDDEV 2.7.1 rev. 3 [15] defined clinical experience data as:

- *"generated through clinical use... outside the conduct of clinical investigations and may relate to either the device in question or equivalent devices. Such types of data may include:*
- *manufacturer-generated post market surveillance reports, registries or cohort studies (which may contain unpublished long-term safety and performance data);*
- *adverse events databases (held by either the manufacturer or Regulatory Authorities);*
- *data for the device in question generated from individual patients under compassionate usage programs prior to marketing of the device;*
- *details of clinically relevant field corrective actions (e.g., recalls, notifications, hazard alerts)"*

The updated guideline, MEDDEV 2.7/1 rev. 4 [1], subsequently listed common

"clinical data generated from risk management activities and the PMS programmes which the manufacturer has implemented...including...

- *PMCF studies, such as post market clinical investigations and any device registries...*
- *PMS reports, including vigilance reports and trend reports*
- *the literature search and evaluation reports for PMS*
- *incident reports... (including the manufacturer's own evaluation and report)*
- *complaints regarding performance and safety sent to the manufacturer, including the manufacturer's own evaluation and report*
- *analysis of explanted devices (as far as available)*
- *details of all field safety corrective actions*
- *use as a custom made device*
- *use under compassionate use/humanitarian exemption programs*
- *other user reports"*

"For clinical experience data it is important that any reports or collations of data (e.g., the manufacturer's PMS reports) contain sufficient information for the evaluators to be able to undertake a rational and objective evaluation of the information and make a conclusion about its significance with respect to the performance and safety of the device in question.

Reports of clinical experience that are not adequately supported by data, such as anecdotal reports or opinions, may contribute to the evaluation, e.g., for the identification of unexpected risks, but should not be used as proof of adequate clinical performance and clinical safety of the device."

Experience data are inherently low quality and are not sufficient clinical data for the CER/PER. Complaints and safety reports may document a single occurrence, and people only file complaints when something goes wrong. Appraising experience data is not really about assessing quality and relevance. Ultimately, all safety reports or device complaints should be included. If the experience data are not about the device under evaluation or the background knowledge or SOTA surrounding the subject device, then these data should be excluded. Similarly, clinical experience data should be excluded if they do not allow the evaluator to draw any conclusions about the device S&P.

Clinical experience data may identify certain DDs, device risks, and failure modes; however, clinical experience data often provide only a tiny amount of uncontrolled and highly biased information regarding the reported AE or device issue. In addition, not every experience is reported and each report documents only a single event without any larger context. Even so, clinical experience data are essential to understanding the full clinical picture to explain what actually happens when the device is being used.

MDCG 2020-6 states:

> *"The use of vigilance data, in general is appropriate for identification of any new risks, events in subpopulations, examining trends in PMS reports etc. With respect to the utilisation of post-market surveillance data for the purpose of conformity assessment, it is important to recognise that uncontrolled sources of clinical data—for example complaint or incident report data—cannot always provide reliable data with respect to the incidence of risks and cannot provide an estimate of uncertainty i.e. a confidence interval. Due to limitations of complaints reporting, the use of estimates such as [number of incidents or complaints] / [number of device sales] cannot generally be considered sufficient to provide proof of safety; their use should be limited to cases where data from pre-market or post-market clinical investigations or PMCF studies are not deemed appropriate"* [5].

In addition, the evaluator must be careful not to double-count events. If the same event (e.g., a particular patient death) is reported in both the internal PMS system and an external public database, such as the US FDA Manufacturer and User Facility Device Experience database (MAUDE), the event should be noted as appearing in both groups and only reported as a single event in the CER/PER. Appraising clinical experience data is about grouping similar reports and assessing what AEs and device malfunctions are occurring.

4.5.1 Manufacturer-held clinical experience data

Manufacturer-held experience data include complaints, surveys, sales data, and other data generated by PMS. Ideally, all PMS data would have already been gathered, appraised, and analyzed in the required PMS report (PMSR) or periodic safety update report (PSUR). Clinical experience appraisal involves simply determining if the PMSR/PSUR provides sufficient clinical experience data details and analyses to include in the CER/PER.

The same standards for quality, relevance, and methodology apply to clinical evidence data as for clinical trial and clinical literature data. The clinical experience data should be comprehensive and should date back to when the device entered the market. If each subsequent PMSR/PSUR updates the previous one in a comprehensive manner, only the most recent PMSR/PSUR may be needed; however, if each PMSR/PSUR only addresses a specific timeframe (e.g., 2017–2020) without including previous data, all PMSRs/PSURs will need to be appraised for inclusion in the CER/PER.

The evaluator must consider if the PMSR/PSUR provides sufficient clinical experience data detail to be appraised and analyzed with a reproducible result. If publicly-available data were included in the PMSR/PSUR, the evaluator should confirm if the numbers align with independent searches and appraisals of the publicly available data.

Similar to the clinical trial and literature data appraisal steps, crieria used to appraise and analyze the clinical experience data must be stated and defined to be reproducible. Relevant clinical experience data are not restricted to device malfunctions or "unavoidable" patient injuries. All device complaints should be included in the CER/PER appraisal process, and all complaints about equipment used with the device (i.e., capital equipment used with multiple devices) should be explained if relevant or excluded if not relevant to the device under evaluation. For example, device "user error" issues should be included. User errors and misuses provide information about the end-user "learning curve" and if user error and/or misuse occur on a regular basis, changes to the design or indication for use may be required.

The evaluator must determine if the clinical experience data are sufficient or if additional clinical experience data are required. Simply stating complaint rates without documenting and analyzing the complaint details is not sufficient. Complaint details about what happened to the patient are required for the CER/PER. For example, if a company receives 10 complaints for every 100,000 units sold, the complaint rate (0.01%) may seem sufficiently low; however, this rate and the individual events must be analyzed more carefully to understand what happened to each patient. A complaint about an incorrect shipment or a damaged package is not the same as a complaint about a patient being injured by the device. In addition, the complaint rate is simply not a valid number since both the numerator (complaint number) and the denominator (units sold) are not accurate (many complaints go unreported in the numerator and many sold devices may be unused in the denominator).

Sometimes, the PMSR only provides details for the "reportable" complaints without specifying the patient details and without documenting anything about the remaining "unreportable" complaints. Consider the following example: a company producing stents and angioplasty balloons received a user complaint about an angioplasty balloon failing to inflate during stent placement. The complaint was listed as reportable for both the balloon and the stent; however, if the stent was being evaluated, the complaint would not be relevant while the complaint woudl be relevant if the balloon was being evaluated. When the user attempted to inflate the balloon a second time, the balloon inflated; however, the complaint was still relevant and reportable even though the balloon eventually worked. Procedural delay, particularly when general anesthesia is involved, is an AE. The longer a patient is sedated, the more risk is incurred. Additionally, a device should work as intended on the first try.

The manufacturer must capture enough clinical experience data details for the evaluator to analyze the clinical experience data and to draw conclusions independently.

4.5.2 Clinical experience databases

Various regulatory agencies maintain AE databases collecting clinical experience data from around the world. For example, the Therapeutic Goods Administration (TGA) in Australia maintains a Database of Adverse Event Notifications (DAEN) to capture AEs and DAEN provides separate search portals for medical devices and drugs used in Australia since July 2012 (https://www.tga.gov.au/safety/safety/safety-monitoring-daen-database-adverse-event-notifications/database-adverse-event-notifications-daen). In addition, Health Canada maintains an AE database covering both drugs and devices (https://cvp-pcv.hc-sc.gc.ca/arq-rei/index-eng.jsp), and the EU is planning to make the EUDAMED medical device AE database available to the public.

Currently, the US FDA maintains MAUDE, the largest English-language medical device experience database with all medical device reports received by the FDA in the past 10 years readily accessible to the public and additional data back to 1991 also available (https://www.accessdata.fda.gov/scripts/cdrh/cfdocs/cfmaude/search.cfm). Any US-based device manufacturer, US agent of a non-US-based manufacturer, or importer of devices into the US must submit a report to the FDA within 30 days after becoming aware of device-related AEs including deaths or serious injuries occurring anywhere in the world. US-based device user facilities must submit reports to the FDA and the manufacturer within 10 business days after becoming aware of a device-related AE. US-based importers must also report device-related AEs and device malfunctions to the manufacturer. Manufacturers, user facilities, and importers may voluntarily report malfunctions to the FDA. The related Total Product Life Cycle (TPLC) database provides aggregate data from medical device reports at the product code level, and the Medical Product Safety Network (MedSun) database provides reports from the user facility network.

The MAUDE search results can be downloaded directly into an Excel spreadsheet; however, the DAEN search results must be input into a spreadsheet manually. These clinical experience data should be appraised and analyzed separately from the manufacturer-held complaints data, which are the most valuable clinical experience data. Because the clinical experience data are such poor quality, the clinical experience data appraisal process differs substantially from the appraisal processes used for clinical trial or literature data. First, group all manufacturer-held complaints data into appropriate groups by separating out all reports with no evaluable clinical data and then grouping all the remaining clinical data into appropriate subgroups based on the subject device risks as well as the S&P acceptance criteria. Then, do the same for all the other clinical experience data deemed worthy of evaluation.

4.5.3 Manufacturer and User Facility Device Experience database appraisal steps: patient problems versus device deficiencies

For all clinical experience data reports collected by searching publicly available databases including MAUDE, DAEN, and other databases, start by deciding if each report describes a subject device use or not. Reports may be mislabeled and searches may return results for other, similar devices, which can be used in the background/SOTA or alternative therapy/device of the CER/PER. Assign I&E codes to each data point. Clinical experience I&E codes will not be as detailed as the codes used for clinical trials and literature because the clinical experience data details are limited. For example, a MAUDE search was completed for the brand name "Honeywell ONE-fit N95 respirator"; however, no results were identified. A separate, broader search for the term "N95" alone identified 40 results on 08 May 2023, and these results were relevant to all N95 devices regardless of manufacturer or design. The appraisal process excluded four reports about reprocessed face masks because the Honeywell ONE-fit N95 respirator is not indicated for "reprocessing," even though the competitor product, the "Battelle system, X1" face mask, is designed for reprocessing (Table 4.10). This note about reprocessing for some N95 face masks may be appropriate for the Background/SOTA/Alternative therapies section of the CER.

TABLE 4.10 N95 example MAUDE search results.

I&E code	Description	Count
2	About similar device	35
1	About subject device (Honeywell ONE-fit N95 respirator)	0
− 1	Not about device	1
− 2	Duplicate	0
− 3	About reprocessed masks (BATTELLE system, X1)	4
	Include	35
	Exclude	5
	Total	40

I&E, inclusion and exclusion.

This inclusion and exclusion (I&E) coding moves unrelated, duplicative and irrelevant clinical experience data out of focus as more analysis is completed on the included data. As the process proceeds to the analysis step, each experience report will be analyzed separately for device deficiencies and patient problems.

4.6 Clinical data appraisal by clinical eevaluation/perfomance evaluation type

Often, novel and new to market or equivalent devices have little or no clinical data and legacy or obsolete devices may not have recent data. Given these situations, clinical data appraisal methods may need to be modified for the various CE/PE types.

4.6.1 Viability clinical evaluation/viability performance evaluation

When appraising clinical data for a viability CER/PER, the background knowledge, SOTA, and alternative therapies must be clear so the relevant clinical trials, literature, and experiences will be included with the appropriate caveats. Some new devices are completely novel while others have novel aspects. The new technology may only have manufacturer bench data available, and a clinical trial may be in the planning stages. These data and plans need to be the focus for novel device S&P and benefit-risk ratio evaluations. Since no clinical data are available from the new, never before used in a human, subject device, the only clinical data available will be from other "similar" devices. Claiming equivalence may not be possible for truly novel devices prior to marketing. In this case, the inclusion codes should be set as broad as possible, but the evalutaor must exclude any irrelevant or poor-quality data to keep the CER/PER focused on the most relevant and highest-quality data.

Ensure the rationales for each inclusion and exclusion criterion should be well described so the CER/PER reader will understand why these data from "similar" devices were used in lieu of the subject device or any equivalent device clinical data. In addition, clinical data weighting activities must not overestimate the value of the clinical data when the vCER/vPER clinical data are not about the subject device. Be clear about this point: no clinical data are included from the subject device; however, "related" clinical data are provided to explain why the subject device is expected to be safe and to perform as intended as required under GSPRs 1, 6, and 8.

4.6.2 Equivalence clinical evaluation/equivalence performance evaluation

Be truthful and not misleading when appraising equivalent device clinical data. These data are not the same as the clinical data for the device itself; however, the equivalent device data should represent a similar S&P level as the subject device. The appraisal parameters will need to include as much S&P data as possible and good explanations will be required for the details about which data are to be included or excluded. For example, perhaps the equivalent device is available in many different sizes, but the subject device is available in only one size. In this situation, the other device sizes may be appropriate for the background/SOTA section but not for the subject device section of the CER/PER.

4.6.3 First-in-human clinical evaluation/first-in-human performance evaluation

As the clinical data begins to accumulate for the subject device, the reliance on equivalent and similar device data should decrease. Unfortunately, sometimes the manufacturer wishes to continue claiming equivalence throughout the device life cycle and may not progress beyond the equivalent device clinical data collection. Because all subject device data have a higher value in the CER/PER than the equivalent or similar device clinical data, the evaluator should take great care to account for and include all subject device experience data especially when no clinical trial or clinical literature data were available for the subject device during the first-in-human (fih) clinical experience data appraisal step. As discussed previously, the equivalent and similar device clinical trial and literature data should also be included in the appraisal steps when no clinical trial or literature data are available from the subject device.

4.6.4 Traditional clinical evaluation/traditional performance evaluation

Often, "traditional" devices have been on the market for some time and may or may not have clinical trials or literature data available for evaluation. A special case was made for CE marked "legacy" devices onthe EU market under the regulations in place before the 2017 MDR and IVDR were enacted. The clinical data provided for these devices during the prior conformity assessment/s will be considered acceptable even though they may lack sufficient clinical evidence as required under MDR [5] and IVDR [16]. The PMS clinical data meeting MDR/IVDR requirements will be specifically considered along with the previously existing clinical data used to gain a CE mark. In this setting, legacy device PMS systems often need enhancement to include specific clinical data and patient experiences in the PMCFP/PMPFP to demonstrate conformity with MDR/IVDR GSPRs.

Legacy (i.e., traditional) devices must still meet all the future update requirements for the CE/PE document cycle (i.e., CEP/PEP, CER/PER, PMCFP/PMPFP, summary of safety and clinical performance (SSCP)/summary of safety and performance (SSP), and postmarket clinical follow-up evaluation report (PMCFER)/postmarket performance follow-up evaluation report (PMPFER) are still required and must be updated as required). For some older devices, research interest may have decreased over time, and the clinical data may be dated. In this case, the evaluator and writers must justify the I&E decisions and include the appropriate data. If clinical data are lacking, this clinical data gap and the appropriate plans to generate the clinical data must be recorded as part of the written record in the CE/PE documentation.

When creating or updating a legacy device CEP/PEP, consider if the premarket studies including fih, feasibility, and pilot studies are still relevant to the legacy device, especially if the device, diagnostic/treatment paradigms, and/or patient population have changed over time. In addition, MDR Article 2(48) limits the clinical data sources definition and no longer includes "unpublished reports" of some clinical experience data. Reports should be carefully appraised and potentially excluded or moved to the background/SOTA/alternative therapy section of the CER/PER with sufficient caveats about the unpublished, speculative nature of the clinical data. On the other hand, identifying the relevant GSPRs and device details (e.g., intended purpose, target groups, indications, contraindications, quantified benefits/outcomes, risks, and S&P acceptance criteria details, benefit-risk ratio acceptability) in the CEP/PEP will be relevant and must align with the labeling (e.g., instructions for use, user manual, promotional materials, risk management documents) and SOTA.

Unlike the vCER/vPER, eqCER/eqPER, and fihCER/fihPER, the traditional trCER/trPER devices have been on the market for some timeeven though they often lack clinical trial or literature data. These "legacy" medical devices are allowed to remain on the EU market as long as PMS data provide sufficient clinical evidence under the new regulations. Unfortunately, if past PMS clinical evidence is insufficient "to confirm safety, performance and the acceptability of the benefit-risk determination in relation to the state-of-the-art for the legacy devices" according to MDCG 2020-6 [5], then new clinical data "may need to be generated prior to CE-marking under the MDR." In addition, the legacy device clinical evaluation "should not rely on new PMCF studies started under MDR to bridge gaps (e.g., indications not supported by clinical evidence)." Existing, historical device indications must be fully supported with existing clinical data, otherwise additional clinical data collection (i.e., beyond a PMCF study) may be required.

This guideline also discussed including and evaluating indirect clinical benefits, risks managed in the risk management system, and the benefit-risk ratio based on the SOTA and alternative therapies as follows:

> *"...while direct clinical benefits should be supported by clinical data, indirect clinical benefits may be demonstrable by other evidence such as: . . .pre-clinical and bench test data (eg compliance to product standards or common specifications); . . .real world data such as registries, information deriving from insurance database records, . . .data from another device that is used with the subject device which does have direct clinical data (eg, data from a stent used to justify safety and performance of a guidewire). . . A determination of the level of clinical evidence required to demonstrate an indirect clinical benefit should be made on the basis of a thorough risk assessment and evaluation of short, medium and long term clinical risks (for example, a guidewire, although used transiently, may have long term clinical risks if it leads to vessel dissection)."* MDCG 2020-6 end of section 6.5 (a) [5]

Legacy devices may no longer have "sufficient clinical data for certification under the MDR" for many reasons (e.g., SOTA changes, new risks identified during PMS, new device sampling, new equivalence requirements, more explicit clinical data definitions which may remove prior data sources such as unpublished literature). Alternatively, the PMS data within a well-functioning QMS may support a WET when the device risk is low, and the device fulfills a standard of care without much SOTA evolution over time [5].

The importance of appropriate appraisal for the trCER/trPER cannot be overstated, especially when the subject device had a CE mark and was on the EU market before the new MDR/IVDR took effect. The trCER/trPER appraisal should include "real world" clinical data and clinical data showing "indirect clinical benefits" as these clinical data may be instrumental to maintain the CE mark for the legacy device.

4.6.5 Obsolete clinical evaluation/obsolete performance evaluation

Once the device has reached the end of the device life cycle, the appraisal will endeavor to include all relevant, high-quality data for patients who are still using the device. This means gathering and including as much clinical data as possible from long-term device users. In this setting, clinical data are not likely to be found in clinical trials; however, clinical literature and clinical experience data may have sufficient clinical evidence to ensure the obsolete device is known to be safe and performs as intended. Alternatively, the device may need to be removed from the EU market if the product is no longer supported with an oCER/oPER by the manufacturer.

4.7 Who should appraise clinical data?

Clinical data appraisal is a complex, scientific process involving many different clinical data inclusion and exclusion decisions with inter-related data ranking and weighting considerations. Appraisal methods need to be scientifically sound, objective, and reproducible. The evaluator completing the clinical data appraisal steps must understand how research is conducted, how to read clinical trial reports, scientific literature and clinical experience reports and how to appraise and weight clinical data for clinical and performance evaluations. In addition, the evaluator must understand MDR/IVDR appraisal requirements, how clinical data appraisal relates to subsequent clinical data analysis, and the overall device evaluation. The evaluator must also understand the appraisal process is ongoing and iterative and the evaluator must be flexible enough to apply the appraisal principles to diverse datasets. Appraisal has no set rules, but has principles used to develop rules specific to each dataset.

4.8 What if the prior appraisal missed or duplicated clinical data?

No CER/PER will include all available clinical data. This means some clinical data will be missed in every single CER/PER.

Clinical data evaluation under the MDR/IVDR is not about documenting every single possible data point. The goal is to identify and analyze enough relevant, high-quality clinical data to determine whether the device meets the S&P acceptance criteria, the clinical data are sufficient to demonstrate conformity to the GSPRs, and if the benefit-risk ratio is acceptable for the device be on the EU market. The clinical evaluation must be comprehensive and complete.

In addition, sometimes data changes over time. Clinical trial data may be updated as the trial progresses, old literature may be added or removed from a database like PubMed or clinical experiences may be followed up with new information added. As a result, newly identified clinical data will need to be identified, documented in the spreadsheet or software system and appraised in each subsequent CER/PER. The evaluator should watch for new safety signals and device performance changes with the potential to change the benefit-risk ratio. Changes to the benefit-risk ratio will likely mean the CER/PER will not have sufficient clinical data any more, because the new safety signal or performance change will need to be evaluated in a new PMCFP/PMPFP to capture these data.

In addition to finding missing and identifying new clinical data, sometimes the clinical data were duplicated and care should be taken to remove duplications. Duplication may occur because the same clinical data were identified in several different clinical data types (e.g., a clinical trial may be available both from a manufacturer-held clinical study report and separately in an article published in a journal article). In addition, sometimes, the same clinical data are reported several times in the literature or experience reports. This data duplication should be documented, excluded, and may need to be discussed in the CER/PER appraisal methods and results sections if the problem was particularly onerous.

4.9 Conclusions

Appraising clinical data is a critical clinical evaluation step. Proper appraisal makes clinical data analysis possible; however, no single, correct appraisal method exists. The appraisal must include sufficient clinical data to demonstrate conformity to GSPRs 1, 6, and 8 related to device suitability, S&P, benefits, risks, benefit-risk ratio, device lifetime, and all undesirable side effects. Each clinical trial, literature, and experience data point is appraised at the individual data point level (i.e., for relevance to the subject device and for quality) and at the group dataset level (i.e., to ensure the data bring value to the clinical evaluation).

Clinical trial, literature, and experience data appraisals each require different appraisal approaches and offer unique challenges; however, across all clinical data types, the central focus should be on clinical data relevance and quality. MDR and IVDR require both background/SOTA/alternative therapy/device data and subject-device-specific data to be identified, appraised, and evaluated. Thus, a good first appraisal step is to determine whether the clinical data are relevant to the subject device or to the background/SOTA/alternative therapy/device or neither. If neither, the clinical data should be excluded.

Three passes through the dataset are common for clinical data appraisal. For example, with clinical literature data: first read the document title, then read the abstract, and finally skim through the entire article to assign I&E coding. This process is meant to help with large volumes of data. The goal is to include or exclude the obvious data in the first pass without needing to read the abstract or full paper (i.e., the title alone made the I&E decision clear), then to exclude more by reading the abstract and finally exclude more after reading the entire article. Potentially, hundreds of articles may be identified for appraisal with potential relevant clinical data points. If the title is relevant to subject device characteristics and intended use or to the background/SOTA/alternative therapy/device, then assign the appropriate include code for the article. Include all questionable articles until the exclusion rationale is clearly justified.

Once relevance is determined, quality must be established. Clinical data quality exists on a spectrum from a well-done clinical trial, clinical practice guideline, metaanalysis, systematic review, or RCT to a "single complaint report." The single most important factor in determining I&E, based on quality, is context. The same articles may be included in one CER and excluded in another depending on the overall clinical data quality. The context is the whole dataset: if sufficient high-quality data are available and included, then the poor-quality data can be excluded as long as no unique safety signals or performance issues are present in the clinical data to be excluded. Sometimes, a CER/PER may have only low quality clinical experience data. In this context, all clinical experience data will be included because no clinical trial or literature data (i.e., higher-quality data) were identified. As the comprehensive data change, so do the individual I&E decisions.

The appraisal must favor good statistical quality and unbiased approaches. The clinical data appraisal process should be methodologically sound with appropriate data checks and reviews before sign-off. In other words, every appraisal decision should be well-documented (e.g., using I&E codes), transparent and defensible. The notified body (NB) assessor should understand why each decision was made and they should be able to reproduce the appraisal if necessary.

Appraising clinical data is a dynamic and multidimensional task focused on clinical data relevance and quality. Within each background/SOTA/alternative therapy and subject/equivalent device section dimension are many detailed layers about what to include or exclude and how the included data relate to each other. Enough data must be included to demonstrate the device is safe and performs as intended with all the S&P caveats considered over the device lifetime.

Once all identified clinical/performance data are appraised, the manufacturer should analyze all included clinical/ performance data (i.e., clinical trials, literature, and experiences) and the results should be documented in the CER/PER.

4.10 Review questions

1. What do the following acronyms mean? CE/PE, GCP, I&E, S&P, NB, RCT, MAUDE, DAEN, WET.
2. What does the term "appraisal" mean during CE/PE?
3. What are the two main considerations when deciding whether to include or exclude clinical data from the CE/PE?
4. How is clinical data relevance determined?
5. At what point should data be excluded?
6. What should the evaluator do if the MAUDE search has no results?
7. Do full-length articles need to be read during appraisal?
8. What types of complaints should be included in a CER/PER?
9. Where can EU-specific device AE data be located?
10. Why is statistical quality important in the CE/PE process?
11. What are the "three passes" when appraising clinical trials or clinical literature?
12. What is a legacy device?

13. How are clinical data ranked?
14. How are clinical data weighed?
15. What are some research tools used to limit bias?
16. What is the difference between quality control (QC) and quality assurance?
17. What are "sufficient" clinical data for the CER/PER?
18. What type of clinical data are found in MAUDE and DAEN?
19. True or false? Rigorous literature searches will identify all published literature.
20. Which is the lowest-quality clinical data available: clinical trial data, clinical literature data, or clinical experience data, and why?

References

[1] European Commission. Medical device directives. Guidelines on medical devices. Clinical evaluation: a guide for manufacturers and notified bodies under Directives 93/42/EEC and 90/385/EEC. Med Dev 2.7/1 Rev 4. June 2016. Accessed May 11, 2024. https://www.medical-device-regulation.eu/wp-content/uploads/2019/05/2_7_1_rev4_en.pdf#page = 20&zoom = 100,27,785.

[2] European Parliament and Council of the European Union. Consolidated text: regulation (EU) 2017/745 of the European Parliament and of the Council of 5 April 2017 on medical devices, amending Directive 2001/83/EC, Regulation (EC) No 178/2002 and Regulation (EC) No 1223/2009 and repealing Council Directives 90/385/EEC and 93/42/EEC. 2023 Accessed May 11, 2024. https://eur-lex.europa.eu/legal-content/EN/TXT/?uri = CELEX%3A02017R0745-20230320.

[3] European Parliament and Council of the European Union. Consolidated text: regulation (EU) 2017/746 of the European Parliament and of the Council of 5 April 2017 on *in vitro* diagnostic medical devices and repealing Directive 98/79/EC and Commission Decision 2010/227/EU. 2023. Accessed May 11, 2024. https://eur-lex.europa.eu/legal-content/EN/TXT/?uri = CELEX%3A02017R0746-20230320.

[4] X. Zeng, Y. Zhang, J.S. Kwong, et al., The methodological quality assessment tools for preclinical and clinical studies, systematic review and meta-analysis, and clinical practice guideline: a systematic review, Journal of Evidence-Based Medicine 8 (1) (2015) 2–10. Available from: https://doi.org/10.1111/jebm.12141Review, https://www.ncbi.nlm.nih.gov/pubmed/25594108.

[5] Medical Device Coordination Group. MDCG 2020-6. Regulation (EU) 2017/745: clinical evidence needed for medical devices previously CE marked under Directives 93/42/EEC or 90/385/EEC: a guide for manufacturers and notified bodies. April 2020. Accessed May 26, 2023. https://health.ec.europa.eu/system/files/2020-09/md_mdcg_2020_6_guidance_sufficient_clinical_evidence_en_0.pdf.

[6] IMDRF/Registry WG/N42. 2017. Methodological principles in the use of International Medical Device Registry Data. Accessed May 26, 2023. http://www.imdrf.org/docs/imdrf/final/technical/imdrf-tech-170316-methodological-principles.pdf.

[7] IMDRF/Registry WG/N33. 2016. Principles of international system of registries linked to other data sources and tools. Accessed May 26, 2023. http://www.imdrf.org/docs/imdrf/final/technical/imdrf-tech-160930-principles-system-registries.pdf.

[8] A.B. Nagella, V.S. Madhugiri, Journal retraction rates and citation metrics: an ouroboric association? Cureus 12 (11) (2020) e11542. Available from: https://doi.org/10.7759/cureus.11542PMID: 33365211; PMCID: PMC7748576. Accessed May 26, 2023., https://www.ncbi.nlm.nih.gov/pmc/articles/PMC7748576/pdf/cureus-0012-00000011542.pdf.

[9] National Institutes of Health (NIH). What is the NIH public access policy? (last updated September 9, 2016). Accessed May 27, 2023. https://www.nih.gov/health-information/nih-clinical-research-trials-you/what-is-nih-public-access-policy.

[10] G. Schillaci, F. Battista, G. Puci, Are observational studies more informative than randomized controlled trials in hypertension, Hypertension 62 (2013) 470–476. Available from: https://doi.org/10.1161/hypertensionaha.113.01501Accessed May 26, 2023., https://www.ahajournals.org/doi/pdf/10.1161/hypertensionaha.113.01501.

[11] OCEBM Levels of Evidence Working Group 2009 Accessed May 26, 2023. https://www.cebm.net/2009/06/oxford-centre-evidence-based-medicine-levels-evidence-march-2009/.

[12] American Academy of Orthopaedic Surgeons (AAOS) Website. AAOS Clinical Practice Guideline Methodology v4.0. Accessed May 11, 2024. https://www.aaos.org/globalassets/quality-and-practice-resources/methodology/cpg-methodology.pdf

[13] Reddy A., Scott J., Checketts J., Fishbeck K., Boose M., Stallings L., et al. Levels of evidence backing the AAOS clinical practice guidelines. *Journal of Orthopaedics, Trauma and Rehabilitation*. (OnlineFirst February 10, 2021). Accessed May 11, 2024. https://journals.sagepub.com/doi/epub/10.1177/2210491721992533.

[14] OCEBM Levels of Evidence Working Group*. The Oxford levels of evidence 2. Oxford Centre for Evidence-Based Medicine. 2024. Accessed May 11, 2024. https://www.cebm.ox.ac.uk/resources/levels-of-evidence/ocebm-levels-of-evidence *OCEBM Levels of Evidence Working Group: Jeremy Howick, Iain Chalmers (James Lind Library), Paul Glasziou, Trish Greenhalgh, Carl Heneghan, Alessandro Liberati, Ivan Moschetti, Bob Phillips, Hazel Thornton, Olive Goddard and Mary Hodgkinson.

[15] European Commission. Medical device directives. Guidelines on medical devices. clinical evaluation: a guide for manufacturers and notified bodies. Med Dev 2.7.1 Rev 3. December 2009. Accessed May 26, 2023. http://meddev.info/_documents/2_7_1rev_3_en.pdf.

[16] Medical Device Coordination Group. MDCG 2022-8. Regulation (EU) 2017/746—application of IVDR requirements to 'legacy devices' and to devices placed on the market prior to 26 May 2022 in accordance with Directive 98/79/EC. May 2022. Accessed May 26, 2023. https://health.ec.europa.eu/system/files/2022-05/mdcg_2022-8_en.pdf.

Chapter 5

Analyzing clinical/performance data

The evaluator, document author, writers, and team members must read and analyze each piece of clinical/performance data. The evaluator must determine if the device is safe enough when used as indicated in the appropriate patient population to be entered into or to stay on the European Union market. This decision is based on a thorough safety and performance evaluation and an unbiased assessment if the device benefits outweigh the device-related risks to the patient, user or any others exposed to the device (i.e., the benefit-risk ratio must be acceptable). A key concept is to ensure the evaluated clinical/performance data are sufficient to draw the required clinical evaluation report (CER)/ performance evaluation report (PER) conclusions. Any inconsistency in the CER/PER may draw a deficiency from the notified body or expert panel when they assess the CER/PER for compliance to the Medical Device Regulation or *In Vitro* Diagnostic Device Regulation.

Scientific substantiation principles require all reasonable clinical data interpretations to be considered during the Clinical Evaluation/Perfromance Evaluation (CE/PE) analysis step. Clinical/performance data analyses and conclusions should not mislead the reader. Both the included individual clinical data points and the clinical data analyses overall need to provide competent and reliable evidence about the device S&P in the context of the background knowledge, state of the art (SOTA), and alternative therapies. The conclusions drawn must be based on the evidence presented and the conclusions must be a reasonable data interpretation as presented.

Clinical/performance data analysis requires professional expertise in relevant areas and objectivity to ensure analytic clinical data interpretations are unbiased by hidden issues. Qualified evaluators are expected to use generally accepted procedures to yield accurate and reliable results and to draw conclusions from their clinical data analysis.

BOX 5.1 Use the "scientific method" to evaluate clinical data

Clinical data evaluation must reveal the relevant, scientifically substantiated conclusions about device conformity with General Safety and Performance Requirements (GSPRs) 1, 6, and 8. Scientific "conformity" substantiation requires valid data analysis using the scientific method and complying with relevant recognized standards and best practices. An example clinical evaluation "scientific method" testable hypothesis might be stated as follows:

"The <subject device> clinical data are sufficient to demonstrate acceptable device S&P and benefit-risk ratio to document conformity to the MDR/IVDR GSPRs 1, 6, and 8 clinical data requirements."

The evaluator must take an unbiased and highly critical view when analyzing all clinical data (e.g., the analysis will examine all details for each included clinical trial, each included clinical literature document and each included clinical experience data point). Appropriate questions must be asked about the clinical data from each data source. The data must be tabulated, compared, and contrasted within the overall data generated after using the subject device and across all clinical data found for the background knowledge, SOTA, and alternative therapies.

Conclusions must be linked directly to the specific clinical data analyzed in order to answer questions about device S&P, benefit-risk ratio and conformity with the GSPRs 1, 6, and 8. All clinical data gaps mustbe documented and addressed in the plans for postmarket clinical follow-up/postmarket performance follow-up (PMCF/PMPF) studies. If clinical data are insufficient to meet the GSPRs, additional remediation is required.

The evaluator is expected to be objective. Objectivity requires a level of independence from the manufacturer and from all the obvious biases, as well as hidden, subtle biases which might come from various conflicts of interest. Data point quality, relevance and the total evidence impact must be carefully measured and weighed when analyzing the data.

Planning, Writing and Reviewing Medical Device Clinical and Performance Evaluation Reports (CERs/PERs). DOI: https://doi.org/10.1016/B978-0-443-22063-0.00007-9

The evaluator must understand what "analysis" means. Previously, some manufacturers presented a simple list of data points (e.g., studies done, articles found, abstracts provided) as though this was an evaluation (i.e., analysis) but this is not analysis. This is just a listing exercise. Analysis happens when the evaluator tries to make sense of the list or the articles found. What do the individual clinical/performance data items contribute, how do they compare to each other, when were they done compared to the device use, did they use the subject device, equivalent device, no device? The evaluator must ask questions and document answers during data analysis (Table 5.1).

TABLE 5.1 Analysis differentiation.

Analysis is not	Analysis is
Listing clinical data from trials, literature (articles or abstracts) or experience data	Creating new knowledge based on data
Listing or citing references	Interpreting tests, studies, scientific research
Repeating author conclusion without comparison	Deciding what data mean
Documenting anecdotal customer stories	Comparing professional expertise/experiences
Copying newspaper or magazine articles	Conducting objective data review by qualified persons
Promoting a product with sales materials	Weighing accurate, accepted procedural information
Citing a number, rate or money back guarantee	Calculating means, deviations, standard errors
Tabulating abstracts from references	Documenting similarities and differences between data

Clinical/performance data analysis seeks to determine if the device was safe and performed as intended under normal use conditions with device benefits outweighing patient risks (i.e., testing the hypothesis). The device and conditions of use should remain "constant" in the clinical/performance dataset if possible. The measurements to judge device S&P and benefit-risk ratio should be specified and carefully controlled in the clinical evaluation plan/performance evaluation plan (CEP/PEP) Acceptance Criteria (i.e., as "dependent" variables) and the methods used to test the device may be allowed to vary within each experimental design (i.e., as "independent" variables). Any deviations from these expectations should be documented and accounted for in the CER/PER.

The evaluator needs to gather, understand, and organize the clinical data to tell the story. In addition, the evaluator needs to make the clinical data readable in a way the reader will understand the story. Specifically, the evaluator must answer the restated hypothesis question: Do the clinical data clearly demonstrate: (1) the device was safe and performed as intended under normal use conditions of use, and (2) the device benefits clearly outweigh the risks to the patient when used as indicated in the device labeling?

Finding a negative result is ok. Answering the question is the point. If the device is not safe or the risks are greater than the benefits, then the device should probably be redesigned before the device is marketed for use in human patients.

The CER/PER is the knowledge repository including all the clinical/performance data analyses over time. CERs/PERs are meant to be updated repeatedly over time. Institutional knowledge is meant to grow and support quality improvements for the device design and for the patient benefit-risk ratio. Verifying and expanding the evaluation over time is an important CER/PER attribute. Others who repeat the analysis should come up with the same acceptable result, repeatedly, until something changes and the device may become no longer acceptable for human use. In addition, a CER/PER clinical data analysis summary must be made public in the summary of safety and clinical performance/summary of safety and performance (SSCP/SSP), so the CE/PE process should plan this public disclosure document from the start.

5.1 Quality management system must define clinical/performance data analysis process

The quality management system (QMS) and specific clinical/performance data evaluation (i.e., analysis) processes must ensure "valid," relevant and high-quality clinical trial, literature, and experience data are analyzed in the CE/PE. The first step is to analyze each individual "included" clinical trial report, published article, and patient experience data point and then analyze the data collectively, as a whole dataset, to formulate results and draw conclusions. Evaluators should:

- specify the clinical/performance data needed to demonstrate regulatory compliance (i.e., S&P acceptance criteria, acceptable benefit-risk ratio, etc.);
- use rigorous, scientific methods to plan, collect and evaluate data, determine results and draw conclusions;
- ensure clinical/performance data analysis follows the plan;
- document clinical/performance data, analyses and outcomes in the workbook and then write up and summarize the details in the CER/PER;
- document useful data abstractions from each clinical data document in the workbook or software system including the following for comparison among data being analyzed:
    - "disease state" (i.e., define clinical setting for each patient in the analysis),
    - "clinical measures" (i.e., specify clinical data and clinical end points measured),
    - "performance results" (i.e., document how device performed),
    - "device issues" (i.e., record device problem list),
    - "safety results" (i.e., document device safety including adverse device events),
    - "safety issues" (i.e., document specific device safety concerns),
    - "conclusions" (i.e., answer all questions asked during the evaluation), and
    - "limitations" (e.g., detail small sample size, author bias, missing control groups, poor quality);
- be comprehensive (i.e., include all relevant clinical/performance data over all relevant times);
- ensure clinical/performance data analysis processes have quality control (QC) and quality assurance (QA) practices functioning during the analyses (i.e., proactively plan QA within the QMS and react to needed updates via QC during the analysis);
- specify clinical data gaps (i.e., list all missing S&P and benefit-risk ratio clinical data);
- recommend PMCF/PMPF studies (i.e., clinical data collection methods) to ensure all missing clinical data gaps are filled; and
- recommend removal from market if device does not meet acceptable S&P and benefit-risk ratio levels.

BOX 5.2 How data are analyzed against S&P acceptance criteria

The CEP/PEP defined the prequantified and prequalified S&P acceptance criteria and the clinical/performance data must meet these device-specific criteria. Often these detailed criteria are based on the most common and the most concerning adverse device effects (ADEs). In practice, the clinical team (led by the chief medical officer) will maintain a list of all device-related adverse events (ADEs) including a reason why each ADE is "acceptable" during device use (e.g., the device benefits must be greater than the individual AE risks), and the benefit-risk ratio should explain why all benefits are greater than all risks when using the device as indicated.

During S&P analysis, the workbook or software should have a specific place to record S&P acceptance criteria measures from each clinical data source. In other words, each row will be a different clinical trial and each column header will list each S&P acceptance criterion in the order presented in the CEP/PEP. A table showing these data will be provided in the CER/PER with a range and a conclusion for each criterion. For example:

Author, year (size, *N*)	<1% infection... (more columns with other S&P criteria)
Smith, 2023 (*N* = 2200)	0.5% (12/2200)
Doe, 2022 (*N* = 100)	*2% (2/100)
James, 2023 (*N* = 5000)	<1.0% (49/5000)
...(more rows with individual data points from clinical trial, literature, experience data)	
Range	0.5%–2%
Criterion Met?	No*

**In this case the criterion stated less than 1% infection; however, one* study *had an infection rate of 2% which was greater than the "< 1% infection" S&P acceptance criterion.*

When analyzing the tabulated data, the important conclusion is about why the prespecified S&P acceptance criteria were or were not met. In the example above, one of the three clinical trials had an infection rate at 2%, so the evaluator will want to explain the similarities and differences between the three trials. Perhaps a rationale will emerge about why this one trial had such a high infection rate and the others did not. Perhaps the S&P acceptance criterion for infection rate was set at an unrealistic value and should be increased. A change like this may be required for the next CEP/PEP provided the change can be justified as "acceptable" given the benefit-risk ratio for patient safety.

The clinical data analysis procedure should be established and practiced in alignment with the company QMS and policies about including all required details in company standard operating procedures (SOPs), work instructions (WIs), and templates. These documents should define how to do a systematic, comprehensive clinical data analysis (e.g., analyze the highest quality data first and include each individual clinical/performance trial, then each individual piece of clinical literature, and finally the lowest-quality data, each individual clinical experience data point). Once the individual data are analyzed, then consider each dataset as a whole for each data type (e.g., clinical trials, literature, experience data) and finally as an overall clinical dataset. All clinical data gaps must be documented. The evaluator must document all clinical data gaps and the new clinical/performance data generation required in future PMCF/PMPF plans (PMCFP/PMPFP) for new data collections to address the clinical data gaps in the past CER/PER.

5.2 *In vitro* diagnostic and other devices have different data analysis needs

Unlike the CER, three specific clinical data analysis reports are required within the PER: the scientific validity report (SVR), the analytical performance report (APR), and the clinical performance report (CPR). Often, the SVR is included within the IVD device verification and validation (V&V) procedures. This is long before the IVD device is actually used for human testing, when the analyte is chosen and formally documented as specifically associated with a clinical condition or physiological state. Once the analyte is chosen and scientifically validated, the IVD device is developed and both the analytical and clinical performance must be analyzed (i.e., do the clinical data support the chosen analyte as a valid disease marker, does the IVD device provide a valid analysis, and does the IVD device align with the clinical end point, as expected?).

The EU IVDR 2017/746 [1] Annex II defines these unique and important terms as follows:

- "(38) 'scientific validity of an analyte' means the association of an analyte with a clinical condition or a physiological state";
- "(40) 'analytical performance' means the ability of a device to correctly detect or measure a particular analyte";
- "(41) 'clinical performance' means the ability of a device to yield results that are correlated with a particular clinical condition or a physiological or pathological process or state in accordance with the target population and intended user";
- "(42) 'performance study' means a study undertaken to establish or confirm the analytical or clinical performance of a device";
- "(44) 'performance evaluation' means an assessment and analysis of data to establish or verify the scientific validity, the analytical and, where applicable, the clinical performance of a device."

In addition, each IVD device requires careful analysis of sensitivity and specificity, predictive value, and likelihood ratio which are not relevant to other medical devices:

- "(49) 'diagnostic specificity' means the ability of a device to recognise the absence of a target marker associated with a particular disease or condition";
- "(50) 'diagnostic sensitivity' means the ability of a device to identify the presence of a target marker associated with a particular disease or condition";
- "(51) 'predictive value' means the probability that a person with a positive device test result has a given condition under investigation, or that a person with a negative device test result does not have a given condition";
- "(52) 'positive predictive value' means the ability of a device to separate true positive results from false positive results for a given attribute in a given population";
- "(53) 'negative predictive value' means the ability of a device to separate true negative results from false negative results for a given attribute in a given population";
- "(54) 'likelihood ratio' means the likelihood of a given result arising in an individual with the target clinical condition or physiological state compared to the likelihood of the same result arising in an individual without that clinical condition or physiological state."

Furthermore, because the IVD device measures the analyte, the PER evaluator must be careful to analyze the IVD device in relation to the calibrators and the control materials used.

- "(55) 'calibrator' means a measurement reference material used in the calibration of a device";
- "(56) 'control material' means a substance, material or article intended by its manufacturer to be used to verify the performance characteristics of a device."

In particular, the PER has some unique performance analysis requirements compared to the CER for items required in the technical files for the device. General Safety and Performance Requirement (GSPR) #9 in the EU MDR 2017/745 Annex I [2] defined risk analysis details for devices without a medical purpose as follows:

9. For the devices... [without a medica purpose], the general safety requirements... shall be understood to mean that the device, when used under the conditions and for the purposes intended, does not present a risk at all or presents a risk that is no more than the maximum acceptable risk related to the product's use which is consistent with a high level of protection for the safety and health of persons.

Meanwhile, GSPR 9 in the EU IVDR 2017/746 Annex 1 [1] defined performance requirements for IVD devices as follows:

9. Performance characteristics

9.1. Devices shall be designed and manufactured in such a way that they are... suitable with regard to the performance they are intended to achieve, taking account of the generally acknowledged state-of-the-art. They shall achieve the performances, as stated by the manufacturer and in particular, where applicable:

the analytical performance, such as, analytical sensitivity, analytical specificity, trueness (bias), precision (repeatability and reproducibility), accuracy (resulting from trueness and precision), limits of detection and quantitation, measuring range, linearity, cut-off, including determination of appropriate criteria for specimen collection and handling and control of known relevant endogenous and exogenous interference, cross-reactions; and

the clinical performance, such as diagnostic sensitivity, diagnostic specificity, positive predictive value, negative predictive value, likelihood ratio, expected values in normal and affected populations.

9.2. The performance characteristics of the device shall be maintained during the lifetime of the device as indicated by the manufacturer.

9.3. Where the performance of devices depends on the use of calibrators and/or control materials, the metrological traceability of values assigned to calibrators and/or control materials shall be assured through suitable reference measurement procedures and/or suitable reference materials of a higher metrological order. Where available, metrological traceability of values assigned to calibrators and control materials shall be assured to certified reference materials or reference measurement procedures.

9.4. The characteristics and performances of the device shall be specifically checked in the event that they may be affected when the device is used for the intended use under normal conditions:

for devices for self-testing, performances obtained by laypersons;

for devices for near-patient testing, performances obtained in relevant environments (for example, patient home, emergency units, ambulances).

These GSPR 9 technical data should be addressed specifically and sufficiently within the PMS and RM systems. In addition, any clinical data in the PMS and RM system reports should be evaluated carefully in the CER/PER.

The CER for a device without a medical purpose, in particular, should analyze the clinical data related to all safety, risks, and safety/health protections details in the technical file or design dossier. The PER should analyze the clinical data related to any and all claimed performance technical details in the technical file or design dossier (i.e., scientific validity, sensitivity, specificity, bias, repeatability, reproducibility, accuracy, limits of detection, limits of quantitation, measuring range, linearity, cut-off, specimen collection, handling, interference, cross-reactions). Although GSPR 9 is about risk assessment or design, manufacturing, maintenance, calibration, and control, sometimes these nonclinical details may overlap with actual clinical data. When available, all clinical data should be identified, appraised, and analyzed in the CER/PER. The evaluator should be clear about what constitutes clinical data for inclusion in the specific clinical evaluation as opposed to the full technical documentation evaluation.

BOX 5.3 Clinical data versus technical documentation

According to the United States (US) National Institutes of Health (NIH), "Clinical data is a collection of data related to patient diagnosis, demographics, exposures, laboratory tests, and family relationships." [3] These clinical data are just one part of the technical documentation required for CE marking in the EU.

The EU MDR and IVDR define technical documentation required for a CE mark as follows:
"ANNEX II: Technical Documentation

1. Device Description
2. Information to be Supplied by the Manufacturer
3. Design and Manufacturing Information
4. General Safety and Performance Requirements
5. Benefit-Risk Analysis and Risk Management
6. Product Verification and Validation (V&V) [*MDR Section 6.1(c) requires 'the clinical evaluation report and its updates...' and IVDR Section 6.2 requires 'Information on clinical performance and clinical evidence. Performance Evaluation Report']

ANNEX III: Technical Documentation on Post-Market Surveillance

1. The post-market surveillance plan...
2. The PSUR... and the post-market surveillance report..."

As illustrated above in bold, the CER/PER is a required Product V&V document within the EU MDR/IVDR technical documentation and the manufacturer must select the right "clinical" experts to develop and review the CER/PER within the technical documentation. Of interest, the EU MDR and IVDR do not use the "technical file" or "design dossier" terms as they are used in the US.

Because PER clinical data analyses are not exactly the same as CER clinical data analyses, the PER will have a slightly different analytical format compared to the CER. The PER must call out the SVR, APR, and CPR analyses specifically. In addition, the PER conclusions must be carefully aligned with the SVR, APR, and CPR clinical data analysis in the background, SOTA, alternative device and the subject/equivalent device sections. These sections must present specific information about the IVD device use including whether the IVD device was used for "self-testing" or "near-patient testing" or as a "companion diagnostic" because the benefits and risks associated with these use types will vary considerably.

5.3 Clinical/performance trial analysis

First, analyze each clinical trial individually to determine if the S&P acceptance criteria were met and if the benefit-risk ratio was acceptable. Then, analyze the clinical trial data as a whole to determine if the overall dataset met the S&P acceptance criteria and if the benefit-risk ratio was acceptable overall. Document the relevant S&P data when the subject device was used as indicated. The clinical/performance evaluation in the CER/PER is not intended to simply restate the trial methods, results, and conclusions from the clinical trial. The point is to analyze the clinical/performance data from the clinical trial to determine if the subject device S&P and benefit-risk ratio were acceptable.

Clinical trial data meet the "gold standard" when well-controlled clinical trials generate high-quality clinical data about the subject device used as indicated. The nubmer of study subjects must be sufficient to define the device S&P and the patient benefit-risk ratio across all indications for use. Many different clinical trials are included and some may be relevant even if they are not about the subject device S&P and benefit-risk ratio. For example, a clinical trial may be done using the subject device to explore a disease-related topic or to test a particular device feature. Sometimes, the clinical data needed for the CER/PER clinical/performance analysis may be entirely unrelated to the clinical trial findings and conclusions. This is expected since the CER/PER purpose may not be the same as the clinical trial objectives.

The evaluator must understand the difference (if any) between the clinical trial and the CER/PER objectives. When the clinical trial data are not aligned with the CER/PER purpose, appropriate caveats must be stated for the data analyses in the CER/PER. For example, subject device safety data are almost *always* available in any clinical trial using the subject device simply because the device was used and no negative comments or safety concerns were reproted about the device. These data can be included and discussed with a caveat stating device safety was not the point of the study.

5.3.1 Trial analyses: single trials

Begin each clinical trial analysis by documenting general trial aspects in the workbook (i.e., include trial objectives, number of subjects, device names/uses, methods used, measurements made, etc.). Record answers to specific clinical data analysis questions: Did the site investigator follow the trial protocol? Does the trial have good data integrity? Was the trial complaint with ISO 14155: 2020, "Clinical investigation of medical devices for human subjects—Good clinical practice"? If not, should the trial be excluded from the CER/PER? Review all clinical study reports (CSRs), including current interim and follow-up reports. Combined data from multiple studies, if performed, should be inspected for accuracy, consistency, reproducible results, and any trending S&P issues.

Data integrity is often accounted for by ensuring the clinical/performance trial data are ALCOA + compliant. This means the data are attributable, legible, contemporaneous, original, accurate, complete, consistent, enduring, and available. In addition, all clinical trial results should be documented in a clinical/performance study report (CSR/PSR) and each CSR\/PSR should follow an internationally acceptable, standard format to present the clinical trial data clearly and completely [4]. A few example questions may help when analyzing the CSR/PSR (Table 5.2).

TABLE 5.2 CSR/PSR analysis example questions.

CSR topic	Analysis questions
Protocol, design	What was the study design? What were the limitations for this study design? Did the protocol require appropriate intended end point measurements? How were the clinical data documented and reported? What statistical analyses were used? Was the study approved and overseen by an institutional review board (IRB) or internationalethics board (IEB)? How well did the study follow the protocol?
Aims, hypothesis	Were the study hypotheses, aims, and methods appropriate? Was adequate and relevant background/SOTA information provided? Did the study hypothesis align with the study aim and design? Did the study groups/arms/comparisons align with the study aim? Were study objectives met or not? If not, why not? Was the protocol/study design adequate to achieve the study aims?
Methods	Was patient enrollment clearly described? Was an appropriate patient population studied? Was the sample size as described in the protocol? Was the sample size sufficient (based on statistical power analysis)? Was the patient treatment schedule/process clearly described and followed? Were all planned treatments completed in all study subjects? Were primary end point measures and was confounder management described? Which device generations or models were used?
Data management, analysis	Were data collected as planned? Were changes and deviations documented sufficiently? Were any additional "ad hoc" evaluations clearly identified? Did collected data align with study end points? Were safety and performance data collected adequately? Did the device perform safely and was the device used as indicated? Did the clinical data support the device intended use and all clinical claims? Was data quality appropriately addressed? Were valid statistical analyses completed?
Conclusions, limitations, bias	Were results correctly stated and not overstated (compared to the data presented)? Were the safety and performance results clear? Did the device benefits outweigh the risks? Were the identified device hazards and patient harms adequately addressed and mitigated? Were the device outcomes comparable to other SOTA device uses or alternative therapies? Were limitations and deviations adequately described? Was the trial registered prior to trial start? Was funding described? Were trial personnel biases and financial/other conflicts of interest described, documented, and mitigated when appropriate? Was the CSR/PSR signed, dated, and version controlled?

The answers to these CSR/PSR questions may be helpful to keep in the spreadsheet or evaluation software. Each clinical trial may have different strengths and weaknesses. If no CSR/PSR is available or the full data are not included in the report, then the clinical/performance trial data may not be "reliable" enough for use in the CER/PER. This is a regrettable, yet all too common problem.

Evaluators should document concerns during individual clinical trial analyses. Major problems such as illegal trial activities, failure to comply with international standards such as GCP, unfounded conclusions, or suspected data errors noted during analysis may require a re-appraisal and exclusion or a data inclusion justification depending on relevance and clinical data volume available for analysis. Less serious concerns may simply require notation in CER/PER.

After analyzing the basics and the device-specific S&P details from each trial, the evaluator must evaluate the trial data impact on the benefit-risk ratio. The evaluator should list, analyze and discuss the overall clinical trial S&P results from the CSRs/PSRs including all AEs, ADEs and device deficiencies (DDs), all ADE treatments and whether any ADEs and/or DDs were not resolved during the clinical trials.

BOX 5.4 Example ADEs and DDs

ADEs in a hysteroscopic morcellator device trial were listed in the CSR, including fluid volume overload, bleeding, and perforation. Each ADE included a specific descriptor for device relatedness (e.g., possible, probable, or definite), occurrence rate (e.g., occurred in 3/231 patients or a 1.3% occurrence rate) and severity (e.g., 1/3 required hospitalization, 2/3 resolved with IV diuretics postprocedurally). DDs were also defined, including specific details about the physical device part with the problem, the dysfunction details, and the DD impact on the study subject. Sometimes a DD was associated with an AE, and this relationship was also defined. The evaluator was required to analyze the ADEs and DDs carefully to see if they were already considered in the S&P acceptance criteria and the benefit-risk ratio. If not, these were added in, and the entire collection of ADEs and DDs were considered in the CER/PER.

Device-related AEs (i.e., ADEs) represent device risks and must be analyzed and documented in the PMS system files, risk management system files, and the CER/PER benefit-risk ratio analyses. For the CER/PER, the evaluator should document the clinical-trial specific safety end point data and their analysis notes in the designated CER/PER spreadsheet (e.g., each trial represents a row in a spreadsheet or software).

ADEs relevant to the safety acceptance criteria as defined in the CEP/PEP should be recorded specifically. For example, a hysteroscopic morcellator device CEP/PEP may define a safety criterion as less than 1% perforations. Perforation, then, should have a specific column in the CER/PER spreadsheet where the total perforation number and percentage are recorded for each trial. ADEs not relevant to the defined safety acceptance criteria should be recorded and compared to the prespecified acceptance criteria. Evaluators should analyze all documented AEs, even nonsevere AEs or AEs indirectly related to subject device use. Only AEs worthy of comment should be documented in the CER/PER spreadsheet or software. Any new ADEs or DDs will need to be added to the PMS and RM system files and reconsidered in the definition of the S&P acceptance criteria during the next CE/PE cycle.

After safety details from each trial are analyzed, the evaluator should analyze efficacy results. Efficacy in a medical device trial represents device performance. ADEs (safety problems) and DDs (performance problems) can sometimes be difficult to isolate from each other and from other issues (e.g., underlying disease, surgical procedures, medications, diet). The predefined S&P acceptance criteria should guide the performance analysis by providing specific, prespecified and prequantified performance criteria to analyze. Trial data relevant to each S&P criterion should be extracted and recorded in the designated spreadsheet along with any additional performance issues noted during the clinical trial evaluation (i.e., new S&P issues may be discovered during the analysis and new S&P acceptance criteria may be needed when these new or changed S&P details are found).

The CE/PE spreadsheet or software should allow separate columns for the ADE and DD analyses (i.e., Table 5.3 is an example S&P analysis from two hysteroscope clinical trials). In addition, the evaluator should ensure all ADEs and DDs are fully considered in the benefit-risk ratio.

TABLE 5.3 Example clinical trial ADE and DD analysis.

Trial no.	End points	ADEs (Safety)	DDs (Performance)
02646498141	Uterine polyp removal success (full polyps removed), patient pain scale.	During procedure pain: mild 6.3% (5/79), moderate 1.3% (1/79), severe 0% (0/79); immediate postprocedure pain: mild 5.1% (4/79), moderate 3.8% (3/79), severe 0% (0/79); 1.3% (1/79) extended procedure time due to device failure; 2.5% (2/79) cervical tear as evidenced by vaginal spotting postprocedure (no treatment provided).	100% (79/79) polyp removal; One device failure: screen black, physician unable to visualize uterine cavity. Device removed, system rebooted, device reinserted and screen worked. No patient complications.

(Continued)

TABLE 5.3 (Continued)

Trial no.	End points	ADEs (Safety)	DDs (Performance)
65498798465	Uterine polyp removal success (full polyps removed), patient satisfaction with procedure scale and quality of life scale at 3 months compared to baseline.	0.6% (1/156) extended procedure time, 0.6% (1/156) piece manually removed from patient with no further reported harm, 3.2% (6/156) cervical perforation, 1.3% (2/156) blood loss requiring transfusion, 0.6% (1/156) fluid volume overload requiring 1 day of hospitalization; 0.6% (1/156) itching abdominal skin in patient with psoriasis.	99.4% (155/156) polyp removal success; 1 disposable speculum package open, determined to be unsterile and discarded (no patient harm), 1 device break in patient, removed with forceps (no patient harm), 1 fluid return measurement inaccurate causing patient fluid volume overload.

The evaluator determined "itching abdominal skin in patient with psoriasis" was not a new device-related risk, because the "itching" was related to the underlying condition (psoriasis) and was not related to the device, device intended use, or condition device was intended to treat. This systematic approach to tracking and recording the S&P data analyses details for individual clinical trials also supported the overall device S&P analysis across all trials. Over both trials, pain, cervical tear/perforation, extended procedure time due to device problems including breakage, and system malfunction were common device-related safety issues, and device failure/breakage as well as compromised sterility and fluid measurement errors were DD concerns. To facilitate a more detailed analysis, the details in each column may need to be separated into multiple columns as needed to make the data discrete enough for analysis. Furthermore, individual data analysis columns in the workbook/software should be dedicated to the specific acceptance criteria.

5.3.2 Trial analyses: grouped and as a whole

After all appraised and included clinical trials are individually analyzed, the evaluator must analyze the clinical trial data as a whole. Did the collected clinical trials adequately represent the device use in the entire indicated population for use? Were all clinical trials scientifically valid? Were ADEs and DDs similar or different across the clinical trials? Were the S&P acceptance criteria are consistently met? When analyzing the data for the S&P acceptance criteria table, the evaluator must document the occurrence percentage as the number of subjects who experienced each predefined ADE or DD (i.e., the numerator) over the number of subjects treated with the device (i.e., the denominator). Then, the evaluator must determine if the percentage for each ADE or DD met the acceptance criteria. If not, the evaluator must determine why not. Is one particular trial contributing a disproportionately large number of ADEs or DDs? If so, the trial data must still be included, but the possible reasons for the difference must also be discussed and the need for additional clinical data generation must be considered with all evaluator conclusions documented.

BOX 5.5 Case study—device changes

Case study situation

A manufacturer sponsored a series of clinical trials for their novel, active, implantable medical device 5 years ago. In the past 5 years, the device was upgraded and technical details were changed to make the device more efficient (e.g., smaller battery and overall device size, longer battery life, fewer sensing/pacing leads, less invasive implantation process). These changes improved the benefit-risk ratio and were docuemnted in the postmarket data.

Case study issues identified

During clinical trial data analysis for the fifth CER, the evaluator noticed the specific device used was not clearly documented in the early pilot and pivotal clinical trials. Based on trial timing, the device used in the clinical trials could not have been the most recent device generation (i.e., the subject device in the current CER). The manufacturer wished to cite the original clinical trial results as sufficient clinical evidence supporting the current device with the upgraded device and related device claims to meet the GSPRs.

Case study outcomes

The evaluator analyzed and documented all clinical trials, literature, and experience data in the CER. This evaluator found many clinical data gaps between the original device models used in the device-specific clinical trials and the current device model. Noclinical trial data were available using this new device model. In particular, the modifications to the current device model had changed the benefit-risk ratio and the evaluator documented in the CER how the S&P acceptance criteria could be improved to

(Continued)

BOX 5.5 (Continued)

represent the current device model more accurately in the next CEP. The evaluator documented all clinical data gaps in the CER and explained the need for additional clinical data to be generated in order to document more clearly device conformity with GSPRs 1, 6, and 8. The specified additional clinical data gathering was proposed in the PMCFP, which followed CER completion and sign-off. In this case, gathering specific clinical data from the "new model" device launch in the postmarket setting was considered sufficient to address the clinical data gaps, and no new clinical trial was required.

A better case study solution enacted

This case study illustrates the importance of the CER/PER "developmental context" section to describe the historical device models. In particular, the evaluator must consider all device changes over time when analyzing the clinical data. Not only must the evaluator identify and document all clinical data gaps for all current marketed device models, but the evaluator must also recommend how to remedy all identified clinical data gaps in the CER/PER. Sometimes a new clinical/performance trial may be required, and the evaluator must apply due diligence when drawing conclusions about the clinical data overall sufficiency including the anticipated sufficiency of proposed *future* clinical data generation.

One of the key considerations for the overall clinical trial analyses is for the evaluator to determine if the clinical trial data are sufficient to support conformity with GSPRs 1, 6, and 8 about the device S&P and benefit-risk ratio over the device lifetime when used as indicated. The evaluator often recommends collecing new clinical trial data to cover every segment of the indication for use (i.e., each age group, each disease state, all possible conditions, etc.), all device features and uses, all benefits and risks including their occurrence rates, etc. to ensure the CER/PER clinical data are accurate, complete, and acceptable. Many options are available for new clinical trials to address specific clinical questions including the gold standard, randomized controlled trial (RCT), retrospective chart reviews, registries and many other types of clinical trials. In addition, the company may want to publish clinical data in the peer-reviewed literature and will continue to collect clinical experience data.

Clinical trials for the CER/PER background/SOTA/alterative therapy and the subject/equivalent device sections should be analyzed in the same way, but separately. Any clinical/performance trials from previous device iterations or different device models with different S&P acceptance criteria and benefit-risk ratios should be analyzed in the background/SOTA section and not pooled with data from the subject device unless adequately justified in the CER/PER. Upcoming, incomplete, orterminated clinical trials should also be documented, analyzed, and explained in the CER/PER.

5.4 Clinical/performance literature analysis

The clinical/performance literature data evaluation should include a critical review of each study design, the trial execution, the pivotal S&P data and the subject/equivalent device benefits and risks. Any potential bias should be documented and fully considered. The evaluator should analyze each individual article first and then the literature as a whole. The analytical details should be recorded in a customized datasheet/workbook or an appropriate software designed for this purpose. Comparisons of totality of clinical literature data from SOTA and subject device literature should be made and significant differences, both favorable and unfavorable, should be described and documented as part of the clinical literature analysis. This in-depth background/SOTA/alternative therapy/device literature analysis may help to identify additional subject device S&P objectives and limitations may help to identify potential gaps to be addressed in the next PMCFP/PMPFP. Variations in clinical settings and device intended uses will confound the literature analysis. The evaluator should document and describe the relevant details so the evaluation can be validated by others and found reliable. The evaluator must not overinterpret the clinical/performance literature data.

5.4.1 Literature analyses: single articles

The evaluator should begin the literature analysis by skimming through and briefly reading the entire article. The methods and results sections are critical to the analysis. The introductions and conclusions typically do not include clinical data presentations and are often restricted to historical background or future predications, respectively. The CER/PER literature analysis must focus on the new clinical data specifically and should not focus on opinions, predictions or unsubstantiated claims. As a starting point, the evaluator should re-read the title and abstract for each individual article to understand what was done and then the relevant clinical/performance data details should be documented in the workbook or software for further inspection and analysis across many articles. Copying over all clinical/performance data from each article is not the point. Extracting only the relevant details is required and takes a good deal of clinical data analytical prowess.

The goal is to find important and relevant details and to document them in the clinical data analysis workbook or software in a useful format for comparison to other clinical data. Sometimes the data presented in a given published article may need to be transformed into an appropriate format for comparison to other data. For example, days of device exposure may be listed in some articles by days, weeks, or months and converting all data in the spreadsheet or software into one unit of time is helpful to allow direct comparisons between individual articles (e.g., 1 week can be represented as 7 days, and 6 months can be represented as 182 days, so the units of time are using the same "per day" timescale for comparison).

5.4.2 Separate the background and subject device clinical literature

When reading articles for the CER/PER background/SOTA/alternative therapy/device section, the goal is to put the subject/equivalent device in the context of all known and relevant details. Providing a comprehensive review of every detail from each background article is not required; however, a thorough and complete analysis and discussion of the relevant background/SOTA/alternative therapy details is required. The evaluator should understand the device and device issues before starting the clinical literature data analysis.

BOX 5.6 Abstract relevant clinical literature data only

When working on a "leadless" cardiac pacemaker CER, 56,000 articles were identified about pacemakers. This clinical literature included more than 6,000 review articles, 600 metaanalysis and systematic review articles, nearly 250 historical articles, and 150 guideline articles. The evaluator decided to review a few historical articles published in the last 5 years before beginning the actual "clinical data" literature analysis. This dataset included 11 "narrative review" articles, which did not report specific clinical data analyses per se; however, they did review the history of pacemaker development and these articles allowed the evaluator to focus on relevant background knowledge/SOTA/alternative therapies. For example, these recent narrative review articles covered topics including details about the lithium battery power source, action potentials, pacemaker programming, changes in the technology over time, and safety concerns about extracting leads after implantation as well as pacing the heart in heart failure patients. Of interest, one review article included a discussion about the leadless pacemaker.

This "extra work" outside the CEP for the CER clinical data analyses allowed the evaluator to become familiar with the topics commonly discussed in the literature. Although these articles did not provide clinical data for analysis in the CER and no data were abstracted for further analysis, these articles helped the evaluator understand the history of cardiac pacemakers including the new "leadless" cardiac pacemaker. Using this "history of cardiac pacemaker" understanding with the detailed subject/equivalent device description and the actual clinical data available for analysis from the clinical data appraisal step, the evaluator was able to structure specific clinical data abstractions to be "on target" and "relevant" for the analyses to address the subject/equivalent device S&P and benefit-risk ratio. In addition, the evaluator understood the basic framework for many relevant issues to be explored in the CER background knowledge/SOTA/alternative therapies section.

In this example, the CER background knowledge/SOTA/alternative therapies section needed to discuss the leadless pacemaker evolution from pacemaker devices which required leads and were still available as alternative therapies; however, because the leadless pacemaker has no leads, the clinical data related to the leads were simply not relevant to the CER subject/equivalent device section. No clinical data needed to be extracted from the clinical literature articles about the leads (i.e., lead tract infections, breakage, dislodgments, etc.), since these were not relevant to the leadless device. In this example, the CER subject/equivalent device section needed to focus on the S&P and benefit-risk ratio specifically for the leadless device. Critical performance details and complications needed to be documented and analyzed to document how the leadless device met the GSPRs 1, 6, and 8 within the context of all pacemakers, as briefly discussed in the CER background knowledge/SOTA/alternative therapies section.

The CER background/SOTA/alternative therapy/device section was not focused on any "best in class" analysis but was focused instead on the S&P as well as the benefit-risk ratio to document how the leadless pacemaker was able to meet the GSPRs for safety, performance over the device lifetime. The evaluator did not have any requirement to identify if the leadless pacemaker was "best in class" and did not need to determine if any pacemakers with leads should be removed from the market because the leadless pacemaker was somehow better than the older pacemaker models. The CER regulations do not require only "best in class" devices to be marketed in the EU. On the other hand, all devices on the EU market must be safe and must perform as indicated with sufficient clinical data to support the S&P, benefit-risk ratio, and all claims made by the manufacturer.

The evaluator is required to search for S&P and benefit-risk details throughout the full-length article because critical details about the subject/equivalent device S&P and the device benefits and risks may be discussed within the article but not mentioned in the title or abstract. The evaluator needs to find and use the relevant clinical data details from each article in the CER.

5.4.3 Use appropriate analytical methods specific to the literature type

Many analytical methods are used when analyzing the clinical literature because no one analytical method will work for every single article type. Some guidelines suggest using PICO (problem, intervention, control, outcome) to analyze the literature. The acronym, PICO, helps the evaluator to focuson the main concerns or clinical problems, the testing/treatments and recommendations, the comparison or control measures, and the plans to improve or affect symptoms in a measurable way. Although simple to use, PICO does not address the validity, reliability, or limitations for each article, and these details are crucial to the clinical literature analyses.

Additionally, the Cochrane Handbook for Systematic Reviews of Interventions suggests using PRISMA (The Preferred Reporting Items for Systematic Reviews and Meta-Analyses Statement and MOOSE Proposal (Meta-analysis of Observational Studies in Epidemiology) to evaluate clinical literature; however, PRISMA and MOOSE are only relevant to specific article types (i.e., metaanalyses and systematic reviews) and may not be applicable to the clinical literature analysis required for other article types including case reports, narrative reviews, retrospective reviews, RCTs, observational studies, etc.

Each article type may have a different objective and may require a different analytic approach. The CER/PER articles for the background knowledge/SOTA/alternative therapies section should be analyzed separately from the articles for the subject/equivalent device sectionand many factors including the treatments used/discussed and any potential author bias should be considered for each article type (Table 5.4).

TABLE 5.4 Example analysis details by article type.

Article type	Article objective	Example details to analyze
CER/PER background knowledge/SOTA/alternative therapies section		
Guideline	To assist care-giver and patient decisions about appropriate care for a specific clinical circumstance.	Who wrote the guideline? When was the guideline written? What areas does the guideline cover? What is the standard of care for the disease or condition? Is the subject device or device class recommended? Under what conditions is the subject device or device class recommended?
Metaanalysis	To assess the strength of evidence for a disease or treatment by examining data from numerous independent studies to determine overall trends.	What question was asked? How was the analysis conducted? How was the literature search conducted? How many studies were evaluated? Were any pivotal studies missing? Was the subject device evaluated? What outcomes were evaluated? What were the conclusions? What was the level of certainty for the conclusions? Were the statistical methods appropriate and was the meta-analysis executed correctly?
Systematic review	To summarize existing evidence using a systematic literature search.	How was the literature search conducted? How many articles were included? Were any pivotal studies missing? Was the subject device evaluated? What were the conclusions?
CER/PER subject/equivalent device section		
RCT	To evaluate intervention effectiveness and establish a causal relationship between the intervention and outcome.	What was the sample size? How was the trial controlled? Was the trial blinded? Was the assignment to treatment or reference device/intervention random? What were the inclusion/exclusion criteria? Were the baseline demographics comparable between groups? How many arms did the study have? What was the study evaluating (e.g., outcomes, event rates)? Was the study single-site or multisite? What were the study end points and were point estimates and variability measures provided for the primary outcome? How many subjects were treated with the subject device? Were all randomized subjects analyzed?
Retrospective review	To review historical data collected in medical records looking back.	Why were the data collected retrospectively? How was the data collection controlled? Were any efforts extended to control selection bias and other forms of bias? What were the inclusion/exclusion criteria? How far back were data collected, and were the oldest data similar enough to the newest data for pooling into one data analysis?
Prospective trial	To collect clinical data generated going forward.	What other study design details were used in addition to prospective data collection? Was a comparison arm included? Was the study blinded, controlled, or randomized? What data were collected and over what time period?

(*Continued*)

TABLE 5.4 (Continued)

Article type	Article objective	Example details to analyze
CER/PER background knowledge/SOTA/alternative therapies section		
Cohort study	To compare outcomes across device treatment groups.	Was the study a retrospective review or a prospective cohort study? Was the study controlled? Blinded? How were subjects selected? Were baseline demographics comparable between cohorts? How many arms did the study have? What was the sample size? What was the study evaluating? Was the study single-site or multisite? What were the study end points? How many subjects were treated with the subject device? Was follow-up long enough to see outcomes for an appropriate cohort proportion? Were dropouts explained and similar across groups?
Case-control study	To compare device experience in a defined outcomes group versus a control group.	How were cases and controls selected? Was the study blinded? What was the disease state? Were controls selected at random from the same case population? How were subjects selected? Were the baseline demographics comparable between groups? What was measured for the response and nonresponse rates? Were statistical analyses appropriate (e.g., not "over" matched)?
Case report/ series	To document interesting and/or related cases without any comparator.	How many cases are documented? What were the inclusion and exclusion criteria? Was an objective criterion or blinding used? Were subjects in the case series similar to each other? Was series representative of a relevant population? What happened? Were novel S&P signals documented? Why was the case documented (i.e., what makes this case unique, what is the article purpose)? Were populations, prognostic factors, techniques, and outcomes described adequately? Was follow-up long enough to see an outcome? Do the authors have potential biases?
Narrative review	To summarize existing evidence without using a systematic approach.	How many articles were reviewed? Were conclusions supported with clinical data? Do the authors have potential biases? What aspects of the article were helpful to understand the historical context of the subject/equivalent device?

RCT, randomized controlled trial; S&P, safety and performance.

The evaluator must use a rigorous, reproducible, and flexible method to address the issues encountered in the various article types and quality levels. For example, low-quality articles like the case studies, narrative reviews and bench/animal study article types should only be included if they contain novel information about S&P not addressed elsewhere in the literature. These low-quality articles are quite valuable to the CE/PE when high-quality data are lacking and when the article includes detailed clinical data about a rare or novel finding not reported elsewhere. The evaluator must be careful not to overinterpret any included low-quality articles. A single case study with a "potential" novel outcome must be carefully considered since the reliability of this data type is quite low. Reference to a case report or case series describing a unique benefit or risk should probably be referred to in the CER/PER as a "potential" benefit or risk which requires further study.

Another good evaluator practice for critical analysis of each article is to answer specific questions about each section of the individual article (Table 5.5).

TABLE 5.5 Example general clinical literature analyses questions by article section.

Article section	Analysis questions
Study design	What was the study design (e.g., metaanalysis, RCT, case report)? Is this study design appropriate to analyze the device S&P or the device benefit-risk ratio?
Objective	What was the study objective and was the study objective relevant to the device S&P? The objective is often found in the abstract or the last part of the introduction and the objective often starts with a "to" statement such as "to evaluate..."
Subjects	How many study subjects participated in the study? How many used the subject/equivalent device? Was study powered appropriately with sufficient study subjects to address the study objective/s?
Population	What population (e.g., age, sex, disease state, type of surgery performed, anatomical location treated) was studied? Did the population reflect the target population for the subject/equivalent device?

(*Continued*)

TABLE 5.5 (Continued)

Article section	Analysis questions
Treatments	What treatment was used? Was treatment compared to standard of care (e.g., per guidelines and for pre- and posttreatment care, if applicable)?
Clinical measures	What clinical measures were evaluated? What were the study end points? Specifically, what clinical data were collected? Were these clinical measures relevant to the subject/equivalent device use?
Follow-up	How long was the study follow-up (e.g., 1 day, 12 months, or 10 years after initial treatment)? Was the follow-up sufficient to address the device S&P outcomes?
Results/outcomes	What were the results/outcomes? What was the treatment performance? (E.g., any issues/problems with the device? Did the device deliver any benefit to the patient?) What was the treatment safety (e.g., any ADEs, complications, safety concerns)?
Conclusions	What was the conclusion? Were the conclusions appropriate and supported by the data?
Limitations	What were the study limitations (e.g., small sample size, conflict of interest, author bias, lacking safety or performance data, conclusions unfounded, inconsistencies within article, missing or inconsistent data, difficult to fully analyze data based on the information provided, no control group, treatment of interest not able to be separated from other treatment or confounder.)?
Additional/custom	Was any other information included but not properly summarized?

Whenever relevant to the CER/PER purpose, the answers to the general clinical literature analysis questions for each article section should be documented and should populate the customized spreadsheet/software database as the CER/PER Essential Clinical Literature Data Listings. When abstracting clinical literature data, the medical reviewer, evaluator, and writer should review each "study-type" reporting guideline. For example, one resource provides a list of many guidelines and is called the EQUATOR network (i.e., the Enhancing the QUAlity and Transparency Of health Research at http://www.equator-network.org) (Fig. 5.1).

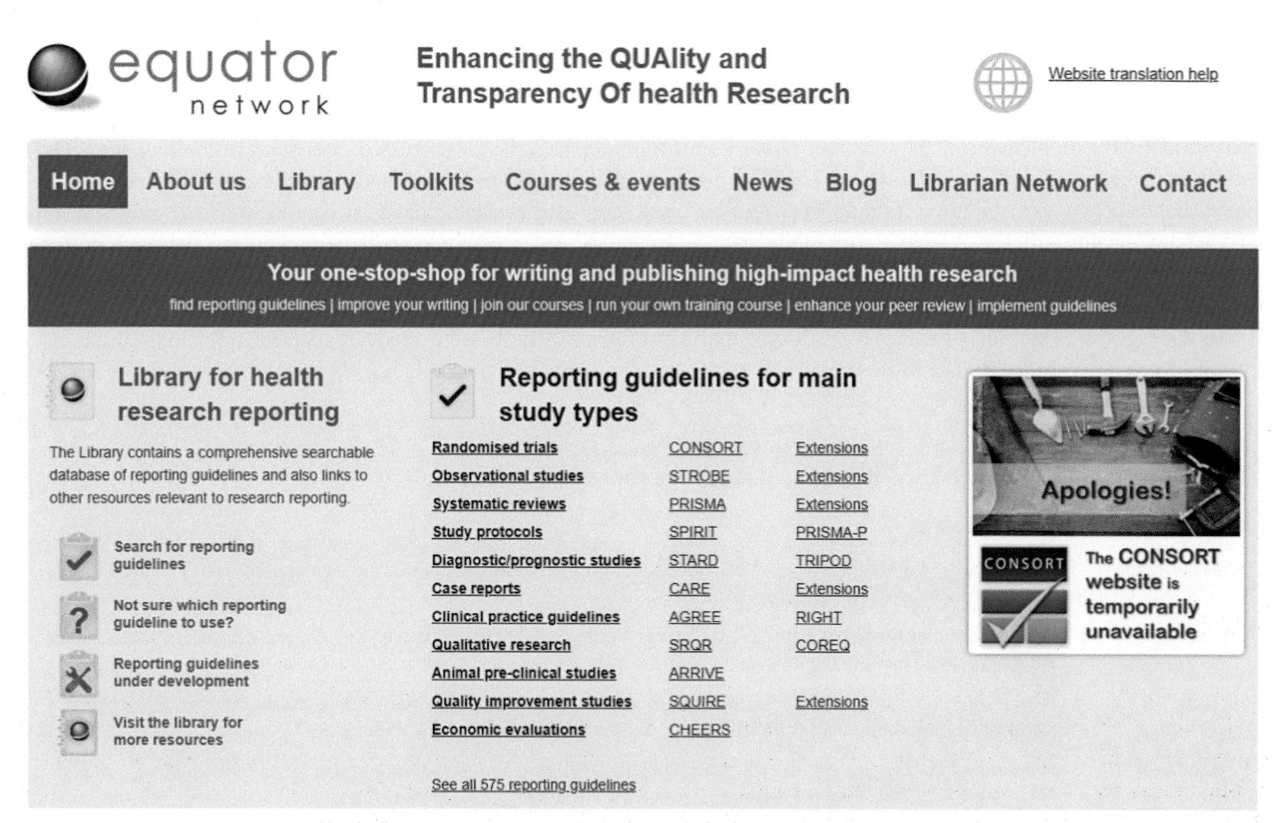

FIGURE 5.1 EQUATOR network reporting guidelines for main study types

Reviewing these guidelines will help while analyzing the clinical data from the literature reporting data from these trial types. Any studies not following good practice guidelines may have quality issues. These guidelines should also prompt good evaluation questions to ask and important clinical data details to report in the CE/PE process.

After clinical data abstraction, the spreadsheet/software should include a listing of all the device-related AEs and DDs from each article as well as documentation related to each predefined and prequantified S&P criterion found in each article. Specifically, the evaluator must document if each criterion was met in each article and if the reported benefits continue to outweigh the risks for using the device as intended in each article. In addition, standardizing the data collection in the spreadsheet or software as much as possible is a good practice. For example, patient age can be a column in the spreadsheet and this data can be standardized to be recorded in years (e.g., a case report about a 1-month-old infant would be documented as "0.08" years in the "age" data column). These analysis results will be tabulated and summarized in the CER/PER, so constructing these tables carefully in the spreadsheet/software during the analysis step is helpful.

5.4.4 Clinical literature data must be relevant to the subject/equivalent device

In addition to understanding the scientific validity of each article and the clinical data within each article, the evaluator should seek to understand the article in the context of the subject/equivalent device. In general, most articles are not written to review the S&P for a specific device unless, for example, the device is new and the clinical trial was designed to evaluate the medical device. In other words, most clinical articles will require the evaluator to identify the relevant device-specific S&P-related clinical data inside an article. This is not easy when the article was not designed to describe the subject/equivalent device S&P clinical data, and this "tricky" data abstraction requires the evaluator to read the methods and results details carefully in order to identify and document the subject/equivalent device-specific S&P data from the article.

Essentially any article reporting a specific medical device use in a patient will have safety information since the device was used. The safety data quality may be low, especially if the article says nothing about the subject/equivalent device other than documenting the device was used. Safety can be assumed if the study design required the device to function in order to generate study results. In these cases, the clinical data in the article was based on the device working without negative impacts, which would have been noted if they had occurred.

The clinical literature typically reports safety data as device-related AEs experienced by the study subject, user, or others exposed to the device. These safety data are often reported as a list of all device-related AEs with a number and percentage of patients experiencing each particular AE (e.g., the AE list may include wide-ranging details from an electrical shock to worsening heart failure; from itching or nausea to death for an active medical device or, for IVD devices, from leakage to missed or incorrect diagnosis).

Performance data are typically related to how the device worked or did not work as experienced by the study subject, user or others exposed to the device as intended (e.g., performance details may include items like: the device operated as intended, had a broken component, the procedure or analysis was abandoned and a different device was used). Performance data may also include device-related benefits (e.g., quality of life index score increased on follow-up documenting the claimed device performance to improve quality of life). Sometimes the S&P clinical data overlap and the evaluator should consistently follow the data analysis practices laid out for the device as specified in the CEP/PEP until a new data analysis method becomes obvious and required.

5.4.5 Evaluators must address complex clinical literature data analysis issues

S&P clinical data may be measured using quantitative and/or qualitative terms. Objective, numerical (i.e., quantitative) clinical data with solid statistical analyses reported in a well-written article are typically considered higher-quality clinical data than subjective (i.e., qualitative) measures without any statistical references reported in a poorly-written article. Often the clinical literature report both quantitative and qualitative data. For example, articles generally document how many patients used the device or how many procedures were completed with the device and many articles will specify how many device-related incidents (e.g., AEs, DDs) occurred. In addition, many articles report opinions about what the data mean, how the patient felt or what happened when using the device. These descriptive data may also be important to document even though these narrative data about the event quality are more difficult to analyze than the numerical data defining the event quantity.

Comparative quantitative data should be abstracted including both compared values and the statistical *p* value for the comparison. A standard format for basic comparisons is: xx versus yy, p<value>. Qualitative data should be abstracted using direct quotes. A standard format for qualitative data is to group the subjective qualities, and give a short descriptor with a rationale for the grouped qualitative data descriptor. For example, "shortness of breath," "difficulty breathing," and "dyspnea" might all be grouped as dyspnea, since dyspnea means difficult or labored respiration. In this way, qualitative descriptions may be able to be assigned numerical values when the same or similar data are reported more than once.

In addition to the individual data point complexities, large variations are found in article quality, study purpose, study objectives, and study end points. This extensive diversity complicates the CER/PER analysis because many included articles might not align fully with the CER/PER objective and purpose. This is to be expected, and the evaluator should develop strategies to mitigate these all-too-common clinical literature data analysis challenges (Table 5.6).

TABLE 5.6 Individual article analysis challenges and possible mitigations.

Challenge	Potential resolution
After full analysis, article should be excluded.	Appraisal and analysis are iterative processes; "recoding" is acceptable with new detailed information. Recode this article as excluded; then, complete QC and analyze the other articles as planned without this one.
No device-related S&P data mentioned.	Device problems may only be reported if they occur or lead to patient complications. Document the number of uses "without reported device problem" or "without reported AEs." Keep this article included if the implied S&P for the device is needed; however, if this article does not add value to the analysis, then exclude this article.
Device mentioned only in methods section with no further information.	Some device risk is determined by the procedure. For example, a surgical article may state a particular clamp was used in a venous surgery; however, the article may only report outcomes for "surgical success" without any specific clamp information. Consider documenting the number of surgeries where the clamp was used, and carefully review the AEs to determine if any AEs could be potentially related to the clamp. For example, was a "torn vessel" AE at the location where the clamp was applied?. Document the findings from the AE analyses and if unable to determine if any AEs were related to device, document the inability to determine device involvement.
Article is missing clinical data.	Recode this article for exclusion unless novel clinical data are present, and document the specific missing clinical data issues (e.g., patient population not defined, outcome details not provided, no safety and or performance data reported).
Trial design not relevant to subject/equivalent device S&P.	Consider this article for the background/SOTA/alternative therapy/device section or recode for exclusion since the trial was not about the subject/equivalent device S&P and does not have relevant info for the background/SOTA/alternative therapy/device section.
Measurements are not standardized.	Transform data into usable measurements for comparisons with other clinical data and document the transformed "standard" value and the transformation method used.
Study includes many devices used together and S&P data cannot be assigned to one device.	Consider this article for the background/SOTA/alternative therapy/device section or recode this article for exclusion unless novel clinical data are present. Include a statement about the inability to assign the data to any specific device.
Study uses subject/equivalent device but does not report device-specific S&P data.	Consider for background/SOTA/alternative therapy/device section only or recode for exclusion since trial does not include relevant S&P data about the subject/ equivalent device.
Study reports only unquantified qualitative clinical data.	Describe data in appropriate terms within tables and text. Recode this article for exclusion if the qualitative clinical data are reported sufficiently in other included articles.
No primary S&P data provided (e.g., narrative review, systematic review, technical description without discrete clinical data).	Document relevant article clinical data in background/SOTA/alternative therapy/ device section; state the data are not primary clinical data; consider listing number and study names referenced in article and determine if those individual studies were also individually identified for analysis in CER/PER process. The data in this secondary source may duplicate primary clinical data reported elsewhere in the literature. If all studies in this review are already present elsewhere in CER/PER dataset, consider using the primary data collection studies instead and recode this article for exclusion. If including both primary and secondary clinical data from this article in the CER/PER, document the data duplication for each specific use to avoid double counting confusion.

(*Continued*)

TABLE 5.6 (Continued)

Challenge	Potential resolution
Authors did not indicate device details (e.g., device type, model, or generation not provided).	Document subject/equivalent device model release dates and device variations over time and explain how the data in the article may or may not be related to the subject/equivalent device based on the known device model release dates.
Author conclusions are unfounded, inaccurate, not valid, or unreliable.	Recode for exclusion, or if this article must be included (e.g., insufficient clinical data are otherwise available), then document the problems in the CER/PER and explain the impact of the problems on the individual and overall clinical data reliability.
Statistical analysis is missing, poor, or inappropriate.	Recode for exclusion unless novel clinical data are present, then include a statement about the statistical issues (e.g., no comparator arm, no sample size calculation, no definition of statistical method used).
Article has potential for significant bias.	Recode for exclusion unless novel clinical data are present, then include with a statement about the bias (e.g., subject selection bias, financial bias, outcomes bias, lack of critical analysis, incorrect statistical method chosen, reports trends, e.g., $p = 0.05$, as though meaningful/significant).
Article is confusing and poorly written.	Recode for exclusion unless novel clinical data are present, then include with a statement about the poor article quality.
Article is about an unrelated topic with only a small portion of the data related to CER/PER purpose.	Recode for exclusion unless novel clinical data are present, then include with a statement about the unrelated article nature. If including this type of article, state the number of data points related to the device as well as the number of data points overall, and explain how the selected data subset was not envisioned by the author or powered for the statistical analyses as was planned for the whole dataset.
Author describes illegal or unethical research methods.	Recode for exclusion. Do not use these data in the CER/PER.
Author describes off-label device use.	Report off-label uses in background/SOTA section. If device is regularly used off-label, identify systematic misuse and recommend corrective and preventive action (CAPA) activity to begin.
Data is reported in a manner incompatible with S&P criteria.	Record data as reported and state the reason why the data are not compatible with the measurements required by the S&P criteria. Article can be included in CER/PER discussion but excluded from detailed S&P analysis, as appropriate.

After reading each article, the evaluator should understand the key clinical S&P data relevant to the background knowledge/SOTA/alternative therapies as well as the subject/equivalent device CER/PER sections. The important clinical data from each clinical literature article should be sufficiently abstracted and detailed in the workbook or software designed for this CER/PER purpose. QC is always required during analysis including the need to double- or triple-check the workbook/software data abstractions and all recoded articles when they are changed from inclusion to exclusion or vice versa. Any concerns about systematic off-label use or other issues that might call for a CAPA should be escalated to management.

BOX 5.7 What is a CAPA?

A Corrective and Preventive Action (CAPA) is taken by a company to improve QMS processes specifically to eliminate nonconformities and other undesirable issues. The process begins when a nonconformity or undesirable issue is identified and the company conducts a systematic root cause analysis (RCA) to suggest actions to prevent their recurrence (corrective) or occurrence (preventive). International standards (ISO 9000, ISO 13485), US FDA regulations (21CFR820.100) and EU MDR/IVDR (field safety corrective actions) define requirements for device manufacturers to use CAPA processes within the quality management system.

Corrective action → Root cause analysis → Fix root cause → Verify fix worked
Preventive action → Lessons learned → Inform organization → Verify no new issue occurs

The US FDA regulations have at least 10 required CAPA objectives for medical devices:

1. APA process must be documented.
2. Quality issues must be identified.

(Continued)

BOX 5.7 (Continued)

3. Unfavorable trends must be identified.
4. Quality/content of data must support ability to identify and act on issues.
5. Statistical process controls (SPCs) should detect quality problems.
6. Level of RCA work is consistent with risk level.
7. Actions address root cause and prevent future occurrences.
8. CAPA actions must be verified/validated before implementation.
9. CAPA actions must be implemented and documented.
10. CAPAs must be reviewed during management review.

CAPA example: A review of manufacturing records identified missing/incomplete data records, lack of computer system control and inadequate written process control procedures. The RCA determined the extent of the data, computer system and procedures problems. A third-party interviewed current/former employees, investigated the root cause and conducted a risk assessment to determine if the failure impacted products, users and/or patients. The management team documented and supported the CAPA including their decisions about the need for a recall, additional testing, communication to customers, etc. Effectiveness checks showed the problems did not recur.

Closing a CAPA should be carefully considered to ensure compliant nonconformity data handling (e.g., RCA, containment steps, CAPA details) must be well-documented including all document revisions. CAPA effectiveness checking is required and the company must (among other details) notify and train staff, complete document management review for all CAPA documents, update technical documentation as specified by the CAPA, monitor CAPAs with additional PMS activities, ensure PMS system identifies the need for new CAPAs, notify distributors and importers if CAPA affects product, document decisions about recalls or field corrections, include CAPAs in PSUR and ensure all CAPA processes are well-documented in the QMS.

In addition to the data abstraction into the workbook (i.e., the CER/PER Essential Clinical Literature Data Listings), all analyzed full-length articles should be stored in an easily retrievable format (i.e., with the other essential CER/PER data records including the CEP/PEP, clinical/performance study reports, clinical literature articles, clinical evidence reports, CER/PER, PMCFP/PMPFP, SSCP/SSP, etc.). The most useful full-length articles will be appended to the CER/PER for source document verification as needed by any reviewer.

5.4.6 Literature analyses: grouped and as a whole

The evaluator should analyze the clinical literature data in appropriate groups and then overall as a best practice. The groups should be customized to meet the specific CER/PER needs. The SOTA/background and subject/equivalent device literature have differing CER/PER purposes and they are analyzed separately. The planned process often starts with analysis of a preidentified data grouping like all data from articles of a particular study design. For example, among all included clinical practice guidelines, the evaluator may analyze the guideline articles to see if the devices simlar to the subject/equivalent device were recommended for use in the target population and if the guidelines universally recommended specific details to be considered when using these devices. Similarly, the metaanalyses, systematic reviews, and individual clinical studies for the similar devices may provide a good background and overview for the subject/equivalent device use. This same approach can be used for the subject/equivalent device analysis (i.e., group articles by article type, then analyze guidelines which include the subject/equivalent device, then metaanalyses, systematic reviews, RCTs, and other clinical trial types, all the way down to the lowest-quality case reports which include the subject/equivalent device).

During this literature analysis process, the evaluator will begin to identify patterns, similarities, and differences among the clinical data and may further group the literature as new patterns emerge. For example, among RCTs, the highest subject enrollment studies may indicate a certain safety problem (e.g., perforation for a catheter or probe device), but smaller studies do not mention this problem. This pattern should be noted and described as a finding in the literature results section. Perforation may be a rare event which did not occur in small, well controlled trials. The evaluator should make sure perforations are listed as a known device risk. If perforation was not listed as a risk for the subject/equivalent device but is in the literature, this is clinical data gap example. Additional clinical data should be gathered to understand this risk better, and the product literature may need to be updated to document this risk.

BOX 5.8 Case study—newly discovered risks

Case study situation

During analysis, an evaluator analyzed five articles reporting eight patients who experienced hypoxia during medical device use including shortness of breath which was relieved by pausing device use temporarily (i.e., the hypoxia was likely to be device related). This safety risk was not described as a potential risk in the device literature including the instructions for use, website, product brochure, or anywhere else in the product labeling.

Case study issues identified

No clinical trials were conducted for this device, and animal testing did not identify this problem. The evaluator identified a new risk and a clinical data gap.

Case study outcomes

The evaluator wrote about this new risk in the CER by first presenting the five individual article findings in a table with descriptions discussed in the results text. In the conclusion, the evaluator identified the issue as eight patients having experienced a new (previously unidentified) and potentially device-related hypoxia risk needing additional analysis. The evaluator emphasized and explained the clinical data gap and the need to collect more clinical data in the PMCFP specifically to explore this safety concern. The overall CER conclusion documented how the specific PMCFP hypoxia data will be evaluated via a physician survey to see if others have seen hypoxia during device use. The CER conclusion also specified the survey results would be documented in the associated PMCF evaluation report (PMCFER) and discussed in the next CER. In addition, for all ongoing clinical data collections for devices of this type, the hypoxia safety concern was highlighted as a critical area of interest needing additional, urgent data collection.

A better case study solution enacted

During discussion with physician users, the company was expected to become more knowledgeable regarding device uses and hypoxia safety concerns. Finding device-related hypoxia in the literature without prior knowledge by the company suggested the quality systems were not working well. As a result, the sales, marketing, PMS and RM teams were retrained and again required to seek out and immediately report all risks and complaints including hypoxia if/when they occur. The post market surveillance (PMS) and risk management (RM) teams were specifically required to learn more about hypoxia including the occurrence rate, the triggers and specific conditions when the hypoxia occurs and all associated clinical findings. If the survey instrument did not yield immediate and comprehensive understanding of this new risk, then a new animal or human clinical registry study was in development to evaluate this potential relationship between the device and the hypoxia risk.

After analyzing every article and recoding each article into the right group, the clinical literature data as a whole should be analyzed. Starting this analysis with the overall background knowledge, SOTA and alternative therapy/device articles will give an excellent introduction to the device, the disease state, the SOTA, and all the therapies available to treat the relevant patient population. Then, separately analyzing the overall subject/equivalent device clinical literature data to determine and document the most important S&P details along with all the reported risks and benefits will allow the evaluator to determine if the S&P acceptance criteria have been met.

Within each of the (1) background knowledge, SOTA and alternative therapy/device and (2) subject/equivalent device article groups, several potential subgroups might include:

- similar study design groups (e.g., metaanalyses, RCTs, case reports);
- similar patient or population characteristic groups (e.g., adults versus children; cancer versus healthy);
- similar anatomic device locations (e.g., shoulder, elbow, finger, hip, knee, toe); and
- similar treatment comparisons (e.g., treatment with and without device, treatment by standard of care or sham treatment compared to device).

Grouping similar articles allows the evaluator to visualize patterns in the data and to identify outliers.

The evaluator can use a grouping strategy to determine how the data are similar or different, relevant or irrelevant, unique or novel. Depending on the clinical data amount available (i.e., number of articles to analyze), the articles may be grouped in a single level (e.g., only group by anatomical device location) or multiple levels (e.g., group first by study type [e.g., RCTs] and then by treatment comparisons [e.g., treatment of interest versus placebo]).

Often, similar studies will vary in size and a large study reporting more subjects and device uses would likely be considered higher value than a small study reporting only a few subjects and device uses. The evaluator will need to decide if all or only some of the articles should be included in the CER/PER. This is often a subjective decision based on the overall weight of the existing clinical data and an assessment if each individual article collectively adds value to the evaluation or not. Small studies with only a few subjects may be excluded if the subject device clinical data are already sufficient without these data and if the excluded article does not have novel clinical data compared to the clinical data already included.

For another example, several case studies may describe a variety of specific device failures resulting in rare "potential" device-related safety complications. Do not exclude these articles. Case studies with novel device-related information (e.g., rare risks and safety concerns) should always be retained and reported in the CER/PER. These device risks and safety concerns will also need to be included in the manufacturer PMS and RM systems to ensure proper accounting and risk mitigation activities occur, respectively, even if the novel device-related risks and safety concerns were documented in similar devices and not specifically in the subject/equivalent devices. For more information about how manufacturers identify, establish and control risks, perform benefit-risk analyses and establish risk acceptability thresholds, see ISO 14971:2019, "Medical devices—Application of risk management to medical devices." To identify if the information is novel or not, the evaluator must analyze the entire dataset and the detailed notes in the CER/PER spreadsheet or software system will be helpful.

Regardless of analysis method, the evaluator must determine if the critical clinical data for the S&P and Benefit-Risk Ratio for the subject device are sufficient to demonstrate conformity with the GSPRs 1, 6 and 8.

5.4.7 Literature analysis: benefits and risks

Benefits typically need to be established in multiple high-quality articles; however, some exceptions to this rule exist. For example, if a case series reported a device successfully treated a rare, fatal or debilitating disease with no other treatment options, the benefit could be allowed. A benefit found in an article evaluating an under-studied population might also be allowed. Simlarly, in a setting where most articles were about adults, but one article (for a device indicated for use in children) reported a benefit in children, then this pediatric benefit might also be allowed. In all of these cases, the manufacturer would need to collect more clinical data to confirm these benefits. The same is true for risks.

5.4.8 Litetature analysis: excluding lower quality articles

The analysis process is iterative and an individual article may have more or less value to the overall analysis as the analysis develops. This change in value may occur because the article had more clinical data in the full text than was obvious by appraising the abstract alone (i.e., the article with more clinical data typically has more weight) or because many other articles had the same type of data in a larger sample size (i.e., as clinical literature data reliability increases, the value of each individual article decreases). In addition, author bias and/or technical errors may not be identified in the abstract, but may become apparent after reading the full article (i.e., this bias will decrease the value of the clinical literature). Sometimes, the abstract may contain inaccurate or incomplete information with even more inconsistencies in the full article text (i.e., this will also decrease the value of the clinical literature). At other times, the abstract may claim a systematic review was performed, but the full text reveals a narrative review without the rigor required of a systematic review. Or, the abstract may not identify the specific device used, so the entire article must be dissected to determine if the subject/equivalent device was used, and the article will have less value if the specific devices used cannot be determined. During analysis, articles previously included during the appraisal step may be excluded as the article weight becomes more accurate and the CER/PER conclusions come into focus.

5.4.9 Identifying whether clinical literature data meet safety and performance acceptance criteria

After each article is analyzed and the literature are appropriately separated into the background/SOTA/alternative therapy/device and subject/equivalent device sections, the clinical data in each group must be analyzed to determine if the S&P acceptance criteria defined in the CEP/PEP were met. To determine if the subject/equivalent device is safe and effective, specific numbers relating to the S&P acceptance criteria need to be tabulated after they are extracted from every article. To do this work, the evaluator may need to calculate the specific S&P data from each article. Once all relevant clinical data are in the table, the evaluator can determine the overall data ranges and can easily identify any outliers in the table. Then, the detailed and pooled data points can be compared to the previously stated S&P acceptance criteria (Table 5.7).

TABLE 5.7 Thrombectomy catheter S&P acceptance criteria table.

Article	Device	*N*	Safety 1 <1% pulmonary embolism	Safety 2 <2% major bleeding	Performance 1 >70% clot removal	Performance 2 <50 mg lytic dose
Smith 2000	Arterial BestCatheter 2.0 F (City, State)	96	**2% (2/96)**[a,b]	0% (0/96)	96.9% (93/96)	28 mg
Johnson 2000	Arterial BestCatheter 2.0 F (City, State)	91	0% (0/91)	1.1% (1/91)	95.6% (87/91)	**55 mg**[b]
Smith 2010	Arterial BestCatheter 2.0 F (City, State)	45	NA	**11.1% (5/45)**[a]	NA	16 mg
Doe 2020	Arterial BestCatheter 2.0 F (City, State)	19	0% (0/19)	**10.5% (2/19)**[a]	94.7% (18/19)	NA
Range		251	0–2%	0–**11.1%**	94.7%–96.9%	16–55 mg
Mean[c]			*0.97% (2/206)*	***3.2% (8/251)***	*95.2% (198/206)*	*33 mg*
Pass?			**Maybe**[b]	**NO**	**YES**	**Maybe**[b]

[a]*Bold numbers are outside the acceptable range and the means are provided in italics*
[b]The outlier study should be carefully evaluated and the acceptance criterion re-evaluated to determine if these slightly out of specification data should continue to be seen as acceptance criteria failures due to the impact on the clinical S&P and the benefit-risk ratio.
[c]The mean of pooled data across diverse studies may not be appropriate.

Care must be taken when calculating overall averages using reported averages from many different studies. Apply the proper statistical methods to control error propagation. Simply averaging reported average values will not give the correct result. The evaluator must do the math. In addition, all outliers should be identified during analysis and discussed in the CER/PER. Failing to meet an S&P acceptance criterion is not a reason for exclusion.

BOX 5.9 Two examples of exclusion coding changes during analysis

Example 1: Two background/SOTA/alternative therapy/device articles studied thrombectomy catheters used to treat deep venous thrombosis in pregnant women. Upon reviewing the IFU, the evaluator noted the subject/equivalent device was contraindicated for use during pregnancy. Because the indication did not include use during pregnancy and because the two articles were not about subject/equivalent device use, the evaluator recoded the articles for exclusion and removed them from the CE. This may not be the best decision, since the background/SOTA/alternative therapy/device section is expected to provide a comprehensive background knowledge discussion including contraindications and off-label uses when appropriate. In this case, many pregnant women may suffer from DVT and other thrombotic events and they may be benefited by thrombectomy catheter use, so the background should explain why this use is contraindicated and report the data from these known off-label uses. On the other hand, if this contraindication and off-label use data were already definedand discussed based on the clinical data in other articles already included in the background/SOTA/alternative therapy/device section, then the exclusion of these articles may be the right choice.

Example 2: An article discusses the inaccurate COVID detection rate reported by an IVD device. After reading the full-length article, the evaluator discovered the device used was not the subject/equivalent device but rather a competitor device. The evaluator recodes the article to move the article from the PER subject/equivalent device section to the background/SOTA/alternative therapy/device section. Once grouped with the rest of the background/SOTA/alternative therapy/device articles, the evaluator learns many other included articles have presented this same conclusion and several of these other studies had much larger sample sizes. The evaluator docuemnted the article did not add value to the background/SOTA/alternative therapy/device section and recoded the article for exclusion.

These two examples describe the flexibility needed when completing the clinical data analysis of the entire data set after the initial clinical data appraisal is completed for the individual articles. Many other situations present themselves, and the evaluator must carefully evaluate each one while applying skill and expertise to decide if the clinical data should be included in a particular CER/PER section with more or less emphasis, if an article should be moved to a different section, or excluded altogether even though the initial appraisal kept the article included for full analysis.

The S&P acceptance criteria were determined and documented in the CEP/PEP process prior to beginning the CER/PER. During evaluation, the S&P clinical data from the background/SOTA/alternative therapy/device group act as a benchmark, allowing direct comparison with the subject/equivalent device S&P clinical data. The background/SOTA/alternative therapy/device analysis helps determine the subject/equivalent device S&P acceptance criteria validity and continued veracity. While S&P acceptance criteria should be based on specific prior subject/equivalent device and general background/SOTA/alternative therapy knowledge, once analyzed, the previously defined S&P acceptance criteria may not represent realistic values for the current clinical data under evaluation. If neither the background/SOTA/alternative therapy/device nor the subject/equivalent device clinical data are meeting a particular S&P acceptance criterion, the current background/SOTA/alternative therapy/device clinical data findings should be used to help explain the problem in the text and should serve as a guide to revise the next CEP/PEP.

CEP/PEP revisions should be carefully version controlled prior to starting the next CER/PER. The CEP/PEP should describe the issues encountered during analysis in the CER/PER rather than changing the plan to fit the evaluation results. Changing the plan during the evaluation may be considered untruthful, misleading, and/or unethical.

5.4.10 Interpreting the clinical literature data

Once the quantitative analysis is complete, the evaluator must determine what the analysis means and the evaluator must draw conclusions about the subject/equivalent device S&P as well as the benefit-risk ratio and the GSPR expectations. All conclusions must be sufficiently supported by clinical data and any outliers or deviations must be discussed with all clinical data gaps identified, listed, and a commitment to address the clinical data gaps specifically in the next PMCFP/PMPFP must be provided. In general, the evaluator should use an organized and systematic approach to produce a consistent, reproducible and reliable clinical literature data analyses (Table 5.8).

TABLE 5.8 Clinical literature analysis summary.

1	Read and analyze all relevant clinical data in each article.
2	Document all relevant clinical data in CER/PER spreadsheet or software.
3	QC all included background/SOTA/alternative therapy/device data in spreadsheet or software.
4	QC all included subject/equivalent device data in spreadsheet or software.
5	Calculate/transform clinical data for S&P acceptance criteria analysis in spreadsheet or software.
6	Compare and document clinical data similarities and differences by study type, population, etc.
7	Determine if clinical data meet prespecified S&P acceptance criteria for each relevant group and overall.
8	QC all data calculations/transformations and S&P decisions.
9	Document all clinical data gaps identified during device S&P.
10	Complete benefit-risk ratio analyses and QC all work: are GSPRs 1, 6, and 8 met?

Clinical literature analysis requires the evaluator to read and understand every included article. The evaluator must identify and abstract all relevant CER/PER clinical data into a spreadsheet or other software system to facilitate data comparison and analysis. The clinical data may also need to be "calculated" or "transformed" if the clinical data in the article do not allow direct comparison to the clinical data details required in the S&P acceptance criteria or the benefit-risk ratio. The evaluator must be vigilant to seek "sufficient" data to demonstrate unequivocal compliance and conformity to GSPRs 1, 6, and 8 regarding device S&P, as well as benefit-risk ratio across the entire device lifetime and for all indicated uses.

5.5 Clinical experience analysis

Clinical experience data are often difficult to analyze because so few actual clinical data are included in these "real-world" experience reports. In addition, clinical experience data typically have no QC, correction, or follow-up possible to improve the clinical data quality or reliability. Furthermore, the clinical data available in the postmarket surveillance report (PMSR)/periodic safety update report (PSUR) may not have the needed details to complete a robust clinical experience data analysis.

The clinical experience data evaluation typically includes both device issues and patient harms, as well as potential harms associated with the subject device. The clinical experience data may include SOTA devices as necessary for a comprehensive review of the device type when used in the field as indicated. Detailed S&P data should be collated and objectively evaluated to document the individual and overall benefits and risks for the subject device as well as similar or alternative device types or therapies. In addition, a comparison to data reported in clinical trials and the clinical literature should be performed, and the similarities and differences between these different data types should be discussed.

5.5.1 Experience analyses: single experiences

Each clinical experience type (i.e., company-held complaint data, Manufacturer And User facility Device Experience (MAUDE) or Database of Adverse Event Notifications (DAEN) database reports, warning letters, Total Product Life Cycle (TPLC), recalls, etc.) requires a separate analysis for patient problems (i.e., AEs) and device problems (i.e., DDs, malfunctions). The evaluator must read and analyze each experience report completely to abstract all relevant S&P and Benefit-Risk clinical data while taking notes to prepare for tabulation and writing the experience data narrative summary.

Some experience data such as complaints may be numerous; however, other experience data, like recalls, may be limited. The evaluator should use spreadsheet and software calculators and formulas to generate accurate accounting when combining experience data from multiple sources. The evaluator should tabulate clinical experience data appraisal groups (e.g., complaints, MAUDE, DAEN) and issue types (e.g., death, infection, rash for AEs and device break, missing component, labeling issue, loss of sterility for DDs) for the CER/PER. Two tables are needed for each dataset: one for patient problems (AEs) and one for device problems/DDs.

Every experience report should be read and assigned a specific patient harm (i.e., AE/complication or report documenting harm or additional risk to the patient) code and a specific device problem (i.e., DD or report documenting device malfunctions or other device-related problems) code representing the highest-risk AE and DD reported. The evaluator should code all AEs and DDs found within each experience report during the first pass through the experience data, and the evaluator should include unique codes for any and all potential AEs and DDs in this coding process. Experience report analysis depends on grouping similar event types together in order to find patterns in the data. When multiple issues occur in the same report, the higher-risk item should be listed first in results tables, discussions, and conclusions. Each report can document an AE, a DD or both. AEs and DDs need to be analyzed separately because they represent different risks. AEs and DDs do not perfectly overlap. Some AEs are caused by DDs, but not all, and some DDs may result in AEs, but not all. Tracking AEs and DDs separately avoids confusion and double counting issues.

For example, a report might describe fluid leaking from a tube. The user inspected the tubing and discovered a crack. Both leak and break (crack) should be listed in the spreadsheet. The evaluator should decide if the DD should be categorized primarily as a leak or primarily as a break or if leaks, breaks, and cracks should always be lumped together for this device.

The process for coding AEs is similar; however, AE appraisal can be challenging from the clinical perspective. For example, a needle stick represents an AE even when no disease transmission occurs. An evaluator analyzing hypodermic needle experience reports may want to consider both the actual needle stick as well as the secondary potential exposure to a pathogen with potential disease transmission (i.e., seroconversion) via the needle stick as separate AE codes (i.e., the disease transmission may require treatment and may be much more serious than a simple needle stick).

For example, surgical revision, device explant and any need for additional surgery or extended surgical/procedure times are all AEs even though surgical revision or device explant at the expected device lifetime end may be expected AEs. The evaluator will want to consider expected AEs separately from unexpected AEs. In addition, the evaluator will want to consider AEs from three different perspectives: relatedness (i.e., to the device), seriousness (i.e., death/hospitalization/congenital anomaly, etc.), and expectedness (i.e., expected AEs will be defined in the product literature along with acceptable mitigations to manage these risks).

BOX 5.10 MAUDE analysis example

A MAUDE report included the following text "an orthopedic surgical screw broke during placement with resulting bone damage." No further data were available. The spreadsheet recorded the DD appraisal coding as SURGICAL SCREW BREAK and the AE appraisal coding was BONE DAMAGE. This report was tabulated in the DD table as a "break" and in the AE table as "bone damage."

(Continued)

BOX 5.10 (Continued)

Note: Another MAUDE report described an orthopedic surgical screw breaking before use in any patient and this would be assigned a "break" DD code and an "NA" (not applicable/available) AE code. This rerport would be accounted for in the DD table but not the AE table. This experience data point may *not be relevant if the broken screw did not have any clinical impact on the patient (i.e., if the surgery was not delayed, etc.) and if not relevant,* this clinical experience report *could probably be recoded for exclusion since no clinical data were present in the report.*

MAUDE search results include "tags" for AEs ("Patient Problems") and DDs ("Device Problem") as selected by the person who reported the issue in the database. One database entry description (Fig. 5.2) for an implanted cardiac defibrillator stated: "This device was explanted due to eri with unexpected battery behavior. No adverse patient events were reported" and the report was tagged with "Device Problem: Premature Elective Replacement Indicator" and "Patient Problem: No Clinical Signs, Symptoms or Conditions." However, this would not be a correct "Patient Problems" analysis since the patient underwent additional surgery (i.e., was exposed to additional risk), and one AE code for the report could be "additional surgery required" or "early explant" depending on which code is more useful for constructing AE (i.e., patient problems) group codes for this device.

BIOTRONIK SE & CO. KG ILIVIA 7 VR-T DF4 PROMRI ICD Back to Search Results

Model Number 404626
Device Problem Premature Elective Replacement Indicator (1483)
Patient Problem No Clinical Signs, Symptoms or Conditions (4582)
Event Date 12/28/2022
Event Type malfunction
Event Description
This device was explanted due to eri with unexpected battery behavior. No adverse patient events were reported. Should additional information become available, this file will be updated.

FIGURE 5.2 MAUDE report summary.

This example illustrates why the evaluator should not rely on these database "tags" when grouping and coding clinical experience data (i.e., because they lack consistency between the individual reports in the database and informed AE analysis for untoward events encountered by patients in this clinical experience dataset).

During the first pass through the clinical experience dataset, all identifiable AE and DDs are extracted and grouped to consolidate multiple specific AEs/DDs for further analysis, discussion, and conclusions (Table 5.9).

TABLE 5.9 Experience report appraisal example.

Event description	AE (Y/N)	AE types	AE group	DD (Y/N)	DD types	DD group
"This device was explanted due to eri with unexpected battery behavior. No adverse patient events were reported..."	Y	Additional surgery required (explant), early device failure	Additional surgery required	Y	Battery malfunction	Malfunction
"Device remains implanted. Device went into premature eri. No adverse patient events reported..."	Y	Early device failure (not explanted)	Early device failure	Y	Battery malfunction	Malfunction
"This system was explanted due to infection. A new system has not been implanted at this time..."	Y	Infection, additional surgery required (explant)	Infection	N	NA	NA
"This device was being preprogrammed before an implant and the device would not interrogate. The analysis revealed damage to the device."	N	NA	NA	Y	Damage, failure to interrogate	Damage

Once all reports have been coded, AEs and DDs are tabulated in the CER/PER separately based on the assigned codes to document the event volumes in each group.

The evaluator can only use the clinical data provided in the experience report. For example, when analyzing clinical experience data for an N95 face mask, the AE for "Allergic Reaction" was noted in a report where the user documented "redness, burning, and itching around my nose, cheeks, and chin where the mask had made contact" and "Potential chemical contamination" was entered for DD (Table 5.10).

TABLE 5.10 Example N95 face mask MAUDE analysis worksheet.

Event description excerpt	Clinical harm/AE	Device problem/ DD
"... I tried the mask on, and within a few minutes developed redness, burning, and itching around my nose, cheeks, and chin where the mask had made contact... dermatologist... advised this is most likely an allergic reaction or chemical burn. I have previously had a sensitivity to nickel, and the reaction and progression to other areas of my body with previous nickel sensitivity from nickel jewelry make me think this mask was contaminated with nickel, and the dermatologist concurs this is likely... I wear other N95 masks with no problem.	Allergic reaction	Potential chemical contaminant
RN was being fit tested by educator and when she went to put on a small N95 mask, one of the straps broke	NA	Strap break
Two employees changed out small duck bill masks at 0700. By 0945, both employees' masks had become unusable due to straps snapping off of the mask, with normal donning and doffing.	NA	Strap break
The respirator valve was not sealing appropriately.	NA	Inadequate seal
Severe shortness of breath...	Dyspnea	NA
While providing patient care in the emergency room, the piece of elastic of the... N95 mask broke causing the mask to fall off... two other nurses reported the same problem, fortunately they were not providing patient care at that time. Three days later, two RNs reported the same problem; one while providing patient care. The nurses describe the elastic ear pieces as "stretching out" and "breaking." nurses on the patient units have made similar complaints... the... n95 face portion loses its seal.	Exposure when mask fell off	Strap break; Inadequate seal

Another MAUDE report for an N95 face mask included the following event description: "Event Description: RN was being fit tested by educator and... one of the straps broke." A possible DD code was "Strap break." The exact code should be descriptive and the same code should be used for all other reports involving broken N95 face mask straps. Consistent coding makes sorting and tabulating results easier during analysis. As the evaluator goes through the coding, the description and code may need to change to better accommodate the clinical data being evaluated. In addition, in this particular report, no patient harm was described. The description specified the break occurred during fit testing and not while treating patients so the clinical harm/AE for this issue was coded as "NA" (not applicable/ available).

One MAUDE report with the description "while providing patient care... the piece of elastic of the... N95 mask broke causing the mask to fall off" used the same DD code "Strap break"; however, the user was treating a patient and could have been exposed to a contagion when the mask failed to function as intended. This "Exposure when mask fell off" was coded as an AE due to the increased risk for either the patient or the user. The evaluator is not attempting to determine if actual harm occurred or to what extent a patient or user was injured. Rather, the AE simply documents the known AE and increased patient or user risk even if the exposure does not result in disease transmission (e.g., infection risk alone is enough to qualify as an AE).

In addition to MAUDE reports, recalls may have useful clinical experience data. The FDA Medical Device Recalls database at https://www.accessdata.fda.gov/scripts/cdrh/cfdocs/cfRES/res.cfm was searched for any "N95" product recalls, and four recalls were identified. Every recall report was read and assigned a specific recall type (Table 5.11).

TABLE 5.11 Example N95 face mask recall analysis worksheet.

Recall number	Product description	Trade name	Manufacturer recall reason	Recall type
Z-0309-2021	"... decontaminating compatible N95 respirators... for multiple-user... reuse by healthcare personnel (HCP)... to prevent exposure to pathogenic biological airborne particulates when there are insufficient supplies of Filtering Facepiece Respirators (FFRs) resulting from the Coronavirus Disease 2019 (COVID-19) pandemic."	Battelle Critical Care Decontamination System (CCDS)	Masks processed at one site were not maintained at levels of condensation during a portion of the decontamination cycle.	Faulty decontamination procedure
Z-3024-2020	JINYINSHAN, KN95 PROTECTIVE MASK P1, 20 pieces	JINYINSHAN, KN95 PROTECTIVE MASK	Test results revealed the KN95 masks failed to filter greater than 95% of particulates.	Failed to filter particulates
Z-1576-2009	Inovel N95 HealthCare Particulate Respirator, Model/Catalog Number: FRN95-SEZ, Expiration Date: 07/13/2010.	INOVEL HEALTH CARE N95 PARTICULATE RESPIRATOR	Display boxes labeled as FRN95-MLEZ, Medium/Large Size Respirators were inadvertently mixed with Display Boxes labeled for FRN95-SEZ Small Size Respirators.	Incorrect labeling
Z-0587-06	Particulate Respirator and Surgical Mask, Inovel 3000 series Healthcare N95 Model Number/s: 3002N95-M, Lot Number 051130-161.	Inovel 3000 series Healthcare N95 Particulate Respirator and Surgical Mask	Mislabeling-A customer reported mislabeled product display boxes. Review of complaint samples and inventory confirmed some product display boxes were labeled with incorrect model/size number.	Incorrect labeling

In this recall example, two recalls were coded for incorrect labeling. This process of assigning descriptions or codes to the clinical experience data facilitates grouping clinical experience data together for overall analysis after the individual clinical data have been analyzed.

The US FDA TPLC database combines data from several FDA databases including the MAUDE and Medical Device Recalls databases, and the TPLC database outputs all data for a particular product code. The TPLC database tabulates all MAUDE device problem and patient problem tags for each product code, but does not provide device specific information. For example, implantable cardioverter defibrillators are assigned the product code LWS and searching TPLC for LWS provides the combined data for more than 10,000 devices since 2008 with product code LWS (Fig. 5.3).

New Search   show TPLC since 2008   Back to Search Results

Device Implantable Cardioverter Defibrillator (Non-crt)
Definition These devices treat tachycardia (fast heartbeats) with RV defibrillation therapy as necessary.
Product Code LWS
Device Class 3

Premarket Approvals (PMA)															
2008	2009	2010	2011	2012	2013	2014	2015	2016	2017	2018	2019	2020	2021	2022	2023
94	80	119	136	129	148	168	100	145	162	142	98	125	111	91	3

MDR Year	MDR Reports	MDR Events
2013	17237	17237
2014	18593	18593
2015	14042	14042
2016	16635	16635
2017	18453	18453
2018	17238	17238
2019	16189	16189
2020	16912	16912
2021	19675	19675
2022	23500	23500

Device Problems	MDRs with this Device Problem	Events in those MDRs
Over-Sensing	40092	40092
High impedance	30518	30518
Adverse Event Without Identified Device or Use Problem	28228	28228
Inappropriate/Inadequate Shock/Stimulation	19999	19999
Premature Discharge of Battery	17500	17500
Signal Artifact/Noise	14732	14732

Patient Problems	MDRs with this Patient Problem	Events in those MDRs
No Known Impact Or Consequence To Patient	73613	73614
No Clinical Signs, Symptoms or Conditions	30386	30386
No Consequences Or Impact To Patient	19635	19635
Unspecified Infection	17697	17697
Shock from Patient Lead(s)	11460	11460
No Code Available	5500	5500
Electric Shock	4531	4531
Death	3085	3086
Sepsis	2838	2838
Pocket Erosion	1519	1519
Syncope	1402	1402

FIGURE 5.3 Example TPLC output.

TPLC breaks reported problems into separate patient and device problems (i.e., MAUDE tags) and collates totals for all reports filed for every device with a particular product code by year. The AE and DD lists within TPLC allow the evaluator to consider if the AEs and DDs from the subject device broadly align with similar devices in this product code.

5.5.2 Experience analyses: grouped and as a whole

The individual clinical experience data analyzed and tabulated in the complaints, AEs, DDs, and recalls tables also need to be grouped together, analyzed, and interpreted overall to draw conclusions. The clinical experience data in each table should be organized in a logical flow pattern (e.g., from most common to least common problems, or from most serious device-related harm to least worrisome device-related harm). The evaluator should document how the clinical experience data impact the PMS and RM systems (e.g., if any clinical experience data are new and not mentioned previously, they will need to be analyzed fully within the PMS and RM systems, as well as within the CE/PE system). In addition, if the clinical experience data documents ongoing treatments or unresolved issues, the clinical experience data will need to be fully considered and documented in the PMCFP/PMPFP.

The AEs from the entire clinical experience data worksheet are grouped by the most common AE code (i.e., injury group) and tabulated for use in the CER/PER. For example, 41 N95 face mask MAUDE reports were analyzed. (Table 5.12).

TABLE 5.12 Example N95 face mask MAUDE AEs table.

Injury group	Count
Insufficient information, NA	19
Exposure due to mask failure	6
Contracted COVID-19 or potential COVID-19 infection	6
Irritation/allergic reaction (e.g., redness, rash, swelling)	4
Nausea, headache, sore throat, SOB, potential chemical exposure after N95 mask decontamination	4
Potential green dye transfer	1
Self-limited cardiac arrest after bradycardia	1
Total	**41**

In addition, the DDs from the entire clinical experience data worksheet were grouped by the most common DD code (i.e., device problem group) and tabulated for use in the CER/PER (Table 5.13).

TABLE 5.13 Example N95 face mask MAUDE clinical experience DD table.

Device problem group	Count
Insufficient information, no device problem, NA	12
Strap break	11
Poor quality (did not meet NIOSH standard, no seal, fogged glasses)/recall (counterfeit product)	9
Potential chemical residue after decontamination/green dye transfer	3
Mislabeled (not FDA cleared)	6
Total	**41**

FDA, Food and Drug Administration; *NIOSH*, National Institute for Occupational Safety and Health.

After the abstracted individual clinical experience data were grouped, put into tables, and analyzed individually and overall, the QC process should verify all data in the worksheet or software and in all tables. If recalls are documented, the overall recall data should be analyzed for the recall reason, effect on patients, device cause, action taken by manufacturer to correct action, quality of recall report, and resolution.

The aggregated data found in the TPLC data should also be grouped and analyzed separately for patient problems (i.e., AEs) and device problems (i.e., DDs). For example, teh evaluator should group and report the following AE terms together since the CE/PE has no need to detail issues without clinical data:

- no known impact or consequence to patient;
- insufficient information;
- no clinical signs, symptoms, or conditions;
- appropriate clinical signs, symptoms, conditions term/code not available;
- no patient involvement;
- no code available; and
- no consequences or impact to patient.

The evaluator should create a code to group each TPLC MDR report about a patient problem (Table 5.14).

TABLE 5.14 Example Coding TPLC patient problems into groups.

Code	Patient problem	MDRpts with this patient problem	Events in MDRpts
A	No known impact or consequence to patient	23	23
A	Insufficient information	15	15
A	No clinical signs, symptoms, or conditions	11	11
B	Rash	9	9
B	Hypersensitivity/allergic reaction	8	8
NA	[a]Etc.	NA	NA
NA	[a]TOTAL	NA	NA

[a]Table abbreviated: additional problems exist but were not listed here; therefore, the total was not calculated here.

In this example, code A has grouped "no clinical data" and B will group "allergic reaction" including symptoms which fit into the "allergic reaction" group. This grouping by code allows tabulated data to be more easily analyzed and summarized in the CER/PER using a key to describe the issue type (Table 5.15).

TABLE 5.15 Example TPLC patient problems table.

Code	Patient problems	Count
A	No known impact/insufficient information/no clinical Signs, symptoms, or conditions	49
B	Rash/hypersensitivity/allergic reaction	17
C	Etc.[a]	NA
	TOTAL[a]	NA

[a]Table abbreviated: additional problems exist but were not listed here; therefore, the total was not calculated.

A separate table should represent the TPLC device problems. For example, the evaluator grouped "no known impact," "no patient complication," and "no adverse event" into one group for analysis. Again, the grouping by code allowed the TPLC data to be easily analyzed and summarized in the CER/PER using a key to describe the device problem type (Table 5.16).

TABLE 5.16 Example TPLC device problems table.

Code	Device problems	MDRpts	Events
A	Insufficient information, appropriate term/code not available	4	4
B	Break/detachment/loose or intermittent connection/material split, cut or torn	27	27
C	Etc.[a]	NA	NA
	TOTAL[a]	NA	NA

[a]*Table abbreviated: additional problems exist but were not listed here; therefore, the total was not calculated.*

Once every TPLC report has been assigned a category, analysis can begin for each group. As with complaint reports, TPLC data needs to be interpreted and discussed in the CER/PER. Questions to consider include: what was the most common patient problem? What were the most serious patient problems (e.g., death)? Were the device problems related to device malfunction?

After the individual and grouped experience reports were analyzed, the experience data should be considered as a whole. Based on MAUDE/DAEN, TPLC, and PMS outputs, what were the most common patient problems and device problems? Were those issues properly accounted for in the benefit-risk ratio? What is being done to mitigate observed issues? How does the PMCF plan address any observed clinical data gaps related to the clinical experience data?

5.6 Identifying benefit-risk ratio and PMCF needs

The benefit-risk ratio should document how the identified benefits outweigh the patient risks when using the device as indicated. This documentation should align with and satisfy the MDR/IVDR GSPRs 1, 6, and 8. The evaluator must consider all device characteristics, AEs, DDs and S&P results from all clinical data sources (trials, literature, and experiences) for all patients and other users/caregivers. All probable benefits and risks should be documented and evaluated individually and then as a whole. An acceptable, overall benefit-risk ratio requires the risks to be mitigated and reduced as low as possible, and the benefits to outweigh the risks to the patient.

The evaluator must also determine if the beneit-risk ratio data are sufficient and must identify all clinical data gaps. Once identified, the CER/PER must document how the lack of sufficient data and/or all clinical data gaps will be addressed in the PMCFP/PMPFP with specific associated clinical data gathering details and deliverable dates.

BOX 5.11 Sufficient clinical data analysis

Most manufacturers struggle to define the "sufficient" amount of clinical/performance data. Often, the manufacturer wants to know the "lower limit" or the minimal clinical data amount required to generate a "sufficient" CER/PER. Even though this lower limit can be discussed, this should not be the goal or an excuse to seek the minimum CE/PE data amount.

Tips

- Do not exclude any medical-device-related clinical data until the exclusion reason for each medical-device-related clinical data point is completely clear and justified.
- Be absolutely certain all medical-device-related S&P data are explicitly defined with no open questions.
- The clinical data must support the medical-device claimed benefits convincingly.
- The clinical data must show the medical device actually works as promised without fail.
- Keep weighing and reweighing medical-device-related benefits and risks during analysis until the balance is clear.
- List each medical-device-related benefit clearly.
- List each medical-device-related risk clearly.
- Explain the relative weights for each medical-device-related benefit and risk.
- Do not settle for a weak story—the medical-device-related clinical data (i.e., evidence) should be compelling.
- Ensure all medical-device-related details are addressed in the clinical data analyses.
- Make the medical-device-related clinical data messages as clear and concise as possible.
- Double-check accuracy and consistency for all clinical data .
- Sufficient clinical data about the medical device will tell the complete story.

Sufficient, high-quality data are required to support device safety and performance for each indication during use).

According to MEDDEV 2.7/1 rev. 4 [5], clinical data used to determine conformity "need to be in line with current knowledge/the state-of-the-art, be scientifically sound, cover all aspects of the intended purpose, and all products/models/sizes/settings foreseen by the manufacturer." If the available clinical data lacks sufficient quality and/or quantity for any indication or subject device use area, the evaluator should document the clinical data gap in the CER/PER and then design new clinical data gathering to address the clinical data gap in the PMCFP/PMPFP. A few clinical data gap examples are listed below (but clinical data gaps are not limited to these examples):

1. Pediatric clinical data were needed: A device indicated for use in adult and pediatric patients had extremely limited clinical data identified to address the S&P or benefits and risks associated with pediatric use (i.e., no clinical trials were completed in children, only a few small, retrospective studies in <50 children were available in the clinical literature and no clinical experiences reported in children). A clinical data gap was identified for device use in children and more clinical data were needed to evaluate the S&P and benefit-risk ratio during the device use in children. The manufacturer considered changing the indication for use to adults only; however,this device had significant benefits for children outweighing all risks, and the manufacturer decided to collect pediatric use data in the PMCFP/PMPFP as soon as possible to preserve the broad indication for use in adults and children.
2. All device models or versions needed clinical data: Sufficient clinical data was not available for each device model or version (this assessment was made after appropriate device groups and samples were considered). The PMCFP/PMPFP needed to collect clinical data for any identified models or versions without sufficient clinical data.
3. Devices with rare or serious side effects needed clinical data: A device with rare or serious device-related side effects may needed larger datasets over a longer time to accumulate sufficient clinical data to fully define the rare or serious AEs. Each device-related AE was evaluated in the context of the S&P acceptance criteria and the benefit-risk ratio and a clinical data gap was recognized. The PMCFP/PMPFP was designed to collect sufficient, high-quality data to examine these rare and serious (e.g., life-threatening) side effects. The PMCFER/PMPFER was able to identify and define all AEs, risks and hazards with a special focus on these specific, identified rare and serious side effects.
4. Devices with device-related complaints needed clinical data: Several new complaints were received about diabetic patients developing infections when using a device unrelated to diabetes treatment. The evaluator searched to identify device-specific clinical data from diabetic patients and from patients who had infections. None of the identified clinical trials, literature and experience data collections reported device uses in diabetic patients and no prior infections were reported, so the evaluator identified this situation as a clinical data gap and the AE for infection specific to diabetic patients was documented needing additional clinical data collection ASAP. The new infection risk for diabetics and plans for risk mitigation were discussed in the CER and the PMCFP was designed to collect the data collection aimed atneeded to determine and mitigate the risk for this diabetes patient population specific to the infection.

The overall CER/PER conclusions must demonstrate how the subject device achieved the expected performance when used as intended, how the device risks were acceptably balanced with patient benefits, and how the continued safe and effective device use was supported by sufficient clinical evidence. The residual risk acceptability must be discussed along with the appropriate PMS and RM procedures and plans. In addition, the executive summary should be an accurate and concise one-page CE/PE summary of the individual detailed document sections. In the best cases, the executive summary will provide clear, final analyses findings with sufficient, top-level data to support all conclusions about the continued safe and effective device use in conformity with all current regulations and standards.

CERs/PERs need to be rewritten every 1–5 years, depending on device risk. Over time, new clinical data collected and analyzed in the PMCFER/PMPFER will fill in the previously identified clinical data gaps until sufficient data exists for all populations and indications. Clinical data analysis is essential to identify and fill in clinical data gaps over time as required under the EU MDR and IVDR.

5.7 Clinical data analysis by Clinical Evaluation/Perfromance Evaluation type

Different CE/PE types will require specific evaluator skills with different types of analyses.

5.7.1 Viability clinical evaluation / viability performance evaluation

When analyzing clinical data in the viability setting (before any device-specific clinical data have been generated), the evaluator must keep the clinical data analysis focused on the critical differences between the future device and the comparator clinical data being evaluated in lieu of the subject/equivalent device.

5.7.2 Equivalence clinical evaluation / equivalence performance evaluation

When an equivalent device is representing the subject device, the evaluator must analyze the clinical impacts of the differences between the subject and equivalent devices in each section of the CER/PER.

5.7.3 First-in-human clinical evaluation / first-in-human performance evaluation

When the first in human clinical data become available, the evaluator must be careful to analyze the clinical data for the subject device separately from the original viability data or the data from the equivalent device.

5.7.4 Traditional clinical evaluation / traditional performance evaluation

When the device has had many human uses, the evaluator must focus the analysis on the clinical data specific to the subject device for the S&P and benefit-risk ratio (if possible).

5.7.5 Obsolescence clinical evaluation / obsolescence performance evaluation

As the device is obsoleted, the evaluator should focus the analysis on the new S&P and the new benefit-risk ratio for the device after the useful life has been exceeded. The clinical data volume will be getting smaller over time as fewer obsoleted devices are being used, and little to no comparable data may be available.

5.8 Evaluator expertise is required

Due to the iterative, complex clinical data analysis requirements, people with postgraduate degrees are typically the most successful evaluators. Specifically, PhD preparation involves training in the research process, reading and analyzing large volumes of literature and creating new information through research and analysis. Thus, a PhD-prepared professional is ideal. People with extensive experience reading and analyzing data such as those who have written systematic literature reviews and metaanalyses in the past may also be readily capable. The challenge with clinical/performance data analysis is having sufficient training and experience to do this work well. Clinical trials, literature, and experience data vary greatly in type, quality, and relevancy to the target medical device. Understanding this variation and studying these differences takes time and dedication to the detailed analysis tasks for both individual clinical data points as well as the clinical data overall.

In particular, the MEDDEV 2.7/1 rev. 4 guideline [5] states:

> *"clinical evaluation should be conducted by a suitably qualified individual or team"* and evaluators should *"possess knowledge of the following:*
>
> - *research methodology (including clinical investigation design and biostatistics);*
> - *information management (e.g., scientific background or librarianship qualification; experience with relevant databases such as Embase and Medline);*
> - *regulatory requirements; and*
> - *medical writing (e.g., post-graduate experience in a relevant science or in medicine; training and experience in medical writing, systematic review and clinical data appraisal)*".

Appraisal decisions for inclusion and exclusion criteria often need to change as the evaluator learns new knowledge related to the device S&P and benefit-risk ratio. Additionally, a large volume of clinical trials, literature, and experiences needs to be read, understood, and analyzed. Clinical evaluation efficiency is only gained through experience with clinical data. This experience is essential and patience is required since the analysis takes a good deal of time (with the effort often measured in years or decades).

The CER/PER is a stand-alone document and should not be dependent on previous CER/PER versions. The conclusions drawn must be specific and therefore are often expected to be different from the prior CER/PER versions, particularly if additional new clinical data were included and analyzed in the evaluation. In addition, an evaluator may conclude clinical data were misinterpreted by a previous evaluator. If the identified misinterpretation alters the S&P acceptance criteria, benefit-risk ratio, or other critical evaluation outcomes, the misinterpretation may affect the CE/PE conclusions and changes to correct the error/s must be addressed in the CER/PER as well as the related PMS and RM systems.

The evaluator should know how to update the clinical data analysis and conclusions. The evaluator should clearly describe the misinterpretation, the methods and all changes made to correct the misinterpretation, the justification/rationale for the new clinical data analyses and updated conclusions and whether a CAPA was initiated to address the clinical data misinterpretation issues in the future. The key to addressing prior clinical data misinterpretations is transparency. If no new data has become available, but the conclusions and benefit-risk ratio have changed, the evaluator needs to clearly state what changed including why and how the changes impacted the CE/PE overall.

5.9 Conclusions

Clinical data are evaluated to draw conclusions about device S&P and benefit-risk ratio. All relevant clinical data must be analyzed in a comprehensive and reproducible manner. Clinical trials, literature, and experience data all need to be analyzed by a competent, trained, and experienced evaluator. First, each individual trial, article, or experience report needs to be analyzed to determine what subject device S&P and benefit-risk ratio clinical data are present in each individual clinical data point. Then, each clinical data group needs to be tabulated, organized, and analyzed as a whole. For trials and literature, this step involves determining whether the S&P acceptance criteria listed in the CEP/PEP are met. If clinical data gaps are identified, the gaps need to be listed, prioritized and assigned to a specific PMCFP/PMPFP activity and timeline. For clinical experience data, analysis involves grouping similar events together, then tabulating/analyzing and drawing conclusions about these real-world use risks.

The analyzed clinical data must be accompanied by narrative discussion in the CER/PER text to explain the data and all conclusions. Additionally, the CER/PER reviewer (i.e., the NB inspector) must be able to follow the analysis and reproduce the work if necessary. CE/PE is fundamentally about analyzing medical device use in humans to determine and to document in writing whether the S&P acceptance criteria were met and to document how the benefits outweigh the risks when the device was used as indicated. Clinical data evaluation requires data to be quantified, contextualized, and explained. In short, the clinical data need to be fully analyzed and the conclusions need to be derived directly from the clinical data analysis documented in the CER/PER.

5.10 Review questions

1. What do the acronyms SVR, APR, and CPR mean?
2. When are CER/PER clinical data sufficient to document conformity with GSPRs 1, 6, and 8?
3. How are the S&P acceptance criteria in the CEP/PEP analyzed to meet the CER/PER requirements?
4. How is the benefit-risk ratio developed during CE/PE clinical data analysis?
5. What "scientific method" steps are needed when analyzing clinical data?
6. What is "analysis" and what are some actions often confused with analysis?
7. What are some useful data abstractions from each clinical data document to put into the workbook or software for analysis?
8. What is the difference between "diagnostic specificity" and "diagnostic sensitivity"?
9. What do the terms "predictive value" and "calibrator" mean?
10. How are the clinical analyses in the CER and PER different?
11. True or false? Only "gold standard" clinical data are analyzed in the CER/PER.
12. What steps are involved when analyzing clinical data?
13. What is the analysis purpose in the CER/PER?
14. Why does the clinical data analysis process need to be "iterative"?
15. What are "safety" and "performance" data?
16. What are some ways to demonstrate analysis in CERs/PERs for readers?
17. What needs to be done if clinical data gaps in CER/PER are found after analysis is completed?
18. What makes analyzing clinical data particularly difficult?
19. What information should be extracted from clinical experience reports?
20. What is a CAPA?

References

[1] European Parliament and Council of the European Union. Consolidated text: Regulation (EU) 2017/746 of the European Parliament and of the Council of 5 April 2017 on *in vitro* diagnostic medical devices and repealing Directive 98/79/EC and Commission Decision 2010/227/EU. Accessed May 11, 2024. https://eur-lex.europa.eu/legal-content/EN/TXT/?uri = CELEX%3A02017R0746-20230320.

[2] European Parliament and Council of the European Union. Consolidated text: Regulation (EU) 2017/745 of the European Parliament and of the Council of 5 April 2017 on medical devices, amending Directive 2001/83/EC, Regulation (EC) No 178/2002 and Regulation (EC) No 1223/2009 and repealing Council Directives 90/385/EEC and 93/42/EEC. Accessed May 15, 2024. https://eur-lex.europa.eu/legal-content/EN/TXT/?uri = CELEX%3A02017R0745-20230320.

[3] National Institutes of Health, National Cancer Institute, Genomic Data Commons Documentation. Clinical Data. Accessed May 11, 2024. https://docs.gdc.cancer.gov/Encyclopedia/pages/Clinical_Data/.

[4] European Medicines Agency (EMEA). ICH Topic E3: Structure and Content of Clinical Study Reports (July 1996, CPMP/ICH/137/95, © EMEA 2006). Accessed May 15, 2024. https://www.ema.europa.eu/en/documents/scientific-guideline/ich-e-3-structure-content-clinical-study-reports-step-5_en.pdf.

[5] European Commission. Medical device directives. Guidelines on medical devices. Clinical evaluation: a guide for manufacturers and notified bodies under Directives 93/42/EEC and 90/385/EEC. Med Dev 2.7/1 rev 4. June 2016. Accessed May 11, 2024. https://www.medical-device-regulation.eu/wp-content/uploads/2019/05/2_7_1_rev4_en.pdf#page = 20&zoom = 100,27,785.

Chapter 6

Establishing clinical benefit-risk ratios

A benefit-risk ratio framework has been used in pharmaceutical development for many decades and general risk management (RM) concepts are not new to medical device manufacturers. Unfortunately, the clinical benefit-risk ratio was simply not required for medical device manufacturers, even those device companies following ISO 14971, the international standard for medical device RM. Most medical device RM systems simply documented and quantified risks without considering any benefits and most medical device companies had never used a formal clinical benefit-risk ratio evaluation in their RM systems before 2017 when the European Union (EU) Medical Device Regulation (EU Reg 2017/745)/In Vitro Diagnostic Device Regulation (EU Reg 2017/746) specifically required an acceptable clinical benefit-risk ratio documented during the clinical evaluation/performance evaluation (CE/PE) process for all medical devices on the market in the EU.

The process for determining the benefit-risk ratio was defined as follows:

> *'benefit-risk determination' means the analysis of all assessments of benefit and risk of possible relevance for the use of the device for the intended purpose, when used in accordance with the intended purpose given by the manufacturer* (MDR [1] Article 2 (24); IVDR [2] Article 2 (17))

After these new regulations were released, ISO 14971 was updated in 2019 [3] to emphasize the need to record and quantify all anticipated benefits and required a balance between the known residual risks and the anticipated benefits. The benefit-risk ratio quantifies the medical device value. This ratio evaluates pros and cons encountered during each device use, and applies this evaluation to the global device view, including all potential device uses.

6.1 Benefit-risk ratio requirements

Benefit-risk ratios are determined by benefit-risk analysis after clinical/performance data have been identified, appraised, and analyzed. Individual and overall device-related benefits and risks must be analyzed further to determine if device clinical/performance benefits outweigh risks (e.g., this might be an assessment of in vitro diagnostic (IVD) device accuracy or other measure of device benefit analyzed in comparisons to risks like missed diagnosis). The clinical evaluation report/performance evaluation report (CER/PER) must include a section to explain fully the benefit-risk ratio and to determine if the ratio is acceptable for the device to be placed on (or to remain on) the EU market.

The MDR and IVDR benefit-risk ratio requirements are remarkably similar including, but not limited to, the following which are required in the technical file or design dossier outside of the CER/PER:

- Conformity assessment is not required when device "modifications do not adversely affect the benefit-risk ratio of the device" (MDR Article 54) or are not "substantial modifications" (IVDR Article 71, 74).
- The benefit-risk ratio shall be "acceptable," "valid" and based on "sufficient" clinical data (MDR Article 61; IVDR Article 56 (1)).
- Postmarket surveillance (PMS) system data shall be used "to update the benefit-risk ratio and improve" RM (MDR Article 83, 3(a); IVDR Article 78, 3(a)).
- The periodic safety update report (PSUR)* will define the benefit-risk determination conclusions (MDR Article 86, 1(a); IVDR Article 81, 1(a)) **PSUR is only for class IIa, class IIb and class III medical devices or class C or class D IVD devices, low-risk devices use the PMS Report (PMSR).*
- Trend reporting shall be done by manufacturers for "any statistically significant increase in the frequency or severity of incidents that are not serious incidents... that could have a significant impact on the benefit-risk analysis... and which... have led or may lead to... risks to the health or safety of patients, users or other persons" (MDR Article 88(1) including any risks considered "unacceptable when weighed against the intended benefits"; IVDR Article 83(1) including "erroneous results").

Planning, Writing and Reviewing Medical Device Clinical and Performance Evaluation Reports (CERs/PERs). DOI: https://doi.org/10.1016/B978-0-443-22063-0.00012-2

- The Commission analyzes vigilance data and the "competent authority... shall inform the manufacturer" about any "previously unknown risk" or changes in "the frequency of an anticipated risk" which "significantly and adversely changes the benefit-risk determination" and the manufacturer "shall then take the necessary corrective actions" (MDR Article 90; IVDR Article 85).
- Risk reduction general safety and performance requirements (GSPRs) 2, 3 (10 for MDR only):
  - The manufacturer must "reduce risks as far as possible... without adversely affecting the benefit-risk ratio" (MDR Annex I, GSPR 2; IVDR Annex I, GSPR 2).
  - The manufacturer must evaluate the *PMS data* impact on the "overall risk, benefit-risk ratio and risk acceptability" (MDR Annex I, GSPR 3(e); IVDR Annex I, GSPR 3(e)).
  - For medical devices only: when justifying specific objectionable substances in the device (e.g., "substances which are carcinogenic, mutagenic or toxic to reproduction" or endocrine-disrupting substances or phthalates), the justification must explain why changes to remove these substances are inappropriate "in relation to maintaining... the benefit-risk ratio" (MDR ONLY Annex I, GSPR 10.4.2(c), 10.4.3).
- The manufacturer must have *technical documentation* including the benefit-risk analysis and RM (MDR Annex II, 5(a); IVDR Annex II, GSPR 5(a)).
- The PMS plan (PMSP) shall include "suitable indicators and threshold values" for benefit-risk analysis and for RM (MDR Annex III, 1 (b); IVDR Annex III, 1 (b)).
- The Notified Body (NB) must verify the clinical evaluation/performance evaluation (CE/PE) are adequate in conformity with the GSPRs including "adequacy of the benefit-risk determination... risk management... instructions for use... user training", PMSP and proposed post market clinical follow up/post market performance follow up (PMCF/PMPF) plan (PMCFP/PMPFP) with appropriate specific milestones considered in the CE/PE process (MDR Annex IX, 4.6 and 4.7; IVDR Annex IX, 4.6 and 4.7 and consider the *"Assessment procedure for certain class III and class IIb devices" in Annex IX, 5* of the *MDR only.)*
- The clinical evaluation plan/performance evaluation plan (CEP/PEP) "shall include... parameters to be used to determine... the acceptability of the benefit-risk ratio for the [various indications and] intended purpose... of the device" and how issues will be addressed (MDR Annex XIV 1(a); IVDR Annex XIII, 1.1).
- The PMCFP/PMPFP shall include a continuous proactive process to collect and evaluate clinical data "with the aim of... ensuring the continued acceptability of the benefit-risk ratio" (MDR Annex XIV, 6.1(d); IVDR Annex XIII, 4 and 5.1(d)).
- The Investigator's brochure (IB) must summarize the benefit-risk analysis and RM details (MDR Annex XV 2.5; IVDR, Annex XIV, 2.5).
- Clinical trials may be required to substantiate the benefit-risk ratio (MDR Annex XV, 2.1) and performance studies are widely discussed in the IVDR to demonstrate the IVD device analytical performance (IVDR Annex XIII, 1.2.2).
- MDR only: NB must assess changes to devices using human cells or tissues to "ensure that the changes have no negative impact on the established benefit-risk ratio of the addition of the tissues or cells of human origin or their derivatives in the device" (MDR Annex IX, 5.3.1(d)).

The listing above was imported into a checklist (**Appendix P: Benefit-Risk Ratio Checklist**), which may be helpful when assessing progress toward completely verified benefit-risk ratio documentation to meet the MDR/IVDR requirements.

The benefit-risk ratio must be updated as new clinical data becomes available through clinical investigations or PMS activities. Benefit-risk ratio changes must be monitored with strict PMS attention to the reporting requirements of various global regulatory authorities. All changes to the benefit-risk ratio generate potential updates to multiple related documents in the CE/PE, PMS, and RM systems.

6.2 Benefit-risk ratio evaluation methods

No single accepted benefit-risk ratio evaluation method exists; however, the main building blocks for evidence-based device benefit-risk analyses include effects, evidence, and uncertainties. All device-related outcomes (i.e., device effects) evaluated in benefit-risk ratio analyses must eventually be fully supported by clinical/performance data (i.e., evidence) in "sufficient" data amounts and qualities (i.e., uncertainties need to be considered).

Many other factors also affect the benefit-risk ratio including, but not limited to:

- therapeutic context (i.e., standard of care, treatment options, alternative therapies, etc.);
- target population (i.e., age, height, weight, etc.);
- disease state or condition (i.e., condition or disease being treated or diagnosed, etc.);
- measures taken to enhance benefits and reduce risks (i.e., HCP training/experience, RM activities, etc.); and
- for IVD devices, other factors may include chemistries which interfere with the analytic detection or performance of the IVD device.

When conducting the benefit-risk ratio analysis, all observed effects (benefits as well as risks) must be demonstrated with clinical data evidence of a sufficient amount and quality to eliminate any uncertainty about the benefit-risk ratio being acceptable for use as indicated in the labeling for the intended patient population. The benefit-risk ratio for the IVD device will probably include an assessment of IVD device accuracy or other measure of device benefit analyzed in comparison to the IVD device issues/risks including missed diagnoses, sensitivity/specificity drift, device hazards, etc.

6.2.1 Indications for use are "benefits"

The starting point for any medical device benefit-risk analysis is the indication for use. The evaluator must consider how the medical device is used to treat a patient with a given disease or how the IVD device delivers an acceptable analytical and clinical performance to reliably detect and analyze a specific analyte, etc. The disease context within the indication for use helps the evaluator to understand the public health context for the device benefit-risk ratio, including how the intended population is affected by the disease to be treated, diagnosed, evaluated, or otherwise benefited by the device use. The evaluator should focus on the disease aspects addressed or investigated by the device. For example, if the subject device alleviates symptoms and reduces long-term morbidity, those aspects should be discussed and clinical data relevant to them should be evaluated to understand the benefit-risk ratio. Any known differences in disease state, progression, or outcomes across subpopulations should also be quantified and evaluated during the benefit-risk ratio evaluation.

Analyzing all available alternative treatments and diagnostic options for each treated indication is important. Clinical data should be used to quantify treatment safety, efficacy, tolerability, and the uncertainties or limitations surrounding each treatment or diagnostic option for comparison to the subject device. The standard of care is typically used for comparison; however, alternative devices, including unapproved or off-label therapies or diagnostics, should also be evaluated to put the subject device in the proper medical context. The limits of detection, sensitivity, and specificity are measures used for IVD devices. The subject device benefit-risk ratio should be directly compared to other benefit-risk ratios for other available treatments.

For both medical and IVD devices, the benefit-risk ratio acceptance determinations must also consider the harm done by the disease or condition being treated or measured. More device risk may be acceptable when treating or diagnosing a terminal condition than when treating a mild, chronic condition. The disease state may also drive the perceived value placed on device treatment benefits or diagnostic/therapeutic device assessments. For example, a patient with metastatic cancer may tolerate more risk for short-term benefits (e.g., a temporary quality of life improvement) than a patient dealing with a mild, chronic condition like eczema. In other words, understanding all aspects of the device indications for use and the benefits/risks involved in treating or not treating the condition (or using vs. not using the IVD device) allows the evaluator to calibrate and contextualize the benefit-risk ratio profile.

6.2.2 Identify/quantify all benefits

The EU MDR Article 2 (53) [1] defined clinical benefit as "the positive impact of a device *on the health of an individual, expressed in terms of a meaningful, measurable, patient-relevant clinical outcome/s, including outcomes/s related to diagnosis*, or a positive impact on patient management or public health." Similarly, EU IVDR Article 2 (17) [2] defined clinical benefit as "the positive impact of a device *related to its function, such as that of screening, monitoring, diagnosis or aid to diagnosis of patients*, or a positive impact on patient management or public health."

For medical devices, the requirement for "meaningful, measurable, patient-relevant clinical outcomes" means the benefits, and the benefit-risk ratio as a whole, must be based on quantitative data and analysis. Each and every benefit must be measurable and supported by sufficient clinical data and the uncertainty around each clinical data source for a benefit must be discussed. For IVD devices, the benefit-risk ratio measurement is more about the device function, rather than patient clinical outcomes. This is an important distinction. Rather than creating or supporting a specific clinical outcome for a patient directly, the IVD device provides the benefit of screening, monitoring, diagnosing, or aiding the diagnosis.

Each beneficial effect should be quantified based on the medical need as well as the likelihood, magnitude, and duration of the beneficial effect.

Medical device benefits include positive outcomes like increasing survival, reducing symptoms, or reducing/ eliminating risks. For example, if the standard of care for a particular indication for use includes known risks and the subject device reduces those risks in a quantifiable manner, the risk reduction is a benefit. These risk reductions must be quantifiable and based on clinical data. Nonclinical evidence, such as *in vitro* or animal evidence, is not sufficient.

Similarly, IVD device benefits include analyte detection, sensitivity, and specificity to screen, monitor, and diagnose a disease or a disease stage accurately (e.g., an IVD device may diagnose cancer or cancer metastasis). Again, the diagnostic utility must be documented and quantified with clinical data.

If a device beneficial effect varies across patient subpopulations (e.g., has a different likelihood or duration in different patient groups), the benefit should be quantified in each subpopulation with supporting clinical data. For example, if a device benefit occurs more frequently in pediatric patients than in adult patients, clinical data must be provided to support the benefit likelihood, magnitude and duration in both adult and pediatric groups and the corresponding uncertainties must be defined for each group separately. In addition, under EU MDR/IVDR, all clinical claims (which are typically benefits) must also be supported by sufficient clinical data with all clinical data uncertainties evaluated.

6.2.3 Identify/quantify all risks

Risk is defined as "the combination of the probability of occurrence of harm and the severity of that harm" (EU MDR Article 2 (23); EU IVDR Article 2 (16)). Risk analysis involves identifying all known device-related harms (i.e., adverse events, AEs) and quantifying the AE likelihood (i.e., probability of occurrence), magnitude (i.e., severity), duration and all associated uncertainties. As with benefits, clinical data needs to support all conclusions regarding risks and all differential effects on known subpopulations. PMS data (i.e., passively-collected complaints) can help to identify potential risks but are usually not sufficient for quantification.

AEs are stratified into the following groups and subgroups:

- serious AEs (SAEs) or nonserious AEs;
- expected or unexpected AEs (expected AEs are generally those found in device labeling);
- device-related or not-device-related AEs; and
- device deficiencies (DDs) with or without reported AEs.

Serious, unexpected and device-related AE will typically require expedited "safety reporting" to regulatory authorities among other interested parties, so all clinical team members must understand these risk definitions.

MDR Article 2 (58) defined SAEs as any AEs "that led to any of the following:

death,
serious deterioration in the health of the subject, that resulted in any of the following:

> *life-threatening illness or injury,*
> *permanent impairment of a body structure or a body function,*
> *hospitalisation or prolongation of patient hospitalisation,*
> *medical or surgical intervention to prevent life-threatening illness or injury or permanent impairment to a body structure or a body function,*
> *chronic disease,*

foetal distress, foetal death or a congenital physical or mental impairment or birth defect."

The SAE definition in IVDR Article 2(58) was identical except for the addition of one statement at the start of the definition, as follows:

"a patient management decision resulting in death or an imminent life-threatening situation for the individual being tested, or in the death of the individual's offspring."

This is an important distinction between the MDR and IVDR since IVD devices are often linked to patient healthcare decisions not direct device benefits. The medical device is directly applied on or in a patient to provide a beneficial effect while the IVD device test result provides the beneficial effect.

EU MDR Article 2 (59) and IVDR Article 2(59) both defined device deficiency as "any inadequacy in the identity, quality, durability, reliability, safety or performance of an investigational device [device for performance study], including malfunction, use errors or inadequacy in information supplied by the manufacturer."

For medical devices, the evaluation of all risks should include the AE treatment, duration, and resolution, or reversibility. In addition, any information about whether an AE or risk can be predicted, prevented, or treated should be included, quantified, and discussed.

Risks associated with "on-label" uses (i.e., uses complying with the labeled indication for use) should be considered in the benefit-risk ratio and separately, risks associated with "off-label" uses (i.e., uses that do not comply with the labeled indication for use) should be considered under the potential for misuse, along with any associated risk mitigation strategies. If the potential for misuse is high and the associated "off-label" risks are serious, these risks should be discussed in the benefit-risk ratio evaluation. In addition, the evaluator must pay careful attention to the potential for systematic misuse wherein a manufacturer is aware of widespread and intentional off-label uses of their devices. In this case, the evaluator should generate the clinical data needed to determine if the indication for use should be changed to incorporate the previously "off-label" use or if actions are required to prevent the misuse, especially if patients are being exposed to unnecessary serious device-related risks during this off-label use.

For IVDs and other diagnostic devices, false positive and false negative results represent major risk factors. The risk is not only the erroneous test result, but the added risk if a healthcare provider treats or does not treat a patient based on the erroneous test result. For false positives, the risk is unnecessary treatment (i.e., patient receives all the treatment risks with no potential benefit) and, for false negatives, the risk is no access to necessary treatment (i.e., patient receives no treatment risks because they are not treated; however, the patient also received no potential benefit as the untreated disease or condition risk is unmitigated without the appropriate treatment).

6.2.4 Assess risk mitigation effectiveness

Evaluate all planned or implemented risk mitigation/management strategies and analyze the clinical data quantifying the remaining residual risk (i.e., risks remaining after applying all risk-mitigations). For each risk, the benefit-risk assessment should document the steps taken by the company to mitigate the risk. Risk mitigation can be a simple as adding specific language in the device labeling, instructions for use (IFU) and product literature or as complex as changing the manufacturing steps in the medical device design controls to prevent misuse. No matter the mitigation strategy, clinical data must be documented to support the mitigation strategy and to define the residual risk after mitigation.

6.2.5 Consider patient/user perspectives

Patient and user perspectives are important to consider when assessing benefits and risks. Ultimately, patients and users are the ones who experience the benefits and risks associated with device use and patient/user perspectives about those benefits and risks may not align with physician or manufacturer perspectives. Patient perspectives can be gathered through validated patient surveys or other data collection tools with a goal to understand how patients perceive the benefit-risk ratio. Sometimes a particular side effect may not be serious or severe, but may be intolerable for certain patients or users, and may impact device use adoption and/or compliance. When considering long-term device use in particular, patient perspectives (including perspectives on adjuvant therapies used to mitigate risks) are essential to understand the clinical benefit-risk ratio. User perspectives are also particularly important to consider when a risk or benefit primarily affects healthcare workers who use the device. For example, an easier-to-use surgical clamp may not directly benefit the patient (although the device may benefit the patient if the clamp changes the benefit-risk ratio associated with a particular surgical procedure), but an improved design may benefit the surgeon.

Each indication for use population may have a separate benefit-risk ratio needing clinical data support and documentation. For example, a device indicated for use in people with two distinct but similar conditions may require a separate benefit-risk analysis for each of the two conditions, because differences in uncertainties, medical needs or other device/patient interaction factors can change the benefit-risk analysis across similar indications for use.

Benefit-risk evaluations are "highly subjective" and require significant effort to "incorporate as many different perspectives into the process as possible" to ensure the conclusions are as unbiased as possible [4]. Unfortunately, objectivity may not be possible because benefits and risks are based on "attitudes and preferences about which reasonable people disagree." Quantifying highly subjective benefits and risks is complicated because people have widely differing attitudes about benefits and risks. What one person considers a risk may be a benefit to another person. In addition, no uniform measurements are available for benefits and risks and rare safety events, long-term harms or events occurring only in subpopulations may be missed.

6.2.6 Explore various benefit-risk evaluation approaches

Although focused on drugs, a report from the European Medicines Agency (EMA) identified 18 distinct quantitative approaches for the benefit-risk analysis, along with additional qualitative approaches [5]. Ultimately, the EMA recommended a comprehensive eight-step framework, PrOACT-URL, as follows:

1. "PrOBLEM. Determine the nature of the problem and its context: what is the medicinal product (e.g., new or marketed chemical or biological entity, device, generic); what sort of decision or recommendation is required (e.g., approve/disapprove, restrict); who are the stakeholders and key players; what factors should be considered in solving the problem (e.g., the therapeutic area, the unmet medical need, severity of condition, affected population, an individual's social context, time frame for outcomes). Then frame the problem (e.g., as mainly a problem of uncertainty, or of multiple conflicting objectives, or as some combination of the two).
2. OBJECTIVES. Identify objectives that indicate the overall purposes to be achieved (e.g., maximise favourable effects, minimise unfavourable effects), and develop criteria against which the alternatives can be evaluated (i.e., what are the favourable and unfavourable effects?).
3. ALTERNATIVES. Identify the options (actions about a medicinal product or the products themselves) to be evaluated against the criteria (e.g., pre-approval: new treatment, placebo, active comparator; post-approval: do nothing, limit duration, restrict indication, suspend).
4. CONSEQUENCES. Based on available data, describe how the alternative would perform on the criteria (e.g., describe the magnitude of possible favourable and unfavourable effects)... consider intermediate outcomes, such as safety and efficacy effects. Consequences describe clinically relevant effects. Create a 'consequence table' with alternatives in rows and criteria in columns. Write descriptions of the consequences in each cell, qualitative and quantitative... record the basis for uncertainties about the consequences...
5. TRADE-OFFS. Assess the balance between favourable and unfavourable effects.

   These five steps are common to all decisions in which the consequences are known with certainty. In approving drugs, regulators typically must face uncertainty and risk, in which case three additional steps are relevant:
6. UNCERTAINTY. Consider how the balance between favourable and unfavourable effects would change by taking account of the uncertainty associated with the consequences.
7. RISK. Judge the relative importance of the Agency's risk attitude for this medicinal product (by considering, e.g., the therapeutic area, the unmet medical need and patients' concerns) and adjust the uncertainty-adjusted balance between favourable and unfavourable effects accordingly. Consider, too, how risks would be perceived by stakeholders (according to their views of risk).
8. LINKED DECISIONS. Consider the consistency of this decision with similar past decisions, and assess whether taking this decision could impact future decisions either favourably or unfavourably (e.g., would it set a precedent or make similar decisions in the future easier or more difficult)."

This EMA report [5] provided a table derived from work by the US Food and Drug Administration (FDA) to identify key issues in benefit-risk determinations (Table 6.1).

TABLE 6.1 FDAs benefit-risk deliberation framework.

Consideration	Favorable benefit-risk	Noncontributory	Unfavorable benefit-risk
Condition severity			
Unmet medical need			
Clinical benefit			
Risk			
Risk management			

The balance between favorable and unfavorable effects (i.e., benefits and risks) is considered along with the uncertainty or risk tolerance associated with the effects. The EMA Effects Table template for implementing the PrOACT-URL framework [6] included the following six steps:

1. Identify only favorable/unfavorable effects relevant to the benefit-risk ratio.
2. Describe the effects.
3. Define the measurement scales.
4. Identify the options.
5. Display the data.
6. Describe remaining effect uncertainties.

The effects table template provides a framework for contextualizing the medical device benefit-risk ratio table (Table 6.2), and can be used to list all favorable and unfavorable effects, descriptions, and supporting data [6,7].

TABLE 6.2 EMA effects table template [6,7].

Name	Description	Best	Worst	Units	Placebo	Device	Uncertainty	Source/s
Effect name	Describe effect	Effect range (#)	Effect range (#)	Effect units	Effect value with standard treatment	Effect value with device treatment	Describe uncertainty	Clinical data sources
Favorable effects								
Unfavorable effects								

The United States (US) food and drug administration (FDA) also published the benefit and risk factors to consider for medical device benefit-risk analysis [8]. Each potential benefit and risk type can be listed and the entire list could then be considered as a whole. Benefit and risk listings included type, magnitude, likelihood, duration of effect, patient perspective, user factors for healthcare professionals or caregivers, and medical necessity. Risk considerations included concerns about distributing nonconforming devices, false-positive or false-negative results (for IVDs and diagnostic devices), etc. (Table 6.3). Although not directly comparable, each benefit and risk should also be considered across multiple dimensions

TABLE 6.3 Benefit and risk factor considerations.

Benefit factor	Benefit definition	Data source	Risk factor	Risk definition	Data source
Type	Benefit grouped by type?	Clinical trials, literature, and experiences	Type	Risk grouped by type?	Clinical trials, literature, and experiences
Likelihood	Chance a given patient will experience benefit (consider different subpopulations within indication)?	Clinical trials, literature	Likelihood	Chance a given patient will experience harm (consider AE and non-AE events as chance a given device use will result in harm)?	Clinical trials, literature
Magnitude	Degree of treatment benefit/effectiveness (assessed against standard of care or gold standard for treatment)?	Clinical trials, literature, and experiences	Magnitude	Is risk serious, unexpected, and device-related (consider all deaths, SAEs, device-related non-serious AEs, and medical-device events without reported harm)?	Clinical trials, literature, and experiences (e.g., MAUDE reports)
Duration	How long does benefit last?	Clinical trials, literature, and experiences	Duration	How long does risk last?	Clinical trials, literature, and experiences
Patient perspective	What value do patients place on the benefit?	Clinical trials, literature, and experiences (e.g., patient surveys)	Patient perspective	How risk-tolerant are patients and how do patients feel about potential harm?	Clinical trials, literature, and experiences (e.g., patient surveys)
User benefits	Does benefit improve patient outcomes, caregiver activities, or clinical practice?	Clinical trials, literature, and experiences (e.g., physician surveys)	User risks	Does risk including a potential adverse impact on patient, caregiver, or clinical practice?	Clinical trials, literature, and experiences (e.g., physician surveys)

(Continued)

TABLE 6.3 (Continued)

Benefit factor	Benefit definition	Data source	Risk factor	Risk definition	Data source
Medical necessity	Is benefit provided by alternative therapies/ devices?	Literature	Device needed OR False-positive or false-negative results	Risks/potential harms associated with device OR diagnostic error (e.g., patient received unnecessary treatment or does not receive needed treatment)	Literature

AE, adverse event; *MAUDE*, Manufacturer and User Facility Device Experience; *SAE*, serious adverse event.

(e.g., uncertainty, mitigation, detectability, failure mode, device issue scope, patient impact and preference for availability, nature of violations/nonconforming product, firm compliance history, and decision-making activities).

Each identified risk or benefit should be sourced from high-quality clinical data. Data quality helps to determine the uncertainty around the benefit or risk. Poor-quality data or disagreements across data sources creates a higher uncertainty level. For example, when multiple RCTs show a benefit from the subject device with a similar effect magnitude and likelihood, but one RCT finds a much shorter effect duration than the others, then uncertainty around effect duration is higher than the uncertainty around the benefit magnitude or likelihood.

The standard of care for the device intended use is a good baseline comparison for the benefit-risk analysis. If the device has the same benefit-risk ratio as the standard of care, then the device benefit-risk ratio is acceptable because the standard of care use was, by definition, widely accepted. On the other hand, if device use requires non-standard-of-care procedures, then the device risk includes the new procedure risk and may not be acceptable. For example, if standard of care for a condition is a medication, then an implantable medical device requiring surgery may not be acceptable, since the implant incurs surgical risks in addition to any device-use risks. In this case, the implanted medical device benefits must be great enough to outweigh the surgical and device-use risks when compared to the medication standard of care. Alternatively, if the standard of care is surgery, then the implantable medical device risk will include any and all surgical risks above and beyond the normal standard-of-care surgery risks. Risk (and benefit) are subjective and relational, not only in comparison to similar devices or devices used for the same indication, but also in comparison to the current standard of care in the industry.

The various device effects in different patient subpopulations must also be considered when analyzing the likelihood of various benefits and risks. Did certain subpopulations have more benefits or higher risks than others? Did the medical device IFU discuss the observed benefit and risk differences observed in all patient subpopulations using the device? Did the clinical trial inclusion/exclusion criteria allow the entire indicated population to be included? A new device meeting an unmet medical need as a medical necessity is a major consideration for the device benefit and the novelty of the new treatment will be a major consideration for the device risk. If no therapies are available for a particular indication and the standard of care is palliative only, then the baseline for both risk and benefit is patient quality of life. More risk is generally acceptable if the device is meeting a significant unmet medical need, even if the unmet medical need is because patients cannot access treatment.

Many benefit-risk ratio evaluation methods can be used; however, but none has been selected as the method of choice. L choosing the best benefit-risk ratio evaluation will depend on the clinical/performance data supporting the medical device benefits and risks during use in humans.

6.3 Benefit-risk ratio documentation

The benefit-risk ratio is an essential CER/PER component and needs to be documented for both the background/state of the art (SOTA) and the subject device. Generally speaking, the CER/PER benefit-risk ratio discussion is aimed at determining device compliance with GSPRs 1, 6, and 8, and the CER/PER is structured to present and discuss all clinical evidence. In other words, the benefit-risk ratio discussion will be built upon the data discussed during the entire clinical data evaluation in the CER/PER. For example, the benefits and risks identified in the background/SOTA evaluation should be listed in the background/SOTA section to serve as a baseline for comparison to the subject device benefits and risks. Additional, missing or different subject device benefits and risks should be documented specifically in the appropriate subject device CER/PER sections. Uncertainty about the clinical data should be discussed in all clinical data sections, and the product literature evaluation should serve to describe the risk mitigations activities listed in the IFU and other product labeling.

BOX 6.1 Example CER benefit and risk lists

Clinical data for the Honeywell ONE-Fit N95 Respirators documented the following benefits and risks:

Benefit list

- Reduced airway inflammation when exposure to traffic pollution.
- Key personal protective equipment specified in many guidelines for protection from pandemic viruses.
- Recommended for immunocompromised patient use when leaving a protected environment.
- Less pandemic virus transmission risk than generic medical or surgical masks.

Risk list

- Rebreathing consequences (blood gas alterations, etc.).
- Increased blood pressure in healthy and pregnant women.
- Uncomfortable.
- Skin irritation.
- AEs when worn during manual labor include: headache, tiredness, giddiness, difficulty breathing, difficulty walking, sweating, itching, uncomfortable seal, exhaustion.

These benefits and risks were documented in the literature and experience data. The device risks and side effects were considered acceptable when weighed against the device benefits when used as indicted in the appropriate patient population (GSPR 1). The device performance issues included patient device incompatibility, packaging issues, damaged or faulty device and device operates unexpectedly, and this side effect profile was also considered acceptable during the device lifetime when weight against the benefits (GSPR 6). In addition, analyses during device manufacturing, PMS, and RM documented the occurrence of each individual clinical risk was rare, and the manufacturer was working to increase the benefit-risk ratio continuously by minimizing all risks encountered during normal conditions of use (GSPR 8).

In conclusion, Honeywell ONE-Fit N95 Respirators provided the benefit of reduced risk of viral transmission or environmental contaminants without creating any overwhelming risk to the patient or healthcare personnel users when used as indicated. Although hotly debated for persons with breathing difficulties, this benefit-risk ratio evaluation concluded the benefits outweighed the risks for the continued safe use of the Honeywell ONE-Fit N95 Respirator as indicated.

Once all included clinical data benefits and risks have been listed and discussed, only "key" benefits and risks need to be presented in the CER/PER (i.e., all benefit and risk details should be collected in the RM files, not in the CER/PER). The CER/PER benefit-risk ratio discussion should focus on the benefits and risks most closely related to GSPR conformity issues. The benefit-risk ratio acceptability must be justified with clinical data showing why any and all differences between the standard of care/background/state-of-the-art device benefit-risk ratio and the subject device benefit-risk ratio were considered acceptable. The device type and CER/PER type both influence this benefit-risk ratio discussion even though all CER/PER types must identify, appraise, and analyze clinical data to support the benefit-risk ratio evaluation.

6.3.1 Viability clinical/performance evaluation reports

Viability clinical/performance evaluation reports (vCERs/vPERs) do not generally have subject device clinical data available because the device has not been used in any human during this early design phase. The subject device benefit-risk ratio evaluation must therefore be based on the background/SOTA benefit-risk ratio evaluation using the known or anticipated benefits and risks for similar devices (if such exist). Since no pre- or postmarket clinical data are available, any and all anticipated differences in benefits and risks between similar devices and the subject device must be discussed and tracked over time during device development.

6.3.2 Equivalent device clinical/performance evaluation reports

Equivalent device clinical/performance evaluation reports (eqCERs/eqPERs) will base benefit-risk ratio evaluations on the benefits and risks associated with the equivalent device. By definition, the subject device must have the same benefit-risk ratio as the equivalent device; however, the eqCER/eqPER must analyze the equivalent device benefits and risks and determine the benefit-risk ratio for the equivalent device as though the equivalent device was the subject device. Since little, if any, pre-or postmarket clinical data are available for the subject device, all anticipated differences in benefits and risks between the equivalent device and the subject device must be discussed and tracked over time during device development.

6.3.3 First-in-human clinical/performance evaluation reports

First-in-human clinical/performance evaluation reports (fihCERs/fihPERs) have clinical data for the subject device; however, the available clinical data are often limited to a single, small clinical trial. The benefit-risk ratio evaluation can be based on available subject device data; however, the clinical data are likely to be quite limited, with large associated uncertainties. As a result, the evaluator may need to supplement the limited available subject-device data with data from similar or equivalent devices. Equivalent device PMS data can help contextualize and expand the subject device benefit-risk ratio evaluation; however, the data from equivalent or similar devices must be clearly identified to prevent confusion. The evaluator must avoid mixing up and misrepresenting the subject device, equivalent device, and similar device benefit-risk ratio data. Since PMS data are not yet available for the subject device, any anticipated differences in benefits and risks between similar or equivalent devices and the subject device must be discussed and tracked over time during device development.

6.3.4 Traditional clinical/performance evaluation report

Benefit-risk ratio evaluations for well-established devices should be based on the specific, subject device benefit-risk data compared to background knowledge/SOTA and alternative therapy benefit-risk data. Clinical data must be properly sourced and evaluated, including a discussion of all associated uncertainties. All available clinical trial, clinical literature and clinical experience data should be used to update the previously established and maintained benefit-risk ratio. Large uncertainties may still be associated with anticipated benefits not yet represented in the clinical data and risks may be more or less severe than initially represented in the benefit-risk ratio. Additionally, the background/SOTA/alternative device benefit-risk ratio must be documented and all uncertainties discussed, even for a well-established device. Each update to the traditional clinical/performance evaluation report (trCER/trPER) must include a fully updated and aligned benefit-risk ratio. The trCER/trPER should specifically state if the benefit-risk ratio has not changed since the previous CER/PER and this statement must be supported by all updated data reported in the trCER/trPER including all the updated CE/PE, PMS, and RM data.

6.3.5 Obsolete clinical/performance evaluation reports

As devices become obsoleted, their benefit-risk ratios may change and PMS data are required to track and trend these changes. Obsolete clinical/performance evaluation reports (oCERs/oPERs) rely heavily on PMS data to determine if benefit-risk ratios have changed. In addition, new/updated SOTA, standards of care and alternative devices and treatments must be considered. As new therapies and new devices become available, the benefits and risks associated with the obsoleted device (or any device) may become less tolerable or acceptable and the benefits may have less weight than previously anticipated. Special attention must be paid to the background/SOTA section in oCER/oPERs to ensure the proper context is applied, especially with regard to the benefit-risk ratio acceptability evaluation.

6.4 Define the benefit-risk ratio process steps

The international RM standard, ISO 14971, is now more closely aligned with actual clinical data than ever before. The RM team begins the RM file with only theoretical risks for the subject device because the device engineering plans often begin before the device is prototyped and used in humans. Once the device is used in humans, the nature of the RM work becomes more clinical in nature and requires clinical team inputs in addition to the engineering team.

The process steps for the clinical benefit-risk ratio evaluation should describe how the transition will be completed from the design engineering team to the clinical data team. Standard operating procedures (SOPs) and work instructions (WIs) are required to ensure the benefit-risk ratio evaluation process is methodologically sound and reduces bias as much as possible. Similar to the other CE/PE steps, the benefit-risk ratio evaluation involves many subjective value judgements. Even the most quantitative, statistics-driven benefit-risk ratio methods require some measures involving basic value determinations and alignments with prevailing customer views. Objectivity is simply not possible because benefits and risks are about user attitudes and preferences, which differ between individual persons. Data-driven benefit-risk ratio evaluation requires a direct evaluator subjectivity assessment, which is difficult to document in a reproducible manner.

Special care must be applied when developing the benefit-risk ratio evaluation SOPs and WIs and specific process steps should be defined with specific, responsible parties. Benefit-risk ratio development is relevant to CE/PE, PMS, and RM systems and the interplay between these three systems must be documented and included in the manufacturer's quality management system.

6.5 Conclusions

Benefit-risk ratio evaluations document the overall medical device value and are fundamental to medical device clinical, regulatory, and quality decisions. Weighing all benefits and risks encountered when using the medical device must start by listing the specific pros and cons for each of the many different patient types. The benefit-risk ratio evaluation is best applied with a global view considering all potential device uses.

The evaluator is expected to develop a valid, scientifically sound and rigorous, evidence-based benefit-risk ratio evaluation. All benefits and risks must be identified from within the available clinical data. The clinical data benefits and risks must be appraised and analyzed for magnitude/severity, likelihood, duration, and uncertainty. In addition, the clinical data after any risk mitigation strategies must be identified and evaluated as well.

The concepts of benefit and risk are highly subjective and extensive patient and user perspectives must be considered, particularly when determining the contribution of each benefit or risk to the overall ratio for a given subpopulation of patients represented in the medical device indication for use. Ultimately, the evaluator must determine if the benefits outweigh the risks for the intended use by each indicated patient population. The clinical data supporting each benefit or risk and the process used to evaluate the clinical data in the benefit-risk ratio determination must be properly documented in the records as required by the QMS.

6.6 Review questions

1. Where did the benefit-risk ratio framework originate?
2. True or false? The EU MDR and IVDR have different definitions for benefit-risk determination.
3. What is required when significant changes to a device benefit-risk ratio are identified?
4. When do device modifications require a new conformity assessment for an updated benefit-risk ratio determination?
5. Which benefits and risks need to be considered during the benefit-risk ratio evaluation?
6. Where should the evaluator start when conducting the benefit-risk ratio evaluation?
7. What are the differences between the MDR and IVDR definitions for clinical benefit?
8. What are some factors involved in assessing clinical benefits?
9. What is uncertainty in the context of benefit-risk analysis, and how does uncertainty inform benefit-risk analysis?
10. How is risk defined in the MDR/IVDR?
11. What additional element was added to the SAE definition for an IVD device?
12. What are some risk mitigation activities used by a manufacturer to reduce risk?
13. What role do clinical context, standard of care, and alternative therapies play in benefit-risk analysis?
14. When should risks associated with device misuse be considered?
15. What are some benefit-risk ratio evaluation method examples?
16. Which CER/PER type has its benefit-risk ratio evaluation most directly challenged as new devices become available?
17. Which international standard is associated with medical risk management?
18. How can objectivity be incorporated into the CER/PER benefit-risk ratio evaluation?
19. Which system is most relevant to the benefit-risk ratio evaluation?
20. What is the progression of various CER/PER types during the device lifetime?

References

[1] European Parliament and Council of the European Union. Consolidated text: regulation (EU) 2017/745 of the European Parliament and of the Council of 5 April 2017 on medical devices, amending Directive 2001/83/EC, Regulation (EC) No 178/2002 and Regulation (EC) No 1223/2009 and repealing Council Directives 90/385/EEC and 93/42/EEC. Accessed May 19, 2024. https://eur-lex.europa.eu/legal-content/EN/TXT/?uri = CELEX%3A02017R0745-20230320.

[2] European Parliament and Council of the European Union. Consolidated text: regulation (EU) 2017/746 of the European Parliament and of the Council of 5 April 2017 on in vitro diagnostic medical devices and repealing Directive 98/79/EC and Commission Decision 2010/227/EU. Accessed May 19, 2024. https://eur-lex.europa.eu/legal-content/EN/TXT/?uri = CELEX%3A02017R0746-20230320.

[3] ISO 14971:2019. Medical devices—application of risk management to medical devices. Accessed May 19, 2024. https://www.iso.org/standard/72704.html.

[4] C.H. Coleman, 13—Risk-benefit analysis, in: G. Laurie, E. Dove, A. Ganguli-Mitra, C. McMillan, E. Postan, N. Sethi, et al. (Eds.), *The Cambridge handbook of health research regulation* (Cambridge Law Handbooks, Cambridge University Press, Cambridge, 2021, pp. 130−138. https://doi.org/10.1017/9781108620024.017. Accessed May 19, 2024. https://www.cambridge.org/core/books/cambridge-handbook-of-health-research-regulation/riskbenefit-analysis/5582ABF8A81BA89D6F72686B3929F0A2.

[5] European Medicines Agency. EMA/549682/2012—revision 1. Human Medicines development and evaluation. Benefit-risk methodology project. Work package 2 report: applicability of current tools and processes for regulatory benefit-risk assessment. (August 31, 2010). Accessed May 19, 2024. https://www.ema.europa.eu/en/documents/report/benefit-risk-methodology-project-work-package-2-report-applicability-current-tools-processes_en.pdf.

[6] European Medicines Agency. EMA/297405/2012—revision 1. Human medicines development and evaluation. Benefit-risk methodology project. Work package 4 report: benefit-risk tools and processes. (May 9, 2012). Accessed May 19, 2024. https://www.ema.europa.eu/en/documents/report/benefit-risk-methodology-project-work-package-4-report-benefit-risk-tools-processes_en.pdf.

[7] EMA/74168/2014. Human medicines development and evaluation. Benefit-risk methodology project. Update on work package 5: effects table pilot (Phase I). (February 6, 2014). Accessed May 19, 2024. https://www.ema.europa.eu/en/documents/report/benefit-risk-methodology-project-update-work-package-5-effects-table-pilot-phase-i_en.pdf.

[8] FDA Guidance. Factors to consider regarding benefit-risk in medical device product availability, compliance, and enforcement decisions: guidance for industry and Food and Drug Administration staff. 2016. Accessed May 19, 2024. https://www.fda.gov/files/medical%20devices/published/Factors-to-Consider-Regarding-Benefit-Risk-in-Medical-Device-Product-Availability--Compliance--and-Enforcement-Decisions---Guidance-for-Industry-and-Food-and-Drug-Administration-Staff.pdf.

Chapter 7

Writing clinical/performance evaluation documents

This chapter provides tips to help the evaluator and clinical evaluation/performance evaluation (CE/PE) document writers stay organized and efficient. In addition, this chapter provides advice to CE/PE writing group managers and discusses CE/PE document writing strategies to support good writing "flows" and "processes" within integrated teams. The goal is to define clearly the path to CE/PE document writing success.

1. Stay organized and focused on the clinical data.
2. Understand why clinical evaluation documents are difficult to write.
3. Assign the right staff to complete the CE/PE documents.
4. Plan enough time to write the clinical evaluation documents.
5. Follow the natural flow when writing individual CE/PE documents.
6. Evaluate statistical quality.
7. Follow good writing tips.
8. Manage CE/PE document writing cycles, to ensure benefits outweigh risks so the device can be marketed in the EU, and review thoroughly to ensure successful PE/CE Documents.

These tips are discussed with a few examples to help illustrate how to do this work.

The evaluator and writing team members should understand the whole process before starting to write any clinical evaluaiton/performance evaluation (CE/PE) documents. If the overall document writing scope is unclear, the evaluator and writer may miss critically important details. In addition, understanding the overall CE/PE document writing scope and completing specific CE/PE document writing steps should make the individual CE/PE document writing easier.

For example, after understanding and completing all three prior clinical evaluation report/performance evaluation report (CER/PER) document development steps (i.e., planning, appraising, analyzing all available clinical data), the CER/PER writing step should be formulaic (Fig. 7.1).

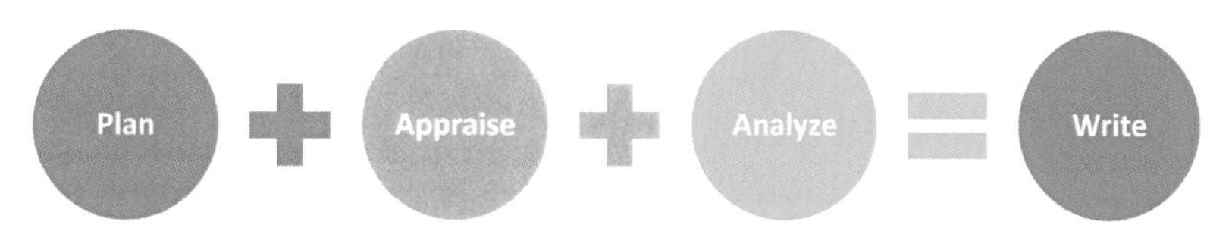

FIGURE 7.1 Basic CE/PE document writing formula.

In other words, the writing simply becomes the story-telling process, detailing what was learned along the way during the planning, appraising, and analyzing steps.

Planning, Writing and Reviewing Medical Device Clinical and Performance Evaluation Reports (CERs/PERs). DOI: https://doi.org/10.1016/B978-0-443-22063-0.00003-1

7.1 Stay organized and focused on clinical data

The CE/PE document writing process must cover the following important topics (at a minimum):

- Clinical data must be relevant to the subject/equivalent device.
- Clinical data must be high quality (while still including novel, low-quality data when appropriate).
- Good clinical practices (GCPs) must be followed when generating clinical data.
- Solid statistical designs are required for all pivotal clinical data analyses.
- Clinical data must be critically analyzed against prespecified and prequantified safety and performance (S&P) acceptance criteria as defined in the clinical evaluation plan/performance evaluation plan (CEP/PEP).
- The benefit-risk ratio must be continuously updated in alignment with the postmarket surveillance (PMS) and risk management (RM) systems.
- All clinical claims must be evaluated and adequate clinical data substantiation must be presented in the CER/PER or the clinical claims should not be used.
- Unlike the CER, the PER must include the scientific validation report (SVR), analytical performance report (APR), and clinical performance report (CPR). These are not required to be separate stand-alone reports, but they must be clearly defined in the PER.
- Conclusions must be drawn from and fully supported by the clinical data evaluation. If the reader cannot follow and easily verify the story, the logic and the clinical data, the CER/PER is deficient by definition.
- Clinical data gaps must be documented in the CER/PER and plans to generate new clinical data are required in the postmarket clinical follow-up plan/postmarket performance follow-up plan (PMCFP/PMPFP).

Typically, the CER/PER background/state of the art (SOTA)/alternative device section is presented before the subject/equivalent device section. This allows the background/SOTA/alternative device section data to justify the S&P acceptance criteria used to evaluate the subject/equivalent device clinical data in the following CER/PER section.

BOX 7.1 The three Cs of CER/PER writing.

The CER/PER is a stand-alone document and must tell a complete, comprehensive and compelling story. The CER/PER must include an introduction explaining the background/SOTA/alternative device clinical data and supporting the S&P acceptance criteria derived from this work. The subject/equivalent device section must clearly describe the methods, results and conclusions from the clinical trial, literature, and experience data evaluation relevant to the device under evaluation.

Complete: The CER/PER story must not leave out any details. The writer must organize the contents so the reader can see all the various clinical device uses first at a high level (i.e., introduce all the uses), then at a detailed level (i.e., justify each use with clear and specific clinical/performance data) and then at a high level again in the conclusion (i.e., explain what the overall device data mean).

Comprehensive: The CER/PER must include all relevant clinical trial, literature, and experience data using appropriate references and citations. All critical clinical data should be appended to the CER/PER.

Note: This does not mean every report should be appended; only the most relevant and compelling trial, literature, and experience data and reports should be appended.

Compelling: The CER/PER binds the company to regulatory decisions allowing the device to be sold in the European Union (EU) market. The CER/PER must accurately and appropriately convey the background/SOTA/alternative device details as well as the subject/equivalent device details in a factual and *convincing* manner. The conclusions must not be overstated; instead, the conclusions must be directly aligned with and drawn from the clinical data. The conclusions should explain how even the smallest clinical data gaps were identified and will be resolved. In essence, this future commitment to complete the PMCFP/PMPFP work on a specific and documented timeline should drive home the message documenting the company commitment to clinical quality and device S&P where the benefits clearly outweigh the risks when using the device as indicated in the appropriate patient population.

After the analytical work to identify, appraise and analyze the available clinical data and to conclude what the clinical data acually mean; the writing takes on an artistic form to lay out the story and to provide an ordered, clear, accurate, and convincing narrative for the reader. The more graphic and illustrative the story, the better. Tell the story with images. "Paint" the picture for the reader with words and related illustrations. Use tables, graphs, and figures. Be truthful and not misleading.

Ultimately, the stand-alone CER/PER must tell a complete, comprehensive, and compelling story.

Staying organized while writing CE/PE documents (including the CER/PER) requires good templates for each docuemnt type. Staying focused requires significant flexibility, especially when the clinical data are complicated. The CE/PE document writer must cover many different topics seamlessly and well. For example, when writing a PER, the writer must migrate the story from defining the association between the analyte and the clinical condition or physiological state (i.e., the scientific validity of the analyte), to the analyte detection and/or measurement (i.e., the analytical performance of the *in vitro* diagnostic (IVD) device) and then to the current data surrounding how well the IVD device yields results correctly correlated with the clinical and pathological condition of the target population and intended user (i.e., the clinical performance of the IVD device). The nuances for these transitions during the story "building" should be the glue holding the "complete, comprehensive, and compelling" story lines together.

In addition, the PER writer needs to know how to address required PER details like the IVD device sensitivity and specificity along with the expected device test results for normal and affected populations, including the predictive values and likelihood ratios. The writing details often differentiate a successful PER from a PER littered with many deficiencies and unanswered questions. Good CE/PE document writing requires discipline and practice.

7.2 Understand why clinical evaluation/performance evaluation documents are difficult to write

CEs/PEs are complex. The clinical data for the CE/PE are highly variable and may come from all medical device and patient types with every possible disease or no disease at all. This data complexity and diversity often creates difficulty when evaluating and writing about the clinical data, because the evaluator has to consider many different data points, not only about the product but also about the patient, their disease, and many alternative treatment strategies. In particular, CE/PE writing is particularly difficult when the CE/PE document writing is:

Fragmented: The evaluator and CE/PE document writers must gather data from many scattered resources. For example, CERs and PERs may separate the systematic literature review (SLR) from other parts of the report, and this can be problematic. This problem may be especially pronounced for the PER where the SVR, APR, and CPR may be in separate reports. These unlinked clinical data are often poorly analyzed and the separated messages and details may be inconsistent or even incorrect. When provided as attachments to the PER, these fragmented reports may also lead the PER to become ridiculously long, unfocused, and sometimes unintelligible. A good goal is to keep the CER/PER as short, simple, and concise as possible to *focus* on the *clinical* data.

Unsupported by management: The evaluator and CE/PE document writers must be supported by senior management for the company. Since CE/PE requirements have changed so dramatically over time, business operations may require additional staff to support CE/PE document writing. Qualified leaders and managers are needed to ensure CE/PE documents meet all regulatory requirements prior to affixing the Conformité Européene (CE) mark to sell medical devices in the EU. The cost for this work is now clear to all companies marketing medical devices and IVD devices in Europe; however, the human resources and budgets allocated to writing and updating CE/PE documents may still be under development. The cost of doing business in the EU has increased and the cost to write the CE/PE documents is not trivial. Failing to support CE/PE is likely to cause a major delay for releasing a device onto the EU market.

Poorly planned: The evaluator and CE/PE document writers must plan appropriately. Both clinical and regulatory affairs team members and managers must effectively communicate the planned CE/PE development timelines (e.g., submission dates, audits, CE recertification deadlines) and costs (e.g., human resources, budgets, outsourcing needs). Without a reasonable timeline and budget, the underresourced evaluator or writer may be overwhelmed by the poor planning as they write the required CE/PE documents. The evaluator should meet with the writing team manager to discuss any roadblocks or lags in timelines and to identify clinical data gaps as early as possible. In addition, the evaluator and writers should not be the only ones working on the CE/PE documents. CE/PE activities are interrelated with other clinical, regulatory, engineering, PMS, and RM activities and planning is required between departments to ensure the CE/PE documents are appropriately resourced.

Subjective: The evaluator and CE/PE document writers must evaluate large amounts of subjective clinical data. Evaluators may differ in their opinions about how best to identify, appraise, analyze and present the clinical data in a concise and accurate manner. Subjective writing decisions often cause confusion. Similarly, the evaluator's expert opinions about how to weigh benefits against risks must be based on clinical evidence, but is also subjective. For example, two well-qualified evaluators may differ in their opinions about exactly when the benefits outweigh risks for using a medical device. CE/PE document regulatory requirements and guidelines contain a lot of "gray area" with interpretation left entirely to the manufacturer for the lowest-risk devices, to the notified body (NB) upon CE/PE document review and to the expert panel for the highest risk, implantable devices. The reviewers must determine if the CE/PE documents adequately evaluated the clinical data as required in the relevant general safety and performance requirements (GSPRs) 1, 6 and 8.

Written by inexperienced evaluators, authors, and reviewers: The evaluator and CE/PE document writers must work with evaluators, other authors and a large variety of reviewers. CE/PE documents are not typically written by one person. Often, review and input from multiple stakeholders is helpful to ensure the device description (i.e., CE/PE document scope) is correct and the clinical evaluation knowledge is integrated with other important activities like PMS, RM and clinical trial development. The clinical development plan is especially important to integrate with the CE/PE documents when a PMCF study is needed to cover the clinical data gaps identified by the evaluation documented in the CER/PER. Assembling a CE/PE document writing team to meet regularly for review requires significant effort, and keeping many individuals focused to meet required timelines can sometimes be an impossible task when team members are inexperienced. Requiring consensus or agreement by all team members is not required, especially if many team members are inexperienced, untrained, and lack sufficient understanding to evaluate the data, and arrive at valid, substantiated conclusions independently. Executive management should understand the risks if the CER/PER is illogical or poorly written: the device may lose the required CE mark and may need to be removed from the EU market. Having a qualified evaluator developing the CE/PE conclusions is required.

Poorly communicated: The evaluator and CE/PE document writers must be skilled communicators. For example, sourcing clinical data to meet minimum CE/PE document requirements requires good communication between the evaluator, writers, and department representatives responsible for the clinical trial, PMS, and RM systems. Clinical trial data may come from the clinical affairs group, clinical literature may come from the field engineering team and clinical experience data may come from the sales and marketing teams who may oversee the call center receiving the PMS and RM clinical data. Team members within these diverse departments may use the same words with different meanings and this can cause consfusion if the teams do not understand how to communicate with each other clearly, efficiently, and effectively. The evaluator and writers must be sensitive to many communication nuances when generating and sharing clinical data during the CE/PE process.

BOX 7.2 Three "difficult communication" examples for CE/PE documents.

1. Gathering clinical data from an ongoing clinical trial requires a careful description to ensure the potentially "unclean" or "interim" data are clearly documented as such. Clinical data will change as more participants are treated in the trial, as the data are "cleaned," and as the statistical analyses are completed. Using interim data is a particular hazard for CE/PE document accuracy and traceability because the clinical data may change from interim to final clinical study report.
2. Gathering clinical data from sales and marketing team members must be free from any promotional activity language or activity. Sometimes personal perceptions may influence actual clinical data reported especially when the person reporting clinical data has a significant bias. Both intentional and unintentional bias must be avoided in CE/PE document development to ensure clinical data validity.
3. Translating each individual experience report into clinical experience data requires PMS and RM team support during clinical data analysis. The PMS and RM team focus on the number of complaints and the risk occurrence rates, respectively, must now focus on the clinical details for each reported complaint or risk for the CE/PE work. Changing PMS and RM team routines to focus on clinical data can be challenging. For the CE/PE work, both the numbers/rates and the clinical data details are important. This is true because a single device-related death or serious injury may be a "game changer" when the device was otherwise considered safe. The PMS and RM systems must be updated to report all the clinical details in every case. Experience data are the largest clinical data resource and having sufficient clinical data detail in this data resource (i.e., for patient benefits and risks as well as overall device S&P) is required to allow clinical experience data relevance and sufficiency evaluations.

Gathering clinical data together from so many different places and teams is an arduous task.

Too diverse: The evaluator and CE/PE document writers must understand the device use because each specific medical device use may have a specific clinical evidence need. In addition, benefits and risks are specific to the device and the patient. For example, some devices may save lives while others may simply meet aesthetic patient preferences. Low-risk, Class I devices may not have much clinical trial or literature data available; however, they often have large amounts of low quality, highly diverse experience data available. Conversely, high-risk, Class III, implanted devices may have well-controlled, high quality clinical trial data with widely diverse literature and experience data (e.g., including many off-label uses and complicated adverse event scenarios to consider). This clinical data diversity means CE/PE document templates are difficult to develop and training persons to write CE/PE documents is complicated by the many variables encountered during the evaluation.

Not diverse enough: The evaluator and CE/PE document writers must also know what to do when the clinical data do not cover all the areas needing clinical data. Having more diverse data makes the CE/PE documents more reliable. All devices require clinical data to demonstrate S&P with sufficient diversity across all patient populations described in the device intended use and indication for use statements. Compared to well-established, lower-risk devices, novel and higher-risk devices often need more clinical data with more diversity. In this setting, the PMCFP/PMPFP must be developed with appropriate clinical data collection activities to supply the needed additional clinical data with sufficient diversity to cover all intended use and indication for use variations.

Not well supported by clinical data: The evaluator and CE/PE document writers must understand how to support the device safety and performance as well as the benefits and risks with clinical data. Sometimes clinical data are simply not sufficient in volume or quality, and too little clinical data are identified to complete the CE/PE documents. Clinical data from trials, literature, and experiences may be generally lacking and the company may refuse to run an expensive clinical trial or performance study. Even though nonclinical bench work and animal studies have been completed and even though safety reports, PMS, RM, and sales data have been gathered and combined to form a device S&P overview, the CE/PE may determine the need for additional human clinical data to be gathered. In this setting, the CER/PER conclusion should inform a new premarket or postmarket clinical follow-up/postmarket performance follow-up (PMCF/PMPF) clinical investigation to gather more clinical data.

Too complex and unrelated: The evaluator and CE/PE document writers must dissect complex clinical data to show how the clinical data support specific S&P concerns related to the medical device. For example, when searching for clinical literature on a broad term like "pacemaker," literally tens of thousands of articles are identified. The challenge is to limit the included clinical data to the most relevant, related, and highest-quality clinical data without missing novel and important clinical data. The CE/PE includes clinical data from many sources and some sources share interconnected details. To sort through this complex clinical data mixture, the evaluator may use the device description and instructions for use to select the most relevant clinical data (e.g., device technical details and specifications, historical device changes over time, regulatory status in different geographies where the device is used, marketing documents, and comparisons to other devices including those devices considered equivalent, benchmark, and similar devices will provide clarity about the specifications needed to guide relatedness).

Missing equivalent, benchmark, and similar devices when needed: The evaluator and CE/PE document writers must understand the generic device types used for comparison to the subject device during CE/PE (Table 7.1).

TABLE 7.1 Equivalent, benchmark, or similar device.

Equivalent device	Benchmark device	Similar device
Clinical, biological, and technical equivalence must be established and the clinical impact for each difference between the subject device and the equivalent device must be documented in writing with a strong rationale to justify using the equivalent device clinical data to represent the clinical data expected from the subject device.	A benchmark device has the same intended use and is a competitor's device used for clinical S&P as well as benefit-risk ratio comparisons with the subject device.	Medical Device Coordination Group (MDCG) 2020-6 [1] defined a "similar device" as a device in the "same generic device group. The MDR defines this as a set of devices having the same or similar intended purposes or a commonality of technology allowing them to be classified in a generic manner not reflecting specific characteristics" (MDR Article 2(7)).

The "equivalent" device must meet specific regulatory requirements because clinical data from the equivalent device will be used to represent the subject device while benchmark, and similar device clinical data are often grouped together with other SOTA device clinical data for comparison to the subject/equivalent device clinical data. Equivalent devices should be used more often during early device development than after the device has been on the market for some time. The benefit-risk ratio is less certain with a new device than an established device in chronic use over a long time. As a result, a new device often needs more clinical data gathered (often using clinical data from an equivalent device) to ensure the new device offers appropriate and sufficient S&P and benefit-risk ratio certainty to each individual patient.

BOX 7.3 Equivalence testing.

Clinical equivalence is the most rigorous of the three tests required for all equivalence claims in the CE/PE process. The device clinical application, intended use, indication for use, and population treated should be identical to demonstrate clinical equivalence between devices.

Biological equivalence allows more diversity since many different construction materials can have quite similar biological effects on the cells and tissues of the human body. For example, medical-grade stainless steel and inert, medical-grade plastic may have similar "inert effects" when used briefly on the external human skin surface.

Technical equivalence is the most diverse equivalence test since devices with many technical differences may still allow the same safety and performance as well as the same benefit-risk profile. For example, technical differences in equipment drape sizes and shapes might be clinically insignificant when considering the S&P and benefit-risk ratio of several different equipment drapes.

When writing the equivalence test justifications, as long as the differences are documented and the clinical impacts of the differences do not change the benefit-risk ratio or the S&P acceptability, the devices may still be considered "equivalent."

As discussed above, the EU MDR/IVDR definition of the word "equivalent" is specific regarding the clinical, biological, and technical similarity between two devices (i.e., the subject and "equivalent" devices); however, "substantial equivalence" has a different meaning when used by the United States (US) Food and Drug Administration (FDA) to define a "predicate" device in a 510(k) submission. The CE/PE and the 510(k) process documents have confusing similarities and differences, since the two word uses for "equivalent" and "equivalence" are not yet harmonized between the EU and US.

Overly specific and unique: Sometimes CE/PE document-specific terms cause confusion because individual words have several different meanings in different locations, contexts, or uses. For example, the word "risk," when used by a device engineer most often relates to device problems (e.g., breakage or mechanical parts not fitting together well, using the wrong materials or a poor engineering design to make the device); however, the same word "risk" when used by a clinician (e.g., physician, medical professional, or caregiver) most often relates to patient problems (e.g., adverse device events or side effects experienced by the patient, patient not receiving intended device benefit). The evaluator and CE/PE document writers should understand certain generic terms like risk, evaluation and data specifically refer to clinical risk, clinical evaluation and clinical data when conducing the CE/PE.

Constantly changing: The evaluator and CE/PE document writers must stay vigilant because the regulations and guidelines, enforcement dates, and the clinical data keep changing. Each CE/PE document requires many steps to complete, and the processes and requirements for this work are evolving. No one person will know all the details, and each CE/PE document will reinterpret the regulations and guidelines for specific details associated with the subject device and the specific document under development. The CE/PE document scope should be set at the kick-off meeting with details about the need to meet the current, required regulations and guidelines.

Unclear and poorly defined: The evaluator and CE/PE document writers must understand the SOTA/background knowledge, including all similar devices and alternative therapies even if they were not previously well defined. The CE/PE must document problems encountered with a similar device even if the problem occurs outside the EU because the same problem might occur with the subject/equivalent device in the EU. Exploring the breadth and depth of the available clinical data requires careful attention to define the details especially related to clinical data relevance and quality.

Confusing: The evaluator and CE/PE document writers are sometimes confused by the NB clinical experts who use regulatory discretion when they inspect and write up their clinical evaluation assessment reports (CEARs) and performance evaluation assessment reports (PEARs). This lack of consistency during CER/PER inspections can raise concerns and the manufacturer may have difficulty deciding when they are required to perform a task in a certain way or when the NB has only suggested they should do a task in a certain way. Different NB reviewers often expressed different opinions and different devices, device groupings and company histories often have different regulatory considerations for detailed clinical data requirements in various CE/PE documents.

Not compliant with the need to be a single, stand-alone report: The evaluator and CE/PE document writers must ensure each CE/PE document is a single, comprehensive, stand-alone document. Like a textbook or autobiography, CE/PE documents summarize many previous layers and details. All clinical data evaluation details are required to be in one location and the manufacturer must keep these data revised and up to date.

These examples are intended to help the reader understand why writing these CE/PE documents is so difficult. Finding device clinical data in many different public databases requires skill, and ensuring all relevant clinical data are captured while filtering out the less relevant or duplicative data is challenging. Assessing clinical data relevance to the device, the level of evidence, and the relative weight for each clinical data point requires the evaluator/writer to process clinical data from documents even when the details may be about unfamiliar topics. The CE/PE analysis and document writing requires not only understanding the contents surrounding each trial, article, and experience report, but also how to group and analyze the relevant concepts and endpoints with logical arguments and data-driven conclusions. A good CE/PE docuemnt writing practice is to “grow” the document as the clinical data become available and are evaluated, piece by piece.

7.3 Specify roles and responsibilities

Company policies and quality management system (QMS) procedures should specify the hiring requirements for for individuals who will hold CE/PE roles and responsibilities. Hiring decisions and work assignments should be based on verified documentation for each qualified individual in the appropriate roleEmployees and consultants working on CE/PE projects must be qualified based on their documented education, training, and experience as requried under the EU regulations, guidelines, and international standards. Team managers and members must ensure the CE/PE documents include all the required contents, analyses, and conclusions. To this end, medical device companies worldwide should have clearly written policies and standard operating procedures (SOPs) defining the hiring practices as well as the roles and responsibilities for CE/PE medical reviewers, clinical evaluators, writers, and other authors.

7.3.1 Medical reviewer role and responsibility

The medical reviewer should be a reputable clinician who uses the device in their medical practice and who has minimal bias, conflicts of interest or other concerns related to the company or device under clinical evaluation. The responsible medical reviewer should participate in all aspects of the CE/PE including, but not limited to:

- defining the CE/PE scope, clinical device applications, appropriate use settings, users and patient populations for each indicated device use;
- comparing the subject device to equivalent, benchmark or similar devices and uses; and
- defining the background knowledge/SOTA/alternative therapies/devices relevant to the subject device use or clinical decisions when no therapy is required.

All medical reviewer corrections, changes, and concerns should be addressed before the medical reviewer signs the CE/PE document/s. The medical reviewer role should not be underestimated, especially if the other evaluator/s, author/s, and reviewer/s do not use the device to treat patients in a clinical/medical practice. The text before the signature should specify the medical reviewer has personally reviewed, substantiated, and agreed with all CE/PE document contents and conclusions. In this way, the medical reviewer signature can validate the CE/PE.

7.3.2 Clinical evaluator role and responsibility

The clinical evaluator is required to identify, read, appraise, and analyze the relevant, high-quality clinical trial, literature, and experience data from individual documents and overall data collections. In other words, the evaluator must be able to read and critically evaluate each clinical trial report, each piece of medical literature, and each individual clinical experience report entirely and completely. In addition, the evaluator should know and understand the historical record for the device development.

Technical writing skills are critical to communicate these clinical/performance data evaluations in a clear and concise format including the reported statistical analyses, adverse events (AEs) and other clinical findings. In addition, the evaluator must determine the statistical quality of each included clinical data report. Poor statistical designs should be excluded or “marginalized” as much as possible to focus on the higher-quality, more reliable data reports (i.e., clinical data reports with solid statistical designs).

Essentially, the evaluator is the one who completes the evaluations. This role is not defined in the MDR or IVDR; however, the evaluator is defined in the MedDev 2.7/1 guideline [2] as follows:

The clinical evaluation should be conducted by a suitably qualified individual or a team. The manufacturer should take the following aspects into consideration:

- *The manufacturer defines requirements for the evaluators that are in line with the nature of the device under evaluation and its clinical performance and risks.*
- *The manufacturer should be able to justify the choice of the evaluators through reference to their qualifications and documented experience, and to present a declaration of interest for each evaluator.*
- *As a general principle, the evaluators should possess knowledge of the following:*
  - *research methodology (including clinical investigation design and biostatistics);*
  - *information management (e.g., scientific background or librarianship qualification; experience with relevant databases such as Embase and Medline);*
  - *regulatory requirements; and*
  - *medical writing (e.g., post-graduate experience in a relevant science or in medicine; training and experience in medical writing, systematic review and clinical data appraisal).*
- *With respect to the particular device under evaluation, the evaluators should in addition have knowledge of:*
  - *the device technology and its application;*
  - *diagnosis and management of the conditions intended to be diagnosed or managed by the device, knowledge of medical alternatives, treatment standards and technology (e.g., specialist clinical expertise in the relevant medical specialty).*
- *The evaluators should have at least the following training and experience in the relevant field:*
  - *a degree from higher education in the respective field and 5 years of documented professional experience; or*
  - *10 years of documented professional experience if a degree is not a prerequisite for a given task.*

There may be circumstances where the level of evaluator expertise may be less or different; this should be documented and duly justified.

The evaluator iteratively applies learning from the evaluation to draw appropriate and well-supported conclusions. Ideally, the evaluator should be involved at every time point throughout the CE/PE document development (at least to understand and approve the steps being taken by the writers and other authors). They must evaluate a large body of evidence while forming independent conclusions based on the evidence evaluated. The evaluator also makes decisions about whether the device benefit-risk ratio is acceptable for patient use as indicated, and if the evaluated and documented clinical data provide sufficient evidence to demonstrate conformity with GSPRs 1, 6, and 8. The evaluator should also drive decisions about which devices should be grouped into large device families or split into smaller groups.

7.3.3 Person responsible for regulatory compliance role and responsibility

Both MDR Article 15 [3] and IVDR Article 15 [4] specify the manufacturer shall have "at least one person responsible for regulatory compliance" (PRRC) with "expertise in the field of... medical devices," including a degree "in law, medicine, pharmacy, engineering or another relevant scientific discipline, and at least one year of professional experience in regulatory affairs or in quality management systems relating to... medical devices" or four years of professional experience. This person should assist the CE/PE evaluator and technical writing team.

7.3.4 Writer role and responsibility

The CE/PE writer must understand the writing process and they must be capable and experienced enough to write clearly about why certain clinical data are ranked or weighted higher than other clinical data. The writer must be talented enough to make the story clear. Specifically, the writer must clearly describe not only the clinical data quantity and quality but also the medical device purposes, S&P results, and benefit-risk ratio determinations made by the medical reviewers and evaluators. The writer must clearly justify in writing why the benefit-risk ratio is "acceptable" for the patient using or depending on the medical device or IVD device and the methods used by the medical reviewers and evaluators to determine and to justify this acceptability decision.

A writer with experience in the medical device industry will often have a basic understanding of the regulatory requirements, device development processes, and device technologies available in the global market; however, they may not have sufficient training or experience to evaluate all of the clinical trials, literature, and experience data. Companies should use caution here because the evaluation work should not be assigned to the writer role.

The writer should assist the medical reviewer, evaluator, and PRRC by providing technical writing support. For example, the writer will fill in tables and text as instructed and they will typically rely on subject matter experts (SMEs) because they may not be SMEs themselves. Like the evaluator role and responsibility, the "writer" role and responsibility were not defined in the MDR/IVDR, guidelines or standards, so companies should clearly articulate the role and responsibility for the CE/PE technical writers since their job is to assist the evaluator and not to do the required evaluation work which may be outside their education, training and experience. Many companies miss this point entirely and expect the technical writer to be the SME. This is a mistake for writers without the appropriate credentials.

7.3.5 Author role and responsibility

The medical reviewer and evaluator should be authors and should sign off on the CEP/PEP (where the CE/PE scope is defined) and on the CER/PER, since these documents drive the development of the other CE/PE documents (i.e., the PMCFP/PMPFP, summary of safety and clinical performance/summary of safety and performance (SSCP/SSP), if required, and the postmarket clinical follow-up evaluation report/postmarket performance follow-up evaluation report (PMCFER/PMPFER)).

Unlike the evaluator or writer roles, the author role is sometimes more about the approval process than the technical writing process. Often the author is equated to the person who will sign off on the CE/PE documents even if they have not written a single word of text in the document. An appropriate signature block should specify the qualified authors including the evaluator, reviewers and approvers. Not everyone who works on a CE/PE document should be listed as an author even if they wrote the document (i.e., under the guidance of the evaluator, reviewers and approvers).

A "gold standard" to follow for authorship is found in the International Committee of Medical Journal Editors (ICJME) definition of authors and contributors. The ICJME clearly defines the author role for peer-reviewed biomedical journal articles, and this definition can be applied to CE/PE document writers in order to differentiate who should be considered authors. Using the ICJME criteria as a best practice, each author must meet all four criteria below:

1. *Substantial contributions to the conception or design of the work; or the acquisition, analysis, or interpretation of data for the work; AND*
2. *Drafting the work or reviewing it critically for important intellectual content; AND*
3. *Final approval of the version to be published; AND*
4. *Agreement to be accountable for all aspects of the work in ensuring that questions related to the accuracy or integrity of any part of the work are appropriately investigated and resolved.*

   ... All those designated as authors should meet all four criteria for authorship, and all who meet the four criteria should be identified as authors. Those who do not meet all four criteria should be acknowledged... [5,6]

In this standard, all authors must be approvers (per criterion #3) and their contributions to the work are not trivial (per criterion #1). Simply drafting and reviewing (per criterion #2) are not sufficient to confer authorship. In addition, the last criterion (i.e., criterion #4) is often problematic for individuals who are company employees since they may not be at liberty to criticize the device or the coauthors, evaluators, medical reviewers, or others for fear of losing their jobs. Independence and integrity of authorship and clarity about the evaluator role are paramount to ensure acceptable CE/PE documents. Only those persons with appropriate education, training, and experience as defined in MedDev 2.7/1, Rev. 4 and meeting all four criteria for authorship as described above should sign off as "authors" and be responsible for the CE/PE document contents. In addition, the manufacturer should have a "disclosure of interest" (DOI) form [7] on file for each person making a significant contribution to the CE/PE. An example ICJME DOI form is available at https://www.icmje.org/disclosure-of-interest and another example is found in **Appendix N Declaration of Interest (DOI) template**.

7.3.6 Justify medical reviewers, evaluators, and authors

Specific assigned medical reviewer, evaluator, and author roles should be justified in writing by the person responsible for CE/PE documentation within the company. For example, the chief medical officer (CMO) should sign and date a statement explaining they have reviewed and selected the medical reviewer, evaluator, and author based on their documented qualifications and experiences related to the specific CE/PE work assigned. In addition, companies would be well advised to include this signed justification statement in a CER/PER appendix along with the individual resumes and declaration of interest statements for each medical reviewer, evaluator and author.

The medical reviewer, evaluator, PRRC, writer, and author roles and responsibilities are often (but not always) distinct for the various, appropriately qualified persons. In low-risk situations where a single person may write the CER/PER or any other CE/PE document/s, this person needs to be fully qualified and should embody all of the required clinical evaluator characteristics. CE/PE roles and responsibilities are often assigned using a risk-based approach.

Companies should strive to avoid assigning roles to conflicted individuals who do not have the training or experience to do the work. The evaluator must be able to read, interpret, and evaluate all the clinical trial, clinical literature, and clinical experience/complaint reports. For high-risk devices, this is a critical point because the NB person reviewing the CE/PE documentation will assess the medical reviewer and evaluator expertise. The medical reviewer, and evaluator documentation must illustrate the required clinical, regulatory, and industry experience with strong analytical, research, and writing skills.

7.3.7 Internal or external staff

Often manufacturers bring in medical reviewers, clinical evaluators, writers, and other editors and reviewers from outside the company to help with CE/PE document writing. In this situation, the QMS documents (i.e., the CE/PE SOPs, work instructions (WIs), and templates) should be written to allow both internal and external team members to evaluate data, write, review, and revise all assigned CE/PE documents. CE/PE team leaders must ensure all team members are aligned. Sometimes internal and external team members see things differently, so team leaders also need to be prepared to intervene and resolve issues as they are escalated to management for action.

Blended internal and external writing teams may have an increased potential to identify and raise previously unidentified biases or data concerns compared to entirely internal writing teams. These newly identified issues should be documented and resolved within the processes controlled by the manufacturer QMS system. Discussing these differences in clinical data evaluations, interpretations and opinions should drive improved communication among the team and higher quality in the written documents; however, sometimes, these discussions may be quite complex and the work scope may become extended which often results in seriously delayed CE/PE document development timelines. As a result, the writing, review, and discussion process timeline should be calculated carefully to ensure enough time is allotted for the draft, review, and revising process especially when resourcing blended internal and external writing teams.

One example of a blended team might be when a manufacturer is writing their first CE/PE document set for a new device and they may be bringing in an experienced, external CE/PE resource to help. In this case, the manufacturer may want to use the CE/PE SOPs, WIs, and templates from the experienced external CE/PE medical reviewers and clinical evaluators. Another example may be when a manufacturer just needs a little extra help getting specific CE/PE documents written under their own well-developed SOPs, WIs, and templates. Either way, the manufacturer QMS needs to be robust enough to handle these staffing variations while getting the CE/PE documents completed.

As a best practice, the CE/PE team manager should take time to orient both internal and external team members and to carefully plan and explain the number of review cycles, the amount of time allotted for each review round and the identity, role and responsibility for each medical reviewer, clinical evaluator, PRRC, writer, and author for each CE/PE document in each round of review. For example, sometimes individuals are assigned to review only certain sections of each CE/PE document (typically aligned with the reviewer's expertise); however, at other times, reviewers are assigned to review the entire document. Either way, a best practice is to have the entire document reviewed by at least one if not two independent reviewers. In particular, the medical reviewer should be asked to review the entire document and to ensure the background, methods, results and conclusions are accurate, adequate, and consistent with the current SOTA medical practice for the patients indicated to use the subject device.

7.4 Plan enough time to write the required documents

The CE/PE document writing timelines may vary from 3 to 19 months for many reasons. One reason may be related to company size. For example, a small start-up company may take even longer than 19 months if they simply do not have sufficient staff or clinical data collected for their device and they keep changing the device design. In this case, their application for CE Mark will be delayed until the CE/PE work is completed. A different problem may arise in a large company when the writing typically takes 12 months to complete because they have large teams working on many inter-related projects; however, the next update is due every 12 months for a high-risk, implantable medical device or a Class D IVD device. Here, the CE/PE document cycle is never ending and this never-ending cycle is simply not a good strategy, since each evaluation cycle is never actually completed.

Other issues affecting the timeline are numerous, including device clinical data complexity, device application variability, device clinical data amount, clinical data quality, etc. Logistics also plays a role in the CE/PE timing. For example, writing timelines are often affected by holidays, delays in clinical trial data availability, and publication delays. The regulatory requirements do not change (e.g., updates are still due annually for high-risk, class III, implantable devices and up to 5 years for low-risk, Class I, nonsterile devices); however, specific CE/PE documents may be dependent on data coming in from other data reporting functions in the company, such as revisions and updates to the postmarket surveillance report (PMSR) or periodic safety update report (PSUR), risk management report (RMR), etc. Careful planning with firm deadlines for document writing is a better strategy than a never-ending cycle of document writing.

7.4.1 Follow document writing process steps

Although the timeline will vary depending on the device and device use complexity as well as the amount of clinical data available, a typical 12-week timeline was developed and used repeatedly with excellent success for all device types. This 12-week timeline used the following time-based CE/PE document writing process steps:

1. Kick-off meeting (KOM): Present all clinical data and all relevant documents to the CE/PE document writing team including any feedback from the NB on the last CER/PER and the PMCFER/PMPFER which was completed since the last CER/PER.

   Note 1: Before the KOM, the PMCFER/PMPFER should already be completed with all required signatures, dates and version controls met, the CEP/PEP should be in rough draft form and the "sandbox" clinical data searches should be completed so these can all be discussed during the KOM.

2. Week 1: Use the version-controlled PMCFER/PMPFER to draft/update the CEP/PEP to include all required components and send CEP/PEP out for review. Use the complete, verified device description, CE/PE scope, and search instructions from the CEP/PEP to start writing the CER/PER and complete all initial clinical data searches.

   Note 2: Initial clinical data searches are typically completed within 1–2 days after the KOM and the search methods are recorded in the draft CER/PER.

3. Week 2: Incorporate all comments and finalize the CEP/PEP by securing all required signatures, dates, and version controls. Then, use the version-controlled CEP/PEP to repeat and document all clinical trial, literature, and experience searches.

   Note 3: All searches should be updated within 1–2 days after CEP/PEP sign-off and the final methods used should be documented in the draft CER/PER including the actual search dates and details (remember the search dates must have occurred after the CEP/PEP was signed off and all internal and external databases must be searched, including Embase for the literature searches).

4. Week 3: Complete all clinical data appraisals and secure all included clinical trial, literature, and experience data. Document appraisal methods and results in the draft CER/PER, then begin the included clinical/performance data evaluations.

   Note 4: all clinical data appraisals should be completed within 1–2 days after the searches are completed, and the methods written in the CER/PER must be easy to understand and repeat.

5. Weeks 4–8: The clinical data evaluation/analysis is the longest part of the process. The clinical/performance data analyses, tabulations and CE/PE document writing should happen concurrently even though the CE/PE document writing generally follows the analytic results tabulation. Having concurrent and iterative processes during document development helps to strengthen the final result, but requires expert time management and collaboration between team members. Here are a few suggestions:
   a. Start with tables and figures, then write narratives around each table and figure.
   b. Take clinical trial data directly from each final clinical study report (CSR) which should include the comprehensive and fully analyzed clinical trial data source. Only the relevant and required CSR data should be abstracted, tabulated, and comprehensively summarized in each CE/PE document.

c. Separate clinical literature data into background/SOTA/alternative therapy/device data and subject device data while weighing relevance and quality. Clinical literature data are often highly variable (e.g., including metaanalyses, systematic reviews, RCTs, case studies, narrative reviews) and require a great deal of work to determine how to use the various literature data most effectively and appropriately in the CE/PE documents. As a result, the clinical literature data analyses often takes longer than the clinical trial or clinical experience data analyses. Plan sufficient time for the qualified medical reviewers and evaluators to discuss the literature.
d. Include all complaints in the clinical experience data taken directly from the PMSR or periodic safety update report (PSUR) (and possibly some insights gathered from the RMR). The PMSR/PSUR results should be compared to the clinical data found in international database search results to ensure accuracy and completeness. The clinical experience data are usually the quickest dataset to analyze because most complaints and adverse event (AE) reports have so little detail. The experience data are typically grouped by AE and device deficiency (DD) type, then tabulated and evaluated with simple summaries regarding complaint numbers by type in the context of number of device units sold.
e. The data from each clinical study, literature and experience report should also be compared and analyzed across all available clinical data.
f. Be sure to use appropriate quality control (QC) checks for all clinical data and keep exploring if the clinical data presentation can be made more concise or better organized to clearly state the analytic findings.

6. Week 9: Send the entire CER/PER draft for QC review (i.e., CER/PER parts should be well-developed; however, double check the executive summary, introduction, device description, methods, clinical trial/literature/ experience data evaluation results, benefit-risk assessments, conformity with GSPRs and overall conclusions). Begin drafting the PMCFP/PMPFP and the SSCP/SSP.
7. Week 10: Send the CER/PER draft for team review including medical/clinical expert review, and require reviewers to return comments within one week. Verify all identified clinical data gaps are documented in the draft PMCFP/PMPFP and the SSCP/SSP has a 6th-grade reading level.
8. Week 11: Revise the CER/PER based on comments received and circulate the PMCFP/PMPFP and SSCP/SSP for review and approval.
9. Week 12: Facilitate sign-off on the CER/PER, PMCFP/PMPFP, and SSCP/SSP.

The CE/PE QMS, SOP, step-by-step WIs, and templates should specify how to approach the contents in each CE/PE document, section by section and each CE/PE document template should offer suggested language as appropriate. The QMS for the CE/PE process should not force all writers into prewritten (i.e., scripted) templates or processes, and cut-and-paste errors should be avoided by training staff to think and write critically, accurately, and concisely using an iterative process with built-in quality checks.

The evaluator must know how to focus the CE/PE document writing on the *new* clinical data, which will not be provided in the suggested template language.

In other words, the evaluator must become an expert at helping the writers explain the clinical data details unhindered by the template whenever necessary. The CE/PE documents must properly describe:

1. the available and fully evaluated/analyzed clinical data from the trials, literature, and experiences in both the background and subject device sections of the CE/PE documents;
2. all clinical data conclusions;
3. all specific, identified, and fully detailed clinical data gaps; and
4. all future plans and the timelines required to generate the additional clinical data to fill the clinical data gaps.

CE/PE document templates cannot be clear on these points, since the data evaluations are new each time the cycle repeats.

7.4.2 Use a draft, review, and revise writing process

Using a clear writing process should help craft the document to fit a "concise and accurate" size and shape. The evaluator must focus the writers, reviewers and editors on areas of particular importance and must ensure each interrelated CE/PE document is an independent, stand-alone document. The proposed draft, review and revise process requires careful attention by the entire team, especially the evaluator and team leader during each round of review. If left unchecked, the writing process may spiral out of control.

BOX 7.4 Revise writing specifically to avoid overstatements and understatements.

During the writing process, overstatement and understatement may occur. These inappropriate remarks should be "fixed" during the draft review and revision process.

Overstatement example: Sometimes, even though detailed clinical/performance data are required in all CE/PE documents, the text gets too focused on tangential information when a simple and clear factual statement about the evaluated data will suffice.

Understatement example: At other times, the team may be pushing too hard to minimize certain required, accurate, and important clinical/performance data details. The text may become too obscure when a simple and clear factual statement about the evaluated data will suffice.

The reviewer should identify these problems, share the appropriate data, and revise the text to accurately and appropriately balance the clinical data presentation. The medical reviewer/clinical evaluator should edit the final text to keep the text factual, well-balanced and concise.

The team must produce a complete, factual, accurate, and not misleading story based on the relevant clinical/performance evidence. A good "rule of thumb" to remember is: "the data are simply the data." The draft, review and revise process should reveal this simple, elegant clinical/performance data clarity within each and every CE/PE document. The evaluation should not overstate or understate the findings in the clinical data.

Many clinical data details need to be carefully evaluated, tabulated, and described in each CE/PE document. The conclusions must be drawn specifically and directly from the clinical data provided in the document. As the documents are drafted, a good practice may be to keep the discussion framed by the "big picture" conclusions being developed within each document and across all CE/PE documents. Each round of review and revision should bring the document closer to a clear and cohesive story while removing all extraneous, irrelevant, and poor-quality prose.

For example, in the CER/PER, both the conclusions at the end and the executive summary at the beginning must state if the clinical benefits outweigh the clinical risks to the patient and the specific clinical evidence regarding the device safety and performance when used by humans must be summarized to support these conclusions. Each reviewer should begin and end with the executive summary and the conclusion. Even when reviewing only a subpart of a CE/PE document, each reviewer should ensure the executive summary and the conclusion clearly represent the actual clinical data reviewed within the document (i.e., an updated CER/PER document should never use the identical executive summary and conclusions from the earlier document).

When drafting the CER/PER executive summary and conclusions, start by writing a clear sentence summarizing the main points from each CER/PER section and then differentiate the text in the two sections by helping the reader to see "what is coming" in the executive summary and "what was clinically evaluated" in the conclusion. These "big picture" sections should carefully describe the actual safety and performance data as well as the benefits and risks reported within each clinical data source (i.e., within the device clinical trials, literature, and experiences/uses). In this way, the evaluator/writer will ensure the most important points are covered from each CER/PER section within these "big picture" summary sections. In addition, this process will ensure the new clinical data are represented accurately and in the appropriate context within these important sections.

Knowing how to summarize actual clinical data accurately and appropriately in the conclusions and summary sections is critically important. When this part is done right, the CER/PER reader is better prepared and able to understand more easily the clinical data components within the CER/PER.

7.4.3 Know when a new clinical evaluaiton/performance evaluation document is required

Normally, the CE/PE cycle will follow the planned update cycle; however, the PMS and RM systems may identify new data and these new data may change the benefit-risk ratio. A change increasing the risk or lowering the benefit to the patient will require an unplanned update to the CE/PE documents and this update will need to start prior to the regularly scheduled CE/PE update.

When a new CER/PER is required due to this benefit-risk ratio change, the process should typically start with the updated PMCFER/PMPFER where the newly discovered data leading to the change in the benefit-risk ratio is fully evaluated by the PMS team. The draft PMCFER/PMPFER with the new clinical data clearly identified should lead directly to the CEP/PEP and the CER/PER with a subsequent PMCFP/PMPFP to gather more clinical/performance data exploring this new change, if needed. Then the SSCP/SSP will also need to be updated (if required for high risk devices only) to incorporate the changed benefit-risk ratio.

If the CE/PE was recorded in a justification (JUS) document, the JUS should be reviewed to determine if a new CER/PER is required due to the changed benefit-risk ratio.

The challenge for the CE/PE team is to stay organized and focused on the CE/PE tasks during the scramble to make sense of the new data. The CE/PE team should help to stabilize the clinical data picture by a rapid and clear response to the newly identified clinical/performance data. This can be difficult when the CE/PE cycle is interrupted. For example, the CE/PE documents require final, version-controlled PMSR/PSUR, RMR, biological evaluation report (BER), and other documents from systems outside the CE/PE process systems within the company, and these may not be available at the time when the updated CE/PE documents are needed. In this setting, the team should focus on the data available and draft CE/PE documents to help the other teams focus on the most important clinical/performance S&P details present in the newly identified data.

A good process is to start by having the medical reviewer and evaluator draft a statement about the clinical impacts of the new data. This statement should include updated S&P acceptance criteria and benefit-risk ratio tables incorporating the new data along with the old data evaluated in the prior CE/PE documents. The new data should be evaluated in the context of all clinical/performance data to determine if S&P acceptance criteria can still be met with this new data included.

After the initial "new data" triage evaluations by the medical reviewer and evaluator, the full CE/PE document development cycle should be completed unless the triage evaluations determine the CE/PE document updates can wait for the planned update cycle (Fig. 7.2).

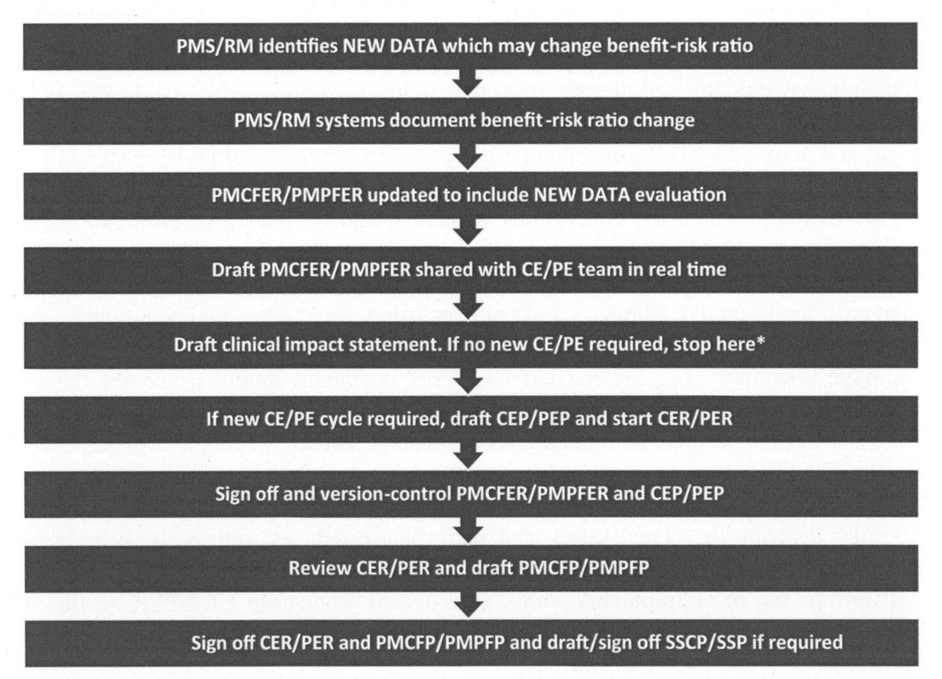

FIGURE 7.2 Restart the CE/PE cycle when new data change the benefit-risk ratio. **If the medical reviewer and evaluator determine the new data do not require an earlier-than-planned CE/PE update, then the CE/PE process can stop with the decision well documented in the RM and PMS files.*

The decision to wait for the previously planned CE/PE document cycle or to implement a new CE/PE document cycle immediately should be documented clearly in writing along with the rationale for the decision and the credentials of the person/s making the decision. These three documents (i.e., the benefit-risk ratio change, draft PMCFER/PMPFER, and CE/PE update decision documents) must be stored in the appropriate PMS, RM, and CE/PE locations within the medical device technical files or design dossiers. These documents must be easy to retrieve when the next CE/PE cycle begins.

7.4.4 Use software to shorten document writing cycle

Computer hardware and software are instrumental in controlling and expediting CE/PE writing. All team members should have access to and be trained on the software storing the QMS documents and all other systems related to writing CE/PE documents [8]. Some companies use Software as a Service (SaaS) products to help with literature reviews and document management (Table 7.2).

TABLE 7.2 CE/PE SaaS solutions.

Name	Manufacturer	Country	Description	Differentiators
CER/PER specific (especially for background literature reviews)				
DistillerSR	DistillerSR Inc.	Canada	Systematic literature reviews (subscription)	Automated screening, extraction and checking (AI enabled)
Covidence	Covidence	Australia/UK	Systematic literature reviews (subscription)	Screening, data extraction (used by Cochrane Reviews)
EPPI-Reviewer 4	Evidence for Policy and Practice Information and Coordinating Centre	UK	Systematic literature reviews (subscription)	All types of literature reviews in health, education, welfare, and other public policy sectors; reference management
SWIFT Active Screener	Sciome	US	Systematic literature reviews (subscription)	Automatically prioritizes articles (ranking)
SUMARI	System for the Unified Management, Assessment and Review of Information, Joanna Briggs Institute, Lippincott	Australia	Evidence review (subscription)	Qualitative, quantitative, economic, textual data review
Rayyan QCRI	Qatar Computing Research Institute, Rayyan Systems	Qatar/ US	Systematic literature reviews (free)	PICO-based searches (AI enabled)
SysRev	Insilica LLC	US	Systematic literature reviews (free)	Screening, data extraction
RevMan	Cochrane	UK	Review manager (subscription)	
Dialog Solutions	Clarivate	UK	Literature review services (subscription)	Search/alert management, PICO-based searches
Captis	Celegence LLC	US/India	MDR/IVDR compliance and CE document writing (subscription)	Report writing about literature and AEs (AI enabled)
Basil	Basil Systems	US	Regulatory file searching online (free)	Data mining regulatory, PMS, clinical trial data (AI enabled)
Generic software for reference and document management				
End Note	Clarivate	UK	Reference manager (subscription)	Format citations, tags
Documentum	OpenText	Canada	Document repository (subscription)	Manage documents, emails, forms, and process-created data
SharePoint	Microsoft	US	Document storage/ management (subscription)	Collaborative platform integrated with Microsoft 365
Grammarly	Grammarly Inc.	US/Kyiv	Effective communication (subscription)	Suggests grammar, spelling, style and tone (AI enabled)
ChatGPT	OpenAI	US	AI chatbot (free)	Generates human-like text
MasterWriter	MasterWriter Inc	US	Creative writing (song writing) (subscription)	Word families; dictionary and thesaurus

AEs, adverse events; *AI*, artificial intelligence; *CE*, clinical evaluation; *IVDR*, *in vitro* diagnostic regulation; *MDR*, medical device regulation; *PICO*, population or patient problem, intervention, comparator, outcomes; *PMS*, postmarket surveillance.

Some websites include systematic review guideline and software search tools such as the Systematic Review Toolbox (systematicreviewtools.com), and statistical software can be used including Stata (https://www.stata.com), R (https://www.r-project.org), and Comprehensive Meta-Analysis (CMA, https://www.meta-analysis.com). In addition, many other SaaS systems have been created, and the CEP/PEP should specify all software and databases used during CE/PE document development.

Of interest, two different author groups evaluated software tools for systematic review (SR) literature searches. The original "benchmark" article [9] compared 16 different tools and reported:

> *DistillerSR, EPPI-Reviewer, Covidence, and SWIFT Active Screener support all mandatory features. These tools are preferred for screening references, but none of them are free... The lowest scoring tools were those not specifically designed for SRs, like Microsoft Word and Endnote. Their use can only be advised for small and simple SRs.*

These authors had no conflicts of interest and they reported three criteria to consider when choosing a SR software tool to use as follows:

> *1) available funding, 2) scope and/or difficulty of the specific SR, and 3) how many SRs are planned. If sufficient funding is available, DistillerSR, EPPI-Reviewer, SWIFT Active Screener, and Covidence are appropriate choices. If the SR is very small and straightforward (dozens of references instead of hundreds), EndNote using Bramer's method or Microsoft Excel using VonVille's method can be used... Rayyan is currently considered the best free option... its only major drawback is that it does not support distinct title/abstract and full-text phases, which may cause some delays... The second-best free scoring tool SysRev... it has a few major drawbacks... it lacks distinct title/abstract and full-text phases... SysRev only supports one type of import file suitable for mass importing references (.xml)... it does not include negative and positive key-word highlighting, which slows down the screening and increases user fatigue...*

The other author group reviewed 53 web-based software tools for systematic literature review searches and features [10,11], and they reported:

> *DistillerSR, Nested Knowledge, and EPPI-Reviewer Web each offer a high density of SR-focused web-based tools. By transparent comparison and discussion regarding SR tool functionality, the medical community can both choose among existing software offerings and note the areas of growth needed, most notably in the support of living reviews.*

Unfortunately, this second report is not an unbiased report, since all five authors had significant conflicts of interest with ownership in Nested Knowledge.

The buyer should consider carefully which SaaS solution is best based on the needs of the team members doing the work and don't forget the free and low-tech solutions like Microsoft Excel and Word. Caution is advised to ensure all data can be exported from the software into a useable format if the company decides to change platforms in the future.

7.4.5 Plan for notified body reviews and unannounced audits

In addition to scheduling time for CE/PE document writing, the timeline should include time to review NB findings from earlier CER/PER assessments. Sometimes, the NB inspector may be assessing and providing feedback on the prior CER/PER version while the team is working on the next CER/PER update. This problem occurred frequently in several companies while this book was being written. In each case, the writing team was distracted by the NB comments about the prior CER/PER version and the team had difficulty separating out the issues related to the past documents under NB review and the draft documents currently being updated. Rather than trying to change the current CER/PER to address the NB comments, a better approach might be to stay focused on the planned evaluation as stated in the CEP/PEP and to await the CEAR/PEAR from the NB for the next CE/PE update as planned.

A similar situation exists for high-risk, Class IIb or III, implantable or Class D IVD devices when the CER/PER is sent to an expert panel for review. Specifically, per IVDR Article 48(6), the NB must send the PER within 5 days of receipt from the manufacturer. In these cases, the NB will wait for the expert panel review results and this may delay the NB reply by weeks or months.

The QMS should define the process to follow while awaiting and then receiving NB and expert panel feedback, if relevant. In general, following a standardized CE/PE process will help teams to stay focused and the NB and expert panel findings should only rarely interrupt the CE/PE process within any company. Once the CE/PE processes are well-established and all clinical data are repeatedly and completely evaluated on a regular basis, the interruptions by the NB or expert panel should become negligible, since the clinical data will be well understood and clearly evaluated in an ongoing manner.

7.5 Follow the natural flow when writing

Writing the device description first helps the evaluator and writer to understand the device before starting the clinical data evaluations. Understanding the device details, mechanism of action, developmental context, and how the device fits among all other therapies and devices is a good foundation for the overall CE/PE process. A good device description sets the evaluation scope and allows relevant clinical data to be identified within clinical trials, literature, and experiences. Once all the relevant, high-quality clinical data are gathered and appraised, then the included data can be tabulated, analyzed and written up.

7.5.1 Document writing flow

The CER/PER involves at least six critical components to write about in individual steps (Fig. 7.3).

1. Device Description

- Gather information from device documents, internet and FDA website searches, etc.
- Compose device component/parts tables and summarize device details
- Support description details with appropriate citations and references
- Specify CE/PE document scope

2. Clinical Trials

- Secure protocol and final clinical study report for each clinical trial
- Compose tables and summarize findings
- Write conclusions related to the subject device

3. Clinical Literature

- Conduct literature searches to identify relevant clinical data in as many databases as possible
- Export data and create master literature database in Excel workbook (or other software solution)
- Appraise literature (title --> abstract --> full article) and code for inclusion/exclusion
- Compose tables and summarize findings
- Write conclusions related to the subject device and background/SOTA/alternative therapy/device

4. Clinical Experiences

- Conduct experience searches to identify relevant clinical data in as many databases as possible
- Export data and create master experience database in Excel workbook (or other software solution)
- Appraise experience data and code for inclusion/exclusion
- Compose tables, narratives and summarize findings
- Write conclusions related to the subject device

5. Benefit-Risk Ratio

- Build benefit-risk ratio table summarizing all clinical risks and benefits to the patient
- Write conclusions related to the subject device

6. Overall Conclusion/Executive Summary

- Summarize each CER/PER section, write executive summary and overall conclusions
- Ensure benefits outweigh risks to the patient
- Verify citations, references, and appendixes are correct
- Check for deficiencies then hand off to independent reviewer to perform QC checks

FIGURE 7.3 Step-by-step CER/PER document writing flow.

This natural writing flow for the CER/PER will accommodate writing about the background and subject device clinical data separately and putting the clinical data for the subject device into the proper perspective given all the background knowledge, SOTA, and alternative therapies/devices. In addition, many manufacturers struggle with the sequence for writing the individual CE/PE documents. Sometimes the manufacturer schedules all five documents to be written with the exact same deadline. This is a mistake.

The plan (CEP/PEP) needs to come before the report (CER/PER) and the report needs to come before the follow-up plan (PMCFP/PMPFP) which needs to come before the public summary (SSCP/SSP). The follow-up plan evaluation report (PMCFER/PMPFER) typically comes after the summary has been released to the public and just before the next CEP/PEP. The sequence here is rather simple; however, scheduling time to get all work done is not simple. For example, linking CE/PE deliverables to other preexisting systems adds scheduling complexity. Specifically, outputs from four systems (i.e., the CE/PE, clinical/performance trial, PMS, and RM systems) are required in the CE/PE processes and vice versa. Unfortunately, these systems are typically managed separately, and intentionally linking these interrelated processes together takes careful planning and maintenance since each system needs to be integrated yet interdependent.Changes in one system must ripple through other systems and back again to improve all four systems. Incorporating these dynamic changes in real-time and within living documents were completely new concepts for most medical device manufacturers in 2017, and confusion remains.

Within each CE/PE document update cycle, the CER/PER must define the data supporting the device S&P story and the device benefit-risk ratio. After reading the CER/PER, the reader should understand exactly why the device is safe and performs as intended, and should be able to point to specific clinical data supporting each and every one of the clinical claims related to the device S&P and the benefit-risk ratio.

7.5.2 Device description

One good CE/PE document writing practice is always to start with a clear device description, to keep all text in focus and to have a firm grasp on the CE/PE scope. The goal is to keep all work on track and in bounds. Here are a few (but not all) important questions to answer in the CE/PE document device description:

- What are the proper names and alternative names for the device?
    - Spell out any device name abbreviations or acronyms.
- Will the CE/PE include multiple device family members?
    - If the device family member list is long, include an appendix to specify the catalog or part numbers with a brief description for each device or part included in scope. Sometimes the declaration of conformity may be used to ensure no devices are missed.
- Does the device have multiple components/accessories which much be used together?
    - If the device component/accessories list is long, include the list in an appendix.
- How is the device packaged for sale to the user?
    - Describe the components/accessories packaged with the device and if anything must be purchased separately (or from other vendors) before the device can be used as intended.
- What are the competitor devices or alternative therapies?
    - If the competitor device list is long, include the list in an appendix.
- What are the critical benefits, risks, and concerns when using the device?
    - If the benefits, risks and concerns list is long, include the list/s as appendix/es.
    - *Note: Do not include theoretical risks recorded in the RM system files; focus the CE/PE list only on benefits, risks, or concerns with documented clinical data.*
- What is the developmental history for this device?
    - How was the device developed? Is the device novel? Does the device contain drugs or cellular tissues?
- Does the device have any obvious S&P concerns?
    - List the S&P concerns and describe how the risks are mitigated.
- Has the benefit-risk ratio been acceptable throughout the device history?
    - If the device has had unacceptable risks in the past, briefly describe the history and carefully justify with clinical data why the new version of the device is considered safe enough for use now.
- What are the clinical claims made about the device?
    - If the claims list is long, include the claims list in an appendix.
    - Alternatively, if available, attach the full-length "claims matrix" as an appendix; each verbatim claim should be documented and linked directly to the clinical data substantiating each individual claim.
    - *Note: The device description should state how the CE/PE process will evaluate all clinical claims and no unsubstantiated claims will be allowed.*
    - *Note: Article 7 in both the MDR and IVDR prohibits the following untruthful or misleading statements:*

(a) *ascribing functions and properties to the device which the device does not have;*
(b) *creating a false impression regarding treatment or diagnosis, functions or properties which the device does not have;*
(c) *failing to inform the user or the patient of a likely risk associated with the use of the device in line with its intended purpose;*
(d) *suggesting uses for the device other than those stated to form part of the intended purpose for which the conformity assessment was carried out.*

- How does the device work?
- Do most users comply with the "indications for use" and avoid all contraindications during device use in the appropriate patient population/s?
- Are all scoping details clearly documented in the CE/PE document device description?

Note how different CE/PE documents require different details; however, the full device description details in all CE/PE documents should be aligned and consistent with the CER/PER device description. In addition, the CER/PER device description must align with the technical file documents (i.e., the labeling, all marketing materials, all claims, etc.). When the evaluator is writing the device description, the writer should assist with specific sequential writer-support steps and tasks (Fig. 7.4).

1. Gather information
2. Log information
3. Draft tables
4. Draft text
5. Verify information
6. Check conclusions
7. Check all citations/references

FIGURE 7.4 Writing the device description: seven writer-support steps.

Each of these steps will be repeated for each of the CE/PE document sections including the clinical trials, literature, and experience sections, as well as the benefit-risk ratio and conclusions/executive summary sections in each CE/PE document (when present).

The details for the device description writer-support steps are as follows:

1. Gather all information required in the CE/PE document template (e.g., gather all device descriptions found in the labeling and marketing information including the instructions for use, user manuals, brochures, etc. and gather all device description information from internet searches including device website information and device-related information available on FDA and other websites).
2. Log all documents into a library and include info found during internet searches including information from the NB and other regulatory agencies.
3. Draft tables and figures including detailed device description information for the CE/PE document device description including device, component, accessory, competitor, alternative therapy, claims lists, etc.
4. Draft text describing the device physical description and mechanism of action.
5. Verify exact wording for intended use, indication for use, claims, etc. as required in the regulatory documents (e.g., this step should always be required in the template and work instructions).
6. Ensure conclusion has statement about device conforming to GSPRs 1, 6, and 8 for safety, performance, benefit-risk ratio, side effect profile, etc.
7. Justify all written descriptions with appropriate citations and references.

The device description holds a significant volume and diversity of information so resist the temptation to just "put in everything" under the weak disguise of being "overly cautious." The device description should not include the entire design history file (DHF). The DHF and the rest of the technical file will be evaluated separately from the stand-alone CE/PE documents, so keep all CE/PE device description details focused on clinical information relevant to the device uses in humans. In addition, since the NB will focus on claim substantiation for all EU advertising and promotion when reviewing CE/PE documents under the MDR/IVDR, the device description must document how all device clinical claims will be evaluated in the CE/PE process, and the CE/PE process must do this claim evaluation work.

BOX 7.5 CE/PE and medical device claims, promotion, and advertising.

One interersting article [12] reviewed EU regulatory medical device frameworks for promotion and advertising, including:

- Directive 2005/29/EC about unfair business-to-consumer commercial practices;
- Directive 2006/114/EC about misleading/comparative advertising;
- individual legislation within EU Member States—even though comparative advertising is lawful when truthful and nondeceptive; the EU Directives require implementation in the EU Member States national laws and will vary between the EU Member States;
- ethical codes of conduct and professional transparency rules between business and healthcare professionals;
- European Advertising Standards Alliance (EASA) best practice for advertising claim substantiation; and
- European Diagnostics Manufacturers Association (EDMA)/European Medical Technology Industry (Eucomed), members of MedTech Europe, Code of Ethical Business Practice with "minimum standards to interact with healthcare professionals or healthcare organizations in reference to the sale, promotion or other activity of medical device products."

Starting with a definition of claims being about "benefits, characteristics, and/or performance of a product or service designed to persuade the customer to make a purchase," the author reported all "statements, results and claims must be verifiable and evidenced by data" and "properly documented." In addition, Article 2 in both MDR and IVDR defines the device intended purpose as "the use for which a device is intended according to the data supplied by the manufacturer on the label, in the instructions for use or in promotional or sales materials or statements and [or] as specified by the manufacturer in the clinical evaluation," and both MDR and IVDR require clinical investigations whenever appropriate to "confirm or disprove" clinical claims.

The author stated "any statements... in the labelling, instructions for use and promotional materials must be included" in the CER/PER, which "plays a fundamental role" in demonstrating conformity to the GSPRs. The author also warned against presenting statements without substantiation, omitting claims from the CE/PE documents and "referring to the literature that does not provide any knowledge in this respect." In other words, the author reports, the practice of filling up the CE/PE documents with irrelevant literature "is unacceptable." Specifically, the NB will assess CEs/PEs including the intended use and all resulting claims per MDR Annex VII, 4.5.5 and IVDR Annex VII, 4.5.4. The author interpreted these regulations to mean the NB "will not permit medical device... claims... unless those claims are substantiated by statistically significant clinical data." In addition, only CE-marked medical devices can be legally marketed and promoted on the EU market, and all "promotion should be restricted to the purposes for which the device has been CE-marked."

The author acknowledged "advertising and claims substantiation are complex and challenging activities for the manufacturers" and "stricter requirements for clinical evaluations will provide regulators with new tools to attack unsubstantiated claims in advertising and promotion." In addition, under MDR/IVDR, manufacturers are now required to inform uses about residual risks, and making these risk claims clear will take some work.

In conclusion, a good practice is to start writing the device definition as soon as possible and to leverage all past device descriptions without simply cutting and pasting the old device description into the new CE/PE documentes. The writer should take the time to update all relevant details and to incorporate all marketing claims into the CE/PE documents. The writer should control the CE/PE scope by clearly stating in the device description what is and is not included in the CE/PE work.

7.5.3 Clinical trial reports

Although clinical trials, literature and experience source data can be evaluated and written about concurrently, the relative clinical data source value is important to prioritize. First and foremost, a robust company-sponsored clinical trial using the subject device as indicated in the appropriate patient population is the gold standard for clinical evidence. This manufacturer-sponsored clinical trial data type will be the strongest clinical evidence in the CE/PE document if the trial is of high quality, directly relevant to the subject device and used as indicated (i.e., the trial is about the subject device used in the appropriate patient population). This clinical trial data should be well represented and emphasized in the CER/PER, and the strength of the evidence should be made clear to the reader, especially if the clinical trial was designed to answer questions about the subject device S&P and benefit-risk ratio.

When writing the clinical trial data evaluation section, the writer should assist the evaluator with the following tasks:

1. Gather all information about the clinical study from each relevant subject device CSR (e.g., clinical trial protocol, statistical analysis plan (SAP), clinical trial database information).
2. Log all documents into a library and include info found in the clinical trial registries and from regulatory agencies about the clinical trial (e.g., investigational device exemption or investigational new drug documentation).
3. Draft tables and figures including detailed clinical data from the CSR subject device clinical data including trial demographics, deviations, all relevant measurements, etc.
4. Draft text describing clinical trial methods (based on the study protocol), summarizing all trial results and conclusions, including any clinical data gaps identified.
5. Verify exact wording for clinical trial data in tables, figures, and text including conclusions, regulatory labeling, etc. by checking all data against the CSR data and text.
6. Ensure conclusion has statement about device conforming to GSPRs 1, 6, and 8 for safety, performance, benefit-risk ratio and side effect profile as identified in the CSR/s.
7. Justify all written descriptions with appropriate citations and references to the relevant clinical trial protocols and CSRs.

The tables, figures, and text should include relevant details about the clinical trial inclusion/exclusion criteria, population demographics and results including all relevant safety results and AE details along with the primary and secondary efficacy end point findings, as appropriate. The results summary should focus on the device S&P measures and benefit-risk ratio, while the conclusions should explain what was learned during the clinical trial and what clinical trial research is still pending.

Note: Detailed clinical trial data are only rarely available for background/SOTA/alternative therapy/device sections because clinical trial data are rarely shared between competing device companies. Most clinical trial data in the background/SOTA/alternative therapy/device sections will come from secondary clinical literature like metaanalyses and systematic reviews or guidelines and not from individual CSR sources.

7.5.4 Clinical literature reports

The clinical literature evaluation typically forms the largest section in each CE/PE document since comprehensive, global literature searches must document the background/SOTA/alternative therapy/device as well as the subject device clinical data. Often, literature searches identify many articles and other times the literature may not be available yet for the background or subject device. Being efficient when evaluating the clinical literature is important. The evaluator should focus on the context immediately surrounding the subject device for the background/SOTA/alternative therapy/devices section. Spending too much time and energy on irrelevant background information is a common pitfall. In addition, for the subject device section, evaluators should focus only on the relevant literature about the subject device S&P and the benefit-risk ratio.

When the evaluator is writing the clinical literature data evaluation section, the writer should assist the evaluator with the following tasks:

1. Gather all information including all included full-length articles.
2. Log all documents into a library including all literature from the manufacturer, all literature fromformal structured searches and all literature from anecdotal sources. Review all clinical literature data in master CER/PER literature database (e.g., Excel workbook).
3. Draft tables and figures including clinical literature details for author year, methods, safety and performance results, along with the author's conclusions (ensure all data in the table is relevant to the subject device).
4. Draft text describing the literature search methods and summarizing the literature evaluation findings including the S&P and benefit-risk results from the literature and the conclusions of the clinical literature evaluation (be careful not to overstate the contribution from each individual literature document; however, at the same time, do not understate the concerns expressed in the literature either).
5. Verify exact wording for clinical literature data in tables, figures, and text by checking the full-length articles (not secondary source data extracted and entered in the workbook or other software system). This is especially important when quoting details from a particular piece of literature (e.g., when author/s are presenting device-specific findings or conclusions relevant to the device S&P and benefit-risk ratio). (This step should always be required in the SOP, work instructions and templates.)
6. Ensure conclusion has statement about device conforming to GSPRs 1, 6, and 8 for safety, performance, benefit-risk ratio, and side effect profile reported in the literature.
7. Justify all written descriptions with appropriate citations and references to the literature evaluated in the CE/PE document/s.

Be sure to rank articles from strongest to weakest, and to describe appropriate cautions when including and analyzing weak clinical data. Describe the inclusion and exclusion criteria and abstract the critical device-related details from each piece of literature into a table with a narrative clinical data summary of the data in the table after analysis.

The literature for the background/SOTA/alternative therapy/device section is often focused on metaanalysis and systematic reviews of many clinical trials using various devices and therapies and leading to global treatment and practice guidelines. The conclusions of this background section should explain how the subject device fits into the global picture.

Sometimes, clinical data about the subject device are also available in metaanalyses or systematic reviews. These articles should be highlighted and discussed in the subject device section before the literature about individual trials, as long as the metaanalyses or systematic reviews are relevant to the subject device intended use and the articles include clinical data about the subject device S&P and benefit-risk ratio.

Summarize the most important S&P signals and benefit-risk ratio details first when writing the clinical literature evaluation. For example, in the CER/PER, the evaluator should document the number and percentage of S&P data analyzed from each article in a table (Table 7.3).

TABLE 7.3 Example CER/PER S&P literature analysis table.

Unique identifier	Method	Device	Safety	Performance	Conclusion
Author year	Population: objective: disease: measures:	Device name (manufacturer, city, state)	List device-related data (e.g., adverse events, clinical risks)	List device-related data (e.g., clinical benefits, device problems, did device work?)	Quote author's conclusions relevant to the device S&P or benefit-risk ratio

Keep generic article information pulled during literature identification and appraisal (e.g., article title, author names, number participants, study aim, link, author conclusion, etc.) on the same spreadsheet line. This example depicts analysis-relevant columns for CER/PER writing.

Some articles including metaanalyses, systematic reviews and narrative reviews do not contain any primary clinical data. These articles analyze and review clinical data published by others (i.e., these are "secondary sources"). In these cases, the available secondary clinical S&P data may be duplicated in the literature provided in the primary data articles and this potential for data duplication in the CE/PE documents must be addressed.

BOX 7.6 Managing duplicated clinical literature data.

Clinical data are often duplicated in published literature in many different ways. For example, an author may attempt to intentionally publish the same data in more than one journal even though the "Ingelfinger rule" applies [13]. Another type of duplication occurs when secondary sources (e.g., metaanalyses and systematic reviews) analyze previously published studies and these studies may also have unpublished, manufacturer-held clinical trial reports on file.. These data may overlap partially or completely, so the evaluator must pay careful attention to the included clinical data to ensure they are not counting the same data more than once. In these cases, the CE/PE process should link these reports together during the appraisal and analysis steps so the evidence clearly and accurately accounts for the clinical trial data.

Sometimes a company may have a publication plan to analyze and publish the data from the same clinical trial report in many bits of literature. The evaluator should analyze the multiple reports from the same study and document one cohesive study result in the CE/PE documents even though the literature had many different reports about the same study. Authors may even plagiarize the work of others as their own or they may duplicate their own work (i.e., "self-plagiarism"). These duplications should be accounted for during the clinical literature appraisal step and these unethical or illegal acts should be excluded from analysis.

The evaluator must understand how to evaluate and write without unnecessary clinical data duplication; however, they must also understand when replication is acceptable and appropriate. For example, multiple publications may be neccessary when a particularly important message needs to reach multiple different audiences or a much larger audience. In these cases, the duplicated articles should include full disclosure and should meet the terms of the journal editors and the ICJME.

The CE/PE document should have each data set are evaluated and presented only one time to avoid skewing the clinical literature evidence and deceiving the reader. If the same clinical data are reported more than once in the CE/PE document, then careful disclosure and accounting must be provided and explained.

During the clinical literature S&P data analysis, the evaluator should review each article in context of the device risks (negative outcomes/side effects/adverse events) and the device benefits (based on adequate device performance). The evaluator will want to identify (in each article) if device use decreased risk or increased benefit for the patients (or vice versa). The evaluator will want to evaluate how the subject device risks (i.e., AEs, safety concerns and device malfunctions) and benefits (i.e., health improvements and performance successes) documented in the article compared to the historical and background knowledge, SOTA, and results from similar devices, and alternative therapies including no therapy (Table 7.4).

TABLE 7.4 Hypothetical example: CER/PER S&P evaluation table (abbreviated[a]).

Article	Device	Safety	Performance
Smith, 2020	Arterial BestCatheter 2.0 F (City, NY)	2.1% (2/96) symptomatic pulmonary embolus	96.9% (93/96) had ≥75% embolus removed
Jones, 2024	Arterial BestCatheter 2.0 F (City, NY)	Direct arterial stenting had shorter hospital length of stay versus staged stenting (4.6 vs. 5.8 days, $p < 0.001$); AEs included 3 recurrent thrombus, 1 minor bleeding at surgical site for direct stenting versus 1 blood transfusion required immediately postsurgery for staged stenting	95.6% (87/91) had >80% embolus removed; immediate symptom improvement better after direct versus staged stenting [91.3% (42/46) vs 68.9% (31/45), $p < 0.001$]; embolus removal better staged than direct (85.7% vs. 81.5%, $p < 0.05$)
Doe, 2019	PigTail Arterial Catheter 2.0 F (City, NY), Arterial BestCatheter 2.0 F (City, NY)	Residual obstruction >50% (25% PigTail vs. 18% BestCatheter, $p = 0.03$); 40%–60% reflux rates; more patients than expected had bleeding complications in both groups	No difference in long-term valve function/reflux rates PigTail versus BestCatheter (58% vs. 42%)
Smith, 2010	PigTail Arterial Catheter 2.0 F (City, NY)	AEs: 57.9% (11/19) insertion point pain, 36.8% (7/19) hypotension, 15.8% (3/19) atrial spasms, 10.5% (2/19) hematoma, 10% (2/19) arterial rupture	94.7% (18/19) had >70% thrombus clearance (1 failed), 1-year primary patency 86.1%, 1 recurrent thrombosis at 1 year postinitial procedure

[a]*Note: Table is abbreviated and should provide methods, conclusions, and other details. The authors are fictitious and not meant to represent any real person/s.*

The evaluator in the hypothetical example above read and evaluated each article, identified how the authors determined safety (embolus, obstruction, and other AEs) and then documented these AEs in the safety column. Article two noted a clinical benefit related to decreased hospital stay with the target catheter used for direct arterial stenting versus the staged stenting technique. The evaluator also identified how the authors determined performance success for the device used (amount of thrombus removed) and recorded these data in the performance column ensuring all comparisons contained *p*-values (statistical significance) when provided. Article two also noted a performance benefit for immediate symptom improvement associated with direct rather than staged stenting. The evaluator should consider if they can make a statement about whether the device performed as intended/indicated, even if this was not explicitly stated by the author. If the author did make such a statement, a conclusion quote from the author/s would be useful to include in the CE/PE document.

A separate table is also typically constructed to document if the specific predefined and prequantified S&P acceptance criteria were met (Table 7.5).

TABLE 7.5 Hypothetical example: S&P acceptance criteria table.

Author year	Safety criterion <1% pulmonary embolism	Performance criterion >70% clot removal
McJones, 2021	2.1% (2/96)[a]	96.9% (93/96)
Johnson, 2024	0% (0/91)	95.6% (87/91)
Carlson, 2015	NA	NA
MacArthur, 2001	0% (0/19)	94.7% (18/19)
Range	0%–2.1%	94.7%–97%
S&P acceptance criteria	<1%	>70%
Were criteria met?	No[a]	Yes

[a]*One study was 2.1%. The authors are fictitious and not meant to represent any real person/s.*

When constructing the S&P acceptance criteria table, the evaluator must pay careful attention to specific details related to the measurement and to the criterion itself. One challenge is related to the differences in measurements used across the literature because different articles rarely use the same measurements. As a result, when extracting data from the literature for S&P evaluation table and the S&P acceptance criteria table, the evaluator must be meticulous about aligning the data using the same, precise calculations anticipated when the criteria were created. This often requires data transformation and calculation.

After all data are placed in the S&P acceptance criteria table, another challenge is evaluating and deciding what to do when these criteria are not met. In the case illustrated hypothetically above, one study failed the safety criterion and is indicated with an asterisk "*" in the table. The evaluator will need to acknowledge this failure in the text and discuss the clinical impacts of this failed acceptance criterion. Should the S&P acceptance criteria be expanded to allow a wider range? Should the device be removed from the market? The evaluator will need to state the facts and make the case with appropriate justifications about what should happen as a result of the identified clinical data gap.

When writing about the clinical literature, the evaluator and writer should be concise and should not include every detail of every article. Irrelevant details will often hide the main takeaway messages. The evaluator should not write a narrative summary for each article; instead, the evaluator should summarize the articles in a table and then describe the tabulated analysis results in the text. One paragraph should typically analyze several studies. For example, if four RCTs described the subject device compared to placebo and three RCTs compared the subject device to a particular treatment, then one paragraph summarizing the S&P results for each comparison (e.g., device vs. placebo and then device vs. treatment) may be appropriate.

BOX 7.7 More guidelines for evaluating clinical literature.

Published guidelines or handbooks are availble to help experts develop clinical evidence reports.

For example, the Cochrane Handbook for Systematic Reviews of Interventions [14] describes standard methods for preparing systematic literature reviews for publication in the Cochrane library including: planning, searching and selecting studies, collecting data, assessing bias, completing statistical analyses, using GRADE (Grading of Recommendations Assessment, Development and Evaluation) to specify levels of certainty, and interpreting results.

The Journal of the American Medical Association (JAMA) has a useful series of articles entitled "Users' Guide to the Medical Literature" available at https://jamanetwork.com/collections/44069/users-guide-to-the-medical-literature. These evidence-based articles describe "how to read" different articles and they include checklists and other useful evaluation tools.

In addition, the American Academy of Orthotists and Prosthetists (AAOP) developed the "State-of-the-Science Evidence Report Guidelines" [15] "to yield a greater understanding of the literature than is possible by simply reading the relevant literature and developing a typical literature review." And, one author group [16] evaluated and systematically reviewed 40 "Appraisal tools for clinical practice guidelines" available from 1995 to 2011. They concluded "trustworthiness of guidelines" and "conflicts of interest" among the guideline developers required more attention. As a result, CE/PE evaluators must understand the details surrounding all appraisal and critical analyis methods and tools used, including how to justify and document all CE/PE decisions and conclusions.

At the end of the clinical literature evaluation section, a conclusion paragraph must be written to describe if the S&P acceptance criteria were met, if the benefits outweighed the risks to the patient during device use and if the product literature appropriately described the risks and benefits. In particular, the conclusions must cover all the clinical literature data evaluated and must document exactly what clinical literature evidence showed the device performed safely and effectively and the device benefits outweighed the risks. If the device did not meet the S&P and benefit-risk ratio acceptability thresholds, then the clinical impact of these unacceptable findings must be summarized and the PMCFP/PMPFP must describe how any and all planned future studies will cover these clinical data gaps or how the device will be removed or temporarily withdrawn from the EU market, as appropriate.

7.5.5 Clinical experiences

The CE/PE clinical experience data analysis should identify the most common clinical complaints received by the manufacturer, the actions taken by the manufacturer to address/resolve the complaints, any currently ongoing issues and what the PMS system is doing to monitor the complaints for recurrent problems.

When the evaluator is writing the clinical experience data evaluation section, the writer should assist the evaluator as they complete the following tasks:

1. Gather all information from the internal complaints, PMSR/PSUR, RM reports, and verify the clinical experience data accurately represent the device-related information available on the international US Manufacturer And User facility Device Experience (MAUDE) database and the Australian Database of Adverse Event Notifications (DAEN).

2. Log all documents into the CE/PElibrary and include clinical data found during internet searches including experience information from the NB and other regulatory agencies.
3. Draft tables and figures with detailed clinical experience data including the report type (patient problem or device issues/deficiencies) and subgroup details.
4. Draft text describing device clinical experience data evaluation methods, clinical experience data findings, and results in context of the overall sales data and all conclusions drawn from these clinical experience data.
5. Verify exact wording when required for complaints and problems described in regulatory documents by checking the data against the source documents.
6. Ensure the CE/PE document conclusion specifies if the subject device conforming to GSPRs 1, 6, and 8 for safety, performance, benefit-risk ratio, and side effect profile.
7. Justify all written descriptions with appropriate citations and references.

One source of data for CE/PE clinical experience data will be the manufacturers own nonpublished customer complaints, surveys or other PMS clinical data (i.e., clinical experience data recorded in prior PMSRs/PSURs, and PMCFERs/PMPFERs). Sometimes clinical experience data may also be found in corrective and preventive action (CAPA) reports, so the evaluator must ensure all prior search, appraise and analyze steps have been completed adequately. The evaluator should verify the experience data found in these reports by completing independent searches of relevant international experience databases like MAUDE and DAEN. When completing these independent searches, the clinical experience data will need to be exported, appraised and analyzed as described earlier; however, if the independent searches substantiate the work done in the PMSRs, PSURs, and PMCFERs/PMPFERs, then the clinical experience data in these "secondary source" documents can be used to populate the CE/PE documents.

Clinical experience data are generally anecdotal observations reported without the rigor of a clinical trial or scientific peer-reviewed article in the medical literature. The CER/PER should describe the methods, inclusion and exclusion criteria and critical points about the device from each piece of experience data. The evaluator should describe the most common type of complaint and the actions taken by the manufacturer to resolve the compliant. The evaluator must also draw device-related conclusions from the experience data, including if the product literature appropriately described the risks and benefits, if any complaint was unresolved and ongoing, and how the PMS system was monitoring recurrent or ongoing issues.

7.5.6 Benefit-risk ratio

After the device is well described and the clinical data from trials, literature, and experiences are analyzed, the next step is to carefully finalize the benefits and risks tabulation and discussion to determine if the benefits outweigh the risks when using the device as indicated in the appropriate patient populations. To write the benefit-risk ratio evaluation, the writer should assist the evaluator as they complete the following tasks:

1. Gather all information required to evaluate the subject device benefits and risksin the context of the background knowledge/SOTA/alternative therapy/device.
2. Log all documents into the CE/PE library and include subject device benefit and risk data found during internet searches or received from the NB or other regulatory agencies.
3. Draft tables and figures summarizing the benefits and risks to the patient from all the clinical data evaluated when the device was used as indicated.
4. Draft text describing the device clinical benefits and clinical risks and conclusions regarding device benefits and risks.
5. Verify exact wording for device benefits and risks, including the benefit-risk ratio definition.
6. Ensure conclusion has statement about the subject device conforming to GSPRs 1, 6, and 8 for device benefits and risks. The device benefits must outweigh the risks or the device will most likely not be allowed on the EU market. Specify PMCF/PMPF details if additional clinical data collection is required.
7. Justify all written descriptions with appropriate citations and references.

The company should maintain a clinical benefits and risks list. Starting with the benefit-risk ratio table from the prior CER/PER, the evaluator should modify the list as new clinical data are evaluated. The benefit/risk table will be the basis to answer the question about whether the benefits continue to outweigh the risks to the users (both the individual patients and their various caregivers) when the device is used as intended based on the clinical data evaluated in the CER/PER. If the benefits do not outweigh the risks, the CER/PER should not be signed off and the device should be returned for additional device development and risk management mitigation activities to validate an acceptable benefit-risk ratio before being released to the EU market. If clinical data gaps are present, the PMCFP/PMPFP should detail the additional clinical data to be collected to address the gaps, and the CER/PER should justify why the device should be able to stay on the EU market or to enter the EU market while this PMCF/PMPF work is ongoing. If no clinical data gaps are found, the clinical data has met the S&P acceptance criteria and the benefit-risk ratio is acceptable, then, the CE/PE documents may be ready for finalization.

7.5.7 Conclusions and executive summary

After the benefit-risk ratio and PMCF/PMPF needs are determined, the conclusions and executive summary sections should be finalized to represent all clinical data in the CE/PE document. This is not a trivial point: the conclusion and executive summary sections must focus on the clinical data. The device is not meant for everyone and each CE/PE document should make this point clear, no matter what the marketing team says. The benefits and risks must be detailed and the details must be explicitly clear about how to use the device appropriately in the appropriate patients.

Often, evaluators and writers get lost in the CE/PE document work instructions and templated language and a focus to get the product on (or keep the product on) the EU market rather than on the actual clinical data evaluation and the conclusions drawn specifically from the clinical data. Each CE/PE document benefits when all CE/PE team members stay focused on evaluating the clinical data.

The CE/PE conclusions and the executive summary are meant to review all evaluated clinical data, to determine the overall subject device S&P, to define the clinical benefit-risk ratio and all residual clinical risks, and to decide if the relevant MDR/IVDR GSPRs were met. Suggested writer-support steps for writing the CE/PE conclusions and executive summary are as follows:

1. Gather all information required in the CE/PE conclusions and executive summary from each prior document-section conclusion, and from the template for the overall specific document conclusions and executive summary sections.
2. Log all documents and ensure references and appendixes are accurate and complete.
3. Draft text describing the conclusions starting with a sentence from each CE/PE document section, as appropriate. Conclusions and executive summary sections do not normally include tables or figures, since these are typically found in the results section.
4. Verify exact wording when required for GSPR conformity, clinical data gaps, and details to be listed in the PMCFP/PMPFP, etc.
5. Ensure conclusion has a well-substantiated statement about device conforming to GSPRs 1, 6, and 8 for safety, performance, benefit-risk ratio, and side effect profile.
6. Justify all written descriptions using appropriate citations and references. The conclusion and executive summary sections should be written by the evaluator and should not require references, since the data have already been referenced in the results and elsewhere, and the conclusions should be the evaluator's opinion after doing all the clinical data evaluation work.

In addition to developing the overall conclusions and executive summary sections based on the clinical data evaluated, the team will need to list all references, compose appendixes, check for deficiencies, and then hand off the full document draft to another staff member for the draft version to be checked and edited for QC. Someone other than the primary author should check the CE/PE document contents. A fresh review by a person with excellent critical thinking capabilities and preferably more training/education/experience than the initial author should help to avoid costly CE/PE mistakes and NB or expert panel deficiency findings.

The work to conclude the CE/PE document writing and to prepare for full draft review involves many steps (Fig. 7.5).

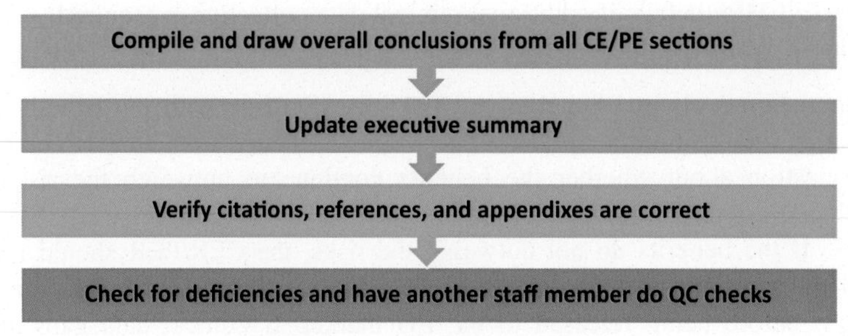

FIGURE 7.5 CER/PER conclusions.

Conclusions drawn from clinical data specifically analyzed within each section (e.g., background/SOTA, clinical trials, clinical literature, clinical experiences, benefit-risk ratio sections) should populate the overall CE/PE conclusions. The process for writing the overall conclusions can start by compiling single-sentence conclusion summaries from each individual section in the CE/PE document. Then, expand the overall conclusions to cover all the salient clinical data points from within each document section. The conclusions should also specify the clinical data gap details to be addressed in the next PMCFP/PMPFP and should provide a timeline for the PMCFER/PMPFER completion and the next CEP/PEP, CER/PER, and SSCP/SSP (if needed) updates. The evaluator must understand what details are acceptable to leave out and what must be kept in.

Once the overall conclusion is drafted, the executive summary can be updated to align with the conclusions. The emphasis in the executive summary should not be a simple numbers listing (e.g., number of trials run, number of articles evaluated, number of experiences described). Instead, like the conclusions, the executive summary should be about the high-quality clinical data within every CE/PE document section describing the clinical details about the device S&P and benefit-risk ratio with specific clinical impacts in human patients and device users.

Working on the conclusions and the executive summary together throughout the writing process is a good idea. In addition, checking the updated conclusions and executive summary contents as each document section is drafted saves time at the end of drafting step and improves QC overall. Similarly, accumulating and cleaning up the references and appendixes in the document during development will also save time at the end.

Once the document draft is completed, a first, full-document QC check for accuracy and completeness is appropriate. This first pass QC with a comprehensive CE/PE document checklist is important to catch high-level issues related to document flow, duplicated information, missing pieces, etc. and the QC reviewer should document any missing, incorrect, or unclear information for revision.

Review the process flow described in the MedDev 2.7/1 guideline to identify, appraise, analyze and then write about the clinical data. Also review the six "writing steps starting with the device description, then the clinical trial, literature, and experience data evaluations. The benefit-risk ratio and the conclusions and executive summary sections should be drafted and updated as the clinical data evaluation sections are being written and then fine-tuned once all the evaluations are completed. Each CE/PE document needs to go through the draft, review, and revise process before sign-off and release.

7.6 Evaluate statistical quality

The evaluator must assess the statistical quality used in each trial, article, and experience report. Far too often, data points and data sources in medical technology reports are based on poor statistical designs and methods. The evaluator should focus on data sources with well-designed, excellent statistical methods. Data collected and analyzed using poor statistical methods are often equivocal and may not be valid. The evaluator should document when a clinical data source lacks a sound statistical design and how this statistical flaw reduces confidence in the clinical data reliability. To complete any clinical data evaluation, the evaluator must consider at least two main statistical questions: (1) were appropriate statistical methods used to evaluate the clinical data? and (2) were appropriate details provided in the reported statistical analysis results and conclusions?

Clinical trial data come in many forms from an individual, anecdotal, case report, experiences to large adaptive clinical trials with multiple arms and multiple-stages. The clinical literature includes, but is not limited to, case studies/series, observational studies, randomized controlled clinical trials, large global systematic reviews and metaanalyses, and international guidelines. Most statistical analysis plans (SAPs) for the clinical data from clinical trials and publications use two main statistical methods: (1) simple descriptive statistics (e.g., mean, median, max, standard deviation) for things like patient demographics and enrollment numbers and (2) inferential statistics to draw conclusions from the data using statistical tests (e.g., testing the probability of a difference between data groups with student's t-text, analysis of variance (ANOVA), Chi square, etc.). Some SAPs describe sophisticated statistical models and computational methods.

Each statistical method choice typically depends on the study aims and objectives, the data types and distributions, and the data relationships (e.g., paired/unpaired data) [17,18]. Although training on statistical methods is outside the scope of this textbook, the following concepts should be evaluated:

- statistical analysis plan (SAP);
- descriptive and inferential statistics with p values;
- hypothesis testing and study objectives;
- statistical power and sample size estimates; and
- statement of clinically significant differences.

At a minimum, evaluators should ensure each clinical data source has all five items clearly stated and addressed. If not, the clinical data source should be ranked and weighed as "low quality" and should be considered for exclusion unless novel S&P data are present. If the novel subject device S&P data does not have a clear and compelling justification to exclude without bias, then keep the data in the evaluation even though the statistical quality is low.

The clinical data report should refer to the SAP. For clinical trial data, the SAP may be in the clinical trial protocol; however, sometimes the SAP is a separate, stand-alone document. For clinical literature, the SAP should be part of the methods. In excellent, high-quality clinical data sources, the SAP is clearly explained; however, in poor-quality clinical data sources, the SAP is confusing, not well explained, not well executed or missing entirely. Clinical trial and clinical literature reports without an SAP defined in the methods section should be considered for exclusion. Unlike the clinical trials and literature data, clinical experience data typically does not have a SAP because, like case studies, no statistics are possible for an N of one (i.e., the number of human subjects was one) study. Even so, case reports and experience data must be evaluated and reported appropriately because low data quality is never an automatic reason for exclusion. Important rare events are often discovered in low quality, experience data and case reports.

All clinical data documents should include descriptive statistics, like the mean, median, maximum, standard deviation, or standard error for the dataset. Most clinical datasets use descriptive statistics to describe the population demographics (e.g., mean age, height, weight, body mass index, number of men vs. women, disease state, prior treatments). Typically, each mean value should be reported with a data variation measure (i.e., the standard deviation or standard error of the mean).

When using inferential statistics to draw conclusions from the data, each clinical data report should state the hypothesis, study objectives, purpose and the methods used to test the hypothesis and each study objective as well as the study end point measures to be compared. The purpose or hypothesis should include all the components required by the acronym PICO (P = population or patient problem, I = intervention, C = comparator, O = outcomes; and sometimes T = time for the acronym PICOT).

The clinical data will be compared with inferential statistics and statistical significance will be calculated (e.g., a p-value <0.05 will demonstrate a significant difference between the two groups). For example, an RCT may compare a treatment group to a placebo control group and the resulting conclusion may be the data show a high probability the datasets are different (greater than 95% of the time is represented as a p-value less than 0.05). The report must specify if the data were paired (meaning the same subjects were measured at different times, etc.) or unpaired (meaning different subjects were compared to each other). Different statistical tests are needed for paired and unpaired comparisons due to the differences in the variation amount within each type of dataset (i.e., paired comparisons have less variability than unpaired comparisons). The statistical calculations may also document the "effect size" which measures the relationship between two variables and how important the difference is between the two variables.

Different statistical methods are used for parametric (i.e., comparing means) and nonparametric (i.e., comparing ranks, proportions, etc.) comparisons and many other specialized statistical methods are used in specific settings. For example, the Kaplan-Meier curve is commonly used to calculate and compare survival times/probabilities between groups, and mathematical methods are used to calculate the area under the curve (AUC). Appropriate statistical methods should be used to calculate and compare two or more IVD device groups for details like diagnostic accuracy, sensitivity, specificity, positive and negative predictive values, etc.

The study methods should define the minimum study population "sample size" and "statistical power" needed to see a difference between the compared groups. For example, the methods should state something like: *a sample of "N" size, will have at least an 80% power, at a minimum of at least a 95% confidence level, to see an "X" difference between the two proposed groups where N is the number of subjects to be evaluated in the clinical study and X is the clinically meaningful difference to be measured between the groups.* Many trials are underpowered, do not have an adequate sample size, or do not yet know the effect size and so cannot reliably draw conclusions. The report should justify and document how the clinically significant (i.e., clinically relevant or clinically meaningful) difference worth measuring was determined.

Each clinical data document should state if the differences found were clinically significant or not. Clinical significance is not the same as statistical significance. A statistically significant difference between two treatment options does not always translate into a clinically significant difference. For example, a large study reported a statistically significant 1-day difference between average recovery times for two groups (e.g., 59 vs. 60 days, $p<0.05$); however, the authors should not report this small difference (i.e., 1 day of recovery) as clinically significant in the larger clinical picture of the overall, often long-term, clinical benefit-risk ratio. This 1-day difference in recovery time is often considered too small to be clinically meaningful, and should not be overstated in the results and conclusions. No clinically significant difference was found between the groups for recovery time, even though, technically, the 1-day difference in recovery days between the two groups was statistically significant.

Prior to performing a study, researchers should establish the magnitude of difference needed for device performance to be considered clinically significant. To review: a few things to look for in the methods for each clinical trial, literature, or experience document or report under clinical evaluation are as follows:

- Statistical analysis details should be documented.
- SAP should be defined prior to clinical data analysis.
- Methods should specify the hypothesis and study objectives to be tested.
- Sample size, power and effect size should be clearly stated.
  - Larger sample sizes result in more statistical power; however, larger samples sizes require more resources.
  - For a result to be meaningful, a statistical power of at least 80% is recommended.
- Descriptive statistics should be used to describe data including demographic data.
- Clinical significance is not the same as statistical significance.
  - The statistical significance of any differences between groups should be clearly stated (e.g., p-value <0.05).
  - The clinical significance of any differences between groups should be clearly stated (e.g., the difference between groups is clinically meaningful or has clinical relevance for a specific reason).

In addition, since statistical significance (i.e., the p-value chosen, typically 0.05 or 0.01), study design and measurement variability are not generally changeable, the balance between the selected power and sample size (N) should be documented. The methods used to determine statistical and clinical significance for the clinical data results and conclusions should be evaluated. Another measure of statistical quality is related to how well the data are presented in the clinical data report under evaluation. The clinical data results should be clearly and completely represented in tables and graphs whenever possible and the correct representation should be chosen for the data comparisons (Fig. 7.6).

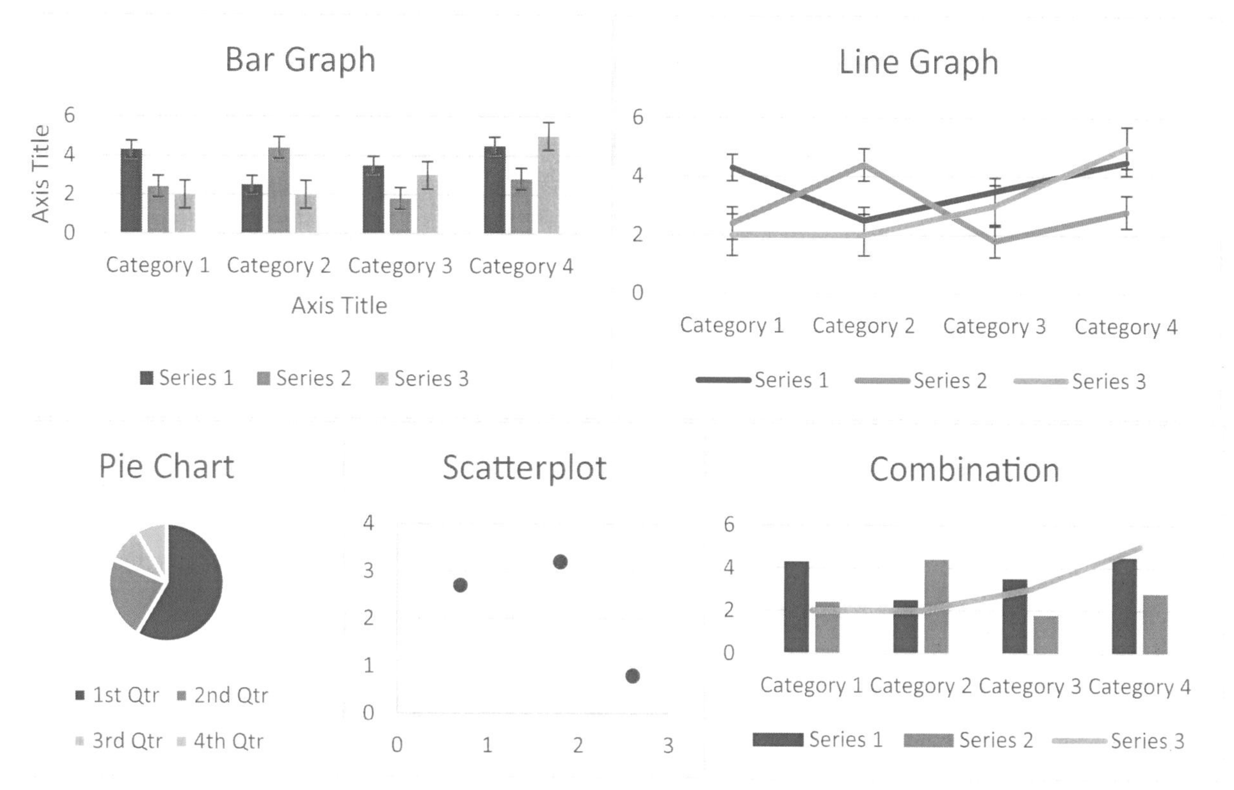

FIGURE 7.6 Clinical data visualization examples.

Key statistical concepts should be obvious in the clinical data analysis results and conclusions. Specific visual aids are useful to help the reader understand how the data points relate, the magnitude of measurements and the size of the differences between the groups. Bar charts or histograms show the counts or percentage of results at given measurements and are good for visualizing how the data are distributed within the sample population. Time series line graphs can show measurements over time. A pie chart can show how a portion of the data relates to the other portions and to the data overall. Scatterplots show one variable is defined by another variable and time series graphs show measurements over time. These are useful for visualizing trends in measurements or results. Box plots and bean or violin plots aid in visualizing the differences between two sets of data, showing how the distribution of results for one treatment compares to the distribution of treatments for another treatment. For longitudinal studies comparing rates of incidence over time or comparing survival between treatments, Kaplan-Meier survival curves create a visual representation of the differences between survival or incidence probabilities over time. Sometimes multiple types of graphs are combined within one visualization, and this may be used to good effect if the data representation is not confusing.

In every case, the results should be characterized with sufficient details (e.g., the axis titles should appear with the units measured, standard error bars should represent the amount of variation around the mean, etc.). In the clinical data report, the confidence interval (CI) is often used to indicate how close to reality the data are based on the statistical theory. This variability can be represented in graphs by standard deviation, standard error, or CI bars as the graphical representation for data variability. Understanding the CIs for specific data points can be very helpful when evaluating the clinical significance/relevance of the reported *p*-values.

Clinical trial, literature, and experience data are highly variable in both the quality of the clinical data collected and the statistical quality of the clinical data analysis and conclusions. For example, an RCT may compare a test treatment to a placebo or sham treatment and the *p*-values may show good statistical significance between the groups. Similarly, an observational uncontrolled, one-arm trial may compare the observed data to prespecified "objective performance criteria" or "performance goals." The evaluator must know the difference between studies like these and the strengths and weaknesses of each for the overall clinical data evaluation.

7.7 Writing tips

The goal is to write excellent CE/PE documents as efficiently as possible. CE/PE document writing excellence requires training, experience, and perseverance while CE/PE document writing efficiency requires speed and due diligence to get the job done right the first time. With these two key writing goals in mind (i.e., CE/PE document writing excellence and efficiency), 10 writing tips are discussed below.

7.7.1 Be rigorous and methodical about each task

The CE/PE evaluator and document writer have many tasks to complete, including (but not limited to) the following:

- Understand the regulatory submission/approval timeline; track all timeline changes.
- Gather all clinical data, workbooks/evaluations, technical reports, marketing, and clinical materials; track all missing materials.
- Start with the device description and document all device name variations; track all name changes.
- Develop the developmental context within the device description section; track all device changes over time.
- Substantiate the equivalent device when required; track all missing clinical impact statements for any clinical, biological, and technical differences.
- List comparator devices within the SOTA; track all comparators or alternative therapies.
- Focus on key S&P data from each clinical data document; track all S&P details.
- Write about similarly grouped data within the background/SOTA/alternative therapy/device section and the subject device section; track all data groups.
- Follow the relevant template, style guide, work instructions, SOPs, guidelines, and regulations; track all clinically-relevant exceptions and address these in the CE/PE document writing.
- Use tables, graphs, and figures (images) whenever possible to illustrate concepts and summarize data; track all data visualization tools.
- Compile benefits and risks while writing; track all changes to the benefit-risk ratio over time.
- Summarize the important points in the conclusion at the end of each section; track and remove all unsupported conclusions.

  The evlauator and writer must develop rigorous methods to handle these tasks and many more. All of these tasks should be accounted for in the CE/PE SOP within the Quality Management System.

7.7.2 Keep good records

The evaluator and writer should have a log to identify the precise clinical data held in each clinical data document. The log should document both positive and negative S&P clinical data to get a comprehensive and accurate device S&P understanding. This record keeping will save a lot of time when looking for specific clinical data details or supporting references. For example, keep track of all potentially useful details from the trials, literature and experiences in the workbook or CE/PE software as the literature is evaluated. Not all data is useful in CE/PE documents: check for bias, limitations, conflicts of interest, poor designs, poor statistical quality, data duplication (e.g., multiple metaanalyses using the same references) and unfounded report conclusions. In addition, tracking all comments after the full draft is sent for review will save a lot of time during the revision process for any given CE/PE document.

7.7.3 Get review feedback for each document

Do not attempt to write the CE/PE documents alone. Often one individual is tasked with the job of writing the CE/PE documents; however, the writer may not realize the importance of having other team members participate during the CE/PE document writing process. The evaluator is a critical "writing team" member and the other members of the clinical evaluation team can include the engineers, clinical trial site investigators, and marketing team members who worked to ensure the end users and patients used the device properly and safely. These voices should be considered when writing CE/PE documents. Even within the smallest companies, the work of developing the medical device CE/PE documents is rarely the work of one person. The knowledge gained during device development and marketing experiences needs to be considered and having the right access to the right people whie writing the CE/PE documents should improve S&P accuracy and should ensure all the right risks and benefits are appropriately evaluated within all the trials, literature, and experience data considered within the CE/PE documents.

7.7.4 Review, revise, reiterate evaluation details, and check data quality often

Addressing feedback and comments in real time is crucial. Do not let feedback pile up. Log comments into a spreadsheet and resolve the issues as quickly as possible. For items with no resolution, create a parking lot to document the item will not be resolved at this time.

7.7.5 Work with the medical officer, reviewer, and clinical personnel

If the company has a CMO, align details of the benefit-risk ratio as managed by the CMO. Discuss issues and concerns with the medical reviewer and clinical personnel while the CE/PE documents are being written. Writers may need to look outside the company if no CMO or clinical team are present within the company. In this case, be careful to seek general help without disclosing any company confidential information, secure appropriate confidentiality/work agreements with appropriate consultants and request additional support through the management team.

7.7.6 Create and follow the plan

Plan each CE/PE document early. For example, gather all the clinical data together while writing the CEP/PEP and layout an appropriate timeline for CE/PE document writing based on the CEP/PEP details (Table 7.6).

TABLE 7.6 Example of CER planning steps and deadlines.

Step	Deadline
Start planning early	About 6 months before CER/PER due date
Write CEP/PEP	About 4 months before CER/PER due date
Hold kick-off meeting	About 3 months before CER/PER due date
Complete all clinical data appraisals	About 2 months before CER/PER due date
Complete all clinical data analyses	About 1 month before CER/PER due date
Complete full draft of CER/PER	About 2 weeks before CER/PER due date
Allow time for sign-off	Before CER/PER due date

The CEP/PEP should house all the clinical data identification details and should describe the processes to be used when appraising and analyzing clinical data. Past CER/PER conclusions and any changes since the last CER/PER should be included in the CEP/PEP. For example, any past CER/PER which failed to meet NB requirements may need a major time investment to bring the CER/PER into compliance and major changes to the device since the last CER/PER may require significant analysis to understand the impact of those changes on the device S&P acceptance criteria and the benefit-risk ratio. These changes must be documented in the next CEP/PEP for the future CER/PER.

7.7.7 Use standard procedures, templates, and style guides

Templates and style guides make the CE/PE document writing job a little easier. The CE/PE document writer should ensure all team members are engaged at the start. For example, introduce these tools during the kick-off meeting if not before and give sufficient instruction so each team member understands how they will need to help the writer get the human clinical data accurately represented within each CE/PE document. Clinical data identification, appraisal, and analyses, including S&P and benefit-risk ratio acceptability decisions, can be subjective at times. QC checks of critical information should be done early in the process and at established checkpoints to ensure an accurate CE/PE document. The work instruction should specify the steps for CE/PE document review and revision, and a designated person should be required to adjudicate differences of opinion during CE/PE document writing.

The writer should use the appropriate CE/PE document template to start writing each CE/PE document, and the reviewer can use the template during the early review and revision cycle because the early CE/PE document will likely be closely aligned with the template. Later CE/PE document versions may not be as closely aligned with the template, so the writer should keep checking back on the template to ensure the template and work instructions are flexible enough to allow the changes occurring during the writing process. The template is just a tool to get started on the process and should not be mandated at the end of the CE/PE document writing, because the evidence accumulated during the clinical data evaluation may require a different flow of information than was envisioned in the template.

All team members should comply with the style guide at all times during the writing and especially when drafting citations in text and references in the reference section. The style guide should define the text font, table format, margin size, etc. One of the writer's jobs is to keep the CE/PE document succinct, and the style guide should help the writer to use a clear and active voice when writing CE/PE documents. The style guide should require an easily identified noun and verb in each sentence and a consistent style across all technical reports in the company including the CE/PE documents. The writer should be familiar with styles defined in the Chicago Manual of Style [19], American Psychological Association (APA) [20], American Medical Association (AMA) [21], or another similar style guide. The company style guide does not need to reproduce the details found in these manuals; however, the style guide should help writers teach the acceptable writing style and lead the team of contributors without a lot of debate about the required writing style.

7.7.8 Keep the team updated

Team members will need frequent updates and additional reminders about why the CE/PE documents are needed and what their role is within the CE/PE document writing process. Providing some basic informational slides about the function and CE/PE document components during the kick-off meeting should support the overall team goal to get the CE/PE documents completed on time. After the clinical data evaluation is underway, additional team meetings may be helpful for the team to learn about the clinical data evaluation discoveries, to understand the identified clinical data gaps and to see if someone within the company can find the missing clinical data to cover the identified gaps. If the clinical data cannot be found to address the gap, then the best course is to begin updating the PMCFP/PMPFP as early as possible to address the identified clinical data gaps. This PMCFP/PMPFP work often involves senior management allocating time and resources to do the work outlined in the updated plan, so the messaging should be clear and concise: what clinical data are needed, why is the clinical data needed, when is the clinical data needed, where and how should the data be collected, etc.

7.7.9 Check data accuracy

The corporate QMS should specify how clinical data will be checked for accuracy during the draft, review, and revise process. The analysis of each clinical data component should be double-checked for accuracy and should ensure the clinical data have been appropriately evaluated. Someone other than the initial evaluator, writer, or author should check the clinical data details, the identification, appraisal, analysis, and evaluation writing steps, especially all evaluation results and the CE/PE conclusions drawn from the data. Do not be afraid to revise the scope and conclusions based on information found or analyses done during the CE/PE document writing process. CE/PE and document writing are iterative processes. Corrections should be made and any areas of disagreement should be reviewed by a third-party adjudicator with the authority, training, and experience to overrule the author and initial quality check person.

7.7.10 Execute a smooth sign-off process

No changes can be made after a person approves/signs a CE/PE document without going through the review process again. Delays during the sign-off process are quite costly. This step usually involves senior staff members and medical professionals who expect the CE/PE document work to be fully completed before they are asked to do the final review and sign-off. This means the clinical data needs to be accurately identified, appraised, and analyzed, and entirely appropriate conclusions need to be supported directly by the clinical data in the CE/PE document. If not, a reviewer may ask for the CE/PE document completion process to be significantly delayed in order to correct the newly identified clinical data errors or omissions and to restart the review process.

7.8 Manager tips

The goal for the CE/PE team manager is to oversee and facilitate the evaluations and the CE/PE document writing. Managers should establish and communicate well with all CE/PE team members to ensure the CE/PE system is well-established, well-heeded, and well-documented to comply with the regulations, standards and the corporate QMS. To begin, the manager should ensure all team members have the appropriate education, training, and experience and 10 manager tips are discussed below.

7.8.1 Assign the right team members to the right roles

Managers who oversee CE/PE team members should make sure they assign "qualified" team members to work on the CE/PE team. The medical reviewer, clinical evaluator, PRRC, writer, and author roles were discussed above; however, other team member types, roles and responsibilities may include other subject matter experts (SMEs) often without CE/PE document sign-off or author responsibilities. These SMEs may include, but are not limited to, device engineers, regulatory team members, and sales and marketing team members; however, the sales and marketing team members should be restricted to only certain types of CE/PE meetings, roles and responsibilities, since they should not be suggesting any promotional work or consideration within the CE/PE operations. In addition, nonclinical experts may be helpful when nonclinical data are evaluated during the CE/PE process (e.g., risks and benefits may oiginate from or be traced back to nonclinical work or specific clinical/performance concerns may require additional non-clinical work). Clinical concerns found in the non-clinical testing must be appropriately documented and mitigated in theRM and PMS processes. Most non-clinical work is housed in the technical file or design dossier and does not need to be reported again in the CE/PE process unless specific clinical concerns require additional attention.

7.8.2 Develop proficiency and accountability within the team

The CE/PE team should include team members with all the required education, training, and experience, and the process should require each evaluation to be integrated and to include all the relevant and high-quality inputs from the various team members while disregarding low-quality, irrelevant inputs, when appropriate to do so. In addition to selecting the right team members, the CE/PE team leader should be responsible for developing proficient CE/PE team members and holding team members accountable.

Having the wrong team members or seriously underdeveloped team members can cause the CE/PE document writing to take years to write and can cost the company (and ultimately the patients who need the device) millions of dollars for a single CE/PE document set. These extraordinary costs were common in dysfunctional, large companies who created teams where neither the team members nor the team leader had CE/PE experience. These companies often had a culture assuming all team members had an equal voice in CE/PE decision-making even when the input was low quality and totally irrelevant to the task at hand. This is simply poor CE/PE management and indicates an absence of any viable CE/PE leadership.

Sometimes team members worked in an isolated department (e.g., clinical, regulatory, engineering) or on one isolated data type (e.g., reviewing and summarizing only clinical trial data, only literature data, only complaints data, only benefit-risk ratios), and these isolated team members and experiences often contributed to the team missing a critical rare event or misinterpreting the clinical data overall. The responsible evaluator needs to be aware of all the evaluated data, not just one small part, and they should have another qualified evaluator review the CE/PE documents especially if all the other reviewers are siloed in isolated departments and only reviewing their assigned portion of the CE/PE documents. Development of fully qualified CE/PE evaluators and team members takes time and effort.

7.8.3 Enforce appropriate timelines and control costs

Normal CE/PE document writing timelines should be about 8–12 weeks (~270–500 hours) and should cost $20,000.00–60,000.00 in 2023 dollars, depending on device classification and complexity among other cost drivers. Higher-risk and more complex devices cost more than lower-risk, less complex devices. Cost drivers include, but are not limited to:

- human resource (HR) costs for time from medical reviewer, evaluator, writer, SME, etc.;
- number of team members and meetings required to get work done;
- CE/PE software costs;
- overhead costs;
- literature database/librarian costs;
- data storage and retrieval costs;
- interim data analysis for clinical trial costs, as needed;
- literature copyright costs;
- timeline delays due to missing data;
- timeline delays due to a need to generate new clinical data; and
- NB findings specifying rework or new/revised clinical data evaluation.

The CE/PE team leader should be responsible for keeping the timeline and costs under control. Often the team leader may be the evaluator to keep costs low.

7.8.4 Chief medical officer is ultimately responsible

The CMO within the company (when present) should be responsible for the CE/PE document writing. Any company without a CMO should carefully consider this potential ethical and legal dilemmas, which could arise if the company has no one with sufficient clinical and medical training and experience to evaluate the CE/PE documents to meet company needs. Any confusion about when the device should be pulled from the EU and other markets for a clinically-related reason would be costly.

Prior to the EU MDR/IVDR, many small to medium-sized medical device companies did not have a CMO or clinical department. These companies hired medical reviewer/advisor consultants to serve in this role. Obviously, the part-time CMO consultant will be less connected to the day-to-day clinical operations of the medical device company than a full time CMO. In addition, the part-time CMO consultant may struggle to have enough time to keep up with all the medical device S&P details coming in about the device. The CMO is responsible to help the company staff to understand and interpret the clinical impacts of all the S&P details, and the CMO should guide the company on the changing benefit-risk ratio for the medical device. For these reasons, many medical device companies in the US began to hire clinical teams and CMOs since the EU MDR/IVDR was enacted in 2017.

Both the chief executive officer (CEO) and the CMO should be aware of their professional as well as personal liabilities during these CE/PE processes. CE/PE documents are critical regulatory and legal documents for every medical device sold in the EU.

7.8.5 Require team members to check in and approve drafts in real time

Managers should make sure the medical reviewers, evaluators, writers, and other team members check in and seek approval for their work often so the timeline does not creep out of scope. These approvals should not require tracking during the drafting process. Each team member should be responsible for suggesting improvements to the CE/PE documents and the qualified author/evaluator should be responsible for the overall CE/PE document contents.

Managers should ensure the style guide is followed and writing quality checks are completed as the CE/PE documents are developed. The manager should strive to keep writers writing well without waste. For example, at a minimum, each written item must be clear and correct with all required components as follows (but not limited to these):

- ✓ Device description: device family, CE/PE document scope, developmental context
- ✓ Clinical trial: protocol method, statistics, results, conclusions
- ✓ Clinical literature: search, appraisal and analysis methods, results, conclusions
- ✓ Clinical experience: data sources, analysis methods, results and individual clinical complaints in context of sales volume
- ✓ Benefit-risk ratio: individual benefits and risks "weighed" against each other and acceptability determined and justified
- ✓ Overall conclusions/executive summary: details from all CE/PE document sections

Managers should check progress through all 10 stage gates and should ensure all CE/PE document writing aligns with the clinical data evaluated (Table 7.7).

TABLE 7.7 Manager should verify stage gate progress.

Stage gate	Description	Manager to verify:
Initiating	Data gathering	CEP and scope are clear
Planning	Kick-off meeting	Team members understand their roles and responsibilities
Searching	Identifying data	Search strategies for trials, literature, and experiences were approved
Coding	Appraisal	Inclusion/exclusion coding and weighting were approved
Analyzing	Tables, lists, and comparisons	Tables, risks and benefits lists, and clinical data gap details were approved, and predetermined and prespecified S&P requirements were met
Writing	Sections and conclusions	Clinical data descriptions and evaluations were sound; style guide was followed; drafts, reviews and revisions were completed
Integrating	PMCFP/PMPFP	Clinical data gaps were covered and draft/review/approve steps completed
Summarizing	SSCP/SSP	Summary documents completed draft/review/approve steps
New data	Generate new data	PMCFP/PMPFP accumulated new clinical data as planned
Follow-up	PMCFER/PMPFER	New clinical data reports from PMCF/PMPF activities completed draft/review/approve steps

The CE/PE document writing process is not always linear (and probably should not be). The process must allow learning as the clinical evaluations and CE/PE document writing proceeds. New clinical data from a trial, a piece of literature or a complaint/experience report may change the clinical data evaluation and interpretation. In addition, the outcome may be different each time the clinical data evaluation is updated with new clinical data. For example, at the beginning of the medical device life cycle, only a small amount of clinical data may be available and all device clinical data may clearly show the benefits outweight the risks; however, at a later time point when more clinical data are available from many different sources and locations, a rare risk may become obvious and the device may need to be redesigned, removed from the market or sold only to a more restricted patient population after a significant change in the device labeling. Alternatively, the new clinical data may change the way the clinical data are appraised and weighted.

Managers need to support the team during all evaluation steps and especially when finalizing each CE/PE document. Managers should schedule time for senior leaders to review and approve each CE/PE document and the manager may need to remind the medical reviewer, evaluator, authors, etc. to finalize the CE/PE document in the version-controlled software system and to sign-off once each document is completed. Once signed off, the manager should ensure each CE/PE document is correctly compiled, published, and archived in the version-controlled software system.

7.8.6 Create a "parking lot" for unresolved issues and move on

Managers need to watch out for situations where the CE/PE document flow is interrupted by inconsistent or unavailable information. One of the keys to efficient CE/PE document writing team management is to focus the team members on directly managing what can be accomplished with the clinical data at hand while escalating requests to clarify inconsistent or missing clinical data and information to more senior management team members. Using a "parking lot" spreadsheet in the Excel workbook to document the inconsistent, missing data or suggested trials which the company is not able to complete, etc. should help CE/PE document evaluators, writers, and authors to work around these issues even when significant items are listed in the parking lot. The team manager will need to review the parking lot often to decide if any listed items will mean the CE/PE documents cannot be completed if the item is not addressed. The team manager should escalate these critical "roadblock" issues to senior management to see if the CE/PE document writing should be stopped or if the risk of receiving unacceptable findings from the NB review (or expert panel review for high-risk implantable devices) is acceptable to the company.

7.8.7 Store and use clinical evaluation/performance evaluation documents as learning tools

The CE/PE document should be stored in a central, easily accessible resource library. Once the CE/PE document is signed off, the evaluator, writer, and the entire CE/PE document writing team collectively have learned a great deal about the device. This institutional knowledge should be shared and amplified without being lost. The manager and the team members should offer a "lessons learned" discusson for interested persons as a good CE/PE practice.

7.8.8 Avoid notified body deficiencies

NB deficiency letters are expensive and wasteful. Writing a CE/PE document to sail through the NB review without a single deficiency is an appropriate goal. Managers should ensure they have a seasoned team including a qualified medical reviewer, an appropriate, experienced evaluator, excellent writers/authors, and sufficient expert team members and SMEs to meet this CE/PE goal. External resources should be brought in to facilitate learning as well as document completion and to ensure the broader (independent and potentially less-biased) views from outside the company are heard.

7.8.9 Ensure benefits outweigh risks

The manager should ensure the team developed the benefit-risk ratio well. The clinical team and the CMO should review and comment on the clinical impacts for all benefits and risks to ensure the benefit-risk ratio is acceptable. If the benefits do not outweigh the risks, the CE/PE documents should probably not be completed and the device should not be placed on the market.

7.8.10 Review thoroughly to create "successful" clinical evaluaiton/performance evaluation documents

A successful CE/PE document has a specific look and feel. For example, the CER/PER must have all available clinical data identified, appraised and analyzed in a clear and succinct manner, the SOTA must be clearly described and properly referenced, and the conclusions must be well-founded and specifically and directly based on the clinical/performance data analyzed in the report. Both the data volume and data diversity must demonstrate clear evidence showing the device S&P were acceptable when the device was used as indicated in the appropriate patient populations. The evaluator and manager should complete a thorough review of all CE/PE elements (Fig. 7.7).

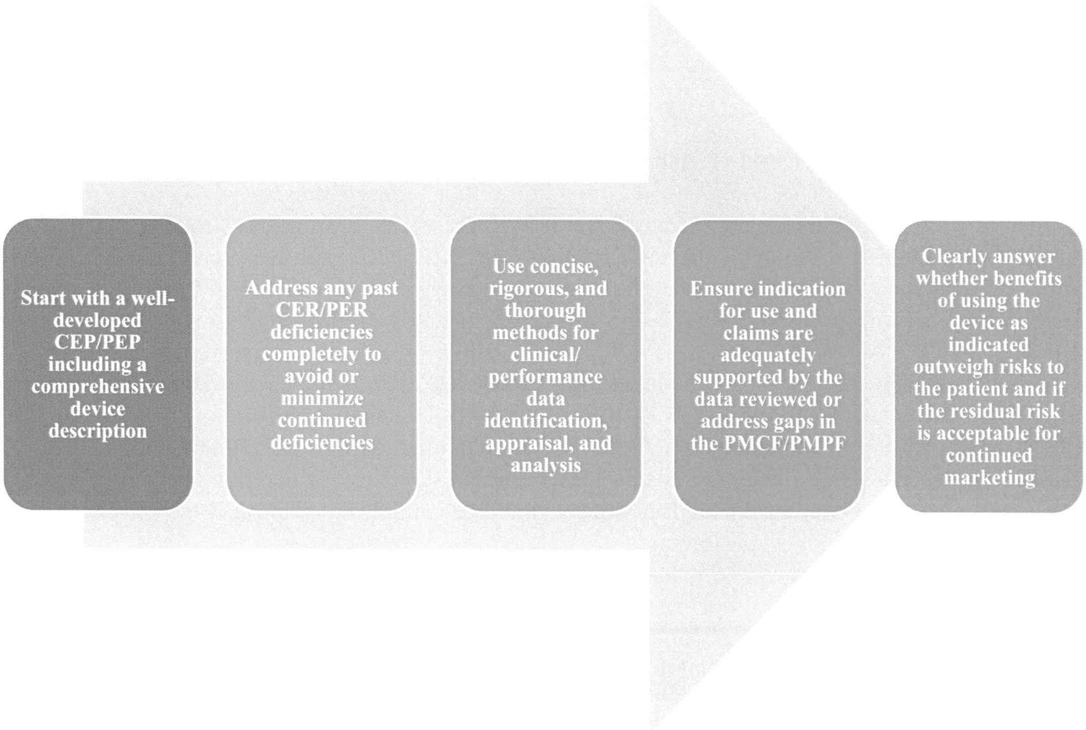

Start with a well-developed CEP/PEP including a comprehensive device description

Address any past CER/PER deficiencies completely to avoid or minimize continued deficiencies

Use concise, rigorous, and thorough methods for clinical/performance data identification, appraisal, and analysis

Ensure indication for use and claims are adequately supported by the data reviewed or address gaps in the PMCF/PMPF

Clearly answer whether benefits of using the device as indicated outweigh risks to the patient and if the residual risk is acceptable for continued marketing

FIGURE 7.7 Successful CER/PER elements.

One key to successful CE/PE document development is having a solid CEP/PEP with clear S&P acceptance criteria, focused clinical data identification, appraisal, and analysis methods, and details about how to address any and all prior CE/PE document deficiencies. Successful CE/PE document writing includes a fair and balanced summary of all clinical/performance data and the benefit-risk ratio. In addition, clinical data gaps must be identified and tracked in the PMCFP/PMPFP with a reasonable timeline for closure. The CE/PE documents must evaluate sufficient clinical/performance data when the subject device was used as indicated in all potential use populations and settings. Another key to successful CE/PE documents is to report all relevant clinical data fairly and completely, without bias. Successful CE/PE document writing involves careful planning, routine reviewing, and rigorous appraisal and analysis steps. Meeting with all stakeholders at the start and regularly throughout the CE/PE document writing process helps to ensure timetables are on track. Essentially, every company department has something at stake in the CE/PE process and should be interested to help ensure the timely and comprehensive CE/PE document completion.

Ultimately, each successful CE/PE document will represent the subject device accurately in a concise, single, stand-alone document (Fig. 7.8).

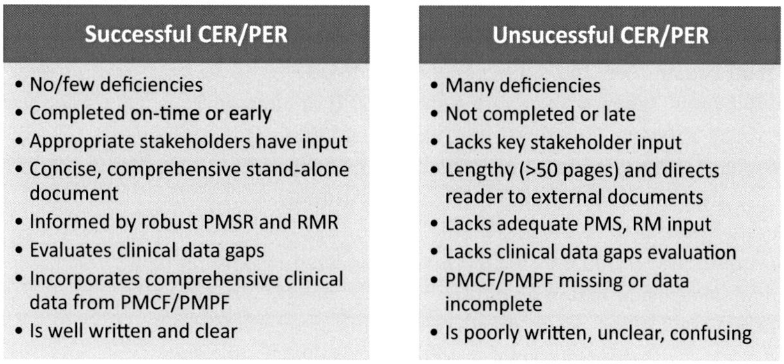

Successful CER/PER	Unsucessful CER/PER
• No/few deficiencies	• Many deficiencies
• Completed on-time or early	• Not completed or late
• Appropriate stakeholders have input	• Lacks key stakeholder input
• Concise, comprehensive stand-alone document	• Lengthy (>50 pages) and directs reader to external documents
• Informed by robust PMSR and RMR	• Lacks adequate PMS, RM input
• Evaluates clinical data gaps	• Lacks clinical data gaps evaluation
• Incorporates comprehensive clinical data from PMCF/PMPF	• PMCF/PMPF missing or data incomplete
• Is well written and clear	• Is poorly written, unclear, confusing

FIGURE 7.8 Successful versus unsuccessful CERs/PERs.

The external CE/PE document success validation will be when the NB CEAR/PEAR documents contain no CE/PE document deficiencies.

7.9 Conclusions

Writing CE/PE documents requires team members to be organized and focused on the clinical data. The CE/PE document writing should be complete, comprehensive, and compelling. Every company should strive to attain a robust and well-organized CE/PE document, even though these CE/PE documents are difficult to write. Clinical evaluation team members must be qualified through education, training, and experience, and many clinical evaluation roles and responsibilities exist, including, but not limited to, the medical reviewer, clinical evaluator, PRRC, writer, and author.

The KOM should provide a working knowledge of the documented clinical evidence presently supporting the acceptable device S&P and benefit-risk profile, as well as the potential gaps in ongoing surveillance and analysis which may tend to hinder CE/PE document development. An appropriate plan with enough time to write the CE/PE documents is required. Understanding and following the natural flow of document writing will help expedite the CE/PE document by starting with the device description and moving through the clinical trials, literature, and experience data and into the benefit-risk ratio, and finally the conclusions/executive summary. In addition, statistical quality must be evaluated to identify high-quality clinical data, and writers should be careful to document if the statistical methods and results were appropriate in each clinical data item.

CE/PE document evaluators, writers and managers actually have a report card showing how well the clinical data were evaluated and how well the CE/PE documents were written to demonstrate conformity with GSPRs 1, 6, and 8 about the safety, performance, residual risks, and benefit-risk ratio for the medical device used as indicated. This report card is the CEAR/PEAR which lists any and all deficiencies identified by the NB reviewer.

To summarize: getting good grades in the "report card" from the NB (i.e., getting few to no CE/PE document deficiencies from the NB) requires each CE/PE document to be accurate as well as complete, comprehensive, and compelling. CE/PE document writing team members including all team managers, medical reviewers, evaluators, writers and authors must complete all training, planning and evaluation steps flawlessly. The CE/PE document writing team manager must understand the strengths and especially the weaknesses of each CE/PE document writing team member. All team members should be held accountable to do their part, and the roles and responsibilities for each team member including the "signing authors" must be clear and enforced. The NB will hold the CEO and CMO professionally and personally liable for the CE/PE document contents.

7.10 Review questions

1. What do the terms N, *p*-value, CI, AUC, and SAP mean?
2. Why are CE/PE documents so difficult to write?
3. What are the "three Cs" of CER/PER writing?
4. Who is qualified to be the "evaluator" as defined in MedDev 2.7/1?
5. How can someone who wants to learn to write CE/PE documents be trained?
6. What are the four prohibited, untruthful or misleading statements defined in Article 7 for both MDR and IVDR?
7. True or false? All clinical claims must be substantiated in the CER/PER.
8. What is often the "strongest clinical evidence" in the CE/PE document?
9. Although the timeline will vary, what is a typical timeline for the CE/PE document writing process steps?
10. How are equivalent, benchmark, and similar devices defined?
11. What is the "natural flow" when writing individual CE/PE documents?
12. What is the difference between statistical significance and clinical significance?
13. What are the two main statistical methods defined in this chapter?
14. List three human study population demographics data examples.
15. What does a Kaplan–Meier curve illustrate?
16. What are the minimum power and confidence levels expected in clinical data?
17. Name three clinical data visualization examples.
18. What three published style guides were mentioned in this chapter?
19. What are at least three "cost drivers" when working on CE/PE documents?
20. How is the "parking lot" helpful to the CE/PE document writing team?

References

[1] Medical Device Coordination Group. MDCG 2020-6. Regulation (EU) 2017/745: clinical evidence needed for medical devices previously CE marked under Directives 93/42/EEC or 90/385/EEC: a guide for manufacturers and notified bodies. April 2020. (Accessed 30 May 2024). https://health.ec.europa.eu/system/files/2020-09/md_mdcg_2020_6_guidance_sufficient_clinical_evidence_en_0.pdf.

[2] European Commission. Medical device directives. Guidelines on medical devices. Clinical evaluation: a guide for manufacturers and notified bodies under Directives 93/42/EEC and 90/385/EEC. Med Dev 2.7/1 Rev 4. June 2016. (Accessed 30 May 2024). https://www.medical-device-regulation.eu/wp-content/uploads/2019/05/2_7_1_rev4_en.pdf#page = 20&zoom = 100,27,785.

[3] European Parliament and Council of the European Union. Consolidated text: regulation (EU) 2017/745 of the European Parliament and of the Council of 5 April 2017 on medical devices, amending Directive 2001/83/EC, Regulation (EC) No 178/2002 and Regulation (EC) No 1223/2009 and repealing Council Directives 90/385/EEC and 93/42/EEC. (Accessed 30 May 2024). https://eur-lex.europa.eu/legal-content/EN/TXT/?uri = CELEX%3A02017R0745-20230320.

[4] European Parliament and Council of the European Union. Consolidated text: regulation (EU) 2017/746 of the European Parliament and of the Council of 5 April 2017 on in vitro diagnostic medical devices and repealing Directive 98/79/EC and Commission Decision 2010/227/EU. (Accessed 30 May 2024). https://eur-lex.europa.eu/legal-content/EN/TXT/?uri = CELEX%3A02017R0746-20230320.

[5] International Committee of Medical Journal Editors (ICJME). Recommendations for the conduct, reporting, editing, and publication of scholarly work in medical journals. Updated in May 2023. (Accessed 30 May 2024). https://www.icmje.org/icmje-recommendations.pdf.

[6] ICJME. Defining the role of authors and contributors. (Accessed 30 May 2024). http://www.icmje.org/recommendations/browse/roles-and-responsibilities/defining-the-role-of-authors-and-contributors.html.

[7] ICJME. Disclosure of financial and non-financial relationships and activities, and conflicts of interest. (Accessed 30 May 2024). https://www.icmje.org/recommendations/browse/roles-and-responsibilities/author-responsibilities--conflicts-of-interest.html .

[8] Center on Knowledge Translation for Disability & Rehabilitation Research (KTDRR) at American Institutes for Research. Software tools for conducting systematic reviews. Updated June 26, 2023. (Accessed 30 May 2024). https://ktdrr.org/resources/sr-resources/tools.html.

[9] S. van der Mierden, K. Tsaioun, A. Bleich, C.H.C. Leenaars, Software tools for literature screening in systematic reviews in biomedical research, ALTEX 36 (3) (2019) 508–517. Available from: https://doi.org/10.14573/altex.1902131, https://www.altex.org/index.php/altex/article/view/1257/1335. (accessed 30.05.24).

[10] K. Cowie, A. Rahmatullah, N. Hardy, K. Holub, K. Kallmes, Web-based software tools for systematic literature review in medicine: systematic search and feature analysis, JMIR Med. Inf. 10 (5) (2022) e33219. Available from: https://doi.org/10.2196/33219, https://www.ncbi.nlm.nih.gov/pmc/articles/PMC9112080/. (accessed 30.05.24). PMID:35499859 PMCID: 9112080.

[11] K. Cowie, A. Rahmatullah, N. Hardy, K. Holub, K. Kallmes, Correction: web-based software tools for systematic literature review in medicine: systematic search and feature analysis, JMIR Med. Inf. 10 (11) (2022) e43520. Available from: https://doi.org/10.2196/43520. Erratum for: JMIR Med Inform. 2022 May 2;10(5):e33219. PMID:36417760; PMCID: PMC9746781. Accessed on 30 May 2024 at https://www.ncbi.nlm.nih.gov/pmc/articles/PMC9746781/.

[12] M. Malinowski, Advertising of medical devices and principles of claim substantiation in European Union, Am. J. Biomed. Sci. Res. 12 (4) (2021) 389–391. Available from: https://biomedgrid.com/pdf/AJBSR.MS.ID.001777.pdf. AJBSR. MS. ID. 001777. Accessed July 3, 2023. https://doi.org/10.34297/AJBSR.2021.12.001777.

[13] C. Johnson, Repetitive, duplicate, and redundant publications: a review for authors and readers, J. Manip. Physiol. Ther. 29 (2006) 505–509. Available from: https://doi.org/10.1016/j.jmpt.2006.07.001, https://www.jmptonline.org/action/showPdf?pii = S0161-4754%2806%2900182-5. (accessed 30.05.24).

[14] Higgins J.P. T., Thomas J., Chandler J., Cumpston M., Li T., Page M.J., et al. (editors). *Cochrane handbook for systematic reviews of interventions*. Version 6.3. Updated February 2022. Cochrane. 2022. (Accessed 30 May 2024). https://training.cochrane.org/handbook/current.

[15] American Academy of Orthotists & Prosthetists (AAOP). State-of-the-science evidence report guidelines. 2008. (Accessed 30 May 2024). https://cdn.ymaws.com/www.oandp.org/resource/resmgr/docs/pdfs/aaop_evidencereportguidlines.pdf.

[16] U. Siering, M. Eikermann, E. Hausner, W. Hoffmann-Eßer, E.A. Neugebauer, Appraisal tools for clinical practice guidelines: a systematic review, PLoS One 8 (12) (2013) e82915. Available from: https://doi.org/10.1371/journal.pone.0082915, https://www.ncbi.nlm.nih.gov/pmc/articles/PMC3857289/. (accessed 30.05.24). PMID:24349397; PMCID: PMC3857289.

[17] P. Mishra, C.M. Pandey, U. Singh, A. Keshri, M. Sabaretnam, Selection of appropriate statistical methods for data analysis, Ann. Card. Anaesth. 22 (2019) 297–301. Available from: https://www.ncbi.nlm.nih.gov/pmc/articles/PMC6639881/pdf/ACA-22-297.pdf. (accessed 30.05.24).

[18] Cleophas T.J., Zwinderman A.H. Understanding clinical data analysis: learning statistical principles from published clinical research. Springer eBook ISBN 978-3-319-39586-9. Published: August 23, 2016. Available from: https://doi.org/10.1007/978-3-319-39586-9.

[19] *The Chicago Manual of Style*, 17th ed. Chicago: University of Chicago Press, 2017. https://doi.org/10.7208/cmos17. Accessed May 30, 2024. https://www.chicagomanualofstyle.org/home.html.

[20] American Psychological Association. *Publication manual of the American Psychological Association*. 7th ed. 2020. Accessed May 30, 2024. https://apastyle.apa.org/products/publication-manual-7th-edition.

[21] The JAMA Network Editors. *AMA manual of style: a guide for authors and editors*, 11th ed. March 2, 2020. Accessed May 30, 2024. https://www.amamanualofstyle.com/.

Chapter 8

Writing safety and performance summary documents

The summary of safety and clinical performance (SSCP) is required in the European Union (EU) Medical Device Regulation (MDR) for high-risk implantable and class III medical devices (MDR Article 32(1)) and the summary of safety and performance (SSP) is required in the EU *In Vitro* Diagnostic (IVD) Regulation (IVDR) for high-risk Class C and D IVD devices (IVDR Article 29(1)). SSCPs/SSPs summarize the relevant clinical data drawn from the clinical evaluation report/performance evaluation report (CER/PER), postmarket clinical follow-up plan (PMCFP) and evaluation report (PMCFER)/postmarket performance follow-up plan (PMPFP) and evaluation report (PMPFER). These clinical data are verified against other technical file documents including the postmarket surveillance (PMS) and risk management (RM) file records, as well as the instructions for use along with all other product labeling.

The summary of safety and clinical performance (SSCP) and the summary of safety and performance (SSP) are designed to highlight clinical data of specific interest and relevance to healthcare professionals (HCPs) users and patients. The documents review subject device clinical evaluation/performance evaluation (CE/PE) data including device safety and performance (S&P) details and benefit-risk ratios in the context of therapeutic and diagnostic alternatives and various use conditions. Device benefits, risks, residual risks, undesirable effects, and positive/negative CE/PE findings are summarized, and the SSCP/SSP is intended to provide a concise and comprehensive device clinical data summary for users and patients.

The SSCP/SSP must be "written in a way that is clear to the intended user and, if relevant, to the patient and shall be made available to the public" [1,2]. In addition, the device labeling (e.g., instructions for use) must indicate where the SSCP/SSP was made publicly available (e.g., on the company website).

8.1 Writing objectives for summary documents

The SSCP/SSP is intended to fulfill MDR/IVDR objectives to enhance information transparency and to provide access to medical device information. SSCPs/SSPs are written using source documents such as technical file documents and the CER/PER. In particular, the SSCP/SSP includes (but is not limited to) the following information:

- device and manufacturer, basic unique device identifier-device identifier (UDI-DI) and single registration number (SRN);
- device intended purpose, indication/s for use, contraindications, and target populations;
- device description, components, reference to previous models or variants;
- description of differences between devices, accessories, or products used in combination;
- possible diagnostic or therapeutic alternatives;
- references to any applied harmonized standards and common specifications;
- suggested user profile and user training and qualifications;
- information on residual risks, undesirable effects, warnings, and precautions;
- clinical/performance evaluation summary with relevant PMCF/PMPF summary;
- summary of clinical data related to equivalent device, if applicable;
- safety, performance, and benefit-risk ratio as well as overall conclusions;
- revision history;
- IVD: summary of scientific device validity, relevant interferences, cross-reactions;
- IVD: metrological traceability; unit of measurement, applied reference materials, and/or reference measurement procedures for device calibration;
- IVD: conditions to use device (e.g., near-patient testing, companion diagnostic); and
- IVD: kit components, as applicable, including component regulatory status.

Planning, Writing and Reviewing Medical Device Clinical and Performance Evaluation Reports (CERs/PERs). DOI: https://doi.org/10.1016/B978-0-443-22063-0.00001-8

Overall, the SSCP/SSP is written after the CER/PER is completed. The notified body (NB) reviews the SSCP/SSP to verify the summarized information against the CER/PER and other technical file documents before entering the SSCP/SSP into the EUDAMED database [3] (once available) for public use. The NB requires the data and conclusions to be accurate, scientifically valid, and fully aligned with all known product information.

8.2 The quality management system must define the summary document writing process

The quality management system (QMS) should incorporate a policy statement about using international SSCP/SSP templates to draft the SSCP/SSP after the CER/PER is completed. Standard operating procedures (SOPs) and work instructions (WIs) should define how to complete all required template sections, including where to find the specific, required SSCP/SSP information in the CER/PER or technical file documents outside the CER/PER. The SOPs and WIs should reference the relevant regulations and guidelines to ensure all required SSCP/SSP information is provided and accurately/reproducibly sourced and summarized.

8.3 Guidance for summary document writing

Two guidance documents (Table 8.1) are available for the SSCP [4] and SSP [5] to guide presentation, content, and SSCP/SSP validation.

TABLE 8.1 SSCP/SSP guidance document templates.

Reference	Title	Link	Date
MDCG 2019-9, Rev 1	Summary of safety and clinical performance	https://health.ec.europa.eu/system/files/2022-03/md_mdcg_2019_9_sscp_en.pdf	March 2022
MDCG 2022-9, Rev 1	Summary of safety and performance template	https://health.ec.europa.eu/document/download/b7cf356f-733f-4dce-9800-0933ff73622a_en?filename = mdcg_2022-9_en.pdf	April 2024

MDCG, Medical Device Coordinating Group.

The SSCP guideline stated: "The manufacturer will assign... an identifier (reference number)... within the manufacturer's management system... unique to that SSCP and [this number] will remain the same for the entire lifetime of the SSCP. In combination with the manufacturer's SRN this will allow... the unique identification of the SSCP in EUDAMED and in EU." A similar statement was not present in the SSP guideline which instead stated the "SSP is associated to one or multiple Basic UDI-DI(s)." Unlike medical devices, IVD devices are often associated with multiple different diagnostic testing uses.

8.3.1 Medical device summary document template (March 2022)

One Medical Device Coordinating Group (MDCG) guidance document (MDCG 2019-9 Rev. 1) [4] provides a template for the SSCP which is required for Class III and implantable medical devices. SSCPs are "intended to provide public access to an updated summary of clinical data and other information about the safety and clinical performance of the medical device" (p. 3) and should be "sourced entirely from the technical documentation (TD) of the device" (p. 4). As with all other CE documents, the SSCP need to be "objective and adequately summarise both favourable and unfavourable data" (p. 5).

SSCPs are intended for the public, and as such, have unique requirements. Information contained in SSCPs needs to be presented in a manner legible to indented users and patients. Additionally, no promotional information can be included. The SSCP is not a marketing document.

Implantable devices with implant cards and Class II devices used directly by patients are required to include an SSCP section "specifically intended for patients" (p. 6) in addition to the "part for intended users/healthcare professionals." Both sections "should be clear and provide information at an appropriate depth to reflect the healthcare professionals' and the patients' different levels of knowledge" (p. 6). Patients should not be assumed to have "any formal education in a medical discipline or any prior knowledge of medical terminology or clinical research."

When, as a part of ongoing CE, the manufacturer prepares a periodic safety update report (PSUR) the SSCP must be updated if "the PSUR contains information rendering any information in the SSCP incorrect or incomplete" (p. 8). SSCPs must also be updated when the CE mark is renewed for Class III and class IIb devices (other than well-established technologies, or WETs). For Class IIa and IIb WET devices, SSCPs only need to be updated when new information makes them incorrect or incomplete.

Specific guidance is provided for SSCP content, including both content for users/healthcare professions and content for patients. Descriptive device information along with information on residual risks and undesirable effects, a clinical evaluation summary and relevant PMCF, possible treatment alternatives, a suggested profile and training for users and references to any harmonized standards and common specifications applied must be included. The MDCG 2019-9 guidance provides a detailed SSCP template and table of contents initially released in 2019 and updated in 2022 to clarify the association between the SSCP and the basic Universal Device Identifier (UDI) in EUDAMED with addition of a manufacturer reference number.

8.3.2 *In vitro* diagnostic device summary document template (April 2024)

Another guidance document (MDCG 2022-9 Rev.1) [5] provides the IVDR SSP template for Class C and D IVD, which is similar to the MDCG 2019-9 SSCP template. One notable exception is in the specific SSP template for devices intended for self-testing is provided since no HCP section will apply. In addition, the SSP includes the following data summary information unique to IVD devices: scientific device validity, relevant interferences, cross-reactions; metrological traceability (unit of measurement, applied reference materials, and/or reference measurement procedures for device calibration), conditions to use device (e.g., near-patient testing, companion diagnostic), and kit components, as applicable, including component regulatory status, etc.

8.4 Summary document anatomy

The SSCP/SSP templates were merged and provided in APPENDIX P: SSCP/SSP Template to help streamline compliance and SSCP/SSP writing across medical devices and IVD devices. Just like the original templates, the merged SSCP/SSP template has two parts: one for intended users (typically HCPs) and the other for patients and laypersons using the device. The table format is intended to house data, not full sentences, so, simply enter the required data into each open box in the table. When writing an SSCP, simply check the box which indicates the data are not available because the item is for a medical device and is not an SSP for an IVD device; otherwise, document the IVD specific data. For example, item 1.7 would require data entered to specify if the IVD is for near-patient testing or a companion diagnostic test and the box should not be checked when writing an SSP. This "check the box or enter the data" style is followed in the rest of the merged template whenever the SSCP and SSP template diverge.

8.4.1 Information for healthcare professionals

Both templates (i.e., SSCP for medical devices and SSP for IVD devices) begin with a specifically worded section about the intended use to "provide public access to an updated summary of the main aspects" of device S&P and the suggested wording specifies the SSCP/SSP was not "to replace the Instructions For Use." Every SSCP/SSP should have similar templated wording beginning the HCP section. In addition, the SSCP/SSP should accurately summarize all CER/PER S&P information. The summary should be shorter and more concise than the CER/PER. The SSCP/SSP should not include every data point but should include accurate positive and negative data with appropriate relative data weights. The document should provide an accurate, truthful, and not misleading picture of the device S&P.

8.4.2 Information for patients

The SSCP/SSP patient section is required whenever "relevant to provide information to patients in lay man's language." A good strategy when writing documents is to know the audience for the written document. Laypersons including patients and their caregivers may lack the required clinical and technical training and knowledge to understand the clinical S&P information and terminology used in the HCP section. As a result, the SSCP/SSP patient information section should briefly summarize the CE/PE data at a level where the patient can understand the clinical S&P. This section is intended to empower patient device-related decision-making.

Unlike other patient/user facing materials, the SSCP/SSP must summarize all applicable favorable and unfavorable clinical data evaluated in the CER/PER. Transforming language from a scientific audience to a lay audience can be particularly challenging. Often, a readability test is helpful to ensure a layperson can comprehend the language used in the SSCP/SSP section intended for them. The readability test often requires the layperson user to read the SSCP/SSP and then to answer questions designed to test their understanding of the information contained within the SSCP/SSP.

8.4.2.1 *Resources available to help write clear and effective summaries*

In the United States, the National Institutes of Health (NIH) found 14% of 18,500 adult Americans could only comprehend "basic, simple text," and they recommended a five-step process [6] to develop materials for persons with such limited health literacy as follows:

1. "Define the Target Audience"
2. "Conduct Target Audience Research"
3. "Develop a Concept for the Product" (including reading level to carry the message)
4. "Develop Content and visual Design Features" (content, layout, visuals, readability)
5. "Pretest and Revise Draft Materials" (comprehension, attractiveness, acceptability)

Many resources recommended a reading grade level below the US 8th-grade level when writing in the English language for a text aimed at the general public in the US. This writing process becomes even more complex when writing in the English language for a document designed to be translated into multiple languages for use with persons who do not speak English.

Another excellent resource when writing the SSCP/SSP layperson section is the "Summaries of Clinical Trial Results for Laypersons" document, which indicated:

> *Based on research across Europe, text for the lay summary should be aimed at a literacy proficiency level of 2–3. The International Adult Literacy Survey (IALS) identifies five levels of proficiency ranging from level 1 (lowest level of proficiency in literacy, that is basic identification of words and numbers) to level 5 (highest level of proficiency in literacy, that is able to understand and verify the sufficiency of the information, synthesize, interpret, analyse and discuss the information. At level 5, the individual demonstrates sophisticated skills in handling information).* [7]

A good practice for writing the SSCP/SSP information in the patient section is to keep it simple.

- Shorten sentences.
- Reduce polysyllabic words.
- Ensure every, single, sentence has a clear noun and verb.
- Organize information to flow logically.
- Avoid acronyms.
- Simplify sentences and structure.
- Include figures, tables, and infographics, defining all medical terms.
- Include quantitative data about adverse device events, side effects, and residual risks.
- Use language appropriate for a 6th-grade reading level.

Getting the data clearly and completely shared in a simple format below the EU recommended IALS level 3 or the US 8th-grade reading level may take a lot more time than summarizing the clinical/performance data for the HCP section.

The process to create the patient section wording is often a transformation from the HCP summary into simpler words and sentences for a lower reader-comprehension level and a removal of all information not relevant to the patient. Ultimately, if no information is relevant to patient users and their caregivers (i.e., laypersons do not operate or use the device directly), then this section can be omitted entirely from the SSCP/SSP (Table 8.2).

TABLE 8.2 **SSCP/SSP "information to patients" section.**

Include information to patients section	Consider excluding information to patients section
Implanted materials/tissues	Hysteroscope
Surgically invasive anatomy or body part fixation products (not tattoos or piercings)	X-ray equipment
Dermal/mucous membrane fillers	Surgical devices and robots
IVD tests completed in the home	Diagnostic or therapeutic test devices and software used only in healthcare settings (e.g., IVD devices used on patient samples without patient involvement)

IVD, in vitro diagnostic.

In particular, the manufacturer must analyze and decide if the information to patients section should be included or excluded in the SSCP/SSP. If excluding this section, the manufacturer should document the rationale for this exclusion decision. This decision rationale is typically based on the analysis of patient/caregiver interaction with the device. Generally, if the patient interacts with the device, either by directly using the device or by having tissues/materials implanted, then the information to patients section seems relevant and will likely be included. Conversely, if the patient/caregiver does not interact with the device (i.e., the patient does not directly "use" the device and has no tissue/materials implanted), then this section seems not particularly relevant and can likely be omitted. For example, the section may not be needed for surgical devices and surgical robots used only by surgeons, IVD/therapeutic devices and software used only in a healthcare setting, and other devices only used by medical personnel.

8.4.3 General identification information

The SSCP/SSP information for HCP and patient sections each begin with general device and manufacturer information and a good way to represent this information is in a table. The SSCP/SSP often use identical device identification and general information types. For example, the first six and the last three items in the "device identification and general information" SSCP and SSP sections are nearly identical, while only item 1.7 about "a device for near-patient testing and/or a companion diagnostic" is entirely unique in the SSP (Table 8.3).

TABLE 8.3 General identification information.

SSCP (MDR)	SSP (IVDR)
1.1 Device trade name/s	1.1 Device trade name/s
1.2 Manufacturer's name and address	1.2 Manufacturer's name and address
1.3 Manufacturer's single registration number (SRN)	1.3 Manufacturer's single registration number (SRN)
1.4 Basic UDI-DI	1.4 Basic UDI-DI
1.5 Medical device nomenclature description/text	1.5 European Medical Device Nomenclature (EMDN) description/text
1.6 Class of device	1.6 Risk class of device
1.7 NA	1.7. Indication whether it is a device for near-patient testing and/or a companion diagnostic
1.8 Year when the first certificate (CE) was issued covering the device	1.8 Year when the first certificate was issued under Regulation (EU) 2017/746 covering the device
1.9 Authorized representative if applicable; name and the SRN	1.9 Authorized representative if applicable; name and the SRN
1.10 NB's name (the NB that will validate the SSCP) and the NB's single identification number	1.10 NB's name (the NB that will validate the SSP) and the NB's single identification number

CE, clinical evaluation; *EU*, European Union; *IVDR*, *In Vitro* Diagnostic Regulation; *MDR*, Medical Device Regulation; *NB*, Notified Body; *SSCP*, summary of safety and clinical performance; *SSP*, summary of safety and performance. Bolded information differs between SSCP and SSP templates.

The general information details to enter into the templated form should be drawn from the CER/PER and verified against source documents.

8.4.4 Device labeling details

Both SSCP/SSP templates require essentially the same labeling details; however, the IVD device SSP includes slightly more descriptive information (Table 8.4).

TABLE 8.4 Device labeling details.

SSCP (MDR)	SSP (IVDR)
2. Intended use of the device	2. Intended use of the device
2.1 Intended purpose	2.1 Intended purpose (including target, specimen required, etc.)
2.2 Indication/s and target population/s	2.2 Indication/s and target population/s
2.3 Contraindications and/or limitations	2.3 Limitations and/or contraindications (e.g., relevant interferences, cross-reactions)

IVDR, In Vitro Diagnostic Regulation; *MDR,* Medical Device Regulation; *SSCP,* summary of safety and clinical performance; *SSP,* summary of safety and performance. Bolded information differs between SSCP and SSP templates.

The SSCP/SSP requires three parts for the intended use to help manufacturers document the *detailed* and specifically required "indication for use" statement, including: (1) intended purpose, (2) indication and target population, and (3) contraindications and limitations (including IVD-device-specific interference and cross-reactivity details). Previously, the intended use was sufficient to focus broadly on device functionality in a general population; however, now, the MDR/IVDR requires the indication for use statement and the specific target populations to be stated clearly. When using the template, align the contents with the relevant "Technical Documentation" which is required to be on file at the manufacturer in a "clear, organised, readily searchable and unambiguous manner" [1,2], including the device intended purpose details (Table 8.5).

TABLE 8.5 Intended purpose within the technical documentation.

MDR EU REG 2017/745, Annex II (1.1(c)) [1]	IVDR EU REG 2017/746, Annex II (1.1(c)) [2]
. . .the intended patient population and medical conditions to be diagnosed, treated and/or monitored and other considerations such as patient selection criteria, indications, contra-indications, warnings. . .	. . .the intended purpose of the device. . . 1. what is to be detected and/or measured; 2. its function such as screening, monitoring, diagnosis or aid to diagnosis, prognosis, prediction, companion diagnostic; 3. the specific disorder, condition or risk factor of interest that it is intended to detect, define or differentiate; 4. whether it is automated or not; 5. whether it is qualitative, semi-quantitative or quantitative; 6. the type of specimen/s required; 7. where applicable, the testing population; 8. the intended user; 9. in addition, for companion diagnostics, the relevant target population and the associated medicinal product/s. [8]

EU, European Union; *IVDR, In Vitro* Diagnostic Regulation; *MDR,* Medical Device Regulation.

The SSCP/SSP must document the CER/PER clinical data gaps which required a PMCFP/PMPFP. The NB pays careful attention to the device indication, target population, contraindications, and limitations when they assess the CE/PE documents including the SSCP/SSP. For the CER/PER, the NB will also independently determine if sufficient clinical/performance data supported the claimed S&P across the entire breadth of the indication/s (e.g., all indicated patient medical conditions, disease states/stages, treatment histories) and patient population/s (e.g., age group, user group, use situation, both sexes, patients with comorbidities). Any findings identifying the need for additional clinical data collection should be documented and discussed briefly in the SSCP/SSP.

8.4.5 Device description

The SSCP/SSP requires a brief device description, and one item (i.e., 3.2 "In case the device is a kit. . .") was added to the SSP template to define any IVD kit components (Table 8.6). The SSP also needs to define specific IVD use conditions in the laboratory or near the patient.

TABLE 8.6 Device description.

SSCP (MDR)	SSP (IVDR)
3. Device description	3. Device description
3.1 Description of the device	3.1. Description of the device, including the conditions to use the device (e.g., laboratory, near-patient testing)
3.2 NA	3.2. In case the device is a kit, description of the components (including regulatory status of components, for example, IVDs, medical devices and any basic UDI-DIs)
3.3 A reference to previous generation/s or variants if such exist, and a description of the differences	3.3. A reference to previous generation/s or variants if such exists, and a description of the differences
3.4 Description of any accessories which are intended to be used in combination with the device	3.4. Description of any accessories which are intended to be used in combination with the device
3.5 Description of any other devices and products which are intended to be used in combination with the device	3.5. Description of any other devices and products which are intended to be used in combination with the device

IVDR, In Vitro Diagnostic Regulation; *MDR,* Medical Device Regulation; *SSCP,* summary of safety and clinical performance; *SSP,* summary of safety and performance. Bolded information differs between SSCP and SSP templates.

The SSCP/SSP summarizes the CER/PER device description including, but not limited to, the operating principles, modes of action, patient-contacting materials, if the device is single use, method of sterilization. Constituent information should be included if the device incorporated a medicinal substance, tissues or cells of human or animal origin (or their derivatives), substances absorbed by the body, carcinogenic, mutagenic, toxic to reproduction (CMR), or endocrine-disrupting substances or sensitizing or allergenic substances.

A labeled device drawing or image can be especially helpful in the SSCP/SSP, especially for devices with multiple parts and operations/directions. Previous device generations, models, and/or variants should be presented in chronological order (including basic UDI-DI information for each variation, if different). The differences between device versions/generations should be described, and the device changes over time should be justified. The SSCP/SSP should specify which generations were used to generate the clinical trial data used for device S&P support and which generations were not yet supported by clinical data, if applicable. The accessories, compatible devices, and products intended to be used in combination with the device should also be listed.

The SSP template was developed about 3 years after the SSCP template and had slightly more detail. For example, both medical devices and IVD devices may be provided in kits, so item 3.2 is relevant to both SSCP and SSP documents even though the two guidance documents do not align on this topic. The SSCP template (MDCG 2019-9) stated:

> *If the device is a system of several components/devices, each device in the system should have a Basic UDI-DI but also one Basic UDI-DI for the system... The device description in the SSCP shall... include all the device/s/device system associated with the same Basic UDI-DI.* [4].

Similarly, the SSP IVD device guidance (MDCG 2022-9) included component details within the template item 3.2 with no further narrative. In other words, put the details in the table if the subject device is part of a kit regardless of whether the product is a medical device or an IVD device and include the regulatory status, UDI-DIs, etc.

8.4.6 Risks, undesirable effects, warnings, and precautions

The SSCP/SSP must summarize CER/PER risk data (Table 8.7).

TABLE 8.7 Risks, undesirable effects, warnings, and precautions.

SSCP (MDR)	SSP (IVDR)
4. Risks and warnings	5. Risks and warnings
4.1. Residual risks and undesirable effects	5.1. Residual risks and undesirable effects
4.2. Warnings and precautions	5.2. Warnings and precautions
4.3. Other relevant aspects of safety, including a summary of any field safety corrective action (FSCA including FSN) if applicable	5.3. Other relevant aspects of safety, including a summary of any field safety corrective action (FSCA including FSN), if applicable

FSCA, field safety corrective action; *FSN,* field safety notice; *IVDR, In Vitro* Diagnostic Regulation; *MDR,* Medical Device Regulation; *SSCP,* summary of safety and clinical performance; *SSP,* summary of safety and performance.

Although the risk section content is identical between SSCP and SSP templates, the numbering differs because the SSCP prioritized reference to harmonized standards and common specifications (MDCG 2022-9) as item 8, while the SSP moved this to item 4. The merged template aligns with the SSP to have the reference section near the end in item 8, not in item 4.

Device risks include both theoretical and improbable clinical and nonclinical risks which may or may not all warrant SSCP/SSP inclusion. Decisions about which risks to include in the SSCP/SSP should be clearly justified. The SSCP template states "'risk' includes both clinical and non-clinical harms" [4]. All risks known or anticipated must be included with an exception to exclude improbable and solely theoretical risks with proper justification (e.g., actual documented patient deaths at least possibly related to the device should be included while purely theoretical "irritation" risks associated with the potential for harm if the device were to break in a particular way may not need to be included). Data quantification with frequency (e.g., potential for a harm to happen over a certain period of time) is also required when describing risks. A good PMS system will quantify and determine risk frequency over time; however, quantification and frequency may be challenging for some devices without long-standing data collection.

The SSCP template also states "The SSCP should contain information on at least the same residual risks and undesirable side-effects as included in the IFU." Warnings and precautions relevant to device "installation/ preparation... or relating to special procedural steps" [4] should be included, but may not need to be discussed in depth if a link to the IFU is provided (e.g., a link to the manufacturer's website can suffice). In addition, safety information related to the disease should be included in this section.

8.4.7 Summary of clinical/performance evaluation including post-market follow up

The SSCP/SSP template requires specific CER/PER and PMCF/PMPF clinical data summaries and specific, additional detail is required to summarize the IVD device scientific validity (Table 8.8) in the SSP.

TABLE 8.8 CER/PER and PMCF/PMPF summary.

SSCP (MDR)	SSP (IVDR)
5. Summary of clinical evaluation and PMCF	6. Summary of performance evaluation and PMPF
5.1 NA	6.1 Summary of scientific validity of the device
5.2 Summary of clinical data related to equivalent device, if applicable	6.2 Summary of performance data from the equivalent device, if applicable
5.3 Summary of clinical data from conducted device investigations before the CE-marking, if applicable	6.3 Summary of performance data from conducted device studies prior to CE-marking
5.4 Summary of clinical data from other sources, if applicable	6.4 Summary of performance data from other sources, if applicable
5.5 An overall summary of the clinical performance and safety	6.5 An overall summary of the performance and safety
5.6 Ongoing or planned PMCF	6.6 Ongoing or planned PMPF

CE, conformité européenne; *IVDR*, *In Vitro* Diagnostic Regulation; *MDR*, Medical Device Regulation; *PMCF*, postmarket clinical follow-up; *PMPF*, postmarket performance follow-up; *SSCP*, summary of safety and clinical performance; *SSP*, summary of safety and performance. Bolded information differs between SSCP and SSP templates.

The SSCP/SSP must clearly summarize the CER/PER, including the clinical data supporting the device S&P and benefit-risk ratio conclusions when the device was used as indicated over the entire device lifetime. The SSCP/SSP must also summarize the plans (PMCFP/PMPFP) and evaluation reports (PMCFER/PMPFER) for any and all postmarket clinical data collections (e.g., clinical trials, chart reviews, surveys) completed to fill in any previously identified CER/PER clinical data gaps.

8.4.8 Possible diagnostic or therapeutic alternatives

The SSCP and SSP templates differ for possible subject medical device diagnostic and treatment alternatives because the medical device template requires "possible diagnostic or therapeutic alternatives"; however, the IVD devices must explain the various units of measurement and reference materials used in the IVD device (Table 8.9).

TABLE 8.9 Possible diagnostic or therapeutic alternatives templates.

SSCP (MDR)	SSP (IVDR)
6. Possible diagnostic or therapeutic alternatives	7. NA
6.1 NA	7.1. Explanation of the unit of measurement, if applicable
6.2 NA	7.2. Identification of applied reference materials and/or reference measurement procedures of higher order used by the manufacturer for the calibration of the device

IVDR, *In Vitro* Diagnostic Regulation; *MDR*, Medical Device Regulation; *SSCP*, summary of safety and clinical performance; *SSP*, summary of safety and performance.

This section requires a standard of care medical practice summary and an alternative therapy/device summary from the CEP/PEP and CER/PER including lifestyle changes, therapies, medications, procedures or other devices used, if available and appropriate for the disease/condition treated. Reference materials used to calibrate the device are important to include for IVD device SSPs.

8.4.9 Suggested user profile and training

The SSCP/SSP should summarize the experience or user training required to ensure users can safely and effectively operate the device (Table 8.10).

TABLE 8.10 Suggested profile and training for users.

SSCP (MDR)	SSP (IVDR)
7. Suggested profile and training for users	8. Suggested profile and training for users

IVDR, *In Vitro* Diagnostic Regulation; *MDR*, Medical Device Regulation; *SSCP*, summary of safety and clinical performance; *SSP*, summary of safety and performance.

For example, a "cardiologist" may be too generic for the suggested user profile for implanting a peripheral stent device and an "interventional cardiologist with experience in peripheral stent placement" may be more appropriate. Keep in mind, the manufacturer must develop, manage, and document any required training programs, and may also need to measure and document training effectiveness.

8.4.10 Reference harmonized standards and common specifications applied

The SSCP/SSP requires lists of harmonized standards and common specifications (Table 8.11).

TABLE 8.11 Harmonized standards and common specifications references.

SSCP (MDR)	SSP (IVDR)
8. Reference to any harmonized standards and common specifications applied	4. Reference to any harmonized standards and common specifications applied

IVDR, *In Vitro* Diagnostic Regulation; *MDR*, Medical Device Regulation; *SSCP*, summary of safety and clinical performance; *SSP*, summary of safety and performance.

Be sure to include a summary description, year, and revision number for each standard and specification. Although not required by the regulations, a brief rationale about why certain standards and specifications were followed, partially followed, or not followed can provide context for users to understand why competitor device SSCP/SSPs may have more or different standards listed.

8.4.11 Revision history

The SSCP/SSP requires a revision history to track versions validated by the NB (Table 8.12).

TABLE 8.12 Revision history.

SSCP (MDR)	SSP (IVDR)
Revision validated by the NB	**Revision validated by the NB**
9. Revision history • Yes, validation language: ___________ • No, only applicable for Class IIa or some IIb implantable devices (MDR, Article 52 (4) 2nd paragraph) for which the SSCP is not yet validated by the NB	9. Revision history • Yes, validation language: _______ • No, only applicable for Class C (IVDR, Article 48 (7)) for which the SSP is not yet validated by the NB

IVDR, *in vitro* diagnostic regulation; *MDR*, medical device regulation; *NB*, notified body; *SSCP*, summary of safety and clinical performance; *SSP*, summary of safety and performance.

The SSCP uses MDR classifications (i.e., Class IIa and IIb) and the SSP uses IVDR classifications (i.e., Class C) to explain exceptions to the NB version validation requirement.

8.5 Strategies for writing safety and performance summaries

Unlike the CER/PER audience which includes the NB and expert panel, if the device is high risk, the SSCP/SSP audience includes the healthcare professional (in the first part) and the device user, patient, and caregiver (in the second part). The SSCP/SSP is not meant to replace the CER/PER, instructions for use, or any other technical file/design dossier documentation. In addition, SSCP/SSP writing strategies and challenges depend, in part, on the current device developmental stage because clinical data varies during viability (v) testing, or when using equivalence (eq) to another device, during first in human (fih) testing, during traditional (tr) use, or during obsolescence (o).

8.5.1 Viability summary of safety and clinical performance/viability summary of safety and performance

The vSSCP/vSSP relies on clinical data from similar devices as well as bench and animal data from the subject device as a proof of concept evaluated in the vCER/vPER. This low-quality evidence may preclude the vSSCP/vSSP from meeting the NB review requirements to be entered into EUDAMED, so this type of document is meant to be kept in the manufacturer technical files during development and is to be replaced once either equivalence is claimed and human data are gathered from equivalent device use or first-in-human data are gathered from subject device use.

8.5.2 Equivalence summary of safety and clinical performance/equivalence summary of safety and performance

The eqSSCP/eqSSP relies on data from equivalent devices evaluated in the eqCER/eqPER before sufficient data collection has occurred for the subject device. As such, the eqSSCP/eqSSP will probably contain little to no clinical data from the subject device. The eqSSCP/eqSSP should make the reliance on data from any equivalent devices clear to both users and patients and must state the subject device was granted CE mark based on equivalence and not on data from the device itself. All general information, along with all long-term S&P clinical data, for the equivalent device must be summarized and the relationship to the subject device must be made clear to the reader.

8.5.3 First-in-human summary of safety and clinical performance/first-in-human summary of safety and performance

The fihSSCP/fihSSP relies on clinical data from the first human patients to use the subject device as evaluated in the fihCER/fihPER. All clinical trial data using the subject device should be summarized in the fihSSCP/fihSSP; however, no subject device clinical literature or experience data are likely to be available because the subject device may not be on the market yet and the PMS system may not be fully functional. This summary must make clear the early use setting for this specific subject device.

8.5.4 Traditional summary of safety and clinical performance/traditional summary of safety and performance

The trSSCP/trSSP summarizes all clinical data from trials, literature, and experiences as evaluated in the trCER/trPER because the product is now available on the market and the PMS system is active. Unless only minimal subject device clinical data are available, equivalent device data should not be the primary focus in trSSCP/trSSP even if the equivalent device path was used earlier in device development.

8.5.5 Obsolete summary of safety and clinical performance/obsolete summary of safety and performance

The oSSCP/oSSP summarizes all clinical data from past clinical trials, literature, and experiences for devices no longer actively marketed or sold. After a device is obsoleted, previously sold devices may be still in use/implanted and benefit-risk ratio changes may occur based on events disclosed in literature and experience data. Clinical trial data are not expected because the device is no longer available for sale on the market; however, clinical trials may be conducted on explantation of a previously implanted device or about other clinical concerns for obsoleted devices. The oSSCP/oSSP should document the obsoleted device status, and the summary should clearly evaluate the clinical data for any humans who may continue to experience clinical effects from the obsoleted device.

As discussed above, the lack of interest on the part of the manufacturer is obvious for obsoleted devices which cannot be sold to cover the costs of this CE/PE work. Also the challenges are obvious when trying to convey and justify the viability, equivalence and first in human situations for the lay public.

8.6 Summary of safety and clinical performance/summary of safety and performance translations

EU documents must be translated. Often, translations are completed for all 24 EU official languages of the Member States where the device will be sold: Bulgarian, Croatian, Czech, Danish, Dutch, English, Estonian, Finnish, French, German, Greek, Hungarian, Irish, Italian, Latvian, Lithuanian, Maltese, Polish, Portuguese, Romanian, Slovak, Slovenian, Spanish, and Swedish [8]. An English SSCP/SSP copy is also required because "English is the most common language used in medical scientific publications and is understood by many healthcare professionals in the EU." Manufacturers should develop "one SSCP document for each language" [4,5] and the manufacturer must ensure all translations are correct. A good translation practice is to translate two times (forward and back) to ensure the translated copy functions as expected. The forward translation takes the document from the original into a foreign language, and the second translation (by a different person/group) translates the document from the foreign back into the original language. This process should identify translation difficulties, and the difficult-to-translate text sections should be edited to clarify the meaning. This forward-and-back process helps to ensure quality; however, the manufacturer should also seek native speaker interpretation, use reviews, etc. to ensure quality. The NB validates the SSCP/SSP in one language and this language must be documented within the SSCP/SSP.

8.7 Summary of safety and clinical performance/summary of safety and performance updates

The SSCP/SSP is updated in response to any updated CER/PER and/or PMCFER/PMPFER. The CER/PER is updated as required under the regulations (i.e., from 1 to 5 years depending on the device risk) or whenever new clinical S&P data become available with the potential to change the S&P thresholds or the benefit-risk ratio. A good practice is to involve clinicians, patients, caregivers, and laypersons to review and revise the SSCP/SSP language prior to NB submission and translation.

8.8 Conclusions

SSCPs/SSPs summarize CE/PE S&P data and benefit-risk ratios for high-risk medical devices and IVD medical devices based on the CER/PER and other technical documentation provided to the NB for review. Each SSCP/SSP includes two separate summaries: one for healthcare professions and one for layperson users. In the layperson user section, the SSCP/SSP summary language used must be understandable by the layperson. This requirement makes the SSCP/SSP considerably different from all the other CE/PE documents. Considerable attention should be paid to native speaker interpretation, use reviews and readability testing to ensure the lay reader can comprehend and use the information in the SSCP/SSP document. The manufacturer is well advised to use the EU Medical Device Coordinating Group (MDCG) SSCP/SSP templates which have a sufficiently detailed structure to help individuals create successful summary documents.

For all SSCP/SSP types (v, eq, fih, tra, o), the clinical data summary in the SSCP/SSP: must align with CER/PER and PMCFER/PMPFER; must be truthful and not misleading; must be scientifically valid; must be written in clear form so providers and patients can use the information to understand the device; and, users must be able to understand and apply the information to decision-making. The SSCP/SSP must be made publicly available because the EU initiative for data transparency requires users, including patients, to have access to the S&P summary with a clear depiction of the benefit-risk ratio. EUDAMED will eventually house SSCPs/SSPs; however, all device manufacturers should make the SSCP/SSP available from all relevant websites.

8.9 Review questions

1. What do the acronyms SSCP and SSP mean?
2. Who is the audience for the SSCP/SSP?
3. Is the SSCP/SSP written before or after the CER/PER is completed?
4. Who enters the SSCP/SSP into EUDAMED?
5. How should the SSCP be uniquely identified by the manufacturer?
6. What are the two main SSCP/SSP parts?
7. How is the SSCP/SSP used in clinical practice?
8. What three parts are specifically required in the SSCP/SSP device labeling details?
9. True or false? SSCP/SSP device descriptions should focus solely on the current model and should not attempt to provide previous device history.
10. What labeling document is used to specify the SSCP/SSP risks and undesirable side effects?
11. How often is the SSCP/SSP updated?
12. What are appropriate SSCP/SSP data sources?
13. How should the manufacturer ensure the SSCP/SSP patient section is written so laypersons can understand the information?
14. How does the SSCP/SSP writer determine which "residual risks" must be included?
15. What are some good practices when writing the layperson/patient SSCP/SSP section?
16. What are some differences between the SSCP and SSP documents?
17. How can a manufacturer ensure SSCP/SSP translations are correct?
18. What SSCP/SSP Revision History details are required?
19. What makes the SSCP/SSP so different from all the other CE/PE documents?
20. What is a "readability test"?

References

[1] European Parliament and Council of the European Union. Consolidated text: Regulation (EU) 2017/745 of the European Parliament and of the Council of 5 April 2017 on medical devices, amending Directive 2001/83/EC, Regulation (EC) No 178/2002 and Regulation (EC) No 1223/2009 and repealing Council Directives 90/385/EEC and 93/42/EEC. Accessed June 1, 2024. https://eur-lex.europa.eu/legal-content/EN/TXT/?uri = CELEX%3A02017R0745-20230320.

[2] European Parliament and Council of the European Union. Consolidated text: Regulation (EU) 2017/746 of the European Parliament and of the Council of 5 April 2017 on in vitro diagnostic medical devices and repealing Directive 98/79/EC and Commission Decision 2010/227/EU. Accessed June 1, 2024. https://eur-lex.europa.eu/legal-content/EN/TXT/?uri = CELEX%3A02017R0746-20230320.

[3] EUDAMED—European Database on Medical Devices. Accessed June 1, 2024. https://ec.europa.eu/tools/eudamed/#/screen/home.

[4] MDCG 2019-9. Rev. 1 Summary of safety and clinical performance. A guide for manufacturers and notified bodies. March 2022. Accessed June 1, 2024. https://health.ec.europa.eu/system/files/2022-03/md_mdcg_2019_9_sscp_en.pdf.

[5] MDCG 2022-9. Rev.1 Summary of safety and clinical performance template. May 2022. Accessed June 1, 2024. https://health.ec.europa.eu/document/download/b7cf356f-733f-4dce-9800-0933ff73622a_en?filename = mdcg_2022-9_en.pdf.

[6] National Institutes of Health. Clear communication: clear and simple website. July 7, 2021. Accessed June 1, 2024. https://www.nih.gov/institutes-nih/nih-office-director/office-communications-public-liaison/clear-communication/clear-simple.

[7] Summaries of Clinical Trial Results for Laypersons: Recommendations of the expert group on clinical trials for the implementation of Regulation (EU) No 536/2014 on clinical trials on medicinal products for human use. Version 2. February 22, 2018. Accessed June 1, 2024. https://health.ec.europa.eu/system/files/2020-02/2017_01_26_summaries_of_ct_results_for_laypersons_0.pdf.

[8] European Union. Languages, multilingualism, language rules. Accessed June 1, 2024. https://european-union.europa.eu/principles-countries-history/languages_en#: ~ :text = The%20EU%20has%2024%20official,%2C%20Slovenian%2C%20Spanish%20and%20Swedish.

Chapter 9

Reviewing clinical/performance evaluation documents

Clinical evaluation (CE)/performance evaluation (PE) document reviewers perform critical review tasks to ensure all the clinical/performance data have been appropriately identified, appraised and analyzed and to ensure CE/PE data gaps have been fully addressed by the postmarket clinical follow-up plan/postmarket performance follow-up plan. The roles and responsibilities for CE/PE document reviewers and approvers first falls to each member of the CE/PE document writing team including internal and external evaluators, writers, design engineers, CE/PE team members, risk management team members, postmarket surveillance team members, clinical, regulatory and quality team members, and others. After the documents are signed off and version controlled, the review cycle often moves on to additional external reviewers and inspectors (e.g., the notified body and/or expert panel members for high-risk devices).

The quality management system (QMS) should define the steps required to review each clinical evaluation/ performance evaluation (CE/PE) document and their key deliverables. These steps must cover the processes required for each individual CE/PE document type:

- CEP/PEP—Is the clinical evaluation plan/performance evaluation plan well developed and appropriate?
- CER/PER—Does the clinical evaluation report/performance evaluation report clearly define the safety and performance (S&P), benefit-risk ratio, and conformity to the General Safety and Performance Requirements (GSPR) 1, 6, and 8 in the European Union (EU) Medical Device Regulation (MDR) (EU Reg. 2017/745)/EU *In Vitro* Diagnostic (IVD) Device Regulation (IVDR) (EU Reg. 2017/746)?
- PMCFP/PMPFP—Does the postmarket clinical follow-up plan/postmarket performance follow-up plan cover all of the clinical data gaps identified in the CER/PER?
- PMCFER/PMPFER—Does the PMCF/PMPF evaluation report provide sufficient results from the planned clinical data generation and gathering activities to provide sufficient clinical data to cover all of the clinical data needed in the next CER/PER update?
- SSCP/SSP—Does the summary of safety and clinical performance/summary of safety and performance clearly articulate an appropriate summary of the CER/PER and PMCFER/PMPFER, and are both parts of the SSCP/SSP readable by the healthcare professionals and laypersons as required?

The manufacturer must understand the clinical data complexity and the requirement for each CE/PE report to be read and understood by each individual in each intended audience in sufficient detail to understand and be able to usethe clinical data, the completed data analyses, and the data-driven conclusions within each document.

Because the length and effort required to generate these plans and reports vary significantly, each manufacturer should take the time to build a strong case for the clinical data type and amount required to ensure the CE/PE process has evaluated sufficient clinical data to clearly support patient safety and device performance as indicated in the appropriate patient population/s.

9.1 Define document review process steps

Although many different methods can be used to complete CE/PE document reviews, each review cycle might include the following initial steps:

Planning, Writing and Reviewing Medical Device Clinical and Performance Evaluation Reports (CERs/PERs). DOI: https://doi.org/10.1016/B978-0-443-22063-0.00011-0

1. Start by reading the executive summary and conclusion to understand the most important CE/PE points. These two sections should summarize all sections in the document. For example, the device should be briefly described including the indications for use and target populations who used the device to generate the clinical data, and the procedures used to determine if the device met the S&P acceptance criteria. In addition, a well-established and acceptable benefit-risk profile should be outlined, demonstrating how the benefits outweighed the risks for continued safe device use.
2. Examine the entire CE/PE document structure to ensure the structure is complete and meets the regulatory requirements with all the required components. Compare the document to others of the same type, to the relevant company policy, standard operating procedure (SOP), work instruction (WI) and template and to the templates provided within available international guidelines and standards.
3. Review the device description and revision history to understand the CE/PE document scope. The device description must explain how the device works (technically and clinically), and should provide the entire developmental history for the device including models, sizes, accessories, software versions, etc. Details about the labeling and clinical S&P claims should also be articulated here.
4. Verify all clinical/performance evaluation stages were completed and are well documented in the CE/PE document:
   a. Stage 0: scope and plan.
   b. Stage 1: clinical/performance data identification (e.g., were all available clinical trials, literature, experiences found and documented for appraisal?).
   c. Stage 2: clinical/performance data appraisal (e.g., were scientific validity, relevance, and quality considered and weighted for each clinical data element?).
   d. Stage 3: clinical/performance data analysis (e.g., does the device meet the S&P acceptance criteria, acceptable benefit-risk ratio, required reduction of residual risks, uncertainties, and unanswered questions?).
   e. Stage 4: final report (e.g., is document written with sufficient clinical data to substantiate the conclusions?).
5. Consider the background knowledge, state of the art (SOTA), and alternative device/therapy clinical data discussions documented in the CE/PE document. Were the clinical data appropriate, on target, and informative? Were appropriate clinical trials, literature, and experience data evaluated to make the background section sufficient for framing and comparing the subject device evaluation? Were sufficient details provided to explain the clinical data sources, search methods, appraisal criteria, analytical results, and methods used to draw conclusions about the background knowledge, SOTA, and alternative device/therapies? How does the device fit into the background, and does the device adhere to appropriate standards and practice guidelines?
6. Determine if the clinical/performance data support and substantiate every aspect of the subject device indications for use and target populations?
7. Compare the conclusions to the results from clinical data evaluations. Were the conclusions appropriate and adequate to ensure device S&P? Were all risks mitigated and reduced enough for the benefit-risk ratio to be acceptable? Were all residual risks acceptable? Were all clinical claims addressed with sufficient supporting clinical data to substantiate the clinical claim? Did the device labeling appropriately address all of the clinical/performance data concerns, benefits and risks identified in the CE/PE documents? Were all clinical data gaps identified, and addressed or slated to be addressed in the PMCFP/PMPFP? What research and safety mitigation or performance improvement activities are ongoing?
8. Examine CE/PE document version control elements to ensure adequate version and document control for the CE/PE document. Were the dates and signatures as expected? Was the date for the next CE/PE document clearly stated? Were the required details about the evaluator and other authors provided (e.g., declaration of interest, resume or curriculum vitae, financial disclosure)? Were the clinical data used in the CE/PE document appropriately archived and made available for the review?

SOPs and WIs should not be too detailed, and yet they should provide sufficient detail to make a meaningful impact on the quality of the work being standardized. In addition, linkages should be created within the QMS between the interrelated device engineering, CE/PE, RM, and post-market surveillance (PMS) systems and between the corporate teams working within these systems to ensure all benefits, risks, and complaints identified in the clinical data are shared and discussed between all the relevant systems. Electronic data storage systems should store the source document providing the actual human clinical data details for all process owners to access. For example, clinical benefits, risks, complaints, and device-related serious adverse events (SAEs), adverse device effects (ADEs), and relevant adverse events (AEs) should be listed in the benefit-risk ratio table (Table 9.1) for evaluation within the interrelated systems and teams.

TABLE 9.1 Example benefit-risk ratio table

Item	Description	Source document
Clinical benefits (e.g., indications for use and marketing claims)		
Reduced pain	*Details here*	*Details here*
Improved range of motion	*Details here*	*Details here*
Clinical risks (e.g., adverse device events)		
Surgical site infection	*Details here*	*Details here*
Myocardial infarction	*Details here*	*Details here*
Death	*Details here*	*Details here*

The art of creating and managing the benefit-risk ratio table is to keep the focus on the human patient and to "roll up" or group all identified, device-related medical events using appropriate medical terms. The benefit-risk ratio table should only list terms considered at least possibly related to the device, based on all the clinical data evaluated in the CER/PER. Once the benefit-risk ratio table is created, events can be analyzed further and compared to each other. The benefit-risk ratio terminology must be standardized and aligned with an appropriate medical dictionary so similar terms can be grouped and dissimilar terms can be distinguished. The QMS has an important function to ensure all clinical teams use standardized terminology to "roll-up" AEs and complaints into logical groups—for example, four events identified as a blood clot, thrombosis, embolism, deep venous thromboembolism (DVT) can be grouped under the broader term: thromboembolism (N = 4).

Many companies use the Medical Dictionary for Regulatory Activities (MedDRA dictionary) [1] or the Common Terminology Criteria for Adverse Events (CTCAE) developed by the National Cancer Institute [2] to help standardize clinical terms. Raw complaint data should be coded using standardized medical terms. For example, all instances of heart attack might be coded as myocardial infarction and grouped together with all instances of myocardial infarction as the preferred CTCAE term. In this way, the data analysis will be more uniform and appropriate by following the recommendations for a standardized medical terminology. The CTCAE incorporates many elements of the MedDRA terminology, and these two resources provide expert guidance about navigating the complex world of human AE reporting.

The CE/PE benefit-risk ratio table should group risks and AEs not only by the standardized or preferred term, but also by the affected system organ class (SOC). After the preferred terms are applied for all the reported benefits, risks, AEs, etc., the data can also be grouped at a higher level by the body system affected. For example, a heart attack event can be coded as a myocardial infarction and then grouped with other AEs occurring within the cardiovascular system (e.g., myocardial infarction and mitral valve replacment can be discussed as alterations occurring within the cardiovascular system). Standardizing the process for coding and grouping AE terms should allow the reviewer to more easily verify the CE/PE document safety analyses.

The decision to list or not list a specific finding is sometimes an area of concern or fear among those responsible for the benefit-risk ratio table. Deciding and documenting what was not to included may be more difficult, but no less important than deciding what to include. Many poor decisions in this area are led by a person saying, "Let's be conservative and just include everything, so we don't miss anything." Unfortunately, this is a mistake for many reasons. First and foremost, this approach fails to address the evaluation required in the CE/PE process. The CE/PE document writing team has a duty to appraise and analyze the clinical and performance data. This means the evaluator is required to decide what clinical data are relevant and of sufficient quality to be included, what cllinical data are to be excluded and why and how the clinical data should be weighted for analysis and presentation. The evaluator must determine if the clinical data are acceptable give the device S&P acceptance criteria and the existing benefit-risk ratio and (among many other considerations) if the device was used as indicated in the appropriate patients by the appropriate device users.

The benefit-risk ratio table is a relatively new creation in CE/PE documentation. This CER/PER table should not be confused with the related but separate risk management report (RMR) or PMS report (PMSR) tables, which exist concurrently. The format used in each table is related to the purposes of each document (i.e., RMR, for design/process/use engineering to mitigate all risks as low as possible without a negative impact on the benefit-risk ratio; PMSR/PSUR, periodic safety update report, for postmarket complaints and trends analyses to ensure proper regulatory AE reporting to various regulatory authorities and others; CER/PER to ensure all clinical data have been properly evaluated to ensure conformity with GSPRs 1, 6, and 8 for S&P, benefit-risk ratio, etc. over the entire device lifetime as required for CE mark). The form, fit, and function for each type of table/report should be defined in the QMS and the CE/PE document writing team should discuss and align all three types of benefit-risk ratio evaluations in the RMR, PMSR/PSUR and CER/PER.

The reviewer should assess if/how all clinical risks and patient harms were initially categorized into AEs, device deficiencies (DDs) or both. The reviewer should understand the rationale used in the CE/PE document to explain how the DD was or was not related to the corresponding AE and how all AE and DD were documented in the RMR and PMSR/PSUR documents. This level of detail is not necessarily required in the CE/PE documents, since the CE/PE should only be looking at a subset of these events focused solely on the device-related benefits and risks. Unfortunately, many CE/PE teams were required to do the extra work when the clinical, device-related events and deficiencies were not yet identified in the RMR and PMSR/PSUR documents as required.

The QMS must also document how internal team members will interact with external CE/PE reviewers. Some of these reviewers will be doing reviews before the document is finalized for the purpose of improving the CE/PE documents, and some will be reviewing final documents for the purpose of legal/regulatory compliance or quality assurance (QA). The external reviewers and consultants may bring significant, independent and hopefully less biased clinical and regulatory expertise to ensure: all CE/PE documents meet the "expert level" for clinical knowledge sharing regarding device use in human patients; regulatory compliance regarding highly-variable end user needs and conformity with the EU regulations.

9.2 Assign appropriate document reviewers and approvers

Although CE/PE documents are often owned and maintained by the clinical affairs team, the review and approval responsibilities are often filled by a cross-functional team to ensure all required contents are accurate, complete, truthful, and not misleading. Prior to initial release of each CE/PE document, the document should be reviewed by all senior team members from each of the various departments associated with the document development to ensure all required inputs were accurately represented. Each reviewer should critique the CE/PE documents from their department perspective, the patient perspective and from the regulatory perspective.

This initial fact-checking review should progress to a CE/PE document preapproval review, which should be comprehensive and should include all persons who will be approving, signing, and dating the document. Approvals are often requested from key CE/PE document stakeholders in the clinical, regulatory and quality departments but not the sales and marketing departments, since this work is not promotional and should not be influenced by promotional activities. This review should include qualified medical personnel with significant medical expertise using the device as indicated. This medical review should include at least one physician or health care practitioner who uses the device in their medical practice, as well as the company medical director/chief medical officer. The NB assessment typically requires the medical reviewer/approver qualifications, so this known NB deficiency can be avoided by including all evaluator, medical reviewer, and author qualifications and declarations of interest as supporting documents (especially for the CER/PER).

The company SOPs and WIs should spell out which functions will draft, revise, and complete the fact checking reviews and the preapproval reviews for each CE/PE document. All CE/PE documents should be reviewed by the company device design engineering, risk management (RM), and PMS departments so the relevent design control, risk mitigation and complaint handling functions should be involved in the document development. Leaders from these departments must ensure these review responsibilities are well-developed in the QMS, well-supported by company leadership and fully functional.

Each reviewer and approver must be educated, trained, and experienced enough to critically evaluate the background knowledge, disease state, SOTA, target population, subject device, similar devices or alternatives, intended purpose, indication for use, regulatory requirements, S&P acceptance criteria, benefit-risk ratios, etc. Each reviewer and approver should also be trained and certified to uphold the CE/PE document regulatory requirements. For example, each reviewer and approver should be able to confirm basic CE/PE document review criteria, not limited to the examples provided in Table 9.2.

TABLE 9.2 CE/PE document review criteria examples

CE/PE document	Review criteria examples
CEP (MDR Annex XIV, 1(a))/PEP (IVDR Annex XIII, 1.1)	- Specific device description - Conformity to GSPRs 1, 6, and 8 - Specific device intended purpose - Specific intended target population/s - Clear, specific intended use/indications for use, limitations, and contraindications for each target population - Detailed descriptions of intended clinical benefits to patients including relevant, specified clinical outcome parameters - Specific methods to examine qualitative and quantitative clinical S&P details with a clear reference to the determination of residual risks and side effects - Specific parameters listed to determine, based on the SOTA in medicine, the benefit-risk ratio acceptability for various indications, and for the intended device purpose/s as well as the IVD device analytical and clinical performance when appropriate - Specific methods to address benefit-risk issues relating to specific components such as pharmaceutical, nonviable animal, or human tissues - Version-controlled, signed, and dated clinical development plan indicating progression from exploratory investigations, such as first-in-human studies, feasibility and pilot studies to confirmatory or pivotal clinical investigations; and for IVD devices, a development phase outline including sequences and ways to determine scientific validity as well as analytical and clinical performance including milestones and potential acceptance criteria - Specify need for PMCFP/PMPFP with milestones and potential acceptance criteria - Specific justification if CEP/PEP deemed not appropriate For IVD devices: - Specific, defined and methodologically sound procedure to demonstrate scientific validity, analytical performance, and clinical performance - Specific device characteristics - Specific analytes or markers to be determined by the device - Specific certified reference materials or reference measurement procedures to allow for metrological traceability - Specific methods, including appropriate statistical tools to examine the device analytical and clinical performance and limitations, as well as information provided by the device - Described the SOTA, including an identification of existing relevant standards, CS, guidance, or best practices documents - For software qualified as a device, identify and specify reference databases and other data sources used as the basis for software decision-making
CER/PER	• Document can be read and understood by intended third party without experience using the device (e.g., NB assessor, expert panel member) • GSPRs 1, 6, and 8 were clearly supported by specific clinical data • Clinical data collection objectives and comparator device/s were clearly stated • Clinical data generated and held by the manufacturer were identified, evaluated and summarized • PMS data were evaluated including PMCFP/PMPFP, PMCFER/PMPFER, and PMSR/PSUR data • All clinical trial, literature, and experience source data were clearly identified, evaluted and described • Clinical data appraisal for suitability to establish device S&P was clearly described • Clinical data included both favorable and unfavorable data • All discrepancies and/or clinical data gaps were clearly identified with required PMCFP/PMPFP coverage specifically documented to correct the clinical data gaps • The clinical development plan addressed outstanding issues and generated new/additional clinical data through "properly designed clinical investigations" when appropriate • Analyzed all relevant clinical data and drew well-supported conclusions about device S&P and the device benefit-risk ratio • Clinical and performance data were relevant and justified to cover all "in scope" device variations (i.e., all models, sizes, settings, patient populations, etc.)

(Continued)

TABLE 9.2 (Continued)

CE/PE document	Review criteria examples

	• Clinical trial protocol numbers and www.clinicaltrial.gov national clinical trial (NCT) numbers were provided to uniquely identify each clinical trial
	• Clinical trial, literature, and experience outcomes were aligned and clearly stated in the objectives and conclusions, with outliers clearly identified
	• If claimed, equivalence was clearly justified in robust and comprehensive but succinct writing about the clinical impacts for all clinical, biological, and/or technical differences between the device under evaluation and the equivalent device
	• Differences between device under evaluation and the equivalent device, when claimed, did not impact the device clinical S&P or benefit-risk ratio
	• Background knowledge, SOTA, and alternative therapy/device section was current, comprehensive, succinct, and relevant to device under evaluation
	• Background knowledge, SOTA, and alternative therapy/device section was adequately referenced and substantiated by clinical trial, literature, and experience data
	• Clinical data supported every indication for use detail as described in the instructions for use (IFU) and was not excluded by the contraindications:
	  ◦ all intended users and user groups including laypersons if indicated;
	  ◦ entire target population from fetus, preterm infants to old age, males and females if indicated;
	  ◦ every stated medical condition form, stage, and severity including most benign to most serious forms, all acute and chronic stages, all levels of severity from mild to most severe as indicated;
	  ◦ the whole device use duration including maximal number of repeated exposures as allowed by the IFU;
	  ◦ entire device lifetime; and
	  ◦ discrepancies and exceptions to any device labeling were identified and justified including comments in the executive summary and conclusions as appropriate
	• Clinical data supported all S&P and benefit-risk claims made about device
	• Benefit-risk ratio and undesirable side effects were deemed "acceptable" in the context of the current knowledge, SOTA, and available alternative therapies and devices
	• Labeling was consistent with the CE/PE documentation and all labeling and claim discrepancies were identified with a corrective action plan specified
	• Clinical data led directly to conclusions stated
	• Conclusions identified all residual risks, uncertainties, and unanswered questions to be addressed in future PMS PMCF/PMPF studies
	• Report was signed, dated, and version controlled
	• Evaluator, author, approver qualifications were included, current and properly selected (e.g., appendixes included current resumes or curricula vitae, declarations of interest, and specific statements about the selection of these "appropriately qualified" persons signed/dated by appropriate senior clinical leader)
	• For IVD devices: described IVD device technology
	• For IVD devices: demonstrated scientific validity as well as analytical and clinical performance

PMCFP/PMPFP	• Clearly described all clinical data gaps identified and documented in the CER/PER or any other source
	• Specified and justified methods to proactively collect and evaluate clinical data from trials, literature, and experiences (user feedback, registry data, PMCF studies) to:
	  ◦ confirm device S&P over device lifetime;
	  ◦ identify and manage side effects and contraindications as well as emergent risks;
	  ◦ ensure benefit-risk ratio remains acceptable;
	  ◦ identify possible systematic misuse or off-label use; and
	  ◦ ensure intended purpose is correct
	• Referred to appropriate CER/PER, PMS, and risk management system documents
	• Stated specific objectives to be addressed by PMCF/PMPF
	• Identified clinical data to be evaluated for equivalent/similar devices when needed
	• Referred to PMCF/PMPF common/harmonized standards and relevant guidance
	• Stated time schedule for PMCF activities to be completed

(*Continued*)

TABLE 9.2 (Continued)

CE/PE document	Review criteria examples
PMCFER/PMPFER	• Collected and evaluated all planned clinical data as defined in the PMCFP/PMPFP • Included in technical documentation • Evaluation and conclusions taken into account within broader PMSR/PSUR, CER/PER and RM reports • Manufacturer implemented all corrective or preventive measures identified in the PMCFP/PMPFP
SSCP/SSP (high risk only)	• All data details were closely, completely, and thoroughly aligned as required in the Medical Device Coordinating Group (MDCG) guideline templates • Reading comprehension study ensured all readers will understand the information provided for them • All appropriate translations were completed and tested for accuracy

Each reviewer must have the knowledge and expertise required to complete their assigned review function. The clinical S&P data, benefit-risk ratio, and conclusions must be critiqued in a constructive manner, with clear instructions for improvement. In addition, the evaluator and senior management team must check and monitor the review process to ensure each reviewer is doing a good job reviewing and communicating their review findings to the CE/PE team.

9.3 Assign reviewer roles and responsibilities

In a small company, the reviewer role and responsibility will likely be assigned to one person; however, in a large company, individuals from many different departments will likely review the CE/PE documents for content, accuracy, gaps, and actions to improve the clinical data story for the device or family of devices. Document review is often an exercise with complicated roles and responsibility. Many departments are involved in making sure the device is safe and performs as indicated with benefits outweighing the risks before the device is sold and used for clinical care.

Even with deailed checklists and guidelines, many people are unclear about what the medical practitioner/physician, clinical, regulatory, quality, or design engineer reviewers are supposed to do when they review a CE/PE documents.

9.3.1 Clinical department reviewers including the evaluator and medical practitioner

Clinical evaluators must ensure the evaluated clinical data were high quality and directly relevant to the device under evaluation, or appropriately justified for inclusion. The evaluator must be involved in all CE/PE development stages including the data analysis and document writing. The evaluator must ensure all clinical data were identified, appraised, analyzed, and reported accurately, completely, and correctly. The evaluator writes and/or guides the CE/PE document development and is ultimately responsible for the CE/PE document quality and accuracy. The medical practitioner reviewer/evaluator may share the evaluator role and responsibilities to ensure the data were clinically relevant, had sufficient volume and quality, and were appropriately interpreted with conclusions directly drawn from the data evaluated.

Clinical writers must ensure all clinical data and the clinical/performance evaluation overall were presented clearly, completely and precisely. The grammar, syntax, tables, figures, technical details, and the CE/PE story as a whole must ring true throughout the entire CE/PE document. The conclusions must flow clearly and directly from the clinical data evaluated. Writing clearly and concisely is an art and the clinical writer must respond quickly and completely to reviewer comments guided by the evaluator to ensure technical accuracy within each document and throughout the entire CE/PE document set.

Clinical reviewers must read and check the whole CE/PE document including every detail, data abstraction, transformation, and/or clinical representation. The goal is to ensure accuracy, relevance, quality, and completeness. Clinical reviewers must conduct detailed clinical quality checks to ensure all known human risks are appropriately identified, represented, grouped, listed, and evaluated. In addition, if equivalence is claimed, the clinical reviewers must ensure all clinical impacts of any differences between the device under evaluation and the equivalent device are clearly stated and supported by appropriate clinical data to justify the equivalence claim from the *clinical* perspective.

The clinical authors are the persons who sign the document from the company clinical department and these persons must include the evaluator and the company leader who assigned the evaluator to the CE/PE work (often this is the chief medical officer). This leader will sign not only the CE/PE document overall but also the written statement about how they reviewed the qualifications and assigned the evaluator and medical reviewer to the CE/PE work. This signed/dated statement should be an integral CER/PER component including the curricula vitae and declarations of interest for the evaluator and medical reviewer.

The clinical department must ensure all individuals working on the CE/PE documents have the appropriate education, training, and experience, especially when completing data analysis. The clinical data analysis component is often lacking, so building in appropriate quality control (QC) to achieve ongoing, appropriate QA is required. The clinical team needs extensive training and hands-on experience to do these jobs well. Measuring competence directly for each clinical team member as they work on each CE/PE document requires careful attention to detail by the manager. Assigning the right people to the right tasks is critical, and the only way to do this is to know each individual's competence level for each task included during the device clinical/performance data analysis.

The clinical department team leader must ensure at least four things during the CE/PE process: (1) the clinical/performance data have been accurately abstracted, analyzed and related to the device under evaluation; (2) the CE/PE document accurately and completely represents the expert clinical/performance analysis; (3) the S&P acceptance criteria have been met and are accurately represented in the CE/PE document; and (4) the clinical benefit-risk ratio conclusion specifically determines if the benefits outweigh the risks so the company can claim the benefit-risk ratio is acceptable for all indicated clinical uses. This leader must help team members focus on the most important clinical/performance data details for the medical practitioners and patients who need the medical device.

9.3.2 Quality department reviewers

The quality team reviewer must ensure all appropriate quality checks were completed for QC across each CE/PE document and QA across the entire CE/PE process. The distinction between QC and QA completed by the clinical reviewer and by the quality reviewer lies in the different expertise and focus provided during the review. Since quality team members may not be clinical experts, they should not be assigned the role to check clinical data accuracy and completeness; instead, the quality team member should be checking to ensure each CE/PE document is structurally sound and the written arguments are clear and convincing. For example, the conclusions must be well-stated and must flow logically from the clinical data described within the CE/PE document. Quality team reviewer comments should reflect the careful review of each CE/PE document as well as the entire CE/PE document set to ensure all QC and QA requirements were met.

9.3.3 Regulatory department reviewers

The regulatory team reviewer must ensure all regulatory requirements were met for each specific CE/PE document and for the entire CE/PE document set. The regulatory team reviewer should have a checklist of all required regulatory details specific to each CE/PE document under review and for the entire CE/PE document set. The regulatory reviewer must verify each required regulatory item/detail was provided, or they must require a rewrite to include the detailed information required by a cited regulation. Regulatory expertise is required since the regulations do not specify exactly how to "check" for regulatory compliance and the regulations do not define the details. For example, the S&P acceptance criteria and benefit-risk profile are required by the regulations and guidance documents; however, the manufacturer must provide the details relevant to the specific device. The regulations are also interpreted based on risk with fewer regulatory requirements for low-risk devices than high-risk devices. The regulatory reviewer must understand regulatory discretion and how to exercise the appropriate level of regulatory "control" for a specific medical device. The regulatory reviewer must be skilled at helping the team to write justifications about missing information typical for a low-risk device while not allowing the same type of information to be missing for a high-risk device.

9.3.4 Engineering department reviewers

The engineering team reviewers may include design engineers to ensure each device had clear and accurate device description and development history supported by appropriate bench tests within the technical files, especially if any equivalence claims were made. The design engineer must ensure all CE/PE documents accounted for all device generations, especially when design changes had clinical impacts on the device S&P or benefit-risk ratio.

Packaging and labeling engineers must ensure all CE/PE documents correctly defined the technical packaging and labeling details and, conversely, all device packaging and labeling were updated to comply with the new clinical data evaluated in the CE/PE process as required by the CER/PER conclusions.

The engineering risk management team members must ensure reciprocal integration of all clinical data to/from the risk management records and the CE/PE documents, especially the CER/PER and the PMCFER/PMPFER. The engineering team must also integrate the CE/PE findings into other engineering records in the technical files including, but not limited to, the benefit-risk ratio list, the inputs/outputs documentation and specifications (e.g., the user requirements must specify the correct population and procedures for use as demonstrated in the CER/PER), verification and validation (V&V) records (e.g., the human factors data and use information collected in the CE/PE process should strengthen the V&V records), claims specifications (e.g., each claim must be supported by clinical data in the CER/PER), and labeling (e.g., the entire indication/s for use statement must be supported by clinical data in the CER/PER).

9.3.5 Postmarket surveillance department reviewers

The PMS team reviewer must ensure all surveillance reporting requirements were met in each CE/PE document. The PMS reviewer should use a checklist of PMS reporting requirements specific to each CE/PE document under review. For example, the CE/PE documents must align with and accurately represent the sales, complaints, and experience/use data documented in the PMS aystem and reported in the PMSR/PSUR or other S&P reports to regulatory authorities worldwide. In addition, the PMS reviewer must ensure any new data identified and analyzed in the CE/PE process are reported appropriately. The PMS team reviewer must help the CE/PE document team navigate the device vigilance system. The PMS team reviewer must also write justifications for clinical data collected in the PMS system when the clinical data details were incomplete or required explanation. The PMS clinical experience data limitations must be clear in all CE/PE documents.

9.3.6 External medical reviewers

The medical reviewer is often an MD who used the device in their medical practice. The medical reviewer must verify the clinical data were represented and evaluated appropriately and the conclusions were related to the medical device when used as indicated. The medical reviewer must be well versed in the difference between on-label and off-label uses and must carefully consider and report the potential for misuse and systematic off-label use in the observed real world data documented in the CE/PE documents. As the CE/PE document "guardian", the medical reviewer must read each CE/PE document and must correct any misinterpretations of the clinical data. The medical reviewer must provide detailed feedback including specific document revision requirements as well as overall interpretive guidance about the practice of medicine and the appropriate medical device use as indicated for the device under evaluation.

9.3.7 External notified body and expert panel regulatory authority assessments

External NB reviewers may be required to assess CE/PE documents for high-risk medical devices; however, many low-risk devices do not require NB review. For example, new Class III and implantable medical devices as well as high risk class D IVD devices typically require NB review, and may also require submission to an expert panel. In addition, NBs assess many types of CE/PE documents as part of their sampling program for other, lower-risk, device types. Relationships with external NB reviewers changed dramatically after the European Commission and Member States set up the Notified Body Operations Group (NBOG) in July 2020 to address the "variable and inconsistent" performance in medical device reviews both by the NBs and "the Designating Authorities responsible for them" (https://www.nbog.eu). Specifically, the NBOG meets twice a year in Brussels:

To improve the overall performance of Notified Bodies in the medical devices sector by primarily identifying and promulgating examples of best practice to be adopted by both Notified Bodies and those organisations responsible for their designation and control.

The NBOG focuses on NB documentation in concert with IMDRF/GHTF (International Medical Device Regulators Forum/Global Harmonization Task Force) activities. The NANDO (New Approach Notified and Designated Organisations) information system provides information related to NBs, including NB lists, at https://webgate.ec.europa.eu/single-market-compliance-space/#/notified-bodies, and the Team NB website provides more NB information at https://www.team-nb.org/about-us.

NBs must meet organizational and general requirements (MDR/IVDR Annex VII) and the NB must complete assigned conformity assessments based on:

- QMS and assessment of technical documentation (MDR/IVDR Annex IX);
- type-examination (MDR/IVDR Annex X)—"the procedure whereby a notified body ascertains and certifies that a device, including its technical documentation and relevant life cycle processes and a corresponding representative sample of the device production envisaged, fulfils the relevant provisions of this Regulation"; and
- production QA (MDR/IVDR Annex XI).

NBs and expert panels employ and train clinical experts to review CE/PE documents, engineering experts to review technical files, and nonclinical experts to assess biocompatibility and other bench test reports. The goal is to ensure the device meets all regulatory requirements before releasing the device under a CE Mark. The CE/PE documents only cover a small portion of the medical device regulatory requirements and the NB will assess all the details required by the MDR/IVDR including all CE/PE and PMCF/PMPF specific requirements in MDR Annex XIV/IVDR Annex XIII. Creating a checklist of these details is a good idea for any manufacturer about to undergo a NB or expert panel assessment. The manufacturer should make sure all the details needed to successfully pass through these assessments are embedded in the company SOPs, WIs, and templates.

NBs are also required to have written SOPs for their clinical evaluation assessment report/performance evaluation assessment report (CEAR/PEAR) processes. The NB must review the CE/PE documents (especially the CER/PER), and the NB reviewer must document any noncompliances, deficiencies, and follow-up actions needed to resolve noncompliance issues. The NB must provide access with a complete audit trail of CEAR/PEAR actions if requested by the NB-designating authority. In addition, the NB will send the CER/PER for expert panel review for any new, high-risk, Class IIb and III, implantable medical devices, per Article 54 and Class D IVD devices, per IVDR Article 48, except for certificate renewal, device modification with no change in benefit-risk ratio, or CE addressed in a common specification. The NB will be required to respond to expert panel requests for information, and will grant a positive assessment only after all findings are closed. Fee are charged for NB and expert panel services to complete the conformity assessment and to issue the CE mark after all regulatory requirements have been met.

In order to discuss NB and/or expert panel findings and to correct all documented CE/PE deficiencies, each manufacturer should have a skilled and clinically knowledgeable negotiator who can interface with the NB clinical reviewer to discuss the CE/PE documents and the details surrounding the claimed deficiencies. Over seven years, from 2017-2024, not all NB deficiencies were found to be accurate or appropriate.

One guidance document, entitled "MDCG 2020-13, Clinical evaluation assessment report template" (July 2020), provided a minimum NB assessment checklist for a Class IIb and Class III device clinical evaluation completed by the manufacturer and "assessments of technical documentations on a sampling basis for class IIa/IIb devices" (p. 3) to support the medical device S&P [3]. This guidance does not apply to the IVD device PEAR; however, the same principles generally apply. For example, CEARs/PEARs are provided to expert panels in a clinical evaluation consultation procedure/performance evaluation consultation procedure (CECP/PECP) whenever expert panel review is required or requested.

Clinical evaluators have used the CEAR template to determine whether CERs meet all requirements before the NB assesses the document [4]. For example, the clinical team can anticipate the NB will assess the CER/PER specifically for GSPR conformity, S&P acceptance criteria decisions, and benefit-risk ratio details based on all of the compiled nonclinical and clinical data, data suitability, and relevance for any claimed equivalent devices, SOTA, and supporting background information as well as alternative therapy/device data and all claims/product labeling including the instructions for use. The NB reviewer will verify all CER/PER conclusions were supported by the CE/PE process and will determine if PMS and PMCF/PMPF milestones were met or if they need further review.

The CEAR template includes 11 sections designed to standardize NB-assessments, and manufacturers have used this template to anticipate key NB questions during NB CE/PE assessments (Fig. 9.1).

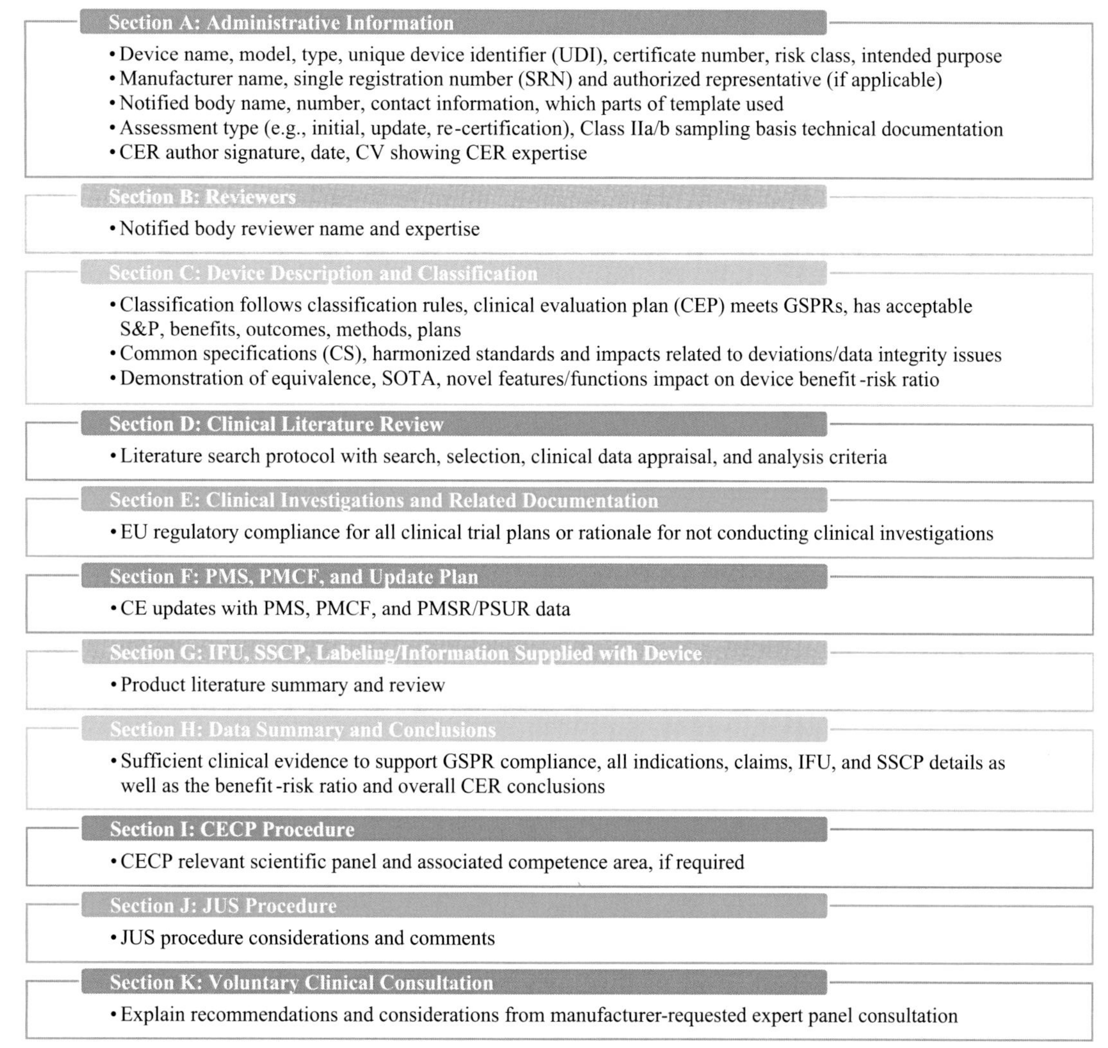

FIGURE 9.1 CEAR template structure.

After assessing the entire CER and all provided supporting CE documentation, the CEAR template requires the NB reviewer to document the CER assessment outcome to serve as the official conformity assessment record for review by the NB-designating authorities. The CEAR template is also used to review JUS documents for "medical devices for which clinical data is not deemed appropriate… and the demonstration of an adequate justification for this" (p. 4). Some NBs interpret this to mean the JUS document should be formatted as a CER/PER; however, not all NBs require the same format for CER/PER and JUS documents.

Section A—Administrative Details includes device, manufacturer, NB information, assessment type, and CER author experience verification. If appropriate, the NB may also include technical file identification numbers and technical documentation assessment report (TDAR) reference if available and the NB will list the specific documents assessed in this section. This section should also specify if an expert panel consultation was required.

Section B—NB Reviewers requires information about all NB reviewers involved in the CER assessment with their name, relevant clinical experience, background knowledge, and expertise identified including the CER aspects they assessed with a specific statement about their relevant competence and experience for their assigned assessment responsibilities as required to justify their role and responsibility to provide an adequate CER assessment.

Section C—Device Description provides background, SOTA, and alternative therapy/device information about how the subject device is used and how and why the device was changed over time. The device description, classification, and brief clinical evaluation plan (CEP) overview should include (but are not limited to) the functional elements, packs, images/diagrams for all sizes, design features, configurations, variants, accessories, compatible/similar devices, previous device generations, reasons for design variations, principles of operation, clinical safety impacts, sales volumes, and time on market in EU and elsewhere. The intended purpose and indications for use and contraindications should contain sufficient detail to specify the precise patient population, disease state, condition, severity, and body part for device treatment. For example, "peripheral vasculature" should specify arterial or venous system, vessel size, or anatomic location, and the target population should specify sex, age, race, disease state, mobility state, etc. to create an unambiguous device description section. In addition, the manufacturer should have direct access to all technical documentation (e.g., a contract may be needed for access to technical data for any high-risk equivalent device manufactured by others) and the applied specifications should be listed. Any equivalence claim should include robust clinical, biological, and technical assessments. The SOTA information should include any benchmark devices used to assess benefits and risks and a description of any alternative treatment options for the same indication, as well as justification for the S&P end points with a clear and concise benefit-risk ratio assessment. Device novelty should be described and used to assess if clinical trial, literature, and experience search parameters were adequate to cover the full range of devices and treatments and to provide a robust evaluation of the benefit-risk ratio.

Section D—Clinical Literature Review summarizes the literature search protocol and the NB will review the search and selection criteria to ensure all sizes, variants, models, accessories and clinical conditions were addressed and the literature selected was related to the device under evaluation, any equivalent or SOTA devices, or any alternative therapies. The literature search strategy must be broad enough to encompass the required background knowledge and SOTA, establish benchmark treatments, find adverse events (AEs) and undesirable side effects, and to adequately describe the databases searched, specific search terms used, inclusion and exclusion criteria applied, and methods used to include both favorable and unfavorable data, assess deviations, and to avoid data duplication and bias. The NB must determine the literature review acceptability after careful consideration of study designs and biases, journal peer-review status, and inclusion of sufficient clinical data with a scientifically valid contribution and weight to support all device sizes, variants, models, accessories, and clinical conditions treated to clearly demonstrate the device S&P and benefit-risk ratio.

Section E—Clinical Investigations and Related Documentation will be assessed by the NB including pre- or postmarket clinical trials conducted by the manufacturer or others or the rationale provided if no clinical investigations were performed. The NB will consider if trials were registered and properly monitored, if required international standards were met, if trial results were published and if conclusions were supported by the study design and resulting clinical data. Clinical investigation plan elements (e.g., scope, design, devices used, study locations, patient population, sample size, objectives, end points and follow-up durations) will be assessed for adequacy to determine device S&P and benefit-risk ratio.

Section F—PMS, PMCF, and Update Plan including PMSR/PSUR, PMCFP/PMPFP, PMCFER/PMPFER or justification for no PMCF/PMPF documentation will be assessed when relevant and available. The CER/PER should state how often and under what conditions CE/PE updates will be required.

Section G—Labeling (e.g., IFUs, SSCPs/SSPs and other information or product literature supplied with the device) should comply with MDR/IVDR GSPRs. The NB will assess if the intended purposes and indications for use were sufficiently supported by clinical data, if the labeling provided adequate direction or training for intended users, and if the labeling fully described the device S&P. The NB will specifically assess if the labelling was clearly presented, understandable, and aligned with the technical documentation, if the warnings and precautions were described appropriately, if all device limitations were listed, and if additional contraindications or warnings were needed in the device labeling.

Section H—Data Summary and Conclusions should discuss whether the clinical data fully supported the device, any equivalence claims and compliance with GSPRs 1, 6, and 8 along with the device S&P acceptance criteria and benefit-risk ratio. The NB will assess if any unanswered device questions remain and if detailed plans for PMS and PMCF activities were appropriately listed to cover all identified clinical data gaps. The overall conclusions should state if the evaluation showed the clinically-relevant benefits outweighed the risks, the clinical evaluation was aligned with RM and all deficiencies and noncompliances were resolved. The CEAR conclusion should include the NB recommendation for granting certification or not.

Sections I, J, and K should describe any expert panel CECP activities, justifications for missing CE documentation and any expert panel voluntary clinical consultation details, respectively.

The NB CEAR/PEAR must summarize how well the CER/PER documents met the MDR/IVDR requirements. The NB must assure the EU Commission the device achieved the claimed intended purpose to benefit the patient and the benefits outweighed the acceptable residual risks and undesirable side effects to the patient when the device was used as indicated. The CER/PER must use adequate qualitative and quantitative S&P evaluation methods and must accurately describe the impact of all complaints, device deficiencies, trends, and vigilance issues. The manufacturer should use the CEAR/PEAR to ensure sufficient and relevant clinical data are on file and the NB assessment and CEAR will allow a CE mark to be applied to the medical device.

9.4 Deficiency letter examples

NB deficiency letters typically identify findings in a variety of documents included in the submission package. The NB reviewer must specify each deficiency clearly and completely and they must identify the document in question and the specific regulation requiring compliance. The NB reviewer must give the manufacturer an opportunity to respond and to correct the deficiency. This process can take multiple rounds of questions and answers as some items will be closed and other items will need follow-up to ensure all details are addressed. Sometimes the NB may issue additional questions if the device manufacturer response was deemed insufficient and the NB may limit the manufacturer to three rounds of questions and answers.

The following is a random sample of anonymized medical device deficiency letter findings issued from 2017 to 2023 across a wide variety of device manufacturers. Each deficiency item includes a suggested process change to help avoid the deficiency in future manufacturing operations. These example deficiency items were modified to avoid release of confidential or proprietary information.

9.4.1 Use current references in all documents

One NB reviewer expected contemporaneous clinical data to support the medical device CE and one company received the following deficiency:

> *Referring to the Instructions for Use ... the referenced publications date back decades (up to more than four decades). To ensure ... the Manuals/Information for Users/Instructions for Use have the most current available data implemented/reflect the most recent applicable data and information, references need to be updated.*

The company response highlighted two different approaches taken in response to this deficiency: (1) instructions for use was updated with more up-to-date references; and (2) the company presented evidence showing how some older publications still represented valid information. The NB closed this deficiency and allowed the company to replace the instructions for use with the updated version over a specific time frame allowing the company to exhaust the IFUs already released into the current manufacturing process.

This deficiency shows how the CER/PER and the device labeling (e.g., the instructions for use in this case) should be closely linked to keep the reference list fully aligned.

In practice, do not allow the literature search to be more than 6 months old when the CER/PER is signed off and version controlled and ensure the instructions for use and product labeling are all updated to align with the current CER/PER.

9.4.2 Do not use ambiguous or open-ended indication for use wording

Another NB reviewer required unambiguous language whenever possible, and they identified the following deficiency:

> *Referring to the Instructions for Use ... The Indications read "... is used in ophthalmic surgical procedures of the anterior segment, including: - Cataract extraction and - Intraocular lens (IOL) implantation" This terminology (..."including"...) leads to the conclusion that there might be other specific indications and thus other claims for the device than the two listed above ... please note that generalizing phrases such as "example given," "such as" and "including" are not acceptable since clinical evidence/data demonstrating performance and safety of the device when used for each specifically claimed indication needs to be provided within the CER. Please ensure consistency throughout the different documents...*

This device manufacturer simply removed the word "including" and the response was accepted.

This deficiency shows a clear focus on limiting the indication for use statement and ensuring consistency across all device documents. The indication for use statement wording must be supported strictly and fully by the CER clinical evidence/data demonstrating device S&P when the device as used as specifically indicated. In other words, the indication for use statement must not imply any device use outside of the clinical S&P data evaluated in the CER/PER. The CER/PER must support every specifically stated indicated use.

In practice, do not use the following "generalizing phrases": "for example," "such as," "including," "example given," etc. in the indication for use statement. The CER/PER must have sufficient clinical data evaluated to support each and every aspect of the indication for use statement.

9.4.3 Differentiate clearly between intended use and indication for use

One company received the following deficiency from their NB reviewer:

> *Referring to the Instructions for Use...: No unambiguously clear description/definition of the intended use for the device under assessment could be found within IFU. Please take into account that it needs to be clearly differentiated between the "Intended Use" and the "Indication/s"; i.e. the term "intended use/purpose" describes the effect of a device on the human body (e.g., the replacement or removal of a body part, etc.) whereas the term "indication" refers to the specific clinical condition that is to be diagnosed, prevented, monitored, treated, alleviated, compensated for, replaced, modified or controlled by the Medical Device. Therefore, please update the document describing the intended use, i.e. clearly describing the effect of the device on the human body.*

This device manufacturer crafted a new intended use statement reflecting the broad device effect on the human body, and kept this statement distinct from the indications for use statement. This was acceptable to the NB reviewer.

This deficiency showed how manufacturer team members must clearly understand the difference between intended use and indication for use statements. The written records and all device labeling must carefully portray the intended use and indication for use statements and the manufacturer should expect focused NB reviewer attention on "key labeling claims." Labeling updates are particularly important for legacy or low-risk devices using a single statement to represent both the intended use and indication for use statements.

In practice, all device labeling must include consistent and separate statements for intended use (i.e., the broad device function) and indication for use (i.e., the specific disease state where the device is to be used). CER/PER team members should be able to identify and fix this deficiency for all legacy and low-risk devices, and the manufacturer should avoid this known NB deficiency in the future.

9.4.4 Remove discrepancies between instructions for use and supporting documentation

Another NB reviewer documented the following discrepancy:

> *Please re-visit the device description as it is provided in the Instructions for Use and in the CER to ensure the information is identical, e.g., with regard to the information related to the molecular weight as given in the unit of Daltons.*

The device manufacturer corrected the technical error and the NB reviewer accepted this response.

This deficiency reinforced the need for a QC reviewer during CER/PER document development. Each CE/PE document must have the correct information exactly as presented in the company supporting documentation.

In practice, the CE/PE document development team should ensure each document is accurateand each team member should be responsible to identify any and all technical discrepancies between the CE/PE documents and the supporting documents.

9.4.5 Increase post-market surveillance data

One company received the following discrepancy:

> *"Please provide ..." the device "Plan for pro-active, prospective, systematic post market surveillance. Please note that it must be comprehensible that any procedures and processes that a manufacturer might have in place as post market surveillance activities do not only rely on passive post market surveillance, but indeed need to focus on pro-active and prospective activities to minimize the risk of possible under-reporting of complaints/complications. Note: The fact that there has only been one complaint within the last certification period does not allow the conclusions related to S&P of the device under assessment, but might rather question the reliability and effectiveness of the post market surveillance system as it is in place so far."*

The company argued with the NB and stated the PMS was accomplished by annual CER updates. The company indicated the clinical literature evaluations were sufficient to monitor the market for this device. The NB rejected this argument and responded as follows:

> *The provided response does not fully address the Deficiency and the underlying reason for raising this issue. As had been explained before, measures need to be in place to allow conclusion that the risk of possible "under-reporting" of complaints/ adverse events is mitigated as much as possible; in addition, it is also necessary to collect data that allow reliable conclusions related to persisting device performance. Annual literature searches might not allow reliable conclusions related to both*

safety and performance since there might be various reasons why no specific recent device-related publications are available; such reasons certainly do not have to be all positive or all negative, it might not be determinable. Also the low number of complaints over the years as reported within the manufacturer's post market surveillance system does not allow reliable conclusions related to ongoing safety and performance; this had also been outlined in the initial Deficiency Report... This is said in light of the regulatory requirements that need to be fulfilled to obtain/keep the CE-Mark, independent of the geographical area where a CE-Marked device might be sold. In consequence, please provide a respectively updated response to this question to allow understanding how the device under assessment is being surveilled pro-actively and prospectively with regard to both safety and performance.

The device manufacturer responded by proposing a PMS survey of users (surgeons) designed to actively seeking their device experience and comments about the device S&P. The NB accepted this plan to complete the survey data collection and analysis prior to the next CER update.

This deficiency focused on the MDR/IVDR requirement for proactive and systematic PMS system activities. The PMS system must require more than passively waiting for a user report or collecting and analyzing previously published literature clinical data. Specific proactive PMS activities and associated outputs must be clearly identified in the PMS plan (PMSP) and the interrelated documents including the CER/PER. The PMS system is designed to collect user medical device experiences in order to identify and report new risks to the regulatory authorizes expeditiously as required. The PMS system also informs the ongoing benefit-risk ratio analysis in the CER/PER and the understanding of specific risks within the risk management system. In this case, the NB reviewer also explained how the lack of reported device adverse events does not support device safety. Rather, this lack of clinical data actually casted doubt on the validity of the device S&P conclusions.

In practice, the medical device manufacturer must gather more device S&P data by proactively identifying new benefit and risk details and user reports. Checking and adding to the risk-benefit listing with proactive PMS activities is required and must be documented in the PMSP.

9.4.6 Ensure clinical evaluation report is a stand-alone document

Another manufacturer received the following deficiency:

Please update the CER to implement any remaining clinically relevant risks including the critical rationale why these can be considered acceptable from a clinical point of view. Note: This is requested since the CER needs to be considered as stand-alone document containing all relevant clinical information and data. Therefore, pure reference to the Risk Management Documentation as done in chapter 4 of the CER is not considered acceptable.

The device manufacturer argued with the NB and stated the CER was submitted with all risk management documentation attached as a "stand-alone" document. The NB reviewer rejected this argument and reiterated the requirement to generate a single, stand-alone CER including an integrated risk analysis. The device manufacturer rewrote the CER benefit-risk section, and this was acceptable to the NB reviewer.

This deficiency required the CER/PER to be a single, stand-alone document which synthesized arguments from supporting documentation into a single, coherent report. The CER/PER is not only a statement of what evidence exists but why the clinical data/clinical evidence are compelling enough and relevant enough to specifically and clearly demonstrate an acceptable device S&P and benefit-risk ratio for device use as indicated. Ideas from many sources are extracted and analyzed together to provide a new, clear, clinical-evidence-based picture of the device S&P and benefit-risk ratio.

In practice, the CE/PE team must create a single, stand-alone, fully analyzed and integrated CER/PER document and NOT an index of reports compiled by the device manufacturer.

9.4.7 Provide sufficient state-of-the-art literature search details

Another company received the following deficiency:

Referring to the State-of-the-Art Analysis...a short summarizing list of current medical practice and other treatments is provided. It is assumed that this list is supposed to cover the State-of-the-Art relative to the intended use and indications. It is not comprehensible how the content of this chapter reflects the outcome of the literature search using the search terms as described ... since these focus on the device under assessment and/or supposedly comparable devices. It is thus not conclusive how the search and the results ... match.

The company responded with a revision to the SOTA section; however, this was rejected by the NB reviewer as follows:

The content of the provided additional document . . . is not fully comprehensible. Chapter 2 of this document . . . reads as: ". . . A search was conducted in the published literature for any publicly published papers addressing the state-of-the-art, history or overview of therapies used in ophthalmic surgical procedures in which a space is created to protect the cornea of the eye and other ocular tissues during such procedures. No time limitation was used, however emphasis was placed on the most current articles in order to capture the most current state-of-the-art." This description explains that a search has been performed, but it is not comprehensibly described how the search has been performed, e.g., search terms. Therefore, it is not possible to understand the basis for the conclusions that have been drawn with regard to the State-of-the-Art.

This deficiency suggested the SOTA search must be specified separately from the device-specific search; however, this is not specifically required under the regulations. Even so, clearly documented search strategies and search terms must be provided (e.g., the methods must define the dates, search terms, results, and search applicability).

In practice, the CER/PER must include a methods section explaining how the background knowledge and SOTA clinical/perfromance data were identified and evaluated, along with details about the alternative therapies and devices available to the patients using the subject device. This search strategy must result in a comprehensive review of the background knowledge and SOTA clinical data including the alternative treatments or devices available to the patient to treat the same condition. A short list of alternative treatments is not sufficient for the CER/PER.

9.4.8 Provide sufficient similar device clinical data

Another company received the following deficiency about their ophthalmic viscosurgical devices (OVDs):

Referring to the literature search and State-of-the-Art: If it understood correctly, the most recent literature search did not result in any publications directly related to the device under assessment. According to the documentation the publications cover supposedly equivalent devices . . . the CER lists different OVDs that are assumed to be "similar." Taking the description of the devices as listed in this Table into consideration, it is obvious that they differ at least with regard to formulations, molecular weight and active ingredient and its origin, respectively. Therefore, it cannot simply be concluded that any data applicable for the other devices are easily transferable to the device under assessment. It needs to be demonstrated why any differences of the devices do not lead to differences with regard to clinical safety and performance when used in patients; i.e. equivalence/comparability needs to be actually shown.

The manufacturer responded with an expert consultant discussion of different categories of products used in the target therapy and how the device under evaluation was similar to others based on similarities in the active ingredient. The company defined how the active ingredients were manufactured and supplied by the device manufacturer so the device manufacturer could confirm the similarity of the active ingredient to a high level of confidence. The NB reviewer rejected this argument, and responded with the following:

The provided response is not completely comprehensible; if it is understood correctly [two OVDs] are considered comparable . . . [and the active ingredient] used in all three products [is the same as made by the manufacturer] and thus data for the other two devices are considered transferable to [the subject device] from both a safety as well as a performance standpoint. In addition [the manufacturer] manufactures the [active ingredient in two additional OVDs] . . . (modified to produce a higher MW) . . . and thus it is concluded that it is reasonable to compare [the subject device] to those two products in the literature from a safety standpoint. Please provide clarification which devices are considered the supposedly comparable ones . . .; please be aware that comparability with a reference device needs to be in place for both safety and performance. In addition, please provide the objective scientific evidence for the explanation . . . "With regard to origin, . . . [the active ingredient], by definition is chemically identical whether extracted from . . . [two different sources]." In case at some time, this issue would be resolved adequately, all respective information would need to be implemented into a final version of the CER (State-of-the-Art section); this is needed to allow also an independent third party reviewer understanding the overall context of the clinical evaluation (as stand-alone document) including understanding why published data reporting on other devices than the one under assessment should nevertheless be considered transferable to device under assessment.

The device manufacturer edited the CER with all pertinent discussions about the differences between subject device and similar devices, and the NB accepted this revision.

This deficiency illustrated a special case where "similar" devices were used to secure an ongoing CE Mark for a legacy device and where similar competitor devices were manufactured using the same active ingredients produced by the device manufacturer of the device in question. As a result, even though (1) the NB clearly stated a listing of products which device manufacturers considered "similar" to their device was insufficient, (2) the CER lacked discussion on how this similarity was established with regards to clinical S&P, and (3) expert consultant statements normally cannot stand alone, this rare case considered similar device clinical S&P data after expert discussions established the clear and reasoned rationale for all clinical data to be used (i.e., including data from "similar" devices not derived from the subject device or from any claimed equivalent device).

In practice, CER authors should ensure equivalence is well established when insufficient clinical data are available for the subject device and the clinical data presented in the CER/PER must be specific to the subject device or a device with established equivalence. As this deficiency illustrated, a "similar" device is an ambiguous term and clinical data from such devices cannot easily be presented as applicable to subject device. Even so, similar device clinical data can sometimes be considered sufficient after expert discussions establish a clear and compelling scientific rationale to determine the device S&P and benefit-risk profile acceptability based on the clinical data from "similar" devices using the identical active ingredient but not derived from the subject device or from any claimed equivalent device.

9.4.9 Avoid multiple deficiencies during low-risk medical device sampling

The NB typically issues a list of findings, and the following example is from a manufacturer making low risk devices where the NB assessor called out the following discrepancies:

- No conformity checklist was submitted.
  - This issue was closed when the manufacturer provided a GSPR checklist.
- "IFUs ..., RM Plan ... RM Report ... are outdated ... The U(Use FMECA and the Useability Engineering File and Example) show no date of issue ..."
  - This issue was closed when the manufacturer provided the updated IFU, risk, uFMECA and usability documents.
- "User group is not consistently defined" (sometimes physicians and trained professionals and sometimes only trained professionals were mentioned). The deficiency stated "The user group must be consistently defined in all pertinent documents" and the manufacturer "must define the requested "training" and his own contribution to this in the IFU. Are courses offered? Instruction videos available ..."
  - This issue was addressed by changing the wording to "healthcare professionals" and stating "no further specific training is estimated necessary for proper use of the devices in question ..." The manufacturer committed to making this update "during the next scheduled IFU and CER update ..." and the NB keep this item as an "Observation" to "be checked during the next assessment" in 5 years.
- "... the patient group and their estimated state/condition ... must be mentioned in all relevant documents ..." "... even if there is no limitation."
  - This issue was also kept as an observation after the manufacturer declared they would update the relevant documents during future updates.
- "... manufacturer must present a PMS/PMCF strategy including a design for a sponsored clinical testing or other applicable measures of pro-active data acquisition (scheduled customer feedback, questionnaires or other) to eliminate the doubts concerning safety and performance of the device in question."
  - "The manufacturer submits ... a plan for conduction of proactive feedback acquisition ... in detail his plan to obtain relevant user data within his PMCF. End users, including direct customers and hospital staff, will be included in the user survey. A suitable rationale for the intended selection of participants is provided. Besides the direct user survey, internal data analysis, as well as review of industrial standards, recommendations and white papers, will be performed. Instructions for completion of the user survey, step by step, are included. This will be followed by a structured data analysis and implementation of the obtained results in other documents. Furthermore, the manufacturer supplies a sound overview of the intended questions to be asked (end users and account representatives), assessing reporting habits, observations of clinical performance, device safety, device contamination/sterile technique, general ease of use and observations of the impact on image quality. This represents a sound, even exemplary system for the acquisition of clinical user data. The aforementioned finding is herewith closed."

All of these deficiencies were addressed, and the relationship with the NB was strengthened through this process.

9.4.10 NB deficiency example conclusions

These NB deficiency findings illustrated just a few common CER/PER challenges as manufacturers worked to comply with the regulatory requirements. These real-world examples of actual deficiencies written for various manufacturers by multiple different NBs is not intended to be a comprehensive listing. These examples included requirements about including current data in CE/PE documents, keeping the indication for use statement free from ambiguous language, keeping the intended use and indication for use statements separate, ensuring documents are current and properly dated, user group and disease state are stated clearly, aligning all product information, proactively collecting and analyzing PMS clinical data, ensuring the CER/PER is a single, stand-alone document, and having sufficient clinical data to determine the device S&P and benefit-risk profile acceptability. The CE/PE team should avoid these basic CE/PE deficiencies.

9.5 Examples of expert panel opinions and guidance

The MDR and IVDR both contain provisions for expert panel reviews to ensure all high-risk medical devices are safe and perform as intended when they "are being placed on the EU market whilst supporting innovation" [5]. The regulations provided more control over high-risk devices with premarket reviews, clinical evidence requirements, and expert panel/expert laboratory provisions including specific EU Reference Laboratories (EURLs) described in Article 106 of the MDR and Article 100 of the IVDR. In particular, MDR Articles 106 and 54 and IVDR Articles 100 and 48(6) require a "Clinical evaluation consultation procedure" including an expert panel for certain high-risk Class III and Class IIb devices (including implantable devices) as well as Class D IVD devices. These expert panels were envisioned to support the scientific assessment of medical devices and IVD medical devices, to provide guidance, and to identify emerging medical device concerns for the European Commission, Member States, NBs, and manufacturers to consider.

In addition, EU Regulation 2019/1396 [6] laid down rules to designate expert panels for medical devices including tasks not limited to:

- provide scientific, technical and clinical assistance for implementation of MDR and IVDR;
- consult for NBs on CERs of high-risk medical devices and PERs for IVD medical devices;
- advise MDCG and European Commission about S&P of medical devices and IVD medical devices;
- advise manufacturers about clinical development strategies and clinical investigations;
- develop and maintain guidance documents, common specifications, and international standards; and
- provide opinions to manufacturers, Member States, and NBs.

If we look back at CER/PER history, many issues evolved over several decades before the 2017 EU regulations came into effect in May 2021. NBs began much greater enforcement activities under the MedDev 2.7/1 guideline in 2016 and the MDR/IVDR in 2017, and several review committees including the MDCG and expert panels started functioning in 2019 and 2020. Expert panel CECPs/PECPs for high-risk devices identified multiple NB failures and frequently advised NBs to require more robust clinical data.

The expert panels started working and the CE/PE evaluator competence requirements increased substantially to require more education, training, and experience for evaluators to be able to read, interpret, and draw conclusions from all clinical data types. Each manufacturer needed to change their operations to comply with the rapidly released MDR/IVDR, MDCG guidelines and templates as well as the new NB and expert panel opinions. Medical device manufacturers also needed to understand how and when to use the guidelines for CE/PE documentation (e.g., MedDev 2.7/1 and MDCG 2022-2) and when not to use the guidelines, even though the details sometimes conflicted with regulatory requirements. These changes occurred while NBs were evaluating whether they could stay in business and how they were going to demonstrate proficiency assessing manufacturers under MDR/IVDR so they could become certified as a NB capable of providing a CE mark for some or all marketed medical devices in Europe.

Regulatory changes and medical device complexities increased rapidly while available clinical evidence continued to expand. Better CE/PE systems were developed while NBs and expert panels provided guidance intended to improve CE/PE quality over time. Expert panel opinions were publicly available and included an overarching view of CE/PE, PMS, and RM.

9.5.1 Expert panel consultation guidance

Expert panels began consulting with NBs about regulatory interpretations for high-risk devices in 2020, and two guidance documents clarified the need for these expert panels consults (Table 9.3).

TABLE 9.3 Expert panel guidance documents

Reference	Title	Date
MDCG 2019-3 rev. 1 [7]	Interpretation of Article 54(2)b	April 2020
MDCG 2021-22 rev. 1 [8]	Clarification on "first certification for that type of device" and corresponding procedures to be followed by notified bodies, in context of the consultation of the expert panel referred to in Article 48(6) of Regulation (EU) 2017/746	September 2022

MDCG 2019-3 rev. 1, a short (2.5 page) document [7], discussed three criteria required to "exempt devices from the pre-market clinical evaluation consultation procedure with the involvement of expert panels" and while Article 54(2)(a) and (c) clearly stated two exemptions; the words "already marketed" and "modifications" in Article 54(2)(b) were unclear.

Article 54 (2). The procedure referred to in paragraph 1 shall not be required for the devices referred to therein:

(a) in the case of renewal of a certificate issued under this Regulation;

(b) where the device has been designed by modifying a device already marketed by the same manufacturer for the same intended purpose, provided that the manufacturer has demonstrated to the satisfaction of the notified body that the modifications do not adversely affect the benefit-risk ratio of the device; or

(c) where the principles of the clinical evaluation of the device type or category have been addressed in a CS [common specification] ... and the notified body confirms that the clinical evaluation of the manufacturer for this device is in compliance with the relevant CS [common specification] for clinical evaluation of that kind of device.

The "already marketed" language was not about "a device already marketed uniquely under the new Regulation" and "modification" was restricted to those changes "needed in order to comply with the new legal requirements introduced by the MDR."

This guidance listed additional items for NB review, including a manufacturer statement about the intended purpose not changing, the certificate history (with a copy of the last certificate), and the device changes needed for MDR compliance. The NB will verify the device has an acceptable benefit-risk ratio and a valid certificate, complies with the GSPRs, has "no pending assessment of changes for the device or outstanding non-compliance" and ensures "Limitations of the intended purpose of the device should not trigger the consultation procedure." The MDR created a "Clinical evaluation consultation procedure for certain class III and class IIb devices" and these exemptions explained when a consultation procedure with an expert panel was not needed.

This guidance did not mention IVD devices; however, the IVDR expert panel consultation was required for any Class D devices with no common specifications at the first certification for the device type (i.e., with "no similar device on the market having the same intended purpose and based on similar technology"), and the NB must submit the PER "to the expert panel within five days" of receipt from the manufacturer (IVDR Article 48 (6)). Although the EC may request scientific advice from the expert panels about S&P of any Class D device (Article 50), all other IVD devices are exempt from expert panel review.

The second expert panel guidance, MDCG 2021-22, rev. 1 [8], clarified the term "first certification for that type of device" to include "the first certification under either Directive 98/79/EC or under Regulation (EU) 2017/746 by any notified body," and offered examples of the same and different device types. Procedures were also discussed to assist the NB in deciding if the review will be the first certification for a particular device type.

These two guidance documents helped to clarify the boundaries for expert panel consultations.

9.5.2 Expert panel findings

MDR/IVDR established expert panels as a mechanism to provide advice and expert scientific opinions on high-risk devices. Twelve panels currently exist, including ten for specific medical specialties (e.g., orthopedics, neurology, obstetrics, and ophthalmology), one for IVDs, and the screening panel responsible for determining whether a scientific opinion is needed. For certain high-risk medical devices, expert panels provided opinions on the NB CEAR, and for IVDs, the expert panel provided views on PERs for NBs.

Under MDR, all Class IIb Rule 12 and Class III medical device CEARs were submitted to the screening panel unless the device has already been granted a CE Mark under MDR, the device was a modification to a certified device with the same intended use and benefit-risk ratio or a common specification was available for the device type. The screening panel then decided whether to seek an expert panel opinion based on:

i. the novelty of the device or of the related clinical procedure involved, and the possible major clinical or health impact thereof;

ii. a significantly adverse change in the benefit-Risk Ratio of a specific category or group of devices due to scientifically valid health concerns in respect of components or source material or in respect of the impact on health in the case of failure of the device;

iii. a significantly increased rate of serious incidents reported in accordance with Article 87 in respect of a specific category or group of devices. (Annex IX 5.1)

If a scientific opinion was sought, the thematic expert panel reviewed the NB CEAR, CER, and PMCF plan, and issued a report with recommendations to the NB. If the expert panel finds:

the level of clinical evidence is not sufficient or otherwise gives rise to serious concerns about the benefit-risk determination, the consistency of that evidence with the intended purpose, including the medical indication/s, and with the PMCF plan, the notified body shall, if necessary, advise the manufacturer to restrict the intended purpose of the device to certain groups of patients or certain medical indications and/or to impose a limit on the duration of validity of the certificate, to undertake specific PMCF studies, to adapt the instructions for use or the summary of safety and performance, or to impose other restrictions in its conformity assessment report, as appropriate. (Annex IX 5.1)

For IVDs under the IVDR, Class D devices (i.e., devices intended for transmissible agent detection) are sent for expert panel review if no common specification is available for the device type and no other devices of the same type have been certified under IVDR. However, NBs do not evaluate PERs until the expert panel provides a scientific opinion on the PER, and the conformity assessment is made after considering the expert panel report. As such, the PECP does not involve a screening panel, and scientific opinions are issued for all qualifying IVDs.

The expert panel reports focused on CER/PER clinical data, and both positive and negative comments were released on public websites for CECP expert opinions [9] and PECP expert views provided to the NBs [10], respectively.

Expert panel review began in April 2021. According to the European Medicines Agency (EMA), during the period from April 2021 to June 2022, 24 device and 15 IVD dossiers were submitted for expert panel review, resulting in 5 CECP scientific opinions and 15 PECP scientific views. As of February 17, 2023, 8 CECP opinions were available along with 16 PECP views.

Among the eight medical devices available from thematic panels for scientific CECP opinions by February 2023, six were sent for scientific opinions due to device or procedure novelty (criterion 1) and three opinions were sought due to "scientifically valid health concerns" (criterion 2). No opinions were sought due to an increase in adverse event reports (criterion 3). All eight devices were Class III implantables, and three were on the market in the EU under MDD/AIMDD certification. Three submissions were trCERs, one was an eqCER, and four were fihCERs.

As of June 9, 2024, a total of 30 expert opinions (10 CECP and 20 PECP) were available for review, and these provided excellent teaching tools to avoid mistakes made by other manufacturers and NBs. A good exercise for learning is to read through these opinions to understand what expert panels require for adequate and acceptable CERs and PERs (Table 9.4).

TABLE 9.4 CECP and PECP expert opinions (as of June 9, 2024, $N = 30$)

Thematic expert panel	CECP opinion report
1. Orthopedics, traumatology, rehabilitation, rheumatology	August 25, 2022, NB0459, CECP-2022-000232
1. Orthopedics, traumatology, rehabilitation, rheumatology	October 22, 2021, NB2797, CECP-2021-000205
2. Circulatory system	November 11, 2022, NB0123, CECP-2022-000235
2. Circulatory system	July 5, 2022, NB0344, CECP-2022-000225
2. Circulatory system	June 27, 2022, NB0344, CECP-2022-000216
2. Circulatory system	May 23, 2022, NB0344, CECP-2022-000213
2. Circulatory system	December 7, 2021, NB0344, CECP-2021-000207
3. Neurology	August 1, 2022, NB0344, CECP-2022-000222
4. Respiratory system, anesthesiology, intensive care	NA
5. Endocrinology and diabetes	NA
6. General and plastic surgery and dentistry	October 6, 2022, NB2797, CECP-2022-000227
6. General and plastic surgery and dentistry	June 15, 2021, NB0483, CECP-2021-000201
7. Obstetrics and gynecology, including reproductive medicine	NA
8. Gastroenterology and hepatology	NA
9. Nephrology and urology	NA
10. Ophthalmology	NA
11. *In vitro* diagnostic medical devices	IVD-2021-000001-view
11. *In vitro* diagnostic medical devices	IVD-2021-000002-view
11. In vitro diagnostic medical devices	IVD-2021-000003-view
11. *In vitro* diagnostic medical devices	IVD-2021-000004-view
11. *In vitro* diagnostic medical devices	IVD-2021-000005-view
11. *In vitro* diagnostic medical devices	IVD-2021-000006-view
11. *In vitro* diagnostic medical devices	IVD-2021-000007-view
11. *In vitro* diagnostic medical devices	IVD-2021-000008-view
11. *In vitro* diagnostic medical devices	IVD-2021-000009-view
11. *In vitro* diagnostic medical devices	IVD-2021-0000010-view
11. *In vitro* diagnostic medical devices	IVD-2021-0000011-view
11. *In vitro* diagnostic medical devices	IVD-2021-0000012-view
11. *In vitro* diagnostic medical devices	IVD-2021-0000013-view
11. *In vitro* diagnostic medical devices	IVD-2021-0000014-view
11. *In vitro* diagnostic medical devices	IVD-2021-0000015-view
11. *In vitro* diagnostic medical devices	IVD-2022-0000016-view
11. *In vitro* diagnostic medical devices	IVD-2023-0000017-view
11. *In vitro* diagnostic medical devices	IVD-2023-0000018-view
11. *In vitro* diagnostic medical devices	IVD-2023-0000019-view
11. *In vitro* diagnostic medical devices	IVD-2024-000020-view

These CECP and PECP reports had a similar structure and appearance with administrative information followed by expert panel opinions or views (Fig. 9.2).

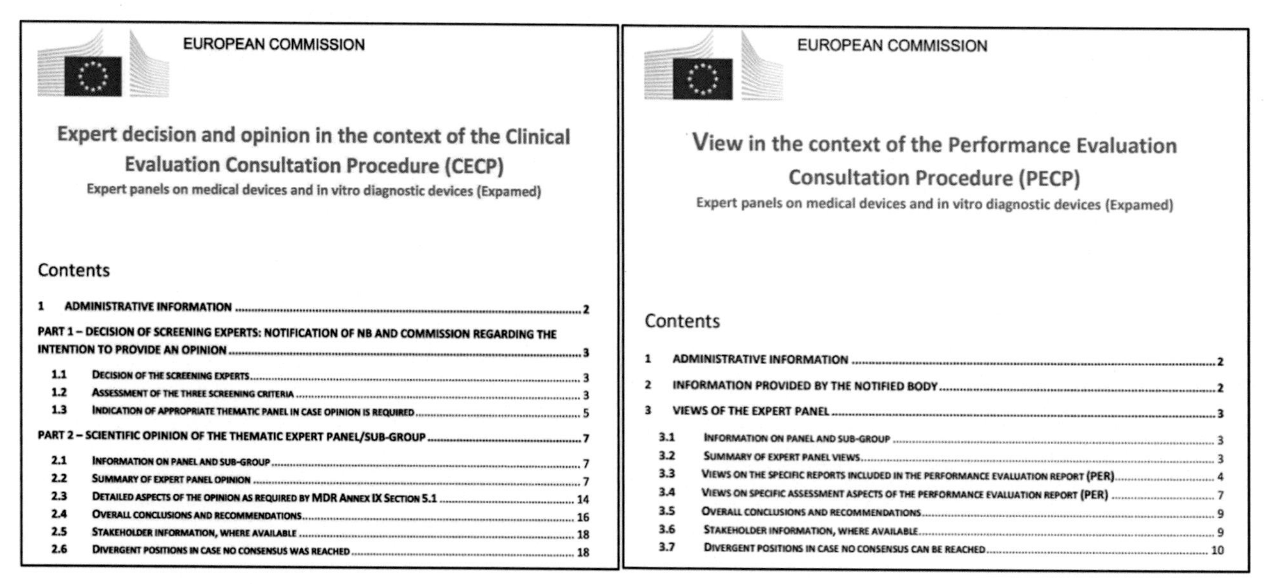

EUROPEAN COMMISSION

Expert decision and opinion in the context of the Clinical Evaluation Consultation Procedure (CECP)

Expert panels on medical devices and in vitro diagnostic devices (Expamed)

Contents

EUROPEAN COMMISSION

View in the context of the Performance Evaluation Consultation Procedure (PECP)

Expert panels on medical devices and in vitro diagnostic devices (Expamed)

Contents

FIGURE 9.2 CECP (Left) and PECP (Right) tables of contents.

The CECP and PECP scope statements within these 30 reports were also quite similar (Table 9.5), with independent experts reviewing NB opinions about the CER (i.e., NB CEAR) or the PER directly as required in the MDR/IVDR.

TABLE 9.5 Typical CECP and PECP scope statements

CECP scope	PECP scope
This scientific opinion reflects the views of independent experts (MDR Article 106) on the clinical evaluation assessment report (CEAR) of the notified body. The advice is provided in the context of the clinical evaluation consultation procedure (CECP), which is an additional element of conformity assessment by notified bodies for specific high-risk devices (MDR Article 54 and Annex IX, Section 5.1). The notified body is obliged to give due consideration to views expressed in the expert panel scientific opinion and in particular in case experts find the level of clinical evidence not sufficient or have serious concerns about the benefit-risk determination, the consistency of the clinical evidence with the intended purpose including the medical indication/s or with the postmarket clinical follow-up (PMCF) plan. Having considered the expert views, the notified body must, if necessary, *advise the manufacturer on possible actions*, such as specific restrictions of the intended purpose, limitations on the duration of the certificate validity, specific PMCF studies, adaption of instructions for use or the summary of safety and clinical performance (SSCP) or may impose other restrictions in its conformity assessment report. In accordance with MDR Annex IX, 5.1.g., the notify body shall provide a full justification where it has not followed the advice of the expert panel in its conformity assessment report.	This scientific view reflects the opinion of independent experts (MDR Article 106.1) on the performance evaluation report (PER) of the manufacturer. The advice is provided in the context of the performance evaluation consultation procedure (PECP), which is an additional element of conformity assessment by notified bodies for specific high-risk *in vitro* diagnostic devices (IVDR Article 48.6). When making its conformity assessment decision, the notified body is obliged to give due consideration to opinions expressed in the scientific view of the expert panel, where applicable (Annex IX, Section 4.9 or, as applicable, Annex X, Section 3, point (j)). For class D devices, the *notified body must provide a full justification in the case of divergent views between the notified body and the experts.* This justification shall be included in the notification to the competent authority (IVDR Article 50; mechanism for scrutiny of Class D devices).

The CECP scope statement indicated the NB advised the manufacturer about "possible actions" and "specific restrictions" while the PECP scope stated the NB must justify any divergent views about the highest-risk devices directly to the competent authority.

As the NB completed their medical device conformity assessment procedure, mandatory expert panel consultation was required for certain high-risk devices and the results of this expert panel consultation were recorded within the CECP/PECP. These publicly-available CECP/PECP expert panel reports illustrated the need for higher quality CERs/PERs.

9.6 Conclusions

When reviewing CE/PE documents, each reviewer should clearly understand their role and responsibility during each review cycle. Most critical are the tasks to ensure clinical and performance data are accurate and complete (a task often assigned to the quality team with full support by the clinical team for clinical/performance data interpretation). Of similar importance is the oversight of this immensely detailed work by the team manager and the medical reviewer. The evaluator must ensure every data component was critically evaluated and the identification, appraisal, analysis, and conclusions were consistent with the clinical data evaluated.

CE/PE documents are reviewed by both internal and external reviewers, so the manufacturer QMS should incorporate the work of the external stakeholders including all contracted writers, reviewers, medical experts, NBs, and expert panels, as appropriate. The processes should allow changes before, during, and after each CE/PE document is signed off and version controlled. When addressing NB and expert panel deficiencies, the negotiation process should be clearly described, including a process for trending and tracking all medical reviewer, NB, and expert panel questions, answers, and company decisions for each deficiency. The CER/PER revisions after initial sign-off should be version-controlled to document all changes implemented to meet the NB/expert panel request/s. These changes may affect the interrelated CE/PE, PMS, engineering, and RM systems and teams. For example, changes to labeling are a common requirement, as NBs often limit the indication for use to reflect the specific, fully evaluated clinical/performance data and these required indication fo use changes need to be recorded in all device labeling, CE/PE, PMS and RM documents.

9.7 Review questions

1. What do the acronyms CTCAE, NBOG, CEAR, PEAR, and MDCG stand for?
2. What eight steps are suggested to complete CE/PE document reviews?
3. What resources are used to help standardize clinical terms in the benefit-risk ratio?
4. How should the evaluator group details in the benefit-risk ratio table?
5. What three key things should the clinical department team leader check during CE/PE document writing?
6. What are the differences between the clinical team reviewer and the quality team reviewer roles and responsibilities?
7. What must the regulatory team reviewer ensure during review of a CE/PE document?
8. True or false? Medical device labeling should never change during a CE/PE.
9. What documents must the PMS reviewer ensure are aligned with the CER/PER?
10. Which devices must be evaluated by expert panels?
11. Which guidance document provided a NB checklist for CER assessment?
12. How many sections are in the CEAR template?
13. True or false? The CEAR includes info about the expert panel.
14. What is an example of ambiguous language used in an indication statement, and why did at least one NB assessor require removal of the ambiguous language from the indication for use statement?
15. What is the difference between "intended use" and "indication" for use?
16. True or false? The CER can simply refer to risk management and other documents in the technical file without evaluating those documents.
17. When updating labeling to address a deficiency, does the change need to be implemented immediately?
18. What is the role of the expert panel?
19. What are the most critical tasks to complete during CE/PE document review?
20. True or false? The NB often limits the medical device indication for use.

References

[1] ICH MedRA Management Committee. Medical dictionary for regulated activities. Accessed June 10, 2024. https://www.meddra.org/.

[2] US Department of Health and Human Services, National Institutes of Health, National Cancer Institute. Common Terminology criteria for adverse events (CTCAE) Version 5.0. Published November 27, 2017. Accessed June 10, 2024. https://ctep.cancer.gov/protocoldevelopment/electronic_applications/docs/ctcae_v5_quick_reference_8.5x11.pdf.

[3] MDCG 2020-13. Clinical evaluation assessment report template. July 2020. Accessed June 10, 2024. https://ec.europa.eu/health/sites/health/files/md_sector/docs/md_2020-13-cea-report-template_en.pdf.

[4] Frestedt J. Clinical evaluation assessment report (CEAR) template (MDCG 2020-13). Accessed June 9, 2024. https://frestedt.com/wp-content/uploads/2020/10/MDCG-CEAR-Training-10-15-20-1626.pdf.

[5] European Commission. Medical devices – expert panels. Accessed June 10,2024. https://ec.europa.eu/health/md_expertpanels/overview_en.

[6] EU Reg. 2019/1936 laying down the rules for the designation of expert panels in the field of medical devices. Accessed June 10, 2024. https://eur-lex.europa.eu/legal-content/EN/TXT/?uri = uriserv:OJ.L_0.2019.234.01.0023.01.ENG&toc = OJ:L:2019:234:TOC.

[7] MDCG 2019-3 rev.1. Interpretation of article 54(2)b (April 20). Accessed June 9,2024. https://health.ec.europa.eu/system/files/2020-09/md_mdcg_2019_3_rev1_cecp_en_0.pdf.

[8] MDCG 2021-22, Rev.1. Clarification on "first certification for that type of device" and corresponding procedures to be followed by notified bodies, in context of the consultation of the expert panel referred to in Article 48(6) of Regulation (EU) 2017/746. September 2022. Accessed June 9, 2024. https://health.ec.europa.eu/system/files/2022-09/mdcg_2021-22_en.pdf.

[9] European Commission. List of opinions provided under the CECP. Accessed June 10, 2024. https://health.ec.europa.eu/medical-devices-expert-panels/experts/list-opinions-provided-under-cecp_en.

[10] European Commission. List of views provided and ongoing consultations under the PECP. Accessed June 10, 2024. https://health.ec.europa.eu/medical-devices-expert-panels/experts/list-views-provided-and-ongoing-consultations-under-pecp_en.

Chapter 10

Integrating clinical evaluation, postmarket surveillance and risk management systems

The manufacturer must monitor medical device safety and performance after product launch into the European Union (EU) market. Specifically, every medical device company with a CE-marked product on the EU market must have a postmarket surveillance (PMS) system including a postmarket clinical follow-up plan (PMCFP)/postmarket performance follow-up plan (PMPFP). The PMS plan must document specific activities and timelines required to produce postmarket evidence or a justification documenting why postmarket clinical follow-up (PMCF)/postmarket performance follow-up (PMPF) was not required. These PMCF/PMPF activities must meet the EU Medical Device Regulation (MDR) (EU Reg. 2017/745)/*In Vitro* Device Regulation (IVDR) (EU Reg. 2017/746) requirements. The MDR/IVDR also requires a risk management (RM) system to be fully integrated with the clinical evaluation/performance evaluation (CE/PE) and PMS systems. This chapter highlights the importance of integrating clinical data from corporate CE/PE, PMS, and RM systems.

Post market clinical follow up (PMCF)/post market performance follow up (PMPF) clinical data are required after a medical device is placed on the EU market because the premarket clinical evaluation (CE)/performance evaluation (PE) used to secure the CE mark was limited in scope or because the clinical evaluation/performance evaluation (CE/PE), postmarket surveillance (PMS), or risk management (RM) systems identified certain safety and performance (S&P) or benefit-risk ratio areas needing additional clinical data. Once the medical device is on the EU market, many more uses and a greater variety of clinical trials, literature and experiences become available to evaluate in the PMS, CE/PE, and RM systems. In addition to the routine PMS required for real time reporting of reportable events, the PMS system is required to include PMCF/PMPF to gather additional clinical data specifically for the CE/PE system.

10.1 Postmarket follow up is required

After the clinical evaluation report/performance evaluation report (CER/PER) is issued, the post market clinical follow up plan (PMCFP)/post market performance follow up plan (PMPFP) must detail the specific activities intended to generate new CE/PE clinical data for at least three different reasons:

1. To address all clinical data gaps identified in the CER/PER.
2. To capture and evaluate additional clinical data about device S&P and the benefit-risk ratio when the device is used as indicated in "real-world" medical practice settings, over the device lifetime, by broad patient populations to include all indication for use possibilities, etc.
3. To add subject-device-specific S&P clinical data for any device approved for CE marking, especially for any device gaining a CE mark based on clinical data solely from equivalent device clinical data.

The PMS system was designed to detect new risk signals continuously and to report these new or changed risks to the regulatory authorities in a timely manner. This continuous PMS process was intended to confirm device S&P throughout the device lifetime and to ensure the identified and emerging risks were "acceptable." In addition, the PMCF/PMPF process within the PMS system should identify potential "systematic misuse" or off-label uses to evaluate and ensure the intended purpose and indications for use were clear and correct for all users.

Planning, Writing and Reviewing Medical Device Clinical and Performance Evaluation Reports (CERs/PERs). DOI: https://doi.org/10.1016/B978-0-443-22063-0.00005-5

PMS PMCF/PMPF activities were designed to create and capture device-specific clinical data not previously available for evaluation. The PMSP and PMCFP/PMPFP need to describe carefully the clinical data collection activities and the specific clinical data collection time frame for each activity. The PMCFP/PMPFP must describe both passive clinical data evaluation and proactive clinical data generation and evaluation activities. After the new clinical data are collected and analyzed, the data are reported in the PMCF/PMPF evaluation report (PMCFER/PMFPER) within the PMS system. One important PMS PMCF/PMPF goal is to ensure sufficient clinical data have been collected and analyzed, to address fully all the identified CER/PER clinical data gaps. These new clinical data must be incorporated into the next CER/PER update.

All marketed medical devices in the EU must address any missing clinical data with appropriate proactive clinical data collection activities within the PMS system. The right clinical data collection tool should be used to address specific types of clinical data gaps. Sometimes, the right clinical data collection tool will be a full-blown prospective, randomized, and controlled trial (RCT); however, more often, the right clinical data collection tool is a less rigorous clinical data collection activity than a large and expensive clinical trial.

BOX 10.1 Trial versus study

As the regulatory focus on clinical data grows in the EU and worldwide, the language we use to describe clinical data is also changing. One such language change is about the term "trial" to mean a rigorous clinical trial (e.g., an interventional RCT, a prospective clinical trial designed to test an interventional treatment or to diagnose a condition or disease) and the term "study" to mean a less rigorous type of clinical data collection (e.g., an observational study, real-world research, genetic study, epidemiological study, registry study, survey study). Part of this language differentiation may be related to historical differences between clinical research done for drugs and for medical devices. For example, clinical trials are required for new drugs; however, clinical trials are not typically required for new medical devices, depending on the device risk level. In addition, the size, cost, and complexity of clinical trials are much greater for drugs than devices.

Previously, the International Conference on Harmonization (ICH) Guideline for pharmaceutical trials stated the terms clinical trial and clinical study were synonymous (ICH E6 R2, Section 1.12) [1]; however, this description was challenged because some types of clinical research should not be called "clinical trials." For example, the US FDA stated: "*Clinical trials are a kind of clinical research designed to evaluate and test new interventions such as psychotherapy or medications. Clinical trials are often conducted in four phases. The trials at each phase have a different purpose and help scientists answer different questions*" [2]. In addition, the US FDA stated a noninterventional observational clinical study design is not considered a clinical trial [3] and the EU Clinical Trials Regulation (EU Reg. 536/2014 Article 2.2(4)) defined "*a non-interventional study is a clinical study other than a clinical trial*" [4].

In this setting, medical device research is evolving to include many alternative types of clinical studies which are not clinical trials, per se. These clinical data should document if each medical device is safe and performs as indicated and if the benefit-risk ratio is acceptable when used as intended. Medical devices do not follow the four phases used for drug development. High-risk medical devices often completed a pilot and a pivotal trial to document medical device S&P; however, most medical devices had no clinical trials prior to marketing. As the need for medical device clinical data is increasing, the terms clinical trial and clinical study are also changing. Many clinical studies are being developed to fill the expanding need for clinical data in the pharmaceutical, medical device and food industries. Evaluating these highly variable clinical data requires training and experience.

A clinical trial will probably continue to mean a rigorous, interventional research program conducted outside of a routine clinical practice. A clinical study will probably incorporate many different, less rigorous human investigation research program types and may include real-world evidence (RWE) and registry studies conducted within a routine clinical practice or without any clinical professional involvement and possibly without any intervention (i.e., a noninterventional study).

Knowing the right clinical trial or clinical study design to use is critical when designing, completing, reporting, and integrating PMCF/PMPF activities into the PMS, CE/PE, and RM system processes to meet the EU MDR/IVDR.

10.2 Postmarket follow up activities are varied

PMCF/PMPF is "a continual process that updates the [clinical/performance] evaluation...with the aim of confirming the safety and performance throughout the expected lifetime of the device [and] ensuring the continued acceptability of identified risks and...detecting emerging risks on the basis of factual evidence" (EU MDR Annex XIV Part B; EU IVDR Annex XIII Part B).

PMCF/PMPF activities are required to generate and gather additional, proactive clinical data during continuous S&P and benefit-risk ratio evaluations for a marketed medical device. PMCF/PMPF generates new clinical/performance data for CE/PE along with other, more traditional, PMSR/PSUR clinical data. PMSR/PSUR, device vigilance, and safety reporting activities were well-established within most medical device companies before the MDR/IVDR. Unless a company is new, the quality management system (QMS) already described the PMSR/PSUR development processes; however, integrating the CER/PER "clinical/performance data" evaluation requirements under the EU MDR/EU IVDR into the PMSR/PSUR processes was a new and interesting dilemma. Most device companies had established processes to evaluate "clinical data"; however, these historical "clinical data" were not quite the same as the more rigorous "clinical data" required under the MDR/IVDR.

The EU Active Implantable Medical Devices Directive (AIMDD)/MDD/IVDD did not define clinical data to include data specifically sourced from PMS systems; instead, these old EU regulations defined experience data as "published and/or unpublished reports on. . .clinical experience of either the device in question or a similar device for which equivalence to the device in question can be demonstrated." In contrast, the EU MDR included "clinically relevant information coming from post-market surveillance, in particular the post-market clinical follow-up." Note the EU IVDR does not specifically define clinical performance data. The inclusion of "clinically relevant information" and the emphasis on PMS and PMCF in the definition of clinical data explicitly expanded the regulatory scope during clinical data evaluations. Experience data must be appraised to determine clinical relevance (i.e., every report must be read in full to determine what safety or performance information it contains). Any clinically relevant information must then be included in the CE/PE and RM system evaluations as well as the PMS system evaluations. The MDR/IVDR requires proactive PMS clinical data collection in addition to the traditional passive PMS clinical data surveillance (Table 10.1).

TABLE 10.1 PMS examples: passive versus proactive activities.

Passive PMS activities	Proactive PMS activities
Reacting to unsolicited clinical data received	Gathering clinical data to cover CER/PER clinical data gaps
Waiting for phone call with complaint	Executing a clinical trial per protocol
Recording complaint while working in the field	Creating a registry or observational study to collect new data
Documenting manufacturing problems	Hosting a focus group or advisory board
Evaluating service records for complaints	Reporting responses from a specific customer survey
Reviewing biocompatibility records	Evaluating specific, solicited user feedback
Analyzing design vigilance records	Analyzing data in patient info form (e.g., returned postcard or "fill in the blank" online from device package)
Hearing about a device misuse case	Collecting real-world clinical experience data in a specific comparative clinical trial
Evaluating device design compatibility with other devices records	Managing PMCF studies
Recording customer satisfaction communications	Designing and executing retrieval studies (evaluation of retrieved devices after removal from patient)

CER, clinical evaluation report; *PER*, performance evaluation report; *PMCF*, postmarket clinical follow-up; *PMS*, postmarket surveillance.

As the manufacturer conducts each CE/PE in preparation for MDR/IVDR conformity assessment in order to receive a CE mark in the EU, the manufacturer should consider all benefits and risks in device S&P and benefit-risk ratio terms. The CE/PE process will identify clinical data gaps needing additional PMCF/PMPF clinical data to complete the S&P and benefit-risk ratio evaluation. The clinical data generation and gathering plans are documented in the PMCFP/PMPFP including the specific methods and procedures along with the rationale defining why these particular measures were considered appropriate to cover the clinical data gaps in the previous CE/PE process. The data generated and gathered during the process are documented in the PMCFER/PMPFER.

The PMCFP/PMPFP is required for the initial CE mark and the PMCF/PMPF is an ongoing PMS process to update the CE/PE and RM systems with newly created or newly identified clinical data. The PMCFP/PMPFP is modified as necessary to confirm S&P and to adequately identify emerging risks throughout the expected lifetime of the device. Specific PMCFP/PMPFP methods typically include clinical trials, clinical studies, clinical literature, and clinical experience data collections from user feedback documents, device registries, advisory boards and other sources. The PMCFP/PMPFP requirements vary greatly, as determined by the specific medical device clinical evidence needed.

The PMS plan (PMSP) must describe how the PMS system will capture and incorporate the required PMCF/PMPF clinical data into the various corporate systems in order to address all the PMS clinical data needs for the company. The PMS report (PMSR) is generated for Class I (low risk) devices while the periodic safety update report (PSUR) is generated for Class IIa, IIb, and III medical devices and Class D IVD devices. All PMS and regulatory reporting details should be explained in the quality management system (QMS).

The MDR/IVDR described a PMS system as a proactive and systematic process to gather, record, and analyze relevant data on device quality, S&P, and benefit-risk ratio throughout the entire device lifetime. This process was intended to be proportionate to the risk class, appropriate for the device type and designed to draw the necessary conclusions and to determine, implement, and monitor any preventive and corrective actions. The PMS process was also intended to compare the subject device to similar products available on the market. The PMSP defined in MDR/IVDR Annex III included the need to correctly characterize device performance, to have effective and appropriate methods to evaluate complaints, field reports, trends, and corrective actions using suitable indicator and threshold values to assess RM and the benefit-risk ratio, to communicate effectively with regulatory authorities, operators, and users, to fulfill manufacturer PMS obligations with an appropriate PMS plan, PSMR/PSUR, and PMCFP/PMPFP, or to justify why a PMCF/PMPF was not applicable.

The PMS system data (including PMCF/PMPF data) were used to:

a. update benefit-risk determinations and improve RM;
b. identify events which require reporting to regulatory authorities);
c. update design and manufacturing information, instructions for use and labelling;
d. update the CE/PE;
e. update the summary of safety and clinical performance (SSCP)/summary of safety and performance (SSP);
f. identify needs for preventive, corrective or field safety corrective action;
g. identify needs to improve device usability and S&P;
h. when relevant, contribute to the PMS of other devices; and
i. detect and report trends.

Using the right PMCF/PMPF activities to generate and gather additional, proactive PMS clinical data supports the required and continuous S&P and benefit-risk ratio evaluations for the marketed medical device.

10.3 Integrate risk management, clinical evaluation and postmarket surveillance processes

Clinical data must ensure regulatory, legal, and manufacturing needs are met. The RM system needs clinical data to validate assumptions before any devices are used in humans and to capture risks and benefits during use in the pre- and post-market settings. The CE/PE system must generate, capture, evaluate and report the clinical data while meeting all relevant international ethical, regulatory, and good clinical practice (GCP) requirements. The PMS system must evaluate clinical data to identify safety and performance signals and to define the benefit-risk profile while ensuring postmarket reports are filed with relevant government bodies and device users as approprate. The sales, marketing and publications teams must share relevant, truthful, and not misleading clinical data with those who may be interested in purchasing and using the device.

Clinical data gathering and evaluation are now required CE/PE activities within many different medical device manufacturer systems. For example, the RM system must gather and evaluate clinical data to estimate risk probability and severity in order to establish residual risk acceptability. The PMS system must gather and evaluate clinical surveillance data to see if any reportable events must be reported to the regulatory authorities. At the same time, but separately, the CE/PE system gathers and evaluates all clinical data to determine if the medical device can establish conformity with the EU MDR/IVDR General Safety and Performance Requirements (GSPRs) 1, 6, and 8. Each system should provide a check and balance to the other systems. Specifically, the PMCF/PMPF process steps must be integrated within the PMS and linked to the other RM and CE/PE systems to provide redundant device S&P control with strong corroborated evidence showing the benefits outweigh the risks for the continued safe medical device use as indicated in the appropriate human patients or patient samples for IVD devices.

The CER/PER must determine and document if all RM activities were conducted as identified in the device risk management plan (RMP) and if all reasonably foreseeable risks and combinations of events were considered in the risk analysis. The RM report (RMR) should evaluate all known risks, both in normal and fault conditions, along with the overall residual risk. Additionally, like the PMSR/PSUR, the RMR should document how adverse events (AEs) or unforeseen device complications were identified in the literature and evaluated to determine the acceptable medical device S&P and benefit-risk ratio. The RMR results should document how the device-related risks were appropriately balanced to ensure the device-use benefits outweighed the risks or the RMR (like the PMSR/PSUR and CER/PER) should recommend mitigations up to and including removal of the medical device from the market.

Knowing the medical or IVD device life cycle stage is helpful to understand how much clinical data may be available for the individual system reports. The QMS should define how clinical data flows both within each system and between the regulatory, RM, clinical, PMS, and sales and marketing systems (Fig. 10.1).

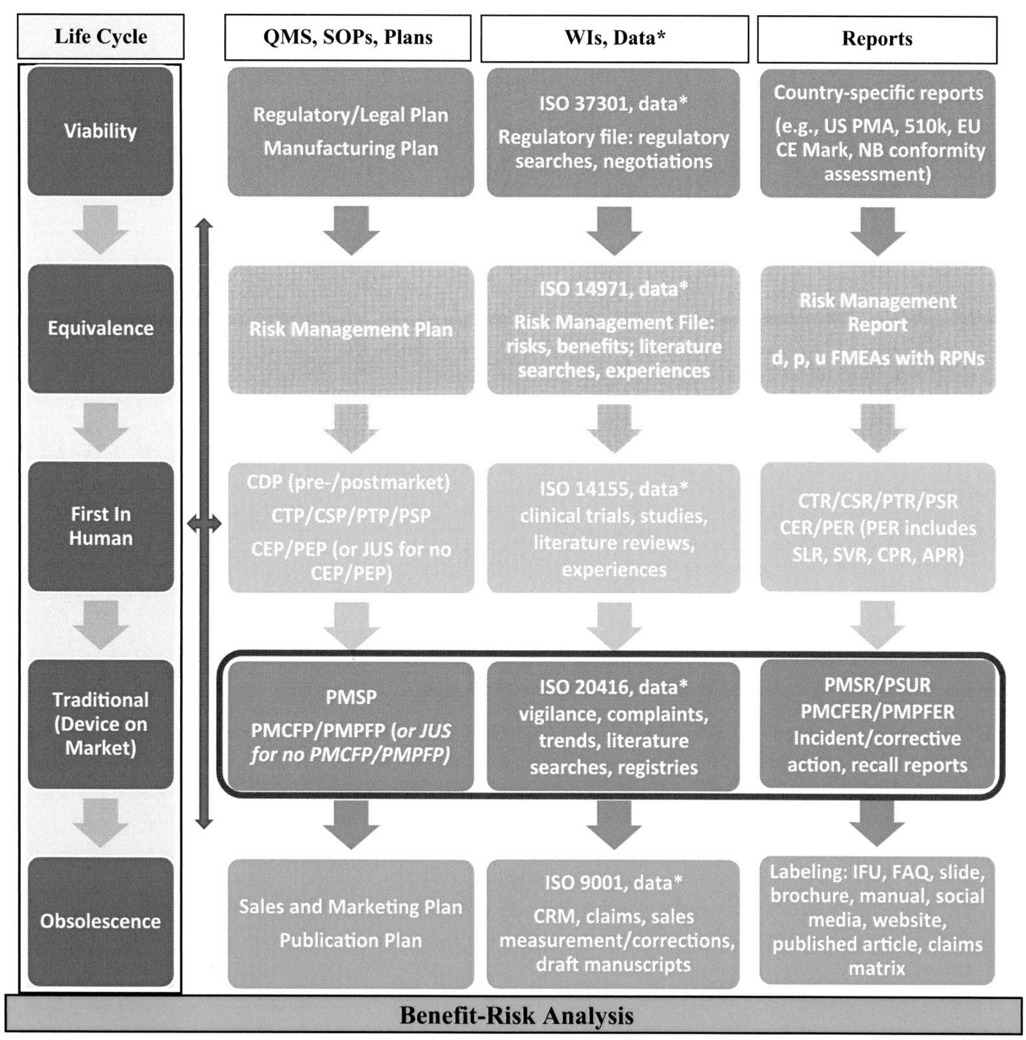

FIGURE 10.1 PMCF/PMPF studies integrated into corporate systems. Information flows from top to bottom for the device life cycle (on the left) and the amount of clinical data collected are least to most during the life cycle. These clinical data are needed for the regulatory/legal/manufacturing, risk management, clinical affairs, PMS and sales and marketing/publication systems (illustrated on the right). Information flows from left to right within each manufacturing system from plans and SOPs to WIs and data generation to data analysis and reports. The red circle indicates the PMS system which must be integrated with all the other systems. *Clinical data includes controlled databases holding subject device, equivalent device and similar device or alternative therapy or device data and data analyses typically require data management and statistical analysis plans (DMP/SAP), etc. *APR*, analytical performance report; *CDP*, clinical development plan; *CEP*, clinical evaluation plan; *CPR*, clinical performance report; *CRM*, customer relationship database; *CSP*, clinical study protocol; *CTP*, clinical trial protocol; *JUS*, justification; *PEP*, performance evaluation plan; *PMCFER*, postmarket clinical follow-up evaluation report; *PMCFP*, postmarket clinical follow-up plan; *PMPFER*, postmarket performance follow-up evaluation report; *PMPFP*, postmarket performance follow-up plan; *PMS*, post market surveillance; *PMSP*, postmarket surveillance plan; *PMSR*, postmarket surveillance report (Class I); *PSUR*, periodic safety update report (Class IIa, IIb, and III, Class D for IVD devices); *QMS*, quality management system; *SLR*, scientific literature review; *SOP*, standard operating procedure; *SSCP*, summary of safety and clinical performance; *SSP*, summary of safety and performance; *SVR*, scientific validation report; *WIs*, work instructions.

All manufacturing activities should be responsive to the regulatory/legal plans and updates since all activities are required (by law) to comply with all applicable local, national, and international laws and regulations as well as all company policies and requirements. In addition, all company systems are required to interact with each other and to contribute to a common and overarching corporate benefit-risk ratio analyses for each medical device which derives specific, benefit and risk elements from within each system (i.e,. including the regulatory, risk, clinical, PMS, and sales and marketing systems). In particular, benefit data collection and tracking has expanded greatly since 2017 to help each manufacturer interpret the device benefit-risk ratio.

BOX 10.2 ISO 14971:2019 now mentions *benefit* 40 times

ISO 14971:2019 "Medical Devices—application of risk management to medical devices" defined newly added terms, including benefit, reasonably foreseeable misuse, and state-of-the-art (SOTA), and this revision also expanded details regarding benefits throughout the updated standard as follows:

Foreword: "Definitions of *benefit, reasonably foreseeable misuse* and *state-of-the-art* have been introduced. More attention is given to the *benefits*... expected from... the medical device. The term *benefit-risk* analysis has been aligned with terminology used in some regulations."

Introduction: "*Risk management* is a complex subject because each stakeholder can place a different value on the acceptability of *risks* in relation to the anticipated *benefits*... The acceptability of a *risk* to a stakeholder is influenced... by the stakeholder's perception of the *risk* and the *benefit*... The decision to use a *medical device* in the context of a particular clinical *procedure* requires the *residual risks* to be balanced against the anticipated *benefits* of the *procedure*. Such decisions... take into account the *intended use*, the circumstances of use, the performance and *risks* associated with the *medical device*, as well as the *risks* and *benefits* associated with the clinical *procedure*. Some of these decisions can be made only by a qualified medical practitioner with knowledge of the state of health of an individual patient or the patient's own opinion."

Section 3.2: "*benefit*" was defined as a "positive impact or desirable outcome of the use of a *medical device*... on the health of an individual, or a positive impact on patient management or public health... *Benefits* can include positive impact on clinical outcome, the patient's quality of life, outcomes related to diagnosis, positive impact from diagnostic devices on clinical outcomes, or positive impact on public health."

Section 4.2: "Management Responsibilities" included "The *manufacturer's* policy for establishing criteria for *risk* acceptability can define the approaches to *risk control*: reducing *risk* as low as reasonably practicable, reducing *risk* as low as reasonably achievable, or reducing *risk* as far as possible without adversely affecting the *benefit-risk* ratio."

Section 7.1: "*Risk control* option analysis" required "If, during *risk control* option analysis, the *manufacturer* determines that *risk* reduction is not practicable, the *manufacturer* shall conduct a *benefit-risk* analysis of the *residual risk*."

Section 7.4: "Benefit-risk analysis" was changed from "Risk/Benefit analysis" in ISO 14971:2007 and was simlarly defined in ISO14971:2019 as: "If a *residual risk* is not judged acceptable using the criteria established in the *risk management* plan and further *risk control* is not practicable, the *manufacturer* may gather and review data and literature to determine if the *benefits* of the *intended use* outweigh this *residual risk*... If this evidence does not support the conclusion that the *benefits* outweigh this *residual risk*, then the *manufacturer* may consider modifying the *medical device* or its *intended use*... Otherwise, this *risk* remains unacceptable... If the *benefits* outweigh the *residual risk*, then proceed to... [the next step]. The results of the *benefit-risk* analysis shall be recorded in the *risk management file*... Compliance is checked by inspection of the *risk management file*." The option to modify the medical device or intended use was new in the 2019 version of this international standard and this nuanced detail was fast becoming a requirement rather than an option.

Section 8: "Evaluation of overall *residual risk*" required, "After all *risk control* measures have been implemented and verified, the *manufacturer* shall evaluate the overall *residual risk* posed by the *medical device*, taking into account the contributions of all *residual risks*, in relation to the *benefits* of the *intended use*, using the method and the criteria for acceptability of the overall *residual risk* defined in the *risk management* plan... If the overall *residual risk* is not judged acceptable in relation to the *benefits* of the *intended use*, the *manufacturer* may consider implementing additional *risk control* measures... or modifying the *medical device* or its *intended use*... Otherwise, the overall *residual risk* remains unacceptable." Again the option to modify the device or intended use was new and this text replaced the 2007 language "If the overall *residual risk* is not judged acceptable..., the manufacturer may gather and review data and literature to determine if the medical benefits of the intended use outweigh the overall residual risk. If this evidence supports the conclusion that the medical benefits outweigh the overall residual risk, then the overall residual risk can be judged acceptable."

Section 10.2: Information review also included new review requirements: "The *manufacturer* shall review the information collected for possible relevance to *safety*, especially whether... the overall *residual risk* is no longer acceptable in relation to the *benefits* of the *intended use*; or the generally acknowledged *state of the art* has changed."

(Continued)

BOX 10.2 (Continued)

Annex A: Rationale for Requirements, A1 General continued as follows: "...When discussions on an International Standard for *risk management* began, crucial features of *risk management* needed to be addressed, such as the *process* of *risk evaluation* as well as the balancing of *risks* and *benefits* for *medical devices*.... More emphasis was put on the *benefits* that are anticipated from the use of the *medical device* and the balance between the (overall) *residual risks* and those *benefits*."

A2.1: Scope explained "The scope of this document does not include clinical decision making, i.e., decisions on the use of a *medical device* in the context of a particular clinical *procedure*. Such decisions require the *residual risks* to be balanced against the anticipated *benefits* of the *procedure* or the *risks* and anticipated *benefits* of alternative *procedures*. Such decisions take into account the *intended use*, performance and *risks* associated with the *medical device* as well as the *risks* and *benefits* associated with the clinical *procedure* or the circumstances of use. Some of these decisions can be made only by a qualified health care professional with knowledge of the state of health of an individual patient and the patient's own opinion."

A2.3: Terms and definitions reported "*benefit* is defined because of the increased emphasis by regulatory bodies on balancing the (*residual*) *risks* against the *benefits* of the *medical device*. For the same reason the phrase "*benefit-risk* analysis" is used."

A2.7.1: Risk control option analysis sugggested "one possible result of the *risk control* option analysis could be... no practicable way of reducing the *risk* to acceptable levels according to the pre-established criteria for *risk* acceptability. For example, it could be impractical to design a life-supporting *medical device* with such an acceptable *residual risk*. In this case, a *benefit-risk* analysis can be carried out... to determine whether the *benefit* of the *medical device*, to the patient, outweighs the *residual risk*."

A.2.7.4: *Benefit-risk* analysis mentioned "...particular *hazardous situations* for which the *risk* exceeds the *manufacturer's* criteria for *risk* acceptability. This subclause enables the *manufacturer* to provide a high-*risk medical device* for which they have done a careful evaluation and can show that the *benefit* of the *medical device* outweighs the *risk*. However, this subclause cannot be used to weigh *residual risks* against economic advantages or business advantages (i.e. for business decision making)."

A.2.8: Evaluation of overall *residual risk* summarized "The method to evaluate the overall *residual risk* as defined in the *risk management* plan includes balancing the overall *residual risk* against the *benefits* of the *medical device*. This is particularly relevant in determining whether a high-*risk*, but highly beneficial, *medical device* should be marketed."

A.2.10: Production and *postproduction* activities stated "the *manufacturer* needs to collect and review production and *postproduction* information and evaluate its relevance to safety. The information can relate to new *hazards* or *hazardous situations*, and/or can affect their *risk* estimates or the balance between *benefit* and overall *residual risk*. Either can impact the *manufacturer's risk management* decisions. The *manufacturer* should also take into account considerations of the generally acknowledged *state-of-the-art*, including new or revised standards" [5].

The term "risk" was mentioned 586 times while benefit was only mentioned 40 times. Similarly, ISO 14971:2019 was also updated to state that the RM file should include a full benefit-risk analysis, as defined in Section 7.4. The RM report including the full benefit-risk ratio should be used to exert "risk controls" and should be reviewed during the CE/PE process. ISO 14971:2019 also stated some decisions about risks and benefits "can be made only by a qualified health care professional with knowledge of the state of health of an individual patient and the patient's own opinion." This clinical complexity suggests persons with clinical expertise (i.e., qualified health care professionals) should be participating in the benefit-risk ratio determinations and discussions.

Harmonization between RM, PMS, and CE/PE systems should follow a carefully constructed strategy, house work in the appropriate systems and include appropriate, clinically trained personnel in the work. Each system should be designed to contribute to the benefit-risk analysis work in the RM system. The benefit-risk ratio is completed in the RM process while the PMS and CE/PE systems provide inputs into the RM system. All benefit-risk ratio work should comply with the international standard, ISO 14971 as well as the appropriate standards and regulations for PMS and CE/PE systems.

Device manufacturers must evaluate benefits and risks and this work is typically done under ISO 14971 from a theoretical device benefit and risk perspective. RM was defined as the "systematic application of management policies, *procedures*..., and practices to the tasks of analyzing, evaluating, controlling and monitoring risk." This includes determining if all identified risks relating to patient treatment, device operational methods or risks relating to usability have been mitigated as far as possible or if questions remain regarding clinical risks to be resolved.

Engineering teams were responsible to assess prevalence/probability for each theoretical clinical "risk" or "hazard" and they were required to assign a "severity" rating to each risk. One challenge to overcome in CE/PE training is to help engineering, clinical, RM, and PMS team members and others understand the term "risk" is different when considered from the device perspective versus the patient perspective. Clinical teams collect and document specific benefits and risks from actual human experiences rather than theoretical device benefits and risks from the design, process, and use failure mode perspectives used when designing, manufacturing and writing the instructions for use.

The RMR and PMSR/PSUR data are inputs to the CER/PER, and the CER/PER data are inputs to the RMR and PMSR/PSUR. Identification of new risks or AEs in the clinical literature, clinical investigation, or other PMS activities evaluated in the CER/PER may result in changes to the RMR and PMSR/PSUR. The evaluator must confirm alignment between the risks in all RM, CE/PE, and PMS documents. All discrepancies, residual risks, and uncertainties or unanswered questions (e.g., about rare complications, long-term S&P under widespread use) should be identified, whether these are acceptable for CE marking, and whether they need to be addressed during PMS. Sometimes, the RM and PMS system reports will need to be updated after the CE/PE documents are completed and released (e.g., to add newly identified benefits or risks or to remove theoretical benefits or risks which were never observed in a clinical setting). The evaluator will need to document gaps in the CER/PER whenever additional clinical data are needed for the CE/PE, RM and/or PMS systems.

This system integration knowledge and understanding is especially important as the CE/PE, RM, and PMS system expand to include more clinical and performance data and especially as the clinical data collected are intended to include more rigorous clinical trial or clinical study data collected by the clinical affairs teams rather than the PMS teams. Companies must decide which team collects which clinical data type and how the clinical data analyses will be done within the company to generate the appropriate reports with sufficient checks and balances between the CE/PE, RM and PMS systems, but without "too much" redundancy and duplication of effort.

For example, the company "claims matrix" should link each sales/marketing claim to the specific claim substantiation data document. The claim substantiation documents often come from the clinical literature, but they may be tangential to (i.e., not directly related to) the device S&P explored in the CER/PER. The sales and marketing teams will probably want to make many clinical claims; however, supporting many claims with clinically-relevant scientific substantiation often proves difficult. Claims substantiation is a notified body (NB) target for CE mark assessments, especially in the CER/PER and PMS activities in the PMCFER/PMPFER. The need for a medical device claims matrix has increased under the EU MDR/IVDR since all claims and product labeling must now be substantiated by clinical data evaluated and documented in the CER/PER.

10.4 Different departments need specific literature searches

The QMS standard operating procedures (SOPs), work instructions (WIs), templates, and forms should define precisely how the CE/PE, RM and PMS systems work together inside the company. For example, medical literature searches must be completed to identify clinical data within all three systems, so the company may have difficulty managing the clinical literature required by and discovered in these diverse systems. Sometimes, manufacturers may mistakenly believe one literature search will serve the needs for all systems; however, this is not the case since each system has unique needs for the clinical data taken from the medical literature (Table 10.2).

TABLE 10.2 Examples: different systems need different literature searches

System	Literature search purpose	Literature search input (search terms)	Literature search output (search results)
Regulatory, legal, manufacturing	To identify specific details about similar or competitor products, etc.	Search terms specific to the regulatory legal or manufacturing issues	Regulatory submissions, legal briefs and manufacturing detail reports
Risk management	To identify benefits and risks, to assess and prioritize benefit and risk likelihood and acceptability, etc.	Search terms specific to individual device-related benefit, risk, harm	Event details to estimate probability of occurrence and magnitude of clinical impact to calculate risk prioritization number and benefit-risk ratio
Clinical	To design clinical research, to address S&P acceptability, etc.	Search terms specific to patient problem, intervention, comparator, outcomes (PICO)	Clinical S&P and benefit-risk ratio data for patients, subject device and other devices, etc.
Postmarket surveillance	To identify reportable device issues, new trends, etc.	Device name, specific device issues needing focused surveillance	PMSR, PSUR, and incident reporting to regulatory authorities and risk management
Sales, marketing, publications	To identify messages, to answer customer questions, to publish new research, etc.	Specific sales, marketing and research topics	Labeling, marketing materials, new literature publications, etc.

PMSR, postmarket surveillance report; *PSUR*, periodic safety update report; *S&P*, safety and performance.

Early in device development, the regulatory department may complete clinical literature searches to identify similar devices or device features so the company can justify the proposed device S&P prior to any human clinical data collection. As the device life cycle progresses, the regulatory literature searches may focus on justifying equivalence to another device and then on comparing the early, first-in-human, clinical device data to equivalent or similar device clinical data already available in the literature. If the device has any safety or performance concerns requiring design changes during routine device use, then, the regulatory literature searches will likely focus on justifying proposed design changes. In addition, during obsolescence, the regulatory literature searches will probably focus on long-term uses to justify the ongoing device S&P for the final few device users. The regulatory department uses clinical data from their literature searches to answer regulatory questions from government and other agencies when preparing to launch and while keeping medical devices in various markets.

By comparison, the PMS department is required to seek safety signals in the postmarket setting. Initially, PMS literature searches are typically focused on safety signals reported for equivalent, similar and first-in-human device uses. Later on, the PMS literature searches will seek to identify every reported device safety or performance concern for every subject device use. The clinical data gathered and analyzed from the literature are used to complete the manufacturer signal identification and PMS reporting activities required by many different intenational regulations. The PMS department is required to surveil the literature and report specific clinical data details collected from the literature to relevent authorities where the device is marketed. This device vigilance requires extensive literature searching with short reporting time frames and the manufacturer is required to act on new safety signals in a timely manner. In other words, waiting to do the literature search during the next CER/PER, 1−5 years later, will be too long to wait if the safety signal reported in the literature involved a serious public health threat, a device-related subject death, or any other unexpected, serious (or potentially serious) AE related to the device.

The literature searching for new PMS safety signals is not the same as the literature searching for RM benefits and risks or the literature searching for CE/PE device S&P clinical data. The clinical department must create a CEP/PEP with specific, predefined, and prequantified S&P acceptance criteria as well as literature search criteria which are then used to search for literature about the background knowledge surrounding the device, the SOTA for the device and any/all alternative therapies for the device. This background literature searching serves to benchmark the available CE/PE clinical literature data. The CEP/PEP also defines the search terms for literature searches to identify and quantify the S&P clincal data available in the literature for the subject device, so the evaluator can determine if the subject device clinical literature data meet the predefined and pre-quantified S&P thresholds stated in the CEP/PEP. Unlike the CE/PE literature searches, the PMS literature searches are completed to identify new safety signals promptly so they can be reported to the authorities in the manufacturer incident report (MIR) form. Serious public health threats must be reported within 2 days, death or serious "deterioration in the state of health" must be reported within 10 days, and events with potential concerns must be reported within 15 days from the date when the manufacturer "became aware" of the event. The MIR must include AEs coded using the International Medical Device Regulators Forum (IMDRF) coding system [6] and must group similar incidents including trend reporting for similar incidents. The MIR must use the single registration number (SRN) assigned to the medical device manufacturer by the EU and the unique device identification (UDI) number assigned to the medical device as required in the EU.

The sales and marketing departments may also do literature searches, but these are often directed at specific topics to answer customer questions or to track company publications adn to use in company marketing messages.

10.5 "Signal detection" is required

Signal detection requires device-specific AE pattern interpretation. The IMDRF [7] defined signal detection as "The process of determining patterns of association or unexpected occurrences that have the potential to impact patient management decisions and/or alter the known benefit-risk profile of a device."

AE patterns identified for further investigation often include a new safety issue, a more frequent AE occurrence, or a greater AE severity among previously identified device-related AEs, a patient subpopulation with a greater overall AE incidence, or an increase in a specific AE type, finding other risks associated with device use, etc. Signals may stem from defects occurring during production (e.g., sterility issues caused by faulty packaging), design issues (e.g., a weak part which breaks during surgery and may harm the patient), use issues (e.g., a user misunderstanding about how to use the device), etc. True safety signals are difficult to identify and differentiating potential, but false, signals from true safety signals can take significant time to investigate. In addition, once identified and coded, the process to verify the signal requires careful case-by-case evaluation followed by careful response prioritization depending on variables like the signal strength, novelty, clinical importance, public health impact, and ability to treat, prevent, or mitigate the identified AE.

One international author group from Spain, the USA, and the Netherlands [8] completed a systematic literature search and selected 24 articles from among 20,819 articles initially identified describing medical device safety (i.e., search terms: "medical device" AND "signal detection" or "postmarketing surveillance" or "risk management"; timeframe: 2004 to 2017; and limited to PMS data sources, signal detection methodologies or coding dictionaries). Data sources and coding were "remarkably heterogeneous" with "no agreement on the preferred methods for signal detection... and no gold standard for signal detection." In addition, the 24 evaluated reports were found in 24 separate journals, and 12 of these described many "spontaneous reporting systems" as follows:

- Food and Drug Administration (FDA) Manufacturer and User Facility Device Experience (MAUDE) database (US);
- Therapeutics Goods Administration (TGA) Database of Adverse Event Notifications (DAEN) database (Australia);
- *Future* European Databank on Medical devices (EUDAMED) (European Union, EU);
- MHRA database (UK);
- MEDSUN database (US);
- Adverse Event Triggered Reporting for Devices (ASTER-D) (US);
- MEdical DEvices VIgilance and Patient Safety (MEDEVIPAS) (Greece); and
- National Electronic Injury Surveillance System (NEISS) (US).

Nine (9/24, 38%) articles discussed the following registries:

- Orthopedic: National Joint Registry (NJR) (England, Wales and Northern Ireland), Canadian Joint Replacement Registry (CJRR) (Canada), Kaiser Permanente Orthopedic Registry (KPOR) (US), Dutch Arthroplasty Registry (LROI) (Netherlands), National Implants Registry (RNI) (Brazil), Australian Orthopaedic Association National Joint Replacement Registry (AOANJRR) (Australia).
- Vascular: Vascular Quality Initiative (VQI) (US), Australasian Vascular Audit (AVA) registry (Australia and New Zealand), National Vascular Registry (NVR) (UK), Japanese Registry of Endovascular Aneurysm Repair, abdominal and thoracic (JREAR) (Japan).
- Cardiac: National Cardiovascular Data Registry (NCDR) (US), Swedish Coronary Angiography and Angioplasty Registry (SCAAR) (Sweden), Massachusetts Angioplasty Registry (US), Percutaneous Coronary Intervention (J-PCI) (Japan), Percutaneous Coronary Intervention (Cath-PCI) (US), US trans-catheter valve therapies (TVT) (US), Japanese trans-catheter valve therapies (TVT) (Japan), Japan Adult Cardiovascular Surgery Database (JACVSD) (Japan), Database of Sprint Fidelis and Quattro Secure implantable cardioverter defibrillator leads (US).
- Ophthalmic: European Registry of Quality Outcomes for Cataract and Refractive Surgery (EUREQUO) (EU).
- General: Data Extraction and Longitudinal Trend Analysis (DELTA) Registry (US), Medicare database (US claims database constituting a person-specific registry of medical histories recording the use of all hospital services that are eligible for payment, including use of medical devices).

One article described an online, diabetes safety network social media app where patients, caregivers, and family members shared case reports of medical device events. These real-world data resources were growing, and four reports discussed a few statistical signal detection methods:

- disproportionate analysis (DPA) methodologies (frequentist and Bayesian);
- multivariate methods (change point analysis and entity matching algorithm); and
- data extraction and longitudinal trend analysis (DELTA) network (to analyze registry data).

Four reports discussed coding dictionaries, and these were not standardized or cross-referenced to each other, so the codes used were difficult to compare to each other:

- FDA codes (including patient problem and device issue);
- International Organization for Standardization (ISO);
- IMDRF codes for product problems and investigation results and Systematized Nomenclature of Medicine—Clinical Terms (SNOMED CT);
- IMDRF patient codes for patient outcomes;
- Medical Dictionary for Regulatory Activities (MedDRA); and
- International Classification of Diseases (ICD).

The MAUDE, DAEN, and the future EUDAMED were described as the main publicly available databases holding spontaneously reported safety data. Unfortunately, these databases lacked: a harmonized reporting standard, an ability to determine root cause for the event, access to the device or patient for further investigation and these databases had problems with missing or incomplete data, underreporting (e.g., due to a lack of time, unclear device causality, difficult forms, insufficient awareness/understanding of the need), and overreporting (e.g., "notoriety bias"). Registries provided some advantages for PMS signal detection because they often contained device information as well as patient disease and treatment information not normally found in publicly available databases holding spontaneously reported safety data. Future worldwide unique device identifier (UDI) requirements were expected provide an opportunity to harmonize PMS data linked to these UDIs for better global signal detection practices. Unfortunately, current medical device signal detection and reporting requirements were seriously underdeveloped compared to pharmacovigilance safety reporting. The need to standardize medical device data sources, signal detection methods, and coding dictionaries was emphasized in this analysis.

The EU MDR Article 90 [5] and IVDR Article 85 [9] required specific "Analysis of vigilance data" by the European Commission and Member States using electronic systems to "*identify trends, patterns or signals in the data that may reveal new risks or safety concerns... Where a previously unknown risk is identified or the frequency of an anticipated risk significantly and adversely changes the benefit-risk determination, the competent authority... shall inform the manufacturer... which shall then take the necessary corrective actions.*" This system requires the manufacturer to report these issues in the "*Electronic system on vigilance and on post-market surveillance.*" Similar to the ongoing work by the US FDA on the Sentinel active surveillance program for medical product signal detection, additional work on medical device vigilance and signal detection are expected once the EUDAMED electronic system becomes available for use by manufacturers and the public.

Signal detection remains a challenging process needing careful attention during both passive and proactive PMS and PMCF/PMPF activities.

10.6 Postmarket follow up guidance documents are available

The PMCF/PMPF rationale and requirements were first outlined in the MDD/IVDD quality annexes in 1990 and 1993, and these became requirements in 2017 under the MDR/IVDR [5,9]. In general, the MDR/IVDR required:

- the clinical evaluation to include PMCF/PMPF (MDR Article 10, 61 and Annex XIV/IVDR Article 10, 61, Annex XIII Part B);
- NBs to impose restrictions on the intended device purpose (e.g., restricting uses to certain groups of patients) or requiring manufacturers to undertake specific PMCF/PMPF studies (MDR Article 56 and Annex XIV, part B/ IVDR Article 56 and Annex XIII Part B);
- confirmation the PMCFP/PMPFP was appropriate including PMS studies demonstrating device S&P when no clinical investigation was provided for conformity confirmation (MDR Article 61, Annex III/IVDR Article 61, Annex III);
- the Sponsor to notify the Member State at least 30 days prior to conducting a PMCF/PMPF study using a CE-marked product in or on subjects with invasive or burdensome procedures in addition to those performed under normal, device-use conditions (MDR Article 74/IVDR Article 70), for example, when outside the scope of its intended purpose (see MDR Articles 62 and 81/IVDR Article 70), and these PMCFER/PMPFER findings should be documented in the CER/PER, PMSP, PMSR/PSUR as well as the RM file.

The NB clinical reviewer was responsible for the clinical data assessment in addition to the NB technical reviewer who was responsible for the technical file, and standard NB assessment of conformity.

As discussed in earlier chapters, two 2020 guidance documents on medical device software under MDR/IVDR and on sufficient clinical evidence under MDR clarified the PMCF/PMPF requirements. The Medical Device Coordinating Group (MDCG) 2020-1 guidance [10] indicated the clinical development plan should specify PMCF/PMPF milestones and acceptance criteria for new data collections including, but not limited to postmarket complaints, AEs, real-world data, end-user feedback, new guidelines, and other literature. In addition, the CER/PER should be updated with the data obtained after implementing the PMCFP/PMPFP to collect additional data regarding the medical device valid clinical association (MDR)/scientific validity (IVDR), technical performance (MDR)/analytical performance (IVDR), and clinical performance.

In addition, MDCG 2020-6 [11] cautioned about a "Lack of clinical data providing sufficient clinical evidence" and the need for PMS data to be incorporated in the CER, including PMCF data required after the MDR requirements gap analysis. Bridging those gaps may require a controlled clinical investigation; however, clinical data may also be gathered from systematic literature reviews or PMCFER results from clinically relevant and scientifically sound questionnaires, registries or other reliable sources. PMCF studies were generally required for devices using equivalence to another device during the CE and for device-related issues. The CEP and PMSP must define the PMCF basis and any postmarket studies required to gather clinical data after "initial CE marking." The CER can use clinical data from similar devices based on well-established technology (WET); however, legacy devices should include clinical data from the subject device (when available) after using equivalent device data in the initial CE marking. In particular, complaints reporting using "*estimates such as [number of incidents or complaints]/[number of device sales] cannot generally be considered sufficient to provide proof of safety*" and these estimates "*should be limited to cases where data from... clinical investigations or PMCF studies are not deemed appropriate.*" This guidance described:

> *When "post-market data on the device itself (including PMCF) is not adequately comprehensive to provide sufficient clinical evidence, and the demonstration of equivalence is no longer possible under the definition of equivalence in the MDR, new data may need to be generated prior to CE-marking under the MDR..." Generally "sufficient clinical evidence" is required "to confirm safety, performance and the acceptability of the benefit-risk determination in relation to the state of the art for the legacy devices prior to CE-marking under the MDR, and such demonstration should not rely on new PMCF studies started under the MDR to bridge gaps (e.g., indications not supported by clinical evidence). Where other evidence, for example results of pre-clinical testing etc. as described in MDR Article 61(10), is used for confirmation of safety and performance, PMCF studies may be undertaken to confirm these conclusions."*

In other words, the demonstration of compliance with the GSPRs can be based on "*reliable, justified and sound analytical methods*," comprehensive clinical data analysis results, identification of missing data/gaps, and additional PMCF data gathering.

MDCG 2020-6 also described "*scientifically-sound studies*" as those with a clear research question, objectives, end points, rationale, data management plan, and statistical analysis plan, and with sources of study distortion or bias evaluated and minimized to improve data integrity and to allow appropriate conclusions to be drawn from the data. This guideline stated:

> *If there is not sufficient supportive clinical evidence with regard to the declared intended purpose... indications and claims... manufacturers shall narrow the intended purpose of the device under evaluation until it is supported by the available clinical evidence.*

This important and much anticipated requirement is often under appreciated by many manufacturers. If the clinical data do not support the indications for use, then change the indications for use to be more truthful and less misleading. In addition, for some medium- to high-risk legacy devices with limited clinical data (i.e., even for well established technologies and devices on the market for decades), the PMCFP/PMPFP is a critical document to generate new clinical data for S&P evaluation in relation to the evolving SOTA and to support ongoing and future conformity assessments and CE marking under MDR/IVDR. Alternatively, for low-risk, standard-of-care, and WET devices with little change in the SOTA and device background knowledge, GSPR conformity may be defended with more limited clinical data. The guidance states:

> *"Stable, well-established technologies that perform as intended and are not associated with safety concerns, and where there has been no innovation, are less likely to be the subject of research, and therefore literature data may be limited or non-existent." In these cases "a lower level of clinical evidence may be justified to be sufficient for the confirmation of conformity with relevant GSPRs. This may be supported by clinical data from the PMS provided that there has been a quality management system in place to systematically collect and analyse any complaints and incident reports, and that the collected data support the safety and performance of the device."*

In other words, the PMS and RM systems must be operating efficiently and appropriately to meet the MDR/IVDR requirements. In addition, manufacturers must consider if the clinical data on file are "sufficient" for certification under MDR/IVDR. For example, since the clinical data were generated, has the SOTA changed? Have the PMS or RM systems identified any new device benefits or risks or new clinical data related to the indications or contraindications? Were enough devices sampled and certified by the NB (e.g., with appropriate CER/PER assessment) to meet the new MDR/IVDR quality system requirements? Does past use of equivalence meet MDR/IVDR requirements? Do the clinical data evaluated in the past meet the new, more explicit MDR/IVDR definition of clinical data (i.e., without using unpublished reports as "clinical data," etc.)? These are just a few questions to help design the PMCF/PMPF for legacy and older devices. In addition, a wide variety of additional MedDev and MDCG guidance documents clarified the PMCF/PMPF medical device requirements over time (Table 10.3).

TABLE 10.3 PMCF/PMPF guidance documents and international standards.

Guidance	Title/online link	Link	Date
MEDDEV 2.12-2	Guidelines on post market clinical follow-up	https://www.team-nb.org/wp-content/uploads/2015/05/documents2012andolders/2_12-2_05-2004.pdf	May 2004
MEDDEV 2.7.1, Appendix 1	Evaluation of clinical data—a guide for manufacturers and notified bodies—Appendix 1: Clinical Evaluation of Coronary Stents	https://www.qarad.com/wp-content/uploads/2022/05/ODM034-27_1-Appendix-1-Clinical-evaluation-of-Coronary-Stents.pdf	December 2008
MEDDEV 2.7.1, rev. 3	Guidelines on medical devices. clinical evaluation: a guide for manufacturers and notified bodies	http://meddev.info/_documents/2_7_1rev_3_en.pdf	December 2009
MEDDEV 2.7/4	Guidelines on clinical investigation[a]: A guide for manufacturers and notified bodies	http://meddev.info/_documents/2_7_4_en.pdf	December 2010
MEDDEV 2.12/2, rev. 2	Post market clinical follow-up studies	http://meddev.info/_documents/2_12_2_ol_en_.pdf	January 2012
MDCG 2020-7	Guidance on PMCF plan template	https://health.ec.europa.eu/system/files/2020-09/md_mdcg_2020_7_guidance_pmcf_plan_template_en_0.pdf	April 2020
MEDDEV 2.7.1, rev. 4	Clinical evaluation: a guide for manufacturers and notified bodies under directives 93/42/EEC and 90/385/EEC	https://www.medical-device-regulation.eu/wp-content/uploads/2019/05/2_7_1_rev4_en.pdf	June 2016
MDCG 2020-8	Guidance on PMCF evaluation report template	https://health.ec.europa.eu/system/files/2020-09/md_mdcg_2020_8_guidance_pmcf_evaluation_report_en_0.pdf	April 2020
ISO/TR 20416:2020	Medical devices—Post market surveillance for manufacturers	https://www.iso.org/standard/67942.html	July 2020

Note how the PMCF has a specific guidance for the PMCFP and PMCFER; however, no similar guidance was provided for the PMPFP or PMPFER. In addition, clinical investigations and PMS systems are discussed in many documents, but these were much bigger topics than PMCF/PMPF which rarely include a formal clinical investigation.
[a]*Although PMS system and clinical trial designs and activities are outside the scope of this CER/PER book, MedDev 2.7/4 is included here to appreciate the PMCF/PMPF, which may require a formal clinical trial/investigation.*

The PMPF concept was made explicit and mandatory in the IVDR (EU Reg. 2017/746, Annex XIII, Part B), and the PMPFER was meant to be considered fully in the RM and PER as well as the PMSR for low-risk devices or the PSUR and SSP for higher-risk Class C and D IVD devices. Similarly, the PMCF was a mandatory requirement in the MDR (EU Reg. 2017/745, Annex XIV, Part B), and the PMCFER was meant to be considered fully in the RM and CER processes as well as the PMSR for low-risk devices or the PSUR and SSCP for higher-risk Class III medical devices. In addition, the need for additional clinical data has been integral to medical device sales and marketing since the first medical device was sold.

Many methods exist to assemble critical data required for the residual S&P risks and clinical data analyses surrounding the benefit-risk ratio. As the manufacturer considers specific device parameters, device groups can be identified which may need a narrower or broader PMCF/PMPF scope. This device grouping activity can help a manufacturer to establish recurring clinical data collection activities by type and to use these clinical data collection activity types across the specific devices needing this particular clinical data collection activity type.

Each individual device should have a checklist and the manufacturing systems must have interrelated checks and balances to ensure PMCF/PMPF needs are being met. NBs are required to pay careful attention to PMCF/PMPF activities due to past NB and manufacturer oversight failures and the negative impacts on many patients using medical devices in the EU. Better medical device S&P evaluation and monitoring through PMS and PMCF/PMPF processes are among the most important MDR/IVDR requirements critically and specifically addressing these past medical device S&P oversight failures. The international standards, laws, regulations, and available guidance documents have consistently required these clinical data evaluations for decades, and enforcement actions have been greatly increased.

10.6.1 MEDDEV 2.12/2 post market clinical follow up studies (May 2004, rev. 2 January 2012)

Prior to the current PMCF guidelines, the Co-ordination of Notified Bodies Medical Devices (NB-MED) on Council Directives 90/385/EEC, 93/42/EEC, and 98/79/EC released a recommendation NB-MED/2.12/Rec1 in March 2000 to meet the MDD, AIMD and IVD Directive requirements [12]. This document defined a "graduated approach," based on the device risk and intended use, to audit/verify PMS existence and effectiveness even in the absence of a QMS. This recommendation offered the example of a "registration card project" where pacemakers were registered from 1980 onward, and stated: "*This registration system, together with a return goods/failure analysis and corrective action loop, is considered to be an adequate basis for a PMS system as required in the AIMD annexes.*" For MDD, the minimum requirements were met simply by "*labeling the device with the manufacturers name and address,*" because this gave "*the user the ability to report back any experience gained in using the device.*" The Annex in this recommendation listed the following:

"Possible achievements of a manufacturer PMS system" including

- *"detection of manufacturing problems*
- *product quality improvement*
- *confirmation (or otherwise) of risk analysis*
- *knowledge of long-term performance/reliability and/or chronic complications*
- *knowledge of changing performance trends*
- *knowledge of performance in different user populations*
- *feedback on indications of use*
- *feedback on instructions for use*
- *feedback on training needed for users*
- *feedback on use with other devices*
- *feedback on customer satisfaction*
- *identification of vigilance reports*
- *knowledge of ways in which the device is misused*
- *feedback on continuing market viability.*"

This Annex also included proactive and reactive "*Sources of PMS information*" including:

- *"expert users groups ('focus groups')*
- *customer surveys*
- *customer complaints and warranty claims*
- *post-CE-market clinical trials*
- *literature reviews*
- *user feed-back other than complaints, either direct to manufacturer or via sales force*
- *device tracking/implant registries*
- *user reactions during training programmes*
- *other bodies (e.g., the CA)*
- *the media*
- *experience with similar devices made by the same or different manufacturer*
- *maintenance/service reports and*
- *retrieval studies on explants or trade-ins*
- *in-house testing*
- *failure analysis."*

Many of these achievements and sources are still considered today for PMS and PMCF planning. MEDDEV 2/12-2 was initially released in 2004 [13] and was updated in MEDDEV 2.12/2, rev. 2 in 2012 [14]. Initially, the guidance described PMS including many strategies "*in addition to complaint handling and vigilance*" including "*active supervision by customer surveys... inquiries of users and patients... literature reviews*" and PMCF, etc., with PMCF through clinical studies and registries having "*great importance among these strategies*." In addition, this guideline listed "*Circumstances that may justify PMCF studies*" as follows:

- *innovation, e.g., where the design of the device, the materials, substances, the principles of operation, the technology or the medical indications are novel;*
- *significant changes to the products or to its intended use for which pre-market clinical evaluation and re-certification has been completed;*
- *high product related risk e.g., based on design, materials, components, invasiveness, clinical procedures;*
- *high risk anatomical locations;*
- *high risk target populations e.g., paediatrics, elderly;*
- *severity of disease/treatment challenges;*
- *questions of ability to generalise clinical investigation results;*
- *unanswered questions of long-term S&P;*
- *results from any previous clinical investigation, including adverse events or from post-market surveillance activities;*
- *identification of previously unstudied subpopulations which may show different Benefit-Risk Ratio (e.g., hip implants in different ethnic populations);*
- *continued validation in cases of discrepancy between reasonable premarket follow-up time scales and the expected life of the product;*
- *risks identified from the literature or other data sources for similar marketed devices;*
- *interaction with other medical products or treatments;*
- *verification of S&P of device when exposed to a larger and more varied population of clinical users;*
- *emergence of new information on safety or performance;*
- *where CE marking was based on equivalence.*

The initial release of MEDDEV 2.12-2 also stated "*PMCF will not be required for products for which the long term clinical performance and safety is already known from previous use of the device.*" In addition, PMCF should always be considered for devices with a CE mark granted using equivalence. Routine PMS activities with complaints and AE data review are required when no PMCF was planned, and a formal protocol was suggested whenever a PMCF was planned. PMCF study suggestions included extending the follow-up time for patients in premarket trials or designing a prospective study of a representative subset of premarket study patients after the device was placed on the market.

The PMCF was clearly stated as not being required when long-term clinical S&P were already known. In other words, companies should probably stop doing PMCF studies when the clinical data are clear and relatively unchanging. Continuing to do this work simply drives the device cost up without any real benefit. The PMS and RM system functions should remain intact and ongoing, but additional clinical data collection is no longer required once the device S&P are clearly understood. A PMCFP will be needed if a risk is discovered with a negative impact on the benefit-risk ratio.

Both pre- and postmarket clinical S&P data and all clinical claims must be evaluated in the clinical evaluation and PMCF. A wide-range of clinical experience data types outside of clinical investigations were to be included in the CE including manufacturer PMS/PMCF registries, cohort studies, AE reports, etc., including ongoing PMCF literature reviews/surveys. PMCF studies were required to address possible residual risks, specific high-risk areas (e.g., biomaterials, anatomical locations, target populations, disease severities/challenges, device interactions), new innovations or device changes after completing the updated CER and recertification, unclear long-term clinical S&P, or to explore events encountered by specific patient subpopulations. A PMCFP or justification was required if a PMCF was not part of the PMS plan.

MEDDEV 2.12/2, rev. 2 described general PMCF study principles and elements, including:

- PMCF study objectives to address the identified residual risks with a formal hypothesis clearly expressed;
- PMCF study design which must address the study objectives; the study plan must spell out typical clinical trial required elements like study population, inclusion/exclusion criteria, site/investigator selection, test methods, end points, statistical considerations, sample size, follow-up duration, etc. (note: retrospective data reviews may not require all of these details); and
- PMCF study implementation plan with adequate controls, appropriate experience required for analyzing data and drawing conclusions related to the original objectives/hypothesis and sufficient details for the final study report.

The PMCF study data are used in the CE process, and may be linked to public health notices, corrective and preventive actions (CAPA), labeling changes (e.g., changes to the instructions for use), or changes to manufacturing or device designs. The NB will assess the PMS system including the PMCFP and PMCFER as part of the conformity assessment (Table 10.4).

TABLE 10.4 NB PMCF assessments listed in MEDDEV 2.12/2, rev. 2 (2012).

No.	Description	Note
1	Verify manufacturer appropriately considered the PMCF need within the PMS system based on device residual risks (e.g., CER clinical data gaps and device characteristics where PMCF is indicated).	PMCF is directly linked to the premarket CE.
2	Verify PMCF is conducted when CE used equivalent devices clinical data exclusively for initial conformity assessment and PMCF must also address residual risks identified for the equivalent devices.	PMCF is needed if equivalence was used in CER.
3	Assess if justification is appropriate when not conducting a specific PMCF plan within the PMS system; if justification is not valid, seek appropriate remediation.	Lack of PMCFP must be justified appropriately.
4	Assess if PMCFP is appropriate to address stated objectives, residual risks and long-term device clinical S&P issues.	PMCFP must address objectives, risks, long-term S&P.
5	Verify PMCF data gathered (favorable or unfavorable) are used to update the CE and the risk management system actively.	PMCF is part of the PMS used to update the CER and RMR.
6	Consider whether specific device PMCF data should be sent to the NB between scheduled assessment activities (e.g., surveillance audit, recertification assessment).	PMCF may identify reportable clinical data, be sure to report.
7	Consider appropriate device certification period and set particular time for NB PMCF data assessment or set specific conditions for subsequent certification follow-up (e.g., based on residual risks, device risk characteristics, and CER presented during initial assessment). NB may require the manufacturer to submit interim PMS and PMCF clinical data reports between certification reviews.	NB may require more frequent PMS/PMCF updates than required for CE Mark, per se.

This PMCF guidelines was obviously due for an update and will likely be aligned more closely with the MDR and IVDR including the PMPF as well.

10.6.2 MEDDEV 2.7.1 Appendix 1: evaluation of clinical data (December 2008)

This guideline specifies a PMCF for all drug-eluting stents (DES) and all bare-metal stents (BMS) in Section VII, starting on p. 8 [15] as follows:

An appropriate post-market clinical follow-up programme in accordance with MEDDEV 2.12/2 shall be performed for all DES and innovative stents and for all BMS unless duly justified.

Such a programme must be planned and can take the form of a clinical investigation (where the CE marked device is used according to its intended use) and/or registry 'All comer' registries, to include those cases treated off-label, should be conducted to better provide clinical safety and performance data in 'real world' clinical practice. Any data gathered from real world usage by manufacturers should be used to feedback directly into device labelling.

A clinical investigation or a registry should include:

- *a clearly stated objective;*
- *a scientifically sound design with an appropriate rationale and statistical analysis plan designed appropriately to address the objectives of the study and scientifically sound to allow for valid conclusions to be drawn;*
- *a study plan which should justify the patient population (to include a representative population with risk factors such as diabetes and hypertension), the selection of sites and investigators, the endpoints and statistical considerations, the number of subjects involved (to ensure capture of the true incidence of late complications), the duration of the study for a minimum of 3 years (taking into account the lifespan of the device and the time of occurrence of late complications), the data to be collected, the analysis planning including any interim reporting, and the procedures for early study termination.*

In other words, the MDR/IVDR discussion of PMCF/PMPF is not a new concept. The April 2003 version of MEDDEV 2.7.1 did not mention PMCF at all [16]; however, MEDDEV 2.7.1 was updated again and MEDDEV 2.7.1, rev. 3 (December 2009) *"Clinical Evaluation Checklist for Notified Bodies"* [17] mentioned the postmarket setting more than 20 times with links to MEDDEV 2.12/2 PMCF Guidelines. Appendix F required the NB to analyze risks, PMS studies, and AE listings, while Section 4 indicated the NB should conduct a PMCFP review as follows:

> 4 Post-market clinical follow up—the notified body should check and review the manufacturer's post market clinical follow up plan:
>
> 4.1 Has the manufacturer presented an appropriate plan for post-market clinical follow up in line with appropriate guidance? Yes ● No ● N/A.
>
> 4.2 If no post-market clinical follow up plan is presented, has this been adequately justified by the manufacturer? Yes ● No ● N/A.
>
> 4.3 Has the manufacturer an adequate post-market surveillance system in place? Yes ● No ● N/A.
>
> 4.4 Has the manufacturer committed to inform the NB of significant updates to their clinical evaluation arising from PMS/PMCF? Yes ● No ● N/A.

This NB review required a clear manufacturer PMCF process well before the MDR/IVDR was released in 2017.

PMCF requirements were explained further in the MEDDEV 2.7.1, rev. 4 (June 2016) guidance [18] with clear PMCFP and PMCF study definitions derived directly from MEDDEV 2.12/2, rev. 2 as follows:

- PMCF plan: the documented, proactive, organized methods and procedures set up by the manufacturer to collect clinical data based on the use of a CE-marked device corresponding to a particular design dossier or on the use of a group of medical devices belonging to the same subcategory or generic device group as defined in Directive 93/42/EEC. The objective is to confirm clinical performance and safety throughout the expected lifetime of the medical device, the acceptability of identified risks and to detect emerging risks on the basis of factual evidence (MEDDEV 2.12/2, rev. 2).
- PMCF study: a study carried out following the CE marking of a device and intended to answer specific questions relating to clinical safety or performance (i.e. residual risks) of a device when used in accordance with its approved labeling (MEDDEV 2.12/2, rev. 2).

The references to Directive 93/42/EEC and the requirement for a "study" were quite confusing at the time, because this guideline predated the MDR, which came out a short time later in 2017, and this guideline was not fully aligned with the MDR. Confusion about what study type constituted a PMCF study continued and included many different clinical data collection types (rigorous clinical trials, registries, surveys, information collection tools, patient reported outcomes, data extracted from healthcare records, database reviews, real-world data collections of individual case reports, patient/physician/manufacturer conversations, etc.). In particular, one guideline item on the proposed checklist for release of the CER asked: "*Do the report's conclusions identify all residual risks and uncertainties or unanswered questions that should be addressed with PMS/PMCF studies?*" The PMCFP needs to include the right PMCF clinical study design to address the specific device S&P residual risks, and needs to be aligned with the GSPRs. To meet these requirements, all device companies needed to update their QMS to ensure proper PMS/PMCF study development and reporting.

10.6.3 MEDDEV 2.7/4 guidelines on clinical investigation (December 2010)

The MEDDEV 2.7/4 guideline [19] discussed clinical investigations intended to support the initial clinical evaluation and conformity assessment but was not intended for IVD devices or combination products. Some parts were reported to apply to postmarket studies, even though the guideline pointed to the more specific MEDDEV 2.12/2 PMCF guideline. In particular, this guidance recommeded *"The clinical evaluation and its documentation must be actively updated with data obtained from the post-market surveillance. New such data as well as considerations for new or changed intended purposes need updating of the clinical evaluation and may indicate necessity of additional clinical investigations."*

Although clinical trial design was outside the scope of this book, this guideline offered general principles for deciding when a clinical trial was needed to assess device S&P. This guideline was based on the GHTF SG5 guidance, ISO 14155 standard, and MEDDEV 2.7.1, rev. 3 guidance. Class III and implantable medical devices required a clinical investigation unless justified by a proper CE. In addition, this guidance clearly stated all medical devices may require a clinical investigation depending on the clinical claims, RM outcomes, and CE results.

Clinical investigations must comply with ethical, legal, regulatory, RM, and administrative requirements including proper study design with appropriate objectives, data integrity management processes, and statistical considerations (e.g., the clinical investigation must have clinically relevant end points, a testable hypothesis, sufficient statistical power, sample size and statistical analysis plan including sensitivity and poolability analyses and the design should ensure a clinically meaningful outcome). This guidance lists many device and design factors to consider when designing the clinical investigation to test device S&P and benefit-risk ratio. In addition, four possible study designs are discussed: RCTs, cohort studies, case-control studies, and case series. The guidance discusses clinical trial conduct and the requirement for a final study report.

10.6.4 Medical Device Coordinating Group 2020-7 postmarket clinical follow up plan template (April 2020)

This PMCFP template [20] provided a guide for MDR compliance and included appropriate MDR citations and definitions about the PMCFP. Manufacturers defined postmarket clinical activities designed to gather more clinical data in this PMCFP. Often, additional data gathering, as defined in the PMCFP, was critical for any CER without sufficient clinical data depth to address the device S&P and/or benefit-risk ratio. This template can be modified for use with an IVD medical device, PER and SSP, when no PMPFP template was available to comply with IVDR requirements.

The MDR required PMCF as an ongoing CE component, and the PMCFP and PMCFER documents were required within the PMSP. The PMCFP must "*specify the methods and procedures set up by the manufacturer, to proactively collect and evaluate clinical data from the use in or on humans of a CE marked device.*" In particular, the PMCF intended to be proactive, and aimed at:

- *confirming the safety and performance, including the clinical benefits if applicable, of the device throughout its expected lifetime;*
- *Identifying previously unknown side-effects and monitor the identified side-effects and contraindications;*
- *Identifying and analysing emergent risks on the basis of factual evidence;*
- *Ensuring the continued acceptability of the benefit-risk ratio...;*
- *Identifying possible systematic misuse or off-label use of the device, with a view of verifying that the intended purpose is correct.*

Template sections A, B, F, and G required straightforward details to be entered into each table; however, sections C, D, and E were more complex. For example,

- "Section C. Activities related to PMCF: general and specific methods and procedures" required the methods for each activity to be described, and the template detailed examples for a device registry, PMCF studies, real-world evidence, and surveys.
- "Section D. Reference to the relevant parts of the technical documentation" required details about how the CER and RM file information were to be analyzed, followed up, and evaluated in this PMCFP.
- "Section E. Evaluation of clinical data relating to equivalent or similar devices" required all clinical data, analyses, and conclusions from equivalent or similar devices to be documented.

Evaluators should use this detailed PMCFP template along with relevant MDR citation and definitions to document the PMCFP.

10.6.5 Medical Device Coordinating Group 2020-8 postmarket clinical follow up evaluation report template (April 2020)

This PMCFER template [21] provided a guide for MDR compliance and included appropriate MDR citations and definitions about the PMCFER. PMCF was considered a continuous process within the PMSP and the PMCFER template included a complete and organized postmarket clinical data evaluation based on the PMCFP. The PMCFER was intended to become part of the CER and technical documentation for the medical device. Specifically, PMCFER conclusions must be considered during the subsequent CER, RM, PMSP, and SSCP updates, if applicable. This template can also be modified for use with an IVD medical device, PER and SSP, when no PMPFER template is available to comply with IVDR requirements.

Template sections A, B, and F required straightforward details to be entered into each table; however, sections C, D, E, and G were more complex. For example,

- "Section C. Activities undertaken related to PMCF: results" required the results from each activity in the PMCFP and a full analysis with potential impact on the CER, RM, and SSCP documents previously reviewed during the conformity assessment. Data quality and justifications for any deviations from the PMCFP must also be described for each activity performed as specified in the PMCFP.
- "Section D. Evaluation of clinical data relating to equivalent or similar devices" required all clinical data, analyses, and conclusions to be documented along with the impact on the device benefit-risk ratio, CER, or PMCFP.
- "Section E. Impact of the results on the technical documentation" required details about how each PMCF activity might impact the CER or RM file.
- "Section G. Conclusions" required an overall conclusion related to the aims of the PMCFP, the need for any preventive or corrective actions and suggestions for the next PMCFP.

Revisions to this template are anticipated to help clarify the data reporting requirements and to make the template more user-friendly.

These MEDDEV and MDCG guidance documents focused primarily on PMCF and did not offer much support for the PMPF needed for IVD devices. In 2020, the ISO/TR 20416:2020(E) was released to guide effective monitoring of S&P as well as usability in everyday device use. This ISO standard briefly mentioned PMPF in a note as follows:

> A PMCF study was defined as a "study carried out following marketing approval intended to answer specific questions relating to clinical safety or performance (i.e. residual risks) of a medical device when used in accordance with its approved labelling.
>
> Note 1 to entry: These may examine issues such as long-term performance, the appearance of clinical events (such as delayed hypersensitivity reactions or thrombosis), events specific to defined patient populations, or the performance of the medical device in a more representative population of providers and patients.

(Source: GHTF/SG5/N4:2010, modified—"device" changed to "medical device.")
Note 2 to entry: For in vitro *diagnostics, a similar type of study exists, e.g., postmarket performance follow-up (PMPF) study in Europe.*"

PMS was defined as a "systematic process to collect and analyse experience gained from medical devices" (source: ISO 13485:2016, 3.14).

10.7 Sufficient postmarket follow up clinical evidence

Similar to the question about "sufficient" CER/PER clinical data, the regulations and guidance documents do not specifically answer the question about "sufficient" PMCF/PMPF clinical data. Rather, the answer must be defined using specific subject device clinical data collections to address specific device clinical data questions not resolved as the CER/PER attempted to demonstrate GSPR conformity. The PMCF/PMPF tools used to gather additional clinical data to address the clinical data gaps should be thoughtfully matched to the questions raised using a least burdensome approach. When insufficient clinical data were found in the CER/PER, the indication for use may need to change to represent only the clinical data documented in the CER/PER and PMCFP/PMPFP.

PMCF/PMPF studies are designed to fill device-specific S&P clinical data gaps identified, qualified, and quantified in the CER/PER. The clinical data generated in the PMCFER/PMPFER are added to the next CER/PER update to define more clearly and completely and to answer more specifically any questions about the device S&P and the clinical benefit-risk ratio. If clinical data gaps are found again when the CER/PER is updated, then another PMCFP/PMPFP is required as the CE/PE cycle repeats.

10.8 Integrating clinical/performance evaluation, postmarket surveillance and postmarket clinical follow up systems

PMS system changes were particularly challenging due to the rapid expansion of EU PMS requirements between 2017 and 2024 and the resulting explosive hiring and training needed to incorporate many new PMS staff. These new hires were expected to ensure appropriate and clinically relevant PMCFP/PMPFP data were collected and analyzed in the PMCFER/PMPFER for the CE/PE process. Often, this QMS expansion and growth work fell to CE/PE evaluators because the existing PMS systems and teams did not analyze clinical data as needed in the CE/PE system and CE/PE evaluators did not analyze data as required by the prior PMS system. This difference in purpose was obscured by the similarity in language, making the problem difficult to address and discuss.

Experience data were primarily evaluated in the PMS system by simply tracking and reporting AEs; however, the MDR and IVDR defined PMS as follows:

> *all activities carried out by manufacturers in cooperation with other economic operators to institute and keep up to date a systematic procedure to proactively collect and review experience gained from devices they place on the market, make available on the market or put into service for the purpose of identifying any need to immediately apply any necessary corrective or preventive actions.* (MDR Article 2, 60; IVDR Article 2, 63)

In addition, MDR Article 83 specifically required the manufacturer PMS system to comply with specific regulatory requirements. The PMS and CE systems needed to work together to determine if the device had sufficient clinical experience data and corrective/preventive actions to keep the product safely on the market and performing as intended. The following PMS system checklist template was developed based on requirements stated for each line item in MDR Article 83/IVDR Article 78 (Table 10.5).

TABLE 10.5 PMS System Checklist per MDR Article 83 and IVDR Article 78.

Line	Description	Source record	Acceptable? (Yes/No)	If no, define problem and remediation plan	Due date
1	Did manufacturer plan, establish, document, implement, maintain, and update an entire PMS system proportionate to device risk and appropriate for the device?				
	Is the PMS system integrated within the QMS?				
2	Does the PMS system gather data actively?				
	Does PMS system gather data systematically?				
	Are relevant PMS system data about device quality, performance, and safety recorded and analyzed?				
	Do PMS system data represent the entire device lifetime?				
	Were appropriate necessary conclusions drawn based on PMS system data?				
	Were appropriate preventive and corrective actions determined, implemented, and monitored based on PMS system data?				
3a	Were PMS system data used to improve the risk management system?				
	Were PMS system data used to update benefit-risk determinations?				
3b	Were PMS system data used to update design and manufacturing information?				
	Were PMS system data used to update the labeling including the instructions for use?				
3c	Were PMS system data used to update the clinical/ performance evaluation?				

(Continued)

TABLE 10.5 (Continued)

Line	Description	Source record	Acceptable? (Yes/No)	If no, define problem and remediation plan	Due date
3d	Were PMS system data used to update the SSCP/SSP?				
3e	Were PMS system data used to identify needs for preventive, corrective, or field safety corrective action?				
3f	Were PMS system data used to identify options to improve the usability and device S&P?				
3g	Were PMS system data used to contribute to the PMS of other devices (if relevant)?				
3h	Were PMS system data used to detect and report PMS data trends?				
	Was the PMS system technical documentation updated when needed?				
4	Were appropriate measures implemented when preventive or corrective action or both were needed during the PMS process?				
	Were appropriate competent authorities and NB notified appropriately when preventive or corrective action or both were needed during the PMS process?				
	Were identified serious incidents and implemented field safety corrective actions reported as required?				

The checklist user should determine if manufacturer had an acceptable source document showing how each requirement and each part of each requirement were met. The source record (i.e., source document) should be listed and the decision about acceptability should be recorded. If not acceptable, then the detailed problem and the plan to fix the problem (i.e., remediation plan) should be documented with a clear due date for resolution.

The PMS system needed to provide data for CE/PE, which, in turn, outputs PMCF activities for the PMS system to address. The PMS system must be flexible enough to adapt to changing PMCF activities. When designing a CE/PE system, the PMS system must be taken into account and mechanisms needed to be in place for the two systems to interact and cooperate effectively.

In particular, three PMS and CE/PE process system integration questions were raised:

- Are relevant PMS system data about device quality, performance, and safety recorded and analyzed?
- Are PMS system data used to update the CE/PE?
- Are PMS system data used to update the SSCP/SP?

If experience data were not collected, analyzed, and reported in a manner useful to the CE/PE process, then the PMS system should be improved to ensure proper integration supporting all needs for proactively gathered clinical experience data within the PMS system.

Similarly, three RM and CE/PE process system integration questions have been raised:

- Were PMS system data used to improve the RM system?
- Were PMS system data used to update benefit-risk determinations?
- Were PMS system data used to update design and manufacturing information?

If experience data were not collected, analyzed, and reported in a manner useful to the RM process, then the PMS system should be improved to ensure proper integration supporting all needs for the proactively gathered clinical experience data within the PMS system.

Well-integrated systems are required to comply with many different international and local regulations and standards. For example, MDR/IVDR specifically required all manufacturers to:

- conform with "relevant harmonized standards" (Article 8) or "common specifications" (Article 9);
- design, manufacture devices and establish RM, CE/PE, and PMS systems (i.e., set-up, implement, maintain and "keep up to date" the systems) within the corporate QMS (Article 10);
- have a "Person responsible for regulatory compliance" (PRRC) to ensure these PMS obligations were met (Article 15);
- draw up a "declaration of conformity" (DoC) to state regulatory requirements have been met (MDR Article 19/ IVDR Article 17);
- ensure CE mark is "affixed before device is placed on the market" (MDR Article 20/IVDR Article 18);
- assign a unique device identifier (UDI) as required (MDR Article 27/IVDR Article 24);
- create an SSCP/SSP (MDR Article 32/IVDR Article 29);
- interact with the electronic system on UDI tracking, vigilance and PMS (MDR Article 33/IVDR Article 30);
- complete a conformity assessment (MDR Article 52/IVDR Article 48);
- plan, conduct, and document a CE/PE (MDR Article 61, Annex XIV Part A/IVDR Article 56, Annex XII Part A);
- conduct PMS activities in accordance with MDR Article 83/IVDR Article 78 and base PMS activities on a PMSP (MDR Article 84/IVDR Article 79);
- prepare a PMSR for class I devices summarizing PMS data analyses, results and conclusions as required in the PMSP "together with a rational and description of any preventive and corrective actions taken... updated when necessary" (MDR Article 85/IVDR Article 80);
- prepare a PSUR annually for class IIa, IIb and III devices or Class C and D IVD devices summarizing PMS data analyses, results and conclusions as required in the PMSP "together with a rational and description of any preventive and corrective actions taken... conclusions of the benefit-risk determination... main findings of the PMCF[PMPF]; and... volume of sales of the device and an estimate evaluation of the size and other characteristics of the population using the device and, where practicable, the usage frequency of the device" (MDR Article 86/IVDR Article 81); and
- report "serious incidents and field safety corrective actions" and trends according to the PMSP (MDR Article 87–90/IVDR Article 82–85).

Note: Many additional requirements apply to specific clinical trials and investigations outside of CE/PE processes.

The MDR, expanded the corporate PMS system scope from data collection to include data generation. A device manufacturer can no long simply wait for complaints and communications to come in; they must actively identify problems/gaps or predict them and take action to mitigate and prevent the problems by generating new clinical data about device usage. The PMCFP/PMPFP is, in essence, a scientific protocol, laying out exactly what data will be collected and how this work will be done. Specifically, the PMCFP/PMPFP is developed in response to CER/PER data gaps and MDR/ IVDR requires manufacturers to "actively" seek PMCF/PMPF clinical/performance data in different ways depending on the clinical data gap. The PMCFP/PMPFP can include (but is not limited to) randomized controlled trials (RCTs), prospective trials (observational studies), retrospective chart reviews, registries, patient information forms, advisory boards, and surveys (e.g., patients, physicians or users) (Table 10.6). Choosing the appropriate clinical data collection and analysis tool is an important part of the PMCFP/PMPFP process within the PMS system. The CER/PER clinical data gaps must be addressed fully and completely and the PMCFER/PMPFER must be fully evaluated in the next CER/PER update.

TABLE 10.6 Examples of clinical data gaps and potential data collection methods.

Potential clinical data gaps	Potential data collection methods
Insufficient clinical use feedback from users	Passive unsolicited clinical data collection (e.g., complaints/ customer satisfaction by phone, email, or during field work)
Insufficient design feedback from users	Passive evaluation of manufacturing problems, design vigilance, biocompatibility, and service records for complaints
Insufficient understanding of misuse, off-label use	Passively overhearing discussion of a device misuse
Insufficient compatibility experiences between devices	Passive design compatibility review between devices

(Continued)

TABLE 10.6 (Continued)

Potential clinical data gaps	Potential data collection methods
Insufficient clinical literature	Proactive clinical data gathering (e.g., literature reviews)
Insufficient observational data	Proactive retrospective chart reviews
Insufficient device-specific S&P data	Proactive RCT
Rare occurrence of specific AE/device problem; safety signal detection	Proactive registry [22] or observational study
Potential misuse or off-label uses	Proactive focus group or advisory board convened and documented
Questions about specific device uses, use conditions, patient populations	Proactive device-specific customer survey
Need for expert advice about device design, S&P or other concern	Proactive device-specific solicited user feedback (e.g., advisory board or physician/user questionnaire)
Missing details about real-world device users/uses (e.g., confirmation of device implantation and surgical details)	Proactive patient info form (e.g., returned postcard or "fill in the blank" online from device package)
Need for comparative data to substantiate comparative claims with a valid and controlled analysis	Proactive real-world clinical experience data in a specific comparative clinical trial
Insufficient PMS data	Proactive PMCF/PMPF studies
Insufficient explant data	Proactive retrieval studies (evaluation of retrieved devices after removal from patient)

PMCF, Postmarket clinical follow-up; *PMS*, postmarket surveillance; *RCT*, randomized controlled trial; *S&P*, safety and performance.

10.9 Postmarket surveillance and periodic safety update report differences

In addition to the PMCF/PMPF requirements, the PMS system must report all surveillance activities in the PMSR or PSUR. The main difference between the PMSR and PSUR is the requirement for higher-risk devices to provide greater attention to S&P details in the PSUR. The PMSR is only for Class I (lowest-risk group of devices), while the PSUR is for higher-risk (above Class I) devices (Table 10.7).

TABLE 10.7 PMSR versus PSUR.

PMSR	PSUR	Same?
MDR Article 85/IVDR Article 80	MDR Article 86/IVDR Article 81	No
Low-risk, Class I medical devices and Class A and B IVD devices only	Moderate/high risk, Class IIa/b and III devices and Class C and D IVD devices	No
Not implantable	Implantable	No
Update "when necessary"	Class IIa devices: update "at least every two years"; Class IIb and III implantable and Class C and D IVD devices: update "at least annually"	No
Fill in technical documentation and make available to NB upon request	Submit to NB through EUDAMED for Class III implantable device and Class D IVD device	No[a]
Summarize PMS results/conclusions	Summarize PMS results/conclusions	Yes
Corrective actions rationale/description	Corrective actions rationale/description	Yes
PMSR only	Extends PMSR to summarize postmarket info, vigilance reporting, current device status in EU	No

EU, European Union; *IVD*, *in vitro* device; *IVDR*, *In Vitro* Device Regulations; *MDR*, Medical Device Regulations; *NB*, notified body; *PMS*, postmarket surveillance, *PMSR*, postmarket surveillance report; *PSUR*, periodic safety update report.
[a]*Similar: Technical documentation available on request for Class II and Class C IVD devices.*

Knowing when to write a PMSR or PSUR is important. The update frequency is less for PMSRs, which are kept on file and only updated "when necessary," while PSURs are updated regularly and submitted to the NB via EUDAMED or made available to the NB during the conformity assessment (depending on device class). Until EUDAMED is available for PSUR submissions, companies must submit each PSUR to their relevant NB directly.

A guidance document, MDCG 2022-21 [23], was available for the medical device PSUR including a template and details about data presentation. Medical device manufacturers were required to submit PSURs for a:

1. new MDR certified device;
2. legacy device becoming MDR certified during the transition period; or
3. legacy device remaining AIMDD or MDD certified until after the transition period.

The PSUR template includes the following sections:

1. PSUR cover page
    - 1.1. Manufacturer information
    - 1.2. Medical device/s covered by PSUR
    - 1.3. Notified body name and organization number
    - 1.4. PSUR reference number assigned by manufacturer
    - 1.5. PSUR version number
    - 1.6. Data collection period covered by PSUR
    - 1.7. Table of contents
2. Executive summary
3. Description of devices covered by PSUR and their intended uses (Article 86.1)
4. Grouping of devices
5. Volume of sales (Article 86.1)
6. Size and other characteristics of population using device (Article 86.1)
7. PMS: vigilance and CAPA information
    - 7.1. Information concerning serious incidents (Article 87, Annex III MDR)
    - 7.2. Information from trend reporting (Article 88, Annex III MDR, nonserious incidents and expected undesirable side effects)
    - 7.3. Information from field safety corrective actions (FSCA) (Article 87, Annex II MDR)
    - 7.4. Preventive and/or corrective actions (CAPA) (Article 83.4 and Article 86 MDR)
8. PMS: information including general PMCF information (Annex III and Annex XIV, Part B, 6.2(a) and (f) MDR)
    - 8.1. Feedbacks and complaints from users, distributors, and importers
    - 8.2. Scientific literature review of relevant specialist or technical literature
    - 8.3. Public databases and/or registry data
    - 8.4. Publicly available information about similar medical devices
    - 8.5. Other data sources
9. Specific PMCF information (Article 86, MDR Annex XIV, Part B, 6.2(b))
10. Summary of findings and conclusions of PSUR
    - 10.1. Collected data validity
    - 10.2. Overall conclusions from collected data analysis
    - 10.3. Actions taken by the manufacturer

MDCG 2022-21 also detailed PSUR objectives, contents, scope, and PSUR requirements druation, device grouping, PSUR preparation, and issuance via EUDAMED, including submission in the absence of EUDAMED. The PSUR must be updated at least annually for Class IIb and III devices and Class C and D IVD devices; however, this is required only once every 2 years for Class IIa devices, and only "when necessary" for Class I medical devices and Class A and B IVD devices. The PSUR must be updated throughout the device lifetime, and the manufacturer is no longer required to update the PSUR after the final manufactured device model on the market outlives the intended lifetime (i.e., no PSUR update required during the obsolescence phase). One PSUR can be written and updated for a "justified" group of relevant devices. A single "leading device" must be designated when submitting a PSUR for a device group, and the "leading device" must be one of the highest risk devices in the device group (see MDR Article 86.1).

The PSUR, PMS, RM, and CE/PE processes must be integrated to ensure all systems stay current with appropriate complaint frequencies, review details, and updates to S&P conclusions whenever required but especially for serious incident reporting or any changes in the benefit-risk ratio. Both passive and proactive feedback must be collected, complaints and recalls databases must be searched, all data must be analyzed, and conclusions must be drawn from the data. The PSUR should describe design changes and anticipated new knowledge potentially contributing to changes in the benefit-risk ratio and future PSUR updates.

10.10 Postmarket surveillance systems must be proactive and integrated

PMS is no longer a passive activity. Companies must proactively collect device S&P data after the device is available for sale on the market through phone calls, surveys, registries, clinical trials, etc. Previously, PMS existed primarily to document complaints, device problems and device-related AEs. For example, EU AIMDD/MDD required CE/PE to be "actively updated with data obtained from the post-market surveillance" (Annex X) and required manufacturers to report

> *any malfunction of deterioration in the characteristics and/or performance of a device, as well as any inadequacy in the instructions for use which might lead to or might have led to the death of a patient or user or to a serious deterioration in his state of health.* (Annex II)

PMS itself was not clearly defined or enforced. Similar regulations in the US led to primarily passive medical device PMS systems. where the manufacturer simply gathered experience data from complaints and AE reports submitted by users. These reports were combined with publicly-available experience data (e.g., MAUDE reports) and sales numbers to generate PMSRs detailing metrics such as complaints per unit sold and total reported AE numbers. In contrast to AIMDD/MDD/IVDD, the EU MDR and EU IVDR each have 10 Articles in each regulation dedicated to PMS and vigilance (MDR Articles 83–92 and IVDR Articles 78–87), and the MDR/IVDR clearly require the corporate PMS systems to be

> *actively and systematically gathering, recording and analysing relevant data on the quality, performance and safety of a device throughout its entire lifetime, and... drawing the necessary conclusions and... determining, implementing and monitoring any preventive and corrective actions.* (MDR Article 83/IVDR Article 78)

This more active and systematic context required major changes in many established corporate PMS systems during the 7 years from 2017 to 2024.

10.10.1 Proactive postmarket surveillance includes postmarket follow-up planning

Under MDR/IVDR, manufactures must be an active participants in gathering and analyzing postmarket clinical data. Active PMS systems must design, execute, and analyze postmarket clinical data generated proactively by surveying users, holding advisory boards, developing clinical trials and more. The PMS system is responsible for the PMCFP/PMPFP and PMCFER/PMPFER, and for integrating these clinical data into the CEP/PEP and CER/PER documents (Fig. 10.2).

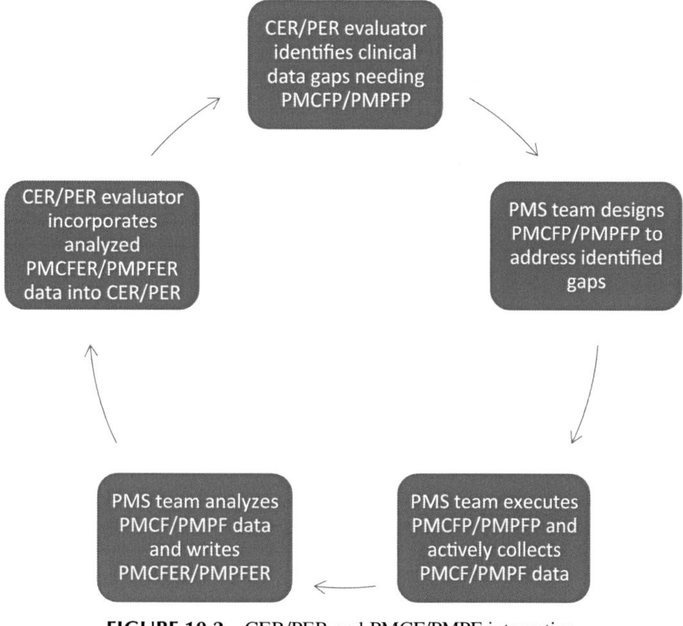

FIGURE 10.2 CER/PER and PMCF/PMPF integration.

Sometimes, PMS data are the only clinical data available for a given device. Device companies were not required to conduct clinical trials before launching the medical device and no clinical literature data were written about the specific device S&P or benefit:risk ratio per se. This was especially true for low-risk devices, and for some higher risk devices when an equivalent (EU)/predicate (US) device regulatory pathway was used. Obviously, once the device was well-established, only limited clinical literature were published about the device because the S&P and benefit-risk profile were well known and more clinical data about the well known devices were not of interest to the scientific community. For devices without clinical trials or literature data, clinical experience/use data was the only way to assess clinical S&P in the CER/PER. PMS data were helpful to determine if the device continued to be safe and if the device continued to perform effectively as intended.

10.10.2 Postmarket surveillance clinical experience data include safety and performance signals

Even when clinical trial and literature data were available, experience data helped to contextualize device use and provided real-world evidence. Trials and studies were highly structured and controlled and often reflected best-case uses rather than average or routine uses. For example, device users in a clinical trial may receive more training and instruction on how to use the device than the average user. Later on, these insufficient training/instruction or design issues in real world uses may only become apparent when analyzing clinical experience data including complaints and real-world clinical data. Similarly, rare risks or risks associated with particular subpopulations may only become apparent after a device is on the market and the PMS data analysis detects such cases.

BOX 10.3 Signal detection requires detailed clinical data

Signal detection is an important concept to understand when building a PMS system. A purely quantitative PMS analysis (i.e., only reporting AE numbers without context) will miss important signals. For example: when a serious AE (SAE) occurs, the PMS system must document sufficient clinical context to provide meaningful data for the PMSR/PSUR, RM, and CE/PE systems.

To illustrate this point, consider the following scenario: a PMSR documented one reported death associated with device use each year for the past 5 years since the device was first introduced into the EU market. Each year, the PMSR concluded no changes were found in the benefit-risk ratio because only one death was reported each year.

This conclusion suggesting "no changes" in the device safety signals was not actually supported by the data without knowing why each of the five patients died. What if all five patients died unexpectedly due to one particular, identical device malfunction and this malfunction could have been easily fixed by a simple design change? This would be a serious safety signal to notice, report, and remediate. Much more clinical detail was needed.

Also, in this example, an unexpected device-related SAE or death is not the same as an expected device-related SAE or death. These details must be included in the PMS triage process: is the report about an AE? is the report/AE serious, related to the device, and unexpected? If yes, the clinical team should be engaged to evaluate the PMS report.

In certain situations, medically-complex, high-risk patients using a high-risk device may experience device-related SAEs, complications or deaths, and knowing whether or not these events are expected will change the reporting urgency.

Collecting and analyzing the proper contextual details in a methodologically sound manner is required to enable appropriate safety signal detection and conclusions about the S&P and the benefit-risk ratio for the device.

Experience data (including complaints and real-world data) have important limitations due to the low quality data, the lack of controls and comparisons and the poor data accuracy. For example, each complaint or MAUDE/DAEN report represents a single data point, and these complaints and reports are not collected in a systematic way. The experience data only contains the information the potentially inexperienced reporter felt necessary to include and these reports are often impossible to verify. In addition, mild or expected side effects and AEs are unlikely to be reported in a passive complaint-based system (i.e., mild/expected AEs are underreported) and sales data do not necessarily equate directly to patient uses because not every unit sold is actually used in a patient (i.e., patient uses are overreported). As a result, experience data should not be used to estimate side-effect or AE frequency in a population, since neither the numerator (i.e., number of AEs/side effects) nor the denominator (i.e., number of uses) can be reliably estimated in the experience data. The PMS team responsible for PMS clinical data analyses and PMSR/PSUR writing, the RM team, and the CER/PER evaluator must all understand how to use experience data and the limitations when evaluating and integrating the PMS data into the CE/PE process properly.

10.10.3 Postmarket surveillance clinical experience data must be current and comprehensive

For the CER/PER, NBs expect the clinical data to be current (i.e., less than 6 months old). In other words, literature and database searches and PMS analyses must be completed no more than 6 months prior to CER/PER sign-off. This NB rule of thumb has not been published in guidelines or regulations, but has been expressed during negotiations with multiple NBs, and this is important for CE/PE, PMS, and RM teams to understand. The clinical data in each CER/PER must be as up to date as possible, since relying on out-of-date information may cause the CE/PE to generate an inaccurate S&P acceptability or benefit-risk ratio analysis. The 6-month time frame helps to determione when searches need to be repeated, especially if the CER/PER or PMSR/PSUR writing or the NB review takes longer than expected.

For devices without clinical trials and with a low-publication and complaint volume, the available clinical data may not change significantly in nature or number over time. In this case, if the CE/PE process is protracted, a practical solution is for the evaluator to write the CER/PER based on available data and then rerun the searches just prior to submission. If no new data were found, the evaluator can document the new search while not changing the CE/PE. In addition, if new data were found but they do not change the S&P analysis or the benefit-risk ratio substantially, then the new data can be included without the need for updated conclusions or discussion. If new data were found and previously unknown risks emerged, then the evaluator and manufacturer must address those risks regardless. Any newly identified data need to be included to ensure the CER/PER is current and comprehensive prior to CER/PER sign-off.

10.10.4 Postmarket surveillance clinical experience data integration by clinical/performance evaluation types

No PMS work is possible for viability (v)CE/PE documents because the device is not yet on the market. In addition, for the equivalence (eq) and first-in-human (fih) CE/PE documents, the PMS data will be largely from the similar, benchmark or equivalent devices since the subject device may not yet be released onto the EU market. After more clinical data are collected in the postmarket setting for traditional (tr) CE/PE documents, the PMS reports will become focused on the subject device rather than the similar, benchmark or equivalent devices. For the first subject-device-specific PMSR/PSUR, "the data analysis may be supported by the device's historical data collected through [PMS] activities as they were conducted prior to Date of Application or MDR Certification data" and subsequent PMSR/PSUR data collection periods should be "contiguous to avoid any gap or overlap of data... the end of data period for one PSUR marks the start data of the next PSUR's data collection period" [23]. All PMSRs/PSURs for a subject device should cover the entire period during which the device has been on the market. Once the device is obsoleted, PMSR/PSUR are no longer required.

10.11 Conclusions

PMS system integration with the CE/PE and RM systems is critical and complex. PMS system PMCF/PMPF activities not only collect missing clinical data for the CE/PE system, but also for the PMS signal detection and safety reporting functions and RM mitigation and control systems. The PMSR/PSUR must clearly document all device-related patient experiences, S&P concerns and impacts of the clinical data on the benefit-risk ratios.

PMS system experience data identification, appraisal, and analysis must be proactive, systematic, and comprehensive. PMSRs/PSURs must include appropriate clinical data analysis for the PMS purpose. The PMS data must be related to and integrated with (but are not the same as) the CE/PE or the RM system outputs. The PMS system is intended to detect new S&P signals, the CE/PE system is required to evaluate all subject device-related clinical data to ensure regulatory requirements were met for S&P and benefit-risk ratio acceptability, and the RM system is required to mitigate all risks to acceptable levels. These divergent purposes were previously not designed to be integrated together with each other, but are now required to do so.

CER/PER clinical experience data must cover the entire device history and CERs/PERs are stand-alone documents. This means a CER/PER update is not meant to be read in conjunction with previous CERs/PERs, and each CER/PER needs to completely capture all relevant clinical data, not just data from the most recent PMSR/PSUR data collection period. This can be a point of confusion and difficulty when CE/PE, PMS and RM teams do not understand the different purposes for CE/PE, PMS and RM system functions.

Comprehensive clinical experience data identification, appraisal, and analysis are required for each subject device CER/PER. However, if high-quality clinical experience (PMS) data are being analyzed by the PMS team at regular intervals for PMSRs/PSURs and/or PMCFERs/PMPFERs, the evaluator may not need to reanalyze the clinical experience data for the CER/PER, provided the PMSR/PSUR clinical data analysis spanned from market introduction to within 6 months of CER/PER sign-off. The specific PMSR/PSUR can be evaluated in the CER/PER rather than having to repeat the raw data evaluations conducted by the PMS team. The PMS team is responsible to ensure the missing clinical data identified in the last CER/PER were captured in the PMCFP/PMPFP and reported PMCFER/PMPFER so the next CER/PER update will have these new clinical data available for clinical evaluation.

Unlike the newer PMCF/PMPF activities, the PMSR/PSUR, device vigilance, and safety reporting activities were typically well-established within most medical device companies. The CE/PE system must also consider the PMSR/PSUR clinical data outside of the PMCF/PMPF processes in the CER/PER. Unless a company is new, the QMS will have detailed steps about developing CE/PE, RM, and PMS systems; however, integrating these systems is a relatively new experience worldwide as required under MDR/IVDR for any devices on the EU market.

GSPR 8 stated "All known and foreseeable risks, and any undesirable side-effects, shall be minimised and be acceptable when weighed against the evaluated benefits to the patient and/or user arising from the achieved performance of the device during normal conditions of use." EU MDR Article 61 stated "the acceptability of the benefit-risk- ratio... shall be based on clinical data providing sufficient clinical evidence." Taken together, this means manufacturers must consider PMS data (if available) when determining benefit-risk ratios. Beyond simply meeting a regulatory requirement, PMS data can provide information on benefit duration and rare risk occurrences. RCTs and other studies may have limited follow-up periods and a limited number of clinical end points; however, PMS systems gather data across the entire device lifetime and can be used to determine if the predetermined clinical end points were aligned with real-world benefits and risks.

10.12 Review questions

1. What do the following acronyms stand for? PMSP, PMSR, PSUR, PMCF, PMPF, PMCFP, PMPFP, PMCFER, PMPFER.
2. Give three reasons why the PMCFP/PMPFP must detail specific activities intended to generate new CE/PE clinical data.
3. List some potential differences between the terms clinical trial and clinical study.
4. List some differences between passive and active PMS activities.
5. What types of devices are required to have a PMSR versus a PSUR?
6. What are some MDR/IVDR Annex III requirements for PMSP elements?
7. How are PMS system data used?
8. True or false? The manufacturer quality management system (QMS) must only define how the clinical data flows within each regulatory, RM, clinical, PMS, and sales and marketing system.
9. What is the function of the company claims matrix?
10. True or false? One well-done literature search is required and can be used for all manufacturer systems.
11. How does the IMDRF define signal detection?
12. What is the value of coding dictionaries such as SNOMED CT, MedDRA, and ICD?
13. In general, for a device without clinical trial or clinical literature data, will clinical experience data (i.e., complaints reporting) alone be considered sufficient proof of safety within the PMSR/PSUR?
14. What are scientifically sound studies?
15. What should a manufacturer do if insufficient supportive clinical evidence is found to support the declared intended purpose, indications, or claims?
16. When is using a lower level of clinical evidence potentially justified?
17. What is a PMCF plan?
18. What is a PMCF study?
19. Which MDR/IVDR Annex specified the PMCF/PMPF is a continual process to update the clinical evaluation?
20. What are the divergent purposes for the PMS, CE/PE, and RM systems?

References

[1] International Conference on Harmonization (ICH E6 R2), Section 1.12. Accessed June 17, 2024. https://database.ich.org/sites/default/files/E6_R2_Addendum.pdf.

[2] US FDA. What are the different types of clinical research? Accessed June 17, 2024. https://www.fda.gov/patients/clinical-trials-what-patients-need-know/what-are-different-types-clinical-research.

[3] US FDA. Framework for FDA's real-world evidence program. December 2018. Accessed June 17, 2024. https://www.fda.gov/media/120060/download.

[4] European Parliament and Council of the European Union. Consolidated text: regulation (EU) No 536/2014 of the European Parliament and of the Council of 16 April 2014 on clinical trials on medicinal products for human use, and repealing Directive 2001/20/EC. (Article 2.2(4)). Accessed June 17, 2024. https://eur-lex.europa.eu/legal-content/EN/TXT/?uri = CELEX%3A02014R0536-20221205.

[5] European Parliament and Council of the European Union. Consolidated text: regulation (EU) 2017/745 of the European Parliament and of the Council of 5 April 2017 on medical devices, amending Directive 2001/83/EC, Regulation (EC) No 178/2002 and Regulation (EC) No 1223/2009 and repealing Council Directives 90/385/EEC and 93/42/EEC. Accessed June 17, 2024. https://eur-lex.europa.eu/legal-content/EN/TXT/?uri = CELEX%3A02017R0745-20230320.

[6] IMDRF/Registry WG/N43. 2020. IMDRF/Registry WG/N42. 2017. Terminologies for categorized adverse event reporting (AER): terms, terminology and codes. Accessed June 17, 2024. https://www.imdrf.org/documents/terminologies-categorized-adverse-event-reporting-aer-terms-terminology-and-codes.

[7] IMDRF/Registry WG/N42. 2017. Methodological principles in the use of international medical device registry data. Accessed June 17, 2024. http://www.imdrf.org/docs/imdrf/final/technical/imdrf-tech-170316-methodological-principles.pdf.

[8] J. Pane, K.M.C. Verhamme, D. Villegas, L. Gamez, I. Rebollo, M.C.J.M. Sturkenboom, Challenges associated with the safety signal detection process for medical devices, Medical Devices (Auckland) 14 (2021) 43–57. Available from: https://doi.org/10.2147/MDER.S278868, https://www.ncbi.nlm.nih.gov/pmc/articles/PMC7917351/pdf/mder-14-43.pdf. Accessed June 17, 2024. PMID: 33658868; PMCID: PMC7917351. February 24.

[9] European Parliament and Council of the European Union. Consolidated text: regulation (EU) 2017/746 of the European Parliament and of the Council of 5 April 2017 on in vitro diagnostic medical devices and repealing Directive 98/79/EC and Commission Decision 2010/227/EU. Accessed June 17, 2024. https://eur-lex.europa.eu/legal-content/EN/TXT/?uri = CELEX%3A02017R0746-20230320.

[10] MDCG 2020-1. Guidance on clinical evaluation (MDR)/performance evaluation (IVDR) of medical device software. March 2020. Accessed June 17, 2024. https://health.ec.europa.eu/system/files/2020-09/md_mdcg_2020_1_guidance_clinic_eva_md_software_en_0.pdf.

[11] MDCG 2020-6. Regulation (EU) 2017/745: Clinical evidence needed for medical devices previously CE marked under Directives 93/42/EEC or 90/385/EEC. April 2020. Accessed June 17, 2024. https://health.ec.europa.eu/system/files/2020-09/md_mdcg_2020_6_guidance_sufficient_clinical_evidence_en_0.pdf.

[12] Co-ordination of Notified Bodies Medical Devices (NB-MED) on Council Directives 90/385/EEC, 93/42/EEC and 98/79/EC. NB-MED/2.12/Rec1. March 2000. Post-marketing surveillance (PMS) post market/production. Accessed June 17, 2024. https://www.team-nb.org/wp-content/uploads/2015/05/documents2012andolders/Recommendation-NB-MED-2_12-1_rev11_Post-Marketing_Surveillance_(PMS).pdf.

[13] European Commission. MEDDEV 2.12/2. Guidelines on medical devices. Post market clinical follow-up studies: a guide for manufacturers and notified bodies. May 2004. Accessed June 17, 2024. https://www.team-nb.org/wp-content/uploads/2015/05/documents2012andolders/2_12-2_05-2004.pdf.

[14] European Commission. MEDDEV 2.12/2 rev2. Guidelines on medical devices. Post market clinical follow-up studies: a guide for manufacturers and notified bodies. January 2012. Accessed June 17, 2024. http://meddev.info/_documents/2_12_2_ol_en_.pdf.

[15] European Commission. MEDDEV 2.7.1 Appendix 1. Guidelines on medical devices. Evaluation of clinical data: a guide for manufacturers and notified bodies. Appendix 1: clinical evaluation on coronary stents. December 2008. Accessed June 17, 2024. https://ec.europa.eu/docsroom/documents/10324/attachments/2/translations.

[16] European Commission. MEDDEV 2.7.1. Guidelines on medical devices. Evaluation of clinical data: a guide for manufacturers and notified bodies. April 2003. Accessed June 17, 2024. http://meddev.info/_documents/2_7.pdf.

[17] European Commission. MEDDEV 2.7.1 Rev. 3. Guidelines on medical devices. Clinical evaluation: a guide for manufacturers and notified bodies. December 2009. Accessed June 17, 2024. http://meddev.info/_documents/2_7_1rev_3_en.pdf.

[18] European Commission. MEDDEV 2.7.1 Rev 4. Guidelines on medical devices. Clinical evaluation: a guide for manufacturers and notified bodies. June 2016. Accessed June 17, 2024. https://www.medical-device-regulation.eu/wp-content/uploads/2019/05/2_7_1_rev4_en.pdf.

[19] European Commission. MEDDEV 2.7/4. Guidelines on clinical investigation: a guide for manufacturers and notified bodies. December 2010. Accessed June 17, 2024. http://meddev.info/_documents/2_7_4_en.pdf.

[20] MDCG 2020-7. Guidance on PMCF plan template. April 2020. Accessed June 17, 2024. https://health.ec.europa.eu/system/files/2020-09/md_mdcg_2020_7_guidance_pmcf_plan_template_en_0.pdf.

[21] MDCG 2020-8. Guidance on PMCF evaluation report template. April 2020. Accessed June 17, 2024. https://health.ec.europa.eu/system/files/2020-09/md_mdcg_2020_8_guidance_pmcf_evaluation_report_en_0.pdf.

[22] IMDRF/Registry. WG/N42FINAL: 2017. March 2017. Methodological principles in the use of international medical device registry data. Accessed June 17, 2024. https://www.imdrf.org/sites/default/files/docs/imdrf/final/technical/imdrf-tech-170316-methodological-principles.pdf.

[23] MDCG 2022-21 Guidance on Periodic Safety Update Report (PSUR) according to regulation (EU) 2017/745. December 2022. Accessed June 17, 2024. https://health.ec.europa.eu/system/files/2023-01/mdcg_2022-21_en.pdf.

Chapter 11

Understanding clinical evaluation regulations outside of Europe

Many countries require clinical data and clinical evidence to support the safety and performance (S&P) of medical devices introduced and sold into their geographies. The European Union (EU) Medical Device Regulation 2017/746 and *In Vitro* Diagnostic Device Regulation 2017/746 are pan-European regulations coveried many European countries (i.e., Member States in the EU) and required clinical evaluation reports (CERs) and performance evaluation reports (PERs). Many other countries around the world also have regulatory requirements for clinical data evaluations to substantiate medical device S&P claims; however, most countries do not require CERs/PERs per se. This chapter describes clinical evidence required in parts of the world outside of Europe.

Throughout this chapter, similarities and differences are highlighed for clinical data evaluations required by specific countries outside of the European Union (EU) including Australia, China, the Association of Southeast Asian Nations (ASEAN), United States (US), Canada, Russia and Eurasian Economic Union (EU), Japan, New Zealand, United Kingdom (UK) and Northern Ireland after Brexit.

11.1 Rest of world clinical evaluations

Many countries had specific clinical evaluation report (CER)/performance evaluation report (PER) regulatory requirements, while other countries required clinical evidence in general without specific CER/PER requirements (Table 11.1).

TABLE 11.1 Rest of world clinical evaluation regulations (outside the EU).

Country	Regulatory authority	Link	CER/PER
Countries with specific CER/PER requirements			
Australia (AUS)	Australian Therapeutic Goods Administration (TGA)	https://www.tga.gov.au	Specific TGA CER requirements for medical [1] and *in vitro* diagnostic (IVD) devices [2].
China (CN)	National Medical Products Administration (NMPA) (translated Chinese name is State Drug Administration)	https://www.nmpa.gov.cn	Medical devices in clinical trial (CT) exemption catalogue (simple report) and CT unexempted (full report, based on predicate, or based on CT) [3].
Member States of the Association of Southeast Asian Nations (ASEAN)	Governments of Brunei Darussalam, Kingdom of Cambodia, Republic of Indonesia, Lao People's Democratic Republic, Malaysia, Republic of the Union of Myanmar, Republic of the Philippines, Republic of Singapore, Kingdom of Thailand and Socialist Republic of Viet Nam	https://asean.org/wp-content/uploads/2016/06/22.-September-2015-ASEAN-Medical-Device-Directive.pdf	Common submission dossier template (CSDT) is used. Clinical evidence must comply with essential principles and CER is same for medical devices and *in vitro* medical devices per the ASEAN Medical Device Directive 2015 [4].
Countries without specific CER/PER requirements			
Canada (CAN)	Health Canada	https://www.canada.ca/en/health-canada.html	CERs/PERs are not required; however, guidance is provided on clinical evidence requirements for medical devices [5].

(*Continued*)

Planning, Writing and Reviewing Medical Device Clinical and Performance Evaluation Reports (CERs/PERs). DOI: https://doi.org/10.1016/B978-0-443-22063-0.00009-2

TABLE 11.1 (Continued)

Country	Regulatory authority	Link	CER/PER
United States (US)	US Food and Drug Administration (FDA)	https://www.fda.gov/medical-devices	CERs/PERs are not required; however, US requires premarket authorization (PMA) with clinical evidence for high-risk devices [6].
Russia	Federal Service for Control of Healthcare and Social Development in the Russian Federation (Roszdravnadzor)	http://www.roszdravnadzor.ru	CERs/PERs are not required. Russia requires safety and efficacy device testing in Russia [7].
New Zealand (NZ)	Ministry of Health and Medsafe (New Zealand Medicines and Medical Devices Safety Authority)	https://www.medsafe.govt.nz	CERs/PERs are not required; however, evidence must be "on file" including "Current certificates from a European Notified Body, Health Canada, TGA or FDA attesting compliance with medical device directives and/or standards" [8].
Japan (JPN)	Pharmaceuticals and Medical Device Agency (PMDA) and Ministry of Health, Labor and Welfare (MHLW)—see Pharmaceuticals and Medical Devices (PMD) Act	https://www.pmda.go.jp/english and https://www.mhlw.go.jp/english/index.html	CERs/PERs are not required; however, MHLW is fully aligned with ISO 13485:2016: "Clinical evaluations and/or evaluation of performance of the medical devices are required to be implemented as part of design and development validation, in the case that the medical device is designated by 23-2-5.3 or 23-2-9.4 of PMD Act" [9].
United Kingdom (UK)	UK Medicines and Healthcare Products Regulatory Agency	https://www.gov.uk/government/organisations/medicines-and-healthcare-products-regulatory-agency	CERs/PERs are not required; however, the UK Medical Devices Regulation 2002 (SI 2002 No. 618) require clinical and performance evaluations [10] and these requirements are changing [11].

Often, regulatory authorities in many countries around the world referred to international standards to harmonize regulations about clinical data evaluations including CERs/PERs (Table 11.2).

TABLE 11.2 International organizations providing CER/PER guidance.

Organization	Link	Description
International Medical Device Regulators Forum (IMDRF)	https://www.imdrf.org	Members included: Australia, Brazil, Canada, China, Europe, Japan Russia, Singapore, South Korea and the US; Formerly Global Harmonization Task Force (GHTF) Working Group (WG) 5 "Clinical Evaluation."
Asian Harmonisation Working Party (AHWP)	http://www.ahwp.info	Members included: Brunei Darussalam, Cambodia, Chile, Chinese Taipei, Hong Kong SAR, China, India, Indonesia, Japan, Jordan, Kazakhstan, Kingdom of Bahrain, Kingdom of Saudi Arabia, Kyrgyz Republic, Laos, Malaysia, Mongolia, Myanmar, Pakistan, People's Republic of China, Philippines, Republic of Kenya, Republic of Korea, Singapore, South Africa, State of Kuwait, Sultanate of Oman, Tanzania, Thailand, United Arab Emirates, USA, Vietnam, Yemen, and Zimbabwe.
International Organization for Standardization (ISO)	https://www.iso.org/home.html	ISO/AWI 18969 is under development to define CER processes at https://www.iso.org/standard/85514.html (PERs are not yet included).
World Health Organization	https://www.who.int/publications/m/item/guidelines-on-clinical-evaluation-of-vaccines-regulatory-expectations	Specific report entitled "Guidelines on clinical evaluation of vaccines: regulatory expectations, Annex 1, TRS No 924" from November 19, 2004.

As illustrated in these global regulatory requirements, many similarities were found; however, several global regions outside the EU, including Australia, China, and the ASEAN countries, had specific CER/PER requirements differing from the EU CER/PER requirements. In addition, several countries did not have specific CER/PER requirements, including Canada, the US, Russia, New Zealand, Japan, and the UK. Many countries are considering global harmonization efforts and some countries (Canada, UK and others) started accepting approvals and clearances from other regulatory authorities and some countries provided expedited pathways for medical devices with a CE Mark which required a CER/PER.

11.2 Australia

The Australian Therapeutic Goods Administration (TGA) regulated medical devices through premarket assessments, postmarket monitoring, and manufacturer inspections. All medical devices were regulated under the Therapeutic Goods Act of 1989 and, similar to the EU MDR/IVDR, the TGA required an established clinical evaluation (CE) process to document how medical devices [1] and *in vitro* devices [2] complied with Australian essential principles (EPs) related to device S&P.

The EU MDR/IVDR, MedDev 2.7/1 and TGA guidelines recommended using PICO (population, intervention, comparator, outcome) to help structure literature search protocols. The common theme is to include a rigorous literature search protocol with appropriate search terms to support objective, unbiased literature searches to identify appropriate literature. Additional suggested methods to structure systematic search and review methods included the Cochrane Handbook for Systematic Reviews of Interventions, the Preferred Reporting Items for Systematic Reviews and Meta-Analyses (PRISMA), and Meta-analysis Of Observational Studies in Epidemiology (MOOSE). In addition, appraisal and analysis must be completed. A variety of appraisal, weighting, and analysis tools can be used. The TGA guideline for medical devices [1] also referred to three relevant international standards recognized many regulatory authorities around the world including ISO 13485 about quality management systems (QMSs), ISO 14971 about risk management (RM) systems, and ISO 14155 about good clinical practices (GCPs) for medical devices.

The TGA guidelines differed significantly from the MedDev 2.7/1 guideline [12], and the differences were not limited to the following examples:

1. EU MedDev 2.7/1 guideline [12] did not cover IVD devices and was being replaced by multiple MDCG guideline documents (e.g., MDCG 2022-2 "Guidance on general principles of clinical evidence for *In Vitro* Diagnostic medical devices (IVDs)" from January 2022) while the TGA guidelines covered both medical devices [1] and *in vitro* devices [2] in separate guidelines which defined key clinical evidence concepts and CER/PER construction details.
2. EU MDR/IVDR regulations and guidelines did not offer device-specific CER recommendations; however, the TGA medical device guidelines [1] included specific recommendations for certain medical devices generally as follows:
   a. "Total and partial joint prostheses"
      i. List commonly combined devices used with joint prostheses.
      ii. Consider patient data from Harris Hip Score (measures functionality) or other scores depending on body part.
      iii. Report revision rates using Australian Orthopaedic Association National Joint Replacement Registry (AOANJRR).
      iv. Report adjunct, surrogate, clinical marker data using radiological findings, radio-stereometric analyses, metal ion concentrations in body fluids (for metal-on-metal devices) to assist in predicting "late" device failures.
      v. Use Short Form-36 Health Survey (SF 36) or other scores over a minimum 2 years postsurgical implant to measure device performance.
   b. "Cardiovascular devices to promote patency or functional flow"
      i. Benchmark against devices in same class to provide comparative data.
      ii. Define anticipated clinical improvement based on patient scores like quality of life (QoL) and exercise stress tests.
      iii. Consider surrogate markers to predict implant failure when *in vivo* time exceeds one year.
      iv. Provide clinical S&P success outcomes by device type.
   c. "Implantable pulse generator systems"
      i. Specify safety measures including: procedural complications (e.g., pneumothorax, hemothorax, pocket haematoma, infection), device pocket erosion, coronary sinus dissection or perforation, pericardial effusion, device migration, toxic or allergic reactions, defibrillation threshold/lead impedances resulting in arrythmias or inappropriate shocks, hazards related to magnetic resonance imaging, etc.
      ii. Provide key performance outcomes using New York Heart Association (NYHA) Classification tool for QoL.
      iii. Include key acceptable peer-reviewed literature.

   d. "Heart valve replacements using a prosthetic valve"
      i. Provide specific postmarket data requirements like distribution numbers by geographical location by year since launch, AEs categories by clinical outcome and type by year, explanted devices with explanations for reach on device failures and corrective measures.
      ii. Ensure clinical outcomes are reported at 30 days posttreatment and include specific outcomes (e.g., all cause reoperation, thrombosis, mortality, kidney injury and stage, hemodialysis need).
      iii. Include information on valve-related dysfunction, prosthetic valve endocarditis, prosthetic valve thrombosis, thromboembolic events and bleeding event reporting with additional guidance provided.
   e. "Supportive devices—meshes, patches and tissue adhesives"
      i. Ensure patient follow-up for clinical trials at a 24-month minimum for permanent and biological meshes.
      ii. Include performance data on anticipated improvement in patient scores after surgery using an internationally recognized QoL or cough stress test.
      iii. Include registries for experience data (e.g., Swedish Hernia Register, European Registry of Abdominal Wall Hernias).
      iv. Review and ensure recurrence rates are within expected recurrence rates provided: mesh hernia repair (15%–25%) and satisfactory mesh application recurrence rate (18% or less for biologic mesh and 12% or less for seroma formation).
   f. "Software as a medical device"
      i. Obtain or collate "clinical information which assists with clinical decision making."
      ii. Support any software device "claim of clinical benefit" with "clinical evidence including safety, effectiveness and performance data."
3. EU IVDR regulations and guidelines considered clinical data for IVD devices a PER [2]; however, the TGA considered CE supporting IVDs a CER, not a PER like the EU IVDR terminology.
4. TGA medical device guideline [1] recommended specific quality appraisal tools for determining evidence level ranks (e.g., Jadad Score, Downs & Black, QUADAS, AMSTAR, SIGN, CEBM, and Cochrane).
5. TGA medical device guideline [1] repeated International Medical Device Regulators Forum (IMDRF) recommendations on appraisal criteria suitability and contribution "to determine suitability and weighting contribution of each dataset"; however, these particular IMDRF recommendations were specifically removed from MedDev 2.7/1 at version 4 and were not included in EU MDR/IVDR or the TGA IVD device guideline. No discussion was provided; however, this suitability and contribution grading system may be missing several critical features for certain types of articles and is difficult to roll up into an inclusion/exclusion decision matrix for individual articles or article groups.
6. TGA medical device guideline [1] stated essential principles or "EPs 1, 2, 3, 4, 6, 13, 13 A and 14 are particularly relevant to clinical evaluation." As noted, the other EPs (5, 7-12 represented in *italics* in the list below) were more fully addressed in other parts of the technical file outside the CER/PER and the EPs 1, 3, 6 and 14 in bold were quite similar to the EU regulations for the CER/PER.
   a. General principles:
      i. Use not to compromise health and safety.
      ii. Design and construction to conform with safety principles.
      iii. Must perform the way the manufacturer intended.
      iv. Must be designed and manufactured for long-term safety.
      v. *Must not be adversely affected by transport or storage.*
      vi. Benefits must outweigh any undesirable effects.
   b. Specific principles:
      vii. *Chemical, physical and biological properties.*
      viii. *Infection and microbial contamination.*
      ix. *Construction and environmental properties.*
      x. *Principles for medical devices with a measuring function.*
      xi. *Protection against radiation.*
      xii. *Medical devices connected to or equipped with an energy source.*
      xiii. Information to be provided with a medical device.
         - Patient implant cards and patient information leaflets.
      xiv. Clinical evidence.
      xv. Principles applying to IVDs only [2].

In Australia, as in Europe, CE was required for every medical device as stated in EP 14, and this includes every IVD device. IVD devices follow the same EPs in Australia [2]. EPs 1, 3, and 6 are particularly relevant (bolded in the list above) regarding safety, performance, and the benefit-risk profile. EPs 2, 4, 13, and 13 A are more relevant to manufacturer practices including design and device information rather than CE; however, safe use of the device in humans must be clearly documented in the risk management plan (EP 2), must be documented for "the length of time appropriate to the intended purpose" (EP 4), and the information provided to the patient includes a clear intended purpose, instructions for use, etc. and all claims about the device are consistent and supported by CE (EPs 13, 13 A). The Australian Therapeutic Goods Act required a "conformity assessment" similar to the EU requirements; however, the TGA "conformity assessment body" issued "TGA conformity assessment certificates" directly to the manufacturer [13] without the involvement of a notified body (NB) as is the EU. The TGA stated the clinical evaluation assessment report (CEAR) from an EU NB "may aid timely clinical review of the submission" [1].

7. The TGA medical device guideline [1] provided a CE checklist including the CER "heading structure" which contained 17 sections (Table 11.3). The checklist was intended to record if "each of the relevant recommended sections has been included, who authored each section, and on which page(s) they can be located within the CER."

TABLE 11.3 Australian clinical evidence checklist.

No.	CER section header
1	General details and device description including Global Medical Device Nomenclature (GMDN), unique device identifier (UDI), lineage, and version (if applicable)
2	Intended purpose, indications and claims
3	Developmental context and state of the art
4	Regulatory status in other countries (including evidence and supporting documents)
5	Summary of relevant preclinical data (if applicable)
6	Demonstration of substantial equivalence or comparability (if applicable)
7	Summary and appraisal of clinical data
8	Data analysis and benefit-risk analysis
9	Conclusions
10	Name, signature and curriculum vitae of clinical expert and date of report
11	Risk assessment and management documents[a]
12	IFU, labeling, and other documents supplied with the device
13	Full clinical investigation reports
14	Literature search and selection strategy
15	Full text of pivotal articles from the literature review
16	Postmarket surveillance reports[a]
17	Additional relevant information on the device (if applicable)

[a] *EU MDR/IVDR contains different requirements.*
Source: Based on Therapeutic Goods Administration. Clinical Evidence Guidelines: Medical Devices. Version 3.1, June 2022. https://www.tga.gov.au/sites/default/files/clinical-evidence-guidelines-medical-devices.pdf.

The EU MedDev 2.7/1 Rev. 4 [12] requirements differed from the TGA CER section requirements above, and these differences were not limited to the following examples:

- No similar "heading checklist" was provided.
- No specific section for risk assessment and management; however, MedDev 2.7/1 Rev. 4 stated risk management documents were in scope for the CER and CE can "result in changes to the manufacturer's risk management documents, instructions for use (IFU) and PMS activities…" [12]. Risk management documents were considered within "data generated and held by the manufacturer" as resources for the EU CER/PER and must be evaluated in the CER/PER.
- No specific section for postmarket surveillance (PMS) reports (PMSRs); however, the EU CER "must be actively updated with data obtained from post-market surveillance" as provided in the MedDev 2.7/1 guideline [12]. PMSR/periodic safety update reports (PSUR) data were integrated into the EU required evaluation processes for the CER/PER as resources within "data generated and held by the manufacturer" and these resources were required to be evaluated in (but not attached to) the CER/PER for the EU.

When considering IVD devices specifically, the regulatory requirements for EU PERs and TGA IVD CERs were also similar and included critically evaluated details about scientific validity, analytical performance, clinical performance, and clinical utility data [2]. The TGA recommended the following structure for IVD CERs, which wass "similar but not identical" to the medical device CER structure:

1. *Device description, information around the lineage and version (if applicable)*
2. Intended purpose, indications for use and claims
3. Regulatory status in other countries
4. Description of the clinical algorithms used to interpret the test (if applicable)
5. Demonstration of substantial equivalence or comparable performance (if applicable)
6. Overview and appraisal of evidence of scientific validity, analytical and clinical performance and clinical utility
7. Summary of relevant analytical and clinical performance data
8. Usability data (if applicable)
9. Post-market data
10. Critical evaluation of clinical evidence
11. Risk/benefit analysis
12. Conclusions
13. Clinical expert's endorsement
14. The name, signature and curriculum vitae... of the clinical expert. [7]

Although the regulatory requirements were similar, the EU did not endorse a specific PER template or section detail like the structure suggested by the guideline in Australia. A couple of high-level similarlities and differences included: the evaluator was responsible to justify the clinical data presented in the CER/PER conformed to the general safety and performance requirements (GSPRs) in the EU or the EPs in Australia and the conformity assessment body [14] issued a certificate after reviewing regulatory conformity details for high-risk devices while allowing a conformity self-assessment for the lowest-risk devices.

For devices where clinical trial data collection were not appropriate (e.g., data collection is unethical or difficult due to limited patient availability or high-risk procedures), the TGA allowed evaluators to "provide a clinical justification for why clinical evidence is either not required or only partially required" [3]. The EU provided similar statements in Article 61(10) for medical devices "where the demonstration of conformity with general safety and performance requirements based on clinical data is not deemed appropriate, adequate justification for any such exception shall be given based on the results of the manufacturer's risk management and on consideration of the specifics of the interaction between the device and the human body, the clinical performance intended and the claims of the manufacturer" [15], and Article 56(4) for IVD devices stated "Clinical performance studies... shall be carried out unless it is duly justified to rely on other sources of clinical performance data" [16]. As a pragmatic exercise, every device manufacturer should explore the boundaries of these particular clauses carefully to limit unnecessary expense during the appropriate due diligence work to protect all medical device users. Misunderstandings and lack of clarity about regulatory discretion in this specific area of the global regulations have seriously hindered medical device innovation and cost containment.

In addition, the TGA website provided additional guidance. For example, the webpage entitled "TGA safety monitoring of medical devices, Database of Adverse Event Notifications" (DAEN) [17] defined signal detection as follows:

Signal detection involves identifying patterns of adverse events associated with a particular device that warrant further investigation.

A medical device safety signal may arise from:

- *a previously unrecognised safety issue*
- *a change in the frequency or severity of a known safety issue*
- *identification of a new at-risk group*
- *use of the device different to that intended by the manufacturer*

The TGA explained the steps required when a potential safety concern was found as follows:

Each detected safety signal "is assessed to determine the nature, magnitude and significance of the concern, and the impact on the overall [device] benefit risk profile... The information in adverse event reports is insufficient to determine the exact level of risk associated with that medical device. Other information is necessary to assess the impact of reported adverse events on the [device] benefit-risk profile...” The TGA must consider “the balance between the benefits offered by any medical device and the potential risks associated with its use for the Australian population as a whole (or individual patient groups where the risks may be higher) before it makes a decision on the response to the signal... a range of actions... can follow when a potential safety issue is identified. These include:

- *informing health professionals and consumers through alerts and articles on the TGA’s website and other publications*
- *requiring changes to product labelling, such as adding warnings, precautions and adverse event information to the Product Information or device label*
- *withdrawing the market approval of the product*
- *compliance testing of device*
- *referral to the Office of Manufacturing Quality (TGA) for follow up audits*
- *recalling the device or particular batches of the device*
- *requiring no further action at this stage, but continuing to monitor the signal.”*

Like regulatory authorities in many countries, the TGA can seek advice from an advisory committee on medical devices when needed for the following types of issues:

- *medical device safety*
- *medical device risk assessment and risk management*
- *medical device performance*
- *consultations with health professionals.*

Of particular assistance when developing the CER/PER, the searchable TGA DAEN database [18] was widely used to document PMS data. In addition, the TGA accepts certification from EU NBs as evidence of conformity assessment compliance along with the TGA conformity assessment certificates and the list of comparable overseas regulators included the US FDA, HealthCanada, Medical Device Single Audit Program (MDSAP), Japan Ministry of Health, Labor and Welfare and Pharmaceutical and Medical Devices, as well as the Singapore Health Sciences Authority (HSA).

11.3 China

The China State Food and Drug Administration (SFDA) was known prior to 2019 as the Chinese Food and Drug Administration (CFDA). The National Medical Products Administration (NMPA) operates under the Ministry of Health/National Health and Family Planning Commission (MOH/NHFPC), and many medical device regulations in China are relevant to clinical evaluations (Table 11.4).

TABLE 11.4 CFDA clinical regulation overview.

Number (No.)	Title	Link (English translation not official)
Level 1: General State Council Order		
State Council Order No. 739 (2021)	Regulation for Supervision and Administration of Medical Devices	https://www.gov.cn/zhengce/content/2021-03/18/content_5593739.htm
Level 2: State Administration for Market Regulation (SAMR) Decree/Annexes		
SAMR Decree No. 47 (2021)	Administrative Measures for Medical Device Registration and Filing	http://www.camdi.org/news/10240
Annex to SAMR Decree No. 47 (2023)	Catalogue of Medical Devices Exempt from Clinical Evaluation	https://www.cmde.org.cn/flfg/fgwj/tz/20230725085258194.html
SAMR Decree No. 48 (2021)	Administrative Measures for In-Vitro Diagnostic Reagent Registration and Filing	https://www.easychinapprov.com/in-vitro-diagnostic-reagents-registration-order-48-in-2021

(Continued)

TABLE 11.4 (Continued)

Number (No.)	Title	Link (English translation not official)
Level 3: Technical guidance documents		
CFDA Circular 61 (2020)	Catalogue of Class III Medical Devices Subject to Clinical Trial Approval (2020 Revision) (No. 2020 of 61)	https://www.nmpa.gov.cn/xxgk/ggtg/qtggtg/20200918103742111.html
CFDA Circular No. 73 (2021)	1. Technical guidelines for clinical evaluation of medical devices 2. Technical guidelines for deciding whether to carry out clinical trials of medical devices 3. Technical guidelines for clinical evaluation of medical device equivalence demonstration 4. Technical guidelines for clinical evaluation reports for medical device registration application 5. Technical guidelines for comparison of products included in the list of medical devices exempt from clinical evaluation	https://www.cmde.org.cn/flfg/zdyz/fbg/fbgqt/20210929092409574.html
CFDA Circular No. 74 (2021)	Technical Guidelines for Clinical Evaluation of In Vitro Diagnostic Reagents Exempt from Clinical Trials (No. 2021 of 74) (cmde.org.cn)	https://topic.echemi.com/a/release-of-technical-guidelines-for-clinical-evaluation-of-in-vitro-diagnostic-reagents-exempt-from-clinical-trials_189910.html#:~:text=This/20guideline/20aims/20to/20provide/20applicants/20with/20technical,the/20data/20by/20the/20drug/20regulatory/20authority./202
CFDA Circular No 77 (2020)	Technical guideline for the use of real-world data for clinical evaluation of medical devices" (draft)	https://www.nmpa.gov.cn/xxgk/ggtg/qtggtg/20201126090030150.html

Note: Please refer to the official version of the regulations; all translations and links to English language documents are not meant to replace a careful reading of the regulation in the native language.

As in other countries, devices were classified from lowest (Class I) to highest risk (Class III), some high-risk devices required premarket clinical trials, and all devices must meet postmarket requirements with a growing interest in real-world evidence (RWE). One team of authors stated "The NMPA assesses data quality from six specific aspects: representativeness, completeness, accuracy, authenticity, consistency, and repetitiveness"; however, data "quality and completeness... vary greatly by hospitals and regions" [19]. Questions of data authenticity may also arise when working with partners in China; for example, the "China Export" symbol (for products produced in China) is not a registered trademark and is strikingly similar to the CE mark, which should not be confused.

11.3.1 China State Council Order 739 (2021)

The earliest regulations about medical device "Supervision and Administration" came into effect in 2000 (State Council Order No. 276) and were updated in 2014 (State Council Order No. 650), 2018 (State Council Order No. 680) and 2021 (State Council Order No. 739). State Council Order No. 739 (2021) included eight chapters discussing general requirements, registration/filing, production, distribution/use, adverse events (AEs)/recalls, supervision/inspection, legal liability, and supplemental information. The marketing authorization holder (MAH) was responsible for device safety and effectiveness during the entire device life cycle, including requirements for a QMS, postmarket follow-up (PMCF) and RM, AE monitoring and reporting with product traceability, and recall system controls.

Certain high-risk devices (i.e., medical X-ray equipment, hemodialysis equipment, hollow-fiber dialyzers, extra-corporeal blood circuits for blood purification equipment, electrocardiographs, implantable cardiac pacemakers, and artificial heart-lung machines) required a China Compulsory Certification since 2002 issued by the China Quality Certification Centre and managed by certification bodies accredited by the China National Accreditation Service for Conformity Assessment (CNAS). Additional regulations were released about medical device registration (2004) and IVD requirements (2007). The regulatory landscape has evolved rapidly in China.

The State Medical Products Administration issued CFDA Circular 61 (2020) "Catalogue of class III medical devices subject to clinical trial approval" was revised and replaced the earlier CFDA Circular No. 14 (2014). The 2020 document stated "Compared with domestic and overseas marketed products, medical devices that adopt new designs, materials or mechanisms, and/or are suitable for a new scope of application and have a higher risk to the human body, should be approved by clinical trials before clinical trials can be carried out in China." Six categories of devices needing clinical trials were listed including: implantable cardiac rhythm management device, implantable ventricular assist system, implantable drug infusion devices, prosthetic heart valves and endovascular stents, tissue engineering medical products containing living cells, and implantable, resorbable polymer/metal material for internal fixation of limb and long bone fractures.

Box 11.1 Cardiac heart valve example

When comparing regulations and guidelines across countries (e.g., the EU, Australia, and China), some ideas and lines of reasoning are more supportive than others. For example, the EU and China describe "prosthetic heart valves" as a type of medical device needing special attention, but only the TGA Guideline provided specific suggestions about how to do this from a clinical evaluation perspective. Specifically, the TGA CE Guideline [1] required "the following outcomes; valve related dysfunction, prosthetic valve endocarditis, prosthetic valve thrombosis, thromboembolic events and bleeding, be reported in a time-related manner," and provided links to a relevant paper entitled "Guidelines for Reporting Mortality and Morbidity after Cardiac Valve Interventions" [20]. The authors (all experts from the American Association for Thoracic Surgery, the Society of Thoracic Surgeons, and the European Association for Cardio-Thoracic Surgery) were part of an "Ad Hoc Liaison Committee for Standardizing Definitions of Prosthetic Heart Valve Morbidity," and they reviewed current clinical practice as they updated and clarified issues across all prosthetic heart valves including an original guideline issued in 1988 then updated in 1996 and again in this reference in 2008.

As a resource for the background/SOTA/alternative therapy of a CER/PER about heart valves, the 2008 guideline takes a comprehensive approach along with the other guidelines and reports recommended in the TGA CE guideline. Manufacturers of heart valves can use this published resource to set standards of care and outcomes measures for a heart valve used anywhere in the world. This type of resource (a published, peer-reviewed, international guideline about device-related mortality and morbidity) is tremendously helpful to undertand the detailed requirements for CER/PER development. Unfortunately, most devices do not have this level of scientific substantiation available, and manufacturers are left to define how to work with physicians in order to develop these resources for each device.

Similar to other countries, China expanded support for innovative devices with a priority review and early market access, and the China NMPA began issuing CE guidance in 2015. In 2018, the NMPA implemented "Technical Guidelines Governing Acceptance of Medical Device Clinical Data from Foreign Studies." Although foreign clinical data had been used in CERs prior to 2018, the changes allowed foreign clinical data to be used also as an alternative to a clinical trial for device registration in China in certain situations as long as the trial: followed ethical, GCP and scientific principles; was consistent with local product-specific regulatory requirements; and was discussed with and encouraged by the Center of Medical Device Evaluation (CMDE) prior to submission.

In 2021, five new medical device CE guidance documents were released based largely on the IMDRF guidance documents. Over time, the CE process in China became more globally aligned with the ongoing, periodic reviews which must be updated throughout the device life cycle. In addition, many devices needed PMS clinical data collection and analysis linked to PM and PMS systems with careful consideration of target user groups, design features and the anticipated design changes needed in response to the CE-identified concerns.

11.3.2 China State Administration for Market Regulation Decree No. 47 (2021)

SAMR Decree No. 47 by the State Administration for Market Regulation (SAMR) explained CE in Article 33 was required for all devices, except those specified in Article 34, which stated:

In any of the following circumstances, clinical evaluation may be exempted:

(1) *The working mechanism is clear, the design is finalized, the production process is mature, the medical device of the same variety that has been on the market has been clinically used for many years without serious adverse event records, and the conventional use is not changed;*

(2) *Other medical devices that can be proved safe and effective through non-clinical evaluation.*

Those who are exempt from clinical evaluation may be exempted from submitting clinical evaluation materials.

The catalogue of medical devices exempted from clinical evaluation shall be formulated, adjusted and published by the State Medical Products Administration. [21]

Decree No. 47, Article 35 specified the need for clinical trials when "existing clinical literature and clinical data are insufficient to confirm the safety and effectiveness of the product" and pointed to State Medical Products Administration guidelines for more information about CER writing. Article 36 required a comparison between the "registration product" and the devices included in the literature and clinical data included in the CER, and appropriate detailed CE materials for any clinical trials completed to support the CER. Article 54 required the CE data to be included in the product registration in China.

The NMPA published a 2023 update to the Annex "Clinical Evaluation Exempt Catalog for Medical Devices" [22]. This annex listed more than 2,000 Class II and III devices (including IVD devices) which were not required to have a clinical trial or a CER. For products listed in this catalog, the manufacturer can submit a comparison document including the product registration describing the catalog information, the predicate already approved in China, and justifying any differences between the product and predicate devices for at least the following 16 parameters:

1. basic principles (performance, mechanism of action/working principle, application scope);
2. components;
3. production technology (including design);
4. manufacturing materials;
5. specifications;
6. safety evaluation;
7. national and industrial standards;
8. intended use;
9. use methods;
10. contraindications;
11. precautions and warnings;
12. delivery conditions (passive device);
13. sterilization and disinfection methods;
14. packaging;
15. label; and
16. instruction for use.

The justification about any differences must explain why the identified difference does not lead to a new risk in product safety or effectiveness. The exemption only applies if the product to be registered is essentially the same as the product in the Clinical Evaluation Exemption Catalog. This catalog assumes similar products were marketed and used in clinical settings for many years, nonclinical and available clinical data indicated the device was safe and effective, and the likelihood of serious adverse events was rare. If the devices were not the same and these assumptions dic not hold true for a given device, then a CER with sufficient clinical evidence to demonstrate safety and effectiveness was required.

11.3.3 China State Administration for Market Regulation Decree No. 48 (2021)

Similarly, SAMR Decree No. 48 for IVD devices defined CE in Article 35 as a "*process of using scientific and reasonable methods to analyze and evaluate clinical data to confirm whether the product meets the requirements for use or intended use, so as to prove the safety and effectiveness of* in vitro *diagnostic reagents*." Article 37 specified "*clinical trials should be conducted to prove the safety and effectiveness of* in vitro *diagnostic reagents*"; however, several conditions were listed for exemptions to the need for clinical trials including the following:

(a) *The reaction principle is clear, the design is stereotyped, the production process is mature, and the* in vitro *diagnostic reagents of the same species that have been marketed have been clinically applied for many years and have no record of serious adverse events and do not change their routine use.*

(b) *The safety and effectiveness of the* in vitro *diagnostic reagents can be proved by means of methodological comparison of the same species.* [23,24]

Article 39 indicated "*in vitro diagnostic reagents included in the catalogue of exempted clinical trials... [will require] comparative analysis with similar products already on the market, methodological comparison data, analysis of relevant literature data and analysis of empirical data, etc.*" This decree clarified "*Calibration products and quality control products do not need to submit clinical evaluation information when applying for registration alone.*" In addition, Article 44 stated "*When conducting clinical evaluation of* in vitro *diagnostic reagents expected to be used by consumers for personal use, the applicant shall also conduct an evaluation of the cognitive ability of consumers without medical background on the product instructions.*"

CE documents were still required for IVD reagents meeting the exemption conditions above, including a comparative analysis between the new IVD product and the same type of approved product, data of methodological comparison, analysis of relevant literature and data, analysis of empirical data, etc. The IVD reagent registration and filing process must use scientific and reasonable methods to analyze and evaluate clinical data, the CER must verify whether the IVD reagent product meets the needs for use or achieves the expected functions, and the CER must demonstrate the safety and effectiveness of the product [24].

11.3.4 Four regulatory paths to the China market

The State Council Order No. 739 and the two Decrees No. 47 and 48 were supported by more than 500 plans, guidelines, and information releases by the NMPA CMDE for medical devices in China. In this setting, medical devices had at least four CE pathways to the China market:

A. No clinical trial: For Class I devices only a basic clinical description was required; for Class II or III devices listed in exemption catalog, manufacture may simply document:

1. Comparison table of exemption catalog listing description and subject device details with differences identified and data provided regarding key difference including, but not limited to, the following device aspects:
   - **a.** device fundamental working principle and mechanism of action;
   - **b.** structural composition;
   - **c.** materials of construction in contact with human body;
   - **d.** performance requirements;
   - **e.** sterilization/disinfection method;
   - **f.** scope of application; and
   - **g.** methods of use.
2. Describe how subject device had no differences compared to listed device (i.e., no differences in intended use, composition, mechanism of action, manufacturing materials, technology, design, functional or performance specifications, active ingredients, sterilization/disinfection or operation method).

   State Council Decree No. 739, Article 14 listed materials to be filed for Class I medical devices (*note these materials* were *also* required *for registration of Class II and III medical devices*):
   - **a.** Product risk analysis data
   - **b.** Product technical requirements
   - **c.** Product inspection report
   - **d.** *Clinical evaluation data (Class I devices just need comparison form)*
   - **e.** Product manuals and label samples (i.e., instruction for use, user manuals, etc.)
   - **f.** Quality management system documents for product development and production
   - **g.** *Other documents showing product safety and effectiveness*

B. For a device not listed in the exemption catalog but with similar or equivalent devices, the CER must document a "predicate comparison" as follows:

1. Comparison table
2. Subject and equivalent device product technical requirements (PTR) and features lists
3. Identified differences lists
4. Data to show no clinically meaningful differences were found in:
   - **a.** Basic principles (working principles/mechanism of action)
   - **b.** Design information (structural composition, technical characteristics)
   - **c.** Manufacturing process

d. Materials (materials in contact with body, energy)
e. Product performance, function, key technical features
f. Biological characteristics (immune response, degradation performance, safety)
  i. Reuse, if applicable
  ii. Sterilization/disinfection methods
g. Application site and method (i.e., indicated device position on body)
h. Indication
i. Stage and severity of applicable disease
j. Indicated population (i.e., age, gender, physiologic info)
k. Body contact type/duration (mucosal, invasive, implant: way/time of contact)
l. Use environment
m. Use method/intended users
n. Contraindications
o. Warnings, precautions and preventive measures
p. Energy
q. Key technical characteristics
r. Industry standards
s. Delivery status
t. Packaging
u. Label
v. Product specification

5. All nonChina-based clinical data and any data from clinical trial/s in China.

   This "simplified" or "predicate comparison" CER is based on similar or equivalent device/s and must conform to the appropriate guidelines (e.g., "Clinical evaluation of Medical Device, No. 73, 2021" or "Clinical evaluation of IVD, No. 74, 2021") which are similar to the EU CER/PER. The following CER table of contents was recommended:

   a. Product description (name, scope, research and development background)
   b. Clinical evaluation scope
   c. Clinical evaluation pathway
   d. Clinical data analysis and evaluation comparing subject and equivalent devices:
      i. Perform clinical evaluation through equivalent device/s clinical data.
      ii. Assess equivalence.
      iii. Summarize equivalent device/s clinical data.
      iv. Analyze equivalent device/s clinical data.
      v. Provide rationale for subject device clinical safety and performance.
      vi. Analyze and evaluate subject device clinical data, if any.
   e. Conclusion
   f. Clinical evaluators (level of expertise/experience with device, research and disease)
   g. Other issues, if applicable

C. For a device not listed in the exempted catalog but with accepted overseas clinical trial data, the manufacturer must provide the clinical study plan, ethical opinion, and clinical study protocol explaining end points, deviations (especially for Asian population), essential study conditions, and a rationale about why no supplemental trial was required in China.

D. For a device not listed in the exempted catalog but needing a clinical study in China, this is the longest and most expensive pathway. This path is typically required for some Class III, high-risk, innovative products without similar products approved in China.

As in the US and many other parts of the world, most Class I and II device do not require a CER/PER in China as they do in the EU. Specifically, China listed all exempt Class I, II, and III devices with a proven safety record, and all novel, high-risk devices including those supporting or maintaining life must be evaluated with clinical data in China.

The CFDA "Technical Guidance on Clinical Evaluation of Medical Devices" included specific requirements and instructions about writing a CER for the CFDA. The medical device CER should follow a verification process to ensure the CFDA application requirements were met, including the following:

- Appropriate medical device application scope.
- CER was thorough and objective.
- Data came from clinical investigations (by applicant), clinical literature, and clinical experiences (often in postmarket setting).
- Both favorable and unfavorable data were considered.
- Data type was appropriate for the device type and the inherent risks.

11.3.5 Filing class II and III medical devices in China

The steps to file a Class II or III device application in China depended on which pathway was followed for the device: (1) no CE, simply file a comparison form because an equivalent device appears on list of devices exempt from CE; (2) do a "predicate comparison" CE and file a CER comparing target and equivalent devices; (3) run a clinical trial outside of China and file clinical study report; and (4) seek permission to run a clinical trial in China.

Specifically, if the subject medical device was on the exempted list and a comparison to the equivalent products on the exempted list showed no differences, then, the comparison form was expected to be sufficient to show the medical device was exempt. If the subject device was not on the exempted list and a comparison to the equivalent products showed no differences having a negative impact on the safety and effectiveness of the subject device, then a CER should be written to compare the subject and equivalent devices. The CER should be based on nonclinical data as well as clinical literature data, clinical experience data, and/or clinical trials data for the subject and equivalent devices from overseas, as well as from trials conducted in China, to address any differences. If the subject device was not on the exempted list and/or had any differences affecting device safety and effectiveness which could not be addressed by a clinical trial including clinical data gathered in China using the equivalent device, then a clinical trial with the subject device should be performed overseas first before submission to the CFDA. The most rigorous pathway is when a clinical trial is needed in China for a novel product (Fig. 11.1).

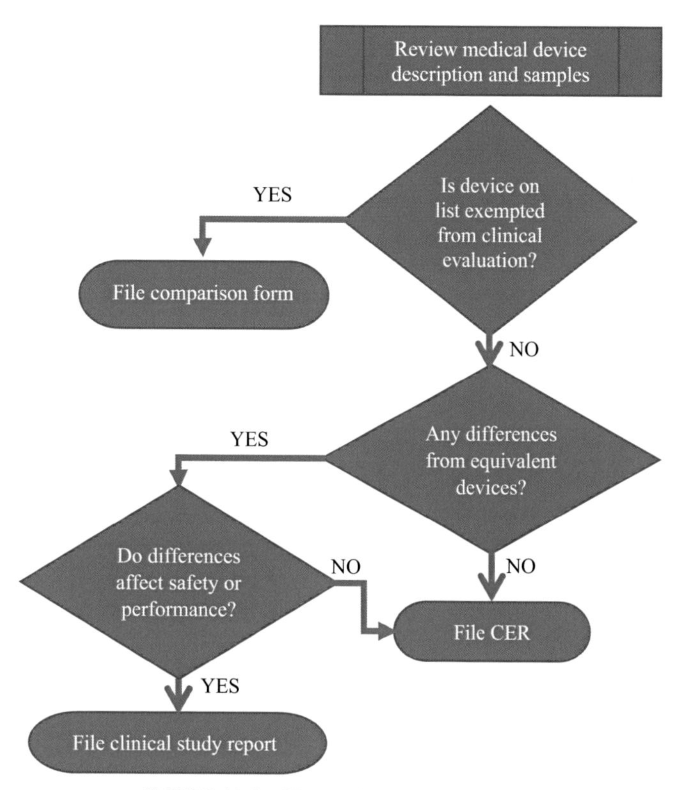

FIGURE 11.1 CFDA clinical evaluation path.

In other words, the CER pathway in China is only available for products with an appropriate equivalent device already approved and used in China (Table 11.5).

TABLE 11.5 Checklist: using a CER in China.

No.	Item	Yes	No[a]	Explanation
1	Class I device?			
2	On exemption list?			
3	On high-risk list?			
4	Unique device? (i.e., no equivalent product in China)			
5	No access to equivalent device info? (E.g., detailed principles of operation, designs, components, materials, processes, performance tests, safety reports.)			
6	Equivalent device differences negatively impact S&P?			
7	Missing or insufficient nonclinical or clinical trials, literature, or experience data to demonstrate equivalent device S&P in China population?			
8	Missing postmarket info in China (e.g., no complaints, sales volumes)?			

[a]*All items must be marked "no" to proceed with CER route.*

For any nonexempt medical device with no equivalent product or any device on a CFDA high-risk device list, a clinical trial is required even if an equivalent product is available and any difference between the subject and equivalent device/s raises a concern about the subject device S&P which might affect the patient benefit-risk profile. In addition, when clinical trials are included, the clinical study report and all associated documents should be provided regardless of whether the trial was conducted in China or overseas.

11.3.6 Clinical equivalence details in China clinical evaluation reports

In China, an equivalent device is defined as a device with domestic registration approval (i.e., in China) and with substantial equivalence to the device under application (i.e., the subject device) for the following:

- expected uses;
- basic principles of operation;
- structural compositions;
- manufacturing materials (especially when contacting the human body or for active devices);
- manufacturing processes;
- performance requirements;
- safety evaluations; and
- conformity with national/industry standards.

The device under application can be considered substantially equivalent only when no adverse effects on the device S&P are potentially associated with differences between the devices.

The CER guidance about equivalence suggested an equivalence table be created to assess equivalence between two devices (Table 11.6). Any deviations from the comparison items (such as not applicable/items do not apply) should be documented and justified.

TABLE 11.6 China equivalence table.

Comparison item		Equivalent device (Name)	Submission device (Name)	Differences and clinical impacts	Summary of supporting data
Clinical scope of application	Indication/s for use				
	Type, stage, and severity of disease				
	Target treatment group (patient population)				
	Site of application in or on the body				
	Contact mode with human body				
	Duration of contact				
	Operational (use) environment				
	Intended user				
	Reuse (number, duration)				
Contraindications					
Precautions and warnings					
Sterilization/disinfection methods					
Packaging					
Labeling					
Instructions/directions for use					
Basic principles	Working principles, method of application			[Same, similar, or different; describe each difference and explain clinical impact of each difference on subject device S&P and justify why equivalent products are expected to have same S&P]	[List relevant test reports, literature, or clinical study reports including page numbers]
	Mechanism of action				
Biological characteristics (biocompatibility, degradation properties, immune responses, tissue integration, etc.)					
Technical structural composition	Product composition				
	Core components				
	Design (dimensions, tolerances, combinations)				
	Software components (core algorithm)				
Manufacturing process					
Manufacturing materials in contact with human body (including material marks, animal-derived/allograft materials, ingredients including pharmaceuticals/bioactives, electrical/radiation safety concerns)					
Conformed/validated national/industry standards for device					
Performance requirements	Performance parameters				
	Function parameters				
Safety evaluation (e.g., injury, clinical, biological, electrical, radiation safety)					
Clinical data, when available	Trial location (China, overseas, multiregional)				
	Registration/cert. no.				
	Outcomes/measures				
	Conclusions				
Equivalence argument					

The CER should describe the differences and must provide an evaluation and determination about why the differences were not expected to indicate a difference between the S&P profiles for the subject and equivalent devices. The rationale must be supported with scientific evidence. Often, these justifications, to address the clinical impacts of the differences between the subject and equivalent devices, were based on the relevant nonclinical study reports, clinical literature, clinical experiences, and/or clinical trials conducted in China. Lengthy discussions for individual differences may be supported by a separate table reviewing the nonclinical/clinical data used to address each individual "difference/clinical impact" question.

11.3.7 Literature search details in China clinical evaluation reports

Comprehensive literature searches shall be conducted on both the equivalent device and the submission device. Like EU CERs, the databases chosen and keywords selected must be carefully considered, robust, and repeatable and search terms/methods/plans/reports should consider the common name, commercial name, manufacturer, basic principles, structural composition, manufacturing materials, design features, key technology, and scope of application (Table 11.7).

TABLE 11.7 Suggested literature search protocol and report template from CFDA guidance.

Required item	Description
Device name	
Model and specification	
Time range of search	
Search date	
Databases/registries searched	
Rationale for databases/registries searched	
Search terms and relationships	
Search methods	
Logical matching of search terms	
Search results (output, list of documents retrieved from each database)	
Description, causes of search deviations, impact on results	
Literature screening criteria (inclusion and exclusion)	
Literature screening procedure	
Excluded literature	
Reasons for exclusions	
Literature screening results (output)	
Description, causes of screening deviations, impact on results	
Note: All literature references need to be in consistent format (e.g., author, title, journal name, year of publication, number of volumes (issues), pages)	
Note: All literature stored in library for retrieval should have consistent naming convention (e.g., first author last name and year with a, b, c... for multiple articles by same first author in a given year)	
Person/signature for literature search and screening activities	
Date	

All database names, links, and search dates must be provided along with the search terms (key words, phrases, and linkages) and inclusion/exclusion criteria. Typical literature databases to search included:

- PubMed at https://www.ncbi.nlm.nih.gov/pubmed;
- Cochrane Library at https://www.cochranelibrary.com;
- Embase at http://www.embase.com (fee required);
- Clinicaltrials.gov at http://clinicaltrials.gov;
- Ovid at http://ovid.com (fee required);
- CNKI (China National Knowledge Infrastructure) at http://cnki.net (in Chinese);
- CQVIP at http://cqvip.com (in Chinese); and
- Wanfangdata at http://www.wanfangdata.com.cn/index.html (in Chinese).

The China CER must include searches of the literature in China. Often these searches and the retrieved literature will need to be translated back and forth into the evaluator's native language. This translation requirement should be considered carefully in the China CER timelines and budgets.

The China CER must complete at least three tasks: (1) evaluate nonclinical and clinical data; (2) evaluate benefit-risk ratio; and (3) draw conclusions about the device S&P and how the device will work in the China population. The China CER should briefly address any regulatory issues regarding the subject and equivalent device comparisons as required by the CFDA, and the China CER Table of Contents should include the following:

A. Equivalent Medical Device Determination
B. Evaluation Path
C. Analysis and Evaluation of Product under Application
    1. Supporting Data
        a. Nonclinical Studies
            i. Risk Management
            ii. Biocompatibility
            iii. Design Verification and Validation
            iv. Preclinical Performance Testing
            v. Packaging Verification
            vi. Sterilization
            vii. Shelf-Life
            viii. Manufacturing Process Validation
        b. Clinical Trials and Data Collection Analysis for Product under Application
            i. Clinical Trial Protocol
            ii. Clinical Trial Results
            iii. Clinical Trial Conclusions
        c. Clinical Literature and Data Collection Analysis for Product under Application
            i. Literature Search Protocol
            ii. Literature Search Results
            iii. Literature Conclusion
        d. Clinical Experience and Data Collection Analysis for Product under Application
            i. Complaints and Adverse Events
            ii. US FDA Manufacturer and User Facility Device Experience (MAUDE)
            iii. Australia TGA Database of Adverse Event Notifications (DAEN)
            iv. British Medical Device Alert (MDA)
            v. "Medical Device Adverse Events Bulletin" and "Alert Newsletter of Medical Devices" issued by China Food and Drug Administration
        e. Chinese Population Data Sets
        f. Conclusions
D. Equivalent Product Analysis and Evaluation
    a. Clinical Trials and Data Collection Analysis for Equivalent Device
        i. Clinical Trial Protocol
        ii. Clinical Trial Results
        iii. Clinical Trial Conclusions

- **b.** Clinical Literature and Analysis of Data Collection for Equivalent Device
  - **i.** Literature Search Protocol
  - **ii.** Literature Search Results
  - **iii.** Literature Conclusion
- **c.** Clinical Experience and Data Collection Analysis for Equivalent Device
  - **i.** Complaints and AEs
  - **ii.** US FDA Manufacturer and User Facility Device Experience (MAUDE)
  - **iii.** Australia TGA Database of Adverse Event Notifications (DAEN)
  - **iv.** British Medical Device Alert (MDA)
  - **v.** "Medical Device Adverse Events Bulletin" and "Alert Newsletter of Medical Devices" issued by China Food and Drug Administration
- **d.** Data Sets of Chinese Population
- **e.** Conclusions

E. Overall Conclusions

Note: The CMDE provided technical guidelines on clinical evaluations for specific devices and device groups (e.g., medical magnetic resonance imaging systems, implantable devices, dental instruments), and more *guidelines for specific devices are being issued.*

11.3.8 Comparisons between China, United States, Eurasian Economic Union (EU), and other clinical evaluation reports

One website, https://www.easychinaapprov.com/clinical-evaluation, stated China exempted "45% of total medical device groups...from clinical evaluation" while "42.5% of medical device need clinical evaluation whereas only 12.5% of total medical device have to undergo clinical study" in 2021. Unlike the EU, where all clinical devices require a CER (with limited exemptions), this high rate of CE exemptions in China was noticeable.

Another publication [25] compared marketing authorizations and reported the CFDA reviewed more Class III devices than the US FDA (23% for China vs. 10% for US) and, although Class I devices were not reviewed in either country, China had fewer Class I devices on the market than the US (36% for China versus 47% for the US). Class II devices were nearly the same in the two countries (41% for China vs. 43% for the US). Previously, Class II and III devices needed to have approval/clearance in the originating country before registering in China, and all medical device registrations (managed through an in-country representative) were valid for 5 years. China regulations are changing; clinical research is now allowed in China before securing approval/clearance in the originating country and China started several special approval programs (e.g., innovative prodcuts registration, priority registration and emergency registration).

Additional differences between China and EU CERs included, but were not limited to, the following:

- China requires clinical data specifically from a China population.
- China does not require a clinical evaluation plan or performance evaluation plan.
- China includes IVD devices with medical devices under the same CER requirements.
- China requires a specific Chinese literature search.
- China requires the comparable/equivalent device to be approved in the country of origin.
- China has more detailed comparison items for clinical and technical and less for biological equivalence (device of a similar kind).
- China has specific input document requirements with a distinct literature/data review and data presentation structure.
- China has specific lists of devices which require a clinical trial and devices which are exempt from a clinical trial or clinical evaluation.

China SAMR Decrees No. 47 and 48 specifically allowed clinical trial/evaluation "*exemptions*" as follows:

- Class I medical devices do not require a clinical trial or clinical evaluation.
- Class II and III devices with clear operating principles, established design records, mature manufacturing processes and without serious adverse events for equivalent devices (devices of the same kind) marketed for years with the same clinical use/purpose do not require a clinical trial or clinical evaluation.

- Class II and III devices with proven safety and effectiveness based on nonclinical evaluation do not require a clinical trial.
- Class II and III devices with safety and effectiveness demonstrated through analysis and evaluation of clinical trials and uses of equivalent devices (devices of the same kind) do not require a clinical trial.

In general, new product registration for Class II and III devices in China typically required 18–48 months. The CFDA pathway evaluated the submitted Application Dossier and Samples at the CFDA Test Center (6–12 months) and, if the device is not on the exemption list, this process required about 1 year for a clinical trial to be developed and completed in China, and 3–6 months for the CER to be developed and completed after the clinical trial data were available for CE. The CMDE conducted a technical review (8–18 months) and the Department of Medical Device Supervision (DMDS) then offered administrative approval (2–3 weeks) by issuing a certificate (2 weeks) through the Administrative Reception Center.

The CFDA conducted clinical trial inspections in China and rejected (or contributed to the withdrawal of) many applications between 2016 and 2018. These CFDA inspections uncovered many problems including untraceable data, poor-quality testing records, unclear AE records, underqualified staff, etc. A clinical trial inspection team was part of the CFDA since 2017 and has increased CFDA emphasis on clinical data authenticity verification especially for the clinical trials required in China.

In summary, the CFDA regulatory framework was established, the regulations were becoming clearer to those outside China, many China CER guidelines were issued, the clinical inspection teams were fully operational and the CFDA was paying more attention to clinical trial quality. CFDA systems were complex and were evolving rapidly to incorporate higher quality standards for clinical trial data. Language barriers and variable quality of in-country representation were problematic; however, international cooperation and regulatory harmonization was improving the quality of all CERs worldwide. Conversely, knowing when a CER is not needed will help greatly to contain costs and here China has been leading the way by listing devices exempt from CE reqiuirements in China.

11.4 Association of Southeast Asian Nations

ASEAN was established in 1967 and included Brunei Darussalam, Cambodia, Indonesia, Lao People's Democratic Republic, Malaysia, Myanmar, Philippines, Singapore, Thailand, and Viet Nam as Member States. This group issued the "ASEAN Agreement on Medical Device Directive" (ASEAN MDD) in 2015 to align with the medical device directives including similar terminology, definitions and "Essential Principles of Safety and Performance of Medical Devices" as defined in Annex 1 [26]. This document recorded the agreement between Member States and included specific terms related to CERs/PERs including (but not limited to) the following:

> *Medical devices shall require clinical evidence, appropriate for the use and classification of the medical device, demonstrating that the medical device complies with the applicable provisions of the essential principles. A clinical evaluation shall be conducted.* (p. 28)

> *4.3.2. Clinical Evidence... should indicate how any applicable requirements of the Essential Principles for clinical evaluation of the medical device have been met. Where applicable, this evaluation may take the form of a systematic review of existing bibliography, clinical experience with the same or similar medical devices, or by clinical investigation. Clinical investigation is most likely to be needed for higher risk class medical devices, or for medical devices where there is little or no clinical experience.* (p. 62)

> *It is appropriate to place a medical device on the market once conformity to the relevant Essential Principles, including a favourable risk/benefit ratio, has been demonstrated. Complete characterization of all risks may not always be possible or practicable in the pre-market phase... questions regarding residual risks... should be answered in the post-market phase through the use of one or more systematic post-market clinical follow-up studies... [to] supplement the pre-market clinical evaluation... [as] one of several options available in a post-market surveillance programme and contribute to the risk management process.* (p. 89)

> *CLINICAL EVALUATION: The assessment and analysis of clinical data pertaining to a medical device to verify the clinical safety and performance of the medical device when used as intended by the product owner.* (p. 89)

> *CLINICAL EVIDENCE: The clinical data and the clinical evaluation report pertaining to a medical device.* (p. 90)

> *When a clinical investigation is conducted, the data obtained is used in the clinical evaluation process and is part of the clinical evidence for the medical device.* (p. 91)

> *Conduct a proper clinical evaluation to demonstrate which clinical data are necessary and can be adequately contributed to by other methods, such as literature searching, prior clinical investigations or clinical experience, and which clinical data remain to be delivered by clinical investigation(s). Available clinical data for comparator medical devices should be carefully examined for comparability and adequacy.* (p. 92)

> *A properly conducted risk analysis is essential in determining what clinical evidence may be needed for a particular medical device. A clinical investigation may be required when the currently available data from preclinical testing, and any prior clinical investigations or other forms of clinical data are insufficient to demonstrate conformity with the Essential Principles. This would be the case when the product owner's risk analysis and the clinical evaluation of a medical device for a particular intended purpose, including new claims, shows that there are residual risks, including aspects of clinical performance, that have not been adequately addressed by the available data and cannot be addressed through other methods.* (p. 92)

> *The outcome of a clinical investigation should be documented in a final study report. The final study report then forms part of the clinical data that is included in the clinical evaluation process and ultimately becomes integrated into the clinical evaluation report for the purposes of conformity assessment.* (p. 96)

Like other regulatory requirements of this type in the EU, Australia, and China, the ASEAN MDD required clinical data from trials, literature, and experiences to be analyzed in a well-developed CER/PER.

In addition, two guidelines were published in a common submission dossier template (CSDT) format for Southeast Asia. The CSDT format was the application standard developed under the ASEAN MDD for the region. The Health Science Authority (HSA) for Singapore was considered the ASEAN "reference" country, as the HSA leads the region with clear guidance and regulations potentially accepted by ASEAN Member States to allow for quick review times throughout the ASEAN region.

The HSA guidance documents were for general medical devices (GN 17) [27] for IVD devices (GN 18) [28] and for clinical evaluations (GN 20) [29]. Similar to all the other CE guidance documents, this HSA guidance introduced the CE concept, general CE principles, and the clinical data sources including literature, experiences, and trials (i.e., clinical investigations) before describing the CER appraisal, analysis, and reporting requirements. Although the guidance was "not intended to cover *In Vitro* Diagnostic (IVD) Medical Devices," this guidance stated:

> *Clinical evaluation should be performed for IVD medical devices as part of conformity assessment to the Essential Principles in a manner similar to other medical devices. The basic principles of objective review of clinical data will apply as described in this guidance document. However, IVD medical devices offer some additional unique challenges that will be addressed in a future document.*

With this obvious oxymoron— IVD guidance was "not intended" but the same "basic principles... will apply"—the HSA guidance documents were fully aligned with the rest of world (ROW) regulatory rhetoric about isolating IVD devices even though the basic principles will apply.

11.5 Examples of countries without specific clinical evaluation report/performance evaluation report requirements

Global regulatory harmonization is evolving on an exponential pace and yet many countries do not have specific "clinical evidence" requirements or guidelines to compare with their EU, Australia, China, or ASEAN counterparts. A few prime examples of countries without CER/PER specific requirements are the US and Canada (CAN) as well as Russia (RUS), Japan (JN), New Zealand (NZ), and the UK including Northern Ireland (NI).

11.5.1 United States

The US does not require a CER/PER. The US medical device regulatory process is different from other countries, and CE is definitely required but the CE presentation used an entirely different format than the CER/PER.

The US Food & Drug Administration (FDA) Center for Devices and Radiological Health (CDRH) medical device regulatory authority [30] oversees the manufacturing, repackaging, relabeling, and/or importing of medical devices marketed in the US under the Federal Food Drug & Cosmetic Act (FD&C Act) [31]. The Act required the FDA to develop and publish regulations in the Federal Register [32] which became the US Code of Federal Regulations (CFR) including parts 800–1299 for Medical Devices under Title 21 [33].

In the US, only certain medical devices and IVD devices were required to submit S&P CE data to the US FDA even though all companies were required to collect and analyze clinical S&P data before, during, and after marketing the devices. Most low-risk devices (Class I and II) were exempt and nothing needed to be filed with the US FDA before marketing these products [34]. Other low-risk Class I and II devices, required a 510(k) or premarket notification (PMN) filed with the FDA [35] to demonstrate the new device had "substantial equivalence" (SE) to another legally marketed medical device already available on the US market with a 510(k) previously cleared by the FDA. A common point of confusion resides in the US FDA SE determination, which is not like the MDR/IVDR equivalence claim in the CER/PER (i.e., the "substantial equivalence" regulatory decision by the US FDA under specific medical device regulations in the CFR and the manufacturer "equivalence" claim under the EU MDR/IVDR use the same word "equivalence," but each country applies a different meaning for the word "equivalence"). In addition, the US FDA specifically uses the term "predicate" to identify the device used to establish SE; however, this term "predicate" is not used in the MDR/IVDR. These terms should not be confused: the US FDA concept of "substantial equivalence" is similar but not the same as the MDR/IVDR concept of an "equivalent device" and the term "prediate" is used by the US FDA but not the EU NB.

Simlarly, premarket 510(k) submissions are NOT the same as CER/PER based on equivalence (i.e., eqCER/eqPERs). In the US, SE means the medical device, when compared specifically to the selected predicate device,:

- *has the same intended use as the predicate; and*
- *has the same technological characteristics as the predicate; or*
- *has the same intended use as the predicate; and*
- *has different technological characteristics and*
- *does not raise different questions of safety and effectiveness; and*
- *the information submitted to FDA demonstrates that the device is as safe and effective as the legally marketed device.* [35]

The FDA determines device safety and efficacy in comparison to the predicate by "*reviewing the scientific methods used to evaluate differences in technological characteristics and performance data.*" The data submitted for the 510(k) review "*can include clinical data and non-clinical bench performance data, including engineering performance testing, sterility, electromagnetic compatibility, software validation, biocompatibility evaluation, among other data.*" The 510k path is not the same as the EU CER/PER process which had specific CE requirements to demonstrate EU-defined equivalency (e.g., same condition, purpose, disease stage, body part, user, expected clinical effect) for the purpose of the CE (i.e., to ensure the clinical data demonstrated conformity with the EU MDR/IVDR GSPRs).

The IVD 510(k) SE process used by the US FDA involved evaluating the submitted (i.e., new) IVD device analytical performance compared to the predicate (potentially without CE) and included:

- *the bias or inaccuracy of the new device;*
- *the imprecision of the new device; and*
- *the analytical specificity and sensitivity* [36].

When no predicate device was available (i.e., when all other devices had different indications for use and/or they had technological differences raising questions about safety and effectiveness, etc.), the device was automatically a Class III device and required a PreMarket Authorization (PMA) decision by the FDA. Manufacturers were allowed to submit a de novo re classification review request to the FDA to reclassify the Class III device into a lower-risk Class I or Class II category. This risk-based device reclassification approach was based on the device description, general and special controls proposed to assure safety and effectives, nonclinical data, biocompatibility data, and available clinical data for medical devices, if any [37]. Manufacturers presented clinical data following the "Acceptance of Clinical Data to Support Medical Device Applications and Submissions" guidance [38]; however, most class I and II devices in the US did not require clinical data. This de novo process in the US was similar to the MDR/IVDR novelty review. In the

EU, devices without equivalent devices were considered novel and typically required clinical trial data to support a regulatory application using fihCER/fihPERs, since eqCER/eqPERs were not possible when no equivalent devices were available.

As was true in most countries around the world, CE was generally required for the highest-risk medical devices or IVDs when they were new or implanted or used to support or sustain life. In the US, these highest-risk, Class III devices required a premarket authorization (PMA) application to the FDA [39] and these PMA applications typically required primary, human clinical data collected during one or more well-designed clinical investigations to document S&P along with the known benefit-risk profile (i.e., clinical data from a pilot and a pivotal trial were common among US PMA approvals). The FDA issued a draft guidance entitled "Demonstrating Substantial Evidence of Effectiveness Based on One Adequate and Well-Controlled Clinical Investigation and Confirmatory Evidence," with a comment period through 18DEC2023 [40].

Although the US FDA does not require a CER/PER, the CER/PER was used to replace two regulatory submission documents in the US including the "report of prior investigations," which iwas a document required in an investigational device exemption (IDE) application as defined in 21CFR812.27 [41] and the required PMA summary of safety and effectiveness data (SSED) document, which reported safety and performance results to the FDA [42]. PMA applications required device and disease details along with scientifically valid clinical trial data about diagnosing and treating the disease with the device and comparisons with alternative therapies similar to the CER/PER background/SOTA/alternative therapy discussion; however, the IDE and PMA regulations and guidelines did not provide specific requirements for planning or conducting literature searches to support the application.

In addition, the US clinical data, just like the EU clinical data, must meet specific requirements. Clinical data used in a regulatory submissions must be scientifically valid and the clinical data provided must have "integrity." In the US, scientifically valid evidence must come from rigorous sources just like the clinical data included in the EU CER/PER:

> *well-controlled investigations, partially controlled studies, studies and objective trials without matched controls, well-documented case histories conducted by qualified experts, and reports of significant human experience with a marketed device, from which it can fairly and responsibly be concluded by qualified experts that there is reasonable assurance of the safety and effectiveness of a device under its conditions of use. The evidence required may vary according to the characteristics of the device, its conditions of use, the existence and adequacy of warnings and other restrictions, and the extent of experience with its use. Isolated case reports, random experience, reports lacking sufficient details to permit scientific evaluation, and unsubstantiated opinions are not regarded as valid scientific evidence to show safety or effectiveness.* [38]

Like the vCER/vPER, when clinical were not yet available, the US FDA has device regulatory pathways including, but not limited to, a humanitarian device exception (HDE) pathway to consider nonclinical data for application support. HDEs are for rare disease applications in the US. HDE applications were often challenged to obtain clinical data; however, benefit-risk assessments were allowed to be supported with nonclinical data in the absence of available clinical data [43].

Similar to the requirements for devices in EU, US device manufacturers must collect clinical experience data on medical devices and IVDs in PMS systems. When close monitoring was required for a cleared or approved device, the FDA often issued a "522 order" and required PMS data to be reported to FDA in a timely manner [44]. Section 522 of the FD&C Act allowed FDA to require device manufactures to collect PMS data for Class II or III devices where safety issues were identified [45]. FDA can also enforce greater safety surveillance by imposing a requirement for manufacturers to conduct a postapproval study (PAS) or a series of studies when a PMA, HDE, or other application is approved or cleared. The PAS allowed continued safety and effectiveness monitoring by the FDA. The PAS was issued as a condition of the approval or clearance and often included clinical or nonclinical study requirements to gather additional data on specific questions about device S&P [46].

Two additional common themes in US regulatory guidance for clinical data evaluations included "Risk Based Monitoring of Clinical Investigations" [47] and specific details surrounding the clinical evaluations of software as a medical device (SaMD) [48]. Clinical risks were differentiated from device risks, and clarity about these different risks was required in the devie RM file. The RM system was required to incorporate traditional risks, as envisioned in ISO 14971, while also incorporating human clinical risks like those envisioned in the risk-based monitoring plans used while collecting human clinical trial data. These clinical risks including "*both risks to participants (e.g., a safety problem) and to data integrity (e.g., incomplete and/or inaccurate data).*" Similarly, the concept of SaMD was expanding exponentially as device makers worldwide explored concepts like artificial intelligence and machine learning. One IMDRF document embedded in the US FDA guidance document about SaMD [48] was helpful to explore SaMD definitions, valid SaMD clinical associations and validation activities required for the CE process. This guidance drew parallels based on the IVD PER steps requiring a scientifically valid clinical association, an analytical validation, and a clinical validation process for SaMD. Specifically, the guidance suggested the SaMD CE process included both verification and validation steps answering specific questions:

- *Is there a valid clinical association between your SaMD output and your SaMD's targeted clinical condition?*
- *Does your SaMD correctly process input data to generate accurate, reliable, and precise output data?*
- *Does your SaMD correctly process input data to generate accurate, reliable, and precise output data?*

This guidance illustrated the "ongoing activities" to be completed during SaMD evaluation of "*clinical safety, effectiveness and performance as intended by the manufacturer in the SaMD's definition statement.*"

Although the US FDA does not require a CER/PER, many features of the regulations requiring clinical data are quite similar worldwide. Most regulators seem to agree: a medical device should only remain on the market when the clinical safety, effectiveness, and performance as indicated are scientifically substantiated in writing by the device manufacturer.

11.5.2 Canada

Like the US, a CER/PER is not required in Canada; however, unlike the US, any CER/PER submitted with a Candian license application is subject to a proactive "Public Release of Clinical Information" requirement.

Health Canada is responsible for regulating medical devices and IVD devices in Canada under the consolidated Medical Device Regulations in SOR/98-282 [49]. Like most countries around the world, Canada determined the regulatory review level based on the device risk level. Devices were classified in groups from I to IV, with Class I representing the lowest risk and Classes II–IV requiring a medical device license for the highest risk devices to be sold in Canada.

In January 2020, the Health Canada Medical Devices Directorate (MDD) was established to review pre- and postmarket safety data. To improve their regulatory system, the Health Canada Action Plan on Medical Devices [50] included a review of "evidence requirements" with a goal to "expand scientific expertise" (especially in areas like "complex software systems"). The review and approve process for medical devices was flexible about the CE type provided to demonstrate medical device safety and effectiveness. Health Canada (like the US) allowed a new medical device to be supported by documenting "similarities in design and performance of an earlier version of the same device." Health Canada intended to strengthen the evidence requirements for higher-risk devices "based on previously authorized versions" and planned to align with international practices.

The Health Canada guidance about CE requirements for Class II, III, and IV medical devices [51] described CE data sources as clinical trials, literature reviews, and clinical experience data including real-world data, usability studies, and simulations. Health Canada used the term "comparator" devices (i.e., not "predicate" or "equivalent" devices) when referring to "an existing device" which was similar to the device submitted for regulatory review. For example, this guidance stated:

> *Certain devices require adequate device-specific clinical data. However, Health Canada understands that many medical devices are developed by making rapid and incremental changes to the design. In some cases, the proposed device may be so similar to an existing device that device-specific clinical data are not required.*
>
> *Where pre-clinical bench testing and side-by-side comparisons can objectively demonstrate that the clinical performance of 2 devices is equivalent, then there is less likely a need for device-specific clinical data. In addition, Health Canada may have already assessed a comparator device to be safe and effective for the same indications for use. This fact alone may be enough evidence of acceptable performance for the subject device, which means that device-specific clinical data may not be required.*

If no similar licensed device can be used as a comparator (does not exist), a manufacturer may use publicly available data sources of comparable devices to assess clinical performance of the subject device. This would avoid the need to generate new clinical data for the device.

Health Canada also issued a guidance with examples of Class II, III, and IV devices and decisions about the need for clinical evidence [52] "*to help manufacturers and regulatory representatives understand when clinical evidence is necessary to demonstrate the safety and effectiveness of a medical device for its intended use.*" Specifically, manufacturers of Class II devices only needed to submit an "attestation" about the objective evidence they had on file demonstrating the device was safe and effective, while manufacturers of Class III devices must summarize all studies and conclusions used to ensure the device was safe and effective, and manufacturers of Class IV devices must submit the actual clinical data to demonstrate safety and efficacy [53].

A CER/PER may be submitted with Class III or IV device applications and was not mandatory. If submitted, the CER/PER may be made publicly available and was expected to demonstrate:

- *the device is effective during normal and expected use conditions*
- *the known and foreseeable risks, and any adverse events, are minimized and acceptable when weighed against the benefits of the intended effectiveness*
- *any claims made about the device's safety and effectiveness (for example, product labelling, marketing material (including website) and indications for use) are supported by suitable clinical evidence* [54].

Canada allowed an equivalence-like process using a comparator device evaluation for regulatory approval similar to an eqCER/eqPER; however, comparators were not equivalent devices by EU definition. As long as objective clinical evidence was provided showing the device was not inferior in regards to S&P compared to the existing device, the device may be allowed on the Canada market after MDD review and licensure.

11.5.3 Russia and Eurasian Economic Union

Russia did not "require" a CER/PER; however, the concept of a CER was not distinguished from a clinical trial which may be needed to support certain Class 2 and 3 devices (or Class 1 devices when no predicate was available). The Ministry of Health or Roszdravnadzor (RZN, Federal Service for Surveillance in Healthcare) regulated medical devices in Russia. Federal Law 323-FZ (21/11/2011) defined medical device details in Article 38, and the Ministry of Health Decree No. 4n (27/12/2012) adopted four risk classes: I (Class I measuring/sterile in other countries and IMDRF), IIa, IIb, and IIIa with separate classification for IVD devices per the GHTF/SG1/N045:2008. Order #1416 (December 27, 2012) defined the registration rules. The Gosstandart (Gost) made sure specific "state" standards were met by all medical devices imported into Russia (GOST-R symbol), and the Rospotrebnadzor (Federal Service for Surveillance on Consumer Rights Protection and Human Wellbeing) ensured products met sanitary and epidemiological requirements for all products in contact with the human body (or those with a potential to negatively impact doctors or patients) [55]. The predicate or comparator device was called the "analog" device in Russia and two pathways were available to register products: (1) Class 1 and 2a device with an analog/predicate device registered in Russia (with same class, application, and efficacy)—report must document the lack of differences between the devices via technical testing/safety evaluation comparing devices; and (2) Class 1 and 2a without an analog/equivalent device. All Class 2b and 3 devices must be tested for product quality, efficacy, and safety in government-approved test centers.

The medical device definition and required elements were similar to other countries and the manufacturer must submit a registration dossier for all product classes according to Ministerial Ordinance No. 11. The documents must be "legalized" ("Notary stamp and corresponding Apostille" [55]), and the registration process was in two phases for all medical devices in Russia: (1) registration dossier with the clinical investigation plan was submitted and put on hold once completed; and (2) clinical investigation was completed and clinical data were submitted so the registration certificate can be issued. Medical device product grouping generally required more applications to cover all models in Russia compared to other markets, and all IVD devices required clinical trials on human samples in Russia. The clinical testing was done at RZN-accredited clinical sites and testing labs, and a locally generated CER with a registered predicate in Russia was used to meet the clinical data requirements. The Russian government did not distinguish between clinical trials and clinical evaluations. An RZN-accredited hospital must provide "evidence of testing" for either the clinical trial or the CER using products already registered in Russia, PMS information from the country of origin, or literature about similar "generic" devices. Russian national standards were unique and all product testing was required in Russia for all products even if they were already CE marked or had FDA 510k clearance, or other national approvals. Core principles were applied in the Eurasian Economic Union (EAEU; including Armenia, Belarus, Kazakhstan, and Kyrgyzstan) to harmonize legislation of Member States, and an EAC (EurAsian Conformity) mark was applied with no expiration once licenses were issued in Russia [56].

The CER/PER was not required as part of the initial submission; however, a clinical trial and/or CER (of sorts) may be required by the RZN after the submitted documents were reviewed for a medical device. Medical device registration rules were expected to change over to EAEU registration rules on December 31, 2025 [57].

11.5.4 Japan

Japan does not have a CER/PER requirement, but used a similar device classification system for medical device and IVD review. This country was the second largest medical device/IVD market after the US [58], and had a rigorous and sophisticated regulatory system developed and managed by the Ministry of Health, Labor and Welfare (MHLW) and the Pharmaceuticals and Medical Devices Agency (PMDA) under the Pharmaceutical and Medical Device Act [59]. Japan used a risk-based regulatory review system classifying devices I–IV. Class I devices, referred to as "Todokede", werere not required to undergo a premarket submission process. Classes II–III were controlled devices, considered low risk to humans, and they required a premarket certification, referred to as "Ninsho", with the need to meet specific certification standards prior to marketing. Class III–IV devices posed medium risk to humans and required premarket approval referred to as "Shonin" [60]. The "Sakigake" system was for innovative devices with minimal clinical evidence and where clinical trials must be started [61]. This system addressed "novel" products, as they are called in the EU. Registered certification bodies (RCBs), similar to EU NBs, were authorized third parties to review Class II controlled medical devices and IVDs. The PMDA reviewed high-risk devices.

All medical devices and IVDs were expected to meet good laboratory practice (GLP), GCP, and good postmarketing study practice (GPSP) standards. The PMDA conducted compliance inspections beginning with GLP prior to premarket reviews. Clinical evidence via clinical trials was expected for Sakigake products following Japanese GCP or determined equivalent standards, but were not required for other levels. The Japanese regulations required regulatory submissions to include "*data concerning the results of clinical studies and other pertinent data to their written applications*" [59], but did not elaborate on evidence levels such as equivalence, literature, or clinical evidence types accepted. Japan does not accept US FDA or EU CER/PER documentation for market approval.

11.5.5 New Zealand

The New Zealand Medicines and Medical Devices Safety Authority governed medical devices under Medsafe, the New Zealand Medicines and Medical Devices Safety Authority (https://www.medsafe.govt.nz). The Medicines Act 1981 [62] required all medical devices created in or supplied to New Zealand to be entered into a database called the Web Assisted Notification of Devices (WAND) 30 days before marketing; however, the Medicines Act does not include a regulatory approval system [63].

Foreign manufactures must appoint a "sponsor" to enter the device into WAND because only New Zealand-based sponsors are allowed to enter information. The database is for notification only and does not represent a regulatory approval; however, New Zealand has a conformity recognition agreement with the EU [64] and Medsafe regulators may ask to review the CE mark documentation provided in a successful CER/PER.

Although clinical evidence was not uploaded to WAND directly, Medsafe can retrospectively request documentation, so manufacturers must have an acceptable clinical evidence certificate confirming S&P "from a European Notified Body, Health Canada, TGA or FDA attesting compliance with medical device directives and/or standards" [65].

Similar to other countries, PMS systems must adequately capture clinical S&P data to meet Medsafe reporting expectations. Safety is monitored by Medsafe and manufacturer-provided safety reports and healthcare provider and device user reports are contained in a regulatory database. Every report is investigated by Medsafe staff and, depending on the event severity, Medsafe may report through publicly accessible websites, request manufacturers change the product or IFU (e.g., warnings, precautions), recall the product from the New Zealand market, provide users additional education and training, or take no action but require continued monitoring [66].

11.5.6 United Kingdom and Northern Ireland after Brexit

The United Kingdom (UK) left the EU on January 1, 2021 before MDR was enforced. The UK Medical Device Regulations (2002, as amended) continued to use MDD under UK law with a new UK Conformity Assessed (UKCA) mark showing regulatory conformity for any device sold in the UK, while the United Kingdom Northern Ireland (UKNI) marking became the appropriate mark to use for devices sold in Northern Ireland. The Medicines and Healthcare products Regulatory Agency (MHRA) was the competent authority in the UK; however, Great Britain and Northern Ireland had different rules. After December 31, 2024, new devices in Great Britain will need UKCA marking including registration with MHRA and representation by a UK-based Authorized Representative. Devices in the UK can be dual marked with CE and UKCA. Unlike in the UK, MDR with CE marking was required in Northern Ireland. Any devices sold in Northern Ireland and the UK, but not elsewhere in the EU, required a UK NB and a UKNI mark with the UKCA mark (UKNI marking is not used independent of the UKCA mark) [67].

The clinical evaluation process for medical devices and IVDs under UK MDR 2002 was similar to EU because clinical data were required and demonstration to set regulatory requirements were necessary for products to obtain certification; however, no CER/PER process existed. Certification was provided by "approved bodies," offering a UKCA mark and legal ability to market devices in the UK. Manufacturers must demonstrate conformity with "essential requirements" (ERs) following Directive 93/42/ECC, and ERs were met when adequate clinical evidence was evaluated to confirm conformity. Medical device clinical evaluations were based on:

1. a critical scientific literature evaluation when relevant data were available relating to device safety, performance, design and intended purpose where device equivalence and data adequately demonstrated essential requirement conformity; or
2. clinical investigations evaluation; or
3. combined clinical data evaluation with literature and clinical investigations [68].

IVD devices under Directive 98/79/EC must meet ERs by providing analytical and diagnostic parameter evidence also "based in particular on clinical data, the adequacy of which is based on the collation of scientific literature or the results of clinical investigations" [69]. Similar to the EU, the UK MDR 2022 medical device and IVD regulations indicated vCER/vPER, eqCER/PERs, fihCER/PER, trCER/trPER, and oCER/PER were acceptable when ERs were adequately supported with valid scientific clinical data.

Medical devices and IVD AEs in the UK were reported to MHRA and were referred to as vigilance reports. Similar to PMS reporting in EU, manufacturers with devices in UK must maintain safe and adequately performing devices, report certain AE types (any event where medical device is suspected to be contributory and the event resulted in, or might have resulted in, death or serious health deterioration to the patient, user or other person), and take field safety corrective actions when required [70].

11.6 Conclusions

This chapter reviewed clinical evidence regulatory requirerments in Australia, China, the ASEAN, US, Canada, Russia and Eurasian Economic Union, Japan, New Zealand, UK and Northern Ireland after Brexit. The EU, Australia, China, and the ASEAN required CERs/PERs for certain medical devices and IVDs; however, they used different methods for CER/PER processes. For example, Australia required additional specific clinical evidence details for CERs/PERs of certain medical device types. Many other countries including Canada, US, Russia, New Zealand, Japan, UK, NI lacked specific CER/PER requirements while all discussed CE as a basic clinical, regulatory and quality requirement for every device.

Typically, the regulations in each country included an established device registration system prior to a device being marketed or imported into the country, a process for submitting documentation for review and a postmarket reporting requirement. Some countries accepted CE-marked products to enter the market and some only agreed to review the EU CER/PER for CE/PE evidence while needing additional or different information for their regulatory submissions (Table 11.8).

TABLE 11.8 CER/PER regulatory differences by country

Country	CER/PER required for market entry	EU CE mark accepted for market entry	EU CER/PER Accepted as CE/PE evidence
EU	Yes	NA	Yes
Australia	Yes	Yes	No
New Zealand	No	Yes	Yes
China	Yes	No	No
US	No	No	Yes
Canada	No	Yes	Yes
Japan	No	No	Yes
UK	No	No	Yes

In addition, some countries (not limited to the list below) have accepted EU CER/PERs for clinical evidence via EU conformity assessment agreements:

- New Zealand;
- Switzerland; and
- Turkey [71].

Many countries used a risk-based device classification approach for regulatory control and oversight and many countries recognized the CE value for all medical devices, but few understood the cost. For comparison, the US FDA 510(k) premarket notification submission allowed the manufacturer to leverage a "substantially equivalent" device without needing to generate clinical data for the device under evaluation. This regulatory path allowed the manufacturer to claim the new device was "SE" to the predicate device without evaluating clinical data per se (a "me too" strategy). In this case, all expectations of medical device S&P would be the same between the two devices, and no new clinical data were needed to ensure the new device S&P. This is a cost-saving pathway to the US market because expensive clinical data collection and evaluation were not required.

The future will determine the risk avoidance cost and how much regulatory oversight should encumber global "free" trade for medical devices. When a medical device is brand new and a viability clinical evaluation (vCER/vPER) is needed, how willing should the public be when deciding to use the device in the open market, given the risk associated with the limited or missing clinical data? Will the costs continue to escalate and drive the best medical devices to be given only to the richest people?

11.7 Review questions

1. What do the following acronyms mean? AHWP, ASEAN MDD, CSDT, EPs, IMDRF, ISO, MHLW, NMPA, PMA, PMDA, SaMD, TGA.
2. Which countries require a CER/PER for products entering market outside EU?
3. List three key differences between EU CER/PER content requirements and Australian CER/PER content requirements.
4. Which EPs are "particularly relevant to clinical evaluation" as described in the TGA medical device guideline?
5. How do the TGA and the EU allow for enforcement discretion when clinical data collection is not appropriate?
6. List three key differences between the EU and the China CERs/PERs.
7. True or False? Searching literature in PubMed and Embase will be sufficient for all CERs/PERs globally.
8. In China, what percentage of all devices were exempted from clinical evaluation in 2021?
9. How does the ASEAN MDD recommend deciding what clinical evidence is needed for a particular medical device?

10. List at least three countries which do not require a CER/PER.
11. How does the US medical device regulatory process differ from other countries when requiring CE and clinical data?
12. Why is the term "equivalence" confusing when considering US FDA regulations and EU MDR/IVDR?
13. True or false? The term "predicate" can be used worldwide to describe a comparator device.
14. What is a "*de novo* classification review" application?
15. What application is generally required by the US FDA to run a medical device clinical trial for a regulatory application?
16. Which two US FDA documents often use CER/PER data?
17. What are a "522 order" and a "PAS"?
18. What are the names of the health authority and the regulation responsible for medical devices in Canada?
19. Which country requires medical device documents to be "legalized"?
20. Which countries have a conformity agreement with EU meaning that CER/PERs are accepted/expected for regulatory purposes?

References

[1] Australian Government, Department of Health, Therapeutic Goods Administration (TGA). Clinical evidence guidelines for medical devices. Version 3.1, June 2022. Accessed September 13, 2023. https://www.tga.gov.au/sites/default/files/clinical-evidence-guidelines-medical-devices.pdf.

[2] Australian Government, Department of Health, Therapeutic Goods Administration. Clinical evidence guidelines supplement In Vitro diagnostic (IVD) medical devices. Version 1.0, March 2020. Accessed September 13, 2023. https://www.tga.gov.au/sites/default/files/clinical-evidence-guidelines-supplement-vitro-diagnostic-ivd-medical-devices.pdf.

[3] China Center for Medical Device Evaluation NMPA (National Medical Products Association). Circular of the State Food and Drug Administration on issuing basic principles for the safety and performance of medical devices (No. 2020 of 18). Accessed September 13, 2023. https://www.cmde.org.cn/flfg/fgwj/tz/20200312141400200.html.

[4] Association of Southeast Asian Nations (ASEAN) Secretariat. Asean Medical Device Directive. Accessed September 13, 2023. 22.-September-2015-ASEAN-Medical-Device-Directive.pdf.

[5] Health Canada. Guidance on clinical evidence requirements for medical devices: overview. Accessed September 13, 2023. https://www.canada.ca/en/health-canada/services/drugs-health-products/medical-devices/application-information/guidance-documents/clinical-evidence-requirements-medical-devices.html.

[6] US FDA. How to study and market your device. Accessed September 13, 2023. https://www.fda.gov/medical-devices/device-advice-comprehensive-regulatory-assistance/how-study-and-market-your-device.

[7] Stepanov A. Medical device regulations in Russia and Eurasian Union (December 30, 2022). Accessed September 13, 2023. https://medicaldevicesinrussia.com/2022/12/30/russian-and-eurasian-medical-device-regulations-highlights-2022/.

[8] M. Hauora. New Zealand Government. Medical Devices: Frequently Asked Questions (July 1, 2014). Accessed September 17, 2023. https://www.medsafe.govt.nz/regulatory/DevicesNew/faqs.asp.

[9] PMDA. Revision of Japanese Medical Device Requirements. Comparison table between ISO13485:2016 and MHLW MO169 Chapter 2, as revised in 2021. [MHLW MO 169 Chapter 2 Basic Requirements Regarding Manufacturing Control and Quality Control of Medical Devices ISO 13485:2016 Medical devices—Quality management systems—Requirements for regulatory purposes]. Accessed September 17, 2023. 000239873.pdf (pmda.go.jp).

[10] Medicines and Healthcare Products Regulatory Agency (MHRA). Consultation outcome. Chapter 7: Clinical investigation/performance studies (June 26, 2022). Accessed September 17, 2023. https://www.gov.uk/government/consultations/consultation-on-the-future-regulation-of-medical-devices-in-the-united-kingdom/chapter-7-clinical-investigation-performance-studies.

[11] Medicines and Healthcare Products Regualtory Agency (MHRA). Consultation outcome. Medicines and Medical Devices Act 2021 Assessment (June 26, 2022). Accessed September 17, 2023. https://www.gov.uk/government/consultations/consultation-on-the-future-regulation-of-medical-devices-in-the-united-kingdom/medicines-and-medical-devices-act-2021-assessment.

[12] European Commission. Guidelines on medical devices; clinical evaluation: a guide for manufacturers and notified bodies under directives 93/42/EEC and 90/385/EEC. MEDDEV 2.7/1, Revision 4 (June 2016). Accessed September 17, 2023. https://ec.europa.eu/docsroom/documents/17522/attachments/1/translations/.

[13] Australian Government, Department of Health, Therapeutic Goods Administration (TGA). The Therapeutic Goods Act 1989. No. 21, 1990. Australian Government. Accessed September 18, 2023. https://www.legislation.gov.au/Details/C2021C00376.

[14] Therapeutic Goods Administration. Medical devices reforms: conformity assessment bodies. Last updated June 30, 2021. Accessed September 18, 2023. https://www.tga.gov.au/medical-devices-reforms-conformity-assessment-bodies.

[15] EU. EU Regulation 2017/745 of the European Parliament and of the Council of 5 April 2017 on in vitro diagnostic medical devices. Accessed September 18, 2023. https://eur-lex.europa.eu/legal-content/EN/TXT/PDF/?uri = CELEX:32017R0745.

[16] EU. EU Regulation 2017/746 of the European Parliament and of the Council of 5 April 2017 on in vitro diagnostic medical devices. Accessed September 18, 2023. https://eur-lex.europa.eu/legal-content/EN/TXT/PDF/?uri = CELEX:32017R0746&from = EN.

[17] Therapeutic Goods Administration. TGA safety monitoring of medical devices, Database of Adverse Event Notifications. Accessed September 18, 2023. https://www.tga.gov.au/resources/resource/guidance/tga-safety-monitoring-medical-devices#: ~ :text = Signal/20detection/20involves/20identifying/20patterns/20of/20adverse/20events,safety/20issue/20identification/20of/20a/20new/20at-risk/20group.

[18] Therapeutic Goods Administration. Database of Adverse Event Notifications (DAEN)—medicines. Accessed September 18, 2023. https://www.tga.gov.au/safety/safety/safety-monitoring-daen-database-adverse-event-notifications/database-adverse-event-notifications-daen-medicines.

[19] J. Li, L. Liu, H. Cao, M. Yang, X. Sun, Use of real-world evidence to support regulatory decisions on medical devices in China and a unique opportunity to gain accelerated approval in "Boao Lecheng Pilot Zone", Cost. Effectiveness Resour. Allocation 21 (7) (2023). Available from: https://doi.org/10.1186/s12962-022-00412-w.

[20] C.W. Kins, D.C. Miller, M.L. Turina, N.T. Kouchoukus, E.H. Blackstone, G.L. Grunkemeier, et al., Guidelines for reporting mortality and morbidity after cardiac valve interventions, J. Thorac. Cardiovasc. Surg. 2008 135 (2008) 732–738. Available from: https://doi.org/10.1016/j.jtcvs.2007.12.002, https://core.ac.uk/reader/82687678?utm_source = linkout.

[21] SAMR Decree No. 47. Administrative measures for medical device registration and filing. 2021. Accessed September 23, 2023. http://www.camdi.org/news/10240.

[22] Annex to SAMR Decree No. 47. Catalogue of medical devices exempt from clinical evaluation. 2023. Accessed September 23, 2023. https://www.cmde.org.cn/flfg/fgwj/tz/20230725085258194.html.

[23] EasyChinapprov. In vitro diagnostic reagents registration and filing management measures, order. 48 in 2021. https://www.easychinapprov.com/in-vitro-diagnostic-reagents-registration-order-48-in-2021.

[24] Wang G. (BaiPharm Website). [Updated] China exempts eligible medical devices and IVD reagents from clinical evaluation or trials. Accessed September 23, 2023. https://baipharm.chemlinked.com/news/china-to-exempt-eligible-medical-devices-and-ivd-reagents-from-clinical-evaluation-or-trials.

[25] Weifan Zhang, R. Lui, C. Chatwin, Marketing authorization of medical devices in China, J. Commercial Biotechnol. 4 (2016). Available from: https://doi.org/10.5912/JCB720, https://core.ac.uk/display/42578571?utm_source = pdf&utm_medium = banner&utm_campaign = pdf-decoration-v1.

[26] ASEAN Secretariat. ASEAN agreement on medical device directive. September 22, 2015 ISBN 978-602-0980-31-7. Accessed September 25, 2023. https://asean.org/wp-content/uploads/2016/06/22.-September-2015-ASEAN-Medical-Device-Directive.pdf.

[27] Health Science Authority (HSA) for Singapore. GN-17: guidance on preparation of a product registration submission for general medical devices using the ASEAN CSDT (July 2023, Revision 2). Accessed September 25, 2023. gn-17-r2-guidance-on-preparation-of-a-product-registration-submission-for-gmd-using-the-asean-csdt-(2023-jul)-pub.pdf (hsa.gov.sg).

[28] Health Science Authority (HSA) for Singapore. GN-18: guidance on preparation of a product registration submission for *in-vitro* diagnostic medical devices using the ASEAN CSDT (JUL 2023, Revision 2). Accessed September 25, 2023. https://www.hsa.gov.sg/docs/default-source/hprg-mdb/guidance-documents-for-medical-devices/gn-18-r2-guidance-on-preparation-of-a-product-registration-submission-for-ivd-md-using-the-asean-csdt-(2023-jul)-pub.pdf.

[29] Health Science Authority (HSA) for Singapore. Regulatory guidance. GN-20: guidance on clinical evaluation (November 2022, Revision 2). Accessed September 25, 2023. https://www.hsa.gov.sg/docs/default-source/hprg-mdb/guidance-documents-for-medical-devices/gn-20-r2-guidance-on-clinical-evaluation(2022-nov)_pub.pdf.

[30] Food & Drug Administration (FDA). Overview of device regulation. Content current as of September 4, 2020. Accessed September 25, 2023. https://www.fda.gov/medical-devices/device-advice-comprehensive-regulatory-assistance/overview-device-regulation.

[31] Federal Food, Drug, And Cosmetic Act. Originally enacted June 25, 1938. Accessed September 25, 2023. https://www.govinfo.gov/content/pkg/COMPS-973/pdf/COMPS-973.pdf.

[32] The Daily Journal of the United States Government. Federal register. Accessed September 25, 2023. https://www.federalregister.gov/.

[33] National Archives. Code of federal regulations. Title 21 food and drugs, chapter I Food and Drug Administration, Department of Health and Human Services, Subchapter H Medical Devices. Last amended March 10, 2023. Accessed September 25, 2023. https://www.ecfr.gov/current/title-21/chapter-I/subchapter-H.

[34] FDA. Class I and Class II device exemptions. Content current as of February 23, 2022. Accessed September 25, 2023. https://www.fda.gov/medical-devices/classify-your-medical-device/class-i-and-class-ii-device-exemptions.

[35] FDA. Premarket Notification 510(k). Content current as of October 3, 2022. Accessed September 25, 2023. https://www.fda.gov/medical-devices/premarket-submissions-selecting-and-preparing-correct-submission/premarket-notification-510k.

[36] FDA. Overview of IVD regulation. Content current as of October 18, 2021. Accessed September 25, 2023. https://www.fda.gov/medical-devices/ivd-regulatory-assistance/overview-ivd-regulation.

[37] FDA. De Novo classification request. Content current as of October 4, 2022. Accessed September 25, 2023. https://www.fda.gov/medical-devices/premarket-submissions-selecting-and-preparing-correct-submission/de-novo-classification-request.

[38] FDA. Acceptance of clinical data to support medical device applications and submissions frequently asked questions. Issued February 21, 2018. Accessed September 25, 2023. https://www.fda.gov/media/111346/download.

[39] FDA. PMA Approvals. Content current as of December 16, 2021. Accessed September 25, 2023. https://www.fda.gov/medical-devices/device-approvals-denials-and-clearances/pma-approvals#GI.

[40] FDA. Demonstrating substantial evidence of effectiveness based on one adequate and well-controlled clinical investigation and confirmatory evidence; draft guidance for industry. Federal register. Accessed September 25, 2023. https://www.federalregister.gov/documents/2023/09/19/2023-20228/demonstrating-substantial-evidence-of-effectiveness-based-on-one-adequate-and-well-controlled.

[41] US Code of Federal Regulations Title 21, Part 812.27 (21CFR812.27) Report of prior investigations. Accessed September 25, 2023. https://www.ecfr.gov/current/title-21/chapter-I/subchapter-H/part-812/subpart-B/section-812.27.

[42] FDA Office of Product Evaluation and Quality. FDA Summary of Safety and Effectiveness Data Template (SEED). TEMPLATE. Accessed September 25, 2023. https://www.fda.gov/media/113810/download.

[43] FDA. Humanitarian Device Exemption (HDE) Program. Issued September 6, 2019. Accessed September 25, 2023. https://www.fda.gov/media/74307/download.

[44] US Code of Federal Regulations Title 21, Part 822 Postmarket Surveillance. Page last updated June 7, 2023. Accessed September 25, 2023. https://www.accessdata.fda.gov/scripts/cdrh/cfdocs/cfcfr/CFRSearch.cfm?CFRPart = 822.

[45] FDA. Postmarket Surveillance Under Section 522 of the Federal Food, Drug, and Cosmetic Act. Issued October 7, 2022. Accessed September 25, 2023. https://www.fda.gov/media/81015/download.

[46] FDA. Procedures for handling post-approval studies imposed by premarket approval application order. Issued October 7, 2022. Accessed September 25, 2023. https://www.fda.gov/media/71327/download.

[47] FDA. A risk-based approach to monitoring of clinical investigations—questions and answers: guidance for industry. April 2023. Accessed September 25, 2023. https://cacmap.fda.gov/media/121479/download.

[48] FDA. Software as a Medical Device (SAMD): clinical evaluation—guidance for industry and food and drug administration staff (December 8, 2018). Accessed September 25, 2023. https://www.fda.gov/files/medical/20devices/published/Software-as-a-Medical-Device-/28SAMD/29-Clinical-Evaluation-Guidance-for-Industry-and-Food-and-Drug-Administration-Staff.pdf.

[49] Health Canada. Medical Devices Regulations. SOR/98-282. Current to September 13, 2023. Accessed September 25, 2023. https://laws-lois.justice.gc.ca/PDF/SOR-98-282.pdf.

[50] Government of Canada. Health Canada's action plan on medical devices: continuously improving safety, effectiveness and quality. Published December 2018. (date modified May 28, 2021). Accessed September 25, 2023. https://www.canada.ca/en/health-canada/services/publications/drugs-health-products/medical-devices-action-plan.html.

[51] Government of Canada. Guidance on clinical evidence requirements for medical devices: Overview. Published November 15, 2022. Accessed March 20, 2023. https://www.canada.ca/en/health-canada/services/drugs-health-products/medical-devices/application-information/guidance-documents/clinical-evidence-requirements-medical-devices.html.

[52] Government of Canada. Examples of clinical evidence requirements for medical devices. November 2022). Accessed September 25, 2023. https://www.canada.ca/content/dam/hc-sc/documents/services/drugs-health-products/medical-devices/application-information/guidance-documents/clinical-evidence-requirements-medical-devices/examples/examples.pdf.

[53] Government of Canada. Guidance on clinical evidence requirements for medical devices: submitting clinical evidence (November 15, 2022). Accessed September 25, 2023. https://www.canada.ca/en/health-canada/services/drugs-health-products/medical-devices/application-information/guidance-documents/clinical-evidence-requirements-medical-devices/submitting-clinical-evidence.html.

[54] Government of Canada. Guidance on clinical evidence requirements for medical devices: clinical data and evaluation. Published November 15, 2022. Accessed September 25, 2023. https://www.canada.ca/en/health-canada/services/drugs-health-products/medical-devices/application-information/guidance-documents/clinical-evidence-requirements-medical-devices/clinical-data-evaluation.html.

[55] Emergo Group. Russia Medical Device Approval Process.pdf (elsmar.com). Accessed September 25, 2023. https://elsmar.com/Cove_Premium/Russia/20Medical/20Device/20Approval/20Process.pdf.

[56] Beawire. Kimes Presentation (2017). Approvals for medical devices: Russian Federation and EAC Med. Accessed September 25, 2023. https://www.google.com/url?sa = t&rct = j&q = &esrc = s&source = web&cd = &ved = 2ahUKEwiWjLHXlseBAxXArokEHdeZAJwQFnoECA0QAQ&url = https/3A/2F/2Fhttp://www.khidi.or.kr/2FfileDownload/3FtitleId/3D160445/26fileId/3D1/26fileDownType/3DC/26paramMenuId/3DMENU01525&usg = AOvVaw23qB-Dc701XNxz1BYbdFQK&opi = 89978449.

[57] KC-PROF. Registration of medical devices according to the Russian Federation rules. Accessed September 25, 2023. https://kc-prof.ru/en/registration-russia/.

[58] Japanese Law Translation. Act on securing quality, efficacy and safety of products including pharmaceuticals and medical devices. Act No. 145 August 10, 1960. Accessed March 21, 2023. https://www.japaneselawtranslation.go.jp/en/laws/view/3213/en.

[59] BSI. Medical Devices. Japan. No Date. Accessed March 21, 2023. https://www.bsigroup.com/en-US/medical-devices/Global-market-access/Japan-market-access/.

[60] Ministry of Health, Labour and Welfare of Japan (MHLW). Strategy of SAKIGAKE. No date. Accessed March 21, 2023. https://www.mhlw.go.jp/english/policy/health-medical/pharmaceuticals/140729-01.html.

[61] PMDA. Reviews and related services. Accessed March 21, 2023. https://www.pmda.go.jp/english/review-services/glp-gcp-gpsp/0002.html.

[62] New Zealand Legislation. Medicines Act 1981. Version as at July 1, 2022. https://www.legislation.govt.nz/act/public/1981/0118/latest/DLM53790.html?search = ts_act_Medicines_resel&sr = 1.

[63] MedSafe. Explanation of the WAND database. Revised May 10, 2011. https://medsafe.govt.nz/regulatory/DevicesNew/3-2Explanation.asp.

[64] Lamph, S. Regulation of medical devices outside the European Union. *J. R. Soc. Med.*, *105*, S12–S21. doi:10.1258/jrsm.2012.120037. https://www.ncbi.nlm.nih.gov/pmc/articles/PMC3326589/

[65] MedSafe. Frequently asked questions. Revised July 1, 2014. Accessed March 17, 2023. https://www.medsafe.govt.nz/regulatory/DevicesNew/faqs.asp.

[66] MedSafe. Safety monitoring of medical devices. Revised May 30, 2017. https://www.medsafe.govt.nz/regulatory/DevicesNew/safety-monitoring.asp.

[67] Gov.UK. Regulating medical devices in the UK. Published December 31, 2020, last updated January 1, 2022. Accessed March 20, 2023 https://www.gov.uk/guidance/regulating-medical-devices-in-the-uk.

[68] Annex X Clinical Evaluation, 1. General Provisions. https://www.legislation.gov.uk/uksi/2002/618/made/data.pdf.

[69] The Medical Devices Regulations 2002 No. 618. 23 (2). Accessed March 20, 2023 https://www.legislation.gov.uk/uksi/2002/618/made/data.pdf.

[70] Gov.UK. Medical devices: guidance for manufacturers on vigilance. Published January 26, 2015, last updated December 5, 2022. Accessed March 20, 2023 https://www.gov.uk/government/collections/medical-devices-guidance-for-manufacturers-on-vigilance#full-publication-update-history.

[71] Titck. Türkiye İlaç ve Tıbbi Cihaz Kurumu (Turkish Medicines and Medical Devices Agency, TMMDA). New Medical Device Regulations. 2021. Accessed September 26, 2023. https://titck.gov.tr/storage/Archive/2021/contentFile/Ek-2.1/20Duyuru/20metni-eng_2e87c6cd-5d45-43d9-9a35-0a94ffee8547.pdf.

Chapter 12

Forecasting clinical evaluation future directions

Predicting the future is sometimes obvious and other times impossible to predict. In the case of clinical evaluations/performance evaluations (CEs/PEs) for medical devices and *in vitro* diagnostic devices, the immediate future is quite clear. More countries will require CE/PE for their medical devices. On the other hand, the mechanism to deliver the required CE/PE details to meet regulatory requirements and to understand the medical device CE/PE impact on the people living in a given geography will probably become more transparent with increasing communication, even though the various political climates around the world continue to evolve.

Predicting the future can be risky, but a few things are clear: 1) regulations tend to increase in both number and focus, 2) people desire new medical discoveries to improve and extend life and 3) clinical data provide insights into medical conditions not avaialble in any other data set. This clarity suggests the future is bright for more and better clinical evaluation work across all drugs, devices and foods globally.

12.1 Introduction

Clinical evaluation reports/performance evaluation reports (CERs/PERs) evolved rapidly since 2017. In addition, the regulatory environment also evolved exponentially as new information became more accessible, as more regulatory authorities harmonized efforts and as notified bodies (NBs) became certified and conscripted to review more and more clinical data.

12.1.1 Clinical data are accessible

Open access (OA) to information online increased information accessibility. In addition, CER/PER-type data were released to the public as summaries of safety and clinical performance (SSCPs)/summaries of safety and performance (SSPs) and published literature became more available as "free to read", "free to reuse" or "free to read online". For example, one article reported the "movement to provide open access (OA) to all research literature" was "over fifteen years old" in 2018 [1]. They found institutions, including the United States (US) National Institutes of Health (NIH), "mandating OA publishing" since 2008 alongside organizations like the European Commission and the Welcome Trust. The ethics and efficiency of "paywall publishing" and "self-archiving" were driving a worldwide discussion of "scholarly communication" systems. Many studies estimated 50% of papers were freely available online and were more likely to be OA if they were more recent; however, the OA rates varied by publisher and discipline. Another article [2] reported OA rates "increased significantly between 2006 and 2010: the OA rate in 2010 (50.2%) was twice that in 2006 (26.3%)." In addition, the number of articles and new medical journals saw "tremendous growth" [3] and included some publishers not following "current best publishing practices". These "predatory publishers" included examples like the OMICS Publishing Group. This increased volume of accessible information and the rise of predatory publishers accentuated the need for expert evaluators to identify relevant and high quality sources of valid clinical data.

12.1.2 Regulations and regulatory bodies are rapidly expanding

At the same time, regulations were expanding and here are just a few of many examples:

Planning, Writing and Reviewing Medical Device Clinical and Performance Evaluation Reports (CERs/PERs). DOI: https://doi.org/10.1016/B978-0-443-22063-0.00010-9

1. NBs and other agents of regulatory authorities worldwide were reviewing more required clinical data globally. Regulatory review groups were expanding to handle the increased regulatory oversight and certification requirements for the medical device regulation (MDR)/*in vitro* diagnostic regulation (IVDR) in the European Union (EU) as well as the similarly growing clinical data requirements in the US, China, Japan, and most other parts of the world.
2. Expert panels were established and growing in the EU to oversee the highest-risk medical devices in Europe.
3. Regulatory authorities in China and the Association of Southeast Asian Nations (ASEAN) countries were releasing many guidelines about how to document the results of clinical trials from all parts of the world in service to the people of the region. For example, China allowed manufacturers of novel devices from overseas to conduct clinical research in China before securing approval/clearance in the originating country as the requirements for CE/PE work in China evolved.

The need for well-trained clinical scientists could not be more obvious; however, the pathway to training them could not be more challenging simply based on the increased information and the rapidly changing regulatory details to digest and understand.

12.1.3 Guidelines, guidelines, and more guidelines

The EU Medical Devices Coordinating Group (MDCG) released many guidelines and, while guidelines help manufacturers understand the MDR/IVDR implementation requirements, the sheer number of MDR/IVDR guidelines was daunting. More than 120 documents in 5 years from 1998 to 2023 was a lot to read, understand an implement, and the diverse regulation and guideline interpretations by the NBs were perplexing. The CE/PE regulations and guidance documents were complex and often difficult to interpret. Reasonable and experienced persons often executed different actions and analyses as they complied with these documents. Efforts were needed to focus on the most important information while allowing a streamlined approach for lower risk issues and devices.

Perhaps the EU MDR/IVDR regulations and guidelines about Class I devices can be rescinded and completely removed. For example, two existing MDCG documents suggested reduced enforcement activity and the risk to removing these seems negligible in the larger picture of EU public health:

- MDCG 2020-2 rev. 1 (March 2020) Class I transitional provisions under Article 120 (3 and 4) (MDR) [4]; and
- MDCG 2019-15 rev. 1 (December 2019) Class I transitional provisions under Article 120 (3 and 4) (MDR) [5].

In essence, the future might hold less regulatory involvement (e.g., less NB oversight) for Class I medical devices and Class A IVD devices as is the case in many countries around the world. If not rescinded and removed, the benefit for this expensive (and potentially excessive) EU MDR/IVDR oversight does not appear to have an acceptable benefit-risk profile. Specifically, the benefits do not appear to outweigh the risks in terms of cost and wasted time specifically when documenting the clinical evidence for such low-risk devices, because the resulting clinical evidence does not offset any clear and compelling risks and does not provide a clear and compelling benefit. Scientific evidence justifying the specific need for such draconian measures is sorely needed, or the regulations and guidelines should be changed to focus on a more pragmatic risk-based approach with a sound scientific rationale. As currently in practice, the NB-enforced EU MDR/IVDR requirements for additional clinical data gathering for low risk and some medium risk, well established devices is debatable and hard to defend.

12.2 Better, more integrated clinical evaluation systems

Company employees struggled to keep pace with increased regulatory changes, medical device complexities and clinical data access points. Employees focused on high-quality, landmark information and discovered they can track and share clinical data for each specific device type, just as the Australian Therapeutic Goods Administration (TGA) provided requirements for specific device types in their clinical evaluation guidelines available at: https://www.tga.gov.au/resources/resource/guidance/clinical-evidence-guidelines. Access to clinical data regulatory findings was made available to all and the clinical data were updated using a planned update frequency to capture all relevant new clinical data. The need for centralized knowledge repositories was quite clear; however, the methods to manage and integrate these repositories lagged behind, with a clear future need to develop better and more integrated CE/PE systems.

As medical device manufacturers worldwide generated CE/PE documents, a system of data sharing was envisioned (at least in Australia). Much like the adverse event (AE) databases allowing direct access to the individual patient report data, clinical data from EU CERs/PERs or from other CE/PE workflows might be made publicly available in an easily and completely analyzable format (e.g., in Eudamed, when available). Sharing common clinical data to support CE/PE work contributed to scientific debate and openned up discussions about device benefit-risk profiles worldwide. Device safety and performance (S&P) data became more available for inspection and training helped evaluators, authors and others understand how to conduct CEs/PEs for every device.

Integrating medical device clinical data evaluations within medical device manufacturing systems (e.g., risk management, post market surveillance (PMS), and clinical/regulatory affairs) started with a common clinical data definition and a clear understanding about why clinical data were required. Creating and using tools to define clinical events (i.e., clinical trial protocols, coding dictionaries, S&P acceptance criteria, specific risks/benefits profiles, regulations, guidelines and templates from global regulatory authorities, etc.) helped to define a common language for CE/PE work and device companies worldwide started including clinical experts (i.e., nurses, doctors, clinical research scientists, healthcare professionals, etc.) in their medical device teams to better integrate CER/PER systems into other well-established company systems.

12.3 Unifying clinical evaluation regulations for medical devices and *in vitro* diagnostic devices

The regulatory landscape will continue to evolve. Pressure to optimize commonality and avoid duplicative documents will make regulations and guidelines more cohesive and less fragmented. By combining all medical devices within one group (as is already done in many countries), the IVD device details can become more organized into appropriate and specific technical file components focused on specific clinical data to be included in the CER to demonstrate S&P (rather than the technical device details currently in the PER which could be moved to the technical files for review similar to all the other device types reviewed in the CER). For example, merging the EU MDR/IVDR makes sense and would simply require calling out specific needs for IVD details in a few sections of the regulation as opposed to having so many identical sections in two separate regulations. Forcing the available guidelines to be updated to include IVD devices will improve the guidance available to IVD manufacturers. The identical sections would not need to be duplicated as they are today, and this might promote support across medical device and IVD device companies for universally better CER writing. This unification might allow future regulatory developments to callout specific details like software as a medical device (SaMD), artificial intelligence (AI), mechanical learning (ML), Chat GPT use in medicine, and more. Again, having a common regulatory framework to reference might facilitate and improve understanding and minimize redundancies.

12.4 Driving international relations and global change

PMS was one area of global exploration. The MDR/IVDR and CE/PE processes improved human clinical data tracking worldwide. This impact is expected to be long lasting and should contribute to a continued evolution in global PMS monitoring. As more databases, registries, and clinical data collections are created and made publicly and freely available, and as language translators allow access and understanding across language barriers worldwide, this part of the MDR/IVDR promises to impart a lasting and significant improvement in medical device clinical data access and evaluation worldwide.

Clinical research is changing. The era of personalized clinical trials and decentralized clinical trial management are stretching the boundaries of clinical medicine. What was "bedside" in the hospital is becoming "bedside" in the home, no matter where the "home" is located.

Individual investigators are able to help and investigator-initiated trials (IITs) expanded as physicians were motivated to help improve the CE/PE available for medical and IVD devices. The burgeoning need for more PMS data caused medical device manufacturers to partner with investigators to capture and analyze missing clinical data and to address all identified clinical data gaps for the moderate and high-risk clinical devices.

Many countries emulated the EU MDR/IVDR, the earlier Therapeutic Goods Administration (TGA) CER guidelines, and the even earlier GHTF (Global Harmonization Task Force) recommendations about CE/PE. Clinical evidence expanded even though many manufacturers obsoleted or removed many products from the EU market because the devices did not generate enough funding to cover the CE/PE costs. With more than 500,000 medical devices on the EU market [6], the EU MDR/IVDR was challenging and some medtech companies have left the EU market entirely due to the compliance costs. In response to negative feedback from manufacturers and the public, and because NBs could not handle the volume of work to get certified and to review these many, many certificate applications, the European Commission delayed the MDR/IVDR transition multiple times. Additional changes are anticipated to make the MDR/IVDR more manageable.

A survey conducted by MedTech Europe analyzed the availability of medical devices in 2022 in response to the MDR [7]. This survey reported EU MDR certificates had not yet been issued for " >85% of the >500,000 devices previously certified under MDD/AIMDD" and the time to certification was "13–18 months on average... double the time historically needed for certification under the Directives." Portfolio reductions were planned for "50% of respondents" and "33% of these companies' medical devices are currently planned for discontinuation... in all product categories." Access to an MDR-designated NB was not possible for 15%–30% of small and medium companies, and "50% of respondents are deprioritizing the EU market," with 20% of respondents identifying "delays in MDR certification to the publication of new or revised MDCG guidance."

12.4.1 Revisiting 2011 European regulatory projections

In an effort to project to the future, sometimes looking at past projections is helpful. In 2011, the European Society of Cardiology Policy Conference Report discussed changes to the Device Directives in the planning stages [8] and they had the following recommendations:

1. *There should be a single, coordinated European system to oversee the evaluation and approval of medical devices*
2. *The NBs should be reorganized as an integrated structure*
3. *The classification of each type of device should be based on a detailed evaluation of risks*
4. *Product standards should be developed for each category of medical device in class II and class III*
5. *Expert professional advice is required*
6. *Adequate transparency is essential*
7. *The concept of conditional approval of a medical device, pending further clinical evaluation, should be developed*
8. *Outcome studies after device implantation should be undertaken as a partnership between physicians, companies, and regulators*
9. *Limits to iterative changes should be defined*
10. *Regulatory systems should retain flexibility for special circumstances*
11. *Manufacturers should be responsible for the clinical evaluation of all class II and class III devices*
12. *Physicians should understand and engage with the regulatory systems for medical devices.*

Note how class I devices were not needing additional clinical data in these recommendations, how the current EU system was clearly aligned to oversee clinical evaluations for medical devices and how NBs have been allowed to issue conformity certificates under MDR/IVDR only if they were reorganized and certified as competent to do so. In practice, the detailed evaluation of risks seemed to be overly focused on the "detail" part and not so much on the "risk" part. Often enforcement discretion appeared to be misunderstood by NB staff. The NB enforcement seemed to have missed the focus on Class II and III devices, as NBs sometimes dive into Class I devices (especially for Class I devices marketed as sterile or with a measuring function) without appropriate risk prioritization of their work. Hiring more NB staff with more clinical experience and better risk management training is an absolute requirement for the future.

Product standards for Class II and III devices were not created, and this might be a good target for future planning. Expert panels started to function, but the conduct of and release of information from those expert panel meetings was exceedingly slow, inefficient and certainly not timely. Insufficient conditional approvals were issued given appropriate PMCF planning and commitments for further CE/PE. Partnership between experts, physicians, companies, and regulators to develop appropriate outcomes studies and device-specific standards should be a welcome future improvement.

NBs and manufacturers worked through the complexities of iterative device and regulatory changes over time while being responsible for device clinical evaluations. In the future, perhaps the CE/PE process will not include Class I devices or Class A IVD devices, as was suggested by this team back in 2011. Additional engagement by physicians and other healthcare professionals (HCP) will be a welcome change in the future at all levels and within all medical device CE/PE, risk management, and PMS systems. Some anticipated benefits from greater HCP involvement will provide a clearer focus on doctor-patient clinical needs with less focus on illogical and clinically trivial details like requiring more clinical data to be generated or implementing engineering "traceability" matrixes for low risk device uses without a clear medical need. Different approaches and better agreement between all parties should be a goal to arrive at more suitable and more functional requirements.

12.4.2 Analyzing medical device safety and performance metadata

Currently, the MDR/IVDR required each SSCP/SSP to be made public via EUDAMED even though this system was not able to hold these data. This requirement was developed solely to fulfill the EU regulatory requirements; however, future generations will likely benefit from a different global clinical data and report sharing vision. Multiple, international centralized, public databases housing global clinical data including the EU SSCPs/SSPs could be linked together. A global reach for clinical data sharing was actually essential for SSCPs/SSPs to serve their initial intended purpose to summarize all evaluated clinical S&P data for high-risk devices. A centralized, global clinical database and SSCP/SSP repository, if adequately granular and searchable for discrete data analysis, would create both a new resource and a new role for SSCPs/SSPs and CERs/PERs to serve the public by providing high-quality medical device S&P and benefit-risk data access for all physicians, patients and caregivers. In this forward-thinking view, all related CE/PE documents (regardless of country of origin) are unique and can be released into one or many public databases. The CE/PE documents would be a helpful resource to make publicly available since no other single document type, regulatory or otherwise, summarized and analyzed all available clinical data relevant to high-risk device S&P and benefits-risks in quite the same way and companies don't share these documents to arrive at the most accurate interpretation possible.

In addition, physicians who regularly use high-risk devices are challenged to keep abreast of all the relevant clinical literature individually. Reading multiple, individual clinical trial reports is taxing, time consuming and may be impossible when the clinical literature volume is high. If the physician must focus solely on review articles or guidelines, even the highest-quality guidelines, metaanalyses and systematic reviews can only evaluate the previously published literature and clinical trial data. CERs/PERs add in all available clinical experience data pieces along with clinical trial details which are not published in the clinical literature.

Currently, only the SSCP/SSP is to be made publicly available, and a well-crafted SSCP/SSP can provide readers with a succinct and comprehensive clinical data summary not available elsewhere. However, the future development of specific CER/PER and/or post market clinical follow up evaluation report/post market performance follow up evaluation report (PMCFER/PMPFER) database/s could allow patients and their HCPs to compare different high-risk devices and potentially make better healthcare decisions, or at least more informed decisions based on more clinical data compiled with a specific intent to understand safety, performance, benefits and risks to the patient.

As the EU and global medical care guidelines evolve, a data sharing movement is likely to take center stage. As clinical data are analyzed and new conclusions are drawn about medical device S&P, expert panels and regulatory authorities should come together to not only oversee the CE/PE work and the approval of medical devices, but also to improve the quality of healthcare information and public understanding worldwide when medical devices are used or as patients consider medical device use for their personal healthcare.

Basic questions should be addressed in the future. For example, how can the global population navigate into a better healthcare future faster, easier and with less cost? Cost simply must enter into the regulatory discussion. Can various regulatory authorities around the world collaborate to provide a "single source" of high-quality medical device S&P clinical data? Can examples like those in the TGA guidelines be developed for more device groups including low-risk devices like tongue depressors, gloves, masks, gowns, and sutures? Can medical device exemption lists such as those in China be created globally to exempt certain low-risk devices from costly CE/PE work globally?

More transparency is also needed in the future related to obsoleted device labeling. For example, what happens to patients with implants when the device is removed from the market? How will patients obtain the latest clinical data and relevant health information when the manufacturer has moved on to other projects? What should we call this obsolete data collection requirement? What should be required of device companies when devices are obsoleted? What if the company no longer exists? Where should the patient go for clinical information about their healthcare related to this obsoleted device? The need for the oCER/oPER and clinical data transparency have increased simultaneously with the MDR/IVDR changes and with additional regulatory changes worldwide. Sharing data and expanded labeling should lead to clearer communication across the entire device lifetime, from first to last human use.

12.4.3 Embracing clinical research changes

Clinical research is evolving to include virtual, at home, remote access, telemedicine, and other clinical trial interactions. Ensuring proper medical care, appropriate communication, and data integrity takes time and diligence.

Clinical trial data are becoming more readily and freely available within systems like AE databases and clinical trial registries; however, the data analyses and reports developed from these data collection tools required experienced review teams. The problems of missing data combined with the high likelihood for inaccurate "clinical" data representation are difficult problems to address, especially in a remote setting when data are collected by persons with no clinical or medical training. This data accuracy challenge is multiplied when typical clinical quality controls such as data monitoring and cleaning activities are simply not possible.

Rapidly growing databases such as http://www.clinicaltrials.gov required the release of clinical research reports resulting from the analysis of the clinical data collected in the listed clinical trials; however, the http://www.clinicaltrials.gov database does not currently have a mechanism for peer-review of the finished report or the data posted into the database. This is a critical quality issue and so is enforcement, since only about half of the studies with results have any results posted in the database.

Clinical trial data recording "structures" should be harmonized globally. Although multiple approaches to harmonization are ongoing, the path to alignment may require a few more decades in the making. Whether AI, ML, or some other information technology (IT) innovation such as Chat GPT might be modified or a new system created to address the specific needs of interpreting CE for specific devices is unclear, but additional computerized advances seem highly likely.

12.5 Conclusions

Future CERs/PERs should evolve to be more useful and should take advantage of ever expanding and more centralized public databases. Regulatory staff should become more proficient in evaluating the quality of CE/PE work. People writing guidelines should become more pragmatic and helpful in guiding future generations to focus on what is most important to keep our medical devices safe and effective.

As global CE/PE conversations continue, the scientific and medical communities should align on the basic ethical message: good medical care is a requirement for a good quality of life. Medical devices are a critical component of good medical care and freely and openly sharing deidentified patient experiences and clinical research data worldwide will deepen our understanding of medical device benefits and risks for all patients.

Patients, physicians, caregivers and medical device manufacturers should encourage global regulatory alignments and convince the authorities to embrace transparency and to facilitate the rapid expansion of information-sharing systems. This sharing should include the open, public, and freely available sharing of CEs/PEs documents worldwide. Just as the PubMed literature database housed more freely available literature, new clinical evidence databases must be built to share more details about individual patient experiences.

In particular, EUDAMED should not only collect safety information including AEs and device vigilance data for regulatory purposes, EUDAMED should share the centralized reports and evaluations publicly. In other words, the EU Parliament should be encouraged to make these data available to the public, similar to the services already provided by the US MAUDE (http://www.accessdata.fda.gov/scripts/cdrh/cfdocs/cfMAUDE/TextSearch.cfm), TPLC (http://www.accessdata.fda.gov/scripts/cdrh/cfdocs/cfTPLC/tplc.cfm), and the Australia DAEN (http://apps.tga.gov.au/prod/DEVICES/daen-entry.aspx) databases. This textbook envisions a future world where clinical data is highly personal and detailed with exquisite clinical quality control, but ethically and intelligently compared to other clinical data around the world. This global time and place is coming soon, not far from here.

12.6 Review questions without answers

1. What is the extent of S&P testing for medical devices?
2. What might a universal agreement about medical device clinical evaluation look like?
3. Where can clinical scientists focus energy to ensure high-quality, medical device clinical evaluations become the norm worldwide?
4. Who should be responsible for medical device clinical data evaluation overall?
5. When should clinicians unite to ensure the best possible medical devices are available worldwide for every patient who might benefit from those medical devices?
6. Why should the regulations for medical devices and *in vitro* devices be merged?
7. How can clinical data be shared and analyzed globally?
8. What resources are needed to expand our knowledge faster and easier?
9. How can the explosive growth of regulations and guidelines be better managed and organized?
10. What are some basic tools to use when a NB assesses a CER/PER?
11. How is a "preditory publisher" identified in a literature search?

12. What process might work to require notified bodies to use the same enforcement actions and clinical evaluation assessment report/performance evaluation assessment report (CEAR/PEAR) findings for specific situations?
13. When should class I medical devices or class A IVD devices be exempted from CER/PER as well as PMCF/PMPF requirements?
14. Where should all regulations and regulatory guidelines be kept so they are better orgnanized and understood?
15. Why should the Australian device-specific clinical data guidelines be extended globally?
16. How can better and more integrated CE/PE systems be developed?
17. What barriers need to be removed before international alignment can be reached for CE/PE processes?
18. How will N of 1 clinical data be analyzed in the future?
19. What happens when physicians and other medical professionals understand and engage with regulatory authorities to improve medical devices globally?
20. How can clinical research evolve to become both less costly, yet more timely and more relevant?

References

[1] H. Piwowar, J. Priem, V. Lariviere, J.P. Alperin, L. Matthias, B. Norlander, et al. The state of OA: a large-scale analysis of the prevalence and impact of Open Access articles, Peer J. 6 (2018) e4375. Available from: https://doi.org/10.7717/peerj.4375. Published online February 13, 2018. Accessed June 20, 2024, https://www.ncbi.nlm.nih.gov/pmc/articles/PMC5815332/.

[2] K. Kurata, T. Morioka, K. Yokoi, M. Matsubayashi. Remarkable growth of open access in the biomedical field: analysis of PubMed articles from 2006 to 2010, PLoS One 8 (5) (2013) e60925. Available from: https://doi.org/10.1371/journal.pone.0060925. Published online May 1, 2013. Accessed June 20, 2024. https://www.ncbi.nlm.nih.gov/pmc/articles/PMC3641021/.

[3] L. Topper, D. Boehr. Publishing trends of journals with manuscripts in PubMed Central: changes from 2008–2009 to 2015–2016, J. Med. Libr. Assoc. 106 (4) (2018) 445–454. Available from: https://doi.org/10.5195/jmla.2018.457Published online October 1, 2018. PMCID: PMC6148616PMID: 30271285. Accessed June 20, 2024, https://www.ncbi.nlm.nih.gov/pmc/articles/PMC6148616/.

[4] MDCG 2020-2 rev.1. Class I transitional provisions under Article 120 (3 and 4)—(MDR). March 2020. Accessed June 20, 2024. https://health.ec.europa.eu/system/files/2020-07/md_transitional-provisions-art-3-and-4_en_0.pdf.

[5] MDCG 2019-15 rev. 1. Class I transitional provisions under Article 120 (3 and 4)—(MDR). December 2019. Accessed June 20, 2024. https://health.ec.europa.eu/system/files/2020-07/md_guidance-manufacturers_en_0.pdf.

[6] I. Rastegayeva. Understanding the impact of the EU Medical Device Regulation (MDR) and its latest evolution. 3DS Blog Dassault Systemes. Accessed June 20, 2024. https://blog.3ds.com/industries/life-sciences-healthcare/impact-of-new-eu-mdr/.

[7] MedTech Europe. MedTech Europe Survey report analyzing he availability of Medical Devices in 2022 in connection to the Medical Device Regulation Implementation. July 14, 2022. Accessed June, 20, 2024. https://www.medtecheurope.org/wp-content/uploads/2022/07/medtech-europe-survey-report-analysing-the-availability-of-medical-devices-in-2022-in-connection-to-the-medical-device-regulation-mdr-implementation.pdf.

[8] A.G. Fraser, J.-C. Daubert, F. Van de Werf, N.A.M. Estes, S.C. Smith, M.W. Krucoff, et al. Clinical evaluation of cardiovascular devices: principles, problems, and proposals for European regulatory reform. Report of a policy conference of the European Society of Cardiologyon behalf of the participants. Eur. Heart J. 32 (2011) 1673–1686. Available from: https://doi.org/10.1093/eurheartj/ehr171Accessed June 20, 2024., https://academic.oup.com/eurheartj/article/32/13/1673/507544.

Appendix A

Example regulatory evolution timeline

1820	US Pharmacopeia
1862	US Department of Agriculture Bureau of Chemistry
1901	First EU National Standards Body (British Standards Institution, or BSI)
1906	US Pure Food and Drugs Act
1938	US Federal Food, Drug, and Cosmetic Act
1950s	EU formed
1965	European Commission (EC) first "standardized" drug approval regulation (65/65/EEC)
1976	US FDA Medical Device Amendments
1984	US FDA Medical Device Reporting (MDR*) regulations
1990	EU Active Implantable Medical Device (AIMD) Directive (90/385/EEC) US FDA Safe Medical Devices Act (SMDA)
1992	Global Harmonization Task Force on Medical Devices (GHTF) (http://www.imdrf.org/ghtf/ghtf-archives.asp) created to unify medical device regulatory systems, to improve patient safety, and to provide "access to safe, effective and clinically beneficial medical technologies around the world." The GHTF had five founding members: the EU, the US, Canada, Australia, and Japan. The International Medical Device Regulators Forum (IMDRF) continues the work of the GHTF.
1993	EU single market completion EU Medical Device Directive (MDD) (93/42/EEC)
1995	European Medicines Agency (EMA) formed
1996	US FDA new Medical Device Reporting (MDR) regulation
1998	EU *In Vitro* Diagnostic (IVD) Medical Device Directive (98/79/EC)
2000	Notified Body Operations Group (NBOG) (https://www.nbog.eu) set up by Member States and European Commission to address poor NB performance and Designating Competent Authority NB oversight when reviewing medical device S&P Earliest medical device regulations in China (State Council Order No. 276)
2001	Clinical Trials Directive (2001/20/EC; repealed and replaced by 526/2014)
2007	MDD/AIMDD/IVDD amendments (2007/47/EC); clinical evaluation for all devices
2011	IMDRF (http://www.imdrf.org) created by representatives from Australia, Brazil, Canada, China, Europe, Japan, the US, and the World Health Organization (WHO) as a voluntary, global forum to discuss medical device regulatory harmonization. Current members also include Russia, Singapore, and South Korea.
2014	Clinical Trials Regulation (EU Reg. 536/2014) (https://ec.europa.eu/health/human-use/clinical-trials/regulation_en) Medical device regulations updated in China (State Council Order No. 650)
2016	MEDDEV 2.7/1, Rev. 4
2017	AIMD/MDD merged into the MDR (EU Reg. 2017/745)
2017	IVD Directive updated (EU Reg. 2017/746) First MDCG meeting

2018	Medical device regulations updated again in China (State Council Order No. 680)
2020	Consolidated MDR
2021	Medical device regulations updated again in China (State Council Order No. 739)
2022	Consolidated IVDR EU 2022/20 laid down the rules to apply EU REG 536/2014 regarding the rules and procedures for Member State cooperation in safety assessments of clinical trials
2023	Template for list of standard fees (by notified bodies)

**Note the confusion when MDR is used to represent Medical Device Reporting in the US under 21CFR803 and Medical Device Regulation (EU Reg. 2017/745) in the EU.*

Further reading

https://www.fda.gov/about-fda/fda-history/milestones-us-food-and-drug-law
https://eumdr.com
https://eur-lex.europa.eu/legal-content/EN/TXT/?uri = uriserv:OJ.L_0.2022.005.01.0014.01.ENG

Appendix B

MEDDEV guidance list

Reproduced from Medical Device Regulations. Accessed August 27, 2024. https://health.ec.europa.eu/system/files/2022-01/md_guidance_meddevs_0.pdf

Acronym	Title	Link
2.1 Scope, field of application, definition		
MEDDEV 2.1/1	Definitions of "medical devices," "accessory," and "manufacturer" (April 1994)	https://ec.europa.eu/docsroom/documents/10278/attachments/1/translations
MEDDEV 2.1/2 rev. 2	Field of application of directive "active implantable medical devices" (April 1994)	https://ec.europa.eu/docsroom/documents/10279/attachments/1/translations
MEDDEV 2.1/2.1	Treatment of computers used to program implantable pulse generators (February 1998)	https://ec.europa.eu/docsroom/documents/10280/attachments/1/translations
MEDDEV 2.1/3 rev. 3	Borderline products, drug-delivery products and medical devices incorporating, as integral part, an ancillary medicinal substance or an ancillary human blood derivative (December 2009)	https://ec.europa.eu/docsroom/documents/10328/attachments/1/translations
MEDDEV 2.1/4	Interface with other directives—medical devices Directive 89/336/EEC relating to electromagnetic compatibility and Directive 89/686/EEC relating to personal protective equipment (March 1994). *For the relation between the MDD and Directive 89/686/EEC concerning personal protective equipment, please see the Commission services interpretative document of August 21, 2009.*	https://ec.europa.eu/docsroom/documents/10281/attachments/1/translations and https://ec.europa.eu/docsroom/documents/10262/attachments/1/translations
MEDDEV 2.1/5	Medical devices with a measuring function (June 1998)	https://ec.europa.eu/docsroom/documents/10283/attachments/1/translations
MEDDEV 2.1/6	Qualification and classification of stand-alone software (July 2016)	https://ec.europa.eu/docsroom/documents/17921/attachments/1/translations
2.2 Essential requirements		
MEDDEV 2.2/1 rev. 1	EMC requirements (February 1998)	https://ec.europa.eu/docsroom/documents/10285/attachments/1/translations
MEDDEV 2.2/3 rev. 3	"Use by" date (June 1998)	https://ec.europa.eu/docsroom/documents/10288/attachments/1/translations
MEDDEV 2.2/4	Conformity assessment of *in vitro* fertilisation (IVF) and assisted reproduction technologies (ART) products (January 2012)	https://ec.europa.eu/docsroom/documents/10340/attachments/1/translations

2.4 Classification of MD		
MEDDEV 2.4/1 rev. 9	Classification of medical devices (June 2010)	https://ec.europa.eu/docsroom/documents/10337/attachments/1/translations
2.5 Conformity assessment procedure—general rules		
NA (see document in the Global Harmonization Task Force, or GHTF)	Quality assurance. Regulatory auditing of quality systems of medical device manufacturers GHTF-Part 4: Multiple Site Auditing GHTF-Part 5: Audits of Manufacturer Control of Supplies	https://www.medical-device-regulation.eu/wp-content/uploads/2019/05/ghtf-sg4-n83-2010-guidelines-for-auditing-qms-part-4-multiple-sites-100827.pdf and https://www.medical-device-regulation.eu/wp-content/uploads/2019/05/ghtf-sg4-n84-2010-guidelines-for-auditing-qms-part-5-control-of-suppliers-100827.pdf
MEDDEV 2.5/3 rev. 2	Subcontracting quality systems related (June 1998)	https://ec.europa.eu/docsroom/documents/10289/attachments/1/translations
MEDDEV 2.5/5 rev. 3	Translation procedure (February 1998)	https://ec.europa.eu/docsroom/documents/10286/attachments/1/translations
MEDDEV 2.5/6 rev. 1	Homogenous batches (verification of manufacturers' products) (February 1998)	https://ec.europa.eu/docsroom/documents/10287/attachments/1/translations
2.5 Conformity assessment procedure—particular groups		
MEDDEV 2.5/7 rev. 1	Conformity assessment of breast implants (July 1998)	https://ec.europa.eu/docsroom/documents/10290/attachments/1/translations
MEDDEV 2.5/9 rev. 1	Evaluation of medical devices incorporating products containing natural rubber latex (February 2004)	https://ec.europa.eu/docsroom/documents/10321/attachments/1/translations
MEDDEV 2.5/10	Guideline for authorised representatives (January 2012)	https://ec.europa.eu/docsroom/documents/10339/attachments/1/translations
2.7 Clinical investigation, clinical evaluation		
MEDDEV 2.7/1 rev. 4	Clinical evaluation: guide for manufacturers and notified bodies (June 2016) Appendix 1: Clinical evaluation on coronary stents (December 2008)	https://ec.europa.eu/docsroom/documents/17522/attachments/1/translations and https://ec.europa.eu/docsroom/documents/10324/attachments/2/translations
MEDDEV 2.7/2 rev. 2	Guidelines for competent authorities for making a validation/assessment of a clinical investigation application under Directives 90/385/EEC and 93/42/EC (September 2015)	https://ec.europa.eu/docsroom/documents/13053/attachments/1/translations
MEDDEV 2.7/3 rev. 3	Clinical investigations: serious adverse reporting under Directives 90/385/EEC and 93/42/EC SAE reporting form (May 2015) *The new SAE reporting form was taken in use by September 1, 2016.*	https://ec.europa.eu/docsroom/documents/16477/attachments/1/translations and https://ec.europa.eu/docsroom/documents/16477/attachments/2/translations
MEDDEV 2.7/4	Guidelines on clinical investigations: a guide for manufacturers and notified bodies (December 2010)	https://ec.europa.eu/docsroom/documents/10336/attachments/1/translations

2.10 Notified bodies		
MEDDEV 2.10/2 rev. 1	Designation and monitoring of notified bodies within the framework of EC directives on medical devices (April 2001) Annex 1 Annex 2 Annex 3 Annex 4	https://ec.europa.eu/docsroom/documents/10291/attachments/1/translations and https://ec.europa.eu/docsroom/documents/10291/attachments/2/translations and https://ec.europa.eu/docsroom/documents/10291/attachments/3/translations and https://ec.europa.eu/docsroom/documents/10291/attachments/4/translations and https://ec.europa.eu/docsroom/documents/10291/attachments/5/translations
2.12 Postmarket surveillance		
MEDDEV 2.12/1 rev. 8	Guidelines on a medical devices vigilance system (January 2013) Additional guidance on MEDDEV 2.12/1 rev. 8 (July 2019) I. MEDDEV 2.12/1 rev. 8 *Latest version form* *MEDDEV 2.12 rev. 7 FSCA is still valid (multiple MIR forms available)*	https://ec.europa.eu/docsroom/documents/32305/attachments/1/translations and https://ec.europa.eu/docsroom/documents/36292 and https://ec.europa.eu/docsroom/documents/41681
MEDDEV 2.12/2 rev. 2	Postmarket clinical follow-up studies (January 2012)	https://ec.europa.eu/docsroom/documents/10334/attachments/1/translations
2.13 Transitional period		
MEDDEV 2.13 rev. 1	Commission communication on the application of transitional provision of Directive 93/42/EEC relating to medical devices (OJ 98/C 242/05) (August 1998)	https://eur-lex.europa.eu/legal-content/EN/TXT/?uri = CELEX:31998Y0801(01)
	As regards the transitional regime of Directive 2007/47/EC see the interpretative document of the Commission's services of (June 5, 2009)	https://ec.europa.eu/docsroom/documents/10264/attachments/1/translations
2.14 IVD		
MEDDEV 2.14/1 rev. 2	Borderline and classification issues. A guide for manufacturers and notified bodies (January 2012)	https://ec.europa.eu/docsroom/documents/10322/attachments/1/translations
MEDDEV 2.14/2 rev. 1	Research use only products (February 2004)	https://ec.europa.eu/docsroom/documents/10292/attachments/1/translations
MEDDEV 2.14/3 rev. 1	Supply of instructions for use (IFU) and other information for *in-vitro* diagnostic (IVD) medical devices (January 2007)	https://ec.europa.eu/docsroom/documents/10293/attachments/1/translations
	Form for the registration of manufacturers and devices *in vitro* diagnostic medical device directive, article 10 (January 2007)	https://ec.europa.eu/docsroom/documents/10294/attachments/1/translations
MEDDEV 2.14/4	CE marking of blood based *in vitro* diagnostic medical devices for vCJD based on detection of abnormal PrP (January 2012)	https://ec.europa.eu/docsroom/documents/10338/attachments/1/translations
2.15 Other guidance		
MEDDEV 2.15 rev. 3	Committees/working groups contributing to the implementation of the medical device directives (December 2008)	https://ec.europa.eu/docsroom/documents/10295/attachments/1/translations

Appendix C

Medical Device Coordinating Group endorsed guidance documents

Reference	Title	Date
Annex XVI products (i.e., without medical purpose)		
MDCG 2023-6	Guidance on demonstration of equivalence for Annex XVI products - A guide for manufacturers and notified bodies	Dec 2023
MDCG 2023-5	Guidance on qualification and classification of Annex XVI products - A guide for manufacturers and notified bodies	Dec 2023
Q&A	Q&A on transitional provisions for products without an intended medical purpose covered by annex XVI of the MDR	Sep-23
Borderline and classification		
Manual on Borderline	Manual on borderline and classification under Regulations (EU) 2017/745 and 2017/746 v3	Sep-23
	Background note on the use of the Manual on borderline and classification for medical devices under the Directives.	Dec-22
MDCG 2022-5	Guidance on borderline between medical devices and medicinal products under Regulation (EU) 2017/745 on medical devices	Apr-22
MDCG 2021-24	Guidance on classification of medical devices	Oct-21
Helsinki Procedure	Helsinki Procedure for borderline and classification under MDR & IVDR	Sep-21
Class I devices		
MDCG 2020-2 rev. 1	Class I transitional provisions under Article 120 (3 and 4) (MDR)	Mar-20
MDCG 2019-15 rev. 1	Guidance notes for manufacturers of class I medical devices	Dec-19
Clinical investigation and evaluation		
MDCG 2024-3	Guidance on content of the Clinical Investigation Plan for clinical investigations of medical devices	Mar-2024
MDCG 2024-3 Appendix A	Clinical Investigation Plan Synopsis Template	Mar-2024
MDCG 2023-7	Guidance on exemptions from the requirement to perform clinical investigations pursuant to Article 61(4)-(6) MDR and on sufficient levels of access' to data needed to justify claims of equivalence	Dec-2023
2023/C 163/06	Commission Guidance on the content and structure of the summary of the clinical investigation report	May-23
MDCG 2021-28	Substantial modification of clinical investigation under Medical Device Regulation	Dec-21
MDCG 2021-20	Instructions for generating CIV-ID for MDR Clinical Investigations	Jul-21

MDCG 2021-8	Clinical investigation application/notification documents	May-21
MDCG 2021-6 - Rev.1	Regulation (EU) 2017/745—Questions & Answers regarding clinical investigation	Dec-2023
MDCG 2020-13	Clinical evaluation assessment report template	Jul-20
MDCG 2020-10/1 rev. 1	Guidance on safety reporting in clinical investigations	Oct-22
MDCG 2020-10/2 rev. 1	Appendix: Clinical investigation summary safety report form	Oct-22
MDCG 2020-8	Guidance on PMCF evaluation report template	Apr-20
MDCG 2020-7	Guidance on PMCF plan template	Apr-20
MDCG 2020-6	Guidance on sufficient clinical evidence for legacy devices. See also "Background note on the relationship between MDCG 2020-6 and MEDDEV 2.7/1 rev. 4 on clinical evaluation" at https://health.ec.europa.eu/system/files/2022-09/md_borderline_bckgr-note-manual-bc-dir_en_1.pdf	Apr-20
MDCG 2020-5	Guidance on clinical evaluation—on Equivalence	Apr-20
MDCG 2019-9 rev. 1	Summary of safety and clinical performance	Mar-22
COVID-19		
MDCG 2021-21 rev. 1	Guidance on performance evaluation of SARS-CoV-2 in vitro diagnostic medical devices	Feb-22
MDCG 2022-1	Notice to 3rd country manufacturers of SARS-CoV-2 in vitro diagnostic medical devices	Jan-22
MDCG 2021-7	Notice to manufacturers and authorised representatives on the impact of genetic variants on SARS-COV-2 in vitro diagnostic medical devices	May-21
MDCG 2021-2	Guidance on state-of-the-art of COVID-19 rapid antibody tests	Mar-21
NA	COVID-19 TESTS: Q&A on in vitro diagnostic medical device conformity assessment and performance in the context of COVID-19	Feb-21
NA	Conformity assessment procedures for protective equipment	Jul-20
NA	How to verify that medical devices and personal protective equipment can be lawfully placed on the EU market and thus purchased and used—also in the COVID-19 context	May-20
NA	Guidance on regulatory requirements for medical face masks	Jun-20
NA	Guidance on medical devices, active implantable medical devices and in vitro diagnostic medical devices in the COVID-19 context	Apr-20
NA	Conformity assessment procedures for 3D printing and 3D printed products to be used in a medical context for COVID-19	Apr-20
MDCG 2020-9	Regulatory requirements for ventilators and related accessories	Apr-20
Custom-Made Devices		
MDCG 2021-3	Questions and Answers on Custom-Made Devices	Mar-21
EUDAMED		
MDCG 2022-12	Guidance on harmonised administrative practices and alternative technical solutions until Eudamed is fully functional (for Regulation (EU) 2017/746 on in vitro diagnostic medical devices)	Jul-22
MDCG 2021-13 Rev. 1	Questions and answers on obligations and related rules for the registration in EUDAMED of actors other than manufacturers, authorised representatives and importers subject to the obligations of Article 31 MDR and Article 28 IVDR	Jul-21
MDCG 2021-1 Rev. 1	Guidance on harmonised administrative practices and alternative technical solutions until EUDAMED is fully functional	May-21

MDCG 2020-15	MDCG Position Paper on the use of the EUDAMED actor registration module and of the Single Registration Number (SRN) in the Member States	Aug-20
MDCG 2019-5	Registration of legacy devices in EUDAMED	Apr-19
MDCG 2019-4	Timelines for registration of device data elements in EUDAMED	Apr-19
European Medical Device Nomenclature (EMDN)		
MDCG 2024-2	Procedures for the updates of the EMDN	Feb-2024
MDCG 2021-12	FAQ on the European Medical Device Nomenclature (EMDN)	Jun-21
NA	The EMDN—The nomenclature of use in EUDAMED	Jan-20
NA	The CND nomenclature—Background and general principles	Jan-20
MDCG 2018-2	Future EU medical device nomenclature—Description of requirements	Mar-18
Implant Cards		
MDCG 2021-11	Guidance on Implant Card—Device types	May-21
MDCG 2019-8 v2	Guidance document implant card on the application of Article 18 Regulation (EU) 2017/745 on medical devices	Mar-20
In-House Devices		
MDCG 2023-1	Guidance on the health institution exemption under Article 5(5) of Regulation (EU) 2017/745 and Regulation (EU) 2017/746	Jan-23
Authorized Representatives, Importers, Distributors		
MDCG 2022-16	Guidance on Authorised Representatives Regulation (EU) 2017/745 and Regulation (EU) 2017/746	Oct-22
MDCG 2021-27 - Rev.1	Questions and Answers on Articles 13 & 14 of Regulation (EU) 2017/745 and Regulation (EU) 2017/746	Dec-23
MDCG 2021-26	Q&A on repackaging & relabelling activities under Article 16 of Regulation (EU) 2017/745 and Regulation (EU) 2017/746	Oct-21
In Vitro Diagnostic (IVD) Medical Devices		
MDCG 2022-20	Substantial modification of performance study under Regulation (EU) 2017/746	Dec-22
MDCG 2022-19	Performance study application/notification documents under Regulation (EU) 2017/746	Dec-22
MDCG 2022-15	Guidance on appropriate surveillance regarding the transitional provisions under Article 110 of the IVDR with regard to devices covered by certificates according to the IVDD	Sep-22
MDCG 2021-22 rev. 1	Clarification on "first certification for that type of device" and corresponding procedures to be followed by notified bodies, in context of the consultation of the expert panel referred to in Article 48(6) of Regulation (EU) 2017/746	Sep-22
MDCG 2022-10	Q&A on the interface between Regulation (EU) 536/2014 on clinical trials for medicinal products for human use (CTR) and Regulation (EU) 2017/746 on in vitro diagnostic medical devices (IVDR)	May-22
MDCG 2022-9	Summary of safety and performance template	May-22
MDCG 2022-8	Regulation (EU) 2017/746—application of IVDR requirements to "legacy devices" and to devices placed on the market prior to 26 May 2022 in accordance with Directive 98/79/EC	May-22
MDCG 2022-6	Guidance on significant changes regarding the transitional provision under Article 110(3) of the IVDR	May-22
MDCG 2022-3	Verification of manufactured class D IVDs by notified bodies	Feb-22

MDCG 2022-2	Guidance on general principles of clinical evidence for In Vitro Diagnostic medical devices (IVDs)	Jan-22
MDCG 2021-4	Application of transitional provisions for certification of class D in vitro diagnostic medical devices according to Regulation (EU) 2017/746	Apr-21
MDCG 2020-16 rev. 2	Guidance on Classification Rules for in vitro Diagnostic Medical Devices under Regulation (EU) 2017/746	Feb-23
New Technologies		
MDCG 2023-4	Medical Device Software (MDSW) – Hardware combinations Guidance on MDSW intended to work in combination with hardware or hardware components	Oct-2023
Infographic	Is your software a Medical Device?	Mar-21
MDCG 2020-1	Guidance on clinical evaluation (MDR)/Performance evaluation (IVDR) of medical device software	Mar-20
MDCG 2019-16 rev. 1	Guidance on cybersecurity for medical devices	Dec-19
MDCG 2019-11	Qualification and classification of software—Regulation (EU) 2017/745 and Regulation (EU) 2017/746	Oct-19
Notified Bodies		
MDCG 2020-3 rev. 1	Guidance on significant changes regarding the transitional provision under Article 120 of the MDR with regard to devices covered by certificates according to MDD or AIMDD	Sep-23
MDCG 2023-2 MDCG 2023-2 MDR form MDCG 2023-2 IVDR form	List of Standard Fees	Jan-23
MDCG 2022-4 rev. 1	Guidance on appropriate surveillance regarding the transitional provisions under Article 120 of the MDR with regard to devices covered by certificates according to the MDD or the AIMDD	Dec-22
MDCG 2022-17	MDCG position paper on "hybrid audits"	Dec-22
MDCG 2019-6 rev. 4	Questions and answers: Requirements relating to notified bodies	Oct-22
MDCG 2022-13	Designation, re-assessment and notification of conformity assessment bodies and notified bodies	Aug-22
MDCG 2021-23	Guidance for notified bodies, distributors and importers on certification activities in accordance with Article 16(4) of Regulation (EU) 2017/745 and Regulation (EU) 2017/746	Aug-21
MDCG 2021-18	Applied-for scope of designation and notification of a conformity assessment body—Regulation (EU) 2017/746 (IVDR)	Jul-21
MDCG 2021-17	Applied-for scope of designation and notification of a conformity assessment body—Regulation (EU) 2017/745 (MDR)	Jul-21
MDCG 2021-16	Application form to be submitted by a conformity assessment body when applying for designation as notified body under the in vitro diagnostic devices regulation (IVDR)	Jul-21
MDCG 2021-15	Application form to be submitted by a conformity assessment body when applying for designation as notified body under the medical devices regulation (MDR)	Jul-21
MDCG 2021-14	Explanatory note on IVDR codes	Jul-21
MDCG 2020-17	Questions and Answers related to MDCG 2020-4: "Guidance on temporary extraordinary measures related to medical device notified body audits during COVID-19 quarantine orders and travel restrictions"	Dec-20

MDCG 2020-14	Guidance for notified bodies on the use of MDSAP audit reports in the context of surveillance audits carried out under the Medical Devices Regulation (MDR)/In Vitro Diagnostic medical devices Regulation (IVDR)	Aug-20
MDCG 2020-12	Guidance on transitional provisions for consultations of authorities on devices incorporating a substance which may be considered a medicinal product and which has action ancillary to that of the device, as well as on devices manufactured using TSE susceptible animal tissues	Jun-20
MDCG 2020-11	Guidance on the renewal of designation and monitoring of notified bodies under Directives 90/385/EEC and 93/42/EEC to be performed in accordance with Commission Implementing Regulation (EU) 2020/666 amending Commission Implementing Regulation (EU) 920/2013	May-20
MDCG 2020-4	Guidance on temporary extraordinary measures related to medical device notified body audits during COVID-19 quarantine orders and travel restrictions	Apr-20
MDCG 2019-14	Explanatory note on MDR codes	Dec-19
MDCG 2019-13	Guidance on sampling of devices for the assessment of the technical documentation	Dec-19
MDCG 2019-12	Designating authority's final assessment form: Key information (EN)	Oct-19
MDCG 2019-10 rev. 1	Application of transitional provisions concerning validity of certificates issued in accordance to the directives	Oct-19
MDCG 2018-8	Guidance on content of the certificates, voluntary certificate transfers	Nov-18
NBOG BPG 2017-2	Best practice guidance on the information required for personnel involved in conformity assessment	Feb-18
NBOG F 2017-8	Review of qualification for the authorisation of personnel (IVDR)	Feb-18
NBOG F 2017-7	Review of qualification for the authorisation of personnel (MDR)	Feb-18
NBOG F 2017-6	Preliminary assessment review template (IVDR)	Feb-18
NBOG F 2017-5	Preliminary assessment review template (MDR)	Feb-18
Person responsible for regulatory compliance (PRRC)		
MDCG 2019-7 - Rev.1	Guidance on article 15 of the medical device regulation (MDR) and in vitro diagnostic device regulation (IVDR) on a 'person responsible for regulatory compliance' (PRRC)	Dec-2023
Postmarket Surveillance Vigilance (PMSV)		
MDCG 2024-1	Device Specific Vigilance Guidance (DSVG) Template	Jan-24
MDCG 2024-1-1	DSVG 01 on Cardiac ablation	Jan-24
MDCG 2024-1-2	DSVG 02 on Coronary stents	Jan-24
MDCG 2024-1-3	DSVG 03 on Cardiac implantable electronic devices (CIEDs)	Jan-24
MDCG 2024-1-4	DSVG 04 on Breast implants	Jan-24
MDCG 2023-3	Questions and Answers on vigilance terms and concepts as outlined in the Regulation (EU) 2017/745 on medical devices	Feb-23
MDCG 2022-21	Guidance on Periodic Safety Update Report (PSUR) according to Regulation (EU) 2017/745	Dec-22
Standards		
MDCG 2021-5	Guidance on standardisation for medical devices	Apr-21
Unique Device Identifier (UDI)		
MDCG 2022-7	Q&A on the Unique Device Identification system under Regulation (EU) 2017/745 and Regulation (EU)	May-22

MDCG 2021-19	Guidance note integration of the UDI within an organisation's quality management system	Jul-21
MDCG 2021-10	The status of Appendixes E-I of IMDRF N48 under the EU regulatory framework for medical devices	Jun-21
MDCG 2021-09	MDCG Position Paper on the Implementation of UDI requirements for contact lenses, spectacle frames, spectacle lenses & ready readers	May-21
MDCG 2018-1 Rev. 4	Guidance on basic UDI-DI and changes to UDI-DI	Apr-21
MDCG 2020-18	MDCG Position Paper on UDI assignment for Spectacle lenses & Ready readers	Dec-20
MDCG 2019-2	Guidance on application of UDI rules to device-part of products referred to in article 1(8), 1(9) and 1(10) of Regulation 745/2017	Feb-19
MDCG 2019-1	MDCG guiding principles for issuing entities rules on basic UDI-DI	Jan-19
MDCG 2018-7	Provisional considerations regarding language issues associated with the UDI database	Oct-18
MDCG 2018-6	Clarifications of UDI related responsibilities in relation to article 16	Oct-18
MDCG 2018-5	UDI assignment to medical device software	Oct-18
MDCG 2018-4	Definitions/descriptions and formats of the UDI core elements for systems or procedure packs	Oct-18
MDCG 2018-3 rev. 1	Guidance on UDI for systems and procedure packs	Jun-20
Other Topics		
Q&A Rev. 1	Q&A on practical aspects related to the implementation of Regulation (EU) 2023/607—Extension of the MDR transitional period and removal of the "sell off" periods	Jul-23
MDCG 2022-18 ADD.1	MDCG Position Paper on the application of Article 97 MDR to legacy devices for which the MDD or AIMDD certificate expires before the issuance of a MDR certificate—Addendum 1	Jun-23
MDCG 2022-18	MDCG Position Paper on the application of Article 97 MDR to legacy devices for which the MDD or AIMDD certificate expires before the issuance of a MDR certificate	Dec-22
MDCG 2022-14	Transition to the MDR and IVDR—Notified body capacity and availability of medical devices and IVDs	Aug-22
MDCG 2022-11 - Rev.1	MDCG Position Paper: Notice to manufacturers to ensure timely compliance with MDR and IVDR requirements	Nov-23
MDCG 2021-25	Application of MDR requirements to "legacy devices" and to devices placed on the market prior to 26 May 2021 in accordance with Directives 90/385/EEC or 93/42/EEC	October 2021
MDCG 2019-7	Guidance on article 15 of the medical device regulation (MDR) and in vitro diagnostic device regulation (IVDR) on a "person responsible for regulatory compliance" (PRRC)	Jun-19
MDCG 2019-3 rev. 1	Clinical evaluation consultation procedure exemptions Interpretation of article 54(2)b	Apr-20
Other Guidance Documents		
MDR/IVDR Language requirements	Overview of language requirements for manufacturers of medical devices for the information and instructions that accompany a device in a specific country	Jan-24
European Medicines Agency (EMA) Guidance	Questions & Answers for applicants, marketing authorisation holders of medicinal products and notified bodies with respect to the implementation of the Medical Devices and In Vitro Diagnostic Medical Devices Regulations ((EU) 2017/745 and (EU) 2017/746)	Jun-21

SCHEER guidelines	Guidelines on the benefit-risk assessment of the presence of phthalates in certain medical devices covering phthalates which are carcinogenic, mutagenic, toxic to reproduction (CMR) or have endocrine-disrupting (ED) properties	Jun-19
CAMD FAQ	CAMD MDR/IVDR Transition Subgroup: FAQ—MDR Transitional provisions	Jan-18

Accessed on 25MAR2024 at https://health.ec.europa.eu/medical-devices-sector/new-regulations/guidance-mdcg-endorsed-documents-and-other-guidance_en—*the following guidance documents were removed from the website between 2018 and 2023 (most were related to NB best practices and applications as well as future nomenclature).*

REMOVED GUIDANCE DOCUMENTS

Reference	Title	Date
NBOG BPG 2017-1	Best practice guidance on designation and notification of conformity assessment bodies	Feb-18
NBOG F 2017-1	Application form to be submitted by a conformity assessment body when applying for designation as notified body under the medical devices regulation (MDR)	Feb-18
NBOG F 2017-2	Application form to be submitted by a conformity assessment body when applying for designation as a notified body under the in vitro diagnostic devices regulation (IVDR)	Feb-18
NBOG F 2017-3	Applied-for scope of designation and notification of a conformity assessment body—Regulation (EU) 2017/745 (MDR)	Feb-18
NBOG F 2017-4	Applied-for scope of designation and notification of a conformity assessment body—Regulation (EU) 2017/746 (IVDR)	Feb-18
MDCG 2018-2	Future EU medical device nomenclature—Description of requirements	Mar-18

Appendix D

Appraisal question examples

**Modified from MDCG SG5/N2R8:2007. Accessed September 29, 2023. https://www.imdrf.org/sites/default/files/docs/ghtf/archived/sg5/technical-docs/ghtf-sg5-n2r8-2007-clinical-evaluation-070501.pdf. This list was previously adapted from "Guidelines for the Assessment of Diagnostic Technologies; Medical Services Advisory Committee 2005" and is also called IMDRF MDCE WB/N56FINAL:2019.*

Check	Question	Comment
Randomized controlled trial: Clinical investigation where subjects are randomized to receive either a test or reference device or intervention and outcomes and event rates are compared for treatment groups.		
	Were inclusion and exclusion criteria specified?	
	Was assignment to treatment groups really random?	
	Was treatment allocation concealed from those responsible for recruiting subjects?	
	Was the prognostic factor distribution sufficiently described for the treatment groups?	
	Were groups comparable at baseline for these factors?	
	Were outcome assessors blinded to treatment allocation?	
	Were care providers blinded?	
	Were subjects blinded?	
	Were all randomized participants included in analysis?	
	Was a point estimate and measure of variability reported for primary outcome?	
Cohort study: Data obtained from groups who have and have not been exposed to device (e.g., historical control) and outcomes compared		
	Were subjects selected prospectively or retrospectively?	
	Was an explicit description of intervention provided?	
	Was subject selection sufficiently described for new intervention and comparison groups?	
	Was the prognostic factor distribution sufficiently described for the new intervention and comparison groups?	
	Were groups comparable for these factors?	
	Did study adequately control for potential confounding factors in design or analysis?	
	Was measurement of outcomes unbiased (i.e., blinded to treatment group and comparable across groups)?	
	Was follow-up long enough for outcomes to occur?	
	What proportion of cohort was followed up and were exclusions from analysis observed?	
	Were drop-out rates and reasons for drop-out similar across intervention and unexposed groups?	

Check	Question	Comment
Case—control study: Patients with a defined outcome and controls without the outcome are selected and information is obtained about whether subjects were exposed to device		
	Was sufficient description about how subjects were defined and selected for case and control groups provided?	
	Was patient disease state reliably assessed and validated?	
	Were controls randomly selected from the source of population of the cases?	
	Was the prognostic factor distribution sufficiently described for the cases and control groups provided?	
	Were groups comparable for these factors?	
	Did study adequately control for potential confounding factors in design or analysis?	
	Was new intervention and other exposures assessed in same way for cases and controls and kept blinded to case/control status?	
	How was response rate defined?	
	Were nonresponse rates and reasons for nonresponse the same in both groups?	
	Was an appropriate statistical analysis used?	
	If matching was used, is it possible cases and controls were matched on factors related to the intervention potentially compromising the analysis due to overmatching?	
	Case series: Device used in a series of patients and results reported, with no control group for comparison	
	Was series based on a representative sample selected from a relevant population?	
	Were criteria for inclusion and exclusion explicit?	
	Did all subjects enter the case series at a similar point in their disease progression?	
	Was follow-up long enough for important events to occur?	
	Were the techniques used adequately described?	
	Were outcomes assessed using objective criteria or was blinding used?	
	If comparisons of subseries were made, was there sufficient description of the series and the distribution of prognostic factors?	

Appendix E

Notified Body clinical evaluation checklist

Modified from MEDDEV 2.7.1 rev. 3.
Available on September 29, 2023 at http://meddev.info/_documents/2_7_1rev_3_en.pdf.

E.1 General details

State proprietary device name and any code names assigned during device development. Identify device manufacturer/s.

E.2 Device description and intended application

Provide concise physical device description and cross-reference relevant sections of manufacturer technical information as appropriate. The description should cover:

- materials, including any medicinal substance (already on the market or new), tissues, or blood products;
- device components, including software and accessories;
- mechanical characteristics; and
- other information (e.g., sterile vs. nonsterile, radioactivity).

State device intended use/application—single use/reusable; invasive/noninvasive; implantable; duration of use or contact with the body and duration of contact along with the organs, tissues or body fluids contacted by the device. Describe how device achieves the intended purpose together with any modifications performed during the investigation.

E.3 Intended therapeutic and/or diagnostic indications and claims

State medical conditions to be treated, including target treatment group and diseases. Outline any specific safety or performance claims made for the device.

E.4 Context of evaluation and choice of clinical data types

Outline device developmental context including whether the device is based on a new technology, a new clinical application of an existing technology, or an incremental change of an existing technology. For a completely new technology, this section should give an overview of the developmental process and points in the development cycle where clinical data exist. For long-standing technology, a shorter description (with appropriate references) could be used. Clearly state if clinical data used in the evaluation are for an equivalent device and identify the equivalent device/s with a justification, cross-referenced to relevant nonclinical documentation supporting the equivalence claim. Equivalence must be defined in three ways:

- Clinical: devices are used for same clinical condition or purpose, at same body site, in similar population (including age, anatomy, and physiology), with similar relevant critical performance according to expected clinical effect for specific intended use.
- Biological: devices are used same materials in contact with same human tissues or body fluids.
- Technical: devices are used under similar conditions of use; have similar specifications and properties: e.g., tensile strength, viscosity, surface characteristics; similar design; similar deployment methods (if relevant); similar principles of operation.

The essential requirements relevant to the device need to be stated, in particular, any special design features posing special performance or safety concerns (e.g., presence of medicinal, human or animal components) identified in the device risk management documentation and needing clinical assessment. Outline how these considerations were used to choose the clinical data used for the evaluation. Where published scientific literature has been used, provide a brief outline of the searching/retrieval process, cross-referenced to the literature search protocol and reports.

E.5 Summary of clinical data and appraisal

Tabulate the clinical data used in the evaluation, categorized according to whether the data address the device performance or safety. (Note many data sets will address both safety and performance.) Within each category, order data according to the importance of their contribution to establishing the device safety and performance and in relation to any specific claims about performance or safety. Provide a brief outline of the data appraisal methods used in the evaluation, including any weighting criteria, and a summary of the key results. Include full citations for literature-based data with titles and investigation codes (if relevant) of any clinical investigation reports. Cross-reference each data entry to the manufacturer's technical documentation location.

E.6 Data analysis

E.6.1 Performance

Provide a description of the analysis used to assess performance. Identify datasets considered most important in-contributing to the demonstration of the overall device performance and, particular performance characteristics, as appropriate. Outline why certain clinical data are "pivotal" and how they demonstrate device performance collectively (e.g., consistency of results, statistical significance, clinical significance of effects).

E.6.2 Safety

Describe total device experience, including patient exposed to device (numbers and characteristics) and duration of follow-up. Provide a summary of device-related adverse events, paying particular attention to serious adverse events. Provide specific comment on whether device safety and intended purpose requires end-user training.

E.6.3 Product literature and instructions for use

State whether manufacturer proposed product literature and instructions for use are consistent with clinical data and cover all hazards and other clinically relevant\information impacting device use.

E.7 Conclusions

Clearly outline the evaluation conclusions about device safety and performance, with respect to the claimed device intended use, indications for use, contraindications, and instructions for use. State whether risks identified in risk management documentation have been addressed by clinical data. For each proposed clinical indication, state whether:

- clinical evidence demonstrates conformity with relevant essential requirements;
- device performance and safety (as claimed) have been established; and
- risks associated with device are acceptable when weighed against patient benefits.

Identify all relevant essential requirements or demonstration of equivalence compliance gaps to be addressed in future, specifically designed clinical investigation/s.

Ref	Requirement	Fulfilled	Comment
0	*Conformity without clinical data*		
0.1	Any demonstration of conformity without clinical data (Annex 7.1.5 of 90/385/EEC and Annex X.1.1d of 93/42/EEC) must be adequately justified and based on the output of the risk management process viewed in the context of the device-body interaction the intended clinical performance the claims of the manufacturer. Adequacy of demonstration of conformity based on performance evaluation, bench testing, and preclinical evaluation in the absence of clinical evaluation must be duly substantiated. NB must review all manufacturer justifications, data adequacy and determine if conformity is demonstrated. Is manufacturer justification adequate? Are performance, bench testing, and preclinical evaluations adequate to demonstrate conformity to the essential requirements?	Yes, no, N/A	
1	*Clinical evaluation, general*		
1.1	Manufacturer should include in the technical documentation a statement on route/s applied to retrieve clinical data used to affix "CE" marking. Statement should make clear whether clinical data were obtained from published literature or results of clinical investigations or both and shall include an adequate justification of route/s selected and a demonstration of equivalency (technical, biological, clinical) and adequacy if clinical data from similar devices have been used.	Yes, no, N/A	Clinical literature, published/unpublished, Equivalence demonstrated, clinical investigation, combination of literature and investigation data
1.2	CER and full clinical data used for CE marking should be included within the technical documentation.	Yes, no, N/A	
1.3	Manufacturer has clearly documented objectives and clinical evaluation scope and specified clinical ERs to be met (e.g., clinical performance/s, safety, risks, favorable benefit-risk ratio related to intended use, target group/s and indication/s).	Yes, no, N/A	
1.4	Manufacturer clearly outlined clinical evaluation steps and procedures completed per this MEDDEV (specifically sections 5–9) with adequate justification for deviations.	Yes, no, N/A	
2	*Clinical investigation route*		
2.1	*Need for clinical investigation*		
2.1.1	Classification of device: is device an implantable or Class III medical device or an active implantable medical device?	Yes, no, N/A	
2.1.2	If clinical investigation is not presented for an implantable or Class III MD or AIMD, has this been adequately justified by the manufacturer in their risk analysis and clinical evaluation?	Yes, no, N/A	
2.1.3	For equivalent devices, are clinical literature data taken together with the available preclinical data sufficient to demonstrate conformity with essential requirements covering device safety and performance under normal conditions of use?	Yes, no, N/A	
2.1.4	For equivalent devices, do any gaps exist in the demonstration of equivalence or the clinical literature demonstration of compliance with each relevant essential requirement to be addressed in specifically designed clinical investigation/s?	Yes, no, N/A	

Ref	Requirement	Fulfilled	Comment
2.1.5	For equivalent devices, are the clinical literature data sufficient to address the clinical hazards identified in the risk analysis? If no, clinical investigation/s will be needed. The objectives of the clinical investigation/s should focus on those aspects not sufficiently addressed by the available data.	Yes, no, N/A	
2.2	*Conduct of clinical investigation*		
2.2.1	Were relevant annexes of the Medical Devices Directives (Annex 7 AIMD, Annex X MDD) and relevant standards (EN ISO 14155-1, 2) taken into account?	Yes, no, N/A	
2.2.2	*Requirements for clinical investigations*		
2.2.3	*The following relevant documentation should be requested and reviewed by the notified body:*		
2.2.4	Protocol submitted to competent authority or other regulatory agency for which no grounds for objection were raised.	Yes, no, N/A	
2.2.5	Letter of "no objection"/approval from competent authority/ authorities (if available) or other approval from relevant regulatory agency/agencies, together with any comments made arising from regulatory review.	Yes, no, N/A	
2.2.6	Ethics committee opinion/s and comments arising from their review or a summary of all ethics committee opinions and any comments/ conditions arising from their reviews.	Yes, no, N/A	
2.2.7	Signed and dated final report.	Yes, no, N/A	
2.3	*Information to be checked—the following information should be checked by the notified body*		
2.3.1	Letter of "no objection" from competent authority/authorities	Yes, no, N/A	
2.3.2	Is clinical investigation plan (CIP) for the clinical investigation, the same as submitted to the competent authority?	Yes, no, N/A	
2.3.3	If parameters are not as set out in original CIP, rationale for nonadherence.	Yes, no, N/A	
2.3.4	Identification of any changes to CIP and rationale for any such changes.	Yes, no, N/A	
2.3.5	Where the clinical investigation/s was performed outside the EU, manufacturer must demonstrate device use (including clinical practice and techniques) and patient population are equivalent to those for which the device will be used within the EU (if relevant).	Yes, no, N/A	
2.3.6	For drug-device combinations, have any issues or concerns raised as part of the clinical assessment of the medicinal substance by the medicinal competent authority or EMEA been considered and/or resolved?	Yes, no, N/A	
2.4	*Final report of investigation—should be reviewed and should include the following*		
2.4.1	Summary—a structured abstract should be provided, presenting study essentials.	Yes, no, N/A	
2.4.2	Introduction—a brief statement placing study in context of device development and identification of guidelines followed in protocol development.	Yes, no, N/A	
2.4.3	Materials and methods.	Yes, no, N/A	
2.4.4	Summary of clinical investigation plan.	Yes, no, N/A	
2.4.5	Results—summary information describing analysis and results.	Yes, no, N/A	
2.4.6	Discussions and conclusions.	Yes, no, N/A	

Ref	Requirement	Fulfilled	Comment
2.4.7	Signatures on final report: sponsor, coordinating clinical investigator (if appointed), principal investigator at each site.	Yes, no, N/A	
2.4.8	Annexes should include clinical investigation plan, amendments, list investigators, institutions, other parties involved, monitors, statisticians (if applicable), ethics committees and their approval letters.	Yes, no, N/A	
2.5	*NB assessment of clinical investigation/s data presented*		
2.5.1	Have identified pass/fail criteria of the investigation/s been met?	Yes, no, N/A	
2.5.2	Have clinical investigation/s results and conclusions demonstrated compliance with the identified relevant essential requirements?	Yes, no, N/A	
2.5.3	Are claims made in device labeling substantiated by clinical data when taken together with relevant preclinical data?	Yes, no, N/A	
2.5.4	Has risk analysis demonstrated risks associated with device use, as set out by the manufacturer, are acceptable when balanced against the patient benefits?	Yes, no, N/A	
2.5.5	Was assessment performed in a critical and objective manner?	Yes, no, N/A	
3	*Clinical literature data: a critical evaluation of relevant scientific literature currently available relating to safety, performance, design characteristics and intended purpose in the form of a written report.*		
3.1	*Methodology*		
3.1.1	A critical evaluation of relevant scientific literature was presented.	Yes, no, N/A	
3.1.2	A search protocol for identification, selection, collation, and review of relevant publications was written.	Yes, no, N/A	
3.1.3	The objective of the literature review was clearly defined.	Yes, no, N/A	
3.1.4	The study types relevant to the objective of the literature review were specified.	Yes, no, N/A	
3.1.5	Data were taken from recognized scientific publications. Unpublished data were taken into account in order to avoid publication bias.	Yes, no, N/A	
3.1.6	*The literature review should state the following*		
3.1.6.1	Data sources, extent of database searches, or other sources of information.	Yes, no, N/A	
3.1.6.2	Rationale for selection/ relevance of published literature.	Yes, no, N/A	
3.1.6.3	Reasons for believing all relevant references, both favorable and unfavorable, were identified.	Yes, no, N/A	
3.1.6.4	Criteria for exclusion of particular references together with a justification for this exclusion.	Yes, no, N/A	
3.1.6.5	Detailed description of different literature search stages including identification, appraisal, analysis, and conclusion.	Yes, no, N/A	
3.2	*Relevance of data presented*		
3.2.1	A literature review should clearly establish the extent to which literature relates to specific characteristics and features of device under consideration.	Yes, no, N/A	
3.2.2	If published studies do not directly refer to device in question, manufacturer must demonstrate equivalence or discuss the similarities between the device under evaluation and the similar devices in the published data.	Yes, no, N/A	

Ref	Requirement	Fulfilled	Comment
3.2.3	To be equivalent, the devices should have similarity with regard to clinical, technical, and biological parameters with special attention to the performance, principles of operation and materials; or if differences are identified, a clinical impact assessment and demonstration of the significance these might have on safety and performance must be set out.	Yes, no, N/A	
3.2.4	The manufacturer must be able to demonstrate the data adequacy to address the aspects of conformity set out in the objective.	Yes, no, N/A	
3.3	*NB assessment of clinical data: literature review should make clear the significance attached to particular references based on a number of factors. These include:*		
3.3.1	relevance of author's background and expertise in relation to the particular device and/or medical procedure involved;	Yes, no, N/A	
3.3.2	whether author's conclusions are substantiated by available data'	Yes, no, N/A	
3.3.3	whether literature reflects current medical practice and generally acknowledged "state of the art" technologies'	Yes, no, N/A	
3.3.4	whether references are taken from recognized scientific publications and whether or not they have been reported in peer-reviewed journals; and	Yes, no, N/A	
3.3.5	the extent to which published literature is the outcome of a study/ studies which have followed scientific principles in relation to design.	Yes, no, N/A	
3.4	*Critical evaluation of the literature: literature review should contain a critical literature evaluation which should:*		
3.4.1	be written by a person suitably qualified in relevant field, reviewed and approved by expert knowledgeable in the "state of the art" and able to demonstrate objectivity;	Yes, no, N/A	
3.4.2	contain a short device description, intended functions, description of intended purpose, and application of use;	Yes, no, N/A	
3.4.3	contain analysis of all available data considered, both favorable and unfavorable;	Yes, no, N/A	
3.4.4	establish the extent to which literature relates to specific device characteristics and features, taking due account of the extent of similarity between device/s covered by the literature and device under assessment;	Yes, no, N/A	
3.4.5	demonstrate device use aspects, including performance, addressed in the clinical risk analysis part are met as claimed by manufacturer, and device fulfils intended purpose as a medical device;	Yes, no, N/A	
3.4.6	analyze identified hazards, associated risks, and appropriate safety measures for patients, medical staff, and third parties involved in study/studies;	Yes, no, N/A	
3.4.7	contain risk analysis relevant to device design, materials, and procedures involved, taking into account any adverse events, results of postmarket surveillance studies, modifications, and recalls (if known);	Yes, no, N/A	
3.4.8	contain a methods description for weighting different papers and statistical analysis methods employed, taking into account assessment methods, study type and duration, and population heterogeneity in study;	Yes, no, N/A	
3.4.9	analyze market experience for the same or similar devices, including results of postmarketing studies, postmarket surveillance, and short- and long-term adverse events;	Yes, no, N/A	
3.4.10	provide a list of publications appropriately cross-referenced in evaluation;	Yes, no, N/A	

Ref	Requirement	Fulfilled	Comment
3.4.11	for equivalent device, describe how equivalence with all relevant characteristics has been demonstrated; and	Yes, no, N/A	
3.4.12	provide a conclusion with justification, including assessment of any probable benefit to health from device use as intended by manufacturer, against probable risks of injury or illness from device use taking account of the "state of the art." The conclusions should make clear how literature review objectives were met and identify any gaps in the evidence necessary to cover all relevant safety and performance aspects.	Yes, no, N/A	
3.4.13	Critical evaluation should be signed and dated by the author.	Yes, no, N/A	
3.5	*NB assessment of critical evaluation of literature presented by manufacturer*		
3.5.1	Are the manufacturer conclusions valid?	Yes, no, N/A	
3.5.2	Are data, taken together with the available preclinical data, sufficient to demonstrate compliance with essential requirements covering device safety and performance under normal conditions of use?	Yes, no, N/A	
3.5.3	Are claims made in device labeling substantiated by clinical data taken together with preclinical data?	Yes, no, N/A	
3.5.4	Was assessment performed in a critical and objective manner?	Yes, no, N/A	
4	*Postmarket clinical follow-up—NB should check and review manufacturer postmarket clinical follow-up plan*		
4.1	Has manufacturer presented an appropriate plan for postmarket clinical follow-up in line with appropriate guidance?	Yes, no, N/A	
4.2	If no postmarket clinical follow-up plan is presented, has this been adequately justified by manufacturer?	Yes, no, N/A	
4.3	Is an adequate postmarket surveillance system in place?	Yes, no, N/A	
4.4	Has manufacturer committed to inform NB of significant updates to clinical evaluation arising from PMS/PMCF?	Yes, no, N/A	
5	*Notified body decision-making*		
5.1	*In reviewing evaluation of clinical data submitted by manufacturer, the NB must decide whether the manufacturer has adequately:*		
5.1.1	described and verified the intended characteristics and performances related to clinical aspects;	Yes, no, N/A	
5.1.2	performed a risk analysis and estimated undesirable side effects; and	Yes, no, N/A	
5.1.3	concluded on the basis of documented justification the risks are acceptable when weighed against the intended benefits.	Yes, no, N/A	
5.2	*NB assessment of benefit-risk ratio presented in clinical evaluation data*		
5.2.1	Listed and characterized device clinical performance intended by manufacturer and expected patient benefit.	Yes, no, N/A	
5.2.2	Used identified hazards list to be addressed through clinical data evaluation.	Yes, no, N/A	
5.2.3	Adequate estimation of associated risks for each identified hazard by: (a) characterizing hazard severity; and (b) estimating and characterizing probability of occurrence of harm (or health impairment or loss of benefit of treatment) (documented with rationale).	Yes, no, N/A	
5.2.4	Decision on acceptability of risks in relation to each identified hazard.	Yes, no, N/A	

Note: Particular attention should be paid to: number of patients entered; objectives of investigation/s (in particular which essential requirements are being addressed); duration of investigation/s and patient follow-up (short- and long-term); end points in terms of diagnostic tools and patient assessment; and inclusion and exclusion criteria.

Include title of investigation/s; identification of medical device/s, device names/models as relevant for complete identification; name of sponsor; statement indicating whether investigation/s was performed in accordance with CEN/ISO Standards; objectives; subjects; methodology; investigation/s initiation and completion dates, including date of early termination, if applicable; results; conclusions; report authors; and report date.

Include clinical investigation objectives; investigation design; type of investigation; investigation end points; ethical considerations; subject population; inclusion/exclusion criteria; sample size; treatment and treatment allocation; investigation variables; concomitant medications/treatments; duration of follow-up; and statistical analysis including investigation hypothesis or pass/fail criteria, sample size calculation, and statistical analysis methods.

Include investigation initiation date; investigation completion/suspension date; disposition of patients/devices; patient demographics; clinical investigation plan compliance; analysis to include safety report, including a summary of all adverse events and adverse device events seen in the investigation, including discussion of severity, treatment required, resolution and assessment by investigator of relation to treatment; performance or efficacy analysis; any subgroup analysis for special population; and description of how missing data, including patients lost to follow-up or withdrawn, were dealt with in the analysis.

Include study performance and safety results; relationship of risks and benefits; clinical relevance and importance of results, particularly in light of other existing data and discussion of comparison with "state of the art"; any specific benefits or special precautions required for individual subjects or at-risk groups; and any implications for conduct of future studies.

For example, having demonstrable and appropriate end points, inclusion and exclusion criteria, an appropriate and validated number of patients submitted, carried out for an appropriate duration, providing evidence and analysis of all adverse incidents, deaths, exclusions, withdrawals, and subjects lost follow-up and identifying an appropriate statistical plan of analysis will help. Ideally, evidence should be generated from a clinical trial (controlled if appropriate), properly designed cohort/case-controlled study, well-documented case histories, or sequential reports conducted by appropriate experienced experts, whether in relation to the device itself or an equivalent device. If unpublished data are being included in the assessment, the literature review will need to weigh the significance attached to each report.

Appendix F

Clinical evaluation plan/performance evaluation plan requirements checklist

Instructions: For each checklist item, document the evidence reviewed to determine if the individual requirement was met, and enter Y/N/NA with relevant notes as needed.

Item	Requirement—CEP/PEP must...	Evidence	Met (Y/N/NA)	Notes
1	establish plan as required to support device intended use			
2	update plan and describe plan procedures and version controls (document version______ and date ________ in spaces provided)			
3	identify general safety and performance requirements (GSPRs) needing support from relevant clinical data including IVD scientific validity, analytical, and clinical performance			
4	identify available common specifications, guidelines, international standards, etc. and whether the CE/PE will comply			
5	justify exceptions/exemptions from clinical/performance requirements (e.g., any elements not deemed appropriate for CE/PE)			
6	include across-the-board SOTA considerations, including existing, current, relevant standards, common specifications, guidance or best practice documents			
7	list and specify prequantified device safety and performance acceptability criteria and parameters used to determine the acceptability of the benefit-risk ratio for each device indication/ intended purpose/use, based on the SOTA in medicine			
8	specify device characteristics, key functional elements (parts, components, software, composition, function, etc.), performance, principles of operation and mechanism/mode of action, and any novel features when used as intended			
9	specify device novelty, degree of innovation, and device classification			
10	specify device intended purpose/use (i.e., specific disorder, medical condition or risk factor of interest to be diagnosed, treated, monitored, etc.)			
11	specify intended user/target population/groups (including disease state, degree of variability for study subject population, prevalence of the clinical state)			
12	specify clear indications, limitations, and contraindications			
13	specify information materials (current product literature) including labeling, instructions for use (where required), and promotional materials			
14	specify clinical claims to be substantiated in the CE/PE			
15	specify device risks to the patient (e.g., resulting directly from device use in a human or from incorrect or delayed result from IVD medical device)			

Item	Requirement—CEP/PEP must...	Evidence	Met (Y/N/NA)	Notes
16	specify device intended clinical benefits to patient, relevant clinical outcome measures and how benefit-risk ratio will be addressed (especially for components including a drug, animal or human tissue, or for an analyte or marker determination needing certified reference materials or reference measurement procedures to allow metrological traceability*)			
17	specify methods used to examine device and generate necessary clinical evidence about device safety (i.e., qualitative and quantitative safety aspects including clear references to document residual risks and side effects), benefit-risk ratio, and clinical/analytical performance (i.e., qualitative and quantitative details, statistical tools, clinical safety, analytical accuracy, information generated by device, residual risks, side effects, device limitations, etc.)			
18	*state (for PEP ONLY) the analyte and assay technology to be evaluated for scientific validity (including certified reference materials and methods to be used and stability of specimens, reagents, etc.)*			
19	*identify and specify (for software as a medical device, SaMD ONLY) the reference databases and other data sources used as the basis for the device decision making*			
20	describe/outline clinical development plan, investigations and performance studies/plans, required regulatory applications, milestones and potential acceptance criteria (i.e., indicate progression from exploratory first-in-human, feasibility and pilot studies, to confirmatory pivotal clinical investigations and sequence and means to determine scientific validity, analytical, and clinical performance, as appropriate)			
21	align with clinical/performance studies using structured and transparent processes to generate reliable and robust data and scientific information			
22	include details about the postmarket follow-up plan (PMCFP/PMPFP) with milestones and potential acceptance criteria			

This checklist was not exhaustive, and other requirements were evaluated using additional evidence as stated below.

Additional notes

* Per IVDR, Annex I, Chapter II, Section 9 and Chapter III, Section 20.4.1(c)).
Requirements drawn from EU MDR 2017/745, Annex XIV and MDCG 2020-13 for CEP, EU IVDR 2017/746, Annex XIII and MDCG 2022-2 for PEP as well as MDCG 2020–1 for SaMD.

My signature below indicates I verified the evidence meets each requirement as specified above.

Signature Date

Appendix G

Clinical evaluation standard operating procedure example 1 (short)

<table>
<tr><td colspan="6">Frestedt incorporated **Frestedt Incorporated Standard Operating Procedure**</td></tr>
<tr><td colspan="5">**Title:** Clinical Evaluation/Performance Evaluation</td><td>**Number**: SOP-001</td></tr>
<tr><td colspan="2">**Revision**: 1</td><td colspan="2">**Supersedes SOP No.**: 0</td><td>**Page:** 1 of 1</td><td>**Effective Date**: 04OCT2024</td></tr>
<tr><td colspan="2">**Author**:

Name/Title

Signature/Date</td><td colspan="3">**Reviewer**:

Name/Title

Signature/Date</td><td>**Approver**:

Name/Title

Signature/Date</td></tr>
<tr><td colspan="6">**Revision History:**</td></tr>
<tr><td>**No.**</td><td>**Section**</td><td colspan="3">**Description**</td><td>**Date**</td></tr>
<tr><td>1</td><td>All</td><td colspan="3">Created</td><td>04OCT2023</td></tr>
</table>

G.1 Policy

Frestedt Incorporated staff will create each clinical evaluation (CE) document in compliance with the associated plan, all applicable regulations, standards, policies, and procedures.

G.2 Purpose

To develop CE documents regarding product safety and performance, to determine if available clinical data are sufficient and establish conformity with international regulations, and to ensure product benefits outweigh product risks to the patient.

G.3 Responsibility

Management will assign an appropriate clinical evaluator and team leader to complete the clinical data evaluation and to write each CE document. Clinical evaluator will ensure the clinical evaluation meets all international regulatory requirements. Project leader will ensure CE documents follow relevant, device-specific plans including required clinical data evaluations and conclusions. All team members will ensure project tasks are completed and deadlines are met.

G.4 Procedures

1. All team members will refer to WI-XXX Writing Clinical Evaluation Documents and WI-XXX Grammar and Style in Science Writing before writing any CE documents.
2. Project leader will ensure KOM discusses the device, CE scope, CE writing process, staff assignments, and due dates.
3. Staff members will collect, appraise, and analyze appropriate clinical data, and will draft assigned CE document sections, as assigned by the clinical evaluator.
4. Project leader will ensure full draft/s are completed to meet due date/s.
5. Clinical evaluator will ensure evaluation is accurate, comprehensive, and complete.

6. Project leader will ensure draft CE documents are reviewed and all CE feedback is received and incorporated.
7. Management will ensure revised and final CE documents are approved, signed, dated (when required), and version controlled.

G.5 References

Number	Document description (title)
WI-XXX	Writing Clinical Evaluation Documents
WI-XXX	Grammar and Style in Science Writing

G.6 Appendices

None.

Appendix H

Clinical evaluation standard operating procedure example 2 (long)

TITLE PAGE*: Clinical evaluation process

Table of Contents

1 Purpose and background

This clinical evaluation (CE) process is ongoing throughout the medical device (including *in vitro* medical device) life cycle. CE documents are initially developed during early design stages, revised during conformity assessment processes leading to device marketing, and then updated according to senior management review decisions (e.g., annually for high-risk devices, every 2–5 years for low-risk/well-established devices, or earlier if new clinical data becomes available which might affect the device safety or performance profile). The CE documents record all clinical data, especially clinical data related to device safety and performance (S&P) when the device is used as indicated (i.e., under normal conditions in appropriate patient populations). Off-label uses are also important to include in the CE in order to assess appropriately the extent of device misuse in the real world.

CE involves the medical device clinical data analysis often related to regulatory compliance in a particular geographical location such as the European Union (EU), China, or Australia. The United States (US) does not require CE documentation in the same way as in other areas around the world. For example, the EU requires CE documentation specifically to illustrate conclusively how the company conforms to General Safety and Performance Requirements (GSPRs) 1, 6, and 8 in the Medical Devices Regulation EU Reg. 2017/745 and in the *In vitro* Diagnostic (IVD) Devices Regulation EU Reg. 2017/746. Different countries have different regulatory requirements, and additional regulations apply to the CE integration with other standard operating processes (SOPs) specifically for clinical trial activities, postmarket surveillance (PMS), and risk management (RM) systems.

**Format tile page per Company QMS Standards/Guidelines*

The CE must include the following evidence substantiated with well-studied clinical data:

- device suitability for intended use/indication for use;
- assessment of all adverse device effects (ADEs) including secondary effects;
- acceptability of safety and performance (S&P), which must be based on prespecified and prequantified "threshold" criteria defined in the CE plan (CEP); and
- ccceptability of benefit-risk ratio.

The manufacturer is expected to have robust PMS and RM systems well-defined within the quality management system (QMS) to monitor the medical device clinical S&P, and changes should be anticipated in the instructions for use and product labeling.

2 Scope

This procedure applies to all devices when company is responsible for design control activities and to all personnel involved in the CE work. External experts can be consulted as necessary, as long as the responsible department verifies all work and issues approval within the QMS.

3 References and associated documents

European Parliament and Council of the European Union. Consolidated text: Regulation (EU) 2017/745 of the European Parliament and of the Council of 5 April 2017 on medical devices, amending Directive 2001/83/EC, Regulation (EC) No 178/2002 and Regulation (EC) No 1223/2009 and repealing Council Directives 90/385/EEC and 93/42/EEC. Accessed on September 28, 2023 at https://eur-lex.europa.eu/legal-content/EN/TXT/?uri = CELEX%3A02017R0745-20230320.

European Parliament and Council of the European Union. Consolidated text: Regulation (EU) 2017/746 of the European Parliament and of the Council of 5 April 2017 on *in vitro* diagnostic medical devices and repealing Directive 98/79/EC and Commission Decision 2010/227/EU. Accessed on September 28, 2023 at https://eur-lex.europa.eu/legal-content/EN/TXT/?uri = CELEX%3A02017R0746-20230320.

MEDDEV 2.7/1 rev. 4 (2016)—Clinical Evaluation: A guide for manufacturers and notified bodies under directives 93/42/EEC and 90/385/EEC.

Additional references from the EU Medical Device Coordinating Group (MDCG) are listed in the following table; however, please note additional regulations, guidelines, and standards may apply. For the most up-to-date information, refer to the current regulations and check the MDCG guidance listings at https://health.ec.europa.eu/medical-devices-sector/new-regulations/guidance-mdcg-endorsed-documents-and-other-guidance.

Reference	Title	Date
MDCG 2020-8	Guidance on PMCF evaluation report template	Apr-20
MDCG 2020-7	Guidance on PMCF plan template	Apr-20
MDCG 2020-6	Guidance on sufficient clinical evidence for legacy devices. See also: "Background note on the relationship between MDCG 2020-6 and MEDDEV 2.7/1 rev. 4 on clinical evaluation" at https://health.ec.europa.eu/system/files/2022-09/md_borderline_bckgr-note-manual-bc-dir_en_1.pdf	Apr-20
MDCG 2020-5	Guidance on clinical evaluation—on equivalence	Apr-20
MDCG 2019-9 rev. 1	Summary of safety and clinical performance	Mar-22
MDCG 2022-9	Summary of safety and performance template	May-22
MDCG 2022-8	Regulation (EU) 2017/746—application of IVDR requirements to "legacy devices" and to devices placed on the market prior to 26 May 2022 in accordance with Directive 98/79/EC	May-22
MDCG 2022-2	Guidance on general principles of clinical evidence for *in vitro* diagnostic (IVD) medical devices	Jan-22
MDCG 2020-1	Guidance on clinical evaluation (MDR)/performance evaluation (IVDR) of medical device software	Mar-20

Associated company QMS documents are listed in the following table.

Medical device	*In vitro* diagnostic device*
SOP XXX Clinical trials	
SOP XXX Postmarket surveillance	
SOP XXX Risk management	
FORM XXXX: Clinical evaluation plan (CEP) template	FORM XXXX: Performance evaluation plan (PEP) template
FORM XXXX: Clinical evaluation report (CER) template	FORM XXXX: Performance evaluation report (TER) template
FORM XXXX: Postmarket clinical follow-up plan (PMCFP) template	FORM XXXX: Postmarket performance follow-up plan (PMPFP) template
FORM XXXX: Postmarket clinical follow-up evaluation report (PMCFER) template	FORM XXXX: Postmarket performance follow-up evaluation report (PMPFER) template
FORM XXXX: Summary of safety and clinical performance (SSCP) template	FORM XXXX: Summary of safety and performance (SSP) template
FORM XXXX: Declaration of interest (DOI)	
REPORT XXXX: Postmarket surveillance plan (PMSP)	
REPORT XXXX: Postmarket surveillance report (PMSR) (low-risk devices)	
REPORT XXXX: Periodic safety update report (PSUR) (high-risk devices)	
REPORT XXXX: Risk management report (RMR)	

**Note: Only a few forms are unique for* in vitro *diagnostic (IVD) device information.*

4 Responsibilities

All company personnel involved in the CE process are responsible to review CE processes and documents and to provide input and feedback to ensure accuracy and completeness.

Quality Engineering will provide the following reports and will ensure all reports are accurate:

1. all PMS reports (PMSRs) including all detailed analyses for all complaints, medical device reports, and sales information relating to the device;
2. all RM reports (RMRs) compliant with the most recent version of ISO 14971; and
3. all needed updates for all reports based on CE document clinical data details.

Clinical Affairs will qualify the evaluator/team members to ensure each CE document is authored and reviewed by appropriately educated, trained, and experienced individuals. Clinical Affairs will also provide all clinical trials, clinical literature, and clinical experience data, and will coordinate the writing, review, and updates for all CE documents.

Regulatory Affairs will provide the device regulatory experience including all approval/clearance information, all recalls, and field safety corrective actions (FCRAs), will ensure the product benefits and risks are documented in the CER, and will verify the frequency of CER updates are completed as required.

Product Management will identify and provide all device data needed in the CE documents, including the device description and all product information including the instructions for use, appropriate labeling, market history, and experience data.

Research & Development/Product Development will identify and provide all data needed in the CE documents relating to the device, background, state of the art (SOTA), alternative therapies, and equivalent device/s (if applicable), including the assessment of alternative products to claim equivalence (as defined in the EU MDR/IVDR) and the detailed technical, biologic, and clinical characteristics for the device itself and the equivalent device to identify specifically any/all differences for assessment of clinical impacts within the CE process.

Senior Management should review the CE documents and should make all appropriate changes to related records and the CE processes as needed.

A qualified and trained clinical evaluators and medical reviewers should evaluate and review the CE processes and all clinical data included the CE document/s to ensure the methods and results are appropriate and have been faithfully applied, and the conclusions are appropriate and fully substantiated by the clinical data evaluated.

5 Procedure

The CE document development process should follow several distinct stages as outlined in the following process flow map.

PROCESS FLOW MAP

5.1 Qualify evaluator and team

Document how team members are each individually qualified by education, training, and experience to complete the CE document development processes assigned to them. Secure and store resumes/curriculum vitas (CVs), training records, and declarations of interest (DOIs) for each individual with evaluator responsibilities, and especially for those drawing any conclusions about device S&P and/or benefit-risk profiles. The evaluator/team must be experienced in research methodology, information management, regulatory requirements, and medical writing, and they must have specific knowledge of the device including the device application as well as diagnosis and management of the conditions related to the device intended use. Evaluators are required to have at least a higher education degree with 5 years documented professional experience or 10 years documented professional experience without a higher education degree. CE documents outsourced to third parties must have documented evidence on file for the same qualifications as discussed above.

5.2 Plan clinical evaluation scope

The clinical evaluation plan/performance evaluation plan (CEP/PEP) must be signed and dated prior to starting the clinical evaluation report/performance evaluation report (CER/PER). The CEP/PEP should include (but is not limited to) the following information:

1. acceptance criteria including prespecified and prequantified S&P thresholds to be evaluated in the CER/PER;
2. subject device or device family description including product codes;
3. marketing history;
4. device design features and specific clinical concerns or target treatment populations which may require specific attention;
5. equivalent devices including clinical, biological, and technical equivalence details;
6. comparable or similar devices which may be used to support the device S&P and the background/SOTA/ alternative therapies related to the device;
7. specific data source/s and type/s to be used including details about available clinical trials, clinical literature, and clinical experience data;
8. current clinical trial documents identifying clinical trial methods, results, and conclusions with all subject level detailed data to evaluate the clinical trial conduct and all ADEs, protocol violations, or other clinical trial details needed to perform the CE;
9. current RM documents identifying the device clinical risks and how these clinical risks have been mitigated;
10. current PMS documents identifying device complaints, safety reports, postmarket surveillance reports/periodic safety update reports (PMSRs/PSURs), and all experience as well as sales volume data;
11. device design changes or intentions to introduce changes in design, materials, labeling/instructions for use, and claims;

12. any new clinical data from PMS postmarket clinical follow-up/postmarket performance follow-up (PMCF/PMPF) studies or any new knowledge from any source about hazards, risks, performance, benefits, or claims; and
13. *for IVD devices: scientific validity, analytical, and clinical performance details.*

5.3 Identify and collect

Pertinent clinical data from clinical trials/investigations (protocols and final study reports), clinical literature, and clinical experiences. Ensure these sources provide suitable clinical data to evaluate the device S&P as well as the benefit-risk ratio for the patient. Clinical data should be separated into two groups:

1. clinical data specifically for the subject device or an equivalent device; or
2. clinical data from a similar device/s specifically for the background/SOTA/alternative therapies sections of the CE documents.

The clinical data from these two groups should be considered separately and as a whole during the CE.

5.4 Appraise

The identified and collected clinical data in order to determine the suitability (merits and limitations) in terms of relevance and quality for each piece of clinical data to ensure the clinical data provide specific details needed to address the device S&P as well as the benefits and risks to the patient. Document the inclusion and exclusion decision details for each clinical data document based on this clinical data appraisal step.

5.5 Analyze

Each piece of included clinical data for clinical S&P details and document the analysis in tabular form especially as related to the prespecified and prequantified S&P acceptance criteria documented in the CEP/PEP for the subject device. In addition, complete a benefit-risk ratio analysis, to document in tabular form if and to what extent the benefits outweighed the risks to the patient using the device as indicated.

5.6 Document, sign, and date

The plans and reports including the CE methods, results, and conclusions of the clinical data evaluations should be in the CER/PER. The CER/PER should include details about the CE scope and context, the available clinical data types and sources, and the rigorous methods used to identify the clinical data evaluated in the CER/PER, as well as the CE appraisal, analysis, and conclusions regarding the subject/equivalent device S&P and benefit-risk profile. The CER/PER should describe any and all clinical data gaps with needs detailed for PMCF/PMPF as appropriate, and the CER/PER should document the timeline commitments for the PMCF/PMPF evaluation report prior to the next planned CER/PER revision (annually for high-risk devices, and every 2–5 years for low-risk/well-established devices), with an appropriate justification for the stated timelines. The completed CE documents should be signed, dated, and version controlled after senior management review and review by a physician using the device in medical practice. Senior management should also provide a signed and dated statement justifying the choice of authors and evaluators and the CVs or resumes, credentials, and declarations of interest for each appropriate author/s and evaluator/s should be attached in an appendix to the CER/PER. Version control must be maintained according to the QMS, and the CE documents must be stored along with all attachments/appendices (e.g., publications, literature references, spreadsheets) in an easily retrievable format.

Appendix I

Clinical evaluation work instruction example

Frestedt incorporated	Title: Clinical Affairs	Doc. Number: WI-003	Rev. Number: 20

PURPOSE: To describe clinical affairs processes including development and monitoring of clinical trials (CTs) and clinical evaluation reports (CERs).

INSTRUCTIONS:

1. Management will assign clinical activities to appropriate staff members as needed.
2. Client history should be reviewed to ensure accurate project scope of work.
3. Clinical staff will interact with internal staff to complete assignments potentially using client-specific instructions.
4. Templates may be used.
   a. To design a new clinical trial (e.g., see protocol template from NIH at https://osp.od.nih.gov/clinical-research/clinical-trials).
      i. F-014 Food Protocol Synopsis
      ii. F-015 Device Protocol Synopsis
      iii. F-016 Drug Protocol Synopsis
   b. To conduct clinical trial monitoring.
      i. F-038 Monitoring Report Template
      ii. F-044 Risk Based Monitoring Plan Template
      iii. F-045 Monitor Letter Template
   c. To develop a clinical evaluation report (CER). *Note the CER may be drafted following the stage gate (SG) process defined in Appendix A of this document.*
      i. F-019 Workbook Template
      ii. F-023 PMCF Report Template
      iii. F-027 SSCP Template
      iv. F-047 Declaration of Interest for CER
      v. F-049 Clinical Evaluation Report Template
      vi. F-072 Clinical Study Report Template
      vii. F-074 MEDDEV 2.7.1 CER Checklist
      viii. F-089 Clinical Evaluation Plan Template
      ix. F-090 PMCF Plan Template
5. An independent person should conduct quality check before closing out each project.
6. Completed projects should be stored in the project file/s.

REFERENCES

Number	Document description (title)
F-014	Food Protocol Synopsis
F-015	Device Protocol Synopsis
F-016	Drug Protocol Synopsis
F-019	Workbook Template

Number	Document description (title)
F-023	PMCF Report Template
F-027	SSCP Template
F-038	Monitoring Report Template
F-044	Risk Based Monitoring Plan Template
F-045	Monitor Letter Template
F-047	Declaration of Interest for CER
F-049	Clinical Evaluation Report Template
F-072	Clinical Study Report Template
F-074	MEDDEV 2.7.1 CER Checklist
F-089	Clinical Evaluation Plan Template
F-090	PMCF Plan Template

Appendix A: Clinical Evaluation Report Process
SG1: Clinical Evaluation Report Kickoff (Initiating)

1. Request documents from client prior to kickoff.
   - a. CER SOP/work instructions/templates
   - b. List of CER reviewers and approvers (names, credentials, titles)
   - c. List of competitor/equivalent devices (device name, manufacturer, city, state)
   - d. CER past comments/deficiencies from notified body or others
   - e. Plans and reports
     - i. Clinical evaluation plan (CEP)
     - ii. Postmarket clinical follow-up plan (PMCF)
     - iii. Prior CERs and summary of safety and clinical performance (SSCP) reports
     - iv. Risk management report (EN ISO 14971:2019 compliant)
     - v. Postmarket surveillance report (PMSR) (e.g., experience data including all prior PMCF data, complaints and issues from Manufacturer and User Facility Device Experience (MAUDE), Database of Adverse Event Notifications (DAEN), SwissMedic)
     - vi. Periodic safety update reports (PSURs)
   - f. Device and use description
   - g. Instructions for use (IFUs), operating manuals, etc.
   - h. Clinical trial data (ISO 14155:2020 compliant clinical study protocols, reports, publications)
   - i. http://www.clinicaltrials.gov listings for each trial
   - j. Literature and literature searches (e.g., published or unpublished articles used during initial or ongoing development especially for safety or performance issues)
   - k. Marketing information
     - i. history and sales numbers (list of countries with sales, date or year first on market in each country/geography with sales potentially broken out by year to align with complaints)
     - ii. marketing materials and claims (including performance and safety claims on website as well as in press releases and promotional documents, etc.)
   - l. EU regulatory documents (CE mark date of approval/renewal, history)
     - i. Notified body certification (CE mark)
     - ii. Audit reports
     - iii. Deficiencies
   - m. US regulatory documents (date of approval/clearance, history)
     - iv. 510k, PMA, De Novo, HDE, etc.
     - v. Audit reports
     - vi. Deficiencies
   - n. ISO certifications

o. GMDN codes and UDI information
p. Design history file and/or technical files (if concise/available)
    i. Labels
    ii. Specifications
    iii. Sterilization information
    iv. Manufacturing information

2. Record each document received from the client in the "Docs Received" tab in the workbook according to the column headers.
3. Store and organize all documents with appropriate names in shared drive without duplication.
    a. < Company Project Name> is main folder for CER and workbook (ensure date and time are used in titles to identify most current versions).
    b. < Library> is main folder for documents received.
    c. < Articles> is Library subfolder (article name: <first author last name> year).
    d. Archive all past revisions, as appropriate.
4. Ensure workbook has appropriate headers and footers on each page (QC check).
    a. Access header/footer by clicking on page layout display setting on bottom right corner of workbook.
    e. Use appropriate header (centered: title of tab without date/time).
    f. Use appropriate footer (left: file name and tab; middle: CONFIDENTIAL; right: page # of ##) on every tab.

SG2: Clinical Evaluation Plan (CEP) (Planning)

1. CEP should include device description as well as predefined and quantified safety and performance end points (with acceptance criteria).
2. Have CEP reviewed (QC check).
3. CEP should be signed and dated by senior management and client (approval).
4. CEP should inform the CER starting with the device description and allowing concurrent development of CER device description, executive summary, and conclusion sections based on the CEP.

SG3: Clinical Data Identification (Searching)

1. For each search, document appropriate information in key and literature/experience worksheets. A good practice is to begin drafting methods and narratives in the CER at the same time searches are being conducted and documented in the workbook.
2. After exploring possibilities, perform appropriate literature and experience searches.
3. Document search number, number of articles, and details in "Literature Key" worksheet.
4. Search PubMed at http://www.ncbi.nlm.nih.gov/pubmed.
    a. Specific device search strategy: subject/equivalent device brand name (if no brand name is/are available, use manufacturer name).
    b. Generic similar device search strategy: similar/competitor device brand names (if no brand names are available, use manufacturer name).
    c. Generic background/state-of-the-art (SOTA) search strategy: type of device, disease state, patient population (focus on what is needed for a robust search to include appropriate alternative therapies). Caution: stay focused on the subject device during these searches (e.g., metaanalyses of rigorous comprehensive research on alternative therapies).
    d. Filter output using two different filter sets.
        i. Metaanalyses: in sequential order, select metaanalysis, systematic review, practice guideline, guideline.
        ii. Clinical trials: in sequential order, select randomized controlled trial, controlled clinical trial, multicenter study, pragmatic clinical trial, clinical trial, clinical study, observational study, comparative study, evaluation studies.
    e. Use "advanced search" feature as needed to remove duplicates.
    f. Download literature search outputs. For example, in PubMed: select "Save," "All Results," "CSV," and "Create file." Cut and paste literature list into literature tab. Remove all line spaces, returns, and extra lines.
5. If additional guideline search is helpful, search National Institute for Health and Care Excellence (NICE) guidance at http://www.nice.org.uk.
6. Search STN Easy at https://stneasy.cas.org/html/english/login1.html?service = STN.

   a. Search strategy: subject device brand name (or company name if needed), search in the Life Sciences databases EMBASE, Biosis, SciSearch. *Note: keep login ID and password secure.*
   b. Narrow search by combing searches with "AND," "OR," "NEAR," or "NOT."
   c. Identify duplicates by selecting "Show Duplicates" to identify other databases with same article.
   d. Do not click "Display answers." Every title clicked results in a bill for the amount listed.
   e. Copy and paste article titles into a separate Word document and remove hyperlinks, being careful not to click on any of these.
   f. Cut and paste titles into "literature" worksheet.
   g. Print "search history" and document client and project name with your initials and date on the printed record for accounting and invoicing.
7. Search Google Scholar at https://scholar.google.com.
   a. Search strategy: subject device brand name (or company name if needed).
   b. Uncheck "include patents" and "include citations" boxes.
   c. Limit to first five pages of results, unless directed otherwise.
   d. Review output for relevant titles, eliminate duplicates, and paste titles into "literature" worksheet.
8. In the "Literature Key," record in the comments column the reason for including or not including the search output (e.g., "Too broad" or "Reviewed all, none our device" or "Keep").
9. Have reviewer check workbook "Literature Key" (including search strategy) and "Required Data List" (with all associated outputs prior to coding) (QC check).
10. Seek management approval and sign-off on the search strategies in "Literature Key and "Required Data List," preferably within 2 days of kick-off (approval).
11. Read PMSR and abstract appropriate experience information.
   a. If helpful, record complaints in "Complaints" worksheet and quantify total number and type of each complaint (e.g., break, separation of pieces, leak, user error, packaging problem, sterility issue). If appropriate, separate out adverse events and device deficiencies for further evaluation.
   b. If not already provided in the PMSR, calculate percent of sales for complaints (number of complaints/number of units sold; numbers must be from same time frame)
   c. If PMSR does not include searches below, consider if a new search is required and conduct the search/es required.
      i. US FDA regulatory filing databases for subject device brand name and indications as appropriate (e.g., 510(k) at http://www.accessdata.fda.gov/scripts/cdrh/cfdocs/cfpmn/pmn.cfm, PMA, HDE, etc.; click "Export to Excel" and copy/paste results into "Regulatory" worksheet. Highlight all included lines.
      ii. MAUDE database for device brand name and indications at https://www.accessdata.fda.gov/scripts/cdrh/cfdocs/cfmaude/search.cfm; click "Export to Excel," then copy/paste results into "MAUDE Data" worksheet using appropriate search and inclusion/exclusion (I/E) codes from "Key" worksheet.
         - Note the "Date Report Received by FDA" defaults to a 1-month span, so ensure the desired date range is correct (most commonly: 01/01/1900 to <current date>). If output "maxes out" at 500 entries, date range may need to be adjusted and explained in "comments."
         - Auto-download: web address, report number, event date, event type, manufacturer, date FDA received report, product code, brand name, device problem, and event text.
      iii. Australian Therapeutic Goods Association (TGA) Database of Adverse Event Notifications (DAEN) at http://apps.tga.gov.au/prod/DEVICES/daen-entry.aspx. *Note: document the report number in the workbook. Click to open each report to obtain tables of data and save a pdf copy of each report on the share drive if needed.*
      iv. MedSun Medical Product Safety Network at http://www.accessdata.fda.gov/scripts/cdrh/cfdocs/Medsun/searchReport.cfm.
      v. US FDA "Recalls" database (limit to relevant and specific terms only) at http://www.accessdata.fda.gov/scripts/cdrh/cfdocs/cfRES/res.cfm.
         - Click "Export to Excel."
         - Copy/paste results into "Recalls" worksheet.
         - Click on Recall to look up "Quantity in Commerce" and enter into worksheet.
      vi. SwissMedic at https://fsca.swissmedic.ch/mep/#.
      vii. Health Canada Recalls and Safety Alerts database at http://healthycanadians.gc.ca/recall-alert-rappel-avis/index-eng.php?_ga = 2.21864760.204481559.1507668614-778876590.1502139133.

viii. Federal Institute for Drugs and Medical Devices (BfArM) field corrective actions database at https://www.bfarm.de/SiteGlobals/Forms/Suche/EN/Expertensuche_Formular.html;
jsessionid = C666D57E79F097877268CB779F575C97.internet562?
nn = 708434&cl2Categories_Format = kundeninfo.

ix. Medicines and Healthcare Products Regulatory Agency (MHRA) Alerts and Recalls for drugs and medical devices database at https://www.gov.uk/drug-device-alerts.

x. FDA Adverse Events Reporting System (FAERS) public dashboard at https://www.fda.gov/drugs/questions-and-answers-fdas-adverse-event-reporting-system-faers/fda-adverse-event-reporting-system-faers-public-dashboard.
- Click on tab for "FAERS Public Dashboard" and accept disclaimer.
- Type in keyword into search bar.
- If results are present, click on "Reaction" then "Listing of Cases."
- Right-click data table, select "Export," then "Export data."
- Select "Click here to download your data file" and copy data into workbook tab.

xi. US FDA Warning Letters database at http://www.fda.gov/ICECI/EnforcementActions/WarningLetters/default.htm.

xii. US FDA Letters to Health Care Providers at https://www.fda.gov/medical-devices/medical-device-safety/letters-health-care-providers.

xiii. US FDA Safety Communications at https://www.fda.gov/medicaldevices/safety/alertsandnotices/default.htm.

xiv. FDA Total Product Life Cycle (TPLC) database at http://www.accessdata.fda.gov/scripts/cdrh/cfdocs/cfTPLC/tplc.cfm. *Note: set "Since" to oldest date available.*

xv. Clinical Trials Databases (as directed).
- International Clinical Trials Registry Platform (ICTRP) maintained by WHO at http://apps.who.int/trialsearch.
- EU Clinical Trials Register maintained by European Commission on Public Health at https://www.clinicaltrialsregister.eu/ctr-search/search.
- NIH ClinicalTrials.gov at https://clinicaltrials.gov; click on "Download" to retrieve "all available columns" in comma-separated values (csv) format, then cut and paste into appropriate places in worksheet.

12. Ensure contents are entered into workbook literature and experience tabs.
13. Highlight in yellow the search outputs to be included in the CER.

SG4: Clinical Appraisal (Coding)

1. Keep articles/experiences included until a definitive inclusion or exclusion criterion is met (i.e., if unknown, keep article/experience in for further review and ensure inclusion and exclusion coding is explicit for each article).
2. To help determine I/E codes, record appropriate data in each worksheet column.
3. For literature, determine I/E code for each article by first reviewing title, then title and abstract, and then full-length article. A good practice is to begin drafting methods and narratives in the CER at the same time the coding is being conducted in the workbook. During coding, be sure to assign a relative weight for each I/E code based on criteria described in the Oxford Center for Evidence Based Medicine (OCEBM) Levels of Evidence, which are available at https://www.cebm.net/wp-content/uploads/2014/06/CEBM-Levels-of-Evidence-2.1.pdf. A best practice when coding literature is to document the following in the literature worksheet:
   a. abstract (be sure to remove all line returns, etc. when pasting abstract into workbook literature tab);
   b. study design (may aid I/E coding based on Oxford criteria):
      i. metaanalysis, systematic reviews, guideline,
      ii. RCT = randomized controlled trial,
      iii. prospective, CT, retrospective review, cohort,
      iv. case report/series (usually <20 patients),
      v. narrative review, editorial/comment/letter, and
      vi. animal/*in vitro*/bench/cadaver/simulation;
   c. study objective, which should start with "to" and should use quotation marks when quoting directly from the abstract—do not plagiarize;
   d. study site (country);

  e. number of subjects;
  f. number of subjects with device;
  g. device names;
  h. whether article is free or will cost money to purchase;
  i. link to article (free or for purchase); and
  j. download article, if article is free and may be helpful to CER.
4. Determine which experiences should be "included" based on relevance to the device/s and indications for use evaluated in the CER.
5. Highlight in yellow each included article or experience finding.
6. Remove any unused tabs from workbook (e.g., MEDDEV TOC and checklist tabs may be removed if not needed) and verify worksheets have appropriate headers, footers, etc.
7. Have reviewer check I/E coding (QC check).
8. Seek management approval and sign-off for I/E coding (approval).
9. After management approval, ensure all free "included" articles have been downloaded and order all "included" articles to be purchased by client or by Frestedt.
10. Secure, document, and store all articles in the "Articles" subfolder using first author's last name and year as the document title. Order all needed articles as soon as possible.

SG5: Clinical Analysis (Analyzing)

1. For literature, analyze appropriate data from each article into the workbook using self-evident column headers to guide development.
   a. Abstract performance results pertaining to device efficacy data.
   b. Summarize specific device problems in device issues column.
   c. Abstract safety results pertaining to patient safety, including adverse events.
   d. Summarize notable safety concerns in safety issues column.
   e. Describe study limitations (e.g., small sample size, author bias as employee, financial COI (conflict of interest), article lacking safety or performance data, unfounded conclusions) in limitations column.
2. For experiences, analyze event text and populate contents in "Experience Data" worksheet: Injury to patient? (Y/N); Injury group (define groups); Device performance issue? (Y/N); and Device problem group (define groups). When developing groups, ensure no overlap between groups is defined.
3. Populate both the workbook and the CER as the data analysis proceeds.
4. Document all device benefits and risks found in clinical trials, literature, or experiences, including a description of each in "Benefit Risk Listing" worksheet.

SG6: Draft and Review (Writing)

1. From the literature, write the methods and summarize metaanalyses, systematic reviews, guidelines, RCTs, and other literature in the CER.
2. From the experience, write the methods and summarize the findings from the PMSR and each experience database search.
3. Indicate whether the general safety and performance requirements (GSPRs) (EU Medical Device Regulation 2017/745) have been fully addressed with clinical data.
4. Evaluate and write the overall clinical evidence conclusions, including benefit/risk profile and PMCF requirements. If data are not sufficient for a positive benefit-risk ratio, explain the PMCF plan to satisfy the gap.
5. Complete internal draft review with senior management.
6. Submit draft to client for review.
7. Incorporate all reviewer comments
8. Ensure critical appraisal (e.g., by physician using device in practice) has been completed, when applicable.
9. Repeat internal and client review and updates until ready for sign-off.

SG7: Sign-Off (Closing)

1. Review entire CER to ensure all requirements are met (may use a checklist).
2. Secure review by an internal CER reviewer and senior management.
3. Critical appraiser and then reviewer should sign and date prior to approver signature and date.

4. Compile CER as a pdf file.
   a. Save CER as pdf.
   b. Save each full-length article (<1Mb each) and merge all articles into one pdf.
   c. Scan signature page/s as pdf/s.
   d. Save declarations of interest (DOIs) and CVs/resumes (author/evaluators) as pdfs.
   e. Compile CER, signatures, DOIs, CVs/resumes, and articles into one pdf.
   f. Embed workbook into CER pdf document.
   g. Ensure all appropriate bookmarks are present in the pdf.
5. Secure senior management approval, then file CER in IP library and send to client (retention is the responsibility of the client).

Appendix J

Kick off meeting agenda/minutes/slides

This 1-hour meeting on <ddMONyyyy> from <xx to yy am/pm> will be conducted to engage all participants in the work to be completed for the *<DEVICE NAME>* *<CER/PER>* activities over the next <*12*> weeks.

I—Introductions (5 minutes); assign scribe[1] to take meeting minutes
II—Review *<DEVICE NAME>* *<CER/PER>* process and process owners[2]
III—Discuss anticipated challenges; assign "parking lot" attendant[3]
IV—Delegate/accept tasks and due dates to meet timeline
V—Schedule next meeting

[1] Scribe is the person who takes and circulates *DRAFT* meeting minutes within 2 working days after each meeting. The meeting minutes should capture all action items and persons assigned to each delegated tasks with due dates. Revisions are due back to the scribe within 2 working days to ensure all updates are captured before the next meeting, which will start with a review of the meeting minutes.

[2] Kick off meeting (KOM) slides are suggested in the following table.

Slide	Slide contents	Slide details	Presenter
1	Title slide	Project name, meeting date/location.	Project Lead
2	Team members	Define each person's role and responsibility, document contact info, and designee if person is not available for any reason.	Project Lead
3	QMS review	Details about how to do the assigned work are written down and doable.	Project Lead
4	Training	Does anyone need additional training? Training sessions will be provided. *(Scribe to record names, dates and times for those needing training.)*	Project Lead
5	Notified Body (NB) plan	Device goals will be shared with NB.	Project Lead
6	Prior guidance	NB, expert panel, deficiencies, and other guidance info received by company in past.	Regulatory Lead
7	Device description	Project scope.	Project Lead
8	Device development	Device developmental history. Is device novel? Does device have CE mark? Past background knowledge, State of the Art (SOTA) details, and alternative therapies list to be provided.	Research and Development (R&D) Lead
9	Manufacturer-held data	Clinical trial data. Literature data. Experience data.	Clinical Affairs Lead / R&D Lead / Post Market Surveillance (PMS) Lead
10	Competitor/other data	Clinical trial data. Literature data. Experience data.	Clinical Affairs Lead / R&D Lead / PMS Lead
11	Equivalence	Prior CER/PER equivalence.	Regulatory Lead
12	Device grouping	Define all SKUs to be included, review Declaration of Conformity (DoC).	Regulatory Lead
13	Document deliverables	Each deliverable document will have an assigned team member with a deadline.	Project Lead

Slide	Slide contents	Slide details	Presenter
14	Document QC/QA	Each document will be reviewed and edited by a more senior writer/ evaluator.	Project Lead
15	CER/PER story line	Provide consistent and clear story line (reviewed at each team meeting).	Project Lead
16	CEP/PEP sign-off date	CEP/PEP will be signed off prior to CE/PE starting (typically 2 weeks after KOM).	Project Lead or CEP/PEP writer
17	Clinical data gaps	Details for PMCFPs/PMPFPs will be anticipated and planned.	PMS Lead and RM Lead
18	Timeline	Review timeline.	Project Lead
19	Review	Ensure all action items are documented and assigned to persons with due dates.	Scribe
20	Next meeting	Reminder: next meeting date, time, and location.	Project Lead

[3] Parking Lot Attendant is person who captures all issues in a spreadsheet with numbered sequence by date. Column headers should include the following.

No.	Date entered	Issue description	Assigned to <Name>	Due date	Notes/comments
1					
2					
3					

Appendix K

Clinical Evaluation Plan Template

Clinical Evaluation Plan
< Product Name >

Approved By:
AUTHOR/EVALUATOR

Print Name/Title Signature Date

MANUFACTURER*

Print Name/Title Signature Date

Revision history

Rev.	Date	Modifications
A	*< DDMONYYYY >*	Created, Initial Release

The file contents were compiled in accordance with European Medical Device Regulation 2017/745 and Medical Device Coordinating Group (MDCG) guidelines. * an asterisk or italic text in this template indicates the term should be replaced with the proper name (e.g., MANUFACTURER* should be replaced with the proper manufacturer name).

Table of Contents

1 Objective and Scope

This clinical evaluation plan (CEP) outlines the process to identify and evaluate clinical data for *Manufacturer* Device** (hereafter collectively referred to as *Product*).*

2 Specifications and Acceptance Criteria

This CEP is a *Manufacturer** quality system component designed to provide a comprehensive analysis of available clinical data regarding the safety and performance of *Product** when used as intended. This CEP will inform a subsequent clinical evaluation report (CER) as required under *Regulation (EU) 2017/745 MDR (e.g., Article 61; Annex XIV, Part A and Annex I, Chapter I)*, and the CER will generally follow the relevant Medical Device Coordinating Group (MDCG) Guidelines.

The CER will document the clinical safety and performance, device lifetime, benefit/risk profile, and acceptability of undesirable side effects to allow *Product** to be marketed in the EU.

Safety

- *<Acceptance criteria, threshold value (XX%)>*

Performance

- *<Acceptance criteria, threshold value (XX%)>*

The prespecified and prequantified threshold values will be reviewed from the background knowledge and state of the art (SOTA) perspective for *<device description>*. The acceptance criteria are expected to guide the clinical evaluation of *<device name>* residual risks and anticipated safety and performance objectives to *<intended use>*.

The CER will discuss all findings including those outside these anticipated acceptance criteria (i.e., threshold values), and the CER will draw conclusions about *<device name>* conformity with the EU MDR 2017/745 General Safety and Performance Requirements (GSPRs) *1, 6, and 8* including clinical safety, benefit/risk profile, performance, lifetime of the device, and acceptability of undesirable side effects. In addition, the CER outcomes from evaluation of these safety and performance objectives and acceptance criteria will be considered in future updates to the CER, risk management (RM), and postmarket surveillance (PMS) plans and activities.

3 Device Description

3.1 Device Name

*Product**

3.2 Device Manufacturer

*Manufacturer** manufactures *Product** at *X* locations globally: *list names* (Table 1).

TABLE 1 Manufacturing locations.

Manufacturer/Distributor	Contract manufacturer	Packaging and sterilization
Manufacturer Name	*Manufacturer Name*	*Manufacturer Name*
City, State Zip Code	*City, State Zip Code*	*City, State Zip Code*

The *Manufacturer** quality management system is certified (*CERT*) under *ISOXXXXX:XXXX* as follows: "*QUOTE*."

3.3 Device Description and Principles of Operation

*Product** is a (*single-use*) device (*describe*) (Table 2).

TABLE 2 *Product* name and auxiliary components.*

Catalog #	Product name	Description	Image
Provide Catalog Number	*Provide Product Name*	*Brief Product Description*	*< image >*

< Describe principles of operations here >

3.4 Lifetime of device

*Product** has an expected lifetime of *<X months or years>* based on *< testing reference >*.

3.5 Intended Use

*Product** devices are intended... *< Intended use >*

3.6 Indications for use

*Product** indications for use are the same in the United States (US) and outside the US to ... *Product** *<Indications for use verbatim in the US> (KXXXXX).*

3.7 Contraindications

In general, *Product** has the following contraindications:

- *List contraindications*

3.8 Safety and Performance Claims

*Product** is associated with several clinical safety and performance claims including:

List claims and include sources

These clinical safety and performance claims will be evaluated in the CER.

3.9 Equivalent Devices

<*Manufacturer Devices* - the comparator products*> were considered equivalent to *Product** based on evaluation of clinical, biological, and technical characteristics when used as intended in the appropriate patient populations. *Manufacturer** had sufficient levels of access to technical files and design history.

Clinical equivalence compared device indications for use and clinical applications, biological equivalence compared the *direct* effect of the device materials on human cells and tissues, and technical equivalence compared physical appearance, design specifications, composition, and operation (Table 3).

TABLE 3 Equivalence Table.

	Device 1 (under clinical evaluation) *Description of characteristics and reference to specifying documents*	***Device 2 (marketed device)*** *Description of characteristics and reference to specifying documents*	**Identified differences or conclusion of no differences in the characteristic**
1. Clinical characteristics			
Same clinical condition or purpose, including similar severity and stage of disease			1.1
Same site in the body			1.2 (Characteristic must be the same for the demonstration of equivalence)
Similar population, including as regards age, anatomy and physiology			1.3
Same kind of user			1.4 (Characteristic must be the same for the demonstration of equivalence)
Similar relevant critical performance in view of the expected clinical effect for a specific intended purpose			1.5
Scientific justification for no clinically significant difference in the safety and clinical performance of *device*, OR a description of the impact on safety and or clinical performance (Use one row for each identified difference in characteristics, and add references to documentation as applicable)			**Clinically significant difference Yes/No**
1.1			
1.2			
1.3			
1.4			
1.5			

(*Continued*)

TABLE 3 (Continued)

	Device 1 (under clinical evaluation) *Description of characteristics and reference to specifying documents*	***Device 2 (marketed device)*** *Description of characteristics and reference to specifying documents*	**Identified differences or conclusion of no differences in the characteristic**
2. Biological characteristics			
Uses same materials or substances in contact with the same human tissues or body fluids			2.1 (Characteristic must be the same for the demonstration of equivalence)
Similar kind and duration of contact with same human tissues or body fluids			2.2
Similar release characteristics of substances including degradation products and leachables			2.3
Scientific justification for no clinically significant difference in the safety and clinical performance of *device*, OR a description of the impact on safety and or clinical performance (Use one row for each identified difference in characteristics, and add references to documentation as applicable)			**Clinically significant difference Yes/No**
2.1			
2.2			
2.3			
3. Technical characteristics			
Device is of similar design			3.1
Used under similar conditions of use			3.2
Similar specifications and properties including physiochemical properties such as intensity of energy, tensile strength, viscosity, surface characteristics, wavelength and software algorithms			3.3
Uses similar deployment methods where relevant			3.4
Has similar principles of operation and critical performance requirements			3.5
Scientific justification for no clinically significant difference in the safety and clinical performance of *device*, OR a description of the impact on safety and or clinical performance (Use one row for each of the identified differences in characteristics, and add references to documentation as applicable)			**Clinically significant difference Yes/No**
3.1			
3.2			
3.3			
3.4			
3.5			
Summary In the circumstance that more than one nonsignificant difference is identified, provide a justification whether the sum of differences may affect the safety and clinical performance of the device.			

These devices have identical clinical applications and indications for use; therefore, they are considered clinically equivalent. < *OR if the devices have different clinical applications or indications for use, etc., describe the differences and justify why the devices can be considered equivalent.* >

Manufacturer Devices are considered biologically equivalent because they are made of <*list of human tissue contacting materials*> and these materials are biologically inert < *OR describe biological impacts on human cells and tissues here* >. Since these materials are identical or have no major differences between the biological impacts of the tissue-contacting materials used, they are considered biologically equivalent. < *OR if the device human tissue contacting materials have any differences in cell/tissue impact, describe the differences and justify why the devices can be considered biologically equivalent.* >

These devices have identical technical features < *describe* >; therefore, they are considered technically equivalent. < *OR if the devices have different clinical applications or indications for use, describe the differences and justify why the devices can be considered equivalent.* >

3.9.1 Justification of Equivalent Devices

*Manufacturer Devices** are considered equivalent to *Product** based on their clinical equivalence (e.g., *clinical applications, intended uses, indications for use, intended populations, environments of use and clinical effect*), biological equivalence (e.g., *interactions between patient-tissue-contacting materials and cells and tissues contacted by the device*; e.g., *patient-tissue-contacting materials are made of biologically inert materials and do not raise any issues of unanticipated safety or efficacy differences*), and technical equivalence (e.g., *device sizes, materials of construction, technical details and filtration capabilities*). < *Give specific details above OR if the devices have differences, describe the differences and justify why the devices can be considered equivalent, including details about clinical, biological, and technical issues contributing to the differences.* >

3.10 Developmental Context

*Products** do not have medicinal, human or animal components (as defined in *Regulation (EU) 2017/745 MDR*) and are based on long standing technologies of <*intended use.*> *Manufacturer** was acquired by *Manufacturer** (formerly *Manufacturer**) in *YYYY*. The level of innovation in the current design is not novel. < *Describe clinical development plan for Product*, including exploratory, first in man, feasibility, pilot and pivotal studies used to secure CE mark.* >

In the EU, *Product** are class *XX* and *Manufacturer** secured a CE mark (*XXXX*) for these devices in *YYYY* (*Notified Body* certificate number: *XXXXX*) under annex I of the European Directive 93/42/EEC (Table 4).

TABLE 4 Devices Regulatory Status.

Device name	EU regulation	US regulation
*Product**	Class: Directive: Annex: Rule:	Class: Regulation Number: 510(k) Number: Product Code:

In the US, *Product** were cleared in *YYYY* as Class *X* devices under FDA regulation *XXX.XXX* and product code *XXX* (*describe*). *Product** are currently available in the following countries: *X, Y, Z.*

*Manufacturer** currently serves as an original equipment manufacturer (OEM) by providing *Product** to <*list other manufacturers* > for use in their private label medical devices. Over *X* devices have been manufactured and shipped to *company* since *YYYY*.

4 Background and State of the Art (SOTA)

< *Describe background/SOTA information here.* >

In addition, < *quantity* > state-of-the-art (SOTA) therapies and *device type** devices comparable to *Manufacturer** *Product** have been identified for further analysis in the CER (Table 5).

TABLE 5 Comparable therapies and devices.

Therapy/product name (Mfr, city, state)	Description	Intended/documented uses

< Discuss similar therapies with details about the clinical data to be included and the relative safety and efficacy to be explored in the CER. >

4.1 Safety and Performance Considerations

The CER will evaluate *Product** conformity with the GSPRs 1, 6, and 8 of the EU Medical Device Regulation 2017/745 (generally including an evaluation of the clinical safety, performance, benefit/risk profile, and acceptability of undesirable side effects).

*Product** has a long history of safe use, and past adverse device events and clinical concerns have included (but are not limited to) *< list adverse device events and clinical concerns here and quantify occurrence/s and include relationship/s to the SOTA as well as the subject and equivalent devices >*.

*Manufacturer** has a robust device development program, and potential failure modes have been identified for *Product**, including:

- *< List performance issues and risks/failure modes and associated effects here and quantify occurrence/s and include relationship/s to the SOTA as well as the subject and equivalent devices. >*
- *Note: include a summary of potential future risk.*

4.2 Benefits and Risks Considerations

*Manufacturer** records and tracks clinical benefits and risks of *Product** when used as indicated in the appropriate patient population (Table 6).

TABLE 6 Benefits vs. Risks.

Benefits
< describe benefit >
< describe benefit >
< describe benefit >
< describe benefit> ...
Risks
< describe risk >
< describe risk >
< describe risk >
< describe risk> ...

The *Manufacturer* Product** prior CER and *risk management reports* have discussed these clinical benefits, risks, misuse conditions and safety concerns for patients and users. *<Describe outcomes to be analyzed in the CER with regard to the list above.>*

5 Clinical Evaluation Process

The process for the identification, appraisal, and analysis of clinical data for the *Product** CER includes the following:

1. Planning, as detailed in this CEP
2. Discovery and data collection to identify clinical data for evaluation
    - Technical documentation
        - Clinical trial data
        - Risk management files
        - Product labels, instructions for use, marketing materials
        - Prior clinical evaluation reports
        - Postmarket surveillance plans and reports
        - Applicable regulations, standards and guidelines
    - Data found by searching clinical literature databases
        - PubMed
        - Embase
        - Internet searches (e.g., Google Scholar), if needed
        - Hand searches, if needed
    - Experience data held by manufacturer and found by searching experience databases
        - Complaints and risk mitigation activities
        - Manufacturer and User Facility Device Experience (MAUDE), and other adverse event database searches
3. Clinical data appraisal
    - Identify inclusion criteria
    - Identify exclusion criteria
    - Apply weighting criteria
4. Clinical data analysis and benefit:risk assessment
    - Analyze each document and develop written summaries of included data
    - Tabulate important benefit:risk info and summarize key findings
    - Assess benefit:risk of product and document findings
5. Writing CER
    - Provide introduction, methods, analyses, results, and conclusions
    - State if identified clinical risks have been mitigated
    - State if benefits outweigh the risks for continued use as intended
    - Compile references and appendices
    - Identify PMCF needs
6. Review and approval by manufacturer-provided clinical expert

5.1 Data Selection Rationale

The clinical data evaluated in the CER will reflect the SOTA therapies specifically related to *device type* when used as indicated. Clinical data will be selected based on applicability to *Product** or similar devices. Clinical data not relevant to *Product** or similar, SOTA devices will be excluded from review.

Efforts will be made to ensure the clinical data included in the CER are fair and balanced concerning the benefits and risks of the *Product**, accessories and/or similar devices and technologies. The clinical data will include both favorable and unfavorable outcomes, and analysis will be conducted by *<name of responsible third party OR if this is an employee of the company, then remove this sentence.>* an educated third party without any current, direct interest in *device type* but with the training and experience required to create the CER and to evaluate the clinical data.

Although extensive clinical data searches will be conducted, some clinical data may be missed due to the specific search strategies used to identify the clinical data. Independent of this CER, *Manufacturer** will continue to refine and verify the clinical data identified in ongoing literature searches and will update risk management activities, postmarket surveillance product reviews, and postmarket clinical follow-up activities related to the *Manufacturer* Product** and accessories for the anticipated life cycle of the device.

5.2 Identification of Clinical Data

Clinical data will be searched in appropriate databases using the following key words alone and in combination with "AND"/"OR" connectors: *<list specific search terms and filters and include references to specific strategies used like PICO, Cochrane, MOOSE or others.>*

5.2.1 Justification of Clinical Literature Database Selection

PubMed is a free resource developed and maintained by the National Center for Biotechnology Information (NCBI), at the U.S. National Library of Medicine (NLM), located at the National Institutes of Health (NIH). PubMed houses more than 37 million records from biomedical, health, life science, behavioral science, chemical science, and bioengineering literature from MEDLINE, life science journals, and online books, including more than 2 million "OLDMEDLINE" articles from 1946 to 1965, and continues to grow with the ongoing conversion of older medical indexes into machine readable formats.

Embase is a biomedical and pharmacological database containing bibliographic records with citations, abstracts, and indexing derived from biomedical articles in peer-reviewed journals. Embase contains more than 37.2 million records spanning 1947 to present, with more than 1 million records added annually. The Embase journal collection is international with more than 8,500 indexed biomedical journals from 90 countries. This international database provides a comprehensive collection of peer-reviewed literature.

Additional published articles may be identified and obtained by reviewing articles identified in the original searches and from other sources (e.g., Internet searches). Using multiple databases improves the likelihood of recovering a greater percentage of the global literature available on specific topics.

5.2.2 Justification of Clinical Experience Database Selection

Clinical experience data from the Australian Therapeutic Goods Administration (TGA) Database of Adverse Event Notifications (DAEN) and the US Food and Drug Administration (FDA) MAUDE, Recall, 510(k), and other clinical experience databases will be searched as part of this CER process to identify experience reports related to *Product** and similar devices. The search will be restricted to 1 year if more than 100 results are available within the 1-year "look back" history. If more than 500 results are found within 1 year, the results will be further restricted to a smaller time frame or more restricted device focus. (The MAUDE database will not release more than 500 results for any MAUDE search even if many thousands of results are present in the database.) In addition, the Total Product Life Cycle (TPLC) database will be searched for the general US FDA Product Code *XXX* (*description of code*).

The DAEN and FDA MAUDE databases maintain medical device reports from a sources including manufacturers, healthcare providers, and the public. DAEN is updated up to 3 months prior to the initiated search, MAUDE is updated monthly, and the search page date range options for both reflect the most recent updates.

The FDA Medical Device Recalls database maintains classified reports since November 2002, and in 2017 began to include correction or removal actions initiated by a firm prior to review by the FDA. The status is updated by the FDA as new violations are classified, and if and when recalls are terminated.

The FDA TPLC database maintains premarket and postmarket data about medical devices including premarket approvals (PMAs), premarket notifications (510[k]), adverse events, and recalls, and is updated as new data are reported from each individual data source. The TPLC data are preferred when MAUDE data are extensive because the FDA has grouped and categorized these data in the TPLC database.

5.3 Appraisal of Clinical Data

Identified clinical data will be appraised for relevance and applicability. Articles will be appraised using a weighted coding process. This appraisal includes an evaluation about the relative weight for each article according to relevance of clinical data to the subject device or similar devices (i.e., whether subject device was used in study), strength of study type (i.e., metaanalyses are given higher weight than narrative reviews and randomized controlled trials are given higher weight than nonrandomized, uncontrolled trials as described by the Oxford Centre for Evidence-Based Medicine (OCEBM) Levels of Evidence, seevOCEBM Reference), and number of study subjects (i.e., studies with more subjects are weighted higher than studies with fewer subjects). In addition, studies using *Product** will be presented before similar devices; meta analyses and systematic reviews will be presented before individual randomized controlled trials; and case studies will be presented last if they include benefits or risks not included in other literature already included in the CER. *<describe specific references here when used>*

5.3.1 Inclusion Criteria

Clinical data will be included in the CER if the data provide clinical outcome parameters associated with use of *Product** or similar devices to treat *disease**. For published literature, clinical data will be taken from peer-reviewed, scientific articles to ensure clinical data were documented and reviewed, and each author had the required background and expertise to support the study. Certain SOTA literature may also explore general treatments for *disease** to clarify device safety and performance when used as intended in the *disease** treatment setting.

5.3.2 Exclusion Criteria

In general, clinical data may be excluded if the data are:

- not about clinical data relevant to *Product** or similar devices;
- without pertinent clinical data regarding safety, performance/efficacy and/or device qualities for *device type** outcomes and/or effects on *disease** data when devices were used as intended to treat *disease**;
- phantom, cadaver, bench, or animal studies;
- not in English and have no relevant, novel clinical data;
- isolated case report or clinical case series including three or fewer cases; or
- proceeding, poster, or abstract of a presentation (even those including clinical data), since they do not typically provide sufficient information for a full clinical data evaluation.

6 Postmarket Clinical Follow-up Planning

A postmarket clinical follow-up (PMCF) plan will be designed by *Manufacturer** to address any gaps in clinical data supporting the safety and performance of the device as documented in the CER.

6.1 Postmarket Surveillance Report

Complaints, vigilance, and occurrence of possible adverse events associated with *Product** as well as other adverse advents published in literature and adverse event databases will be investigated within the PMSR.

6.2 Device Changes

Since the last CER, *Manufacturer** indicated no substantial changes were made to the design, materials, manufacturing procedures or *Product** product claims or labeling supplied by the manufacturer (e.g., the IFU, product brochures, and websites). *<If changes have occurred, describe changes here with particular attention to any safety and performance issues related to the changes. >*

6.3 Newly Emerged Clinical Concerns

*Product** had no new emerging clinical concerns identified. If emerging concerns were present,describe concerns here with particular attention to any safety and performance issues related to the device.

6.4 Summary of Safety and Clinical Performance

<Remove this section if not applicable>

For this *<Class III> OR <implantable>* device, the device description (including background, developmental context and SOTA) developed in this CEP and subsequent CER, along with a summary of the clinical evaluation with information on residual risks, undesirable effects, warnings, precautions, and suggested profile and training for users will be used to inform a publically available summary of safety and clinical performance pursuant to Article 52 of Regulation (EU) 2017/745 MDR.

7 Clinical evolution report Conclusions

The CER will identify, appraise, and analyze clinical data from clinical trials, clinical literature, and clinical experiences pertaining to *Product** and similar therapies within the SOTA using these devices. Benefits and risks associated with *device type** will be properly specified and tabulated to allow a device acceptability evaluation in actual clinical use. The CER conclusion will ensure the evaluation included sufficient clinical evidence to confirm compliance with GSPRs 1, 6 and 8 when *Product** was used as indicated.

8 References

OCEBM Levels of Evidence Working Group *(Jeremy Howick, Iain Chalmers (James Lind Library), Paul Glasziou, Trish Greenhalgh, Carl Heneghan, Alessandro Liberati, Ivan Moschetti, Bob Phillips, Hazel Thornton, Olive Goddard and Mary Hodgkinson)*. The Oxford Levels of Evidence 2. Oxford Centre for Evidence-Based Medicine. Accessed on 21FEB2021 at https://www.cebm.ox.ac.uk/resources/levels-of-evidence/ocebm-levels-of-evidence

Appendix L

Clinical evaluation report template

Logo
Clinical evaluation report
Product Name

Authored by:	**Reviewed by:**	**Approved by:**
Name, Title, Date, Signature	Name, Title, Date, Signature	Name, Title, Date, Signature
Name, Credentials Title Company	**Name, Credentials** Title Company	**Name, Credentials** Title Company
Name, Credentials Title Company	**Name, Credentials** Title Company	

Medical Reviewer
I have reviewed the available clinical trial, literature, and experience data, and reviewed the specific, detailed contents of this report. I have drawn the same conclusions documented in this report.

Name: Name, Credentials
Title: Title, Company/Affiliation

Date: DATE

Revision History

Rev.	Date	Modifications
XX		Initial Release.
Date of next Clinical Evaluation Report		<DDMMMYYYY>

The contents of this file were compiled in accordance with European Medical Device Directives 93/42/EEC and 2007/47/EC, Medical Device Regulation 2017/745 and Guidelines on Medical Devices in MEDDEV 2.7.1 rev. 4 entitled "Clinical Evaluation: A Guide for Manufacturers and Notified Bodies Under Directives 93/42/EEC and 90/385/EEC" (June 2016).

Table of Contents

Table of Abbreviations and Definitions of Terms

Term	Definition
AE	Adverse event
ANOVA	Analysis of variance
CDRH	Center for Devices and Radiological Health
CE (mark)	Conformité Européene
CEP	Clinical evaluation plan
CER	Clinical evaluation report
CFR	Code of Federal Regulations
CGMP	Current good manufacturing practice
DAEN	Database of Adverse Events Notifications
EU	European Union
FDA	Food and Drug Administration
GSPR	General safety and performance requirement
ISO	International Organization for Standardization
MAUDE	Manufacturer and User Facility Device Experience
MEDDEV	Medical device documents
MFR	Manufacturer
OCEBM	Oxford Centre for Evidence-Based Medicine

Term	Definition
OEM	Original equipment manufacturer
PMCF	Postmarket Clinical Follow-Up Plan
PMS	Postmarket surveillance
QS	Quality system
SD	Standard deviation
SE	Standard error
SEM	Standard error of mean
SOTA	State of the art
TGA	Therapeutic goods administration
TPLC	Total product life cycle
US	United States

1 Executive summary

This clinical evaluation report (CER) is focused on the safety and performance of *Manufacturer* Device** (hereafter collectively referred to as *Product**) using clinical data from clinical trials, scientific literature, and postmarket clinical uses. This <*single*-use, *sterile* and *nonsterile*> *Product* has been used as a standard practice <for many years>*. *Product** is not novel and has been CE marked (CE *XXXX*) since *XXXX* in the European Union (EU). In the United States (US), *Product** was cleared for market by the Food and Drug Administration (FDA) in *XXXX*.

No clinical investigations were required and none was conducted by Manufacturer prior to marketing Product**. This CER evaluated clinical data in *XX* published articles including *XX* metaanalyses, *XX* review articles, and *XX* clinical study articles reporting *Product** safety and performance data in *XXXX* patients, and *XX* clinical study articles reporting safety and performance of similar state-of-the-art (SOTA) devices in *XXXX* patients. These *X* articles provide appropriate clinical data to evaluate safety and performance of *Product** when used *<provide function>*. Reported benefits include *<list benefits>*. These benefits outweighed the risks of *<state risks>*.

Clinical experience data includes manufacturer-held complaint and sales data for *Product** and publicly available experience data from *<list databases from PMSR or PSUR>* *<the Australian Therapeutic Goods Administration (TGA) Database of Adverse Events Notifications (DAEN), Manufacturer and User Facility Device Experience (MAUDE)>*, recall, clinical trial, warning letter, and *Total Product Life Cycle (TPLC)* databases. These experience data reflect similar safety and performance as reported in the literature. *Manufacturer** reported *X* complaints including *<list adverse events>*. The *DAEN* and *MAUDE* databases documented *X Product** and *X* similar, SOTA device problems including: *<describe device issues>*. The *TPLC* database showed *X* device problems for *device category*. *Manufacturer** received *X* warning letters and *X* recalls for *Product**. Product literature was found to discuss the benefits and risks appropriately.

Overall, this evaluation of clinical investigations, literature and experience data for *Product** and similar SOTA devices as well as ongoing *Product** risk management processes provided a favorable benefit/risk profile supporting safe use of *Product** as indicated in appropriate patient populations in conformity with EU Medical Device Regulation 2017/745 General Safety and Performance Requirements (GSPRs). This broad review of the global literature and ongoing clinical use data was considered sufficient for this CER, and no new clinical investigation was required or performed.

2 Purpose, specifications, and acceptance criteria (safety and performance criteria)

This CER evaluates clinical data from clinical trials, literature, and experiences about the safety and performance of *Manufacturer* Product**.

2.1 Safety

- *<Acceptance criteria, threshold value (XX%)>*

2.2 Performance

- *<Acceptance criteria, threshold value (XX%)>*

To ensure a robust clinical evaluation and to guide the clinical evaluation of *Product** residual risks, this *Product** CER also evaluated clinical data from background knowledge and similar SOTA *<device name>* or alternatives, like using *<device name>* at all. Specifically, this CER evaluated conformance to EU MDR (Reg. 2017/745) GSPRs 1, 6, and 8 throughout the device lifetime when *<device name>* was used to provide *<indication for use>*.

3 Device description

3.1 Device name and manufacturer

Name: *Product**
*Manufacturer** manufactures *Product** at *X* locations globally: *list names* (Table 1).

TABLE 1 **Manufacturing locations.**

Manufacturer/Distributor	Contract manufacturer	Packaging and sterilization
Manufacturer Name *City, State Zip Code*	*Manufacturer Name* *City, State Zip Code*	*Manufacturer Name* *City, State Zip Code*

*Manufacturer** quality management system is certified (*CERT*) under *ISOXXXXX:XXXX* as follows: "*QUOTE*."

3.2 Device description and principles of operation

*Product** is a (*single-use*) device (*describe*) (Table 2).

TABLE 2 ***Product* name*** **and auxiliary components.**

Catalog #	Product name	Description	Image
Provide Catalog Number	*Provide Product Name*	*Brief Product Description*	*<image>*

<Describe principle of operations here>

3.3 Lifetime of device

* *Product** has an expected lifetime of *X months or years*.

3.4 Device intended use

*Product** devices are intended... *<intended use>*

3.5 Device indications for use

*Product** indications for use are the same in the US and outside the US to ...

- *Product* <indications for use verbatim in the U.S.> (KXXX)*

3.6 Contraindications

In general, *Product** has the following contraindications:

- *List contraindications*

3.7 Safety and performance claims

*Product** is associated with several clinical safety and performance claims, including:

- *List claims and include source*

3.8 Description of equivalent devices

*Product** are considered equivalent due to their clinical, biological, and technical similarities. Clinical equivalence was determined by comparing indications for use and clinical applications for the device, biological equivalence was evaluated by comparing the *direct* effect of the device materials on human cells and tissues, and technical equivalence was evaluated by comparing physical appearance, design specifications, composition, and operation (Table 3).

TABLE 3 Equivalence table.

	Device 1 (under clinical evaluation) Description of characteristics and reference to specifying documents	**Device 2 (marketed device)** Description of characteristics and reference to specifying documents	**Identified differences or conclusion of no differences in the characteristic**
1. Clinical characteristics			
Same clinical condition or purpose, including similar severity and stage of disease			1.1
Same site in the body			(Characteristic must be the same for the demonstration of equivalence) 1.2
Similar population, including as regards age, anatomy and physiology			1.3
Same kind of user			(Characteristic must be the same for the demonstration of equivalence)1.4
Similar relevant critical performance in view of the expected clinical effect for a specific intended purpose			1.5
Scientific justification for no clinically significant difference in the safety and clinical performance of device, OR a description of the impact on safety and or clinical performance (Use one row for each identified difference in characteristics, and add references to documentation as applicable)			**Clinically significant difference Yes/No**
1.1			
1.2			
1.3			
1.4			
1.5			

(Continued)

TABLE 3 (Continued)

	Device 1 (under clinical evaluation) Description of characteristics and reference to specifying documents	**Device 2 (marketed device)** Description of characteristics and reference to specifying documents	**Identified differences or conclusion of no differences in the characteristic**
2. Biological characteristics			
Uses same materials or substances in contact with the same human tissues or body fluids			(Characteristic must be the same for the demonstration of equivalence) 2.1
Similar kind and duration of contact with same human tissues or body fluids			2.2
Similar release characteristics of substances including degradation products and leachables			2.3
Scientific justification for no clinically significant difference in the safety and clinical performance of device, OR a description of the impact on safety and or clinical performance (Use one row for each identified difference in characteristics, and add references to documentation as applicable)			**Clinically significant differenceYes/No**
2.1			
2.2			
2.3			
3. Technical characteristics			
Device is of similar design			3.1
Used under similar conditions of use			3.2
Similar specifications and properties including physiochemical properties such as intensity of energy, tensile strength, viscosity, surface characteristics, wavelength, and software algorithms			3.3
Uses similar deployment methods where relevant			3.4
Has similar principles of operation and critical performance requirements			3.5
Scientific justification for no clinically significant difference in the safety and clinical performance of device, OR a description of the impact on safety and or clinical performance (Use one row for each of the identified differences in characteristics, and add references to documentation as applicable)			**Clinically significant differenceYes/No**
3.1			
3.2			
3.3			
3.4			
3.5			
Summary In the circumstance that more than one nonsignificant difference is identified, provide a justification whether the sum of differences may affect the safety and clinical performance of the device.			

Manufacturer Products** and *Marketed Device* are considered clinically equivalent based on clinical, biological, and technical characteristics because . . ..

Although all *X* products are indicated for *X*, the disease state for each device is slightly different *<list>*. These disease states are still within the same class of *<type of disease>* and the devices are still considered equivalent.

Biomaterials used for *X* devices are not significantly different; however, *<describe any differences>*. These *differences* do not impact the biological interactions of the device (*reason*), and *devices* are still considered equivalent.

Technical differences <Describe any differences> do not alter crucial technical specifications and main technical device functions are identical and principles of operation and technical features related to safety and performance of *device* when used as intended are considered equivalent for *Manufacturer* Products**.

Manufacturer Products** are considered equivalent to each other because clinical applications and intended uses, materials and design specifications are the *same*.

3.9 Developmental context

*Product** contain no medicinal, human or animal components and are based on long standing technologies of *<intended use.> Manufacturer** was acquired by *Manufacturer** (formerly *Manufacturer**) in *yyyy*. The level of innovation in the current design is not novel.

In the EU, *Product** is Class *XX* and *Manufacturer** secured a CE mark (*XXXX*) for this device in *YEAR* (*Notified Body* certificate number: *XXXXX*) under annex I of the European Directive 93/42/EEC (Table 4). In the US, *Product** was cleared in year *XXXX* as a Class *X* device under FDA regulation *XXX.XXX* and product code *XXX* (*describe*). *Product** is currently available in the following countries: *X, Y, Z*.

TABLE 4 Devices regulatory status.

Device name	EU regulation	US regulation
*Product**	Class: Directive: Annex: Rule:	Class: Regulation Number: 510(k) Number: Product Code:

*Manufacturer** currently serves as an original equipment manufacturer (OEM) by providing *Manufacturer* Product** to *<list other manufacturers>* for use in their private label medical devices. More than *X* devices have been manufactured and shipped to *company* since *yyyy*.

3.10 Background and state of the art

<Describe background/SOTA information here.>

X guidelines, X metaanalyses and X systematic reviews describe the background and SOTA treatment options for <intended use.> Describe metaanalyses and systematic reviews.

In addition, *<quantity>* SOTA therapies and *device type** devices comparable to *Manufacturer* Product** have been identified for further analysis in the CER (Table 5).

TABLE 5 Comparable therapies and devices.

Therapy/Product name (Mfr, city, state)	Description	Intended/documented uses

<Discuss similar therapies with details about the clinical data to be included and the relative safety and efficacy to be explored in the CER.>

These background/SOTA articles identified appropriate safety and performance objectives for clinical evaluation (Table 6)

TABLE 6 Safety and performance data for SOTA.

Author (year)	N	*Outcome (measurement)*	*Outcome (measurement)*	*Outcome (measurement)*	*Outcome (measurement)*
Total range	NA				

Consider separate safety/performance tables or bulleted list commensurate with volume of data. *<Summarize safety and performance objectives identified for evaluation of Product* in this CER>*

In addition, benefits and risks associated with *device type** used for *intended use* included:

3.10.1 Benefits

- *<List>*

3.10.2 Risks

- *<List>*

This broad review of SOTA for *device type** in addition to specific safety and performance and benefit/risk evaluations for *Product** is intended to create a robust CER. This provides a wider perspective on possible safety and performance issues as well as benefits and risks for these devices.

3.11 Justification of clinical data type

Manufacturer Product** are well-established medical devices based on existing technologies with a long history of use. This CER evaluates clinical trial, clinical literature, and clinical experience data about the safety and performance of *Product** and similar devices. In particular, *Product**-specific data included... No additional *Product**-specific clinical data was available or required to complete this CER.

4 Clinical investigation review

4.1 Summary of clinical investigations

The manufacturers have not conducted any clinical investigations or postmarket registry studies designed to evaluate the safety and performance of *Manufacturer* Product**. <Or describe clinical trial data in this section with methods pointing to Appendix 9.1 CEP>

4.2 Critical evaluation/analysis of clinical investigations

No company sponsored clinical investigations of safety and performance were available, so no evaluation or analysis was possible.

4.3 Conclusions of clinical investigation review

Although no clinical investigations were conducted by *manufacturer*, others have published results of clinical studies using *Product** and other <*product type*> reported in the following section, "Clinical Literature Review."

5 Clinical literature review

5.1 Methodology

5.1.1 Objectives

This clinical literature review evaluates safety and performance of *Manufacturer* Product** and similar devices in a variety of patient populations (Appendix 9.1). This review includes literature about similar devices made and marketed by other manufacturers to obtain a broad view of safety and performance issues for use of *Manufacturer* Product**.

5.1.2 Search criteria to identify literature

Available literature was gathered from the *prior CER*, manufacturer, the Internet, and independent literature searches designed to identify published review articles and clinical study reports about *Manufacturer* Product**, or more broadly, <*type of devices*>. Independent literature searches were conducted on *DATE* were updated and verified on *DATE* and *DATE* using the US National Library of Medicine PubMed database to identify articles (Appendix 9.2). Search terms included: <*list all search terms and filters*>. Additional searches for the terms <*list all search terms*> on *DATE* were conducted in the Biosis, Embase, and SciSearch databases.

5.1.3 Selection of relevant clinical literature

Independent authors and reviewers with appropriate qualifications and experience (Appendix 9.3) were selected (Appendix 9.4) to evaluate appropriate literature and write this CER analyzing performance and safety of *Product** and similar devices. Articles were coded and excluded from further review for the following reasons: not about performance and safety of <*type of devices*>, not in English, not about a similar device or used a different <*type of device*>, a cadaver/*in vitro*/animal study, no abstract available, or a duplicate study. Full-length articles are attached at the end of this CER (Appendix 9.5).

5.1.4 Appraisal of relevant clinical literature

Articles were appraised using a weighted coding process. This appraisal included an evaluation about the relative weight for each article according to relevance of clinical data to the subject device or similar devices (i.e., whether subject device was used in study), strength of study type (i.e., metaanalyses are given higher weight than narrative reviews and randomized control trials are given higher weight than nonrandomized, uncontrolled trials as described by the Oxford Centre for Evidence-Based Medicine (OCEBM) Levels of Evidence), and number of study subjects (i.e., studies with more subjects were weighted higher than studies with fewer subjects). In addition, clinical study reports are presented in tables and text according to their relative weights (e.g., subject device articles are presented first in tables and text, randomized controlled trials are presented first and case studies are only presented to report a rare safety signal).

5.1.5 Critical evaluation/analysis of clinical literature

Analysis of selected literature was performed by reading and analyzing the following information from each article (prioritized primary sources and scientifically valid methods):

- Title and Abstract
- Patient Population
- Methods
- Performance and Safety Results
- Limitations
- Conclusions/Comments

Safety was evaluated based on adverse events in humans (*examples*) and other complications associated with <*type of device*>. Performance was evaluated based on how *device* performed in <*intended use, desired outcome*> as well as any device malfunctions (*examples*).

5.2 Results of clinical literature review

5.2.1 Summary of clinical literature

This CER initially evaluated *X* articles, and *X* were selected for further analysis after excluding *X* articles (Fig. 1).

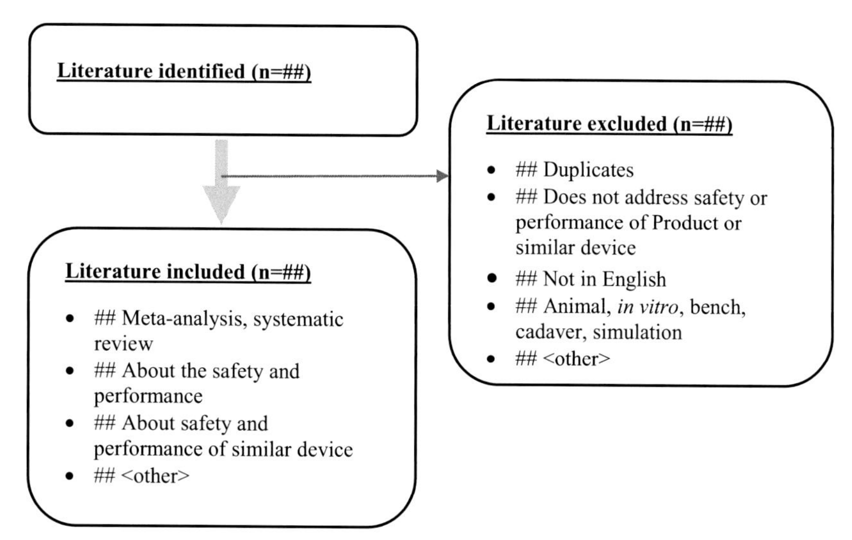

FIGURE 1 Flow chart showing inclusion and exclusion of relevant literature.

X included publications evaluated the safety and performance of <*type of device*> in *X* subjects as follows:

- *List how each included study was conducted and patient population*

X of the included *X* articles are review articles, while *X* of these articles evaluated safety and performance of *Product**in *X* patients.

5.2.2 Metaanalyses and systematic reviews summary

X metaanalyses and *X* systematic reviews included evaluations of *Product** (Table 7). *Describe metaanalyses and systematic reviews.*

TABLE 7 Metaanalyses and systematic reviews.

Author (year)	Method	Safety	Performance	Conclusion
First Author last name (year)	*Brief description*	*State the safety data provided in the study*	*State the Performance results of the study*	The authors state: *"Provide the conclusion quote stated in the study"If not conclusion quote, paraphrase the conclusion in the publication*

The metaanalyses and systematic reviews discussed. . .

5.2.3 Safety and Performance Publications

X articles evaluated safety and performance of *Product** (Table 8).

TABLE 8 Safety and Performance Data.

Author (year)	N	Method	Device (Mfr., City, State)	Safety	Performance	Conclusion
Author (year)		*Population: (age etc.) Objective: Disease State: Measures:*	*Device Name Manufacture Name, City, State of Manufacturing facility*	*State the safety data provided in the study*	*State the Performance results of the study*	The authors state: "*Provide the conclusion quote stated in the study*" *If not conclusion quote, paraphrase the conclusion in the publication*

List of abbreviations for table.

< Alternative option for Table 8 >

Author (year)	N	*Outcome (measurement)*	*Outcome (measurement)*	*Outcome (measurement)*	*Outcome (measurement)*
Total Range	NA				

Consider separate safety/performance tables commensurate with volume of data.

Continued narrative for table above.

X studies evaluated clinical measures including (but not limited to) < *list measures: e.g., lengths of hospital stay, pain, healing complications* >. The safety of *Product**in *X* patients was demonstrated by < *describe safety results* >, and the most common adverse events included: < *list AEs* >. Overall, studies reviewed in this CER reported that *Product** performed safely as intended <*describe performance results*>.

5.3 Conclusions of clinical literature review

This CER evaluated *X* articles (*X* reviews and *X Product** clinical studies) for the safety and performance of *Product** in a clinical setting. The safety of *Product** was acceptable in all studies evaluated. Safety concerns included: *list*. These clinical literature articles were assessed against the prespecified safety and performance criteria (Table 9). Performance was also acceptable in all studies and *Product** performed as intended with highly reliable intended use and with few clinical risks (e.g., *list*). These studies explored variability in *intended use* with many *types of devices* in various situations (e.g., *list how the devices were used in the different studies*).

TABLE 9 Safety and performance acceptance criteria evaluation.

Author (year)	N	Performance *criteria* (%)(n/N)	Safety *criteria* (%)(n/N)	Safety *criteria* (%)(n/N)
Background, SOTA probe covers (by study size)				
Total/Range	*x–y*	*% (xx/yy)*	*% (xx/yy)*	*% (xx/yy)*
Requirement		*X%*	*x%*	*x%*
Meets requirement?		Yes/No	Yes/No	Yes/No

(Continued)

TABLE 9 (Continued)

Author (year)	N	Performance *criteria* (%)(n/N)	Safety *criteria* (%)(n/N)	Safety *criteria* (%)(n/N)
Subject company devices				
Total/Range	*x–y*	*% (xx/yy)*	*% (xx/yy)*	*% (xx/yy)*
RequirementMeets requirement?		*X%* Yes/No	*x%* Yes/No	*x%* Yes/No

<Safety and performance summary (e.g., criteria met/not met)>

Overall, clinical experience data and benefit/risk analyses indicate that *Product** performs safely and effectively, and this CER supports the continued use of these devices as indicated in appropriate patient populations in compliance with the EU MDR.

6 Clinical experience review

6.1 Methodology

6.1.1 Objectives

To evaluate the clinical safety and performance of *Manufacturer* Product** and similar state-of-the-art devices using available clinical experience data from the *Manufacturer* Product** PMS plan and postmarket surveillance report (PMSR) <or periodic safety update report (PSUR)>.

6.1.2 Postmarket data sources and tools

As part of ongoing, continuous PMS for *Product**, *Manufacturer** reviews key sources of PMS data including complaints, feedback, AEs, and case reports received through the *Manufacturer** vigilance system, published literature and scientific session presentations, international issues databases including *MAUDE*, and data from PMCF studies and risk analysis.

If Frestedt is conducting or supplementing experience data analysis outside PMSR, consider the following.

Identification

All available product literature and complaint data was gathered from *Manufacturer** and the Internet. Searches of the Australian Therapeutic Goods Administration (TGA) Database of Adverse Event Notifications (DAEN) for medical devices, and four publicly available US FDA databases including the MAUDE, Warning Letters, TPLC, and medical device recall databases were searched on *DATE*. *<insert other OUS databases searched>* were also used. Additionally, the publicly available clinical trials database (http://www.clinicaltrials.gov) was searched on *DATE* to identify clinical trials related to *Manufacturer* Product**. DAEN, MAUDE, Warning Letters, recalls, and clinical trials databases were searched using the following search terms for specific manufacturers: *<list search terms>*. Additional searches were conducted for the following brand names: *list*. TPLC searches used product codes *XXX* and *XXX*.

Appraisal

Reports were included if they pertained to *Product** or similar devices. MAUDE reports were excluded if they detailed events regarding a nonsimilar device or did not provide sufficient adverse events data. Warning letters and recalls were similarly excluded if they did not relate to a *<type of device>*. Clinical trials were excluded if they were not about safety and performance of *Product** or similar devices. In addition, complaints and product literature for *Manufacturer* Product** were reviewed for reports of clinical uses and design validation activities not published in medical literature.

Analysis

Each clinical experience document was analyzed to determine whether an injury occurred and to define the type of injury of mechanical failure, if any. The number of patient injuries, adverse events, and product problems were documented, and general information about each event (report identifier, date, model/catalog number, brand name) as well as a summary of the event was recorded for each report.

6.2 Results of clinical experience review

6.2.1 Manufacturer* PMSR complaints, experience, and risk management

*Manufacturer** reported *XX* complaints from *Date* to *Date*. The most common complaints included *X* (Table 10).

TABLE 10 Complaints (*date–date*).

Complaint	Number
Problem description	##
Total	*n*

More than *X Manufacturer* Product** were sold from *yyyy* through *yyyy*, giving an estimated complaint rate of # per cent (# complaints among # *units sold*). This small number of complaints relative to the large number of devices on the market underscores the robust and well-established reliability of *Manufacturer* Product**.

6.2.2 Experience database reports

The *DAEN* and *MAUDE* databases included *X* reports about *Manufacturer* Product** and *X* reports about similar devices. The most common *DAEN/MAUDE* issues included *X* (n = *X*), *X* (n = *X*), *X* (n = *X*) (Table 11).

TABLE 11 *DAEN/MAUDE* reports for all related products.

Device Issue	*DAEN*	*MAUDE*	Total
Device issue			##
Device issue			##
Total			##

<*Describe any patient death and/or injuries.*> <*No*> follow-up was available for these patient injury/adverse event reports.

6.2.3 Recalls

X recalls were issued to *Manufacturer** on *date* for *Product** due to <*reason for recall*>. The recall involved *###* units and was terminated on *date*.

6.2.4 Clinical trials

X clinical trials were found related to *Product** as follows:

- "*Title*" (NCT*XXXXXXX*) <consider adding status here as well as linking to any published works above>

< summarize list here >

6.2.5 Warning letters

Manufacturer received an FDA warning letter on *date* for *reason for letter*. This warning letter and close-out letter issued on *date* are available at: *links*

6.2.6 Total product life cycle

The TPLC database reported *X* premarket reviews, *X* recalls, and *X* device problems listed for product code *XXX* (*define*) under regulation number <*regulation #*> (Table 12).

TABLE 12 Device problems for *XXX* product code.

Device problems	<XXX>	<YYY>	Total
Problem			#
Total			#

The most common device problem was *X*. These findings are generally consistent with clinical trials, literature, complaints, and other clinical data evaluated in this CER.

6.2.7 Product literature

Manufacturer Product** product literature (e.g., data sheets, product inserts, and instructions for use (IFU)) addresses safety and performance issues discussed in this CER. Data sheets provide statements about *specs* (e.g., "*list*") and *<other spec>* (e.g., "*list*"). Product literature and IFU provides statements about product specifications as well as warnings and precautions:

- *List*

Manufacturer Product** product literature adequately addresses known clinical risks and product performance issues associated with *Product**.

6.3 Clinical experience review conclusions

Manufacturer Product* PMSR <Or PSUR> concluded* complaints and other experience reports indicated a low level of device malfunctions and manufacturing issues and product literature adequately documented safety and performance issues. *X* complaints reported and *X MAUDE* reports included only *<rare injuries>* in the setting of more than *X* uses. *<Add conclusions about other experience data here.>* Product literature documented safety and performance issues adequately and complaints, *MAUDE* reports, and *TPLC problems* indicated a low level of device malfunctions and manufacturing issues. *<Describe any planned clinical trials, TPLC issues, recalls, or WL issued to manufacturer.>* Safety and performance acceptance criteria evaluation *<did/did not>* meet the prespecified acceptance criteria.

7 Benefit/risk discussion (GSPR 1, 6, and 8 alignment)

Clinical data demonstrate the safety and performance of *Product** and benefits include *list* (General Safety and Performance Requirement 1) (Table 13). Clinical risks include *list*. These benefits and risks have been

TABLE 13 Benefit versus risk.

Benefits
<Describe benefit>
<Describe benefit>
<Describe benefit>
<Describe benefit> ...
Risks
<Describe risk>
<Describe risk>
<Describe risk>
<Describe risk> ...

well-documented in the clinical trial, literature, and experience data including customer complaints and subsequent investigations by *Manufacturer** for issues specifically related to *Manufacturer* Product**. Issues with performance of the product include *<list>* and are well-addressed in *Product** product literature over the lifetime of the device (General Safety and Performance Requirement 6). The side effects are considered acceptable when weighed against the benefits of device use as indicted in the appropriate patient population (General Safety and Performance Requirement 8).

Analyses during production and as part of postmarket surveillance/risk management processes have shown that occurrences of each individual clinical risk are rare, and *Manufacturer** is working to continuously increase the benefit/risk ratio.

The benefit/risk analysis of *Product** (based on literature surveyed and use information reviewed herein) demonstrated benefits outweigh risks for continued use of *Product** as intended. *Product** provide benefits of *<intended use>* without creating any overwhelming risk to patients or healthcare personnel.

8 Conclusions

This CER is based on details in the *<insert title of the clinical evaluation plan (CEP)>*. This CER evaluates safety and performance of *Product**. *No clinical trials were conducted by Manufacturer* and no clinical trials for Product* were listed in the* http://www.clinicaltrials.gov *database*; however, published literature included *X* studies using *Product** and *X* other studies evaluated similar devices among *X* (*categorization, i.e. infants, children and adult*) patients. These articles documented significant benefits (e.g., *list*) and infrequent risks (e.g., *lists risks*). *MAUDE* and *TPLC* reports documented infrequent occurrence of device issues including *list issues*. *<Describe any recalls or WL issued to manufacturer and how they have addressed them. >*

This CER created need for specific details to be tracked in a postmarket clinical follow-up plan (PMCFP). *<Insert details.>*

This CER documents *Product** perform effectively and safely with a small number of product complaints, device malfunctions, or adverse device effects within the limits of the specified safety and performance objectives. Based on clinical studies, published literature, and growing experience data as well as ongoing risk mitigation activities underway at *Manufacturer** and because *Product** is considered to be a standard of care, this CER concludes that *Product** has a well-understood and acceptable benefit/risk profile supporting continued use of this device as indicated in the appropriate patient population in conformity with general safety and performance requirements.

9 References

1. *<Last name of primary author first initial, subsequent authors>*. *<Title of the publication>*. *<Name of the Journal>*. *<year of publication; volume number: Page Numbers>*. Available at: *<Website publication can be found>*.
2. "The Oxford 2011 Levels of Evidence (http://www.cebm.net/index.aspx?o = 5653)"

10 Appendices

10.1 *Product** clinical evaluation plan

The CEP is an attachment (represented in vertical navigation panel by the paperclip symbol).

10.2 *Product** workbook

The attached Excel workbook includes:

Documents Received
Literature Key
Literature
Required Data List
Regulatory
Experience Key
Experience Data
Recall Key

Recalls
FAERS Key
FAERS
Medical Device Safety Key
Medical Device Safety
TPLC Report Key
TPLC Reports
Complaints Analysis
Clinicaltrials.gov Key
Clinicaltrials.gov
Benefit Risk Listing
MEDDEV TOC
Checklist

View the Workbook in vertical navigation panel symbolized by a paperclip.

10.3 Qualifications and Experience of the Evaluators

Resumes and declarations of interest are attached for the authors and reviewers:
Author: *Author Name, Credentials*
Reviewer/Senior Author: *Reviewer Name, Credentials*
Medical Reviewer: *Reviewer Name, Credentials*

10.4 Justification of the choice of author/reviewer

I have reviewed the qualifications and documented experience of *Senior Author*, *Subsequent Authors*, and *Medical Reviewer*, and I have concluded they have sufficient knowledge of the device technology and its application, the research methodology and the diagnosis and management of the conditions intended to be treated or diagnosed by the device to complete a thorough and unbiased review of the clinical evidence related to *Product**.

Approval:

Name: Date:

Title:

10.5 Full length articles

The full-length articles used in the clinical evaluation report are attachments (represented in vertical navigation panel by the paperclip symbol).

Appendix M

Clinical data requirements checklists

UID	CER/PER Development Requirements	Yes	No	NA	If yes, where documented (page #)	If no, enter justification
1	CEP/PEP Written prior to search dates and CER/PER writing including: • identification of GSPRs supported by clinical data • intended purpose • intended target groups with clear indications and contraindications • description of intended clinical benefits with clinical outcome parameters • methods for examination of qualitative and quantitative aspects of clinical safety including determination of residual risks and side effects • specific parameters used to determine acceptable benefit–risk ratio based on the state-of-the-art for various indications and intended purpose or purposes • indication addressing how benefit–risk issues of specific components such as use of pharmaceutical, nonviable animal or human tissues will be addressed • clinical development plan indicating progression from exploratory investigations to confirmatory investigations and a PMCFP/PMPFP with layout of milestones and a description of potential acceptance criteria					
2	Specify appropriate GSPRs; Essential Requirements; Essential Principles covered in CER/PER and document considerations given to expert panel or former NB views expressed to the manufacturer (especially for high-risk devices)					
3	Device name listed					
4	Manufacturer(s) identified					
5	Lifetime of device listed (verify whether clinical evidence is appropriate throughout the device lifecycle)					
6	Intended use defined					
7	Indication for use defined (must be aligned with IFU and labeling, specifically)					
8	Contraindications listed					
9	Device description (including images, main components, and device classification) complete (additional sections included if medicinal substance and human or animal cells/tissues) are incorporated in device or combination product, etc.					
10	Description of principles of operation					

UID	CER/PER Development Requirements	Yes	No	NA	If yes, where documented (page #)	If no, enter justification
11	Safety claims listed (including sterility, MR compatibility, etc., if claimed)					
12	Performance claims listed					
13	Safety and performance acceptance criteria stated (in CER/PER as defined in CEP/PEP)					
14	PER SPECIFIC: IVD scientific validity					
15	PER SPECIFIC: IVD analytical performance					
16	PER SPECIFIC: IVD clinical performance					
17	Equivalent device/s identified					
18	Equivalent devices justified based on technical, biological and clinical equivalence criteria (Table includes all differences and justifications for any differences with clinical impacts) as well as scientific criteria for demonstrating equivalence and sufficient access to technical and design data: • Technical—similar design, condition of use, specifications and properties, deployment methods, principles of operation, and critical performance requirements • Biological—same human-contacting materials (tissues or body fluids) and similar kind/duration of contact and release characteristics (e.g., degradation and leachables) • Clinical—same clinical condition (i.e., severity/ stage), body site, user; similar population (i.e., age, anatomy/physiology); and expected clinical effect					
19	Developmental (historical) context described					
20	SOTA, background, benchmark, and/or similar products identified					
21	(Sponsor-held) Clinical investigations identified/ described with outcome measures and relevant clinical data results for subject device included (The clinical expert should demonstrate and document how any clinical, technical, and biological differences between the devices do not affect the evaluation of safety or performance of the device under assessment.)					
22	Clinical investigations appraised including: completed, ongoing, terminated, and future trials					
23	Lack of clinical investigations justified, if applicable (e. g., verify MDR article 61(10) justification provided, assess decision based on risk management process, device/body interaction, and claims made; determine if performance evaluation, bench testing, and preclinical evaluation are adequate to demonstrate conformity)					
24	Clinical investigations analyzed adequately to summarize clinical results of the trials (verify whether all trials are compliant with all applicable regulatory and ethical requirements)					
25	Clinical investigation conclusion: accurate, logical, and directly based on clinical trial results; discusses limitations					

UID	CER/PER Development Requirements	Yes	No	NA	If yes, where documented (page #)	If no, enter justification
26	Literature search methods/criteria fully described (sources, search terms, dates, duplicate removals, etc., justified) for subject device as well as the background, SOTA, and alternative therapy/device evaluation—The clinical expert should demonstrate and document how any clinical, technical, and biological differences between the devices do not affect the evaluation of safety or performance of the device under assessment.					
27	Literature appraisal described (inclusion, exclusion) and justified (PICO, Cochrane, and MOOSE are recommended)					
28	Literature results summarized (total included; meta-analyses, RCT, case studies, etc.) and verify whether these literature show safety and performance characteristics of the device under normal conditions of use					
29	Meta-analyses and systematic reviews analyzed and described					
30	Articles on safety adequately analyzed and described					
31	Articles on performance adequately analyzed and described					
32	Background articles and SOTA summarized					
33	Literature contains both favorable and unfavorable data					
34	Literature conclusions drawn; limitations discussed; safety and performance conclusions follow logically from clinical data analyses					
35	Safety and performance acceptance criteria evaluated in all literature (do the background literature and subject device literature meet the criteria?)					
36	Experience search criteria described and manufacturer held data identified including past CER/PER, PMSR/PSUR, and PMCF/PMPF data, as well as all suspensions, market withdrawals, cancellations in any jurisdiction, etc.)					
37	Experience appraisal/data selection described					
38	Experience analysis methods described					
39	Manufacturer sales and complaints (AEs, device dysfunctions) listed and summarized					
40	Public complaint database results listed and summarized					
41	Recalls fully described with results					
42	Warning letters described with close out explanation					
43	(Investigator-initiated) Clinical trials listed and summarized from experience communications					
44	Product literature (IFU and labeling) described with specs and warnings described and determine if IFU contents are fully supported/aligned with clinical evidence in the CER/PER and the risk analyses					
45	Labeling claims evaluated for clinical data substantiation and documented in CER/PER					

UID	CER/PER Development Requirements	Yes	No	NA	If yes, where documented (page #)	If no, enter justification
46	Clinical experience safety and performance conclusions follow logically from analysis; limitations discussed					
47	Total benefit–risk discussion takes into account entire clinical evaluation and discusses undesirable side effects and contraindications					
48	All benefits and risks related to device are identified (i. e., risk management and postmarket surveillance must be evaluated including all identified risks and how these are to be mitigated)					
49	Verify benefit–risk analysis under the normal conditions of use (considering other SOTA devices, technologies, and procedures) and demonstrate risks are acceptable when balanced against benefits					
50	Overall conclusion summarizes clinical evaluation including benefits, risks, and data gaps					
51	Conclusion is truthful and not misleading as to whether benefits outweigh risks					
52	PMCFP/PMPFP was described, appropriate, adequate, or justified if not planned in the PMSP					
53	PMCFP/PMPFP adequately addressed clinical data gaps identified and fully described clinical data to be gathered to cover clinical data gaps identified in the CER/PER					
54	PMCFER/PMPFER were fully evaluated in the CER/PER					
55	SSCP/SSP were fully evaluated and found acceptable in the CER/PER					
56	Complete executive summary with manufacturer, product, data results, and "benefits outweigh the risks"					
57	All in-text citations are in correct format					
58	All sources are in the references list (either by occurrence in text or in alphabetical order)					
59	Workbook is appended					
60	Justification for choice (nomination) of author and medical reviewer is signed and dated					
61	Qualifications (CVs) of evaluator and medical reviewer are provided					
62	DOIs are attached for evaluator and medical reviewer					
63	Full-length "key" articles and any additional appendixes are attached					

UID	CER/PER Development Requirements	Yes	No	NA	If yes, where documented (page #)	If no, enter justification
64	Verify whether clinical evaluation followed a defined and methodologically sound procedure and was completed appropriately as part of the device technical file by following all relevant company Standard Operating Procedures, policies, and quality objectives (specifically as documented in the quality manual/ quality management system/SOPs for clinical evaluation); also verify whether appropriate, relevant harmonized/consensus standards were followed					
65	Company logo is on the title page					
66	Revision history is complete including revision number, date issued, and description of changes					
67	Date of next CER established and documented					
68	Page numbers, headers, and footers are accurate and appropriate on all pages					
69	Statement/declaration of conformity/compliance is attached					
70	Table of contents, table of tables, table of figures updated					
71	Table of abbreviations/definitions complete and accurate					
72	Evaluator, reviewer, and approval signatures and dates acquired (CER/PER must be endorsed and signed/dated by a competent clinical expert with sufficient, relevant clinical qualifications and clinical experience)					
73	Full CER/PER pdf compiled with all attachments (i.e., single, stand-alone, and version-controlled copy is archived and easily retrieve)					
74	Other (specify):_______________					
75	Other (specify):_______________					
76	Other (specify):_______________					
	Reviewer Signature	**Date**			**Type or Print Name**	**Title**
	Approver Signature	**Date**			**Type or Print Name**	**Title**

Note: Investigation is synonymous with "clinical trial" or "clinical study" as systematic clinical research in or on one or more human subjects. *AE*, Adverse event; *CE*, clinical evaluation; *CEP*, clinical evaluation plan; *CER*, clinical evaluation report; *CV*, curriculum vitae; *DoC*, declaration of conformity; *DOI*, declaration of interest; *EU*, European Union; *FMEA*, failure modes and effects analysis; *GSPR*, general safety and performance requirement; *IFU*, instructions for use; *IVD*, in vitro diagnostic (device); *MDCG*, medical device coordinating group; *MDR*, medical device regulation; *MOOSE*, meta-analysis of observational studies in epidemiology; *MR*, magnetic resonance; *NB*, notified body; *NBOG*, notified body operations group; *PEP*, performance evaluation plan; *PER*, performance evaluation report; *PICO*, population, indication, control, outcome; *PMCF*, post market clinical follow up; *PMCFER*, post market clinical follow up evaluation report; *PMCFP*, post market clinical follow up plan; *PMPF*, post market performance follow up; *PMPFER*, post market performance follow up evaluation report; *PMPFP*, post market performance follow up plan; *PMS*, post market surveillance; *PMSR*, Post market surveillance report; *PSUR*, periodic safety update report; *QMS*, quality management system; *RCT*, randomized controlled trial; *SOP*, standard operating procedure; *SOTA*, state of the art; *SSCP*, summary of safety and clinical performance; *SSP*, summary of safety and performance.

References:

MEDDEV 2.7.1 revision 3, guideline (Dec 2009)
MEDDEV 2.7/1 revision 4, guideline (June 2016)
European (EU) Medical Device Regulation 2017|745 and 2017|746
NBOG CL 2010-1 NBOG 2010 - Checklist for audit of Notified Body's review of Clinical Data/Clinical Evaluation
MDCG 2020-13 Clinical evaluation assessment report template
MDCG 2022-2 Guidance on general principles of clinical evidence for In Vitro Diagnostic medical devices (IVDs)

DOUBLE CHECK to verify the following do NOT occur. A mark in the "no" column confirms accuracy and completeness of CER.

UID	24 Common Errors to Double Check	Yes	No	N/A	If yes, corrective action	If no, where documented
77	Absence of ANY required components of the CER, more than one CER is provided, poorly presented information and data, illegible components, or sections not in English					
78	Intended purpose(s), indication, and claims inconsistent between documents, that is, submission, IFU, and CER list different intended purpose(s)					
79	Intended purpose(s), indication, and claims not supported by clinical data					
80	Lack of information about the regulatory history of the device in other countries (e.g., approvals for MR conditional use for implantable devices; recalls, withdrawals, removals from market, and suspensions and cancellations in any jurisdiction)					
81	Information on predicate or previous related devices not included and/or substantial equivalence not demonstrated (if relevant)					
82	Insufficient or incomplete clinical investigation(s) data and/or literature review and/or postmarket data (clinical experience) with the device (or predicate/similar marketed device if relevant).					
83	No demonstrated comprehensive literature review					
84	Failure to provide specific literature search and appraisal/selection methods with detailed search results (i.e., did not provide: databases searched, search terms used, and inclusion and exclusion criteria to a level of detail so the search can be reproduced).					
85	Failure to include full-length articles for key points in the CER/PER					
86	Insufficient information and/or poor-quality search protocol resulting in inability to reproduce or understand the literature review strategy					
87	Provision of a multitude of citations or publications with little or no explanation as to why they are of relevance					
88	Little or no critical evaluation of clinical investigation data and/or literature and/or experience data and inadequate synthesis of the data					
89	No discussion of relative strengths of the data (e.g., RCTs, case control studies, and case series)					
90	Substantial equivalence comparing technical characteristics, biological characteristics, and clinical use not established to validate data for a different device (i.e., similar marketed device) to the device under review					
91	Lack of discussion of the validity or otherwise of outcome measures used					

UID	24 Common Errors to Double Check	Yes	No	N/A	If yes, corrective action	If no, where documented
92	Inadequate critique and summary of the totality of evidence provided for the device					
93	Unclear or missing statement on MRI status and, if claiming MR conditional, conditions are incomplete in IFU					
94	No postmarket data including adverse events, complaints, failures in cases where this information is available, lack of Failure Mode Effect Analysis (FMEA) data					
95	Author of CER/PER not included, totality of clinical data not evaluated by competent clinical expert, CER/PER not endorsed/signed by clinical expert					
96	CER/PER is not dated or is outdated or is not version controlled					
97	Inappropriate selection of clinical experts. The clinical expert who critically evaluates the clinical data and endorses/signs and dates the CER/PER is expected to have direct clinical experience in the relevant field using similar devices or performing similar procedures.					
98	CV of clinical expert(s) is not provided					
99	DOI of clinical expert(s) is not provided					
100	The clinical expert failed to provide a critical evaluation of the totality of all clinical data. Provision of a perfunctory conclusion or affirmation that the device is safe and fit for the intended purpose(s) is not sufficient.					
	QC Reviewer Signature	**Date**			**Type or Print Name**	**Title**
	QC Approver Signature	**Date**			**Type or Print Name**	**Title**

Note: Investigation is synonymous with "clinical trial" or "clinical study" as systematic clinical research in or on one or more human subjects. *AE*, Adverse event; *CE*, clinical evaluation; *CEP*, clinical evaluation plan; *CER*, clinical evaluation report; *CV*, curriculum vitae; *DoC*, declaration of conformity; *DOI*, declaration of interest; *EU*, European Union; *FMEA*, failure modes and effects analysis; *GSPR*, general safety and performance requirement; *IFU*, instructions for use; *IVD*, in vitro diagnostic (device); *MDCG*, medical device coordinating group; *MDR*, medical device regulation; *MOOSE*, meta-analysis of observational studies in epidemiology; *MR*, magnetic resonance; *NB*, notified body; *NBOG*, notified body operations group; *PEP*, performance evaluation plan; *PER*, performance evaluation report; *PICO*, population, indication, control, outcome; *PMCF*, post market clinical follow up; *PMCFER*, post market clinical follow up evaluation report; *PMCFP*, post market clinical follow up plan; *PMPF*, post market performance follow up; *PMPFER*, post market performance follow up evaluation report; *PMPFP*, post market performance follow up plan; *PMS*, post market surveillance; *PMSR*, post market surveillance report; *PSUR*, periodic safety update report; *QMS*, quality management system; *RCT*, randomized controlled trial; *SOP*, standard operating procedure; *SOTA*, state of the art; *SSCP*, summary of safety and clinical performance; *SSP*, summary of safety and performance.

References:

MEDDEV 2.7.1 revision 3, guideline (Dec 2009)
MEDDEV 2.7/1 revision 4, guideline (June 2016)
European (EU) Medical Device Regulation 2017|745 and 2017|746
NBOG CL 2010-1 NBOG 2010 - Checklist for audit of Notified Body's review of Clinical Data/Clinical Evaluation
MDCG 2020-13 Clinical evaluation assessment report template
MDCG 2022-2 Guidance on general principles of clinical evidence for In Vitro Diagnostic medical devices (IVDs)

Appendix N

Declaration of Interest template

Declaration of Interest form

Clinical evaluation report (CER) evaluators, reviewers and approvers shall complete this form to show all Declarations of Interest (DOI) as required in MEDDEV 2.7.1 revision 4 (CLINICAL EVALUATION: A GUIDE FOR MANUFACTURERS AND NOTIFIED BODIES UNDER DIRECTIVES 93/42/EEC and 90/385/EEC). This DOI covers relevant financial interests for persons contributing to CER work. In this document: "<company name>" means <company name>, any "Affiliate of <company name>" means that any entity, directly or indirectly, through one or more intermediaries, controls, or is controlled by, or is under common control with <company name>, and "control" means the power to direct or cause the direction of, the management, governance, or policies of an entity, directly or indirectly, through any applicable means including legal, beneficial, or equitable ownership, partnership, or by contract.

Evaluator Name	
Role/Title	
Site/Company	

During the last 36 months, have you, your spouse, partner or any children or adults for whom you are legally responsible:

1. Been employed as a full-time or part-time employee of <*company name*>? *NOTE: contractors are not considered employees*	**YES** ☐	**NO** ☐	***If yes, brief description***
2. Participated as an investigator in the clinical or preclinical testing of the CER subject device?	**YES** ☐	**NO** ☐	***If yes, brief description***
3. Had any equity interest in <*company name*> (e.g., ownership interest, stock options or other financial interest) possibly affected by the CER outcome (e.g., to receive greater compensation for favorable results)?	**YES** ☐	**NO** ☐	***If yes, brief description***
4. Had any grants sponsored by <*company name*>?	**YES** ☐	**NO** ☐	***If yes, brief description***
5. Received any benefits (e.g., travel or hospitality) beyond what is reasonably expected and necessary for the work completed?	**YES** ☐	**NO** ☐	***If yes, brief description***
6. Participated in manufacturing the subject device or constituents of the device?	**YES** ☐	**NO** ☐	***If yes, brief description***
7. Had any intellectual property interest possibly affected by the CER outcome (e.g., royalties, patents, trademarks, copyrights or licensing agreements, whether pending, issued or licensed)?	**YES** ☐	**NO** ☐	***If yes, brief description***
8. Had any other interests or sources of revenues possibly affected by the CER results (e.g., honoraria, consulting fees, grant support for laboratory activities or ongoing research, actual equipment for the laboratory/clinic, or monetary payments of other sorts)?	**YES** ☐	**NO** ☐	***If yes, brief description***

I certify by my signature below that the above information accurately represents my financial interests in the sponsor of the Clinical Evaluation Report and associated product.

Signature	**Date (DD/MMM/YYYY)**

Appendix O

Summary of safety and clinical performance/summary of safety and performance template

This template was modified and merged from original text in two templates: MDCG 2019-9 rev. 1 (March 2022) (accessed May 20, 2024 at https://health.ec.europa.eu/system/files/2022-03/md_mdcg_2019_9_sscp_en.pdf) and MDCG 2022-9 rev. 1 (April 2024) (accessed May 20, 2024 at https://health.ec.europa.eu/document/download/b7cf356f-733f-4dce-9800-0933ff73622a_en?filename=mdcg_2022-9_en.pdf). MDCG suggested that SSCP/SSP general information is italicized at the start of each section. The first part is dedicated to users/healthcare professionals for all implantable and Class III devices and for all Class C and D IVD devices (including self-testing devices) and, when relevant, the second part, for patients/laypersons, should be added.

Summary of safety and clinical performance

This updated summary provides public access to important device safety and performance details. This document does not replace the instructions for use as the main document to ensure safe device use, nor is this document intended to provide diagnostic or therapeutic suggestions to intended users or patients. When this document is provided in two parts, the first part is intended for healthcare professionals and the second part is intended for patients and laypersons.

Manufacturer's SSCP/SSP reference number:________________________

1. Device identification and general information	
1.1. Device trade name/s	
1.2. Manufacturer's name and address	
1.3. Manufacturer's single registration number (SRN)	
1.4. Basic UDI-DI	
1.5. European Medical Device Nomenclature (EMDN) description / text	
1.6. Risk class of device	
1.7. Year when first CE certificate was issued covering the device	
1.8. Authorized representative, if applicable; name and the SRN	
1.9. NB's name (NB that will validate the SSCP) and NB's single identification number	
2. Intended use of the device	
2.1. Intended purpose	
2.2. Indication/s and target population/s	
2.3. *[IVDR only] Indication whether it is a device for near-patient testing and/or a companion diagnostic*	☐ *Not IVDR*
2.4. Contraindications and/or limitations *[IVDR (e.g., relevant interferences, cross-reactions)]*	
3. Device description	
3.1. Device description *[IVDR: including conditions to use device (e.g., laboratory, near-patient testing)]*	

3.2. *[IVDR only] In case device is a kit, description of components (including regulatory status of components, e.g., IVDs, medical devices, and any basic UDI-DIs)*	▯ *Not IVDR*
3.3. A reference to previous generation/s or variants if such exist, and a description of the differences [Self-testing devices] describe how device is achieving its intended purpose	
3.4. Description of any accessories that are intended to be used in combination with the device	
3.5. Description of any other devices and products that are intended to be used in combination with the device	
4. Risks and warnings	
4.1. Residual risks and undesirable effects [For Self-testing] include how potential risk have been controlled or managed	
4.2. Warnings and precautions	
4.3. Other relevant aspects of safety, including a summary of any field safety corrective action (FSCA including FSN) if applicable	
5. Summary of clinical/*performance* evaluation and PMCF/*PMPF*	
5.1. Summary of clinical data related to equivalent device, if applicable *[IVDR: summary of device scientific validity]*	
5.2. *[IVDR only]: summary of performance data from the equivalent device, if applicable]*	▯ *Not IVDR*
5.3. Summary of clinical/*performance* data from conducted investigations of the device before the CE marking, if applicable	
5.4. Summary of clinical/*performance* data from other sources, if applicable	
5.5. An overall summary of clinical performance and safety	
5.6. Ongoing or planned postmarket clinical/*performance* follow-up	
6. Possible diagnostic or therapeutic alternatives or *[IVDR only]: metrological traceability of assigned values* *6.1. Explanation of the unit of measurement, if applicable* *6.2. Identification of applied reference materials and/or reference measurement procedures of higher order used by the manufacturer for the calibration of the device*	
7. Suggested profile and training for users	
8. Reference to any harmonized standards and common specifications applied *(noted as Section 4 in MDCG 2022-9 IVDR)*	

9. Revision History

SSCP/SSP revision #	**Date issued**	**Change description**	**Revision validated by the notified body**
			▯ Yes Validation language: __________ ▯ No: only applicable for Class IIa or some IIb implantable devices for which SSCP is not yet validated by the NB (MDR, Article 52 (4) 2nd paragraph) OR only applicable for Class C for which SSP is not yet validated by the NB (IVDR, Article 48(7))
			▯ Yes Validation language: __________ ▯ No

A clinical safety and performance summary, intended for patients, is given below.

Summary for Patients/Lay Persons

This updated summary provides public access to important device safety and performance information intended for patients or laypersons. A more extensive device safety and performance summary for healthcare professionals is found in the first part of this document. This document is not intended to give advice on treating a medical condition. Please contact your healthcare professional if you have questions about your medical condition or about the device use in your situation. This document is not intended to replace an implant card or the instructions for use to provide information on safe device use. [IVDR only: this document is not intended for self-testing devices.]

Manufacturer's SSCP/SSP revision:_____________ and date_____________

1. Device identification and general information	
1.1. Device trade name	
1.2. Manufacturer; name and address	
1.3. Basic UDI-DI	
1.4. *[IVDR only]: risk class of device*	▯ *Not IVDR*
1.5. Year when the device was first CE marked	
2. Intended use of the device	
2.1. Intended purpose *[IVDR: including intended patient groups]*	
2.2. Indications and intended patient groups	
2.3. Contraindications *[IVDR: and /or limitations]*	
3. Device description	
3.1. Device description *[MDR only: and material/substances in contact with patient tissues]*	
3.2. *[MDR only] Information about medicinal substances in the device, if any*	▯ *Not MDR*
3.3. Description of how device achieves intended mode of action	
3.4. Description of accessories, *[IVDR: or accessories or other devices/equipment need to use the device in question,]* if any	

4. Risks and warnings

Contact your healthcare professional if you believe that you are experiencing side effects related to the device or its use, or if you are concerned about risks. This document is not intended to replace a consultation with your healthcare professional if needed.

4.1. How potential risks have been controlled or managed	
4.2. Remaining risks and undesirable effects	
4.3. Warnings and precautions	
4.4. Summary of any field safety corrective action (FSCA including FSN) if applicable	
5. Summary of clinical/performance and postmarket clinical/performance follow-up	
5.1. Clinical background of the device [IVDR: Summary of scientific validity]	
5.2. [MDR] The clinical evidence for the CE marking [IVDR] summary of performance data	
5.3. [MDR] Safety [IVDR: ongoing or planned post market performance follow-up]	

6. [MDR only] Possible diagnostic or therapeutic alternatives ***When considering alternative treatments, it is recommended to contact your healthcare professional, who can take into account your individual situation.***	▯ *Not MDR*
6.1. [MDR only] General description of therapeutic alternatives	▯ *Not MDR*
7. Suggested profile and training for users	

Appendix P

Benefit-risk ratio checklist

Use this checklist when considering the Medical Device Regulation/In Vitro Diagnostic device Regulation (MDR/IVDR) requirements for benefit-risk ratio determinations across all manufacturing systems. A check in the "Yes" column means that the regulatory requirement is fully verified and no issues remain. A check in the "No" column means that the regulatory requirement is not fully verified. A check in the "NA" column means that the regulatory requirement is not applicable. The "" for any No and NA column entries means that this entry requires a justification in the Comments column.*

UID	Regulation	Citation	Yes	No*	NA*	*Comment
1	Conformity assessment is not required when device "modifications do not adversely affect the benefit-risk ratio of the device" or are not "substantial modifications."	MDR Article 54 IVDR Article 71, 74				
2	The benefit-risk ratio shall be "acceptable," "valid" and based on "sufficient" clinical data.	MDR Article 61 IVDR Article 56 (1)				
3	Postmarket surveillance (PMS) system data shall be used "to update the benefit-risk ratio and improve" Risk Management (RM).	MDR Article 83, 3(a) IVDR Article 78, 3(a)				
4	The periodic safety update report (PSUR)* will define the benefit-risk determination conclusions. **PSUR is only for "class IIa, class IIb and class III medical devices or class C or class D IVD devices, low-risk devices use the PMS Report (PMSR)."*	MDR Article 86, 1(a) IVDR Article 81, 1(a)				
5	Trend reporting shall be done by manufacturers for "any statistically significant increase in the frequency or severity of incidents that are not serious incidents... that could have a significant impact on the benefit-risk analysis... and which... have led or may lead to... risks to the health or safety of patients, users or other persons."	MDR Article 88(1) includes "unacceptable" risks "when weighed against the intended benefits" IVDR Article 83(1) including "erroneous results"				
6	The Commission analyzes vigilance data and the "competent authority... shall inform the manufacturer" about any "previously unknown risk" or changes in "the frequency of an anticipated risk" which "significantly and adversely changes the benefit-risk determination" and the manufacturer "shall then take the necessary corrective actions."	MDR Article 90 IVDR Article 85				
7	The manufacturer must "reduce risks as far as possible... without adversely affecting the benefit-risk ratio."	MDR Annex I, GSPR 2 IVDR Annex I, GSPR 2				
8	The manufacturer must evaluate the PMS data impact on the "overall risk, benefit-risk ratio and risk acceptability."	MDR Annex I, GSPR 3(e) IVDR Annex I, GSPR 3(e))				

UID	Regulation	Citation	Yes	No*	NA*	*Comment
9	For medical devices only: when justifying specific objectionable substances in the device (e.g., "substances which are carcinogenic, mutagenic or toxic to reproduction" or endocrine-disrupting substances or phthalates), the justification must explain why changes to remove these substances are inappropriate "in relation to maintaining... the benefit-risk ratio."	MDR ONLY Annex I, GSPR 10.4.2(c), 10.4.3				
10	The manufacturer must have technical documentation including the benefit-risk analysis and RM.	MDR Annex II, 5(a); IVDR Annex II, GSPR 5(a)				
11	The PMS plan (PMSP) shall include "suitable indicators and threshold values" for benefit-risk analysis and for RM.	MDR Annex III, 1 (b) IVDR Annex III, 1 (b)				
12	Notified Body (NB) to verify that clinical evaluation/performance evaluation (CE/PE) are adequately in conformity with the general safety and performance requirements (GSPRs) including "adequacy of the benefit-risk determination... risk management... instructions for use... user training." The PMSP and proposed PMCF/PMPF plan (PMCFP/PMPFP) with appropriate specific milestones are considered in the CE/PE process. *(Note "Assessment procedure for certain class III and class IIb devices" in Annex IX, 5 in MDR only.)*	MDR Annex IX, 4.6 and 4.7 IVDR Annex IX, 4.6 and 4.7				
13	The clinical evaluation plan/performance evaluation plan (CEP/PEP) "shall include... parameters to be used to determine... the acceptability of the benefit-risk ratio for the [various indications and] intended purpose... of the device" and how issues will be addressed.	MDR Annex XIV 1(a) IVDR Annex XIII, 1.1				
14	The PMCFP/PMPFP shall include a continuous process to proactively collect and evaluate clinical data "with the aim of... ensuring the continued acceptability of the benefit-risk ratio."	MDR Annex XIV, 6.1(d) IVDR Annex XIII, 4 and 5.1(d)				
15	Investigator's brochure (IB) must summarize benefit-risk analysis and RM.	MDR Annex XV, 2.5 IVDR, Annex XIV, 2.5				
16	MDR: clinical trials to substantiate benefit-risk ratio. IVDR: performance studies to demonstrate analytical performance of IVD devices.	MDR Annex XV, 2.1 *IVDR Annex XIII, 1.2.2*				
17	MDR only: NB conformity assessment conditions when assessing changes to devices using human cells or tissues to "ensure that the changes have no negative impact on the established benefit-risk ratio of the addition of the tissues or cells of human origin or their derivatives in the device."	MDR Annex IX, 5.3.1(d) *IVDR NA*				

Appendix Q

Clinical evaluation/performance evaluation strategy checklist

No.	Description	Assigned to	Date due	√	Note/Comment
1	Devices needing CE/PE (i.e., "book of business"—how many devices? What type?)				
2	Past NB deficiencies for each CE/PE process				
3	Team members who can work on CE/PE processes going forward				
4	NB required CE/PE document delivery timeline for each document: CEP/PEP, CER/PER, PMCFP/PMPFP, PMCFER/PMPFER, SSCP/SSP				
5	Meeting plan (i.e., number of meetings, specific topics, detailed agendas, meeting minutes for each meeting over time)				
6	Assignments (i.e., who is doing which CE/PE process parts?)				
7	Due dates (i.e., when are deliverables required to be delivered from each team member?)				
8	Coordinator function (i.e., who is collecting deliverables and updating team on progress?)				
9	Data sourcing details (i.e., where are data housed and are all needed clinical data in house?)				
10	Roadblocks (i.e., what is missing? Where have problems arisen in the past?)				
11	Who can be called for help (i.e., when, why and how soon should outside support be called?)				
12	Links to staff who work on PMSR/PSUR, RMR, senior management reviews, scientific, medical affairs, statistics department				
13	What clinical data should be requested on a regular basis (i.e., can we streamline and make these requests more efficient?)				
14	What will happen if required clinical data are not found?				
15	What will happen if CE/PE concludes INSUFFICIENT clinical data were present?				
16	What is the PMCFP/PMPFP overview for all products in this CE/PE process?				
17	Design details				
18	Labeling				

19	NB requirements—how NB requirements will be addressed if company feels the product is safe and performs (i.e., benefit-risk ratio is appropriate) but NB may not agree? How does NB weigh risks/benefits? *Note some NBs suggest the device must be "superior" to all other devices on the market or data must include RCTs as opposed to other data types. The company CE/PE strategy should be clear if these issues arise.*				
20	The "parking lot" as a place to write down future CE/PE negotiations and needs				

Appendix R

Clinical data requirements matrix

Regulation, guidance, and standard	Chapter											
	1	2	3	4	5	6	7	8	9	10	11	12
MDR (EU REG 2017/745)	X	X	X	X	X	X	X	X	X	X	X	X
IVDR (EU REG 2017/746)	X	X	X	X	X	X	X	X	X	X	X	X
MDCG	X	X	X	X			X	X	X	X	X	X
MED DEV	X	X	X	X	X		X		X	X	X	
GHTF	X	X	X				X		X	X	X	X
IMDRF	X	X		X					X	X	X	
ICH	X				X							
ISO	X	X	X	X	X	X	X	X	X	X	X	
Decree (China)											X	
FDA	X	X	X	X	X	X	X			X	X	X
21CFR (US FDA)	X	X			X						X	
Guideline/guidance	X	X	X	X	X	X	X	X	X	X	X	X

21CFR, Title 21 code of federal regulations; *EU*, European Union; *FDA*, Food and drug administration; *GHTF*, global harmonization task force; *ICH*, international conference on harmonization; *IMDRF*, international medical device regulators forum; *ISO*, international standards organization; *IVDR*, in vitro diagnostic device regulation; *MDCG*, medical device coordinating group; *MDR*, medical device regulation; *Med Dev*, Medical Device guideline; *US*, United States.

Appendix S

Reference list of regulatory jurisdictions

- Australia—Therapeutic Goods Administration (TGA)
- Brazil—National Health Surveillance Agency (ANVISA)
- Canada—Health Canada (HCan)
- China—National Medical Products Administration (NMPA)
- European Union—European Commission Directorate-General for Internal Market, Industry, Entrepreneurship and SMEs
- Ireland—Health Products Regulatory Authority (HPRA)
- Japan—Pharmaceuticals and Medical Devices Agency (PMDA) and Ministry of Health, Labour and Welfare
- Russia—Russian Ministry of Health
- Singapore—Health Sciences Authority
- South Korea—Ministry of Food and Drug Safety
- United Kingdom—Medicines and Healthcare products Regaultory Agency (MHRA)
- United States—US Food and Drug Administration (FDA)

Appendix T

Answers to review questions

Chapter 1

1. Clinical evaluation plan/performance evaluation plan, clinical evaluation report/performance evaluation report, medical device regulation/*in vitro* diagnostic device regulation, postmarket clinical follow-up/postmarket performance follow-up, summary of safety and clinical performance, summary of safety and performance.
2. All devices require a CER in the EU.
3. CERs are required for all devices seeking market approval for use in EU Member States. vCERs/vPERs can be an effective tool during development stage companies to understand and educate the company about the current state of literature evidence for their device. Literature searches and market research during device development can be used to inform the eqCER/eqPER when an equivalent device is chosen and the fihCER/fihPER once human clinical data become available for the device.
4. Prelaunch or at time of product launch, respectively.
5. CERs are needed to demonstrate the medical device's S&P in conformity with the GSPRs under the MDR (EU Reg. 2017/745) in order to receive a CE mark and for the medical device to be placed on the EU market for sale.
6. The CER tells the full story about how a device meets the EU Reg. 2071/745 requirements to be safe and to perform as indicated throughout the device lifetime. The CER must determine if the device benefits outweigh the risks and if sufficient clinical data are available to meet these requirements.
7. No, a CER is only one part of an interrelated system of important technical documentation (e.g., CER, PMSR, RMR, technical file with bench testing).
8. Planning (define the scope), selecting (identify the clinical data), appraising (determine what clinical data to include), analyzing (document safety, performance, benefits, and risks), and writing/compiling the CER.
9. The conclusion, because the CER needs to specify if the device is safe, performs as intended, has benefits outweighing the risks to the patient, and is in conformity with the GSPRs when the device is used as intended in the appropriate patient populations.
10. The clinical/performance data must be scientifically valid and must cover all parts of the indication for use. The CER/PER must plan and execute a complete and thorough evaluation of all the available clinical/performance data. The evaluator must be relatively independent and unbiased to offer a scientifically valid evaluation process (from start to finish, including data identification, appraisal, analysis, and reporting). Any missing clinical/performance data must be identified and documented in the CER/PER for the PMCFP/PMPFP to collect using the appropriate data collection tools, e.g., a full clinical investigation or a bench test or literature review.
11. The CER/PER is used in conjunction with a host of documentation by regulators to inform the decision about whether sufficient evidence exists for a device to be marketed in the EU. CERs/PERs and supporting documents can have deficiencies noted by the NB which require answers or edits. Even if the CER/PER is considered up to MDR/IVDR standards, the device manufacturer may still be required to conduct additional work, such as more clinical studies or increased surveillance. CERs/PERs require updates if new clinical data come to light, especially if the new clinical data impact the clinical benefit–risk ratio.
12. False—the update cadence for each CERs depends on multiple factors. According to current guidelines, Class III implantable and Class IIb active devices should be reevaluated annually, Class IIa devices at least every 2 years, and Class I devices no less than every 5 years. In addition, CERs must be updated when new clinical, safety, or performance data become available or when device changes affect the benefit–risk ratio.
13. Viability, equivalent, first in human, traditional, and obsolete device CERs. This CER classification system is helpful because different evaluation expectations are managed within each of these different CER/PER types. For example, the vCER/vPER will not have clinical data but will have significant amounts of nonclinical data

designed to support the specific device S&P. The eqCER/eqPER will rely on another device, which may not be exactly the same as the device under evaluation, and the fihCER/fihPER will incorporate the initial human clinical trial data. The trCER/trPER will have the challenge of integrating many years of accumulating clinical data as each subsequent trCER/trPER is developed. The oCER/oPER will need to continue evaluating clinical data even though the device itself may no longer be marketed by the manufacturer, as this is important for individuals still using this device.

14. A number of factors should be considered when choosing a NB, including NB costs (https://www.qualitiso.com/en/comparison-of-notified-body-fees), expertise, procedures for reviewing data, and audit stringency (https://www.mddionline.com/business/how-select-notified-body, https://www.greenlight.guru/blog/selecting-notified-body).
15. 39 + 10 = 49 as of October 1, 2023. For more current data, see https://ec.europa.eu/growth/tools-databases/nando/index.cfm?fuseaction = directive.notifiedbody&dir_id = 34 and https://ec.europa.eu/growth/tools-databases/nando/index.cfm?fuseaction = directive.notifiedbody&dir_id = 35.
16. EU certification is established for each NB under national law in the EU Member State or under third-country law (e.g., Switzerland) and is a documented legal status.
17. 6–9 months was a common time frame in 2023.
18. The MDCG provided guidance about the process for evaluating a CER in the CEAR and the factors considered by regulators during this process. High-risk and implantable devices also require an independent assessment by the expert panel. CEARs provide leeway for regulators to require certain PMCF types and to ensure adequate device PMS by the manufacturer.
19. CEP/PEP, CER/PER, PMCFP/PMPFP, SSCP/SSP, and PMCFER/PMPFER.
20. Write the CEP/PEP, then the CER/PER (based on the CEP/PEP), then the PMCFP/PMPFP (based on the CER/PER), then the SSCP/SSP (based on the CER/PER and PMCFP/PMPFP), then the PMCFER/PMPFER (based on the PMCFP/PMPFP).

Chapter 2

1. Clinical trial, kick-off meeting, quality management system, standard operating procedure, work instruction.
2. Review medical device and *in vitro* diagnostic device regulatory definitions in MDR and IVDR and document* company decisions about device definitions (especially if company decides the product is not a medical device). Document* the rationale for each company decision (i.e., why is the product considered a medical device or not). Comply with all regulations relevant to the device. Seek an independent regulatory expert to help make decisions for borderline devices. *Remember all documentation needs to be in writing and ALCOA + compliant.
3. The best CE/PE strategies include comprehensive planning, monitoring, auditing, and learning. The CE/PE leader should know exactly what to expect based on past experience, the KOM should discuss all past CE/PE problems including missing or weak clinical data, and the process should integrate responsible individuals from other work areas including those working on clinical trials, publications, postmarket surveillance, and risk management. The CE/PE document writing process should not be an endless pursuit. The CE/PE leader must set boundaries for time and money to complete the CE/PE work.
4.

SG #	SG name	SG deliverable
SG1	Initiating	CER/PER KOM
SG2	Planning	CEP/PEP
SG3	Searching	All clinical/performance data identified
SG4	Coding	All clinical/performance data appraised
SG5	Analyzing	All clinical/performance data analyzed
SG6	Writing	CER/PER
SG7	Integrating	PMCFP/PMPFP
SG8	Summarizing	SSCP/SSP
SG9	New data	New data is collected
SG10	Follow-up	PMCFER/PMPFER

5. Setting up the strategy should always come before the KOM because having a successful KOM requires careful planning and strategy. Without the road map developed in the strategy meeting, the KOM will take longer, cost more and the result is often directly linked to lower-quality outputs.
6. a. S&P acceptance criteria (specific details must be clear and measurable in the plan).
   b. Benefit-risk ratio (benefits must outweigh the risks in the plan).
   c. Progress to complete SG1 and SG2 (time to complete, output quality, should be measured and all past and present NB deficiencies relevant to each stage gate item along with all the "parking lot" items generated during the process should be consdiered).
7. SG4 is the coding step where clinical data are appraised and weighted, including all inclusion and exclusion decisions.
8. First, all clinical data are gathered from all the existing documents (typically, these data are reviewed at the KOM), then the CEP/PEP defines the device S&P acceptance criteria and past benefit-risk ratio, and the CE/PE process determines if the clinical data are sufficient to demonstrate conformity with the relevant GSPR (e.g., GSPR 1, 6, and 8). The CER/PER must document this decision and explain what will be done if the clinical data amount or type are not sufficient. The required clinical data amount and type will vary directly with the device risk. More risk requires more clinical data especially for novel devices, implantable devices, or devices with an unclear benefit-risk ratio. Conversely, less risk requires less clinical data. Often, NBs have difficulty "drawing the line" for how much clinical data are required for the lower risk devices. In this setting, a clinical expert should be involved in the NB decision-making since other nonclinical experts frequently err on the side of requiring too much unnecessary data collection. Documenting device clinical data is an expensive operation, and the benefit for gathering this data needs to outweigh the cost/risk to the patient (should patients bear the extra cost for these additional clinical data gathering activities if they do not offer any additional benefit?).
9. New clinical data are required to address any and all clinical data gaps documented during the CE/PE process. Typically, the clinical data gathering is completed in a CT protocol or structured within the PMS activities defined in the PMCFP/PMPFP.
10. False. Version control is required for final documents but not for draft documents (unless required by a particular, potentially unwise, company policy).
11. True. All clinical claims require substantiation in the CER/PER.
12. Grouping devices into a "device family" is easily done and the MDR/IVDR rules allow manufacturers to prepare a single "device family" CER/PER. Although no specific guidance is available for this low-risk device grouping, manufacturers commonly group similar devices into individual device families in order to manage technical files and CE/PE documentation. These device groups can be quite large due to the included device technical details (e.g., different device colors, shapes, and sizes, subtle differences in device manufacturing materials, processes, and uses). MDR and IVDR define "generic" device groups with the same or similar intended purposes or common technology including devices in Class I/IIa/IIb WET/IIb Rule 12 devices and Class A/B/C IVDs.
13. No. NBs do not typically assess low-risk, Class I devices except for sterility or measuring functions. In addition, NBs use an annual sampling strategy to assess a single representative device for conformity per device "category" for Class IIa and Class B devices or per "generic device group" in Class IIb and Class C devices.
14. Clinical, biological, and technical equivalence must be defined and documented in the plan and report.
15. Although not defined in the regulations or guidelines, this textbook strongly recommends using a "Developmental context" section in the CEP/PEP and CER/PER to help ensure the device is properly placed within the important background, SOTA, alternative therapy/device, S&P, and benefit-risk ratio developments. The goal is to define clearly the device development story from a clinical (not engineering) perspective. This is the place to describe the global regulatory history (including CE mark and clearances or approvals in all other geographies), device novelty, innovation, and risk levels, any medicinal substances, cells, or tissues used on or in or with the device, any unmet medical needs served by the device, sales numbers (EU and worldwide), future marketing plans, different device versions, and all alternative therapies for the device. The developmental context section should explain how the prototype and all predecessor devices evolved into the product currently on or proposed for the EU market. It is also necessary to describe briefly any ongoing clinical data or other relevant data collection plans here. The evaluator must leave no confusion about how this device fits into the SOTA, is safe and performs as intended, while making clear the device is not required by the regulations to be "best in class." Explaining the variety of devices on the market and how they are used by healthcare professionals and all other users is critical to the developmental context. Taking this approach should avoid many NB deficiencies.

16. The term "sandbox" is used to describe a means to practice searching for clinical data before finalizing details in the CEP/PEP (e.g., identifying the best individual search terms, Boolean connectors and filters as well as the best search strings by practicing in the various databases before putting the search terms into the CEP/PEP). This process requires the evaluator to be intimately familiar with the device, SOTA, and alternative therapies before defining the process in the CEP/PEP. This process also requires leaving room for changes during execution as the evaluator learns more as the searches are executed and the appraisals and analyses are completed during the CE/PE process. The best process is iterative and this flexibility needs to be clear in the CEP/PEP.
17. Immediate issue escalation is required and the issue reporting system should ensure all PMS and RM system owners are alerted. Detailed reporting is required and short timelines are enforced (sometimes reporting is required within 24 hours). If confirmed, an event may require a device recall, user retraining, and/or device labeling changes (even if no actual harm resulted). The CER/PER should be completed as planned, and any clinical data gaps identified for the specific new risk should be documented. The future CE/PE process must incorporate all details learned over time in the PMS and RM systems.
18. Each manufacturer system needs to be integrated within the QMS and the documentation within each system should describe how the related systems are integrated. For example, clinical trials completed within the CT system will have data evaluated in the CE/PE process, the PMS system will track all PMCF/PMPF work, and the RM system will mitigate all risks, including newly identified risks which may be discovered during the CE/PE process.
19. The CEP/PEP needs to be well-defined and the QMS needs to detail all plan steps to ensure all required documents and analyses are completed in compliance with all regulatory requirements.
20. vCEP/vPEP: rationale for safety based on nonclinical and similar device data, plan to collect clinical data and re-establish S&P with benefit-risk ratio during fih uses (for IVD, ensure plan to establish scientific validity, analytical performance and clinical performance are clearly defined).

    eqCEP/eqPEP: tabulate clinical, biological, and technical equivalence details side-by-side including indications, provide rationale for equivalence and describe nonclinical data about subject device with future plan to collect subject device clinical data (similarly, IVD devices must have clear scientific validity, analytical performance and clinical performance plans for both subject and equivalent devices).

    fihCEP/fihPEP: clear details about how emerging clinical data will be collected starting with first human use and the plan must describe how to merge new clinical data into historical CE/PE process documents with plans to evolve into trCE/trPE type while updating to more specific S&P acceptance criteria with reduced reliance on prior nonclinical or equivalent device clinical data (similarly IVD devices must plan to use new subject device clinical data to update scientific validity as well as clinical and analytical performance).

    trCEP/trPEP: define how clinical data are changing over time, how planning for obsolescence is progressing, how benefit-risk ratio and S&P acceptance criteria are changing (IVD devices must explain if clinical data are changing in any way which might impact scientific validity or clinical and analytical performance).

    oCEP/oPEP: explain why device was obsoleted, specify if users need to do anything and how they should interact with the obsoleted device, explain any alternatives and define remaining user status and when the CE/PE process will end.

Chapter 3

1. The PER must evaluate scientific validity, analytical performance, and clinical performance.
2. Clinical data are sourced from clinical trial, clinical literature, and clinical experience reports. Here, the word "clinical" means *human*, not animal or bench data.
3. Clinical data needs to include a person (or relevant analyte for IVD devices), a device, and a clinical effect.
4. The chemical of interest in an investigation or the item to be identified or measured in an IVD test.
5. The "association of an analyte with a clinical condition or physiological state" (IVDR Article 2 (38)).
6. The "ability of a device to correctly detect or measure a particular analyte" (IVDR Article 2 (40)).
7. The "ability of a device to yield results... correlated with a particular clinical condition or a physiological or pathological process or state in accordance with the target population and intended user" (IVDR Article 2 (41)).
8. The "clinical data and performance evaluation results, pertaining to a device of a sufficient amount and quality to allow a qualified assessment of whether the device is safe and achieves the intended clinical benefit/s, when used as intended by the manufacturer" (IVDR Article 2 (38)).

9. A clinical evaluation process with good procedures, careful observations, and skeptical evaluations, often starting with a scientific question or observation phrased as a testable hypothesis.
10. a. Follow the CEP/PEP to identify clinical data.
    b. Search for, gather, and document/list clinical data, covering all benefits and risks as well as all S&P measures.
    c. Identify and separate subject device data from current knowledge/SOTA and alternative therapy/device data.
11. The workbook allows all identified data to be stored and later appraised as excluded. Keeping all data in the Word CER/PER template is unwieldy and difficult to manage. The data are more easily moved, tabulated, formatted, and analyzed in Excel or other appropriate software. Having the Word document separate from the Excel (or other software) allows data manipulation and visualization separate from the writing process.
12. Enough "scientifically valid" clinical data must be identified and collected as stated in the CEP/PEP to allow the CE/PE to substantiate each of the GSPRs 1, 6, and 8, the prespecified and prequantified S&P acceptance criteria, all clinical claims, the benefit-risk ratio and the CER/PER conclusions must document if the benefits outweigh the risks when using the device as indicated.
13. A JUS document is a special type of CER/PER when no clinical data exist for the device or when collecting clinical data is not appropriate. A JUS may be appropriate for low-risk and well-established devices, as well as those where no clinical data are possible or appropriate to collect. The regulations require the manufacturer to "specify and justify the level of clinical evidence necessary to demonstrate conformity with the relevant general safety and performance requirements. That level of clinical evidence shall be appropriate in view of the characteristics of the device and its intended purpose" (MDR Article 61(1) and IVDR Article 56(1), identical).
14. Clinical trial data are incomplete (e.g., data may not be available, accurate, or updated).
15. The literature must be either 1) directly relevant to and focused on the subject/equivalent device used as indicated, or 2) relevant to and focused on the relevant, high-quality background/SOTA/alternative therapies or off-label uses of the subject/equivalent device.
16. The process helps to identify most relevant, manageably sized, and useful search terms for each specific database search considering the terms used in the relevant clinical field and using appropriate limiting or expanding Boolean operators. When executing the plan, additional sandboxing may be required if the database or the contents have changed since the CEP/PEP was written.
17. Ideally, the literature search should not be limited by time and no timespan limits should be used. The CE/PE process should comprehensively cover all literature over all time in some meaningful manner. One possible exception to this rule may be for the specific subject device searches, when a time limit for the date of first use may be appropriate.
18. Experience searches are tailored to the subject/equivalent device as appropriate, and background, SOTA, and alternative therapy/device experience data are generally not considered relevant. In addition, many clinical experience database search outputs may require manual data extraction of the search results into the working Excel spreadsheet or software, while most literature database search outputs can be directly exported in an electronic format.
19. The SOP should provide a high-level standard method to identify clinical data and the WI should provide the "how to" step-by-step details starting with: (1) where to identify clinical data (e.g., specific databases); (2) how to gather data from each clinical data source (clinical trials, literature, and experience data); and (3) how to store the gathered clinical data source information (e.g., in a spreadsheet or specific software) and full documents (e.g., in an electronic library of some sort). Details about the CEP/PEP sandboxing and search term development as well as identification steps should be included along with the detailed QC checks to be conducted during the identification step. The SOP and WI should be extremely clear about how to document each search result properly and how to store clinical data for the next steps in the process (i.e., the appraisal and analysis steps).
20. Legacy devices were CE marked in the EU market before MDR and IVDR were established and effective. This designation is important because transitional provisions were provided in MDR/IVDR for legacy devices, and MDR/IVDR requirements are generally not applicable to "old" devices (i.e., devices placed on the market before May 26, 2021 for medical devices or May 26, 2022 for IVD devices). This means the decision about "sufficient" clinical data may differ for legacy devices and newer devices on the EU market. This "legacy" distinction does not apply to the PMS requirements.

Chapter 4

1. Clinical evaluation/performance evaluation, good clinical practice, inclusion and exclusion, safety and performance, notified body, randomized controlled trial, Manufacturer and User Facility Device Experience, Database of Adverse Event Notifications, well-established technology.
2. Appraisal is the process of assessing identified clinical data to determine whether data will be included or excluded in the CE/PE and how much the data will contribute to the overall benefit-risk ratio.
3. Relevance to subject/equivalent device S&P or background/SOTA/alternative therapies and data quality (including scientific validity) are the two primary considerations. Relevance and quality must be considered for each individual clinical data point as well as the overall clinical trial, literature, and experience dataset.
4. Clinical data are considered relevant if the data are about the subject/equivalent device or the background/SOTA/alternative therapies. Typically the devices are used for the same clinical application, indication for use, disease state, and standard of care. In addition, "off-label" subject/equivalent device uses may be relevant and are typically included in the CER/PER background/SOTA/alternative therapy section.
5. Data should be excluded after the evaluator has a documented, a methodologically sound reason for exclusion. This may be done by applying a standardized numerical code identifying the specific, applicable, well-defined exclusion reason.
6. The search term/s should be modified to capture a broader dataset. For example, if a MAUDE search using the brand name "Honeywell ONE-fit N95 respirator" identified no results in the MAUDE database, then a separate, broader search for the term "N95" would be appropriate. Although these results should not be used in the subject/equivalent device section of the CER, the results are relevant to all N95 face mask devices regardless of manufacturer or design and should be reported in the CER background/SOTA/alternative therapy section.
7. The evaluator must obtain and read full-length articles (i.e., the full-length article must be purchased if not available for free) to determine article relevance and quality but only if the article cannot be clearly be included or excluded based on the title and abstract alone. After appraisal, all included full-length articles must be obtained for the analysis step, which will require reading the full-length articles. Copyright fees will need to be paid at this point.
8. All complaints about the subject/equivalent device should be appraised and all relevant complaints of sufficient quality should be included while all irrelevant complaints or complaints with insufficient quality (e.g., without details about the device problem or clinical impact) should be excluded during the appraisal step.
9. EU MDR specifies PMS vigilance data (e.g., including AEs) will be compiled and made available via EUDAMED. As of 11 May 2024, EUDAMED does not yet share AE data.
10. Statistical quality is required for all included clinical data points in order to ensure results and comparisons are as accurate as possible and to provide a quantified uncertainty measurement (i.e., to estimate variability in the detailed measures). Mathematical accuracy allows researchers to draw reasonable conclusions from well-defined data with sound statistical quality. CE/PE conclusions based on clinical data with good mathematical/statistical quality are considered more reliable when used to inform decisions about device use in patients. Removing clinical data with poor or improper statistical methods helps to avoid misunderstandings about the data. Statistical quality is required to guide decisions about sample size, effect size, differences between groups and the overall mathematical result significance.
11. First pass—make appraisal decisions based on title alone, if possible.
    Second pass—make appraisal decisions based on title and abstract, if possible.
    Third pass—make appraisal decisions based on skimming the entire report.
12. A device previously CE marked under regulations existing before the 2017 MDR/IVDR is considered a "legacy" or a traditional (tr) devices with presumed clinical data showing S&P as required under the prior EU essential requirements.
13. Clinical data are ranked in groups from highest to lowest relevance and quality in order to specify and justify the clinical evidence level required to demonstrate conformity with GSPRs 1, 6, and 8. The evaluator assigns a clinical data rank (e.g., a numerical inclusion or exclusion code) which allows clinical data below a certain rank (e.g., negative numbers below zero) to be excluded from further evaluation. The clinical data "rank" is used to characterize the level of evidence (LOE) showing the relative strength of the clinical data identified, appraised, and analyzed during the CE/PE process.

14. Clinical data are weighed based on their relative value or quality for use in the CE/PE. A good example of clinical data weighing is during the benefit-risk ratio evaluation. The evaluator must determine if the individual and overall clinical data points provide sufficient evidence about the benefits and risks to allow a reasonable conclusion about whether the device should be used or not. The evaluator must "weigh" the clinical data pros and cons (i.e., benefits and risks) to determine if the benefits outweigh the risks and the device is safe enough for use.
15. Statements about potential conflicts of interest should be included with the clinical data. The evaluator must review the conflicts to determine if the concerns should disqualify and exclude the clinical data from further analysis. Excluding poor-quality data will limit unintended bias from unclear data points. Iterative appraisals should continue to improve the clinical data quality as the evaluator team increases the device-specific clinical data knowledge and reduces the level of bias allowed in the included clinical data points. Standards and conventions should be followed in the clinical data collection to avoid and limit bias. For example, RCTs are randomized and controlled to limit bias while other trial types may have significant selection, performance, and author bias when they are not controlled in these ways. Blinding the clinical data helps to avoid detection bias and intent-to-treat clinical data analyses should help to limit attrition bias. Reporting bias may be limited by requirements to report all of the clinical data.
16. QC is the process to check and ensure product quality requirements are met (i.e., product quality, after production, is verified using the QC process). Quality assurance is the overarching instruction and enforcement of processes to give confidence the quality requirements will be met (i.e., standard procedures are followed throughout product development and manufacturing in the quality assurance process). Checking data every day is part of the QC function while auditing to ensure quality is present is part of the quality assurance function.
17. The evaluator must determine and then explicitly and clearly document the appropriate amount of high quality, relevant clinical data required to demonstrate conformity to GSPRs 1, 6, and 8 regarding the device S&P, benefit-risk ratio, and overall CER/PER clinical data conclusions about the device. The level of clinical evidence (i.e., the evaluated clinical data amount, type, and quality) must be specified and justified by the evaluator in the CER/PER since the clinical evidence varies both across devices and within the evaluation of a single device. Lower-risk devices will require a lower level of clinical evidence than higher-risk devices. In order to be "sufficient," the clinical data must be scientifically sound and robust enough to document clearly the S&P and benefit-risk ratio for the specific subject/equivalent device when used as intended (i.e., the clinical data must cover all intended use aspects including all patient populations, device sizes, settings, etc.), and these clinical data must be aligned with the background knowledge/SOTA/alternative therapy/device uses. Insufficient clinical data must be documented in the CER/PER and the PMCFP/PMPFP must document specifically how and when the missing clinical data will be generated to fill the clinical data gaps. Alternatively, the indication for use should be narrowed to indicate only uses where the available clinical data are sufficient.
18. Clinical experience data.
19. False. No matter how rigorous the search, literature can still be missed.
20. Clinical experience data are the lowest-quality data because the clinical experience data collection is anecdotal, highly biased by the reporter, and uncontrolled. These data are collected one report at a time and can be reported in a mandatory or voluntary setting by healthcare professionals or the lay public.

Chapter 5

1. Scientific validity report, analytical performance report, and clinical performance report.
2. When the clinical data analysis results in no clinical data gaps (i.e., all clinical questions have been fully and completely answered with documented and specific clinical data). Specifically, the S&P acceptance criteria must be met, the benefit-risk ratio must show the benefits outweigh the risks and all clinical claims must be substantiated with valid and comprehensive clinical data (e.g., in a claims matrix listing specific clinical data sources for claim substantiation data, reviewed and verified by clinical experts, and version controlled in the technical file).
3. The S&P acceptance criteria from the CEP/PEP are compared directly to clinical data from each clinical data source (clinical trial, literature, and experience reports). If the occurrence rate for each designated S&P criterion is at or below the acceptance level documented, then the S&P acceptance criteria have been met; however, if the occurrence rate for any designated S&P acceptance criterion is exceeded, then the S&P acceptance criteria have not been met. In this case, the CER/PER needs to justify why the device should be allowed to be on the EU market even though the S&P acceptance criteria were not met and the PMCFP/PMPFP is ongoing. The next CEP/PEP should detail what happened and the justification/rationale if the S&P acceptance criteria have changed over time.

4. A list of benefits and risks is developed with appropriate details about the rate of occurrence in the clinical data. The clinical data in the lists are weighed and compared to see if the type and rate of occurrence are acceptable to the patient using the device. The clinical data are iteratively analyzed to determine whether the benefits continue to outweigh the risks to the patient. As new data are analyzed, new benefits and risks are added to the lists and weighed to ensure the benefit-risk ratio remains acceptable.
5. First, the evaluator defines a "testable hypothesis" (e.g., "The < subject device > clinical data are sufficient to demonstrate acceptable device S&P and benefit-risk ratio to document conformity to the MDR/IVDR GSPRs 1, 6, and 8 clinical data requirements."), then the evaluator critically reviews all background/SOTA/alternative device and subject/equivalent device clinical data separately. The evaluator uses appropriate guiding questions and documents the clinical data details in spreadsheets, software, and Word document tables to compare and contrast the clinical data in groups and subgroups. Once the individual clinical data have been analyzed, then the clinical data are also critically evaluated overall, and the results are documented for inclusion in the CER/PER. Clinical data conclusions are linked directly to the specific clinical data analyzed in order to answer the questions about device S&P, benefit-risk ratio and conformity with GSPRs 1, 6, and 8. All clinical data gaps are documented and addressed in the PMCFP/PMPFP. If clinical data are insufficient to meet the GSPRs, additional remediation is required.
6.

Analysis is not	Analysis is
Listing clinical data from trials, literature (articles or abstracts) or experience data	Creating new knowledge based on data
Listing or citing references	Interpreting tests, studies, scientific research
Repeating author conclusions without comparison	Deciding what data mean
Documenting anecdotal customer stories	Comparing professional expertise/experiences
Copying newspaper or magazine articles	Conducting objective data review
Promoting a product with sales materials	Weighing accurate, accepted procedural information
Citing a number, rate or money back guarantee	Calculating means, deviations, standard errors
Tabulating abstracts from references	Documenting similarities and differences between data

7. 1. "disease state" (i.e., define clinical setting for each patient in the analysis);
   2. "clinical measures" (i.e., specify clinical data and clinical end points measured);
   3. "performance results" (i.e., document how device performed);
   4. "device issues" (i.e., record device problem list);
   5. "safety results" (i.e., document device safety including AEs and ADEs);
   6. "safety issues" (i.e., document specific device safety concerns);
   7. "conclusions" (i.e., answer all questions asked during the evaluation); and
   8. "limitations" (e.g., detail data quality problems including small sample size, author bias, missing control groups, poor quality).
8. "(49) 'diagnostic specificity' means the ability of a device to recognise the absence of a target marker associated with a particular disease or condition."
   "(50) 'diagnostic sensitivity' means the ability of a device to identify the presence of a target marker associated with a particular disease or condition."
9. "(51) 'predictive value' means the probability that a person with a positive device test result has a given condition under investigation, or that a person with a negative device test result does not have a given condition."
   "(55) 'calibrator' means a measurement reference material used in the calibration of a device."
10. The PER must carefully align the SVR, APR, and CPR clinical data analyses and conclusions in both the background/SOTA/alternative device and the subject/equivalent device sections. In addition, these sections must present specific clinical data supporting the IVD device intended use, especially when used "self-testing" or "near-patient testing" or as a "companion diagnostic," because the benefits and risks associated with these use types vary considerably. The CER does not require SVR, APR or CPR.

11. False. All clinical data must be identified, appraised, analyzed, and reported in the CER/PER. Included and excluded data must be carefully justified to ensure S&P signals are not missed and the included data analysis is a truthful representation of the entire available clinical dataset at the time of the evaluation. Specifically, clinical data should be kept included regardless of data quality when the clinical data are novel and relevant to the subject/equivalent device.
12. Each trial, article, or experience report is evaluated individually to understand all clinical data relevant to the background/SOTA/alternative therapy/device or subject/equivalent device S&P and to understand the specific clinical data related to subject/equivalent device benefits and risks. Then, the clinical data reports are grouped and analyzed overall (i.e., all clinical trials, literature or experience reports together). Specific information about the subject/equivalent device S&P should be extracted and recorded in a worksheet or specific software for analysis in order to determine whether the S&P acceptance criteria were met. The poor quality clinical experience reports need special coding to ensure all similar events are pooled together during analysis to determine S&P in real-world settings. The evaluator must draw conclusions directly from the clinical data analyzed and discussed in the CER/PER text. All identified clinical data gaps must be defined and listed in the CER/PER with specific details about the future activities to be documented in the next PMCFP/PMPFP.
13. To determine if the subject/equivalent device clinical data clearly demonstrate device compliance with GSPRs 1, 6, and 8, to identify clinical data gaps and to document specific clinical data to be collected by the next PMCFP.
14. Multiple analysis passes are required to analyze each piece of clinical data fully. Analyzing a particular trial, article, or experience data point may change the specific appraisal decision for the data point, or the detailed analysis may change the overall clinical data analysis conclusions. Multiple subgroupings may be added and analyzed separately or combined and reanalyzed. Analysis depends on both the individual trial/article/experience report and the overall dataset and often changes as the process develops and more knowledge and understanding are gained. Having an iterative process allows this learning and application of the learning to occur in real time.
15. Data relevant to GSPRs 1, 6, and 8 and the benefit-risk ratio, and data addressing the how effective a device is and what risks are present. Safety data are commonly collected as device-related AEs or ADEs and performance data are commonly collected as successful device uses or DDs.
16. Analysis is typically documented in tables, narrative discussion in the text, detailed documentation of methodology, and data-based conclusions.
17. Document the clinical data gaps in the CER/PER text and explain how these will be addressed in the PMCFP/PMPFP.
18. The clinical data ofen have variable scientific quality, differing end points (i.e., the study addresses something other than device use), poorly documented methods, and dozens of articles to read and analyze in depth while interpreting and extracting the clinical data whic are presented in many different ways.
19. The clinical data abstracted should explain what happened to the patient (i.e., did the patient experience a complication or increased risk)? What happened to the device (i.e., did the device malfunction in some way)?
20. A corrective and preventive action (CAPA) is a process to collect and analyze information related to product and quality problems. The goal of this process is to document and take action to correct and/or prevent problems appropriately and effectively. The CAPA system is a critical QMS element, and the proper implementation requires careful and intentional verification and validation documenting how the problem was "fixed" and the process used to ensure the problem will not recur. A good CAPA system requires communicating well with all involved persons, having strong management review and guidance, and thorough follow-up by specific responsible parties to effectively prevent or minimize device failures.

Chapter 6

1. The Benefit-Risk Ratio framework began during pharmaceutical development for many decades before use during device development.
2. False. "The 'benefit-risk determination' means "the analysis of all assessments of benefit and risk of possible relevance for the use of the device for the intended purpose, when used in accordance with the intended purpose given by the manufacturer" (MDR [1] Article 2 (24); IVDR [2] Article 2 (17)).
3. Relevant EU authorities (i.e., the NB) need to be informed, the CER/PER needs to be updated with the new clinical data and the new benefit-risk ratio evaluation and an updated PSUR needs to be issued documenting the changes.
4. Conformity assessment is required when device modifications "adversely" affect the medical device benefit-risk ratio (MDR Article 54) or when the IVD device has "substantial modifications" (IVDR Article 71, 74).
5. All benefits and risks relevant to the device used as indicated by the manufacturer and all user benefit and risk perceptions should be represented in the benefit-risk ratio evaluation.

6. A good place to start is by considering clinical benefits associated with the indication for use.
7. The EU MDR defined clinical benefit as "the positive impact of a device *on the health of an individual, expressed in terms of a meaningful, measurable, patient-relevant clinical outcome/s, including outcomes/s related to diagnosis*, or a positive impact on patient management or public health." The EU IVDR defined clinical benefit as "the positive impact of a device *related to its function, such as that of screening, monitoring, diagnosis or aid to diagnosis of patients*, or a positive impact on patient management or public health."
8. Magnitude, duration, likelihood, medical necessity/clinical context, patient perspectives, user benefits and perspectives.
9. Uncertainty is associated with the clinical data strength/confidence level. Each benefit-risk ratio factor has an associated uncertainty level due to the clinical data quality. High uncertainty gives a benefit or risk less weight in the overall benefit-risk ratio evaluation.
10. Risk is defined as "the combination of the probability of occurrence of harm and the severity of that harm."
11. The IVD device SAE definition includes all of the elements from the standard medical device SAE definition, but the IVD device SAE definition also includes "(a) a patient management decision resulting in death or an imminent life-threatening situation for the individual being tested, or in the death of the individual's offspring."
12. Risk mitigations may include adding specific language in the device labeling, instructions for use (IFU) and product literature or manufacturing changes to prevent misuse.
13. They inform overall benefit-risk ratio weighting.
14. Risks from device misuse should be documented and mitigated whenever possible. Systematic misuse must be identified and mitigated, and should be carefully considered and documented during the CER/PER benefit-risk analysis.
15. EMA: PrOACT-URL; FDA: benefit-risk deliberation framework; EMA Effects Table.
16. The obsoleted device oCER/oPER
17. ISO 14971.
18. Objectivity is not possible because benefits and risks are about user attitudes and preferences
19. CE/PE, PMS and RM systems are all relevant and involve benefit-risk evaluations
20. From viability (v) to equivalence (eq) to first in human (fih) to traditional (tr) and finally to obsolete (o).

Chapter 7

1. N = sample size; p-value = statistical significance; CI = confidence interval; AUC = area under the curve; SAP = statistical analysis plan.
2. CE/PE documents include the history, regulatory status, contemporary clinical S&P data, and how the device compares to alternative therapies and fits into the background and SOTA medical practice. The CE/PE documents are kept current and used by EU regulators to determine whether the medical devices are safe enough to be on the EU market. Often medical reviewers, evaluators, writers, and authors are overwhelmed by the amount of data required in the CE/PE documents or the amount of training and experience needed to perform well as a CE/PE document writing team member.
3. The CER/PER story must be complete, comprehensive, and compelling.
4. The evaluator needs a graduate degree and 5 years of experience, or no graduate degree and 10 years of experience. The evaluator needs training and experience using the device, reviewing scientific literature, understanding research methodology and analyzing clinical data including statistical analyses, etc.
5. The first step is to obtain education in biological/medical sciences (e.g., college- and graduate-level education in biology, chemistry, clinical medicine, biomedical engineering). Then, the trainee should find a mentor to advise them about the CE/PE document writing process and to review their work. As the trainee gains experience in writing CE/PE documents over 5–10 years, they will become competent in CE/PE document writing. Reading the CE/PE document regulations and guidance documents and applying this learning to multiple CE/PE documents is required.
6. "(a) ascribing functions and properties to the device which the device does not have;
   (b) creating a false impression regarding treatment or diagnosis, functions or properties which the device does not have;
   (c) failing to inform the user or the patient of a likely risk associated with the use of the device in line with its intended purpose;
   (d) suggesting uses for the device other than those stated to form part of the intended purpose for which the conformity assessment was carried out."
7. True. Clinical data must support each clinical claim and the clinical data must be fully evaluated in the CER/PER.

8. A robust company-sponsored clinical trial using the subject device as indicated in the appropriate patient population will be the strongest clinical evidence in the CE/PE document if the trial was high quality and relevant (i.e., about the study device used in the appropriate patient population).
9. 12 weeks.
10.

Equivalent device	Benchmark device	Similar device
Clinical, biological, and technical equivalence must be clearly established. The clinical impacts for each difference between the subject device and the equivalent device must be documented in writing and must provide a strong rationale to justify using the equivalent device clinical data to represent the clinical data expected from the subject device.	A benchmark device has the same intended use and is a competitor's device used for clinical S&P as well as benefit-risk ratio comparisons with the subject device.	MDCG 2020-6 [1] defines a "similar device" as a device in the "same generic device group. The MDR defines this as a set of devices having the same or similar intended purposes or a commonality of technology allowing them to be classified in a generic manner not reflecting specific characteristics" (MDR Article 2(7)).

11. The natural flow includes six writing steps, beginning with writing the device description, then the clinical trial section, followed by the clinical literature and clinical experience sections before the benefit-risk ratio and the conclusions/executive summary.
12. Statistical significance is the mathematical comparison of the data between two datasets (i.e., do the datasets differ?) while clinical significance is the practical importance of the treatment effect (does the difference matter?).
13. Simple descriptive statistics (e.g., means, medians, standard deviations, etc.) and inferential statistics (e.g., p values < 0.05 identifying differences between groups, etc.).
14. Answers may include three of many examples not limited to: sex, age, height, weight, body mass index, number of patients with a particular disease state, number of prior treatments given, etc.
15. The Kaplan–Meier curve is used to calculate and compare changes over time for survival or different probabilities between groups, etc.
16. The statistical calculations should have at least an 80% power and at least a 95% confidence level.
17. Answers may include three of many examples not limited to: bar chart, line graph, pie chart, scattergram, combined bar chart, etc.
18. Chicago Manual of Style [19], APA [20], and AMA [21].
19.
    - Human resource (HR) costs
    - Number of team members and meetings required to get work done
    - CE/PE software costs
    - Overhead costs
    - Literature database/librarian costs
    - Data storage and retrieval costs
    - Interim data analysis for clinical trial costs, as needed
    - Literature copyright costs
    - Timeline delays due to missing data
    - Timeline delays due to a need to generate new clinical data
    - NB findings specifying rework or new/revised clinical data evaluation
20. The "parking lot" is a document holding details about issues not yet resolved during the CE/PE document writing process. Putting issues into the parking lot allows the team to move on even though those issues are not resolved.

Chapter 8

1. An SSCP is a summary of safety and clinical performance document for a medical device and the SSP is a summary of safety and performance for an IVD device. This regulatory document summarizes all safety and performance data in a short format, documenting how the device benefits outweigh risks when the device is used as indicated in the appropriate patient population.
2. Although NBs review the SSCP/SSP for compliance and scientific validity, the audience includes healthcare and medical providers, and device users and patients when the device is used by patients.
3. The SSCP/SSP are summaries written after the CER/PER is completed.
4. The Notified Body (NB).

5. The manufacturer must assign a unique SSCP reference number/identifier to be used for the entire SSCP lifetime. This SSCP reference number/identifier often includes the manufacturer SRN to uniquely identify the SSCP in EUDAMED.
6. The two main SSCP/SSP sections are intended for different user audiences: one for healthcare professionals and the other for laypersons and patients.
7. The SSCP/SSP provides a safety and performance summary including a review of all available alternative treatment information so healthcare providers can compare treatment options and make informed decisions on the device appropriateness for their patients. In addition, layperson language is provided so patients and caregivers using the device can also understand the clinical data summaries to make informed treatment decisions.
8. The SSCP/SSP must include the following three specific labeling details: 1) intended purpose, 2) indication/target population and 3) contraindications/limitations (e.g., interference and cross-reactivity details for IVD devices).
9. False. The SSCP/SSP should discuss previous device variants and the differences between the prior device models/versions/generations and the device changes over time should be described and justified. The specific device model/s used in testing to generate the clinical trial data should be identified esp. when used to support device S&P. Any device models not yet supported by clinical data should be identified.
10. The IFU.
11. The SSCP/SSP is updated whenever new clinical and/or safety information becomes available with the potential to change the benefit-risk ratio. In this situation, all the other CE/PE documents would also need to be updated.
12. Clinical data sources for the SSCP/SSP are typically secondary sources including (but not limited to) the clinical data analyses completed and documented in the CER/PER, PMCFER/PMPFER, PMSR/PSUR, the IFU, and other product labeling, and in other technical file documents.
13. Writers should use language at a 6th-grade level without much technical or medical terminology (e.g., writers may use a medical term in brackets behind layperson translation for the medical term). A use review process and comprehension/readability testing should ensure laypersons are able to read and answer questions about the SSCP/SSP prior to finalization, submission to the NB, and translation.
14. Not all theoretical risks need to be included, but justification as to why risks included and not included is required even though the decision about what to include or exclude is subjective. The MDGG guideline on SSCPs recommends "The SSCP should contain information on at least the same residual risks and undesirable side-effects as included in the IFU." The IFU does not include every single residual risk and the challenge is to ensure the most important benefits and risks are summarized.
15. The writer should keep the SSCP/SSP patient user section SIMPLE and this chapter suggested the following writing tips:
    - Shorten sentences
    - Reduce polysyllabic words
    - Ensure every, single, sentence has a clear noun and verb
    - Organize information to flow logically
    - Avoid acronyms
    - Simplify sentences and structure
    - Include figures, tables, and infographics, defining all medical terms
    - Include quantitative data about adverse device events, side effects, and residual risks
    - Use language appropriate for a 6th grade reading level.
16. Several data points differed between the templates for the SSCP and the SSP, not limited to the following:
    1.7 Indication - is device for near-patient testing and/or a companion diagnostic
    2.3 Limitations and/or contra-indications (e.g., relevant interferences, cross-reactions)
    3.1 Device description, device use conditions (e.g., laboratory, near-patient testing)
    3.2 If device is a kit, describe components including regulatory status, and any basic UDI-DIs
    6.1 Summary of device scientific validity
    7.1 Explanation of the unit of measurement, if applicable
    7.2 Identify applied reference materials/measurements for device calibration
17. Several good translation practices include forward-and-back translations, native speaker interpretations, use reviews, and editing to make particularly difficult to translate text areas clearer for all readers.
18. The manufacturer must check "Yes" or "No" about whether the NB reviewed the document. If yes, the checkbox requires the validation language reviewed to be documented; if no, the checkbox requires justification for this situation.

19. The SSCP/SSP is unique in part due to the inclusion of a section for the lay public (i.e., patient users and caregivers who are not healthcare professions)
20. A readability test ensures a lay person can read and understand the written document. For eample, the person will read the SSCP/SSP and answer questions about their understanding of the data presented in the written document.

Chapter 9

1. Common Terminology Criteria for Adverse Events, Notified Body Operations Group, clinical evaluation assessment report, performance evaluation assessment report, Medical Device Coordinating Group.
2. a. Read the executive summary and conclusion to understand the most important CE/PE points... the indications for use and target populations who used the device to generate the clinical data, the procedures used to determine if the device met the safety and performance (S&P) acceptance criteria with a well-established and acceptable benefit-risk profile.
   b. Ensure CE/PE document structure is complete and meets regulatory requirements by comparing documents and templates from international guidelines and standards.
   c. Review device description and revision history to fully understand the device description, developmental history, all clinical S&P claims and how the device works (technically and clinically).
   d. Verify all five clinical/performance evaluation stages 0-4 were completed and well documented in the CE/PE document:
      i. Stage 0: scope and plan
      ii. Stage 1: clinical/performance data identification (trials, literature, experiences)
      iii. Stage 2: clinical/performance data appraisal (scientific validity, relevance, quality)
      iv. Stage 3: clinical/performance data analysis (S&P acceptance criteria, residual risks, Benefit-Risk Ratio, uncertainty and unanswered questions)
      v. Stage 4: final report is written with sufficient clinical data to fully substantiate the conclusions.
   e. Consider all background knowledge, SOTA and alternative device/therapy clinical data discussions. Were the clinical data appropriate, on target and informative? Were appropriate clinical trials, literature and experience data evaluated? Were sufficient details provided to explain the clinical data sources used, search method details, selection and appraisal criteria, analytical results and methods used to draw conclusions? How does the device fit into this background and does the device adhere to appropriate standards and practice guidelines?
   f. Evaluate the indication for use and target population. Do the clinical/performance data support and substantiate each and every aspect of the indication for use and target populations?
   g. Test conclusions against the data. Are conclusions appropriate and adequate to ensure device S&P? Were risks mitigated and reduced enough so Benefit-Risk Ratio was acceptable? Were residual risks acceptable? Were all manufacturer claims addressed? Did the device labeling appropriately address all clinical/performance data concerns? Were all clinical data gaps identified and addressed or slated to be addressed? What research and safety mitigation or performance improvement activities were ongoing?
   h. Examine the CE/PE document version control elements. Were dates and signatures as expected? Was the next CE/PE document due are clearly stated? Were the required evaluator details provided (e.g., declaration of interest, resume or curriculum vitae, financial disclosure, etc.)? Were all CE/PE document data appropriately archived?
3. MedRA dictionary or Common Terminology Criteria for Adverse Events (CTCAE)
4. A CE/PE Benefit-Risk Ratio Table should group benefits/indications and risks/AEs based on standardized/preferred term, System Organ Class (SOC) and body system affected. For example, heart attack can be coded as a myocardial infarction and grouped with other cardiovascular system AEs.
5. 1) clinical data were accurately abstracted and analyzed in direct relationship to the device under evaluation, 2) CE/PE document accurately and completely represented the expert analysis by the company clinical team and 3) clinical Benefit-Risk Ratio determined the benefits outweighed the risks
6. The clinical team reviewer and the quality team reviewer have different *expertise* and *focus* areas during the review. The clinical team reviewer should be checking clinical data accuracy while quality team reviewers *should* be checking to ensure each CE/PE document is structurally sound and the written arguments are clear and convincing (i.e., conclusions are well-stated and logically flow from the clinical data described within the CE/PE document)
7. Required regulatory items/details
8. False. Device labeling must be updated to comply with CER/PER conclusions when required

9. The PMSP, PMSR/PSUR, PMCFP/PMPFP, and PMCFER/PMPFER
10. The CER/PER for any new, high-risk, class IIb and III, implantable medical devices, and Class D IVD devices; except for certificate renewal, device modification with no change in Benefit-Risk Ratio or CE addressed in a common specification
11. MDCG 2020-13, Clinical evaluation assessment report template (July 2020)
12. 11 sections
13. True. CEAR Sections I, J, K described Expert Panel CECP activities, justifications for missing CE documentation and voluntary clinical consultation details
14. Ambiguous language included "example given," "such as" and "including" in the indication statement and these generalizing phrases were not acceptable because this languaged suggested other specific indications and claims may exist for the devies
15. "intended use/purpose" describes the effect of a device on the human body and "indication" refers to the specific clinical condition to be diagnosed, prevented, monitored, treated, alleviated, compensated for, replaced, modified or controlled by the Medical Device
16. False. The CER is a stand-alone document containing all critically analyzed, relevant clinical data.
17. No, the NB may keep certain items as observations until the next scheduled IFU and CER update
18. The expert panel provides scientific, technical and clinical assistance, consults with NBs on high-risk medical devices and IVD devices, advises MDCG and European Commission and manufacturers about device S&P, clinical development strategies, clinical investigations, guidance documents, common specifications and standards, and provides opinions to manufacturers, Member States and NBs.
19. Tasks to ensure clinical and performance data were accurate and complete, the evaluator critically evaluated each and every data component and the identification, appraisal, analysis and conclusions were consistent with the clinical data
20. True. NBs often limit the indication for use to the specific, fully evaluated clinical/performance data.

Chapter 10

1. Postmarket surveillance plan, postmarket surveillance report, periodic safety update report, postmarket clinical follow-up, postmarket performance follow-up, postmarket clinical follow-up plan, postmarket performance follow-up plan, postmarket clinical follow-up evaluation report, postmarket performance follow-up evaluation report.
2. a. To address all clinical data gaps identified in the CER/PER.
   b. To capture and evaluate additional clinical data about device S&P and Benefit-Risk Ratio acceptability when the device is used as indicated in "real-world" medical practice settings, over the device lifetime, by broad patient populations to include all indication for use possibilities, etc.
   c. To add subject-device-specific S&P clinical data for any device approved for CE Marking; especially for any device gaining CE-Mark based on clinical data solely from equivalent device clinical data.
3. The term "trial" was intended to mean a rigorous interventional RCT, a prospective clinical trial designed to test an interventional treatment or to diagnose a condition or disease outside of routine clinical practice, while a "study" was intended to mean a less rigorous clinical data collection including an observational study, real-world research, registry and survey studies, etc. The EU Clinical Trials Regulation (EU Reg. 536/2014 Article 2.2(4)) defined "a non-interventional study" as "a clinical study other than a clinical trial."
4.

Passive PMS activities	Proactive PMS activities
Reacting to unsolicited clinical data received	Gathering clinical data to cover CER/PER clinical data gaps
Waiting for phone call with complaint	Executing a clinical trial per protocol
Recording complaint while working in the field	Creating a registry or observational study to collect new data
Documenting manufacturing problems	Hosting a focus group or advisory board
Evaluating service records for complaints	Reporting responses from a specific customer survey
Reviewing biocompatibility records	Evaluating specific, solicited user feedback
Analyzing design vigilance records	Analyzing data in patient info form (e.g., returned postcard or "fill in the blank" online from device package)
Hearing about a device misuse case	Collecting real-world clinical experience data in a specific comparative clinical trial
Evaluating device design compatibility with other devices records	Managing PMCF studies
Recording customer satisfaction communications	Designing and executing retrieval studies (evaluation of retrieved devices after removal from patient)

5. PMSR is generated for Class I (low risk) devices while the PSUR is generated for Class IIa, IIb and III medical devices and Class D IVD devices.
6. The need to correctly characterize device performance, to have effective and appropriate methods for collected data assessments including complaints, field reports, trends analyses and corrective actions using suitable indicator and threshold values to assess RM and the Benefit-Risk Ratio, to communicate effectively with regulatory authorities, operators and users, to fulfil manufacturers obligations for PMS system, PMS plan, PSMR/PSUR and PMCFP/PMPFP or to justify why a PMCF/PMPF is not applicable.
7. PMS System data including PMCF/PMPF data are used to:
   a. update benefit-risk determinations and improve RM
   b. update design and manufacturing information, instructions for use and labelling;
   c. update CE/PE;
   d. update SSCP/SSP
   e. identify needs for preventive, corrective or field safety corrective action;
   f. identify needs to improve device usability, S&P;
   g. to contribute to the PMS of other devices; and
   h. to detect and report trends
8. False. "The QMS needs to define how the data flows both within each system and between the Regulatory, RM, Clinical, PMS and Sales and Marketing Systems."
9. The claims matrix links each sales/marketing claim to the specific claim substantiation data document and the CE/PE and PMS teams will use the claims literature in the CER/PER potentially from the PMCFER/PMPFER to support of the product labeling.
10. False, clinical data from litereature searches are required in all three systems (CE/PE, RM and PMS) and each system has different needs for clinical data searched and gathered from the medical literature.
11. Signal detection determines patterns of association or unexpected occurrences potentially impacting patient management decisions and/or altering the device S&P or benefit-risk profile.
12. These dictionaries provide a standard language for documenting and evaluatnig clinical events.
13. No. complaints use estimates like [number of incidents or complaints]/[number of device sales] are not considered sufficient due to the poor data quality (under and over reporting are common);however, these estimates may be used in cases where data from clinical investigations or PMCF studies are not deemed appropriate.
14. Studies with a clear research question, objectives, endpoints, rationale, data management plan, statistical analysis plan and with sources of study distortion or bias evaluated and minimized to improve data integrity and to allow appropriate conclusions to be drawn from the data.
15. Narrow the device intended purpose and indications for use until fully supported by the available clinical evidence.
16. For low-risk, standard-of-care and WET devices with little change in the state-of-the-art and background knowledge for the device, not associated with safety concerns, and with no innovation. These devices are less likely to be the subject of research, and literature data are limited or non-existent. In these cases, a lower level of clinical evidence may be justified as sufficient and may be supported by clinical data from the PMS and RM systems provided the QMS was in place to systematically collect and analyse any complaints and incident reports, and the collected data support the device S&P. The PMS and RM systems must be operating efficiently and appropriately.
17. A documented, proactive, organised procedure to collect clinical data from CE-marked device uses to confirm clinical S&P throughout the expected device lifetime, the acceptability of the benefit-risk ratio and to detect emerging risks. Typically, the PMCFP is designed to address the clinical data gaps identified in the CER.
18. Study carried out following device CE marking and intended to answer specific questions relating to device clinical S&P (i.e. residual risks) when used as indicated in the approved labelling.
19. EU MDR Annex XIV Part B; EU IVDR Annex XIII Part B.
20. PMS is intended to detect new S&P signals, CE/PE is required to analyze ALL subject-device-related clinical data and RM is required to mitigate all risks to acceptable levels.

Chapter 11

1. AHWP = Asian Harmonisation Working Party, ASEAN MDD = Association of Southeast Asian Nations Medical Device Directive, CSDT = common submission dossier template, EPs = essential principles, IMDRF = International Medical Device Regulators Forum, ISO = International Organization for Standardization, MHLW = Ministry of Health, Labor and Welfare, NMPA = National Medical Products Administration, PMA = premarket authorization, PMDA = Pharmaceuticals and Medical Device Agency, SaMD = software as a medical device, TGA = Therapeutic Goods Administration.

2. Australia, China, and ASEAN countries.
3. The Australian guidance used essential principles not GSPR, offered a CE checklist, and included specific examples of literature and clinical data to consider for specific types of devices among other differences.
4. EPs 1, 2, 3, 4, 6, 13, 13 A and 14 were considered particularly relevant to clinical evaluation.
5. Both TGA and the EU allow justifications when clinical data collection was deemed not appropriate (e.g., data collection was unethical or difficult due to limited patient availability or high-risk procedures) and little to no clinical data were collected. Adequate justification was required based on the risk management, the interaction between the device and the human body, the intended clinical performance and the claims.. IVD devices may also be duly justified to rely on other sources of clinical performance data.
6. Unlike the EU CER/PER, China listed high-risk devices requiring clinical trials in CFDA Circular 61 (2020) "Catalogue of class III medical devices subject to clinical trial approval" and SAMR Decree No. 47 (medical devices) and 48 (IVD devices) specifically listed exemptions for clinical trials and CE. China regulations allowed four unique pathways to the China market: (1) Class I (no clinical data needed); (2) predicate comparison; (3) clinical trial data from overseas for a device already on the market overseas; and (4) clinical data from a clinical trial within China for a device not on the market in any geography. The China CER pathway included both medical devices and IVD devices, required an appropriate equivalent device already approved and used in China, required more technical information iwith 16 specific product technical requirements (PTR), and China had more specific details about the literature search protocol and report, among other difference.
7. False. China required searches of the literature in China including the CNKI and/or other specific China databases.
8. In 2021, a website at https://www.easychinapprov.com/clinical-evaluation reported "45% of total medical device groups are exempted from clinical evaluation, 42.5% of medical device need clinical evaluation whereas only 12.5% of total medical device have to undergo clinical study."
9. The ASEAN MDD suggested doing a risk analysis to document residual risks and to determine the clinical evidence needed. A clinical investigation is needed when preclinical testing and prior clinical data were insufficient to demonstrate Essential Principles conformity.
10. Canada, the US, Russia, New Zealand, Japan, UK and Northern Ireland.
11. CE is presented in a different format than the CER/PER and only certain devices are required to submit S&P CE to the US FDA. All device companies must evaluate clinical S&P data; however, the US FDA provide exemptions formost low-risk devices (Class I and II) otherwise a510(k) must demonstrate substantial equivalence (SE) between the new device and a chosen "predicate" which is a medical device legally marketed in the US market with a previously cleared 510(k). Although CERs/PERs were not required as such, the FDA has considered CERs/PERs to meet the "report of priors" requirement for IND/IDE applications.
12. The US FDA SE determination differs when compared to the MDR/IVDR equivalence claim for the CER/PER. The same word 'equivalence' is used in both the US FDA 'substantial equivalence' *regulatory decision* under the 510(k) path for medical device clearance in the US and the 'equivalence' *claim* made in the CER/PER under the EU MDR/IVDR; however, each country applies a different meaning for this word 'equivalence'. The 510k path and EU CER/PER process differ.
13. False. US FDA uses the term 'predicate' for the device chosen by the manufacturer to establish SE for US FDA clearance under a 510(k); however, the MDR/IVDR does not use the term 'predicate' so manufacturers should not use this terminology globally.
14. De novo is a regualtory pathway created by the US FDA to expedite regulatory review when no predicate device is available for a 510(k) (i.e., all other devices have different indications for use, different technologies or other differences raising questions about the device safety and effectiveness). When no predicate is available, the US FDA automatically classifies the device into the Class III group and requires a PreMarket Authorization (PMA). A de novo classification review by the FDA is designed to reclassify the device from Class III into a lower-risk category using a risk-based approach based on the description, non-clinical data, clinical data and controls in place to assure the device is safe and effective.
15. An investigational device exemption (IDE) application.
16. the 'report of prior investigations' provided with the Investigational Device Exemption (IDE) application and the Summary of Safety and Effectiveness Data (SSED) provided with the premarket authorization (PMA) applicaiton to document the device safety and performance results for the FDA.

17. The '522 order' is issued by the US FDA to require timely reporting of PMS data under Section 522 of the FD&C Act typically for class II or III devices with identifed safety issues. The postapproval study (PAS) can be ordered by the US FDA as a condition of approval or clearance for a PMA, HDE or other application to allow more safety and effectivenss monitoriong by the FDA.
18. Health Canada is responsible for regulating medical devices and IVD devices in Canada under the consolidated Medical Device Regulations in SOR/98-282.
19. Russia requires a Notary stamp and corresponding Apostille to meet the requirement to be "legalized".
20. New Zealand, Switzerland, and Turkey.

Index

Note: Page numbers followed by "*f*," "*t*," and "*b*" refer to figures, tables, and boxes, respectively.

C

F

G

H

I

J

K

L

M

N

O

P

Q

R

T

U

V

W

Printed in the United States
by Baker & Taylor Publisher Services